T_H17 Cells in Health and Disease

Shuiping Jiang
Editor

T_H17 Cells in Health and Disease

Springer

Editor
Shuiping Jiang
Department of Nephrology and Transplantation
Guy's Hospital, King's College London
London, UK
sj4774@hotmail.com

ISBN 978-1-4419-9370-0 e-ISBN 978-1-4419-9371-7
DOI 10.1007/978-1-4419-9371-7
Springer New York Dordrecht Heidelberg London

Library of Congress Control Number: 2011930755

Printed on acid-free paper

Springer is part of Springer Science+Business Media (www.springer.com)

Contents

Contributors

Eleftherios I. Agorogiannis Sir William Dunn School of Pathology, University of Oxford, South Parks Road, Oxford, OX1 3RE, UK

Pornpimon Angkasekwinai Department of Medical Technology, Faculty of Allied Health Sciences, Thammasat University, Pathum-thani 12121, Thailand

Francesco Annunziato Department of Internal Medicine, University of Florence, Viale Morgagni 85, Firenze 50134, Italy

Amit Awasthi Center for Neurologic Diseases, Harvard Medical School, Brigham and Women's Hospital, HIM785, 77 Avenue Louis Pasteur, Boston, MA 02115, USA

M. Kemal Aydintug Integrated Department of Immunology, National Jewish Health, 1400 Jackson Street Denver, CO 80206, USA; University of Colorado Denver, Aurora, CO 80045, USA

Jeremy Bastid OREGA Biotech, 15 chemin du Saquin, F-69130 Ecully, France

Alexandre Belot Université de Lyon, France; Centre de Référence des Maladies Rénales Rares, Service de Néphrologie et Rhumatologie Pédiatriques, Hôpital Femme Mère Enfant, Bron F-69677, France

J. Magarian Blander Immunology Institute, Department of Medicine, Mount Sinai School of Medicine, 1425 Madison Avenue, New York, NY 10029, USA

Bruce R. Blazar Department of Pediatrics, Division of Blood and Marrow Transplantation and University of Minnesota Cancer Center, Minneapolis, MN 55455, USA

Nathalie Bonnefoy-Berard Université de Lyon, Lyon, France; INSERM, U851, 21 Avenue Tony Garnier, F-69007 Lyon, France; Université Lyon1, IFR128, Lyon, France

Willi K. Born Integrated Department of Immunology, National Jewish Health, 1400 Jackson Street Denver, CO 80206, USA;
University of Colorado Denver, Aurora, CO 80045, USA

Jennifer L. Chain Integrated Department of Immunology, National Jewish Health, 1400 Jackson Street Denver, CO 80206, USA;
University of Colorado Denver, Aurora, CO 80045, USA

Xi Chen Department of Neuroimmunology, GlaxoSmithKline Research and Development Center, No 3 Building, 898 Halei Road, Zhangjiang, Hi-teck Park, Pudong, Shanghai 201203, China

Stephen P. Cobbold Sir William Dunn School of Pathology, University of Oxford, South Parks Road, Oxford, OX1 3RE, UK

Lorenzo Cosmi Department of Internal Medicine, University of Florence, Viale Morgagni 85, Firenze 50134, Italy

José C. Crispín Division of Rheumatology, Beth Israel Deaconess Medical Center, Harvard Medical School, Boston, MA 02115, USA

Batya Davidovici Laboratory for Investigative Dermatology, The Rockefeller University, 1230 York Avenue, New York, NY 10065, USA

Rosemarie H. DeKruyff Division of Immunology and Allergy, Children's Hospital, Harvard Medical School, Boston, MA 02115, USA

Chen Dong Department of Immunology, University of Texas and MD Anderson Cancer Center, Houston, TX 77030, USA

Agnès Doreau-Bastid Université de Lyon, Lyon, France; Université Lyon1, IFR128, Lyon, France; INSERM, U851, 21 Avenue Tony Garnier, Lyon, F-69007, France

Jean-François Eliaou Service d'Immunologie, CHU de Montpellier; Faculté de Médecine, Université Montpellier 1, F-34295, Montpellier, France

Enric Esplugues Section of Immunobiology and Howard Hughes Medical Institute, Yale University School of Medicine, New Haven, CT 06520, USA

Lei Fang Department of Neuroimmunology, GlaxoSmithKline Research and Development Center, No 3 Building, 898 Halei Road, Zhangjiang Hi-teck Park, Pudong, Shanghai 201203, China

Richard A. Flavell Section of Immunobiology and Howard Hughes Medical Institute, Yale University School of Medicine, New Haven, CT 06520, USA

Andrew P. Fontenot Department of Medicine, University of Colorado Denver, 12700 E. 19th Avenue, Aurora, CO 80045, USA

Sarah L. Gaffen Department of Medicine, Division of Rheumatology & Clinical Immunology, University of Pittsburgh, Pittsburgh, PA 15261, USA

Kamran Ghoreschi Lymphocyte Cell Biology Section, Molecular Immunology and Inflammation Branch, National Institutes of Arthritis and Musculoskeletal and Skin Diseases, National Institutes of Health, Bethesda, MD 20892, USA

Taylor Guo Department of Neuroimmunology, GlaxoSmithKline Research and Development Center, No 3 Building, 898 Halei Road, Zhangjiang Hi-teck Park, Pudong, Shanghai 201203, China

Yafei Huang Integrated Department of Immunology, National Jewish Health, 1400 Jackson Street Denver, CO 80206, USA;
University of Colorado Denver, Aurora, CO 80045, USA

Samuel Huber Section of Immunobiology and Howard Hughes Medical Institute, Yale University School of Medicine, New Haven, CT 06520, USA

Harumichi Ishigame Center for Experimental Medicine and Systems Biology, The Institute of Medical Science, The University of Tokyo, 4-6-1 Shirokanedai, Minato-ku, Tokyo 108-8639, Japan

Yoichiro Iwakura Center for Experimental Medicine and Systems Biology, The Institute of Medical Science, The University of Tokyo, 4-6-1 Shirokanedai, Minato-ku, Tokyo 108-8639, Japan;
Core Research for Evolutional Science and Technology (CREST), Japan Science and Technology Agency, Saitama 332-0012, Japan

Shuiping Jiang Department of Nephrology and Transplantation, Guy's Hospital, King's College London, London SE1 9RT, UK

Niyun Jin Integrated Department of Immunology, National Jewish Health, 1400 Jackson Street, Denver, CO 80206, USA;
University of Colorado Denver, Aurora, CO 80045, USA

Shabaana A. Khader Division of Infectious Diseases, Children's Hospital of Pittsburgh, Pittsburgh, PA 15224, USA

Chang H. Kim Laboratory of Immunology and Hematopoiesis, Department of Comparative Pathobiology, Purdue Cancer Center, Purdue University, 725 Harrison Street, West Lafayette, IN 47907, USA

Akihiro Kimura Laboratory of Immune Regulation, Osaka University Graduate School of Frontier Biosciences, Osaka, Japan

Tadamitsu Kishimoto Laboratory of Immune Regulation, Osaka University Graduate School of Frontier Biosciences, Osaka, Japan

Jay K. Kolls Department of Genetics, Louisiana State University Health Sciences Center, CSRB 657, 533 Bolivar St., New Orleans, LA 70112, USA

Sergei Koralov Department of Pathology New York University, MSB 531A, 550 1st Ave, New York, NY10016, USA

James G. Krueger Laboratory for Investigative Dermatology, The Rockefeller University, 1230 York Avenue, New York, NY 10065, USA

Vijay K. Kuchroo Center for Neurologic Diseases, Harvard Medical School, Brigham and Women's Hospital, HIM785, 77 Avenue Louis Pasteur, Boston, MA 02115, USA

Vasileios C. Kyttaris Division of Rheumatology, Harvard Medical School, Beth Israel Deaconess Medical Center, Boston, MA 02115, USA

Stephen J. Lalor School of Biochemistry and Immunology, Trinity College Dublin, Dublin 2, Ireland

Arian Laurence Lymphocyte Cell Biology Section, Molecular Immunology and Inflammation Branch, National Institutes of Arthritis and Musculoskeletal and Skin Diseases, National Institutes of Health, Bethesda, MD 20892, USA

Stewart Leung Department of Neuroimmunology, GlaxoSmithKline Research and Development Center, No 3 Building, 898 Halei Road, Zhangjiang Hi-teck Park, Pudong, Shanghai 201203, China

Francesco Liotta Department of Internal Medicine, University of Florence, Viale Morgagni 85, Firenze 50134, Italy

Xuebin Liu Department of Neuroimmunology, GlaxoSmithKline Research and Development Center, No 3 Building, 898 Halei Road, Zhangjiang Hi-teck Park, Pudong, Shanghai 201203, China

Ling Lu Department of Liver Transplantation, the First Affiliated Hospital of Nanjing Medical University, Nanjing, China

Erik Lubberts Departments of Rheumatology and Immunology, Erasmus MC, University Medical Center, s-Gravendijkwal 230, 3015 CE, Rotterdam, The Netherlands

Antonella De Luca Microbiology Section, Department of Experimental Medicine and Biochemical Sciences, University of Perugia, Via del Giochetto, Perugia 06122, Italy

Thomas T. MacDonald Blizard Institute of Cell and Molecular Science, Barts and the London School of Medicine and Dentistry, London E1 4NS, UK

Enrico Maggi Department of Internal Medicine, University of Florence, Viale Morgagni 85, Firenze 50134, Italy

Kingston H.G. Mills School of Biochemistry and Immunology, Trinity College Dublin, Dublin 2, Ireland

Pierre Miossec Department of Immunology and Rheumatology, Hôpital Edouard Herriot, 69437 Lyon, Cedex 03, France

Giovanni Monteleone Department of Internal Medicine, University "Tor Vergata" of Rome, Via Montpellier, 100133 Rome, Italy

Timothy P. Moran Department of Microbiology and Immunology, and Department of Medicine, University of North Carolina, Chapel Hill, Chapel Hill, NC 27599, USA

Tetsuji Naka Laboratory for Immune Signal, National Institute of Biomedical Innovation, Osaka, Japan

Susumu Nakae Frontier Research Initiative, The Institute of Medical Science, The University of Tokyo, 4-6-1 Shirokanedai, Minato-ku, Tokyo 108-8639, Japan

Kristine E. Nograles Laboratory for Investigative Dermatology, The Rockefeller University, 1230 York Avenue, New York, NY 10065, USA

Rebecca L. O'Brien Integrated Department of Immunology, National Jewish Health, 1400 Jackson Street, Denver, CO 80206, USA; University of Colorado Denver, Aurora, CO 80045, USA

John J. O'Shea Lymphocyte Cell Biology Section, Molecular Immunology and Inflammation Branch, National Institutes of Arthritis and Musculoskeletal and Skin Diseases, National Institutes of Health, Bethesda, MD 20892, USA

Wenjun Ouyang Immunology Department, Genentech, Inc., South San Francisco, CA 94080, USA

Francesco Pallone Department of Internal Medicine, University "Tor Vergata" of Rome, Via Montpellier, 100133 Rome, Italy

Christina L. Roark Beacon Biotechnology, Aurora, CO 80045, USA

Sergio Romagnani Department of Internal Medicine, University of Florence, Viale Morgagni 85, Firenze 50134, Italy

Luigina Romani Microbiology Section, Department of Experimental Medicine and Biochemical Sciences, University of Perugia, Via del Giochetto, Perugia 06122, Italy

Massimiliano Sarra Department of Internal Medicine, University "Tor Vergata" of Rome, Via Montpellier, 100133 Rome, Italy

Jonathan S. Serody Department of Microbiology and Immunology, and Department of Medicine, University of North Carolina, Chapel Hill, Chapel Hill, NC 27599, USA

Philip L. Simonian Department of Medicine, University of Colorado Denver, 12700 E. 19th Avenue, Aurora, CO 80045, USA

Mark S. Sundrud Department of Pathology, Harvard Medical School, Immune Disease Institute and the Program in Cellular and Molecular Medicine at Children's Hospital Boston, Boston, MA, USA; Tempero pharmaceuticals, 200 Technology Square, Suite 602, Cambridge, MA 02139, USA

Caroline E. Sutton School of Biochemistry and Immunology, Trinity College Dublin, Dublin 2, Ireland

Julia Tsang Department of Anatomical and Cellular pathology, Faculty of Medicine, The Chinese University of Hong Kong, Hong Kong, China

Michael Tsang Department of Developmental Biology, Division of Rheumatology & Clinical Immunology, University of Pittsburgh, Pittsburgh, PA 15261, USA

George C. Tsokos Division of Rheumatology, Beth Israel Deaconess Medical Center, Harvard Medical School, 330 Brookline Avenue, CLS-937, Boston, MA 02115, USA

Dale T. Umetsu Division of Immunology and Allergy, Children's Hospital, Harvard Medical School, Boston, MA 02115, USA

Patricia Valdez Laboratory of Clinical Infectious Diseases, National Institute of Allergy and Infectious Diseases, National Institutes of Health, Bethesda, MD 20892, USA

Herman Waldmann Sir William Dunn School of Pathology, University of Oxford, South Parks Road, Oxford, OX1 3RE, UK

Cailin Moira Wilke Department of Surgery, University of Michigan, C560B MSRB II, 1150 W. Medical Center Dr., Ann Arbor, MI 48109, USA

Xiang-Ping Yang Lymphocyte Cell Biology Section, Molecular Immunology and Inflammation Branch, National Institutes of Arthritis and Musculoskeletal and Skin Diseases, National Institutes of Health, Bethesda, MD 20892, USA

Teresa Zelante Microbiology Section, Department of Experimental Medicine and Biochemical Sciences, University of Perugia, Via del Giochetto, Perugia 06122, Italy

Jingwu Zhang Department of Neuroimmunology, GlaxoSmithKline Research and Development Center, 898 Halei Road, Building 3, Zhangjiang Pudong, Shanghai 201203, China

Weiping Zou Department of Surgery, University of Michigan, C560B MSRB II, 1150 W. Medical Center Dr., Ann Arbor, MI 48109-0669, USA

Part I
The Immunology of IL-17 Family and T_H17 Cells

From T_H1/T_H2 Paradigm to T_H17 Cells: Le Roi Est Mort, Vive Le Roi

Amit Awasthi and Vijay K. Kuchroo

Abstract Upon activation, naïve $CD4^+$ T cells differentiate into distinct T helper cell subsets with specific cytokine profiles and distinct effector functions. Until recently, effector T cells were classified into T_H1 or T_H2 subtypes depending on the cytokines they produced. However, this paradigm had to be revised with the discovery of a third subset of effector T cells called T_H17 cells. IL-17-producing T_H17 cells play an important role in clearing extracellular pathogens and tissue inflammation. TGF-β and IL-6 are the factors that induce differentiation of naïve T cells into T_H17 cells. Differentiated T_H17 cells produce IL-17, IL-17F, IL-21 and IL-22, and thereby mediate distinct effector functions compared to T_H1 and T_H2 cells. While IL-17, IL-17F and IL-22 induce tissue reaction, IL-21 produced by T_H17 cells is essential for amplification of T_H17 cells and B cell function. Whereas the maturation and stabilization of differentiated T_H17 cells are mediated by IL-23; the transcription factors (STAT3, RORγt, RORα and c-Maf) are involved in the development and transcription of various molecules expressed by T_H17 cells. The requirement of TGF-β in generation of both T_H17 cells and $CD4^+CD25^+Foxp3^+$ regulatory T cells (Tregs) suggest a close developmental relationship between these two cell types. Here, we summarize the current information on the differentiation and effector functions of the T_H17 lineage and their interplay with other T helper subsets during tissue inflammation.

V.K. Kuchroo (✉)
Center for Neurologic Diseases, Harvard Medical School, Brigham and Women's Hospital, HIM785, 77 Avenue Louis Pasteur, Boston, MA 02115, USA
e-mail: vkuchroo@rics.bwh.harvard.edu

S. Jiang (ed.), *T_H17 Cells in Health and Disease*,
DOI 10.1007/978-1-4419-9371-7_1, © Springer Science+Business Media, LLC 2011

1 Introduction

Antigenic stimulation of naïve T cells in the presence of specific cytokines produced by innate cells induces activation, expansion and differentiation of T cells into distinct effector T cells (Abbas et al. 1996). About 25 years ago, Mosmann and Coffman introduced the T_H1/T_H2 paradigm of T helper cell differentiation, which helped explain many aspects of adaptive immunity from eliminating intracellular vs. extracellular pathogens to induction of different types of tissue inflammation (Mosmann et al. 1986; Mosmann and Coffman 1989). IFN-γ and IL-12 induce the differentiation of naïve T cells into a T cell subset that predominantly produces IFN-γ and classified as T_H1 cells (Hsieh et al. 1993). T_H1 cells are indispensable for clearing intracellular pathogens by activating effector functions of macrophages and inducing antibody class switching to IgG2a (Mosmann and Coffman 1989). The presence of IFN-γ during T_H1 differentiation induces STAT-1 dependent expression of T-bet, a master transcription factor required for IFN-γ production (Szabo et al. 2000). Both STAT-1 and T-bet deficient mice lack IFN-γ induction and are susceptible to intracellular pathogens such as *Leishmania* and *Mycobacteria* (Szabo et al. 2003). During the T_H1 development, T-bet induces the expression of IL-12Rβ2, which makes developing T_H1 cells responsive to IL-12 in a STAT-4 dependent manner (Mullen et al. 2001). IL-12 exposure to developing T_H1 cells stabilizes their T_H1 phenotypes, and this loss of STAT-4 in STAT-4 deficient mice enhances the susceptibility against intracellular pathogens (Kaplan et al. 1996b). Altogether, these observations suggest the sequential transcriptional requirements (STAT-1, T-bet, STAT-4) for the development of effector T_H1 cells (Szabo et al. 2003). In contrast, IL-4 inhibits the functions of T_H1 cells and induces the differentiation of T cell subsets that predominantly produce IL-4, IL-5, IL-13 and IL-25. The effector functions of this subset are opposite to that of T_H1 cells in that they eliminate extracellular pathogens (like helminthes) and promote antibody class switching to IgG1 and IgE. IL-4 initiates T_H2 development by inducing phospho-STAT-6, which induces GATA-3, a T_H2 specific transcription factor that binds and trans-activates IL-4 promoter to initiate T_H2 developmental program (Ansel et al. 2006; Takeda et al. 1996; Zheng and Flavell 1997). STAT-6 deficient mice are defective in generating T_H2 cells and are susceptible to extracellular pathogens (Kaplan et al. 1996a). Similarly, loss of GATA-3 in mice results in defective T_H2 development with enhanced susceptibility to T_H2 associated pathogens (Zhu et al. 2004). Moreover, overexpression of GATA-3 induces T_H2 development and inhibits the generation of T_H1 cells (Zheng and Flavell 1997). These observations suggest that both STAT-6 and GATA-3 are the transcription factors essential for the development of effector T_H2 cells (Ansel et al. 2006). One of the striking features of T_H1 and T_H2 cells is that they cross-regulate functions of each other by producing antagonizing cytokines (Mosmann and Coffman 1989). For instance, IFN-γ, a T_H1 cytokine, regulates the differentiation of T_H2 cells, while IL-4 produced by T_H2 cells inhibits the generation of T_H1 cells (Mosmann and Coffman 1989). Both T_H1 and T_H2 cells are indispensable for eliminating both intracellular and extracellular pathogens. Therefore, lack of appropriate differentiation in response to infection would result in dissemination of infection (Mosmann and Coffman 1989). However, deregulated T_H1 and T_H2 cells can mediate tissue inflammation and immunopathology (Kuchroo et al. 2002). T_H1 cells have been associated with multiple auto-immune diseases such as multiple sclerosis, rheumatoid arthritis, type-1 diabetes, and inflammatory bowel disease (IBD). Similarly, a deregulated T_H2 response promotes asthma, allergies, and atopy. Although T_H1 cells with specificity of self antigens were considered to be the pathogenic T cells involved in inducing auto-immune diseases, loss of IFN-γ or T_H1 differentiation factors, IL-12, did not inhibit auto-immune diseases, but paradoxically enhanced the incidence and severity of disease (Cua et al. 2003; Ferber et al. 1996). Similarly, T_H2 cells or cytokines were not able to induce auto-immunity (Kuchroo et al. 1995) and this raised the issue of whether there were other cell types involved in inducing auto-immune tissue reactions. With the discovery of IL-23, a cytokine that shares a common chain with IL-12, it became clear that the loss of IL-23 and not IL-12 made mice resistant to the development of multiple auto-immune diseases (Cua et al. 2003; Oppmann et al. 2000). IL-23 induced IL-17 from T cells and this led to the discovery of IL-17-producing T cells, which were shown to induce massive tissue inflammation and auto-immunity (see below) (Korn et al. 2009). T_H17 cells are a subset of T cells that have differentiation and transcription factors that distinguish them from T_H1 and T_H2 cells (Bettelli et al. 2008; Korn et al. 2009). The discovery of T_H17 cells has expanded the T_H1–T_H2 paradigm, and the integration of T_H17 cells with T_H1 and T_H2 effector T cells is beginning to explain the underlying mechanisms of tissue inflammation in a number of infections and auto-immune disease settings. The highly pathogenic nature of T_H17 cells makes them the primary effector cells type in eliminating specific pathogens that are not adequately handled by T_H1 or T_H2 cells. Besides their role in clearing infections, T_H17 cells are potent inducers of tissue inflammation since IL-17 receptors are present on all parenchymal cells including endothelium, fibroblasts, and epithelial cells. Plus, activation with IL-17 results in the production of IL-1, IL-6, TNF-α, matrix metalloproteinases and chemokines, including IL-8 which makes tissue susceptible to infiltration to other pro-inflammatory cells and propagate tissue inflammation (Bettelli et al. 2008). IL-17 and T_H17 cells have been associated with the pathogenesis of many auto-immune diseases both in mice and man. In the last 5 years, cytokines that are required for differentiation, amplification and expansion of T_H17 cells have been identified and transcription factors that induce T_H17 cells differentiation have been elucidated.

In this chapter, we summarize the findings that led to the identification of T_H17 cells, their differentiation factors, and we describe their helper and effector functions.

2 T_H17 Cells Differentiation

Identification of the differentiation and transcription factors for T_H17 cells has classified T_H17 cells into a unique subset of helper T cells, distinct from T_H1 or T_H2 cells (Korn et al. 2009). The discovery of IL-23 was instrumental in identifying T_H17 cells and studying their role in inducing tissue inflammation and auto-immunity (Cua et al. 2003). IL-23, a member of the IL-12 family of cytokines, was discovered in 2000 and is composed of a unique sub-unit called IL-23p19 which associates with the p40 sub-unit of IL-12 to form a hetero dimeric cytokine (Oppmann et al. 2000). Cua et al. showed that p19 deficient mice, in contrast to p35 deficient mice, were resistant to the development of experimental auto-immune encephalomyelitis (EAE) (Cua et al. 2003). Further analysis showed that the central nervous system (CNS) of IL-23p19 deficient mice harbors less T_H17 cells and therefore suggested that IL-23 may be the differentiation factor for the generation of pathogenic T_H17 cells (Cua et al. 2003). In fact, IL-23 was able to expand IL-17 producing T cells from immunized mice further supporting the interpretation that IL-23 may indeed induce differentiation of T_H17 cells (Awasthi et al. 2009). As IL-23R is predominantly expressed on activated/memory T cells, it was possible that IL-23 might act on previously differentiated T_H17 cells rather than promoting de novo differentiation of naïve T cells (Awasthi et al. 2009; McGeachy et al. 2009; Parham et al. 2002). Furthermore, addition of IL-23 to sorted naïve T cells was not able to induce differentiation of T_H17 cells, confirming that IL-23 is not the diversity factor for the differentiation of T_H17 cells (Bettelli et al. 2006). In 2006, three independent studies reported that instead of IL-23, a combination of the immuno-regulatory cytokine TGF-β and a pro-inflammatory cytokine IL-6 is required to induce the differentiation of T_H17 cells from naïve T cells (Bettelli et al. 2006; Mangan et al. 2006; Veldhoen et al. 2006a). Interestingly, TGF-β induces the generation of $Foxp3^+$ iTregs cells from naïve T cells which are able to suppress immune responses in vitro and in vivo (Fig. 1) (Bettelli et al. 2006).

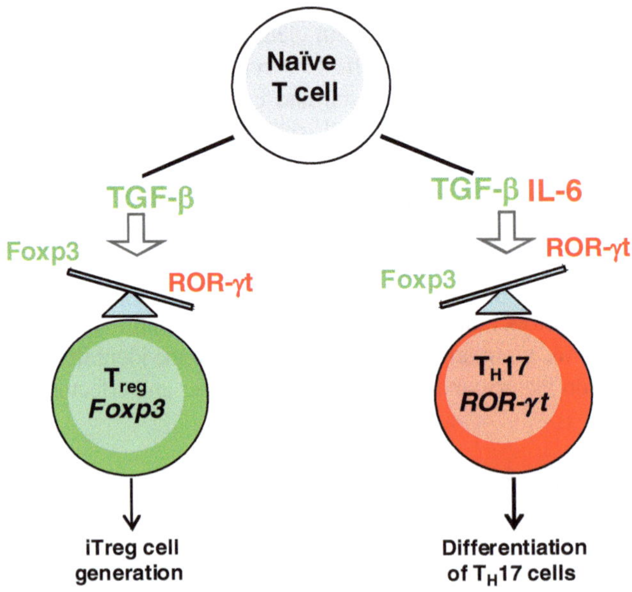

Fig. 1 Reciprocal generation Tregs and T_H17 cells. Activation of naïve T cells with TGF-β induces the generation of Foxp3+ Tregs cells while the addition of IL-6 not only inhibited the induction of TGF-β-induced Foxp3 but also concomitantly induced the generation of T_H17 cells

However, the addition of IL-6 to TGF-β abrogated the induction of Foxp3 expression and resulted in the induction of T cells that predominantly produced IL-17 (Bettelli et al. 2006). The function of TGF-β in differentiation of T_H17 cells was confirmed using TGF-β transgenic mice, where TGF-β overexpression was controlled under IL-2 promoter. Immunization of TGF-β transgenic mice with myelin antigen emulsified in CFA induced severe inflammation in the CNS with the development of overly aggressive EAE (Bettelli et al. 2006). Further analysis of CNS of these mice revealed a higher frequency of T_H17 cells and reduction in frequency $FoxP3^+$ Treg cells (Bettelli et al. 2006). Conversely, when T cells are deficient in a functional receptor of TGF-β and cannot signal to TGF-β, the generation of T_H17 cells was abolished and mice were completely protected from development of EAE (Veldhoen et al. 2006b). Immunization with CFA induces IL-6 from innate immune cells, which in combination with TGF-β induces the generation of T_H17 cells. This data confirmed the in vitro data that TGF-β and IL-6 are essential in generating T_H17 cells and indicated that IL-6 is the key switch factor in determining whether an immune response is dominated by $Foxp3^+$ Tregs or by T_H17 cells (Bettelli et al. 2008). TGF-β is a pleiotropic cytokine produced by multiple lineages of leukocytes and stromal cells (Li and Flavell 2008). However, the cellular source of TGF-β for the induction of T_H17 cells has not been identified. Deletion of the TGF-β1 gene specifically in T cells is detrimental in that the mice develop lethal immunopathology associated with uncontrolled T_H1 and T_H2 responses in multiple organs (Li et al. 2007). TGF-β-deficient $Foxp3^+$ T cells, in contrast to wild type $Foxp3^+$ T cells, fail to control T_H1 mediated inflammation in a T cell transfer model of colitis (Li et al. 2007). In addition, ablation of TGF-β1 production in T cells resulted a defective generation of T_H17 cells and these mice were protected from developing EAE, implying that T cells which produced TGF-β contributed to the differentiation of T_H17 cells (Li et al. 2007). In addition to T cells, DCs express sufficient amounts of TGF-β to drive the differentiation T_H17 cells (Veldhoen et al. 2006b). In fact, conditional deletion of αvβ8 (an integrin required for processing inactive forms of TGF-β into active forms) in DCs resulted in lymphoproliferation and tissue inflammation that was reminiscent of TGF-β1 deficiency in T cells (Travis et al. 2007). These observations indicate that DCs play a role in generating active TGF-β1 locally to support the generation of T_H17 cells in vivo (Veldhoen et al. 2006b). Furthermore, DCs from lamina propria produce TGF-β and retinoic acid induces de novo conversion of T cells into $Foxp3^+$ Tregs cells (Denning et al. 2007). T_H17 cells are also highly enriched in the lamina propria, which raises the possibility that DC-produced TGF-β in the gut in combination with IL-6 might drive T_H17 differentiation in this context.

Similar to TGF-β, IL-6 is also a pleiotropic cytokine induced by infection, inflammation or injury, and mediates a variety of functions in both immune and non-immune compartments (Kishimoto 2005). IL-6 is mainly produced by activated DCs, monocytes, and macrophages in response to TLRs activation. Different approaches led to the identification of IL-6 as a critical player in the differentiation of T_H17 cells: (1) LPS stimulated DC/T cell co-cultures induced the de novo generation of T_H17 cells when TGF-β was present and neutralization of IL-6 by an IL-6 antibody abolished the generation of T_H17 cells (Veldhoen et al. 2006a), (2) the addition of IL-6 together with TGF-β inhibited the expression of $Foxp3^+$ iTregs and resulted in the generation of T_H17 cells (Bettelli et al. 2006). The in vivo function of IL-6 in the generation of T_H17 cells was emphasized in both IL-6 deficient and gp130, a signaling sub-unit of IL-6 receptor complex, and conditional "knock-out" mice (Korn et al. 2007, 2008). These mice were completely resistant to EAE, and in fact, the peripheral repertoire in these mice was dominated by a high frequency of $Foxp3^+$ T_{regs} (Korn et al. 2007, 2008). The reciprocal developmental pathways of T_H17 and iT_{regs} cells suggest that these cells share a common precursor and depending on the cytokines present at the time of their activation, T cells can differentiate into Tregs or T_H17 cells depending on the availability of IL-6 (Fig. 1). It suggests that TGF-β and IL-6 are essential in the differentiation of T_H17 cells both in vitro and in vivo.

3 T_H17 Amplification and IL-21

IL-21, a member of the IL-2 family of cytokines, uses a common γ chain of IL-2 receptor expressed on all T and B cells (Leonard et al. 2008). T_H17 cells produce high amount of IL-21 in addition to IL-17A, IL-17F, and IL-22 (Korn et al. 2007; Nurieva et al. 2007). IL-21, a pleiotropic cytokine, induces a variety of functions on $CD4^+$, $CD8^+$, and B cells (Leonard et al. 2008). Analysis of differentiation in IL-21R deficient T_H17 cells showed that IL-21 is essential to amplify T_H17 differentiation (Korn et al. 2007; Nurieva et al. 2007). This observation is consistent with findings that IL-21, in combination with TGF-β, induces de novo T_H17 differentiation (Korn et al. 2007). These findings imply that T_H17 differentiation can be induced in the absence of IL-6, and cells such as NKT and NK cells that produce IL-21 can support T_H17 differentiation in the absence of IL-6. However, IL-6 still remains the dominant factor in the differentiation of T_H17 cells such that IL-6 deficient mice are completely resistant to EAE development with a defect in T_H17 generation (Korn et al. 2007; Okuda et al. 1998). IL-6 induces IL-21 in T_H17 cells in a STAT-3 dependent manner, which further amplifies the generation of T_H17 cells. IL-21 also helps T_H17 cells in stabilizing their T_H17 phenotype by inducing the expression of IL-23R, and makes T_H17 cells responsive to IL-23 exposure to attain their effector phenotype (Korn et al. 2007; Nurieva et al. 2007). IL-6 also induces IL-23R expression on T_H17 cells sequentially by inducing IL-21 in differentiating T_H17 cells (Littman and Rudensky 2010). These in vitro observations were further confirmed using IL-21 or IL-21R deficient mice showing a diminished expression of IL-23R, resulting in the blunt T_H17 generation. Therefore, IL-21 not only amplifies T_H17 differentiation, but also helps T_H17 cells to attain a mature T_H17 phenotype (Coquet et al. 2008). An initial study showed that IL-21-deficient mice are resistant to EAE and two subsequent studies showed development of a more severe EAE than the wild type cohorts (Coquet et al. 2008). These studies questioned whether IL-21 is critical in vivo for the generation of T_H17 cells. Using Complete Freund's Adjuvant (CFA) for immunization to induce EAE in IL-21 or IL-21R deficient mice might

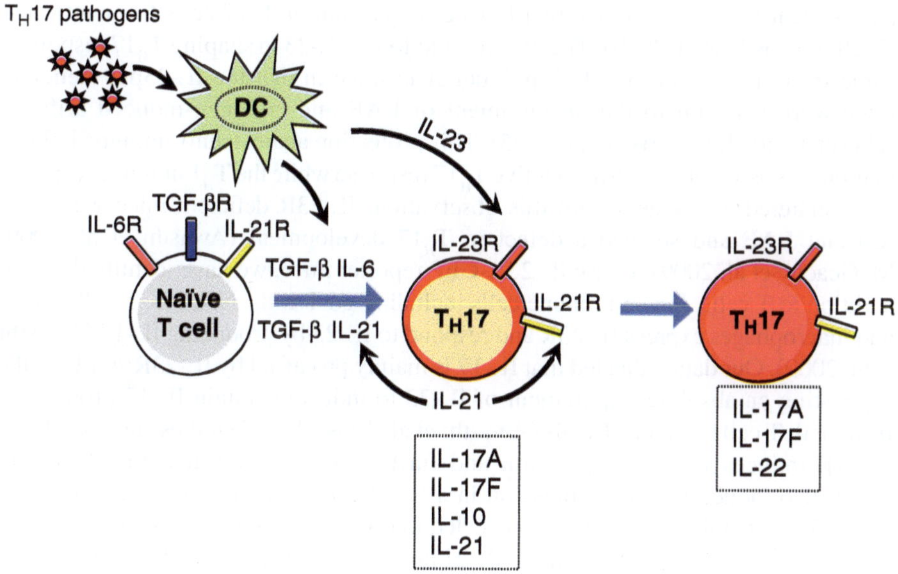

Fig. 2 Developmental pathway of T_H17 cells. Activated DC provide IL-6 that induce T_H17 differentiation together with TGF-β. Differentiated T_H17 cells produce IL-21, which in combination of TGF-β, further enhances the generation of T_H17 cells. IL-21 also induces the expression of IL-23R on T_H17 cells. Innate immune cells (DC and macrophages) produced IL-23 stabilizes T_H17 differentiation

override the need for the induction of IL-21 because massive amounts of IL-6 produced in vivo by immunization with CFA most likely compensates for IL-21 deficiency. This argument is further supported by the observation that NOD mice are completely resistant to the development of type 1 diabetes; if they are deficient in IL-21 or IL-21R induction of auto-immunity occurs spontaneously without any immunization with CFA (Spolski et al. 2008) (Sutherland et al. 2009). In addition to the generation of T_H17 cells, it has been observed that IL-21 is an essential growth factor for IL-27-induced IL-10 producing Tr1 cells (Awasthi et al. 2007; Pot et al. 2009) (Spolski et al. 2009). Therefore, IL-21, or lack of protection against EAE observed in IL-21R deficient mice may be partly due to loss of IL-10 producing Tr1 cells compensating for a defect in T_H17 cells in IL-21 or IL-21R deficient mice. Altogether, it suggests that IL-21 is an essential feed forward loop in self-amplification of T_H17 cells (Fig. 2).

4 T_H17 Stabilization and IL-23

IL-23, a member of the IL-12 cytokine family, is a heterodimer consisting of the p40 sub-unit shared with IL-12 and the IL-23 specific p19 sub-unit (Oppmann et al. 2000). It is expressed by cells of the myeloid lineage, including DCs and macrophages, and is identified as a cytokine that induces expansion of T_H17 cells from activated T cells (Awasthi et al. 2009). The in vivo function of IL-23 in shaping T_H17 responses came from the analysis of IL-23p19 deficient mice in that the IL-23p19 deficient mice were resistant to the development of EAE and collagen-induced arthritis (Cua et al. 2003; Langrish et al. 2005). This protection against auto-immune inflammation was associated with a defective T_H17 response while the T_H1 immune response was unaltered. Consistent with this observation, IL-23R deficient mice were also resistant EAE and showed a defect in T_H17 development (Awasthi et al. 2009; McGeachy et al. 2009). Using IL-23R-GFP reporter mice, we have identified that in addition to T cells, innate immune cells including gd T cells, NK, NK T cells, DCs, and macrophages express IL-23R and respond to IL-23 by producing IL-17 (Awasthi et al. 2009). Our data indicated that IL-17 is mainly produced by IL-23R$^+$/GFP$^+$ cells, suggesting an absolute requirement of IL-23 to induce/maintain IL-17 production from both T cells and non T cells (Awasthi et al. 2009). Initial studies suggested that TGF-β induces IL-23R expression and further analysis revealed that IL-6 and IL-21 also induce IL-23R expression in a STAT-3 dependent manner (Zhou et al. 2007). IL-23R-GFP reporter mice revealed that IL-23 is the best inducer of its own receptor, even more so than TGF-β, IL-21 and IL-6 (Awasthi et al. 2009). However, activation of naïve T cells requires exposure of IL-6 or IL-21, instead of IL-23 to induce initial expression of IL-23R, as IL-23 does not induce IL-23R expression on naïve T cells (Zhou et al. 2007). Once activated in the presence of TGF-β, IL-6 and/or IL-21, T cells express IL-23R and IL-23 further enhances IL-23R expression to induce T_H17 generation (Awasthi et al. 2009; Zhou et al. 2007). Studies using IL-23R-deficient mice revealed that while IL-23R deficient T cells are able to differentiate into T_H17 cells, they fail to maintain their effector functions (Awasthi et al. 2009; McGeachy et al. 2009). Taken together, these observations suggest that the exposure of T_H17 cells to IL-23 matures/stabilizes their pro-inflammatory phenotype. Initial studies suggest that IL-23 not only upregulates expression of IL-23R, but also enhances the expression of T_H17-lineage specific cytokines IL-17A, IL-17F, IL-21, and IL-22 (Awasthi et al. 2009; McGeachy et al. 2009). It also suppresses production of cytokines like IL-10, which do not belong to T_H17 lineage (McGeachy et al. 2007) (Fig. 2). How IL-23 mediates this function at a molecular level is not well understood.

The role of IL-23/IL-23R in human auto-immune diseases is being increasingly recognized. A genome-wide association scan (GWAS) recently identified the IL-23R gene as a susceptibility gene for human inflammatory bowel disease. A single nucleotide polymorphism of arginine 381 to glutamine confers protection against Crohn's disease (Duerr et al. 2006), which is consistent with the mouse model that indicates a pathogenic role of IL-23 in intestinal inflammation (Buonocore et al. 2010; Kullberg et al. 2006). IL-23p19 deficient CD4$^+$ T cells failed to transfer colitis in a T cells transfer model. IL-23 treatment enhances the development of colitis with the increased expression of IL-17 and IL-6 (Yen et al. 2006). These observations indicate that IL-23–T_H17 pathway plays an important role in development of gut inflammation (Wu et al. 2009). Similarly, other genome-wide association studies revealed an association of SNPs in IL-23R with ankylosing spondylitis and psoriasis, and treatment with a monoclonal antibody specific for IL-12/IL-23p40 or IL-17 showing promising results in psoriasis patients, further strengthening the idea that IL-23 and T_H17 cells may be involved in inducing human auto-immune tissue inflammation (Rahman et al. 2009).

5 Transcriptional Regulation of T_H17 Cells

Recently, T_H17 lineage was specified by the expression of a specific factor, ROR-γt, a retinoids-related orphan transcription factor identified as an early element required for T cell and lymphoid organ development (Ivanov et al. 2008; Korn et al. 2009). The gene profiling analysis of T_H17 cells identified ROR-γt as a highly expressed transcription factor in this T cell lineage. Distinction of naïve T cells with TGF-β and IL-6 or IL-21 induces the manifestation of ROR-γt in developing T_H17 cells (Ivanov et al. 2006). Neither T_H1 nor T_H2 cells showed any expression of this transcription factor, suggesting a lineage specific appearance of ROR-γt induces T_H17 differentiation. T_H17 cells are highly enriched in the lamina propria in naïve mice and further analysis revealed that ROR-γt$^+$, but not ROR-γt$^-$ T cells produce IL-17, and forced expression of ROR-γt is sufficient to encourage IL-17 expression without the addition of exogenous cytokines (TGF-β plus IL-6) (Ivanov et al. 2006). These observations confirm the ability of ROR-γt to initiate the development of T_H17 cells. The analysis of ROR-γt-GFP reporter mice revealed the co-expression of T_H17 cytokines with ROR-γt (Lochner et al. 2008). In line with its critical functions in differentiation of T_H17 cells, ROR-γt-deficient T cells were defective in T_H17 cell differentiation and developed less severe auto-immune diseases (Ivanov et al. 2006).

The two cytokines (TGF-β and IL-6) required for the differentiation of T_H17 cells have antagonizing effects on one another, e.g. TGF-β strongly upregulates the appearance of Foxp3, while addition of IL-6 or IL-21 inhibits the expression of Foxp3 and induces the expression of IL-17 by upregulation of ROR-γt expression (Bettelli et al. 2006; Korn et al. 2007; Zhou et al. 2008). TGF-β induces both ROR-γt and Foxp3 without expressing IL-17 unless combined with the pro-inflammatory cytokines require for T_H17 differentiation (IL-6, IL-21 or IL-23) (Zhou et al. 2008). In fact, Foxp3 was shown to associate with ROR-γt, and thereby inhibit its transcriptional activity. Addition of pro-inflammatory cytokines suppresses induction of Foxp3, and relieves ROR-γt from the ROR-γt-Foxp3 complex to induce IL-17 transcription (Zhou et al. 2008). Further insight between Foxp3 and ROR-γt came from the analysis of a truncated form of Foxp3, which lacks exon 2 (Zhou et al. 2008). The truncated Foxp3 was unable to interact with ROR-γt, and thus ROR-γt was functional and able to prompt the expression of IL-17 (Zhou et al. 2008). This data clearly demonstrates the role of Foxp3 in inhibiting the development of T_H17 cells. However, the consequences and relevance of this interaction in vivo are not clear. Runx1, a transcription factor required for both T_H17 cell and Treg functions, interacts with both Foxp3 and ROR-γt (Ono et al. 2007; Zhang et al. 2008).

Recently, it has been shown that ROR-γt and Runx1 together bind to the *il-17a* promoter and lead to increased expression of IL-17 (Zhang et al. 2008) (Fig. 3).

In addition to ROR-γt, differentiation of T_H17 cells also requires ROR-α, another ROR family of transcription factors. ROR-α is also upregulated during T_H17 development. Forced manifestation of ROR-α is able to induce IL-17. On the other hand, overexpression of ROR-γt in ROR-α deficient T cells was not able to induce optimal expression of IL-17 in comparison to control T cells (Yang et al. 2008). The role of ROR-α in T_H17 differentiation is further emphasized by the fact that ROR-α deficient T cells are unreliable in making the development of T_H17 cells (Yang et al. 2008). These observations suggest that ROR-α synergizes the functions of ROR-γt in T_H17 cells differentiation (Yang et al. 2008).

Cytokines act through Jak-Stat pathways to initiate signaling events necessary to induce effector functions. The T_H17 inducing cytokines, IL-6, IL-21, and IL-23 activate STAT-3 to initiate T_H17 cells differentiation (Mathur et al. 2007; Yang et al. 2008). The central role of STAT-3 in T_H17 differentiation was revealed by the observation that STAT-3 deficient $CD4^+$ T cells were flawed in inducing T_H17 differentiation (Liu et al. 2008). Further analysis of STAT-3 deficient T cells showed a defect in the expression of IL-21 and IL-23R, both essential for optimal development of T_H17 cells (Liu et al. 2008). Furthermore, mice lacking STAT-3 in T cells were protected from the development of EAE as they generated lower frequencies of T_H17 cells (Liu et al. 2008). The role of STAT-3 in T_H17 differentiation was also confirmed on a molecular level as chip-based analysis showed that STAT-3 directly binds to the *IL-17a* and *IL-17f* promoter (Chen et al. 2006). Taken together, this data suggests that STAT-3 is essential for the creation of T_H17 cells (Fig. 3).

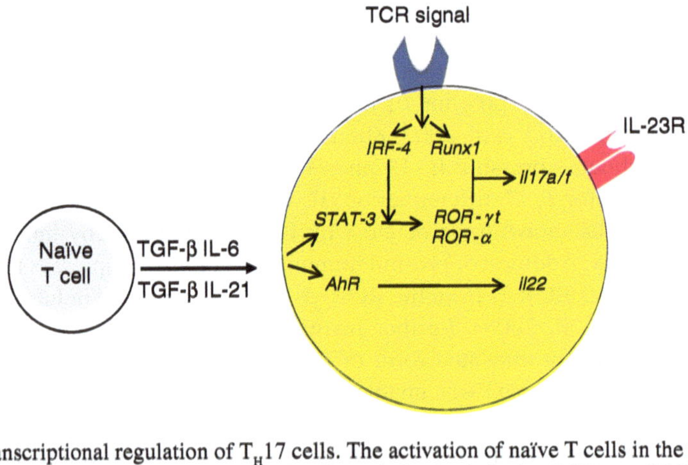

Fig. 3 Transcriptional regulation of T_H17 cells. The activation of naïve T cells in the presence of TGF-β and IL-6 or IL-21 initiate the T_H17 differentiation by inducing RORγt and RORα expression in a STAT-3 dependent manner. T_H17 cells produce IL-21. T_H17 cells also induces AhR expression, which enhances the production of IL-17 and IL-22. Both IRF-4 and Runx1 induced by TCR activation, which further promotes T_H17 differentiation

Suppressor of cytokine signaling- (SOCS) 3 is a major negative feedback regulator of STAT-3 activation and inhibits the generation of T_H17 cells. SOCS-3 negatively regulates IL-23 signaling and thus adjusts the generation of T_H17 cells (Chen et al. 2006). SOCS-3 deficient T cells showed an enhanced generation of T_H17 cells, which might be due to heightened STAT-3 signaling necessary for the development of T_H17 cells (Qin et al. 2009; Taleb et al. 2009). Consistent with the idea, STAT-3 also plays a crucial role in human T_H17 differentiation; a genetic mutation in STAT-3 in Job's syndrome leads to a hyper IgE production, defective T_H17 differentiation with recurrent *Candida albicans* and *Staphylococcus aureus* infection in the lungs of the affected patients (Ma et al. 2008).

Although ROR-γt is the master transcription factor of the T_H17 cells lineage, other aspects cooperate with RORγt for optimal distinction of T_H17 cells and their development. IRF-4, a factor previously shown to enhance T_H2 development was recently shown to also play an important role in the generation of T_H17 cells (Rengarajan et al. 2002). IRF-4 deficient mice were protected from EAE, and T cells from IRF-4-deficient mice failed to separate into T_H17 cells with an attenuated expression of ROR-γt and ROR-α, suggesting that IRF-4 is upstream of these nuclear receptors (Brustle et al. 2007; Huber et al. 2008). Studies using IRF-4 deficient T cells also revealed a shortcoming in induction of IL-21 and IL-23R, factors necessary for T_H17 cell development (Huber et al. 2008; Xiao et al. 2008). Interestingly, IRF-4-deficient T cells showed enhanced expression of Foxp3, suggesting that IRF-4 might be essential in regulating the IL-6 or IL-21 mediated inhibition of Foxp3, and thus improve the development of T_H17. cMaf, a proto-oncogene previously described to be a T_H2-specific transcription factor, is highly expressed in T_H17 cells (Ho et al. 1998). Comparatively, T_H17 cells express cMaf at higher levels than T_H2 cells (Bauquet et al. 2009). Whereas loss of cMaf does not result in a deficiency in T_H17 differentiation, c-Maf lacking T cells are unable to make IL-21 and sustain expression of IL-23R (Bauquet et al. 2009). Therefore, there is a gradual loss of T_H17 cells in cMaf deficient mice. Whereas cMaf is initially induced by IL-6 during T_H17 differentiation, later on the inducible stimulatory molecule (ICOS), which is expressed at higher levels on T_H17 cells, maintains its appearance throughout the life-span of T_H17 cells (Bauquet et al. 2009). Loss of ICOS in ICOS-deficient mice, results in defective T_H17 differentiation by virtue of lower cMaf, IL-21 and IL-23R expression (Bauquet et al. 2009). cMaf has been shown to bind to promoter elements in both IL-21 and IL-23R and specifically trans activate both of these genes (Pot et al. 2009).

Further analysis of T_H17 cells revealed the involvement of aryl hydrocarbon receptors (AHR), a ligand-dependent transcription factor required to mediate the effects of environmental toxins such as 2,3,7,8-Tetrachlorodibenzodioxin (known as TCDD or dioxin). Upon binding to its ligands (FICZ, TCDD), AHR transfers from the cytosol to the nucleus and binds to DNA in order to trans-activate AHR-dependent genes. Activation of AHR with a high affinity natural ligand FICZ during T_H17 differentiation enhances the development of T_H17 cells, but AHR seems to specifically regulate production of IL-22 in T_H17 cells (Quintana et al. 2008; Veldhoen et al. 2008).

Taken together, multiple transcription factors facilitate separation of T_H17 cells and production of various cytokines of T_H17 lineage, but of these, RORγt is the more crucial lineage specific transcription factor required for the development of T_H17 cells (Fig. 3).

6 Inhibition of T_H17 Cells

T_H17 cells are protective against certain extracellular pathogens and a deregulated T_H17 response can induce severe immunopathology of host tissues. Therefore, the regulation of T_H17 cells is essential to prevent tissue damage. Here we described the cytokines that antagonize the differentiation of T_H17 cells.

6.1 IL-27 and Inhibition of T_H17 Response

IL-27, a member of IL-12 cytokine family, is produced by innate immune cells and induces IFN-γ production from T cells (Pflanz et al. 2002). IL-27R consists of the gp130 sub-unit of the IL-6 receptor with a unique chain called WSX-1 and is predominantly expressed on T cells (Pflanz et al. 2004). Initial work suggested that IL-27 is essential to initiate T_H1 response by upregulating IL-12Rβ2, a sub-unit require for IL-12 signaling in a T-bet manner (Hibbert et al. 2003; Takeda et al. 2003). However, soon after its discovery, it become evident that by using different T_H1 and T_H2 associated pathogens, IL-27 negatively regulates both T_H1 and T_H2 responses (Holscher et al. 2005; Rosas et al. 2006; Villarino et al. 2003; Yoshimoto et al. 2007). Toxoplasmosis is a parasitic infection that requires T_H1 mediated effector functions for inhibiting parasite growth within the host. Mice deficient in the T_H1 cells inducing cytokines IL-12 and IFN-γ, die because of high parasitic burden (Liu et al. 2006; Scharton-Kersten et al. 1995, 1996). IL-27 was shown to clear parasites efficiently by mounting a massive T_H1 response that can eliminate them. However, the mice died due to fatal and severe immunopathology (Villarino et al. 2003). This observation was further extended for other intracellular pathogens such as *Leishmania* and *Mycobacterium* (Holscher et al. 2005; Rosas et al. 2006). After the discovery of pathogenic T_H17 cells, the severe immunopathology associated with *Toxoplasma gondii* infection in IL-27R deficient mice was revisited in the context of T_H17 cells. Analysis of *T. gondii* infected IL-27R deficient mice revealed an exaggerated T_H17 response, suggesting that IL-27 adversely adjusts the generation of T_H17 cells (Stumhofer et al. 2006). Furthermore, immunization of IL-27R deficient mice with MOG peptide in CFA induced EAE associated with exaggerated T_H17 response (Batten et al. 2006). An increased number of T_H17 cells were found in the CNS in both models of the IL-27R deficient mice (Batten et al. 2006; Stumhofer et al. 2006). Similarly, addition of IL-27 together with TGF-β and IL-6 also inhibited the variation of T_H17 cells in vitro (Batten et al. 2006; Stumhofer et al. 2006). Since IL-27 significantly induces production of IFN-γ, it is possible that

IL-27 hinders the differentiation of T_H17 cells by enhancing IFN-γ production. Neutralization of IFN-γ in the T_H17 cells cultures did not reverse the IL-27 induced inhibition of T_H17 cells (Batten et al. 2006). IL-27 directly inhibits the differentiation of T_H17 cells in a STAT-1 dependent pathway. STAT-1 deficient mice failed to deter T_H17 differentiation induced by IL-27 (Batten et al. 2006; Stumhofer et al. 2006). Recent data suggests that IL-27 inhibits T_H17 differentiation at the transcriptional level by impeding the expression of ROR-γt (Diveu et al. 2009). IL-6, which inhibits the induction of TGF-β induced iT_{reg} generation with a concomitant, increases the induction of T_H17 cells, and IL-27 inhibits the generation of both T_H17 and iT_{regs} induction (Awasthi et al. 2007; Batten et al. 2006; Stumhofer et al. 2006). The identification of IL-27 as a destructive regulator for the generation of T_H1, T_H2, and T_H17 cells suggested a possibility that it might induce other factors or T cell variation that might directly or indirectly control other effector T cells responses. Recently, a series of papers identified that IL-27 is the differentiation factor for IL-10 producing T cells, raising the possibility that IL-27 might curbT_H17 generation and other effector T cells by inducing IL-10 producing Tr1 cells (Awasthi et al. 2007; Fitzgerald et al. 2007; Stumhofer et al. 2007). These observations might also explain the mechanism of uncontrolled immunopathology in IL-27R deficient mice and is perhaps due to a lack of IL-10-producing T cells. Consistent with this observation, in vivo administration of IL-27 was shown to suppress development of EAE with a significant decrease in the generation of pro-inflammatory effector T cells.

7 T_H1/T_H2 Paradigm and T_H17 Plasticity

Before the discovery of T_H17 cells, most of the inflammatory responses in various infections and auto-immune diseases were described to be associated with either T_H1 or T_H2 cells. In fact past 25 years, T_H1–T_H2 paradigm revolutionized our understanding of adaptive immune responses to various infections and allergic diseases. T_H1 cells are essential to eliminate intracellular pathogens such as *Leishmania*, *Mycobacteria* etc., and were described as the pathogenic effector T cells in many auto-immune diseases such as multiple sclerosis, type-1 diabetes, rheumatoid arthritis, and IBD. Similarly, T_H2 cells that eliminate extracellular pathogens are the effector T cells that induce allergic inflammation. Here we provide evidence that integrates T_H17 cells into the existing T_H1/T_H2 paradigm and explains the underlying mechanisms of tissue inflammation.

7.1 Integration of T_H17 Cells with T_H1/T_H2 Cells in Tissues Inflammation

It is well established that the T_H1 cells with specificity for the self-antigen can transfer auto-immunity. T_H1 clones specific for myelin antigens, MBP, MOG, and PLP were shown to transfer EAE in adoptive transfer mode (Kuchroo et al. 1992). However, genetic deficiencies of molecules associated with T_H1 cells such as IL-12p35, IL-12Rβ2, STAT-1, IFN-γ, and IFN-γR did not abrogate EAE (Korn et al. 2009; McGeachy and Cua 2008). In fact, the incidence and severity of the disease in these genetically deficient mice was enhanced. It is well documented that IL-23, instead of IL-12, is essential for tissue inflammation in EAE (Awasthi et al. 2009; Cua et al. 2003). This observation challenged the notion of association of T_H1 cells with EAE. Further work revealed a clear association of IL-23–T_H17 axis in inducing organ-specific tissue inflammation. T_H1 cells have always been found at the target tissue together with T_H17 cells in EAE. Analysis of CNS infiltrating T cells revealed a population of T cells that co-expressed both IFN-γ and IL-17, suggesting a possibility that Th1 cells synergize the functions of T_H17 cells or vice versa in inducing inflammation. A recent report suggests that T_H17 cells breach the blood brain barrier to allow infiltration of other cell types including T_H1 cells in CNS (Kebir et al. 2007). In addition to a synergy between T_H17 and T_H1 cells, T_H17 cells may convert into IFN-γ-producing T_H1 cells in the target tissue and these IL-17$^+$ IFN-γ$^+$ T cells might be the most potent inducer of tissue inflammation.

Similar to EAE, T_H1 cells play a dominant role in gut inflammation. Both Crohn's disease and the animal model of IBD were shown to be associated with IFN-γ and TNF-α (Neurath et al. 2002a). The association of T_H1 cells in IBD was further evident with the detection of high amounts of IL-12 in patients with Crohn's disease and also in the animal model. Neutralization of IL-12 by anti-IL-12 antibodies enhances intestinal inflammation, presumably by preventing the generation of T_H1 cells (Neurath et al. 1995). Failure of STAT-4 deficient T cells in transferring T cells mediated colitis and the development of overly aggressive colitis in STAT-4 transgenic mice suggested that an IL-12 dependent pathway is dominant in inducing colitogenic T cells at the mucosal surface (Neurath et al. 1995) (Wirtz et al. 1999). Further evidence about the association of T_H1 cells in colitis came with the experiment that the loss of T-bet, a master transcription factor that induces IFN-γ, failed to transfer colitis in T cells (Neurath et al. 2002b). Discovery of a genetic association of IL-23R polymorphism with Crohn's disease in GWAS studies suggested the involvement of IL-23–T_H17 pathways in gut inflammation (Duerr et al. 2006). Consistent with human genetic studies, neutralization of IL-23 ameliorates gut inflammation in a number of mouse models of IBD including colitis induced by transfer of naïve T cells. These observations were further supported with the fact that T_H17 cells are highly enriched in gut associated lymphoid tissues at steady state (Ivanov et al. 2006), suggesting a possibility of their involvement in inducing gut inflammation in certain specific conditions. These observations suggest that both T_H17 and T_H1 cells are essential for inducing tissue inflammation in the gut. Exactly how these T_H1 and T_H17 cells generated in the gut is not clear. New emerging data suggests that the transfer of highly purified T_H17 cells from IL-17F reporter into a lymphopenic host induces T_H1 cells associated colitis, which clearly indicated the conversion of T_H17 cells into T_H1 cells in the gut environment (Lee et al. 2009). However, whether T_H17 cells can induce disease without conversion into IFN-γ producing cells or whether IFN-γ deficient T_H17 cells can also transfer colitis has not been addressed. Altogether, it suggests that T_H17 can readily convert into T_H1 cells under certain conditions or tissue inflammation.

7.2 *Mechanism of T_H17–T_H1 Plasticity*

As discussed above, the co-existence of T_H1 and T_H17 cells in auto-immune inflammation in mouse models such as EAE, rheumatoid arthritis and gut inflammation suggests that both effector T cells are essential for tissue inflammation. In fact, a sizable number of IL-17/IFN-γ double producing $CD4^+$ T cells are always found in the CNS of the EAE affected mice (Korn et al. 2007). The size of this population is such that it has been argued that IL-17/IFN-γ double producing $CD4^+$ T cells are the main effector cells in inducing inflammation in EAE. Similar observations were also found in mycobacterial infection in which T_H17 cells infiltrate lungs in the infection to stimulate permeation of T_H1 cells (Khader et al. 2007). Moreover, IFN-γ/IL-17 double producing $CD4^+$ T cells were also observed in the vaccinated mice after being re-challenged with *M tuberculosis* infection (Khader et al. 2007). Whether IFN-γ/IL-17 double producing $CD4^+$ T cells are generated by the plasticity of T_H17 cells or IFN-γ/IL-17 double producers are induced in the peripheral immune compartment and then infiltrate the target tissue for inducing inflammation is not clear. The cytokines and transcription factors that are required for the differentiation of T_H17 and T_H1 cells are quite different, indicating a divergent differentiation of T_H1 and T_H17 cells (Korn et al. 2009; Szabo et al. 2003). In fact, the differentiation factors for T_H1 and T_H17 cells reciprocally inhibit the differentiation of each other's. The conversion of T_H17 cells into T_H1 cells proposes a convergence of these two distinct pathways at some point in their generation. Analysis of late development of T_H1 and T_H17 cells suggested a link of their convergence. Exposure of T_H1 and T_H17 cells, with IL-12 and IL-23 respectively, are essential and critical for their stable development (Korn et al. 2009; Szabo et al. 2003). The heterodimers of IL-12 and IL-23 share a common sub-unit, IL-12/IL-23p40, which pair either with IL-12p35 or IL-23p19 to provide a functional IL-12 or IL-23 (Oppmann et al. 2000). Similar to IL-12 and IL-23, their receptors also share a common chain; IL-12Rβ1 (Parham et al. 2002). IL-12Rβ2 predominantly expressed on T_H1 cells, but not on T_H17 cells. The T_H1 transcription factor T-bet provides the specificity to IL-12 induced functions on T_H1 cells (Szabo et al. 2003). Similarly, IL-23R, downstream of ROR-γt and specifically expressed on differentiated T_H17 cells, pairs with IL-12Rβ1 to enhance IL-23 induced stabilization of T_H17 phenotypes. A functional requirement for IL-12β1 receptors for both T_H1 and T_H17 cells increases their promiscuity and plasticity. Although downregulation of IL-12Rβ2 sub-unit during T_H17 differentiation is essential to prevent IL-12 mediated deviation of T_H17 differentiation pathway into T_H1 pathway, any increase in the expression of IL-12Rβ2 on differentiated T_H17 cells would increase their responsiveness to IL-12 and induce IFN-γ in T_H17 cells. Identification of both IFN-γ /IL-17 double producing $CD4^+$ T cells in mucosal surface of intestines at steady state and also during inflammatory conditions suggested an overlap or convergence of T_H1 and T_H17 developmental pathways. Recent data supports the idea of late developmental plasticity of highly purified T_H17 cells upon adoptive transfer in vivo (Lee et al. 2009; Lexberg et al. 2008). Lexberg et al. suggested that purified T_H17 precursor cells can be repolarized into T_H1 or T_H2 cells by exposing them to IL-12 or IL-4 (Lexberg et al. 2008). Similarly, other studies have suggested that the transfer of islet-specific BDC2.5 TCR transgenic T_H17 cells into an adoptive transfer model system induces insulitis and diabetes associated with acquisition of IFN-γ phenotype, indicating that transferred T_H17 cells give rise to T_H1 cells (Bending et al. 2009; Martin-Orozco et al. 2009). In another T cells transfer model of colitis, transferring of sorted T_H17 cells from IL-17F reporter into an immuno-deficient host induced inflammation associated with the appearance of T_H1 cells (Lee et al. 2009). On the other hand, transfers of T_H17 cells into an immuno-sufficient host retained their phenotype and were less susceptible to T_H1 conversion (Nurieva et al. 2009). The appearance of T_H1 cells from highly purified T_H17 cells into an immuno-deficient host raises the possibility of contributions from other immune cell types such as macrophages, DCs, and NK cells or cytokines produced by these cell types and late developmental plasticity of T_H17 into T_H1 cells. However, ex vivo isolated T_H17 memory cells retain their T_H17 phenotype even in the presence of IL-12 or IL-4, T_H1, or T_H2 polarizing cytokines (Lexberg et al. 2008). These observations clearly indicate that an essential maturation step is required to promote a precursor T_H17 cells into a stable T_H17 lineage, the mechanism(s) of which has not been yet identified. The in vitro mechanism of late developmental plasticity of T_H17 cells was partially identified. Exposure of TGF-β to T_H17 cells maintains the sustained expression of both IL-17A and IL-17F. However, repetitive stimulation of TGF-β induces the expression of IFN-γ in IL-17 expressing T cells. The mechanism of such IFN-γ/IL-17 co-expression induced by TGF-β is largely unknown, as TGF-β treatment antagonizes IFN-γ production by repressing T-bet expression (Lee et al. 2009). One possibility of such IFN-γ appearance in the presence of TGF-β is that chronic stimulation might make T_H17 cells insensitive to TGF-β mediated inhibition of T_H1 gene-program, making T_H17 cells express IFN-γ. Taken together, T_H17 cells have propensity to convert into T_H1 cells, however, their exposure to IL-23 stabilizes their Th17 phenotypes by promoting their terminal differentiation (Awasthi et al. 2009; McGeachy et al. 2009).

8 Concluding Remarks

The discovery of T_H1/T_H2 paradigm laid the foundation for understanding various immunological and inflammatory conditions including infections and auto-immune diseases. However, the T_H1/T_H2 hypothesis could not adequately explain development of certain inflammatory responses which provided impetus for the discovery of a new subset of T cells called T_H17 cells. After the discovery of differentiation and transcription factors for T_H17 cells, it was clear that T_H17 cells represent an independent subset of T cells with specific functions in eliminating certain extracellular pathogens, presumed to be inadequately handled by T_H1 or T_H2 cells. The major role of T_H17 cells has been described in inducing auto-immune tissue inflammation such as multiple sclerosis, rheumatoid arthritis, and IBD. IL-6 produced by the innate immune system is a major inducer of T_H17 cells. IL-6 in combination with TGF-β induces the differentiation of naïve T cells into T_H17 cells factors that are distinct from those involved in the differentiation of T_H1 or T_H2 cells. Similar to T_H1 and T_H2 cells, T_H17 cells require amplification factors for their expansion and this amplification loop of T_H17 cells is mediated by IL-21, which is produced by T_H17 cells during differentiation. IL-21 not only amplifies the generation of T_H17 cells as a feed forward loop, but also helps T_H17 cells to attain mature phenotype by enhancing the expression of IL-23R. Altogether, it suggests that the generation of T_H17 cells require three essential steps: (1) differentiation, (2) amplification, and (3) stabilization. The cytokines involved in each step are different and some of the steps may overlap depending on the available cytokine in the milieu. In the past 5 years, T_H17 cells have been identified as a crucial cell type involved in the induction of tissue inflammation. However, the presence of T_H1 cells and the co-expression of IFN-γ in T_H17 cells raises a possibility that T_H17 cells may be plastic and IFN-γ and IL-17 may act together to induce and sustain tissue inflammation. The requirement of IL-12β1 receptor as a part of IL-23R makes these cells inherently responsive to IL-12 and increases their promiscuity to produce IFN-γ. Co-expression of IFN-γ in T_H17 cells is beginning to shed new light in the understanding of tissue inflammation during infection and auto-immunity.

Acknowledgements We would like to acknowledge Dr. Nicole Joller and Dr. Gopal Murugaiyan for their critical comments on this review. This work was supported by grants from National Institutes of Health, National Multiple Sclerosis Society, and the Juvenile Diabetes Research Foundation Center for Immunological Tolerance at Harvard Medical School. V.K.K. is a recipient of the Javits Neuroscience Investigator Award from the US National Institutes of Health. AA is supported by a post-doctoral fellowship from the National Multiple Sclerosis Society (NMSS), New York.

References

Abbas, A.K., Murphy, K.M., and Sher, A. (1996). Functional diversity of helper T lymphocytes. Nature *383*, 787–793.

Ansel, K.M., Djuretic, I., Tanasa, B., and Rao, A. (2006). Regulation of Th2 differentiation and Il4 locus accessibility. Annu Rev Immunol *24*, 607–656.

Awasthi, A., Carrier, Y., Peron, J.P., Bettelli, E., Kamanaka, M., Flavell, R.A., Kuchroo, V.K., Oukka, M., and Weiner, H.L. (2007). A dominant function for interleukin 27 in generating interleukin 10-producing anti-inflammatory T cells. Nat Immunol *8*, 1380–1389.

Awasthi, A., Riol-Blanco, L., Jager, A., Korn, T., Pot, C., Galileos, G., Bettelli, E., Kuchroo, V.K., and Oukka, M. (2009). Cutting edge: IL-23 receptor gfp reporter mice reveal distinct populations of IL-17-producing cells. J Immunol *182*, 5904–5908.

Batten, M., Li, J., Yi, S., Kljavin, N.M., Danilenko, D.M., Lucas, S., Lee, J., de Sauvage, F.J., and Ghilardi, N. (2006). Interleukin 27 limits auto-immune encephalomyelitis by suppressing the development of interleukin 17-producing T cells. Nat Immunol *7*, 929–936.

Bauquet, A.T., Jin, H., Paterson, A.M., Mitsdoerffer, M., Ho, I.C., Sharpe, A.H., and Kuchroo, V.K. (2009). The costimulatory molecule ICOS regulates the expression of c-Maf and IL-21 in the development of follicular T helper cells and T(H)-17 cells. Nat Immunol *10*, 167–175.

Bending, D., De La Pena, H., Veldhoen, M., Phillips, J.M., Uyttenhove, C., Stockinger, B., and Cooke, A. (2009). Highly purified Th17 cells from BDC2.5NOD mice convert into Th1-like cells in NOD/SCID recipient mice. J Clin Invest *119*, 565–572.

Bettelli, E., Carrier, Y., Gao, W., Korn, T., Strom, T.B., Oukka, M., Weiner, H.L., and Kuchroo, V.K. (2006). Reciprocal developmental pathways for the generation of pathogenic effector TH17 and regulatory T cells. Nature *441*, 235–238.

Bettelli, E., Korn, T., Oukka, M., and Kuchroo, V.K. (2008). Induction and effector functions of T(H)17 cells. Nature *453*, 1051–1057.

Brustle, A., Heink, S., Huber, M., Rosenplanter, C., Stadelmann, C., Yu, P., Arpaia, E., Mak, T.W., Kamradt, T., and Lohoff, M. (2007). The development of inflammatory T(H)-17 cells requires interferon-regulatory factor 4. Nat Immunol *8*, 958–966.

Buonocore, S., Ahern, P.P., Uhlig, H.H., Ivanov, II, Littman, D.R., Maloy, K.J., and Powrie, F. (2010). Innate lymphoid cells drive interleukin-23-dependent innate intestinal pathology. Nature *464*, 1371–1375.

Chen, Z., Laurence, A., Kanno, Y., Pacher-Zavisin, M., Zhu, B.M., Tato, C., Yoshimura, A., Hennighausen, L., and O'Shea, J.J. (2006). Selective regulatory function of Socs3 in the formation of IL-17-secreting T cells. Proc Natl Acad Sci USA *103*, 8137–8142.

Coquet, J.M., Chakravarti, S., Smyth, M.J., and Godfrey, D.I. (2008). Cutting edge: IL-21 is not essential for Th17 differentiation or experimental auto-immune encephalomyelitis. J Immunol *180*, 7097–7101.

Cua, D.J., Sherlock, J., Chen, Y., Murphy, C.A., Joyce, B., Seymour, B., Lucian, L., To, W., Kwan, S., Churakova, T., *et al.* (2003). Interleukin-23 rather than interleukin-12 is the critical cytokine for auto-immune inflammation of the brain. Nature *421*, 744–748.

Denning, T.L., Wang, Y.C., Patel, S.R., Williams, I.R., and Pulendran, B. (2007). Lamina propria macrophages and dendritic cells differentially induce regulatory and interleukin 17-producing T cell responses. Nat Immunol *8*, 1086–1094.

Diveu, C., McGeachy, M.J., Boniface, K., Stumhofer, J.S., Sathe, M., Joyce-Shaikh, B., Chen, Y., Tato, C.M., McClanahan, T.K., de Waal Malefyt, R., *et al.* (2009). IL-27 blocks RORc expression to inhibit lineage commitment of Th17 cells. J Immunol *182*, 5748–5756.

Duerr, R.H., Taylor, K.D., Brant, S.R., Rioux, J.D., Silverberg, M.S., Daly, M.J., Steinhart, A.H., Abraham, C., Regueiro, M., Griffiths, A., *et al.* (2006). A genome-wide association study identifies IL23R as an inflammatory bowel disease gene. Science *314*, 1461–1463.

Ferber, I.A., Brocke, S., Taylor-Edwards, C., Ridgway, W., Dinisco, C., Steinman, L., Dalton, D., and Fathman, C.G. (1996). Mice with a disrupted IFN-gamma gene are susceptible to the induction of experimental auto-immune encephalomyelitis (EAE). J Immunol *156*, 5–7.

Fitzgerald, D.C., Zhang, G.X., El-Behi, M., Fonseca-Kelly, Z., Li, H., Yu, S., Saris, C.J., Gran, B., Ciric, B., and Rostami, A. (2007). Suppression of auto-immune inflammation of the central nervous system by interleukin 10 secreted by interleukin 27-stimulated T cells. Nat Immunol *8*, 1372–1379.

Hibbert, L., Pflanz, S., De Waal Malefyt, R., and Kastelein, R.A. (2003). IL-27 and IFN-alpha signal via Stat1 and Stat3 and induce T-Bet and IL-12Rbeta2 in naive T cells. J Interferon Cytokine Res *23*, 513–522.

Ho, I.C., Lo, D., Glimcher, L.H. (1998). c-maf promotes T helper cell type 2 (Th2) and attenuates Th1 differentiation by both interleukin 4-dependent and -independent mechanisms. J Exp Med. *188*, 1859–66.

Holscher, C., Holscher, A., Ruckerl, D., Yoshimoto, T., Yoshida, H., Mak, T., Saris, C., and Ehlers, S. (2005). The IL-27 receptor chain WSX-1 differentially regulates antibacterial immunity and survival during experimental tuberculosis. J Immunol *174*, 3534–3544.

Hsieh, C.S., Macatonia, S.E., Tripp, C.S., Wolf, S.F., O'Garra, A., and Murphy, K.M. (1993). Development of TH1 CD4+ T cells through IL-12 produced by Listeria-induced macrophages. Science *260*, 547–549.

Huber, M., Brustle, A., Reinhard, K., Guralnik, A., Walter, G., Mahiny, A., von Low, E., and Lohoff, M. (2008). IRF4 is essential for IL-21-mediated induction, amplification, and stabilization of the Th17 phenotype. Proc Natl Acad Sci USA *105*, 20846–20851.

Ivanov, II, Frutos Rde, L., Manel, N., Yoshinaga, K., Rifkin, D.B., Sartor, R.B., Finlay, B.B., and Littman, D.R. (2008). Specific microbiota direct the differentiation of IL-17-producing T-helper cells in the mucosa of the small intestine. Cell Host Microbe *4*, 337–349.

Ivanov, II, McKenzie, B.S., Zhou, L., Tadokoro, C.E., Lepelley, A., Lafaille, J.J., Cua, D.J., and Littman, D.R. (2006). The orphan nuclear receptor RORgammat directs the differentiation program of pro-inflammatory IL-17+ T helper cells. Cell *126*, 1121–1133.

Kaplan, M.H., Schindler, U., Smiley, S.T., and Grusby, M.J. (1996a). Stat6 is required for mediating responses to IL-4 and for development of Th2 cells. Immunity *4*, 313–319.

Kaplan, M.H., Sun, Y.L., Hoey, T., and Grusby, M.J. (1996b). Impaired IL-12 responses and enhanced development of Th2 cells in Stat4-deficient mice. Nature *382*, 174–177.

Kebir, H., Kreymborg, K., Ifergan, I., Dodelet-Devillers, A., Cayrol, R., Bernard, M., Giuliani, F., Arbour, N., Becher, B., and Prat, A. (2007). Human TH17 lymphocytes promote blood-brain barrier disruption and central nervous system inflammation. Nat Med *13*, 1173–1175.

Khader, S.A., Bell, G.K., Pearl, J.E., Fountain, J.J., Rangel-Moreno, J., Cilley, G.E., Shen, F., Eaton, S.M., Gaffen, S.L., Swain, S.L., *et al.* (2007). IL-23 and IL-17 in the establishment of protective pulmonary CD4+ T cell responses after vaccination and during Mycobacterium tuberculosis challenge. Nat Immunol *8*, 369–377.

Kishimoto, T. (2005). Interleukin-6: from basic science to medicine--40 years in immunology. Annu Rev Immunol *23*, 1–21.

Korn, T., Bettelli, E., Gao, W., Awasthi, A., Jager, A., Strom, T.B., Oukka, M., and Kuchroo, V.K. (2007). IL-21 initiates an alternative pathway to induce pro-inflammatory T(H)17 cells. Nature *448*, 484–487.

Korn, T., Bettelli, E., Oukka, M., and Kuchroo, V.K. (2009). IL-17 and Th17 Cells. Annu Rev Immunol *27*, 485–517.

Korn, T., Mitsdoerffer, M., Croxford, A.L., Awasthi, A., Dardalhon, V.A., Galileos, G., Vollmar, P., Stritesky, G.L., Kaplan, M.H., Waisman, A., *et al.* (2008). IL-6 controls Th17 immunity in vivo by inhibiting the conversion of conventional T cells into Foxp3+ regulatory T cells. Proc Natl Acad Sci USA *105*, 18460–18465.

Kuchroo, V.K., Anderson, A.C., Waldner, H., Munder, M., Bettelli, E., and Nicholson, L.B. (2002). T cell response in experimental auto-immune encephalomyelitis (EAE): role of self and cross-reactive antigens in shaping, tuning, and regulating the autopathogenic T cell repertoire. Annu Rev Immunol *20*, 101–123.

Kuchroo, V.K., Das, M.P., Brown, J.A., Ranger, A.M., Zamvil, S.S., Sobel, R.A., Weiner, H.L., Nabavi, N., and Glimcher, L.H. (1995). B7-1 and B7-2 costimulatory molecules activate differentially the Th1/Th2 developmental pathways: application to auto-immune disease therapy. Cell *80*, 707–718.

Kuchroo, V.K., Sobel, R.A., Laning, J.C., Martin, C.A., Greenfield, E., Dorf, M.E., and Lees, M.B. (1992). Experimental allergic encephalomyelitis mediated by cloned T cells specific for a synthetic peptide of myelin proteolipid protein. Fine specificity and T cell receptor V beta usage. J Immunol *148*, 3776–3782.

Kullberg, M.C., Jankovic, D., Feng, C.G., Hue, S., Gorelick, P.L., McKenzie, B.S., Cua, D.J., Powrie, F., Cheever, A.W., Maloy, K.J., and Sher, A. (2006). IL-23 plays a key role in Helicobacter hepaticus-induced T cell-dependent colitis. J Exp Med *203*, 2485–2494.

Langrish, C.L., Chen, Y., Blumenschein, W.M., Mattson, J., Basham, B., Sedgwick, J.D., McClanahan, T., Kastelein, R.A., and Cua, D.J. (2005). IL-23 drives a pathogenic T cell population that induces auto-immune inflammation. J Exp Med *201*, 233–240.

Lee, Y.K., Turner, H., Maynard, C.L., Oliver, J.R., Chen, D., Elson, C.O., and Weaver, C.T. (2009). Late developmental plasticity in the T helper 17 lineage. Immunity *30*, 92–107.

Leonard, W.J., Zeng, R., and Spolski, R. (2008). Interleukin 21: a cytokine/cytokine receptor system that has come of age. J Leukoc Biol *84*, 348–356.

Lexberg, M.H., Taubner, A., Forster, A., Albrecht, I., Richter, A., Kamradt, T., Radbruch, A., and Chang, H.D. (2008). Th memory for interleukin-17 expression is stable in vivo. Eur J Immunol *38*, 2654–2664.

Li, M.O., and Flavell, R.A. (2008). TGF-beta: a master of all T cell trades. Cell *134*, 392–404.

Li, M.O., Wan, Y.Y., and Flavell, R.A. (2007). T cell-produced transforming growth factor-beta1 controls T cell tolerance and regulates Th1- and Th17-cell differentiation. Immunity *26*, 579–591.

Littman, D.R., and Rudensky, A.Y. (2010). Th17 and regulatory T cells in mediating and restraining inflammation. Cell *140*, 845–858.

Liu, C.H., Fan, Y.T., Dias, A., Esper, L., Corn, R.A., Bafica, A., Machado, F.S., and Aliberti, J. (2006). Cutting edge: dendritic cells are essential for in vivo IL-12 production and development of resistance against Toxoplasma gondii infection in mice. J Immunol *177*, 31–35.

Liu, X., Lee, Y.S., Yu, C.R., and Egwuagu, C.E. (2008). Loss of STAT3 in CD4+ T cells prevents development of experimental auto-immune diseases. J Immunol *180*, 6070–6076.

Lochner, M., Peduto, L., Cherrier, M., Sawa, S., Langa, F., Varona, R., Riethmacher, D., Si-Tahar, M., Di Santo, J.P., and Eberl, G. (2008). In vivo equilibrium of pro-inflammatory IL-17+ and regulatory IL-10+ Foxp3+ RORgamma t+T cells. J Exp Med *205*, 1381–1393.

Ma, C.S., Chew, G.Y., Simpson, N., Priyadarshi, A., Wong, M., Grimbacher, B., Fulcher, D.A., Tangye, S.G., and Cook, M.C. (2008). Deficiency of Th17 cells in hyper IgE syndrome due to mutations in STAT3. J Exp Med *205*, 1551–1557.

Mangan, P.R., Harrington, L.E., O'Quinn, D.B., Helms, W.S., Bullard, D.C., Elson, C.O., Hatton, R.D., Wahl, S.M., Schoeb, T.R., and Weaver, C.T. (2006). Transforming growth factor-beta induces development of the T(H)17 lineage. Nature *441*, 231–234.

Martin-Orozco, N., Chung, Y., Chang, S.H., Wang, Y.H., and Dong, C. (2009). Th17 cells promote pancreatic inflammation but only induce diabetes efficiently in lymphopenic hosts after conversion into Th1 cells. Eur J Immunol *39*, 216–224.

Mathur, A.N., Chang, H.C., Zisoulis, D.G., Stritesky, G.L., Yu, Q., O'Malley, J.T., Kapur, R., Levy, D.E., Kansas, G.S., and Kaplan, M.H. (2007). Stat3 and Stat4 direct development of IL-17-secreting Th cells. J Immunol *178*, 4901–4907.

McGeachy, M.J., Bak-Jensen, K.S., Chen, Y., Tato, C.M., Blumenschein, W., McClanahan, T., and Cua, D.J. (2007). TGF-beta and IL-6 drive the production of IL-17 and IL-10 by T cells and restrain T(H)-17 cell-mediated pathology. Nat Immunol *8*, 1390–1397.

McGeachy, M.J., Chen, Y., Tato, C.M., Laurence, A., Joyce-Shaikh, B., Blumenschein, W.M., McClanahan, T.K., O'Shea, J.J., and Cua, D.J. (2009). The interleukin 23 receptor is essential for the terminal differentiation of interleukin 17-producing effector T helper cells in vivo. Nat Immunol *10*, 314–324.

McGeachy, M.J., and Cua, D.J. (2008). Th17 cell differentiation: the long and winding road. Immunity *28*, 445–453.

Mosmann, T.R., Cherwinski, H., Bond, M.W., Giedlin, M.A., and Coffman, R.L. (1986). Two types of murine helper T cell clone. I. Definition according to profiles of lymphokine activities and secreted proteins. J Immunol *136*, 2348–2357.

Mosmann, T.R., and Coffman, R.L. (1989). TH1 and TH2 cells: different patterns of lymphokine secretion lead to different functional properties. Annu Rev Immunol *7*, 145–173.

Mullen, A.C., High, F.A., Hutchins, A.S., Lee, H.W., Villarino, A.V., Livingston, D.M., Kung, A.L., Cereb, N., Yao, T.P., Yang, S.Y., and Reiner, S.L. (2001). Role of T-bet in commitment of TH1 cells before IL-12-dependent selection. Science *292*, 1907–1910.

Neurath, M.F., Finotto, S., and Glimcher, L.H. (2002a). The role of Th1/Th2 polarization in mucosal immunity. Nat Med *8*, 567–573.

Neurath, M.F., Fuss, I., Kelsall, B.L., Stuber, E., and Strober, W. (1995). Antibodies to interleukin 12 abrogate established experimental colitis in mice. J Exp Med *182*, 1281–1290.

Neurath, M.F., Weigmann, B., Finotto, S., Glickman, J., Nieuwenhuis, E., Iijima, H., Mizoguchi, A., Mizoguchi, E., Mudter, J., Galle, P.R., *et al.* (2002b). The transcription factor T-bet regulates mucosal T cell activation in experimental colitis and Crohn's disease. J Exp Med *195*, 1129–1143.

Nurieva, R., Yang, X.O., Chung, Y., and Dong, C. (2009). Cutting edge: in vitro generated Th17 cells maintain their cytokine expression program in normal but not lymphopenic hosts. J Immunol *182*, 2565–2568.

Nurieva, R., Yang, X.O., Martinez, G., Zhang, Y., Panopoulos, A.D., Ma, L., Schluns, K., Tian, Q., Watowich, S.S., Jetten, A.M., and Dong, C. (2007). Essential autocrine regulation by IL-21 in the generation of inflammatory T cells. Nature *448*, 480–483.

Okuda, Y., Sakoda, S., Bernard, C.C., Fujimura, H., Saeki, Y., Kishimoto, T., and Yanagihara, T. (1998). IL-6-deficient mice are resistant to the induction of experimental auto-immune encephalomyelitis provoked by myelin oligodendrocyte glycoprotein. Int Immunol *10*, 703–708.

Ono, M., Yaguchi, H., Ohkura, N., Kitabayashi, I., Nagamura, Y., Nomura, T., Miyachi, Y., Tsukada, T., and Sakaguchi, S. (2007). Foxp3 controls regulatory T-cell function by interacting with AML1/Runx1. Nature *446*, 685–689.

Oppmann, B., Lesley, R., Blom, B., Timans, J.C., Xu, Y., Hunte, B., Vega, F., Yu, N., Wang, J., Singh, K., *et al.* (2000). Novel p19 protein engages IL-12p40 to form a cytokine, IL-23, with biological activities similar as well as distinct from IL-12. Immunity *13*, 715–725.

Parham, C., Chirica, M., Timans, J., Vaisberg, E., Travis, M., Cheung, J., Pflanz, S., Zhang, R., Singh, K.P., Vega, F., *et al.* (2002). A receptor for the heterodimeric cytokine IL-23 is composed of IL-12Rbeta1 and a novel cytokine receptor sub-unit, IL-23R. J Immunol *168*, 5699–5708.

Pflanz, S., Hibbert, L., Mattson, J., Rosales, R., Vaisberg, E., Bazan, J.F., Phillips, J.H., McClanahan, T.K., de Waal Malefyt, R., and Kastelein, R.A. (2004). WSX-1 and glycoprotein 130 constitute a signal-transducing receptor for IL-27. J Immunol *172*, 2225–2231.

Pflanz, S., Timans, J.C., Cheung, J., Rosales, R., Kanzler, H., Gilbert, J., Hibbert, L., Churakova, T., Travis, M., Vaisberg, E., *et al.* (2002). IL-27, a heterodimeric cytokine composed of EBI3 and p28 protein, induces proliferation of naive CD4(+) T cells. Immunity *16*, 779–790.

Pot, C., Jin, H., Awasthi, A., Liu, S.M., Lai, C.Y., Madan, R., Sharpe, A.H., Karp, C.L., Miaw, S.C., Ho, I.C., and Kuchroo, V.K. (2009). Cutting edge: IL-27 induces the transcription factor c-Maf, cytokine IL-21, and the costimulatory receptor ICOS that coordinately act together to promote differentiation of IL-10-producing Tr1 cells. J Immunol *183*, 797–801.

Qin, H., Wang, L., Feng, T., Elson, C.O., Niyongere, S.A., Lee, S.J., Reynolds, S.L., Weaver, C.T., Roarty, K., Serra, R., *et al.* (2009). TGF-beta promotes Th17 cell development through inhibition of SOCS3. J Immunol *183*, 97–105.

Quintana, F.J., Basso, A.S., Iglesias, A.H., Korn, T., Farez, M.F., Bettelli, E., Caccamo, M., Oukka, M., and Weiner, H.L. (2008). Control of T(reg) and T(H)17 cell differentiation by the aryl hydrocarbon receptor. Nature *453*, 65–71.

Rahman, P., Inman, R.D., Maksymowych, W.P., Reeve, J.P., Peddle, L., and Gladman, D.D. (2009). Association of Interleukin 23 Receptor Variants with Psoriatic Arthritis. J Rheumatol. *36*, 137–140.

Rengarajan, J., Mowen, K.A., McBride, K.D., Smith, E.D., Singh, H., and Glimcher, L.H. (2002). Interferon regulatory factor 4 (IRF4) interacts with NFATc2 to modulate interleukin 4 gene expression. J Exp Med *195*, 1003–1012.

Rosas, L.E., Satoskar, A.A., Roth, K.M., Keiser, T.L., Barbi, J., Hunter, C., de Sauvage, F.J., and Satoskar, A.R. (2006). Interleukin-27R (WSX-1/T-cell cytokine receptor) gene-deficient mice display enhanced resistance to leishmania donovani infection but develop severe liver immunopathology. Am J Pathol *168*, 158–169.

Scharton-Kersten, T., Denkers, E.Y., Gazzinelli, R., and Sher, A. (1995). Role of IL12 in induction of cell-mediated immunity to Toxoplasma gondii. Res Immunol *146*, 539–545.

Scharton-Kersten, T.M., Wynn, T.A., Denkers, E.Y., Bala, S., Grunvald, E., Hieny, S., Gazzinelli, R.T., and Sher, A. (1996). In the absence of endogenous IFN-gamma, mice develop unimpaired IL-12 responses to Toxoplasma gondii while failing to control acute infection. J Immunol *157*, 4045–4054.

Spolski, R., Kashyap, M., Robinson, C., Yu, Z., and Leonard, W.J. (2008). IL-21 signaling is critical for the development of type I diabetes in the NOD mouse. Proc Natl Acad Sci USA *105*, 14028–14033.

Spolski, R., Kim, H.P., Zhu, W., Levy, D.E., and Leonard, W.J. (2009). IL-21 mediates suppressive effects via its induction of IL-10. J Immunol *182*, 2859–2867.

Stumhofer, J.S., Laurence, A., Wilson, E.H., Huang, E., Tato, C.M., Johnson, L.M., Villarino, A.V., Huang, Q., Yoshimura, A., Sehy, D., *et al.* (2006). Interleukin 27 negatively regulates the development of interleukin 17-producing T helper cells during chronic inflammation of the central nervous system. Nat Immunol *7*, 937–945.

Stumhofer, J.S., Silver, J.S., Laurence, A., Porrett, P.M., Harris, T.H., Turka, L.A., Ernst, M., Saris, C.J., O'Shea, J.J., and Hunter, C.A. (2007). Interleukins 27 and 6 induce STAT3-mediated T cell production of interleukin 10. Nat Immunol *8*, 1363–1371.

Sutherland, A.P., Van Belle, T., Wurster, A.L., Suto, A., Michaud, M., Zhang, D., Grusby, M.J., and von Herrath, M. (2009). Interleukin-21 is required for the development of type 1 diabetes in NOD mice. Diabetes *58*, 1144–1155.

Szabo, S.J., Kim, S.T., Costa, G.L., Zhang, X., Fathman, C.G., and Glimcher, L.H. (2000). A novel transcription factor, T-bet, directs Th1 lineage commitment. Cell *100*, 655–669.

Szabo, S.J., Sullivan, B.M., Peng, S.L., and Glimcher, L.H. (2003). Molecular mechanisms regulating Th1 immune responses. Annu Rev Immunol *21*, 713–758.

Takeda, A., Hamano, S., Yamanaka, A., Hanada, T., Ishibashi, T., Mak, T.W., Yoshimura, A., and Yoshida, H. (2003). Cutting edge: role of IL-27/WSX-1 signaling for induction of T-bet through activation of STAT1 during initial Th1 commitment. J Immunol *170*, 4886–4890.

Takeda, K., Tanaka, T., Shi, W., Matsumoto, M., Minami, M., Kashiwamura, S., Nakanishi, K., Yoshida, N., Kishimoto, T., and Akira, S. (1996). Essential role of Stat6 in IL-4 signalling. Nature *380*, 627–630.

Taleb, S., Romain, M., Ramkhelawon, B., Uyttenhove, C., Pasterkamp, G., Herbin, O., Esposito, B., Perez, N., Yasukawa, H., Van Snick, J., *et al.* (2009). Loss of SOCS3 expression in T cells reveals a regulatory role for interleukin-17 in atherosclerosis. J Exp Med *206*, 2067–2077.

Travis, M.A., Reizis, B., Melton, A.C., Masteller, E., Tang, Q., Proctor, J.M., Wang, Y., Bernstein, X., Huang, X., Reichardt, L.F., *et al.* (2007). Loss of integrin alpha(v)beta8 on dendritic cells causes auto-immunity and colitis in mice. Nature *449*, 361–365.

Veldhoen, M., Hirota, K., Westendorf, A.M., Buer, J., Dumoutier, L., Renauld, J.C., and Stockinger, B. (2008). The aryl hydrocarbon receptor links TH17-cell-mediated auto-immunity to environmental toxins. Nature *453*, 106–109.

Veldhoen, M., Hocking, R.J., Atkins, C.J., Locksley, R.M., and Stockinger, B. (2006a). TGFbeta in the context of an inflammatory cytokine milieu supports de novo differentiation of IL-17-producing T cells. Immunity *24*, 179–189.

Veldhoen, M., Hocking, R.J., Flavell, R.A., and Stockinger, B. (2006b). Signals mediated by transforming growth factor-beta initiate auto-immune encephalomyelitis, but chronic inflammation is needed to sustain disease. Nat Immunol *7*, 1151–1156.

Villarino, A., Hibbert, L., Lieberman, L., Wilson, E., Mak, T., Yoshida, H., Kastelein, R.A., Saris, C., and Hunter, C.A. (2003). The IL-27R (WSX-1) is required to suppress T cell hyperactivity during infection. Immunity *19*, 645–655.

Wirtz, S., Finotto, S., Kanzler, S., Lohse, A.W., Blessing, M., Lehr, H.A., Galle, P.R., and Neurath, M.F. (1999). Cutting edge: chronic intestinal inflammation in STAT-4 transgenic mice: characterization of disease and adoptive transfer by TNF- plus IFN-gamma-producing CD4+ T cells that respond to bacterial antigens. J Immunol *162*, 1884–1888.

Wu, S., Rhee, K.J., Albesiano, E., Rabizadeh, S., Wu, X., Yen, H.R., Huso, D.L., Brancati, F.L., Wick, E., McAllister, F., *et al.* (2009). A human colonic commensal promotes colon tumorigenesis via activation of T helper type 17 T cell responses. Nat Med *15*, 1016–1022.

Xiao, S., Jin, H., Korn, T., Liu, S.M., Oukka, M., Lim, B., and Kuchroo, V.K. (2008). Retinoic acid increases Foxp3+ regulatory T cells and inhibits development of Th17 cells by enhancing TGF-beta-driven Smad3 signaling and inhibiting IL-6 and IL-23 receptor expression. J Immunol *181*, 2277–2284.

Yang, X.O., Pappu, B.P., Nurieva, R., Akimzhanov, A., Kang, H.S., Chung, Y., Ma, L., Shah, B., Panopoulos, A.D., Schluns, K.S., *et al.* (2008). T helper 17 lineage differentiation is programmed by orphan nuclear receptors ROR alpha and ROR gamma. Immunity *28*, 29–39.

Yen, D., Cheung, J., Scheerens, H., Poulet, F., McClanahan, T., McKenzie, B., Kleinschek, M.A., Owyang, A., Mattson, J., Blumenschein, W., *et al.* (2006). IL-23 is essential for T cell-mediated colitis and promotes inflammation via IL-17 and IL-6. J Clin Invest *116*, 1310–1316.

Yoshimoto, T., Yasuda, K., Mizuguchi, J., and Nakanishi, K. (2007). IL-27 suppresses Th2 cell development and Th2 cytokines production from polarized Th2 cells: a novel therapeutic way for Th2-mediated allergic inflammation. J Immunol *179*, 4415–4423.

Zhang, F., Meng, G., and Strober, W. (2008). Interactions among the transcription factors Runx1, RORgammat and Foxp3 regulate the differentiation of interleukin 17-producing T cells. Nat Immunol *9*, 1297–1306.

Zheng, W., and Flavell, R.A. (1997). The transcription factor GATA-3 is necessary and sufficient for Th2 cytokine gene expression in CD4 T cells. Cell *89*, 587–596.

Zhou, L., Ivanov, II, Spolski, R., Min, R., Shenderov, K., Egawa, T., Levy, D.E., Leonard, W.J., and Littman, D.R. (2007). IL-6 programs T(H)-17 cell differentiation by promoting sequential engagement of the IL-21 and IL-23 pathways. Nat Immunol *8*, 967–974.

Zhou, L., Lopes, J.E., Chong, M.M., Ivanov, II, Min, R., Victora, G.D., Shen, Y., Du, J., Rubtsov, Y.P., Rudensky, A.Y., *et al.* (2008). TGF-beta-induced Foxp3 inhibits T(H)17 cell differentiation by antagonizing RORgammat function. Nature *453*, 236–240.

Zhu, J., Min, B., Hu-Li, J., Watson, C.J., Grinberg, A., Wang, Q., Killeen, N., Urban, J.F., Jr., Guo, L., and Paul, W.E. (2004). Conditional deletion of Gata3 shows its essential function in T(H)1-T(H)2 responses. Nat Immunol *5*, 1157–1165.

T_H17 Cytokines: Characteristics, Regulation, and Biological Function

Pornpimon Angkasekwinai and Chen Dong

Abstract The recently identified Interleukin 17 (IL-17) cytokine family, which comprises of six members, contributes to immunity in infectious and chronic inflammatory diseases. IL-17 and the most structurally related cytokine, IL-17F, are produced by T_H17 cells, a novel subset of CD4+ helper T cells. Although IL-17 and IL-17F have similar regulation and functions in vitro, they are involved differently in chronic inflammatory diseases and bacterial infection. The aim of this article is to summarize recent work in understanding the function and regulation of T_H17 cytokines, focusing mainly on IL-17 and IL-17F. Elucidation of the function and regulation of these cytokines may yield immuno-therapeutic strategies for the prevention and treatment of inflammatory diseases.

1 Introduction

Cytokines are key players in the regulation of the development and function of immune cells. Several cytokines such as TNF-α, IL-1β and IL-6 are recognized as pro-inflammatory cytokines with functions in promoting inflammation, while others such as IL-27 and IL-10 are involved in inhibiting inflammation. In recent years, extensive attention has been drawn to the recently identified IL-17 cytokine family, especially the founding member IL-17 that becomes an important pro-inflammatory cytokine in auto-immune diseases and in immunity to certain bacterial and fungal infections. The IL-17 cytokine family is comprised of six members, including IL-17 (also called IL-17A), IL-17B, IL-17C, IL-17D, IL-17E (also called IL-25), and

C. Dong (✉)
Department of Immunology, University of Texas and MD Anderson Cancer Center, Houston, TX 77030, USA
e-mail: cdong@mdanderson.org

S. Jiang (ed.), *T_H17 Cells in Health and Disease*,
DOI 10.1007/978-1-4419-9371-7_2,

IL-17F. Among the IL-17 family members, IL-17 shares the highest homology to IL-17F, and IL-25 is the least related; they are the most investigated cytokines in this family. We will discuss the structure, function and regulation of IL-17 and IL-17F in this article.

2 Structure of IL-17 Family Cytokines

IL-17 was initially cloned by Rouvier et al. by subtractive hybridization of activated T cell-specific library and was originally named CTLA-8 (cytotoxic T lymphocyte associated antigen 8) (Rouvier et al. 1993; Yao et al. 1995a). Murine IL-17 displayed 57% identity in amino acid sequence to the ORF13 gene of T lymphotropic herpes virus, *Herpesvirus saimiri* (Rouvier et al. 1993). Subsequently, the human counterpart of murine IL-17 was cloned based on alignment of nucleotide sequences of HSV13 and mIL-17 (Fossiez et al. 1996; Yao et al. 1995b). The IL-17 gene is located on mouse chromosome 1A and human chromosome 2q31. Sequence homology searches for IL-17-related cytokines led to the identification of five additional family members, IL-17B, IL-17C, IL-17D, IL-25, and IL-17F (Fort et al. 2001; Hymowitz et al. 2001; Lee et al. 2001; Li et al. 2000a, b; Shi et al. 2000; Starnes et al. 2001, 2002). All members of the IL-17 family have a similar protein structure with considerable sequence divergence at the N termini. They share four highly conserved cysteine residues, all of which have been shown to form a cystine knot in the crystal structure of IL-17F (Hymowitz et al. 2001). IL-17 family cytokine is therefore recognized as a member of the cystine knot fold superfamily and dimerizes similarly to members of the NGF subfamily. In both human and mouse, IL-17A and IL-17F are closely related, with approximately 50% sequence identity (Hymowitz et al. 2001; Starnes et al. 2001). Structural features of IL-17 suggest that each of members is likely produced as a homodimer (Fossiez et al. 1996). However, our group and others reported that IL-17A and IL-17F were not only secreted as homodimeric proteins but can also form heterodimers in both human and mouse (Chang and Dong 2007; Liang et al. 2007; Wright et al. 2007). Compared with the IL-17A and IL-17F homodimer, the IL-17A and IL-17F heterodimer has intermediate biological activity (Chang and Dong 2007; Liang et al. 2007; Wright et al. 2007). Gene encoding human IL-17F is located adjacent to IL-17A and transcribed in opposite direction, suggesting that both cytokine genes may have been derived during evolution through gene duplication and thus may share same regulatory elements. Multiple non-coding sequences within the IL-17 and IL-17F locus were found to be conserved across species (Akimzhanov et al. 2007). These elements were associated with acetylated histone 3 in a lineage-specific manner and may serve as potential regulatory regions (Akimzhanov et al. 2007).

IL-25 diverges significantly from IL-17A and F with less than 17% homology to IL-17 (Lee et al. 2001). IL-25 cDNA predicts a secreted protein with 177 amino acids with overall conserved characteristics of the IL-17 family, including N-linked glycosylation site and conserved cystine residues (Lee et al. 2001). Several lines of evidence so far indicate that IL-25 possesses different activities compared to other IL-17 family cytokines.

3 Expression and Regulation of IL-17 and IL-17F

3.1 T_H17 Cells as a Novel Effector T_H Lineage

IL-17 was originally found to be mainly expressed by an activated CD4+ T cells, predominantly in memory subset, but not by resting T cells (Yao et al. 1995a). The expression of IL-17 is increased in many chronic inflammatory diseases such as rheumatoid arthritis and multiple sclerosis (Aarvak et al. 1999; Albanesi et al. 2000; Lenarczyk et al. 2000). Naïve T helper cells after activation can differentiate into T_H1 and T_H2 lineage depending on exogenous cytokine IL-12 and IL-4, respectively (Mosmann et al. 1986). T_H1 and T_H2 cells secrete specific cytokines and function differently. Earlier characterization does not provide clear classification of IL-17 in neither T_H1 nor T_H2 subsets (Aarvak et al. 1999; Albanesi et al. 2000; Lenarczyk et al. 2000).

Co-stimulatory molecules are important in determining T helper cell differentiation. Unexpectedly, our initial characterization of mice lacking inducible co-stimulatory molecules (ICOS) showed reduced IL-17 expression in these mice (Dong and Nurieva 2003). The reduction of IL-17 but not IFN-γ production in ICOS-deficient mice was found to be associated with their resistance to collagen-induced arthritis (CIA) (Dong and Nurieva 2003). Moreover, IL-23-deficient mice showed reduced IL-17 expression and alleviated experimental auto-immune encephalomyelitis (EAE) symptoms (Cua et al. 2003). Further studies indicated that IL-23 is critical for expanding pathogenic IL-17-producing T cells that are distinct from T_H1 or T_H2 (Langrish et al. 2005; Murphy et al. 2003). Later, Park et al. and Harrington et al. provided more conclusive evidence indicating that IL-17 was indeed produced by a distinct cell lineage (Harrington et al. 2005; Park et al. 2005). The T_H17 lineage is identified to be a third lineage that plays prominent roles in regulating tissue inflammation. Antigen-specific IL-17-producing cells were shown to be efficiently generated in the absence of T_H1 and T_H2 cell development (Park et al. 2005). The generation of IL-17-producing T cells did not require a master regulator for T_H1 or T_H2, including T-bet, GATA3, STAT1, STAT4, and STAT6 (Harrington et al. 2005; Park et al. 2005). Thus, these studies provide direct evidence that IL-17 producing T cells are a novel T lineage that possess different transcriptional program.

Further studies have demonstrated that several other cell types, including γδ-T cells, NKT cells, NK cells, Paneth cells, and neutrophils are capable of producing IL-17 (Li et al. 2010; Lockhart et al. 2006; Michel et al. 2007; Takahashi et al. 2008). It shows that IL-17 can participate in both innate and the adaptive immune response. IL-17 and IL-17F produced by innate immune cells may contribute to early stage immunity following infection but T_H17 cells are the major cell type producing both cytokines during chronic inflammatory diseases.

3.2 *The Regulation of T_H17 Cells*

3.2.1 Cytokine Regulation of T_H17 Differentiation

Cytokines are important in determining the fate of Th lineage differentiation. IL-12 and IL-4 are critical for driving T_H1 and T_H2 differentiation, respectively (Mosmann et al. 1986). Studies by three independent groups found that a combination of immunoregulatory cytokine, TGF-β and pro-inflammatory cytokine, IL-6 potently induced the differentiation of T_H17 cells from naïve T cells both in vitro and in vivo (Bettelli et al. 2006; Mangan et al. 2006; Veldhoen et al. 2006a). IL-6 was found to be critical for the inhibition of Foxp3 expression induced by TGF-β, while promoted the differentiation of IL-17-producing T cells (Bettelli et al. 2006; Mangan et al. 2006; Veldhoen et al. 2006a). Furthermore, mice lacking a functional receptor for TGF-β or mice with a deleted TGF-β only in T cells had defect in T_H17 generation and were resistant to EAE (Li et al. 2007; Veldhoen et al. 2006b). These data indicate the critical roles of both cytokines in regulating the differentiation of T_H17 lineage. IFN-γ and IL-4, produced by T_H1 and T_H2 cells, respectively, auto-amplify the generation of its own lineage; IL-17 could not act as a growth factor of T_H17 cells. We and others identified that T_H17 cells produced an autocrine IL-21 (Korn et al. 2007; Nurieva et al. 2007; Zhou et al. 2007). IL-21 produced from T_H17 cells can amplify its own expression, and function in promoting and sustaining the differentiation of T_H17. IL-21 was induced by IL-6 through STAT3; both TGF-β together with IL-6 or IL-21 suppressed FOXP3 expression and induced the differentiation of IL-17 (Korn et al. 2007; Nurieva et al. 2007; Zhou et al. 2007).

During T_H17 differentiation initiated by TGF-β and IL-6, pro-inflammatory cytokines such as TNF-α and IL-1β seem to play important roles in enhancing the production of IL-17 (Veldhoen et al. 2006a). IL-1 was found to be critical for the induction of T_H17 cells in humans (Acosta-Rodriguez et al. 2007). More recently, the role of IL-1 in directly regulating T_H17 cells was reported (Ben-Sasson et al. 2009; Chung et al. 2009). IL-1R1 was upregulated in T_H17 cells and IL-6 appeared to be the most important inducer for IL-1R1 expression (Chung et al. 2009). Further characterization by using mixed bone-marrow (BM) chimeras of CD45.1+ WT and CD45.2+ IL-1R1-KO BM cells indicated that IL-1 signaling in CD4+ T cells is necessary to promote the early T_H17 differentiation and for proper T_H17 cell differentiation in vivo (Chung et al. 2009). Moreover, IL-1 appeared to play important roles in maintaining differentiated T_H17 cells even without TCR stimulation (Chung et al. 2009). Similarly, IL-1R1 expression on CD4+ T cells in humans was also shown to be important for both initiation and maintenance of human T_H17 cells (Lee et al. 2010).

TNF receptors expressed on T cells play distinct roles in co-stimulation. We showed that a tumor necrosis factor receptor family member, death receptor 3 (DR3; also known as TNFRSF25), is selectively elevated in T_H17 cells (Pappu et al. 2008). Indeed, TL1A, its cognate receptor was required for the optimal differentiation as well as effector function of T_H17 cells (Pappu et al. 2008).

Besides the mentioned positive regulators, some cytokines can inhibit the differentiation of T_H17 lineages. IL-27 was identified to have an antagonizing effect against IL-17 expression (Stumhofer et al. 2006). The inhibitory activity of IL-27 was dependent on the intracellular signaling molecule STAT1 (Stumhofer et al. 2006). Mice deficient in IL-27 showed increased susceptibility to EAE and generated more IL-17 producing T cells (Stumhofer et al. 2006). Another well-described negative regulator for T_H17 differentiation is IL-2 (Laurence et al. 2007). IL-2 inhibits the production of IL-17 through STAT5 and its deficiency resulted in enhanced IL-17 production (Laurence et al. 2007). In addition, IFN-β was recently found to suppress T_H17 generation through STAT1 activation (Guo et al. 2008). Thus, therapeutics intervention by applying these inhibitory cytokines is important in treating chronic inflammatory diseases mediated by T_H17 cells.

3.2.2 Transcriptional Regulation of T_H17 Differentiation

Members of the signal transducer and activator of transcription (STAT) gene family, which are important in cytokine receptor signaling pathway, play crucial roles in determining T_H lineage commitment. While STAT1 appears to inhibit T_H17 differentiation, STAT4 and 6, which are critical for T_H1 and T_H2 differentiation, are not involved (Harrington et al. 2005; Park et al. 2005). It has been shown that Socs3 (suppressor of cytokine signaling 3), a negative regulator of STAT3 was found to negatively regulate the expression of IL-17 and its deficiency enhanced IL-17 expression (Chen et al. 2006; Pappu et al. 2008; Wong et al. 2006). We subsequently found that STAT3 is indeed a critical factor for T_H17 differentiation (Yang et al. 2007). Overexpression of a hyperactive form of STAT3 promoted the expression of IL-17 and IL-17F, whereas STAT3-deficient T cells showed greatly reduced expression of both cytokines (Yang et al. 2007). Following the identification of cytokines mediating T_H17 differentiation, STAT3 was found to be activated by IL-6, IL-21, and IL-23 in regulating IL-17 expression (Nurieva et al. 2007; Yang et al. 2007).

In a DNA microarray analysis, retinoic-acid-related orphan receptor RORγt was identified as a first T_H17-lineage specific transcription factor (Ivanov et al. 2006). RORγt is selectively expressed in T_H17 cells generated in vitro and in a subset of lamina propria IL-17$^+$ T cells (Ivanov et al. 2006). RORγt$^{-/-}$ mice had impaired T_H17 differentiation and reduced EAE (Ivanov et al. 2006). In these mice, however, T_H17 cytokine expression was not completely abolished. A subsequent study showed us that another member of the retinoid nuclear receptor family, RORα plays a role in T_H17 differentiation and commitment (Yang et al. 2008c). T_H17 cells selectively expressed not only RORγt, but also RORα (Pappu et al. 2008). Co-expression of RORα and RORγt synergistically regulates T_H17 differentiation (Yang et al. 2008a, b, c). Furthermore, mice with T cells deficient in both transcription factors exhibited completely abolished T_H17 generation and were completely protected from EAE (Yang et al. 2008c). Thus, the expression of both RORα and RORγt are essential for the determination of T_H17 cells. Treatment with TGF-β and IL-6 or IL-21 induced the expression of RORγt and RORα, while treatment solely with TGF-β drove the generation of FOXP3$^+$ T cells, suggesting a reciprocal relationship between T_H17 and regulatory T cell development. By utilizing an RFP-IL-17F reporter mouse that we generated together with GFP-FOXP3 reporter mouse, we found the presence of a RFP$^+$GFP$^+$ transient phase upon T cell activation in vitro and in vivo (Yang et al. 2008b). Induction of Foxp3 by TGF-β inhibited T_H17 differentiation by antagonizing the function of RORγt and RORα, possibly by blocking their binding of a co-activator (Yang et al. 2008b). In contrast, IL-6 overcame this suppressive effect of Foxp3 and, together with IL-1, induced genetic reprogramming in Foxp3$^+$ regulatory T cells (Yang et al. 2008b). Consistent with our data, Zhou et al. demonstrated that RORγt and Foxp3 are co-expressed in TGF-β-treated naive CD4+T cells and in a subset of T cells in the lamina propria of the mouse (Zhou et al. 2008). Adding increased concentration of TGF-β can augment Foxp3 expression and reduced IL-23R expression (Zhou et al. 2008). Thus, cytokine-regulated balance of RORγt determines the decision of regulatory T cell and T_H17 cell development.

Besides the transcription factors mentioned above, several other factors are shown to regulate T_H17 cells. The aryl hydrocarbon receptor (AHR), a ligand-dependent transcription factor known for mediating the toxicity of dioxin, was selectively expressed in T_H17 cells (Quintana et al. 2008; Veldhoen et al. 2008). AHR ligation induced the production of the T_H17 cytokine IL-22 and caused accelerated onset of EAE (Veldhoen et al. 2008). Furthermore, the interferon regulatory factor-4 (IRF4) was identified as an important transcription factor necessary for T_H17 lineage differentiation (Brustle et al. 2007). The deficiency of IRF4 was associated with the decreased RORγt expression and increased FoxP3 expression that may negatively impact T_H17 differentiation (Brustle et al. 2007). In addition, it has been shown that Runt-related transcription factor 1 (Runx1) also regulates the expression of IL-17 through a complex that can affect RORγt inhibition by FOXP3 (Zhang et al. 2008). More recently, a basic leucine zipper transcription factor, ATF-like (BATF) belonging to the AP-1 protein family, was identified to be essential for the T_H17 differentiation (Schraml et al. 2009). BATF was highly expressed in T_H1, T_H2, and T_H17 cells compared to naïve T cells. BATF-deficient T cells displayed normal T_H1 and T_H2 differentiation but had a defect in T_H17 generation (Schraml et al. 2009). In contrast, overexpression of BATF in T cells showed increased IL-17 production in both CD4+ and CD8+ T cells. Further in vivo analysis indicated that mice deficient in BATF were completely resistant to EAE (Schraml et al. 2009). Although the production of IL-21 and the expression of RORγt was found to be reduced in BATF-deficient mice, addition of IL-21 and overexpression of RORγt failed to fully restore T_H17 development in BATF$^{-/-}$ T cells, suggesting that BATF may possess distinct interaction with T_H17 specific factors in regulating IL-17 production. Several genes were identified to regulate T_H17 development in a BATF-dependent manner, including RORγt, RORα, the aryl hydrocarbon receptor, IL-22, and IL-17A (Schraml et al. 2009). BATF binds conserved intergenic elements in the IL-17A/F locus and to the IL-17, IL-21, and IL-22 promoters (Schraml et al. 2009). Analysis of the composition of the BATF-containing complex using supershift analysis indicated that BATF preferentially heterodimerized with JunB during T_H17 differentiation (Schraml et al. 2009).

Numerous transcription factors have been identified to regulate the expression of IL-17 and IL-17F, but the mechanisms by which these factors function to establish T_H17 gene expression programs remain unclear. In our study, IL-17 and IL-17F gene promoters undergo lineage-specific chromatin remodeling, providing insights into the regulation of T_H17 differentiation at epigenetic level (Akimzhanov et al. 2007). Several non-coding conserved sites are identified in the IL-17-IL-17F locus and can undergo coordinated chromatin modifications such as histone acetylation and methylation in differentiating T_H17 cells (Akimzhanov et al. 2007). Further analysis of epigenetic changes in the IL-17 and IL-17F locus and the DNA binding specificity will provide better understanding in the regulation of T_H17 lineage differentiation.

3.3 Biological Function of IL-17 and IL-17F and Its Receptors

IL-17 and IL-17F are associated with several immune regulatory functions. Most notably, they are involved in the inflammatory process during infection and in autoimmune diseases. Non-immune cells such as fibroblasts and epithelial cells or hematopoietic cells such as macrophages are known targets for IL-17, which upregulate the expression of many pro-inflammatory cytokines and chemokines in response to IL-17 treatment. As a result, IL-17 mediates the recruitment of neutrophils and macrophages into non-lymphoid tissues. In a microarray analysis on IL-17-treated fibroblasts, we found upregulation of several cytokines including CXCL1 (Gro1), CCL2, CCL7, CCL20, and matrix metalloproteinases (MMP) 3 and 13 (Park et al. 2005). Treatment with IL-17 in lung epithelial cell line resulted in similar up-regulation of above genes (Park et al. 2005). IL-17 blockade led to the reduced severity of EAE associated with decreased expression of several chemokines, while overexpression of IL-17 in the lung epithelial cells caused airway inflammation with the induction of several chemokine genes (Park et al. 2005).

Besides infection, auto-immune diseases, and asthma, IL-17 also participates in tumor immunity. IL-17 was found to be over-expressed in several types of tumors (Kato et al. 2001; Numasaki et al. 2003; Sfanos et al. 2008). Whether this cytokine functions in promoting or inhibiting tumor progression remains controversial. In lymphopenic environment, T_H17 cells were found to mediate protection against skin melanoma after conversion to T_H1 cells (Muranski et al. 2008). By using IL-17-deficient mice, we found that IL-17 is involved in tumor protection (Martin-Orozco et al. 2009). IL-17-deficient mice were more susceptible to B16-F10 melanoma development in the lung that was associated with reduction of CCL20 and CCL2 expression in lungs (Martin-Orozco et al. 2009). Transferring T_H17 cells into tumor-bearing mice elicited the greatest infiltration of granulocytes, macrophages, and DC, whereas T_H1 cell treatment showed slightly increased DC but reduced macrophage numbers. T_H17 cells but not T_H1 cells can induce lung cells to produce CCL2 and CCL20, resulting in DC and activated T cell recruitment, associated with more effective anti-tumor responses (Martin-Orozco et al. 2009). In addition, T_H17 cells can promote

$CD8^+$ T cell proliferation, sustain their cytokine expression, and activate the endogenous anti-tumor $CD8^+$ cells at the tumor sites (Martin-Orozco et al. 2009).

Because IL-17 and IL-17F share the strongest homology, there is a considerable overlap in their biological functions. IL-17F also can induce the production of anti-microbial peptides (defensins), cytokines (IL-6, G-CSF, GM-CSF), and chemokines (CXCL1, CXCL2, CXCL5), as well as enhance granulopoiesis and neutrophil recruitment (Kawaguchi et al. 2004; Kolls and Linden 2004), although its activity seem to be less potent than that of IL-17. Overexpression of IL-17F in the lungs resulted in increased pro-inflammatory cytokine and chemokine expression, and airway inflammation predominantly infiltrated with neutrophil and macrophage (Oda et al. 2005; Yang et al. 2008a). Similar to IL-17, IL-17F had a synergistic effect with TNF-α on enhancing the expression of pro-inflammatory cytokines (Fossiez et al. 1996). As previously mentioned, IL-17 and IL-17F can be secreted as homodimeric protein (IL-17A/A, IL-17F/F) and heterodimeric protein (IL-17A/F). Fully differentiated T_H17 cells secreted IL-17A/F and homodimeric IL-17A and IL-17F (Chang and Dong 2007). However, the potency of heterodimeric cytokine on IL-6 expression was shown to be intermediate (Chang and Dong 2007).

Although IL-17 and IL-17F share the strongest homology and overlap functions, several reports suggest their distinct function in certain cases (Ishigame et al. 2009; Yang et al. 2008a). We generated IL-17- and IL-17F-deficient mice to compare their biological functions. They seem to have distinct functions in the development of inflammatory responses in EAE and asthma. IL-17, but not IL-17F, was required for the initiation of EAE, and while IL-17 contributed positively, IL-17F had a negative effect in allergic asthma (Chang et al. 2008). IL-17$F^{-/-}$ mice showed greater T_H2 cytokine expression and enhanced eosinophil function (Yang et al. 2008a). In a mouse model of dextran sulfurate sodium (DSS)-induced acute colitis, IL-17 was shown to play protective role while IL-17F exacerbated inflammation (Yang et al. 2008a). Studies by Ishigame et al. using a similar approach suggest that both cytokines play distinct functions in immune responses against bacterial infection (Ishigame et al. 2009). They showed that IL-17 played a major role in T cell-dependent auto-immune, but IL-17F only marginally contributed to these responses. Both IL-17A and IL-17F are critically important to protect the mice against mucocutaneous *S. aureus* infections, even though cellular source of both cytokines seem to be different (Ishigame et al. 2009). IL-17A was produced mainly in T cells, whereas IL-17F was produced in T cells, innate immune cells, and epithelial cells (Ishigame et al. 2009).

The reason why IL-17 and IL-17F possess different functions may be explained by their receptors. Both cytokines function through the receptors that belong to the IL-17 receptor family. The IL-17R family consists of 5 members, IL-17 receptor A (IL-17RA or IL-17R), IL-17 receptor B (IL-17RB or IL-17BR), IL-17 receptor C (IL-17RC), IL-17 receptor D (IL-17RD), and IL-17 receptor E (IL-17RE) (Haudenschild et al. 2002; Moseley et al. 2003; Shi et al. 2000; Tian et al. 2000; Yao et al. 1995a, 1997). IL-17, IL-17F and IL-17A/F heterodimer mediates their function through a heterodimeric complex of IL-17RA and IL-17RC (Kuestner et al. 2007; Toy et al. 2006; Wright et al. 2008). Lack of either IL-17RA or IL-17RC completely abrogates the inflammatory function of IL-17 and IL-17F. However, the binding affinities of IL-17A and IL-17F for these receptors are different (Kuestner et al. 2007; Toy et al. 2006; Wright et al. 2008) and different activities may affect cytokine activities. In humans, IL-17 activity can be inhibited by IL-17RA, while IL-17F is inhibited by IL-17RC, and a combination of soluble IL-17RA/IL-17RC receptors is required for inhibition of the IL-17F/IL-17A activity (Wright et al. 2008). In our study, several isoforms of these receptors were detected, suggesting a large number of spice variants in transcripts encoding receptors. Additionally, the distribution of IL-17RA and IL-17RC in tissues seems to be different. IL-17RA mRNA was highly expressed in lymphoid tissues, while IL-17RC mRNA was expressed at high amounts in non-hematopoietic tissues as the colon, small intestine, and lung (Ishigame et al. 2009). T cells expressed IL-17RA but not IL-17RC (Ishigame et al. 2009). Thus, the differential expression and their alternative spices may contribute to different roles of IL-17 and IL-17F.

Early studies showed that IL-17 and IL-17F induced inflammatory cytokines in mouse embryonic fibroblasts (MEFs) through the activation of NF-κB and MAP kinase pathways (Awane et al. 1999; Shalom-Barak et al. 1998). Further studies showed that a tumor necrosis factor receptor-associated factor (TRAF6), an E3 ubiquitin ligase, was required for this activity (Schwandner et al. 2000). Because no TRAF6 binding domain was found in IL-17 receptor, the existence of other adaptors was proposed. All members of the IL-17 receptor family contain a conserved sequence segment that shares similar residues to the conserved motifs of Toll-like receptors (TIR)/IL-1R domain. This domain was named as STIR SEFIR (similar expression to fibroblast growth factor genes and IL-17Rs) and TIR (Novatchkova et al. 2003). The SEFIR domain is also observed in one cytoplasmic protein named Act1 (known as CIKS) with a connection to IκB kinase and stress-activated kinase (Leonardi et al. 2000; Li et al. 2000a, b). Further characterization by our group showed that Act1 physically associates with IL-17RA through the SEFIR domain (Chang et al. 2006). The deficiency of Act1 resulted in the defect of IL-17-mediated function in fibroblast (Chang et al. 2006). IL-17 does not utilize MyD88 and IRAK4 for cytokine induction (Chang et al. 2006). A follow-up study confirmed that Act1 is essential in IL-17-dependent signaling in auto-immune and inflammatory disease (Qian et al. 2007). Because both IL-17RA and IL-17RC are required for IL-17 function, it remains elusive whether Act1 and TRAF6 were recruited through IL-17RA or IL-17RC or both in mediating IL-17 and IL-17F function.

4 Conclusion

In summary, extensive analyses of the IL-17 cytokine family have revealed crucial roles of individual IL-17 family members in immune regulation of infectious and inflammatory diseases. A novel identified T_H17 effector subset of T cells that expresses both IL-17 and IL-17F appears as a central regulator for auto-immune diseases and host defenses to bacterial and fungal infection. These two cytokines may function as homodimeric or heterodimeric secreted proteins, which exert similar activities. Recent characterizations of mice lacking either IL-17 or IL-17F reveal distinct function of individual cytokines. They utilize similar downstream signals but cell-specific receptors and isoforms of their receptors may contribute to distinct function. Further studies on the regulation and function of this important cytokine family may provide better understanding on the roles of the IL-17 family in immune-mediated diseases and such knowledge may lead to the development of immuno-therapeutic strategies for treatment of several inflammatory diseases.

Acknowledgement Our work is supported by research grants from the National Institutes of Health and Leukemia and Lymphoma Society (NIH; to C.D.). C.D. is a Leukemia & Lymphoma Society Scholar and a Trust Fellow and Olga and Harry Wiess Distinguished University Chair in Cancer Research of the MD Anderson Cancer Center.

References

Aarvak, T., Chabaud, M., Miossec, P., & Natvig, J. B. (1999). IL-17 is produced by some pro-inflammatory Th1/Th0 cells but not by Th2 cells. *J Immunol, 162*(3), 1246–1251.

Acosta-Rodriguez, E. V., Napolitani, G., Lanzavecchia, A., & Sallusto, F. (2007). Interleukins 1beta and 6 but not transforming growth factor-beta are essential for the differentiation of interleukin 17-producing human T helper cells. *Nat Immunol, 8*(9), 942–949.

Akimzhanov, A. M., Yang, X. O., & Dong, C. (2007). Chromatin remodeling of interleukin-17 (IL-17)-IL-17F cytokine gene locus during inflammatory helper T cell differentiation. *J Biol Chem, 282*(9), 5969–5972.

Albanesi, C., Scarponi, C., Cavani, A., Federici, M., Nasorri, F., & Girolomoni, G. (2000). Interleukin-17 is produced by both Th1 and Th2 lymphocytes, and modulates interferon-gamma- and interleukin-4-induced activation of human keratinocytes. *J Invest Dermatol, 115*(1), 81–87.

Awane, M., Andres, P. G., Li, D. J., & Reinecker, H. C. (1999). NF-kappa B-inducing kinase is a common mediator of IL-17-, TNF-alpha-, and IL-1 beta-induced chemokine promoter activation in intestinal epithelial cells. *J Immunol, 162*(9), 5337–5344.

Ben-Sasson, S. Z., Hu-Li, J., Quiel, J., Cauchetaux, S., Ratner, M., Shapira, I., et al. (2009). IL-1 acts directly on CD4 T cells to enhance their antigen-driven expansion and differentiation. *Proc Natl Acad Sci USA, 106*(17), 7119–7124.

Bettelli, E., Carrier, Y., Gao, W., Korn, T., Strom, T. B., Oukka, M., et al. (2006). Reciprocal developmental pathways for the generation of pathogenic effector TH17 and regulatory T cells. *Nature, 441*(7090), 235–238.

Brustle, A., Heink, S., Huber, M., Rosenplanter, C., Stadelmann, C., Yu, P., et al. (2007). The development of inflammatory T(H)-17 cells requires interferon-regulatory factor 4. *Nat Immunol, 8*(9), 958–966.

Chang, S. H., & Dong, C. (2007). A novel heterodimeric cytokine consisting of IL-17 and IL-17F regulates inflammatory responses. *Cell Res, 17*(5), 435–440.

Chang, S. H., Park, H., & Dong, C. (2006). Act1 adaptor protein is an immediate and essential signaling component of interleukin-17 receptor. *J Biol Chem, 281*(47), 35603–35607.

Chen, Z., Laurence, A., Kanno, Y., Pacher-Zavisin, M., Zhu, B. M., Tato, C., et al. (2006). Selective regulatory function of Socs3 in the formation of IL-17-secreting T cells. *Proc Natl Acad Sci USA, 103*(21), 8137–8142.

Chung, Y., Chang, S. H., Martinez, G. J., Yang, X. O., Nurieva, R., Kang, H. S., et al. (2009). Critical regulation of early Th17 cell differentiation by interleukin-1 signaling. *Immunity, 30*(4), 576–587.

Cua, D. J., Sherlock, J., Chen, Y., Murphy, C. A., Joyce, B., Seymour, B., et al. (2003). Interleukin-23 rather than interleukin-12 is the critical cytokine for auto-immune inflammation of the brain. *Nature, 421*(6924), 744–748.

Dong, C., & Nurieva, R. I. (2003). Regulation of immune and auto-immune responses by ICOS. *J Autoimmun, 21*(3), 255–260.

Fort, M. M., Cheung, J., Yen, D., Li, J., Zurawski, S. M., Lo, S., et al. (2001). IL-25 induces IL-4, IL-5, and IL-13 and Th2-associated pathologies in vivo. *Immunity, 15*(6), 985–995.

Fossiez, F., Djossou, O., Chomarat, P., Flores-Romo, L., Ait-Yahia, S., Maat, C., et al. (1996). T cell interleukin-17 induces stromal cells to produce pro-inflammatory and hematopoietic cytokines. *J Exp Med, 183*(6), 2593–2603.

Guo, B., Chang, E. Y., & Cheng, G. (2008). The type I IFN induction pathway constrains Th17-mediated auto-immune inflammation in mice. *J Clin Invest, 118*(5), 1680–1690.

Harrington, L. E., Hatton, R. D., Mangan, P. R., Turner, H., Murphy, T. L., Murphy, K. M., et al. (2005). Interleukin 17-producing CD4+ effector T cells develop via a lineage distinct from the T helper type 1 and 2 lineages. *Nat Immunol, 6*(11), 1123–1132.

Haudenschild, D., Moseley, T., Rose, L., & Reddi, A. H. (2002). Soluble and transmembrane isoforms of novel interleukin-17 receptor-like protein by RNA splicing and expression in prostate cancer. *J Biol Chem, 277*(6), 4309–4316.

Hymowitz, S. G., Filvaroff, E. H., Yin, J. P., Lee, J., Cai, L., Risser, P., et al. (2001). IL-17s adopt a cystine knot fold: structure and activity of a novel cytokine, IL-17F, and implications for receptor binding. *Embo J, 20*(19), 5332–5341.

Ishigame, H., Kakuta, S., Nagai, T., Kadoki, M., Nambu, A., Komiyama, Y., et al. (2009). Differential roles of interleukin-17A and -17F in host defense against mucoepithelial bacterial infection and allergic responses. *Immunity, 30*(1), 108–119.

Ivanov, II, McKenzie, B. S., Zhou, L., Tadokoro, C. E., Lepelley, A., Lafaille, J. J., et al. (2006). The orphan nuclear receptor RORgammat directs the differentiation program of pro-inflammatory IL-17+ T helper cells. *Cell, 126*(6), 1121–1133.

Kato, T., Furumoto, H., Ogura, T., Onishi, Y., Irahara, M., Yamano, S., et al. (2001). Expression of IL-17 mRNA in ovarian cancer. *Biochem Biophys Res Commun, 282*(3), 735–738.

Kawaguchi, M., Kokubu, F., Odaka, M., Watanabe, S., Suzuki, S., Ieki, K., et al. (2004). Induction of granulocyte-macrophage colony-stimulating factor by a new cytokine, ML-1 (IL-17F), via Raf I-MEK-ERK pathway. *J Allergy Clin Immunol, 114*(2), 444–450.

Kolls, J. K., & Linden, A. (2004). Interleukin-17 family members and inflammation. *Immunity, 21*(4), 467–476.

Korn, T., Bettelli, E., Gao, W., Awasthi, A., Jager, A., Strom, T. B., et al. (2007). IL-21 initiates an alternative pathway to induce pro-inflammatory T(H)17 cells. *Nature, 448*(7152), 484–487.

Kuestner, R. E., Taft, D. W., Haran, A., Brandt, C. S., Brender, T., Lum, K., et al. (2007). Identification of the IL-17 receptor related molecule IL-17RC as the receptor for IL-17F. *J Immunol, 179*(8), 5462–5473.

Langrish, C. L., Chen, Y., Blumenschein, W. M., Mattson, J., Basham, B., Sedgwick, J. D., et al. (2005). IL-23 drives a pathogenic T cell population that induces auto-immune inflammation. *J Exp Med, 201*(2), 233–240.

Laurence, A., Tato, C. M., Davidson, T. S., Kanno, Y., Chen, Z., Yao, Z., et al. (2007). Interleukin-2 signaling via STAT5 constrains T helper 17 cell generation. *Immunity, 26*(3), 371–381.

Lee, J., Ho, W. H., Maruoka, M., Corpuz, R. T., Baldwin, D. T., Foster, J. S., et al. (2001). IL-17E, a novel pro-inflammatory ligand for the IL-17 receptor homolog IL-17Rh1. *J Biol Chem, 276*(2), 1660–1664.

Lee, W. W., Kang, S. W., Choi, J., Lee, S. H., Shah, K., Eynon, E. E., et al. (2010). Regulating human Th17 cells via differential expression of IL-1 receptor. *Blood, 115*(3), 530–540.

Lenarczyk, A., Helsloot, J., Farmer, K., Peters, L., Sturgess, A., & Kirkham, B. (2000). Antigen-induced IL-17 response in the peripheral blood mononuclear cells (PBMC) of healthy controls. *Clin Exp Immunol, 122*(1), 41–48.

Leonardi, A., Chariot, A., Claudio, E., Cunningham, K., & Siebenlist, U. (2000). CIKS, a connection to Ikappa B kinase and stress-activated protein kinase. *Proc Natl Acad Sci USA, 97*(19), 10494–10499.

Li, H., Chen, J., Huang, A., Stinson, J., Heldens, S., Foster, J., et al. (2000a). Cloning and characterization of IL-17B and IL-17C, two new members of the IL-17 cytokine family. *Proc Natl Acad Sci USA, 97*(2), 773–778.

Li, L., Huang, L., Vergis, A. L., Ye, H., Bajwa, A., Narayan, V., et al. (2010). IL-17 produced by neutrophils regulates IFN-gamma-mediated neutrophil migration in mouse kidney ischemia-reperfusion injury. *J Clin Invest, 120*(1), 331–342.

Li, M. O., Wan, Y. Y., & Flavell, R. A. (2007). T cell-produced transforming growth factor-beta1 controls T cell tolerance and regulates Th1- and Th17-cell differentiation. *Immunity, 26*(5), 579–591.

Li, X., Commane, M., Nie, H., Hua, X., Chatterjee-Kishore, M., Wald, D., et al. (2000b). Act1, an NF-kappa B-activating protein. *Proc Natl Acad Sci USA, 97*(19), 10489–10493.

Liang, S. C., Long, A. J., Bennett, F., Whitters, M. J., Karim, R., Collins, M., et al. (2007). An IL-17F/A heterodimer protein is produced by mouse Th17 cells and induces airway neutrophil recruitment. *J Immunol, 179*(11), 7791–7799.

Lockhart, E., Green, A. M., & Flynn, J. L. (2006). IL-17 production is dominated by gammadelta T cells rather than CD4 T cells during Mycobacterium tuberculosis infection. *J Immunol, 177*(7), 4662–4669.

Mangan, P. R., Harrington, L. E., O'Quinn, D. B., Helms, W. S., Bullard, D. C., Elson, C. O., et al. (2006). Transforming growth factor-beta induces development of the T(H)17 lineage. *Nature, 441*(7090), 231–234.

Martin-Orozco, N., Muranski, P., Chung, Y., Yang, X. O., Yamazaki, T., Lu, S., et al. (2009). T helper 17 cells promote cytotoxic T cell activation in tumor immunity. *Immunity, 31*(5), 787–798.

Michel, M. L., Keller, A. C., Paget, C., Fujio, M., Trottein, F., Savage, P. B., et al. (2007). Identification of an IL-17-producing NK1.1(neg) iNKT cell population involved in airway neutrophilia. *J Exp Med, 204*(5), 995–1001.

Moseley, T. A., Haudenschild, D. R., Rose, L., & Reddi, A. H. (2003). Interleukin-17 family and IL-17 receptors. *Cytokine Growth Factor Rev, 14*(2), 155–174.

Mosmann, T. R., Cherwinski, H., Bond, M. W., Giedlin, M. A., & Coffman, R. L. (1986). Two types of murine helper T cell clone. I. Definition according to profiles of lymphokine activities and secreted proteins. *J Immunol, 136*(7), 2348–2357.

Muranski, P., Boni, A., Antony, P. A., Cassard, L., Irvine, K. R., Kaiser, A., et al. (2008). Tumor-specific Th17-polarized cells eradicate large established melanoma. *Blood, 112*(2), 362–373.

Murphy, C. A., Langrish, C. L., Chen, Y., Blumenschein, W., McClanahan, T., Kastelein, R. A., et al. (2003). Divergent pro and anti-inflammatory roles for IL-23 and IL-12 in joint autoimmune inflammation. *J Exp Med, 198*(12), 1951–1957.

Novatchkova, M., Leibbrandt, A., Werzowa, J., Neubuser, A., & Eisenhaber, F. (2003). The STIR-domain superfamily in signal transduction, development and immunity. *Trends Biochem Sci, 28*(5), 226–229.

Numasaki, M., Fukushi, J., Ono, M., Narula, S. K., Zavodny, P. J., Kudo, T., et al. (2003). Interleukin-17 promotes angiogenesis and tumor growth. *Blood, 101*(7), 2620–2627.

Nurieva, R., Yang, X. O., Martinez, G., Zhang, Y., Panopoulos, A. D., Ma, L., et al. (2007). Essential autocrine regulation by IL-21 in the generation of inflammatory T cells. *Nature, 448*(7152), 480–483.

Oda, N., Canelos, P. B., Essayan, D. M., Plunkett, B. A., Myers, A. C., & Huang, S. K. (2005). Interleukin-17F induces pulmonary neutrophilia and amplifies antigen-induced allergic response. *Am J Respir Crit Care Med, 171*(1), 12–18.

Pappu, B. P., Borodovsky, A., Zheng, T. S., Yang, X., Wu, P., Dong, X., et al. (2008). TL1A-DR3 interaction regulates Th17 cell function and Th17-mediated auto-immune disease. *J Exp Med, 205*(5), 1049–1062.

Park, H., Li, Z., Yang, X. O., Chang, S. H., Nurieva, R., Wang, Y. H., et al. (2005). A distinct lineage of CD4 T cells regulates tissue inflammation by producing interleukin 17. *Nat Immunol, 6*(11), 1133–1141.

Qian, Y., Liu, C., Hartupee, J., Altuntas, C. Z., Gulen, M. F., Jane-Wit, D., et al. (2007). The adaptor Act1 is required for interleukin 17-dependent signaling associated with auto-immune and inflammatory disease. *Nat Immunol, 8*(3), 247–256.

Quintana, F. J., Basso, A. S., Iglesias, A. H., Korn, T., Farez, M. F., Bettelli, E., et al. (2008). Control of T(reg) and T(H)17 cell differentiation by the aryl hydrocarbon receptor. *Nature, 453*(7191), 65–71.

Rouvier, E., Luciani, M. F., Mattei, M. G., Denizot, F., & Golstein, P. (1993). CTLA-8, cloned from an activated T cell, bearing AU-rich messenger RNA instability sequences, and homologous to a herpesvirus saimiri gene. *J Immunol, 150*(12), 5445–5456.

Schraml, B. U., Hildner, K., Ise, W., Lee, W. L., Smith, W. A., Solomon, B., et al. (2009). The AP-1 transcription factor Batf controls T(H)17 differentiation. *Nature, 460*(7253), 405–409.

Schwandner, R., Yamaguchi, K., & Cao, Z. (2000). Requirement of tumor necrosis factor receptor-associated factor (TRAF)6 in interleukin 17 signal transduction. *J Exp Med, 191*(7), 1233–1240.

Sfanos, K. S., Bruno, T. C., Maris, C. H., Xu, L., Thoburn, C. J., DeMarzo, A. M., et al. (2008). Phenotypic analysis of prostate-infiltrating lymphocytes reveals TH17 and Treg skewing. *Clin Cancer Res, 14*(11), 3254–3261.

Shalom-Barak, T., Quach, J., & Lotz, M. (1998). Interleukin-17-induced gene expression in articular chondrocytes is associated with activation of mitogen-activated protein kinases and NF-kappaB. *J Biol Chem, 273*(42), 27467–27473.

Shi, Y., Ullrich, S. J., Zhang, J., Connolly, K., Grzegorzewski, K. J., Barber, M. C., et al. (2000). A novel cytokine receptor-ligand pair. Identification, molecular characterization, and in vivo immuno-modulatory activity. *J Biol Chem, 275*(25), 19167–19176.

Starnes, T., Broxmeyer, H. E., Robertson, M. J., & Hromas, R. (2002). Cutting edge: IL-17D, a novel member of the IL-17 family, stimulates cytokine production and inhibits hemopoiesis. *J Immunol, 169*(2), 642–646.

Starnes, T., Robertson, M. J., Sledge, G., Kelich, S., Nakshatri, H., Broxmeyer, H. E., et al. (2001). Cutting edge: IL-17F, a novel cytokine selectively expressed in activated T cells and monocytes, regulates angiogenesis and endothelial cell cytokine production. *J Immunol, 167*(8), 4137–4140.

Stumhofer, J. S., Laurence, A., Wilson, E. H., Huang, E., Tato, C. M., Johnson, L. M., et al. (2006). Interleukin 27 negatively regulates the development of interleukin 17-producing T helper cells during chronic inflammation of the central nervous system. *Nat Immunol, 7*(9), 937–945.

Takahashi, N., Vanlaere, I., de Rycke, R., Cauwels, A., Joosten, L. A., Lubberts, E., et al. (2008). IL-17 produced by Paneth cells drives TNF-induced shock. *J Exp Med, 205*(8), 1755–1761.

Tian, E., Sawyer, J. R., Largaespada, D. A., Jenkins, N. A., Copeland, N. G., & Shaughnessy, J. D., Jr. (2000). Evi27 encodes a novel membrane protein with homology to the IL17 receptor. *Oncogene, 19*(17), 2098–2109.

Toy, D., Kugler, D., Wolfson, M., Vanden Bos, T., Gurgel, J., Derry, J., et al. (2006). Cutting edge: interleukin 17 signals through a heteromeric receptor complex. *J Immunol, 177*(1), 36–39.

Veldhoen, M., Hirota, K., Westendorf, A. M., Buer, J., Dumoutier, L., Renauld, J. C., et al. (2008). The aryl hydrocarbon receptor links TH17-cell-mediated autoimmunity to environmental toxins. *Nature, 453*(7191), 106–109.

Veldhoen, M., Hocking, R. J., Atkins, C. J., Locksley, R. M., & Stockinger, B. (2006a). TGFbeta in the context of an inflammatory cytokine milieu supports de novo differentiation of IL-17-producing T cells. *Immunity, 24*(2), 179–189.

Veldhoen, M., Hocking, R. J., Flavell, R. A., & Stockinger, B. (2006b). Signals mediated by transforming growth factor-beta initiate auto-immune encephalomyelitis, but chronic inflammation is needed to sustain disease. *Nat Immunol, 7*(11), 1151–1156.

Wong, P. K., Egan, P. J., Croker, B. A., O'Donnell, K., Sims, N. A., Drake, S., et al. (2006). SOCS-3 negatively regulates innate and adaptive immune mechanisms in acute IL-1-dependent inflammatory arthritis. *J Clin Invest, 116*(6), 1571–1581.

Wright, J. F., Bennett, F., Li, B., Brooks, J., Luxenberg, D. P., Whitters, M. J., et al. (2008). The human IL-17F/IL-17A heterodimeric cytokine signals through the IL-17RA/IL-17RC receptor complex. *J Immunol, 181*(4), 2799–2805.

Wright, J. F., Guo, Y., Quazi, A., Luxenberg, D. P., Bennett, F., Ross, J. F., et al. (2007). Identification of an interleukin 17F/17A heterodimer in activated human CD4+ T cells. *J Biol Chem, 282*(18), 13447–13455.

Yang, X. O., Chang, S. H., Park, H., Nurieva, R., Shah, B., Acero, L., et al. (2008a). Regulation of inflammatory responses by IL-17F. *J Exp Med, 205*(5), 1063–1075.

Yang, X. O., Nurieva, R., Martinez, G. J., Kang, H. S., Chung, Y., Pappu, B. P., et al. (2008b). Molecular antagonism and plasticity of regulatory and inflammatory T cell programs. *Immunity, 29*(1), 44–56.

Yang, X. O., Panopoulos, A. D., Nurieva, R., Chang, S. H., Wang, D., Watowich, S. S., et al. (2007). STAT3 regulates cytokine-mediated generation of inflammatory helper T cells. *J Biol Chem, 282*(13), 9358–9363.

Yang, X. O., Pappu, B. P., Nurieva, R., Akimzhanov, A., Kang, H. S., Chung, Y., et al. (2008c). T helper 17 lineage differentiation is programmed by orphan nuclear receptors ROR alpha and ROR gamma. *Immunity, 28*(1), 29–39.

Yao, Z., Fanslow, W. C., Seldin, M. F., Rousseau, A. M., Painter, S. L., Comeau, M. R., et al. (1995a). Herpesvirus Saimiri encodes a new cytokine, IL-17, which binds to a novel cytokine receptor. *Immunity, 3*(6), 811–821.

Yao, Z., Painter, S. L., Fanslow, W. C., Ulrich, D., Macduff, B. M., Spriggs, M. K., et al. (1995b). Human IL-17: a novel cytokine derived from T cells. *J Immunol, 155*(12), 5483–5486.

Yao, Z., Spriggs, M. K., Derry, J. M., Strockbine, L., Park, L. S., VandenBos, T., et al. (1997). Molecular characterization of the human interleukin (IL)-17 receptor. *Cytokine, 9*(11), 794–800.

Zhang, F., Meng, G., & Strober, W. (2008). Interactions among the transcription factors Runx1, RORgammat and Foxp3 regulate the differentiation of interleukin 17-producing T cells. *Nat Immunol, 9*(11), 1297–1306.

Zhou, L., Ivanov, II, Spolski, R., Min, R., Shenderov, K., Egawa, T., et al. (2007). IL-6 programs T(H)-17 cell differentiation by promoting sequential engagement of the IL-21 and IL-23 pathways. *Nat Immunol, 8*(9), 967–974.

Zhou, L., Lopes, J. E., Chong, M. M., Ivanov, II, Min, R., Victora, G. D., et al. (2008). TGF-beta-induced Foxp3 inhibits T(H)17 cell differentiation by antagonizing RORgammat function. *Nature, 453*(7192), 236–240.

TGF-beta and T_H17 Cells

Samuel Huber, Enric Esplugues, and Richard A. Flavell

Abstract TGF-beta plays an important role in the regulation of T helper (T_H) cell differentiation. The presence or absence of TGF-beta in combination with other cytokines (IL-6, IL-1, or IL-23) is critical for the developmental program of regulatory and effector T helper cells and for the final outcome in terms of anti-infective response, pathogenicity, or suppressive capacity. TGF-beta inhibits T_H1 and T_H2 differentiation. By contrast, TGF-beta promotes the development of induced Foxp3+ regulatory T cells and together with IL-6, that of T_H17 cells.

TGF-beta is a pleiotropic cytokine with multiple context dependent functions on different cell types, including immune cells. Three TGF-beta members (TGF-beta1, TGF-beta2, and TGF-beta3) are present in mammals; TGF-beta1 is the major form expressed in the immune system. TGF-beta1 is secreted as a latent form and has to be processed into the active form in order to exert his functions. Active TGF-beta1 binds to the TGF-beta type II receptor (TβRII), which then recruits the TGF-beta type 1 receptor (TβRI) to form a heterodimeric receptor complex. The major TGF-beta signal is then transmitted by Smad proteins. Upon receptor ligation, Smad 2 and/or Smad 3 are phosphorylated and form a complex with Smad 4, which then translocates to the nucleus in order to modulate the expression of TGF-beta regulated genes. Alternatively, TGF-beta can signal through activation of MAP kinases in a Smad-independent signaling pathway (Gorelik and Flavell 2002).

The pivotal function of TGF-beta1 in immune homeostasis was originally revealed in studies of TGF-beta1-deficient mice. These mice develop an early and fatal multi-focal inflammatory disease, leading to death at the age of about 4 weeks (Kulkarni et al. 1993; Shull et al. 1992). T cells are essential mediators of the disease, which is considered to be of auto-immune origin (Kobayashi et al. 1999;

R.A. Flavell (✉)
Section of Immunobiology and Howard Hughes Medical Institute, Yale University School of Medicine, New Haven, CT 06520, USA
e-mail: Richard.Flavell@yale.edu

S. Jiang (ed.), *T_H17 Cells in Health and Disease*,
DOI 10.1007/978-1-4419-9371-7_3, © Springer Science+Business Media, LLC 2011

Letterio et al. 1996). However, since TGF-beta can act on multiple immune cells, it was not clear from these early studies, whether TGF-beta directly controlled T cells. Reports with T-cell specific blockade of TGF-beta signaling then established T cells as a direct target of TGF-beta in vivo (Gorelik and Flavell 2000; Huber et al. 2004; Li et al. 2006; Lucas et al. 2000). It was found that TGF-β blocks both T_H1 and T_H2 differentiation by inhibiting the expression of T-bet and Gata-3, the apparent master regulator of T_H1 and T_H2 differentiation (Gorelik et al. 2000, 2002).

1 The Role of TGF-beta in Treg and T_H17 Cell Differentiation

One essential pathway of peripheral T cell tolerance is mediated by CD4+ regulatory T cells (Treg), which express the transcription factor Foxp3. Treg can be either generated in the thymus (nTreg) or alternatively induced in peripheral lymphoid organs (iTreg) (Sakaguchi 2005). TGF-beta1 plays an essential role in the generation of iTreg (Chen et al. 2003; Fantini et al. 2004; Kretschmer et al. 2005), while nTreg generation is not dependent on TGF-beta1 (Huber et al. 2004; Li et al. 2006; Marie et al. 2005). Therefore, TGF-beta1 inhibits T_H1 and T_H2 effector T cells while promoting Treg cells, which collectively contributes to the maintenance of tolerance. However it was shown that TGF-beta1 induction of iTreg is inhibited in the presence of the proinflammatory cytokine IL-6 (Bettelli et al. 2006). Instead, TGF-beta1 plus IL-6 lead to the generation of T_H17 cells, suggesting a reciprocal development pathway between Treg and T_H17 cells (Bettelli et al. 2006; Veldhoen et al. 2006a) (Fig. 1).

Initial data suggested that the cytokine IL-23 was playing a major role in driving the generation of T_H17 cells (Aggarwal et al. 2003; Harrington et al. 2005; Park et al. 2005; Weaver et al. 2007), but more recent work has shown the importance of IL-23 in preserving the pathogenicity and promoting the expansion of T_H17 cells. It has been argued that this is its major function, rather than playing a key role in the differentiation of T_H17 cells (McGeachy et al. 2007, 2009). TGF-beta in conjunction with IL-6 has been thought to be the key signal for the initial induction and differentiation of T_H17 cells. But recent data showed that the generation of pathogenic T_H17 cells could occur in the absence of TGF-beta signaling (Ghoreschi et al. 2010). The observation of the presence of T_H17 cells in the intestinal lamina propria of mice lacking TGF-beta signaling (Liu et al. 2008; Veldhoen et al. 2006b) supported the idea that in vivo T_H17 differentiation is possible even when the TGF-beta signaling is absent. In line with this observation, $6^{-/-}$ T-bet$^{-/-}$ mice, which are unable to T_H1 and T_H2 cells, IL-6 alone was sufficient to induce T_H17 differentiation. In contrast, TGF-beta1 was dispensable for T_H17 cell differentiation (Das et al. 2009). The combination of IL-6, IL-23, and the IL-1b in the absence of TGF-beta has recently been shown to be sufficient to induce transcription and epigenetic modification of the *il17a/il17f* locus (Ghoreschi et al. 2010), and to increase the expression of T-bet, IL-18R1, and CXCR3; each of which play an important role in the development of EAE and in the trafficking of CD4+ T cells to sites of inflammation (Bettelli et al. 2004; Gutcher et al. 2006; Koch et al. 2009; Yang et al. 2009). This data showed that TGF-beta1 is not essential for the differentiation of T_H17 cells.

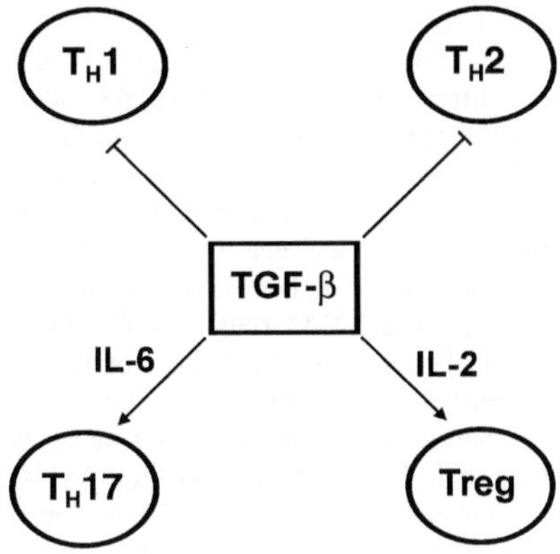

Fig. 1 The role of TGF-beta for different T helper cell lineages. TGF-beta inhibits T_H1 and T_H2 differentiation. In contrast, TGF-beta plus IL-2 leads to the differentiation of Foxp3+ Treg, while TGF-beta plus IL-6 inhibits Treg development and promotes the differentiation of T_H17 cells

Interestingly, T_H17 cells generated in the presence or absence of TGF-beta have different pathogenic potential. T_H17 cells generated in vitro in the presence of TGF-beta and IL-6 are less pathogenic than the T_H17 cells generated in the presence of IL-6, IL-1beta, and IL-23 in a murine model of multiple sclerosis (EAE) (McGeachy et al. 2007).

Although TGF-beta is not essential for the differentiation of T_H17 cells, it seems that TGF-beta is present in most in vivo situations. Therefore, the reciprocal developmental pathway between T_H17 cells and Treg plays an important role in the balance of immune homeostasis and also for defense against pathogens. One physiological stimulus triggering the combination of TGF-beta1 and IL-6 is the phagocytosis of infected apoptotic cells, which then led to the induction of T_H17 cells. Interestingly, the blockade of apoptosis during a gastrointestinal infection impairs the characteristic T_H17 response. In contrast, phagocytosis of apoptotic cells in the absence of microbial signals induces the differentiation of Treg cells (Torchinsky et al. 2009) (Fig. 1).

References

Aggarwal S, Ghilardi N, Xie M H, de Sauvage F J, and Gurney A L (2003) Interleukin-23 promotes a distinct CD4 T cell activation state characterized by the production of interleukin-17. J Biol Chem 278: 1910–1914.

Bettelli E, Carrier Y, Gao W, Korn T, Strom T B, Oukka M, Weiner H L, and Kuchroo V K (2006) Reciprocal developmental pathways for the generation of pathogenic effector TH17 and regulatory T cells. Nature 441: 235–238.

Bettelli E, Sullivan B, Szabo S J, Sobel R A, Glimcher L H, and Kuchroo V K (2004) Loss of T-bet, but not STAT1, prevents the development of experimental auto-immune encephalomyelitis. J Exp Med 200: 79–87.

Chen W, Jin W, Hardegen N, Lei K J, Li L, Marinos N, McGrady G, and Wahl S M (2003) Conversion of peripheral CD4+CD25- naive T cells to CD4+CD25+ regulatory T cells by TGF-beta induction of transcription factor Foxp3. J Exp Med 198: 1875–1886.

Das J, Ren G, Zhang L, Roberts A I, Zhao X, Bothwell A L, Van Kaer L, Shi Y, and Das G (2009) Transforming growth factor beta is dispensable for the molecular orchestration of Th17 cell differentiation. J Exp Med 206: 2407–2416.
Fantini M C, Becker C, Monteleone G, Pallone F, Galle P R, and Neurath M F (2004) Cutting edge: TGF-beta induces a regulatory phenotype in CD4+CD25- T cells through Foxp3 induction and down-regulation of Smad7. J Immunol 172: 5149–5153.
Ghoreschi K, Laurence A, Yang X P, Tato C M, McGeachy M J, Konkel J E, Ramos H L, Wei L, Davidson T S, Bouladoux N, Grainger J R, Chen Q, Kanno Y, Watford W T, Sun H W, Eberl G, Shevach E M, Belkaid Y, Cua D J, Chen W, and O'Shea J J (2010) Generation of pathogenic T(H)17 cells in the absence of TGF-beta signaling. Nature 467: 967–971.
Gorelik L, Constant S, and Flavell R A (2002) Mechanism of transforming growth factor beta-induced inhibition of T helper type 1 differentiation. J Exp Med 195: 1499–1505.
Gorelik L, Fields P E, and Flavell R A (2000) Cutting edge: TGF-beta inhibits Th type 2 development through inhibition of GATA-3 expression. J Immunol 165: 4773–4777.
Gorelik L, and Flavell R A (2000) Abrogation of TGFbeta signaling in T cells leads to spontaneous T cell differentiation and auto-immune disease. Immunity 12: 171–181.
Gorelik L, and Flavell R A (2002) Transforming growth factor-beta in T-cell biology. Nat Rev Immunol 2: 46–53.
Gutcher I, Urich E, Wolter K, Prinz M, and Becher B (2006) Interleukin 18-independent engagement of interleukin 18 receptor-alpha is required for auto-immune inflammation. Nat Immunol 7: 946–953.
Harrington L E, Hatton R D, Mangan P R, Turner H, Murphy T L, Murphy K M, and Weaver C T (2005) Interleukin 17-producing CD4+ effector T cells develop via a lineage distinct from the T helper type 1 and 2 lineages. Nat Immunol 6: 1123–1132.
Huber S, Schramm C, Lehr H A, Mann A, Schmitt S, Becker C, Protschka M, Galle P R, Neurath M F, and Blessing M (2004) Cutting edge: TGF-beta signaling is required for the in vivo expansion and immuno-suppressive capacity of regulatory CD4+CD25+ T cells. J Immunol 173: 6526–6531.
Kobayashi S, Yoshida K, Ward J M, Letterio J J, Longenecker G, Yaswen L, Mittleman B, Mozes E, Roberts A B, Karlsson S, and Kulkarni A B (1999) Beta 2-microglobulin-deficient background ameliorates lethal phenotype of the TGF-beta 1 null mouse. J Immunol 163: 4013–4019.
Koch M A, Tucker-Heard G, Perdue N R, Killebrew J R, Urdahl K B, and Campbell D J (2009) The transcription factor T-bet controls regulatory T cell homeostasis and function during type 1 inflammation. Nat Immunol 10: 595–602.
Kretschmer K, Apostolou I, Hawiger D, Khazaie K, Nussenzweig M C, and von Boehmer H (2005) Inducing and expanding regulatory T cell populations by foreign antigen. Nat Immunol 6: 1219–1227.
Kulkarni A B, Huh C G, Becker D, Geiser A, Lyght M, Flanders K C, Roberts A B, Sporn M B, Ward J M, and Karlsson S (1993) Transforming growth factor beta 1 null mutation in mice causes excessive inflammatory response and early death. Proc Natl Acad Sci USA 90: 770–774.
Letterio J J, Geiser A G, Kulkarni A B, Dang H, Kong L, Nakabayashi T, Mackall C L, Gress R E, and Roberts A B (1996) Auto-immunity associated with TGF-beta1-deficiency in mice is dependent on MHC class II antigen expression. J Clin Invest 98: 2109–2119.
Li M O, Sanjabi S, and Flavell R A (2006) Transforming growth factor-beta controls development, homeostasis, and tolerance of T cells by regulatory T cell-dependent and -independent mechanisms. Immunity 25: 455–471.
Liu Y, Zhang P, Li J, Kulkarni A B, Perruche S, and Chen W (2008) A critical function for TGF-beta signaling in the development of natural CD4+CD25+Foxp3+ regulatory T cells. Nat Immunol 9: 632–640.
Lucas P J, Kim S J, Melby S J, and Gress R E (2000) Disruption of T cell homeostasis in mice expressing a T cell-specific dominant negative transforming growth factor beta II receptor. J Exp Med 191: 1187–1196.

Marie J C, Letterio J J, Gavin M, and Rudensky A Y (2005) TGF-beta1 maintains suppressor function and Foxp3 expression in CD4+CD25+ regulatory T cells. J Exp Med 201: 1061–1067.

McGeachy M J, Bak-Jensen K S, Chen Y, Tato C M, Blumenschein W, McClanahan T, and Cua D J (2007) TGF-beta and IL-6 drive the production of IL-17 and IL-10 by T cells and restrain T(H)-17 cell-mediated pathology. Nat Immunol 8: 1390–1397.

McGeachy M J, Chen Y, Tato C M, Laurence A, Joyce-Shaikh B, Blumenschein W M, McClanahan T K, O'Shea J J, and Cua D J (2009) The interleukin 23 receptor is essential for the terminal differentiation of interleukin 17-producing effector T helper cells in vivo. Nat Immunol 10: 314–324.

Park H, Li Z, Yang X O, Chang S H, Nurieva R, Wang Y H, Wang Y, Hood L, Zhu Z, Tian Q, and Dong C (2005) A distinct lineage of CD4 T cells regulates tissue inflammation by producing interleukin 17. Nat Immunol 6: 1133–1141.

Sakaguchi S (2005) Naturally arising Foxp3-expressing CD25+CD4+ regulatory T cells in immunological tolerance to self and non-self. Nat Immunol 6: 345–352.

Shull M M, Ormsby I, Kier A B, Pawlowski S, Diebold R J, Yin M, Allen R, Sidman C, Proetzel G, Calvin D, and et al. (1992) Targeted disruption of the mouse transforming growth factor-beta 1 gene results in multi-focal inflammatory disease. Nature 359: 693–699.

Torchinsky M B, Garaude J, Martin A P, and Blander J M (2009) Innate immune recognition of infected apoptotic cells directs T(H)17 cell differentiation. Nature 458: 78–82.

Veldhoen M, Hocking R J, Atkins C J, Locksley R M, and Stockinger B (2006a) TGFbeta in the context of an inflammatory cytokine milieu supports de novo differentiation of IL-17-producing T cells. Immunity 24: 179–189.

Veldhoen M, Hocking R J, Flavell R A, and Stockinger B (2006b) Signals mediated by transforming growth factor-beta initiate auto-immune encephalomyelitis, but chronic inflammation is needed to sustain disease. Nat Immunol 7: 1151–1156.

Weaver C T, Hatton R D, Mangan P R, and Harrington L E (2007) IL-17 family cytokines and the expanding diversity of effector T cell lineages. Annu Rev Immunol 25: 821–852.

Yang Y, Weiner J, Liu Y, Smith A J, Huss D J, Winger R, Peng H, Cravens P D, Racke M K, and Lovett-Racke A E (2009) T-bet is essential for encephalitogenicity of both Th1 and Th17 cells. J Exp Med 206: 1549–1564.

IL-6: A Pleiotropic Cytokine and T_H17 Cells

Akihiro Kimura, Tetsuji Naka, and Tadamitsu Kishimoto

Abstract Naïve T cells are multipotential precursors that differentiate into various effector subsets, such as T helper type 1 (T_H1) and T_H2 cells, and are characterized by their distinct functions. IL-17-producing T helper (T_H17) cell has been recently identified as a new subset of T helper cell and a mediator of inflammation associated with various auto-immune diseases. IL-6 induces the generation of T_H17 cells from naïve T cells together with TGF-β and inhibits TGF-β-induced regulatory T (Treg) cells, which suppress adaptive T cell responses and prevent auto-immunity. Because IL-6 plays an important role in directing the balance between Treg and T_H17 cell development, controlling IL-6 activities is an effective approach in the treatment of various auto-immune and inflammatory diseases. Here, we review the recent progress in the field of T_H17 cell differentiation and regulation and describe the critical functions of IL-6 and T_H17 in immunity and diseases.

1 Introduction

$CD4^+$ T cells (T_H) are essential regulators of immune responses and inflammatory diseases. They can be divided into different subsets such as T_H1, T_H2 and regulatory T (Treg) cells, whose development is specified by the transcription factors T-bet, GATA3, and fork head box p3 (Foxp3) (Fig. 1) (Fontenot et al. 2005; Szabo et al. 2000; Zheng and Flavell 1997). The development of T_H1 cells, which activate macrophages and are highly effective in clearing intracellular pathogens, is coupled with the sequential actions of interferon-γ (IFN-γ) and interleukin-12 (IL-12)

T. Kishimoto (✉)
Laboratory of Immune Regulation, Osaka University Graduate School of Frontier Biosciences, Osaka, Japan
e-mail: kishimot@fbs.osaka-u.ac.jp

S. Jiang (ed.), *T_H17 Cells in Health and Disease*,
DOI 10.1007/978-1-4419-9371-7_4,

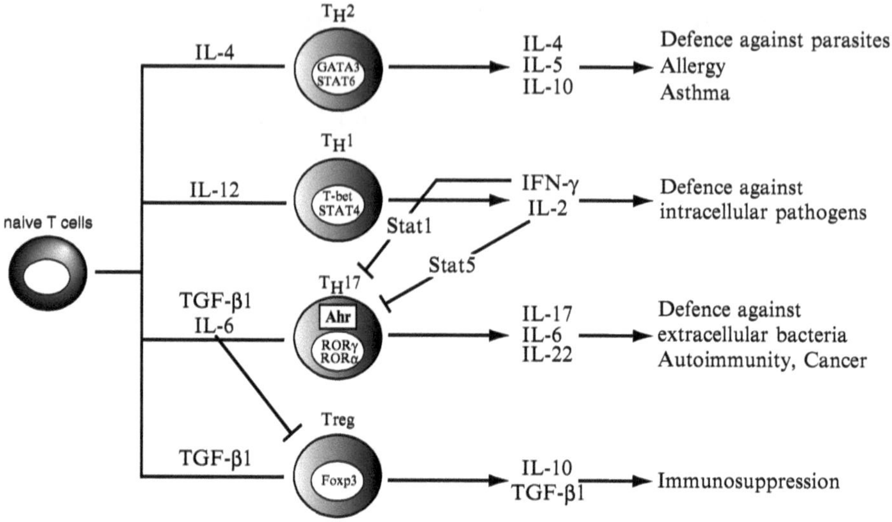

Fig. 1 Th cell differentiation. Naïve T cells can differentiate into several subsets. T_H17 cells were recently identified as a novel $CD4^+$ lineage. They are induced by TGF-β plus IL-6. RORγt and RORα act as the master transcriptional factors for T_H17 cells. (See text for details about their differentiation)

(Hsieh et al. 1993; Scharton and Scott 1993). The T_H2 cells, which differentiations are driven by IL-4, are important for the production of immunoglobulin E and the clearance of extracellular organisms (Bottomly 1988; Min et al. 2004; Shinkai et al. 2002). In addition to these effector subsets, $CD4^+$ T cells can differentiate into distinct regulatory subsets (Treg) which express the fork head/winged helix transcription factor Foxp3. Transforming growth factor-β1 (TGF-β) promotes the differentiation of Treg cells, which suppress adaptive T cell responses and prevent auto-immunity (Chen et al. 2003; Sakaguchi 2000).

Until recent years, it has been believed that T_H1 cells mainly dominate the induction and progression of many auto-immune diseases. However, IFN-γ deficient (KO) mice do not show resistance to auto-immunity. On the contrary, they are even more susceptible to auto-immunity (Tran et al. 2000) which led us to hypothesize that there may be an additional T_H subset that is distinct from T_H1 cells. Recently, a new subset of T_H cells that produces IL-17 (T_H17) has been identified and was shown to have a crucial role in the induction of auto-immune diseases, such as rheumatoid arthritis, experimental auto-immune encephalomyelitis (EAE), and allergen-specific responses (Kolls and Linden 2004; Langrish et al. 2005; Nakae et al. 2002). The differentiation of T_H17 cells from naïve T cells requires stimulation by IL-6 and TGF-β, and induces the master transcriptional factors of the T_H17 subset such as retinoid-related orphan receptor γt (RORγt) and RORα (Fig. 1) (Bettelli et al. 2006; Ivanov et al. 2006; Yang et al. 2008). In this review, we discuss the novel function of the pleiotropic cytokine IL-6 in T_H17 cell development and their relationships in auto-immune and inflammatory diseases.

2 Interleukin-6 (IL-6)

IL-6 was first cloned as a B cell stimulatory factor-2 (BSF-2) in 1986 (Hirano et al. 1986). This molecule has various biological activities, such as a strong stimulatory effect on growth of murine plasmacytoma and human myeloma (Kawano et al. 1988; Suematsu et al. 1992), a hepatocyte stimulating factor (Andus et al. 1987; Gauldie et al. 1987), and the induction of acute phase reaction (Kopf et al. 1994). IL-6 knockout (KO) mice revealed the inhibition of the antiviral antibody response after immunization with a vesicular stomatitis virus, compared with wild-type (WT) mice (Kopf et al. 1994). Thus, IL-6 is a pleiotropic cytokine that is involved in the physiology of virtually every organ system.

2.1 *IL-6 Receptors and Signaling*

The IL-6 receptor (IL-6R) system consists of two polypeptide chains: an 80 kDa IL-6 receptor and a 130 kDa signal transducer (gp130) (Hibi et al. 1990; Kishimoto et al. 1992). IL-6R (80 kDa) exists in both a transmembrane form and a soluble form. IL-6 binds to both of these forms, which can then interact with gp130 and trigger signal transduction and gene expression. gp130 is expressed ubiquitously in tissues. In addition to being a component of the IL-6R, gp130 is a component of receptors for other cytokines, such as leukemia inhibitory factor (LIF), IL-11, oncostatin M (OM), and cardiotropin-1 (CT-1), (Gearing et al. 1992; Ip et al. 1992; Pennica et al. 1995; Yin et al. 1993); which explains the functional redundancy of IL-6 superfamily cytokines.

Although gp130 has no intrinsic kinase domain, members of the Janus kinase (JAK) family, such as JAK1, JAK2, and tyrosine kinase 2 (TYK2) are constitutively associated with gp130 (Stahl et al. 1994). Complexes of IL-6, IL-6R, and gp130 phosphorylate these kinases and then activate the cytoplasmic transcriptional factors, signal transducers, and activators of transcription 1 (STAT1) and STAT3 (Fig. 2) (Ihle and Kerr 1995). In addition, IL-6 activates mitogen-activated protein kinase (MAPK), which phosphorylates the nuclear factor for IL-6 (NF-IL6) (Nakajima et al. 1993; Zhong et al. 1994). Thus, IL-6 activates these kinases and transcriptional factors through IL-6R/gp130 complexes, which leads to gene expression.

2.2 *Negative Feedback Regulation of IL-6 Signaling*

Although IL-6 is essential for the regulation of the immune process, overproduction of the cytokine causes inflammation and auto-immune diseases such as rheumatoid arthritis, systemic juvenile arthritis, and Crohn's disease. Therefore, negative feedback regulation of IL-6 signaling is required for immune homeostasis.

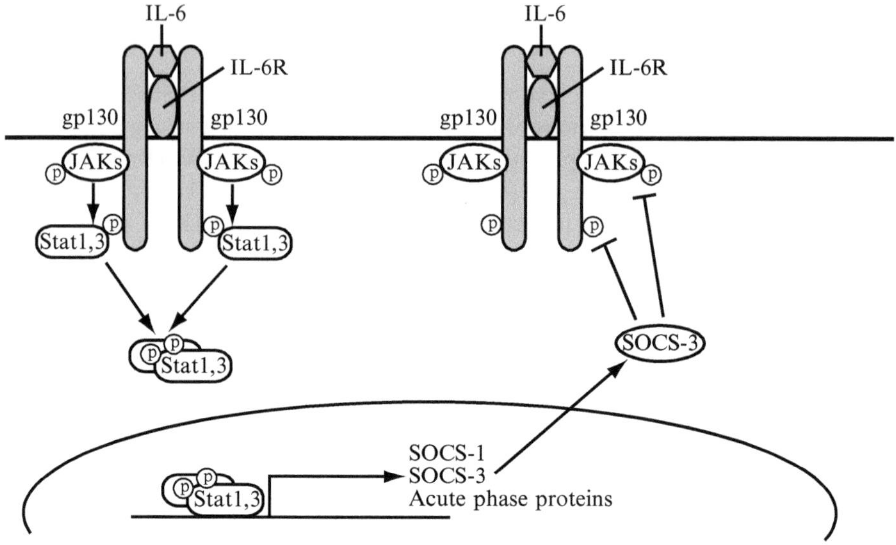

Fig. 2 IL-6 signaling and negative feedback regulation by SOCS. Upon IL-6 interaction, the IL-6R/gp130 complex phosphorylates JAKs and then activates STAT1 and STAT3. Activation of STAT1 and STAT3 induces SOCS1 and SOCS3 gene expression. SOCS1 binds primarily to phosphorylated JAK proteins through its SH2 domain, whereas the SH2 domain of SOCS3 binds to phosphorylated tyrosine residues in the cytoplasmic domain of receptors. These interactions terminate STAT activation and suppress downstream gene expression

Cytokine signaling, such as IL-6 signaling, is negatively regulated by the suppressor of cytokine signaling (SOCS) and the protein inhibitor of activated STATs (PIAS). The SOCS family is composed of eight members: cytokine inducible SRC homology 2 (SH2)-domain-containing protein, (CIS) SOCS1 to SOCS7. SOCS-1, also called STAT-induced STAT inhibitor-1 (SSI-1), and JAK-binding protein (JAB), was initially identified as an intracellular negative-feedback molecule that inhibits JAK-STAT signaling initiated by various stimuli, including IFN-g, IL-4, IL-6, and leukemia-inhibitory factor (LIF) (Endo et al. 1997; Naka et al. 1997; Starr et al. 1997). SOCS-1 mainly inhibits IFN-γ signaling in vivo by binding to JAKs to inhibit its following signal transduction. However, SOCS-1 also negatively regulates innate immune responses such as lipopolysaccharide (LPS) – Toll-like receptor 4 (TLR4) signaling (Kinjyo et al. 2002; Nakagawa et al. 2002). Thus, SOCS-1 acts as an essential negative regulator in not only cytokine signaling, but also TLR signaling.

Although SOCS-3 can be induced by cytokines such as IFN-γ and IL-6, and it can inhibit JAK activation as well as SOCS-1, SOCS-3 binds the cytokine receptors through its SH2 domain (Fig. 2). The activation of Stat1 and Stat3 induced by IL-6 is prolonged in SOCS-3-deficient tissues and cells, but not in SOCS-1-deficient tissues and cells (Croker et al. 2003), which indicates that SOCS-3 is a pivotal regulator of IL-6 signaling in vivo.

2.3 *IL-6 and Immune Diseases*

IL-6 is involved in many diseases such as rheumatoid arthritis, systemic-onset juvenile idiopathic arthritis, systemic lupus erythematosus, Crohn's disease, and inflammatory bowel disease (Gustot et al. 2005; Maini et al. 2006; Tackey et al. 2004; Yamamoto et al. 2006; Yokota et al. 2005). In addition, IL-6 is involved in multiple sclerosis (MS), which is a chronic inflammatory disease affecting the central nervous system (CNS) white matter (Sospedra and Martin 2005). Patients with MS exhibit higher mean levels of IL-6 in their cerebrospinal fluid than normal controls, and the treatment with an anti-IL-6 receptor monoclonal antibody (anti-IL-6R mAb) inhibits the development of EAE, which is a murine model of human MS that shares many pathological and histological characteristics with human MS (Serada et al. 2008).

IL-6 blockade seems to be an innovative treatment strategy for the numerous immune diseases that are impacted by IL-6 overproduction. Actually, blocking IL-6 signaling with a humanized anti IL-6R monoclonal antibody (tocilizumab) is an effective treatment for patients with auto-immune diseases such as Castleman's disease and systemic onset juvenile arthritis (Choy et al. 2002; Nishimoto et al. 2004, 2005; Yokota et al. 2005, 2008). Although many pro-inflammatory cytokines including IL-6, TNF-α and IL-1β are increased in patients with auto-immune diseases such as rheumatoid arthritis and MS, the discovery of T_H17 cells has defined IL-6 blockade as the potent and dominant treatment for these diseases. In Sect. 5, we discuss the details of the relationship between IL-6 and T_H17 cells in auto-immune diseases.

3 IL-17-Producing Helper T Cell (T_H17)

T_H17 cells produce IL-17A (IL-17), IL-17F, IL-22, IL-6 and TNF-α. The IL-17 family is composed of IL-17A, IL-17B, IL-17C, IL-17D, IL-17E (IL-25), and IL-17F. IL-17 is a potent inflammatory cytokine (Kolls and Linden 2004; Moseley et al. 2003). As stated above, auto-immune diseases were previously assumed to be associated with dysregulated T_H1 responses. However, IFN-γ deficiency did not diminish some models of auto-immune diseases like EAE; on the contrary, IFN-γ deficiency worsened the disease. It was recently demonstrated that T_H17 cells are dominantly associated with human and mouse auto-immune diseases such as rheumatoid arthritis, MS and inflammatory bowel disease (Fujino et al. 2003; Komiyama et al. 2006; Nakae et al. 2003; Tzartos et al. 2008). In fact, IL-17 KO mice are resistant to the development of collagen-induced arthritis (CIA) and EAE, and IL-17 blockade by IL-17-blocking antibody prevents the development of EAE (Hofstetter et al. 2005; Komiyama et al. 2006; Nakae et al. 2003).

3.1 T_H17 Cell Differentiation

Although initial reports claimed that IL-23 is required for the generation of T_H17 cells from naïve T cells (Harrington et al. 2005; Langrish et al. 2005), it was subsequently demonstrated that IL-23R is not expressed on naïve T cells and that IL-23 acts as a survival signal for T_H17 cells (Parham et al. 2002; Veldhoen et al. 2006). At present, it is believed that T_H17 cell differentiation is driven by the combination of IL-6 and TGF-β (Bettelli et al. 2006; Mangan et al. 2006). The orphan nuclear receptors, RORγt and RORα, are the key transcription factors that determine the differentiation of the T_H17 lineage. IL-6 together with TGF-β induces these transcription factors, whereas IL-6 inhibits TGF-β-induced expression of Foxp3, a master transcriptional factor for Treg. The levels of RORγt and RORα are significantly reduced in Stat3-deficient T cells, but not in Stat1-deficient T cells under T_H17-polarizing conditions (Diveu et al. 2009; Mathur et al. 2007), which indicates that T_H17 cell differentiation is dependent on Stat3. In contrast to Stat3 activation, Stat1 activation inhibits the development of T_H17 cells (discussed below). Although IL-6 activates both Stat3 and Stat1, Stat3 activation is maintained while Stat1 activation is suppressed in T_H17 cells (Kimura et al. 2007).

Conditioned medium from LPS-stimulated bone marrow-derived dendritic cells (DCCM) can induce the production of IL-17 in naïve T cells. Interestingly, IL-17 was produced by DCCM even with the addition of anti-gp130 antibody or DCCM from IL-6 KO mice, which indicates that there is an IL-6-independent pathway in T_H17 commitment (Kimura et al. 2007). Although several cytokines including TNF-α and IL-23 participate in T_H17 cell development, they are not required for the initiation of T_H17 differentiation. What then, is required for T_H17 differentiation besides IL-6? It has been demonstrated that IL-21 acts as an initiator for T_H17 commitment independent of IL-6 (Korn et al. 2007). IL-21 is a novel cytokine produced by activated T cells and natural killer T (NKT) cells, and its receptor complex is composed of the common IL-2 receptor γ chain (γc) and IL-21 receptor (IL-21R) (Davis et al. 2007). IL-21 and IL-6 inhibit TGF-β-induced Foxp3 expression and induce RORγt and RORα in a Stat3-dependent manner such that naïve T cells differentiate into T_H17 cells (Zhou et al. 2007).

3.2 Other Factors Involved in T_H17 Cell Differentiation

The combination of IL-6 and TGF-β is unable to sustain the activation of Stat1 in T_H17 cells, although it can sustain Stat3 activation. On the other hand, both Stat1 and Stat3 remained activated in T_H17 cells induced by DCCM (Kimura et al. 2007). These findings provide a novel and unknown basis for T_H17 cell differentiation from naïve T cells. In fact, transcriptional factors such as Interferon-regulatory factor 4 (IRF4) and T-bet act as the positive and negative regulator for T_H17 commitment, respectively (Brüstle et al. 2007; Rangachari et al. 2006). It has been also reported that retinoic acid inhibits T_H17 cell development, and dioxin, a ligand of Aryl hydrocarbon receptor, promotes the generation of T_H17 cells (Kimura et al. 2008; Mucida et al. 2007; Quintana et al. 2008; Veldhoen et al. 2008). Thus, IL-6 plays a central role in T_H17 cell differentiation, whereas various factors regulate T_H17 cell development.

3.3 *Negative Regulation of T_H17 Cell Differentiation*

There is a negative regulatory system for T_H17 cell differentiation. IL-27 and IFN-γ are responsible for the inhibition of its development in a Stat1-dependent manner (Batten et al. 2006; Cruz et al. 2006; Stumhofer et al. 2006). IL-27, another IL-12 family member, uses a receptor complex composed of IL-27R and gp130 to transduce its signal and activates both Stat1 and Stat3 (Hunter 2005). Although both IL-6 and IL-27 transduce their signals via the gp130-JAK-STAT axis, IL-6 initiates T_H17 commitment dependent on Stat3 activation, and IL-27 inhibits its development dependent on Stat1 activation. Additionally, IL-27 augments T-bet, the master transcriptional factor for T_H1 cells (Hunter 2005), which indicates that IL-27 has both pro and anti-inflammatory properties in Th cell differentiation.

Laurence et al. demonstrated that IL-2 also inhibits T_H17 cell development (Laurence et al. 2007). They found that IL-2 cannot inhibit T_H17 cell differentiation in Stat5-deficient T cells, and that there are Stat5 binding sites in the IL-17 promoter region. These findings suggest that Stat5 serves as a repressor. Thus, STAT family members activated by various cytokines provide positive and negative regulation of T_H17 cell differentiation. However, the mechanisms of the regulation have not been elucidated.

4 Aryl Hydrocarbon Receptor: A New Player in T_H17 Cells

4.1 *Aryl Hydrocarbon Receptor*

Aryl hydrocarbon receptor (Ahr), also known as dioxin receptor, is a ligand-activated transcription factor that belongs to the basic-helix-loop-helix-PER-ARNT-SIM family (Burbach et al. 1992; Ema et al. 1992). Ahr is present in the cytoplasm where it forms a complex with heat shock protein (HSP) 90, Ahr-interacting protein (AIP), and p23 (Bell and Poland 2000; Kazlauskas et al. 1999; Perdew 1988). Upon binding with a ligand, Ahr undergoes a conformational change, translocates to the nucleus, and dimerizes with Ahr nuclear translocator (Arnt). Within the nucleus, the Ahr/Arnt heterodimer binds to a specific sequence, the xenobiotic responsive element (XRE), which causes a variety of toxicological effects (Dragan and Schrenk 2000; Fujii-Kuriyama et al. 1994; Ohtake et al. 2003; Puga et al. 2005).

Interestingly, it has been recently reported that Ahr is a ligand-dependent E3 ubiquitin ligase (Ohtake et al. 2007), which implies that Ahr has dual functions in controlling intracellular protein levels, serving both as a transcriptional factor to promote the induction of target proteins and as a ligand-dependent E3 ubiquitin ligase to regulate selective protein degradation. Ahr activated by ligands, such as 2,3,7, and 8-tetrachlorodibenzo-*p*-dioxin (TCDD) normalizes the generation of regulatory T cells (Tregs) and modulates the T_H1/T_H2 balance (Funatake et al. 2005; Negishi et al. 2005). In addition, three groups have recently demonstrated that Ahr participates in T_H17 cell differentiation (Kimura et al. 2008; Quintana et al. 2008; Veldhoen et al. 2008). These data collectively demonstrate the importance of Ahr in the differentiation of T cell subsets.

4.2 Ahr Functions in T_H17 Cells

Ahr is specifically induced in naïve T cells under T_H17-polarizing conditions such as TGF-β plus IL-6 or TGF-β plus IL-21. Although the molecular mechanism of Ahr expression in T_H17 development is not known, it is possible that its induction may be regulated downstream of Stat3 by IL-6 and TGF-β because Ahr expression was induced by TGF-β plus IL-6 in Stat1-deficient naïve T cells (Kimura et al. 2008).

As stated above, T_H17 differentiation is positively regulated by IL-6 or IL-21 in combination with TGF-β and negatively regulated by IFN-γ or IL-27. The positive regulation is controlled by Stat3, while the negative regulation is controlled by Stat1. Kimura et al. 2008 found that Ahr binds to Stat1 and Stat5, but not to other members of the Stat family in T_H17 cells and this result suggests that Ahr may regulate the generation of T_H17 cells by modifying the activation of Stat1 and Stat5, which negatively regulate T_H17 generation (Kimura et al. 2008). Indeed, Ahr deficiency prolonged Stat1 activation 24 h after stimulation with TGF-β plus IL-6, whereas its activation was relatively transient and returned to the basal level in WT naïve T cells during the same period. On the other hand, Stat3 activation was equally maintained in both Ahr WT and KO naïve T cells. The mechanism by which Ahr interacts with Stat1 and Stat5 and negatively regulates their activation in T_H17 cell differentiation is not yet understood. Given that Ahr serves both as a transcriptional factor and as a ligand-dependent E3 ubiquitin ligase (Ohtake et al. 2007), it is possible that Ahr marks activated Stat1 for degradation via its ubiquitin ligase function in T_H17 cells (Fig. 3).

5 IL-6 and T_H17 Cells in Auto-immune Diseases

The overproduction of IL-6 or abnormalities in its signal transduction are causative factors in auto-immune disorders including rheumatoid arthritis (Ishihara and Hirano 2002). At present, there is increasing clarification of the role of IL-6 in diseases such as RA in which T_H17 cells are considered to be the primary cause of

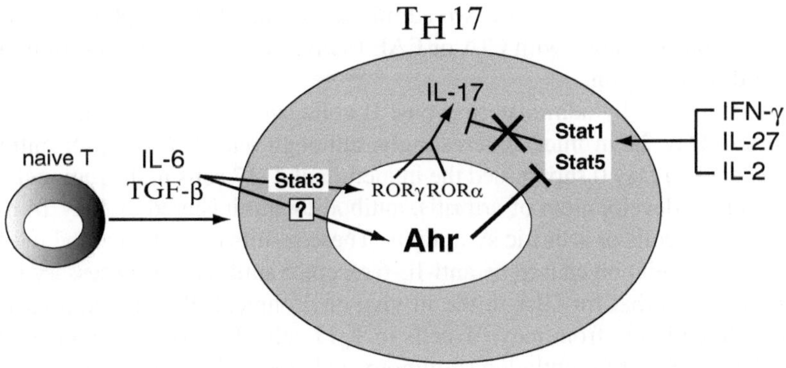

Fig. 3 Ahr functions in T_H17 cells. Ahr is induced under T_H17-polarizing conditions. Ahr participates in T_H17 cell differentiation by regulating Stat1 activation, which suppresses the development of T_H17 cells. Ahr may regulate Stat1 activation by functioning as a ligand-dependent E3 ubiquitin ligase that degrades activated Stat1

pathology. Humanized anti-IL-6 receptor antibodies (Tocilizumab) are currently being used clinically as IL-6-blocking therapy for several auto-immune diseases. This section focuses on the relationship between IL-6 and T_H17 cells in the pathogenesis of inflammatory diseases and an effective approach by Tocilizumab for the treatment of several auto-immune diseases.

5.1 IL-6 and T_H17 Cells in Mouse Auto-immune Disease Models (CIA, EAE)

Previously, T_H1 cells were considered to play a major role in pathogenesis of CIA and EAE. However, in IFN-γ-deficient mice and IFN-γ receptor-deficient mice, CIA and EAE symptoms are not ameliorated, but rather exacerbated (Ferber et al. 1996; Manoury-Schwartz et al. 1997). In contrast, these diseases are suppressed in IL-17-deficient mice (Hofstetter et al. 2005; Komiyama et al. 2006; Nakae et al. 2003), and are alleviated by treatment with IL-17-neutralizing antibodies. These results indicate that in diseases such as CIA and EAE, T_H17 cells are actually the major population in their pathogenesis and the in vivo differentiation and propagation of T_H17 cells can be used as an index of these disease models.

Previous analyses revealed that IL-6 deficient mice are resistant to CIA and EAE (Alonzi et al. 1998; Okuda et al. 1998; Samoilova et al. 1998). However, the reason for this resistance has been poorly understood. In recent years, it has been clarified that T_H17 cells are induced from naïve CD4 T cells by TGF-β and IL-6 in vitro; therefore, it is conceivable that the impaired T_H17 cell differentiation in these mice is the major cause for the resistance to CIA and EAE. However, in genetically deficient mice, their susceptibility to diseases such as CIA and EAE may be influenced by possible intrinsic defects in immune cells or non-uniform genetic backgrounds. Therefore, we administered anti-IL-6 receptor antibodies to congenic wild type mice with CIA or EAE to investigate the in vivo action of IL-6 in T_H17 differentiation.

CIA is induced by administering type II collagen together with an adjuvant on Day 0 and Day 21 in mice. Interestingly, although anti-IL-6 receptor antibodies administered on Day 0 suppressed the induction of T_H17 cells in the regional lymph nodes and the development of arthritis, antibodies administered on Day 14 did not suppress T_H17 cells or arthritic symptoms. These results indicate that the inhibition of T_H17 differentiation caused by anti-IL-6 receptor antibodies is necessary for CIA suppression and that for CIA in the in vivo environment, IL-6 is required for the initial differentiation from naïve T cells to T_H17 cells, but not for the maintenance of T_H17 cells after differentiation (Fujimoto et al. 2008). Moreover, we investigated whether a suppressing effect of T_H17 cell development is observed with TNF inhibitor therapy in CIA. When a TNF-soluble receptor (TNFR-Fc) was administered during the initial CIA induction period (Days 0–14), arthritis and T_H17 differentiation could not be suppressed. However, when TNFR-Fc was administered after Day 21, arthritis is substantially suppressed without any effects on T_H17 cell development (Fujimoto et al. 2008). These results suggest that IL-6 inhibitor treatment in CIA acts primarily on initial CD4 T cell responses including T_H17 cell differentiation, rather than on the effector phase including angiogenesis and osteoclast differentiation. By contrast, it is suggested that the main point of action in TNF inhibitor therapy is different from that in IL-6 inhibitor therapy; it does not play a role in initial T_H17 differentiation but it does act in the effector phase.

EAE is induced by administering the myelin sheath framework protein myelin oligodendrocyte glycoprotein (MOG) peptide together with an adjuvant and pertussis toxin. When anti-IL-6 receptor antibodies were administered immediately after antigen stimulation, the occurrence of EAE could be suppressed in the same manner as for CIA. In EAE models treated with anti-IL-6 receptor antibodies, no T_H17 cells were found in draining lymph nodes or the spinal cord. Immune cells such as T cells, B cells, and macrophages were hardly observed in the lesion of spinal cord (Serada et al. 2008). On the other hand, the effect of anti-IL-6 receptor antibodies on T_H17 cells and the disease onset was abolished when their administration was delayed. This proves that IL-6 is required for the initial differentiation phase for T_H17 in the EAE model, and it appears that IL-6 also acts on cells other than T_H17 cells. These results show that IL-6 inhibitor therapy is highly effective in suppressing the occurrence of EAE (Serada et al. 2008). Analyses of CIA and EAE indicate that initial T_H17 differentiation phase in these auto-immune diseases is fundamentally dependent on IL-6, which suggests that IL-6 is a promising therapeutic target for auto-immune diseases involving T_H17 cell inflammatory functions.

5.2 *IL-6-Blocking Therapy in Human Auto-immune Diseases*

In humans, the therapy by Tocilizumab has become a novel therapeutic strategy for some inflammatory and auto-immune diseases, including RA, systemic-onset juvenile idiopathic arthritis (JIA), Crohn's disease (CD), Castleman's disease, multiple myeloma and systemic lupus erythematosus (SLE) (Illei et al. 2010; Ito et al. 2004; Nishimoto et al. 2005; Yokota et al. 2008). Tocilizumab can block the IL-6 signals induced by the interaction of IL-6 and IL-6R. In the RA patients, Tocilizumab significantly improved the symptoms and ACR (American College of Rheumatology) improvement scores 20, 50, and 70 were 89, 70, and 47%, respectively, and normalized CRP and SAA in the patients within 6 weeks (Nishimoto et al. 2004). Although a role of T_H17 in RA is less clear, it has been reported that IL-17 is detected in the synovial fluid from RA patients and acts as a potent stimulator of osteoclastogenesis (Kotake et al. 1999). Given that IL-6 is important for T_H17 cell differentiation not only in mice, but also in humans, Tocilizumab may improve the symptoms of RA through regulating the development of T_H17 cells. As described above, it has been shown that IL-6 blockade by Tocilizumab is therapeutically effective for other inflammatory diseases such as Castleman's disease, JIA and CD (Ito et al. 2004; Nishimoto et al. 2005; Yokota et al. 2008). However, it is also still controversial whether Tocilitumab can inhibit T_H17 cell differentiation in the improvement of these auto-immune diseases. It is required to demonstrate how Tocilizumab contributes to the treatment for above auto-immune diseases and whether Tocilizumab can also bring the therapeutic benefits for other auto-immune disorders.

TNF-inhibitor therapies (infliximab, etanercept), as well as anti-IL-6 receptor antibodies, are also clinically effective for the treatment of auto-immune diseases such as RA, JIA and Crohn's disease. Although both TNF-α and IL-6 are conventional inflammatory cytokines, our experimental results suggest that they have different roles in T_H17 differentiation and that their therapeutic effects are different in CIA (Fig. 4). Detailed elucidation of the relationship between T_H17 and

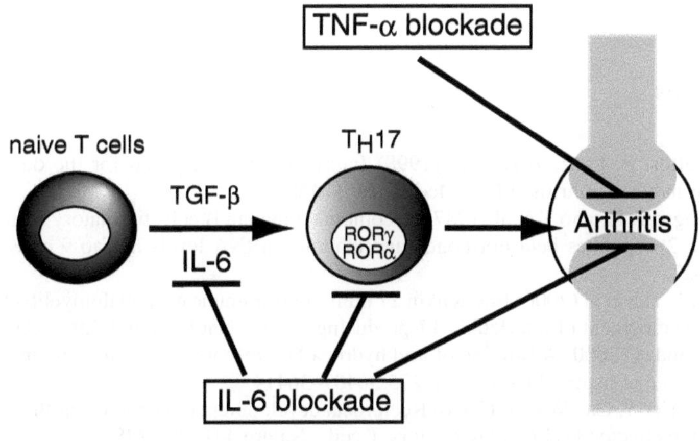

Fig. 4 Different mechanisms between anti-IL-6 receptor antibodies and TNFR-Fc therapies in RA. Anti-IL-6 receptor antibodies inhibit IL-6 at inflammation sites, although they mainly act to subdue the onset of disease by suppressing the initial differentiation phase of T_H17 cells. In contrast, TNF inhibition does not suppress the initial differentiation phase of T_H17, but it is believed to inhibit TNF-a at inflammation sites

pro-inflammatory cytokines such as IL-6 and TNF-α in human auto-immune disorders will provide important information when considering the proper use and switchover of biological agents.

6 Conclusion

In the past two decades, the knowledge on IL-6 has advanced from basic science to medicine. IL-6 is a pleiotropic cytokine that plays a major role in immune response, inflammation, and hematopoiesis, plus its levels are increased in various auto-immune diseases. Tocilizumab is a humanized anti-human IL-6R antibody that inhibits the biological activities of IL-6 by blocking the binding of IL-6 to IL-6R. IL-6 blockade holds therapeutic value in auto-immune diseases, including Castleman's disease, JIA and RA. However, the precise reason why IL-6 blockade leads to the improvement of RA and other human auto-immune disorders is not well understood. The discovery of T_H17 cells sheds light on the novel function of IL-6 and helps to address the above question. T_H17 cells are IL-17-producing helper T cells and IL-17 is involved in the development of several auto-immune diseases. IL-6 is an essential factor for T_H17 cell development, which is one of the reasons why targeting IL-6 activities is an effective approach. In the next decade, we anticipate that the potential of IL-6-targeting therapies in the treatment of various auto-immune diseases will be fully clarified.

Acknowledgments We thank our colleagues in our laboratory for helpful discussion. This work was supported by the Program for Promotion of Fundamental Studies in Health Sciences of the National Institute of Biomedical Innovation and Chugai-Roche Pharmaceutical Co. Ltd, Tokyo, Japan.

References

Alonzi T, Fattori E, Lazzaro D et al (1998) Interleukin 6 is required for the development of collagen-induced arthritis. J Exp Med 187:461–468.

Andus T, Geiger T, Hirano T et al (1987) Recombinant human B cell stimulatory factor 2 (BSF-2/IFN-beta 2) regulates beta-fibrinogen and albumin mRNA levels in Fao-9 cells. FEBS Lett 221:18–22.

Batten M, Li J, Yi S et al (2006) Interleukin 27 limits auto-immune encephalomyelitis by suppressing the development of interleukin 17-producing T cells. Nat Immunol 7:929–936.

Bell DR, Poland A (2000) A Binding of aryl hydrocarbon receptor (AhR) to AhR-interacting protein. The role of hsp90. J Biol Chem 275:36407–36414.

Bettelli E, Carrier Y, Gao W et al (2006) Reciprocal developmental pathways for the generation of pathogenic effector TH17 and regulatory T cells. Nature 441:235–238.

Bottomly K (1988) A function dichotomy in $CD4^+$ T lymphocytes. Immunol Today 9:268–274.

Brüstle A, Heink S, Huber M et al (2007) The development of inflammatory T(H)-17 cells requires interferon-regulatory factor 4. Nat Immunol 8:958–966.

Burbach KM, Poland A, Bradfield CA (1992) Cloning of the Ah-receptor cDNA reveals a distinctive ligand-activated transcription factor. Proc Natl Acad Sci USA 89:8185–8189.

Chen W, Jin W, Hardegen N et al (2003) Conversion of peripheral CD4+ CD25- naïve T cells to CD4+ CD25+ regulatory T cells by TGF-beta induction of transcription factor Foxp3. J Exp Med 198:1875–1886.

Choy EH, Isenberg DA, Garrood T et al (2002) Therapeutic benefit of blocking interleukin-6 activity with an anti-interleukin-6 receptor monoclonal antibody in rheumatoid arthritis: a randomized, double-blind, placebo-controlled, dose-escalation trial. Arthritis Rheum 46:3143–3150.

Croker BA, Krebs DL, Zhang JG et al (2003) SOCS3 negatively regulates IL-6 signaling in vivo. Nat Immunol 4:540–545.

Cruz A, Khader SA, Torrado E et al (2006) Cutting edge: IFN-gamma regulates the induction and expansion of IL-17-producing CD4 T cells during mycobacterial infection. J Immunol 177: 1416–1420.

Davis ID, Skak K, Smyth MJ et al. (2007) Interleukin-21 signaling: functions in cancer and autoimmunity. Clin Cancer Res 13:6926–6932.

Diveu C, McGeachy MJ, Boniface K et al. (2009) IL-27 blocks RORc expression to inhibit lineage commitment of Th17 cells. J Immunol 182:5748–5756.

Dragan YP, Schrenk D (2000) Animal studies addressing the carcinogenicity of TCDD (or related compounds) with an emphasis on tumor promotion. Food Addit Contam 17:289–302.

Ema M, Sogawa K, Watanabe N et al (1992) cDNA cloning and structure of mouse putative Ah receptor. Biochem Biophys Res Commun 184:246–253.

Endo TA, Masuhara M, Yokouchi, M (1997) A new protein containing an SH2 domain that inhibits JAK kinases. Nature 387:921–924.

Ferber IA, Brocke S, Taylor-Edwards C et al (1996) Mice with a disrupted IFN-gamma genes are susceptible to the induction of experimental auto-immune encephalomyelitis (EAE). J Immunol 156:5–7.

Fontenot JD, Rasmussen JP, Williams LM et al (2005) Regulatory T cell lineage specification by the forkhead transcription factor foxp3. Immunity 22:329–341.

Fujii-Kuriyama Y, Ema M, Miura J et al (1994) Ah receptor: a novel ligand-activated transcription factor. Exp Clin Immunogenet 1:65–74.

Fujimoto M, Serada S, Mihara M et al (2008) IL-6 blockade suppresses auto-immune arthritis by inhibition of inflammatory Th17 responses. Arthritis Rheum 58:3710–3719.

Fujino S, Andoh A, Bamba S et al (2003) Increased expression of interleukin 17 in inflammatory bowel disease. Gut 52:65–70.

Funatake CJ, Marshall NB, Steppan LB et al (2005) Cutting edge: activation of the aryl hydrocarbon receptor by 2,3,7,8-tetrachlorodibenzo-p-dioxin generates a population of CD4+ CD25+ cells with characteristics of regulatory T cells. J Immunol 175:4184–4188.

Gauldie J, Richards C, Harnish D et al (1987) Interferon beta 2/B-cell stimulatory factor type 2 shares identity with monocyte-derived hepatocyte-stimulating factor and regulates the major acute phase protein response in liver cells. Proc Natl Acad Sci USA 84:7251–7255.

Gearing DP, Comeau MR, Friend DJ (1992) The IL-6 signal transducer, gp130: an oncostatin M receptor and affinity converter for the LIF receptor. Science 255:1434–1437.

Gustot T, Lemmers A, Louis E et al (2005) Profile of soluble cytokine receptors in Crohn's disease. Gut 54:488–495.

Harrington LE, Hatton RD, Mangan PR et al (2005) Interleukin 17–producing $CD4^+$ effector T cells develop via a lineage distinct from the T helper type 1 and 2 lineages. Nat Immunol 6:1123–1132.

Hibi M, Murakami M, Saito M et al (1990) Molecular cloning and expression of an IL-6 signal transducer, gp130. Cell 63:1149–1157.

Hirano T, Yasukawa K, Harada H et al (1986) Complementary DNA for a novel human interleukin (BSF-2) that induces B lymphocytes to produce immunoglobulin. Nature 324:73–76.

Hofstetter HH, Ibrahim SM, Koczan D et al (2005) Therapeutic efficacy of IL-17 neutralization in murine experimental auto-immune encephalomyelitis. Cell Immunol 237:123–130.

Hunter CA (2005) New IL-12-family members: IL-23 and IL-27, cytokines with divergent functions. Nat Rev Immunol 5:521–531.

Hsieh CS, Macatonia SE, Tripp CS et al (1993) Development of TH1 CD4+ T cells through IL-12 produced by Listeria-induced macrophages. Science 260:547–549.

Ihle JN, Kerr IM (1995) Jaks and Stats in signaling by the cytokine receptor superfamily. Trends Genet 11:69–74.

Illei GG, Shirota Y, Yarboro CH et al (2010) Tocilizumab in systemic lupus erythematosus: Data on safety, preliminary efficacy, and impact on circulating plasma cells from an open-label phase I dosage-escalation study. Arthritis Rheum 62:542–552.

Ip NY, Nye SH, Boulton TG (1992) CNTF and LIF act on neuronal cells via shared signaling pathways that involve the IL-6 signal transducing receptor component gp130. Cell 69:1121–1132.

Ishihara K, Hirano T (2002) IL-6 in auto-immune disease and chronic inflammatory proliferative disease. Cytokine Growth Factor Rev 13:357–368.

Ito H, Takazoe M, Fukuda Y et al (2004) A pilot randomized trial of a human anti-interleukin-6 receptor monoclonal antibody in active Crohn's disease. Gastroenterology 126:989–96.

Ivanov II, McKenzie BS, Zhou L et al (2006) The orphan nuclear receptor RORγt directs the differentiation program of pro-inflammatory IL-17+ T helper cells. Cell 126:1121–1133.

Kawano M, Hirano T, Matsuda T et al (1988) Autocrine generation and requirement of BSF-2/IL-6 for human multiple myelomas. Nature 332:83–85.

Kazlauskas A, Poellinger L, Pongratz I (1999) Evidence that the co-chaperone p23 regulates ligand responsiveness of the dioxin (Aryl hydrocarbon) receptor. J Biol Chem 274:13519–13524.

Kimura A, Naka T, Kishimoto T (2007) IL-6-dependent and -independent pathways in the development of interleukin 17-producing T helper cells. Proc Natl Acad Sci USA 104:12099–12104.

Kimura A, Naka T, Nohara K et al. (2008) Aryl hydrocarbon receptor regulates Stat1 activation and participates in the development of Th17 cells. Proc Natl Acad Sci USA 105:9721–9726.

Kinjyo I, Hanada T, Inagaki-Ohara K (2002) SOCS1/JAB is a negative regulator of LPS-induced macrophage activation. Immunity 17:583–591.

Kishimoto T, Akira S, Taga T (1992) Interleukin-6 and its receptor: a paradigm for cytokines. Science 258:593–597.

Kolls JK, Linden A (2004) Interleukin-17 family members and inflammation. Immunity 21: 467–476.

Komiyama Y, Nakae S, Matsuki T et al (2006) IL-17 plays an important role in the development of experimental auto-immune encephalomyelitis. J Immunol 177:566–573.

Kopf M, Baumann H, Freer G (1994) Impaired immune and acute-phase responses in interleukin-6-deficient mice. Nature 368:339–342.

Korn T, Bettelli E, Gao W et al (2007) IL-21 initiates an alternative pathway to induce pro-inflammatory T(H)17 cells. Nature 448:484–487.

Kotake S, Udagawa N, Takahashi N et al (1999) IL-17 in synovial fluids from patients with rheumatoid arthritis is a potent stimulator of osteoclastogenesis. *J. Clin. Invest.* 103:1345–1352.

Langrish CL, Chen Y, Blumenschein WM et al (2005) IL-23 drives a pathogenic T cell population that induces auto-immune inflammation. J Exp Med 201:233–240.

Laurence A, Tato CM, Davidson TS et al (2007) Interleukin-2 signaling via STAT5 constrains T helper 17 cell generation. Immunity 26:371–381.

Maini RN, Taylor PC, Szechinski J et al (2006) Double-blind randomized controlled clinical trial of the interleukin-6 receptor antagonist, tocilizumab, in European patients with rheumatoid arthritis who had an incomplete response to methotrexate. Arthritis Rheum 54:2817–2829.

Mangan PR, Harrington LE, O'Quinn DB et al (2006) Transforming growth factor-beta induces development of the T(H)17 lineage. Nature 441:231–234.

Manoury-Schwartz B, Chiocchia G, Bessis N et al (1997) High susceptibility to collagen-induced arthritis in mice lacking IFN-gamma receptors. J Immunol 158:5501–5506.

Mathur AN, Chang HC, Zisoulis DG et al (2007) Stat3 and Stat4 direct development of IL-17-secreting Th cells. J Immunol 178:4901–4907.

Min B, Prout M, Hu-Li J et al (2004) Basophils produce IL-4 and accumulate in tissues after infection with a Th2-inducing parasite. J Exp Med 200:507–517.

Moseley TA, Haudenschild DR, Rose L et al (2003) Interleukin-17 family and IL-17 receptors. Cytokine Growth Factor Rev 14:155–174.

Mucida D, Park Y, Kim G et al (2007) Reciprocal TH17 and regulatory T cell differentiation mediated by retinoic acid. Science 317:256–260.

Naka T, Narazaki M, Hirata M (1997) Structure and function of a new STAT-induced STAT inhibitor. Nature 387:924–929.

Nakae S, Komiyama Y, Nambu A et al (2002) Antigen-specific T cell sensitization is impaired in IL-17-deficient mice, causing suppression of allergic cellular and humoral responses. Immunity 17:375–387.

Nakae S, Nambu A, Sudo K et al (2003) Suppression of immune induction of collagen-induced arthritis in IL-17-deficient mice. J Immunol 171:6173–6177.

Nakagawa R, Naka T, Tsutsui H et al (2002) SOCS-1 participates in negative regulation of LPS responses. Immunity 17:677–687.

Nakajima T, Kinoshita S, Sasagawa T et al (1993) Phosphorylation at threonine-235 by a ras-dependent mitogen-activated protein kinase cascade is essential for transcription factor NF-IL6. Proc Natl Acad Sci USA 90:2207–2211.

Negishi T, Kato Y, Ooneda O et al (2005) Effects of aryl hydrocarbon receptor signaling on the modulation of TH1/TH2 balance. J Immunol 175:7348–7356.

Nishimoto N, Yoshizaki K, Miyasaka N et al (2004) Treatment of rheumatoid arthritis with humanized anti-interleukin-6 receptor antibody: a multicenter, double-blind, placebo-controlled trial. Arthritis Rheum 50:1761–1769.

Nishimoto N, Kanakura Y, Aozasa K et al (2005) Humanized anti-interleukin-6 receptor antibody treatment of multicentric Castleman disease. Blood 106:2627–2632.

Ohtake F, Takeyama K, Matsumoto T et al (2003) Modulation of oestrogen receptor signalling by association with the activated dioxin receptor. Nature 423:545–550.

Ohtake F, Baba A, Takada I et al (2007) Dioxin receptor is a ligand-dependent E3 ubiquitin ligase. Nature 446:562–566.

Okuda Y, Sakoda S, Bernard CC et al (1998) IL-6-deficient mice are resistant to the induction of experimental auto-immune encephalomyelitis provoked by myelin oligodendrocyte glycoprotein. Int Immunol 10:703–708.

Parham C, Chirica M, Timans J et al (2002) A receptor for the heterodimeric cytokine IL-23 is composed of IL-12Rbeta1 and a novel cytokine receptor subunit, IL-23R. J Immunol 168: 5699–5708.

Pennica D, Shaw KJ, Swanson TA (1995) Cardiotrophin-1. Biological activities and binding to the leukemia inhibitory factor receptor/gp130 signaling complex. J Biol Chem 270:10915–10922.

Perdew GH (1988) Association of the Ah receptor with the 90-kDa heat shock protein. J Biol Chem 263:13802–13805.

Puga A, Tomlinson CR, Xia Y (2005) Ah receptor signals cross-talk with multiple developmental pathways. Biochem Pharmacol 69:199–207.

Quintana FJ, Basso AS, Iglesias AH et al (2008) Control of T(reg) and T(H)17 cell differentiation by the aryl hydrocarbon receptor. Nature 453:65–71.

Rangachari M, Mauermann N, Marty RR (2006) T-bet negatively regulates auto-immune myocarditis by suppressing local production of interleukin 17. J Exp Med 203:2009–2019.

Sakaguchi S (2000) Regulatory T cells: Key controllers of immunologic self-tolerance. Cell 101:455–458.

Samoilova EB, Horton JL, Hilliard B et al (1998) IL-6-deficient mice are resistant to experimental auto-immune encephalomyelitis: roles of IL-6 in the activation and differentiation of autoreactive T cells. J Immunol 161:6480–6486.

Scharton TM, Scott P (1993) Natural killer cells are a source of interferon gamma that derives differentiation of CD4+ T cell subsets and induces early resistance to Leishmania major in mice. J Exp Med 178:567–577.

Serada S, Fujimoto M, Mihara M et al (2008) IL-6 blockade inhibits the induction of myelin antigen-specific Th17 cells and Th1 cells in experimental auto-immune encephalomyelitis. Proc Natl Acad Sci USA 105:9041–9046.

Shinkai K, Mohrs M, Locksley RM et al (2002) Helper T cells regulate type-2 innate immunity in vivo. Nature 420:825–829.

Sospedra M, Martin R (2005) Immunology of multiple sclerosis. Annu Rev Immunol 23: 683–747.

Stahl N, Boulton TG, Farruggella T et al (1994) Association and activation of Jak-Tyk kinases by CNTF-LIF-OSM-IL-6 beta receptor components. Science 263:92–95.

Starr RTA, Willson EM, Viney LJL et al (1997) A family of cytokine-inducible inhibitors of signaling. Nature 387:917–921.

Stumhofer JS, Laurence A, Wilson EH et al (2006) Interleukin 27 negatively regulates the development of interleukin 17-producing T helper cells during chronic inflammation of the central nervous system. Nat Immunol 7:937–945.

Suematsu S, Matsusaka T, Matsuda T et al (1992) Generation of plasmacytomas with the chromosomal translocation t(12;15) in interleukin 6 transgenic mice. Proc Natl Acad Sci USA 89:232–235.

Szabo SJ, Kim ST, Costa GL et al (2000) A novel transcription factor, T-bet, directs Th1 lineage commitment. Cell 100:655–669.

Tackey E, Lipsky PE, Illei GG (2004) Rationale for interleukin-6 blockade in systemic lupus erythematosus. Lupus 13:339–343.

Tran EH, Prince EN, Owens T (2000) IFN-gamma shapes immune invasion of the central nervous system via regulation of chemokines. J Immunol 164:2759–2768.

Tzartos JS, Friese MA, Craner MJ et al (2008) Interleukin-17 production in central nervous system-infiltrating T cells and glial cells is associated with active disease in multiple sclerosis. Am J Pathol 172:146–155.

Veldhoen M, Hocking RJ, Atkins CJ et al (2006) TGFbeta in the context of an inflammatory cytokine milieu supports de novo differentiation of IL-17-producing T cells. Immunity 24:179–189.

Veldhoen M, Hirota K, Westendorf AM et al (2008) The aryl hydrocarbon receptor links TH17-cell-mediated auto-immunity to environmental toxins. Nature 453:106–109.

Yamamoto T, Saniabadi AR, Umegae S et al (2006) Impact of selective leukocytapheresis on mucosal inflammation and ulcerative colitis: cytokine profiles and endoscopic findings. Inflamm Bowel Dis 12:719–726.

Yang XO, Pappu BP, Nurieva R et al (2008) T Helper 17 Lineage Differentiation Is Programmed by Orphan Nuclear Receptors RORα and RORγ. Immunity 28:29–39.

Yin T, Taga T, Tsang ML (1993) Involvement of IL-6 signal transducer gp130 in IL-11-mediated signal transduction. J Immunol 151:2555–2561.

Yokota S, Miyamae T, Imagawa T et al (2005) Therapeutic efficacy of humanized recombinant anti-interleukin-6 receptor antibody in children with systemic-onset juvenile idiopathic arthritis. Arthritis Rheum 52:818–825.

Yokota S, Imagawa T, Mori M (2008) Efficacy and safety of tocilizumab in patients with systemic-onset juvenile idiopathic arthritis: a randomised, double-blind, placebo-controlled, withdrawal phase III trial. Lancet 371:998–1006.

Zheng W, Flavell RA (1997) The transcription factor GATA-3 is necessary and sufficient for Th2 cytokine gene expression in CD4 T cells. Cell 89:587–596.

Zhong Z, Wen Z, Darnell JE Jr. (1994) Stat3: a STAT family member activated by tyrosine phosphorylation in response to epidermal growth factor and interleukin-6. Science 264:95–98.

Zhou L, Ivanov II, Spolski R et al (2007) IL-6 programs T(H)-17 cell differentiation by promoting sequential engagement of the IL-21 and IL-23 pathways. Nat Immunol 8:967–974.

Apoptosis and T_H17 Cell Differentiation

J. Magarian Blander

Abstract Activation of naive CD4 T cells in culture in the presence of transforming growth factor-β (TGF-β) and interleukin-6 (IL-6) induces their differentiation into T_H17 cells. We discovered that innate recognition of apoptotic cells in the presence of Toll-like receptor engagement directs the simultaneous synthesis of TGF-β and IL-6 by innate immune cells, and provides a cytokine milieu that favors differentiation of naïve CD4 T cells into T_H17 cells. This finding implied that bacterial infections associated with significant apoptosis of host tissues were preferentially the ones that induced T_H17 immunity. The critical role that apoptosis plays in development of T_H17 cells was also evident in vivo where induction of a T_H17 response against an intestinal infection with the rodent pathogen *Citrobacter rodentium* was dependent on the ability of Citrobacter to induce apoptosis of intestinal epithelial cells. Thus, the long known immuno-suppressive nature of apoptotic cell clearance can co-exist with the inflammatory nature of infection by inducing development of a T_H17 response. Other bacterial infections are known to induce T_H17 responses, and the role of apoptosis in directing T_H17 immunity to these infections remains to be investigated. This chapter discusses our original findings and further examines the pathogenesis of T_H17 and IL-17 inducing bacteria with particular emphasis on the expression of virulence factors that mediate host cell apoptosis.

1 Introduction

The innate immune system senses and responds to microbial pathogens and apoptotic host cells through a process that involves the recognition of molecular structures unique to microbial pathogens or to apoptotic cells, respectively. Recognition begins

J.M. Blander (✉)
Immunology Institute, Department of Medicine, Mount Sinai School of Medicine, 1425 Madison Avenue, New York, NY 10029, USA
e-mail: julie.blander@mssm.edu

S. Jiang (ed.), *T_H17 Cells in Health and Disease*,
DOI 10.1007/978-1-4419-9371-7_5, © Springer Science+Business Media, LLC 2011

when these molecular structures engage pattern recognition receptors (PRRs) present on the surface of professional phagocytes such as macrophages and dendritic cells (DCs). Recognition can occur within the context of phagocytosis, a process whereby the professional phagocyte internalizes the microbial pathogen or apoptotic cell into a subcellular compartment called the phagosome (Aderem and Underhill 1999; Kinchen and Ravichandran 2008; Stuart and Ezekowitz 2005). The sequence of events that ensues involves fusion of the phagosome with various endocytic compartments, ultimately leading to its maturation into a phagolysosome characterized by an acidic pH and active hydrolytic enzymes, which initiate the process of processing internalized cargo proteins into small peptides and their presentation within major histo-compatibility complex (MHC) molecules (Savina and Amigorena 2007; Trombetta and Mellman 2005). Peptide-loaded MHC molecules travel to the plasma membrane of the professional phagocyte where they can become engaged by the T cell receptor (TCR) of naïve peptide-specific T lymphocytes. The professional phagocyte becomes an antigen-presenting cell (APC), and within the context of infection-induced expression of co-stimulatory molecules, mediates activation and proliferation of those T lymphocytes by recognizing their cognate ligands on its cell surface (Iwasaki and Medzhitov 2010).

T cell activation is followed by cellular differentiation, and for CD4 T cells, this can be one of many fates such as T helper 1 (T_H1), T helper 2 (T_H2), T helper 17 (T_H17), or regulatory T cells (T_{reg}), each of which is controlled by lineage specific transcription factors and characterized by a unique cytokine expression profile and effector function (Wilson et al. 2009). Other possible T helper cell fates have also been recently described such as T helper 9 (T_H9) (Dardalhon et al. 2008; Veldhoen et al. 2008), T helper 22 (T_H22) (Duhen et al. 2009; Trifari et al. 2009), and T follicular helper (TF_H) cells (King 2009; King et al. 2008; Nurieva et al. 2008; Vogelzang et al. 2008). An important determinant of T cell differentiation fate is the nature of the cytokines within the micro-environment, particularly those made by the priming APC following engagement of its PRRs. For example, inflammatory cytokines such as interleukin-12 (IL-12) and IFN-γ skew CD4 T cell differentiation into T_H1 cells (Wilson et al. 2009). For T_H2 differentiation, the presence of IL-4 in the vicinity of the activated T cell is important (Wilson et al. 2009). Also, DCs can be induced to express the Jagged family of Notch ligands which engages Notch receptor on T cells to initiate *Il4* gene transcription in T cells (Amsen et al. 2009). For T_{reg} cell differentiation, TGF-β is crucial (Wan and Flavell 2007).

Our foray into the physiological trigger for T_H17 cell differentiation is a result of our ongoing interest in how PRR-ligand interactions formed at the beginning of phagocytosis impact the quality of the resultant adaptive immune response (Blander 2007a, b; Blander and Medzhitov 2006a). While phagocytosis of apoptotic cells leads to immune tolerance, phagocytosis of microbial pathogens leads to immunity and host defense against the pathogen (Blander and Medzhitov 2006a). Central to understanding how APCs like DCs induce immunity versus tolerance in response to internalized cargo is examination of the signaling pathways triggered during phagocytosis. The idea that different signaling pathways could be triggered under different conditions of phagocytosis entailed that phagocytosis is not a constitutive process whereby all phagocytic cargo are automatically engulfed and delivered along the same route of phagosome to lysosome, but that an APC could discriminate between a bacterium and an apoptotic cell by virtue of the types of PRRs that were engaged during phagocytosis (Blander and Medzhitov 2004, 2006a). As a result, the fate of the phagosome and ensuing immune response could be tailored accordingly (Blander 2007a, b; Blander and Medzhitov 2004, 2006a, b).

Indeed, each phagocytic cargo has its own unique surface biochemical composition, a composition that would dictate the types of receptors engaged at the plasma membrane of the phagocytic cell. Pathogen associated molecular patterns (PAMPs) present on microbial pathogens engage inflammatory PRRs such as Toll-like receptors (TLRs) and initiate inflammatory phagocytosis (Janeway and Medzhitov 2002; Medzhitov and Janeway 2002). In contrast, an apoptotic cell would engage different kinds of PRRs that recognize 'eat me' signals displayed on the apoptotic cells (Kinchen and Ravichandran 2008). This process results in transduction of a different kind of signal, the nature of which is not clearly defined at this point. Phagocytosis of apoptotic cells is accompanied by the induction of anti-inflammatory genes and likely also a wide variety of tissue repair genes (Erwig and Henson 2008; Li et al. 2010a). This type of phagocytosis is termed as non-inflammatory phagocytosis.

2 Toll-Like Receptor Mediated Innate Recognition of Microbial Pathogens

The signal transduction pathways that have remarkable impact on the function of DCs are those downstream of TLRs. There are currently 13 known TLRs in mice and 10 in humans. TLRs are type I integral membrane glycoproteins. These proteins can form homodimers or heterodimers. They have leucine rich repeats in their extracellular domain and a conserved region in their cytoplasmic domain, known as the Toll/IL-1R (TIR) domain that is crucial for signaling. Remarkably, despite the conservation among LRR domains, different TLRs recognize different PAMPs (Kumar et al. 2009). For example, TLR4 homodimers mediate recognition of lipopolysaccharide, while heterodimers of TLR4 with TLR6 together with the participation of CD36, mediate recognition of oxidized low density lipoproteins and amyloid-β (Stewart et al. 2010). TLR2 forms heterodimers with TLR6 or TLR1 to mediate recognition of triacyl or diacyl lipopeptides, peptidoglycan, lipoteichoic acid, zymosan, and mycobacterial lipoarabinomannan. TLR5 mediates recognition of flagellin and TLRs 3, 7, 8 and 9 mediate recognition of microbial nucleic acids while TLR11 mediates recognition of uropathogenic bacteria and profilin-like protein expressed by *Toxoplasma gondii.*

Much is known about TLR signaling pathways and their control of adaptive immunity and the reader is referred to several recent reviews on this topic (Iwasaki and Medzhitov 2010; Kumar et al. 2009; Manicassamy and Pulendran 2009; O'Neill and Bowie 2007; Takeuchi and Akira 2010). TLRs are expressed on the plasma membrane and along the endocytic pathway, placing them at the proper locations for detecting PAMPs. TLRs 1, 2, 4, 5 and 6 are primarily expressed on the plasma membrane. However, TLR2 was also shown to be present on the membranes of nascent phagosomes forming around *Saccharomyces cerevisiae* (Underhill et al. 1999). Similarly, TLR4 expression is restricted predominantly to plasma membrane and early endosomes. Presumably, all surface TLRs would be expected to recycle between plasma membrane and endosomal compartments. Signaling from the TLRs at the plasma membrane or from phagosomes results in the transcriptional activation of inflammatory cytokines such as IL-12 and IL-6. TLR3, 7, 8, and 9, on the other hand, are confined to late endocytic compartments. The ligands recognized by intracellular TLRs are all microbial constituents that are exposed only after the microorganisms have entered the endocytic compartment and have been subjected to some degree of degradation. For example, TLR3 recognizes double-stranded RNA, TLR7 and 8 recognize single-stranded RNA, and TLR9 recognizes CpG DNA motifs. The expression of TLR3, 7, 8 and 9 primarily in endocytic compartments is ideally suited for the detection of viruses, which at low pH of endocytic compartments uncoat and release nucleic acids. Similarly, delivery of bacteria to lysosomes results in the degradation of their double-stranded DNA genome into shorter single-stranded unmethylated CpG-containing sequences that engage TLR9. TLR activation by microbial nucleic acids triggers the production of type I interferons that have potent antimicrobial activities. Signaling pathways initiated by TLRs involve the adaptor proteins myeloid differentiation factor 88 (MyD88), TIRAP/MyD88-adaptor-like (MAL), TIR-domain containing adaptor protein inducing IFN-β (TRIF)/TIR-domain containing molecule 1 (TICAM-1), and TRIF-related adaptor molecule (TRAM) (O'Neill and Bowie 2007). TIRAP and TRAM are sorting adaptors; they help recruit the signaling adaptors MyD88 and TRIF, respectively, to the TLR TIR domain at the plasma membrane (Barton and Kagan 2009; O'Neill and Bowie 2007). TRIF is involved in signaling downstream of TLR3, MyD88 lies downstream of all TLRs, while TLR4 uses both adaptors. The result of signal transduction is activation of mitogen activated protein kinases (MAPK) p38, ERK and JNK, and transcription of NF-κB and IRF-responsive genes pivotal for the transcriptional initiation of a number of inflammatory, antimicrobial defense, and immune response genes.

3 Innate Recognition of Apoptotic Cells

3.1 'Find Me' and 'Eat Me' Signals

Recognition and phagocytosis of apoptotic cells proceeds through a series of steps. Excellent reviews have been written on this topic and the reader is referred to these for greater detail (Ravichandran and Lorenz 2007; Savill et al. 2002).The first step in apoptotic cell recognition involves the recruitment of professional phagocytes like DCs, monocytes, and macrophages to sites where apoptotic cells are present, and this recruitment is mediated in response to 'find me' signals released by apoptotic cells. Two major 'find me' signals have been identified so far and these include the lipid lysophosphatidylcholine (Lauber et al. 2003) and nucleotides such as ATP and UTP (Elliott et al. 2009). The second step in apoptotic cell recognition involves engagement of diverse receptors present on phagocytes by various 'eat me' signals displayed on apoptotic cells. 'Eat me' signals include molecules that newly appear on the apoptotic cell surface such as externalized phosphatidyl serine (PS) (Fadok et al. 1992) or intercellular adhesion molecule-3 (ICAM-3) (Gregory et al. 1998; Moffatt et al. 1999) and existing molecules that are altered by oxidation, such as oxidized low density lipoprotein-like moieties (Chang et al. 1999; Oka et al. 1998; Sambrano and Steinberg 1995). These molecules either directly engage receptors on phagocytes or do so indirectly via bridging molecules. Examples of indirect recognition include milk fat globule epidermal growth factor 8 (MFG-E8), which links PS to $\alpha_v\beta_3$ integrin on phagocytes (Hanayama et al. 2002; Savill et al. 1990), growth-arrest-specific 6 (GAS6), which links PS to the receptor tyrosine kinases AXL and MER (Nagata et al. 1996), and Thrombospondin, which bridges the apoptotic cell to the vitronectin receptor (Savill et al. 1990). On the other hand, the major receptor involved in direct recognition of PS was for several years thought to be the PS receptor (PSR) (Fadok et al. 2000). Although this receptor appears to be critical in vertebrate development, it seems not to have a role in apoptotic cell uptake (Bose et al. 2004). The PSR knock out (KO) made by Andreas Lengeling's group did not show defects in clearance of apoptotic cells (Bose et al. 2004) and results with KO mice made by other groups were not as clear (Kunisaki et al. 2004; Li et al. 2003). PS is now shown to be recognized by a T-cell immunoglobulin domain and mucin domain (TIM), 4 molecule on macrophages and DC (Kobayashi et al. 2007; Miyanishi et al. 2007), brain-specific angiogenesis inhibitor 1 (BAI1), a seven trans-membrane protein belonging to the adhesion-type G-protein coupled receptor (GPCR) family (Park et al. 2007), and stabilin-2, a multifunctional scavenger receptor with a large extracellular domain containing four epidermal growth factor (EGF) like domains (Park et al. 2008a). Both BAI1 and stabilin-2 have been shown to mediate apoptotic cell uptake via recognition of PS.

Once the apoptotic cell delivers its 'find me' and 'eat me' signals to the phagocyte, the process of apoptotic cell phagocytosis begins. The small Rho-GTPase Rac plays a central role in the induction of membrane ruffling and cytoskeletal changes that mediate phagocytosis of apoptotic cells and two evolutionarily conserved signaling pathways have been described to participate in the orchestration of this process (Ravichandran and Lorenz 2007). In the first pathway, CED-12 engulfment and migration (ELMO) and CED-5/180 kDa protein downstream of chicken tumor virus no. 10 (Dock180) act together as a guanine nucleotide exchange factor for Rac. In the second pathway, the cytoplasmic adaptor protein and CED-6 homolog GULP functions upstream of Rac. While BAI1 uses the ELMO/DOCK180/Rac signaling module (Park et al. 2007), stabilin-2 uses GULP to mediate Rac activation (Park et al. 2008b). TIM-4 does not appear to engage either of the two pathways above and the dispensability of its transmembrane and cytoplasmic domains for engulfment suggests the existence of another membrane protein that might interact with the extracellular domain of TIM-4 and coordinate signaling (Park et al. 2009).

3.2 *Anti-inflammatory Phagocytosis of Apoptotic Cells*

Apoptotic cell uptake by phagocytes has immuno-suppressive effects, first described as reduced inflammatory cytokine production by human monocytes in response to the TLR ligand lipopolysaccharide (LPS) and the upregulation of IL-10 (Voll et al. 1997). Apoptotic cells have also been known to induce the synthesis of anti-inflammatory mediators such as TGF-β, prostaglandin E2 and platelet activating factor by macrophages (Fadok et al. 1998). Recognition of apoptotic cell 'eat me' signals by phagocyte receptors without physical internalization has been reported to be sufficient for inducing the anti-inflammatory response from macrophages (Cvetanovic and Ucker 2004). These studies suggest that signals for the internalization of apoptotic cells and for the synthesis of anti-inflammatory cytokines are distinct and likely diverge downstream of receptor activation at the plasma membrane. Of the PS specific PRRs, given that these are the ones known to primarily mediate TGF-β synthesis in response to apoptotic cells so far, TIM-4 lacks signaling capability (Park et al. 2009). Stabilin-2 activation or agonist anti-stabilin-2 antibody induced TGF-β, suggesting that this receptor may be involved not only in clearance of apoptotic cells, but activation of downstream signaling pathways leading to the release of TGF-β (Park et al. 2008a). The ability of BAI1 to induce TGF-β has not yet been studied.

Although it is still not clear which receptor mediates TGF-β synthesis in response to apoptotic cell recognition, transcriptional and translational regulation appears to be involved. Upregulation of TGF-β mRNA is sensitive to inhibition of p38 MAPK, ERK, or JNK, and translation of TGF-β in response to apoptotic cell uptake required activity of Rho kinase, phosphorylation of PI3K/Akt and mTOR, and subsequent activation of eukaryotic initiation factor 4E (eIF4E) (Xiao et al. 2008). Recent studies have shown that calcium influx is necessary for TGF-β synthesis (Gronski et al. 2009). Phagocytosis of apoptotic cells by macrophages in the absence of extracellular calcium and in the presence of inhibitors of Ca^{2+} release showed that both extracellular and intracellular stores of calcium were important for TGF-β secretion. Cytochalasin D, an inhibitor of actin polymerization had no effect on TGF-β production.

Little is known about the intracellular signaling pathways that connect receptor engagement at the phagocyte plasma membrane during apoptotic cell phagocytosis to the initiation of TGF-β gene transcription. Because many constituents of apoptotic cell membranes are modified fatty acids and sterols, a role was postulated for the peroxisome-proliferator-activated receptors (PPARs) and liver X receptors (LXRs), which function as sensors of modified fatty acids and sterols (Castrillo and Tontonoz 2004). It was shown that phagocytosis of apoptotic cells induced activation or expression of PPAR-γ (Johann et al. 2006; Majai et al. 2007), PPAR-δ (Mukundan et al. 2009), and LXR in macrophages (N et al. 2009). Macrophage specific deletion of PPAR-γ led to downregulation of genes involved in the phagocytosis of apoptotic cells (Majai et al. 2007). PPAR-δ deficient macrophages showed decreased expression of bridge molecules such as GAS6 and Thrombsopondin and a resultant decrease in phagocytosis of apoptotic cells (Mukundan et al. 2009). Similarly, macrophages deficient in both LXRα and LXRβ showed impaired phagocytosis of apoptotic cells and LXR activation-increased expression of the MER receptor tyrosine kinase (N et al. 2009). Consistent with these defects in vitro, global or macrophage-specific deletion of PPAR-δ or deficiency in LXRα/β led to defects in apoptotic cell clearance in vivo and consequent increases in serum levels of auto-antibodies and progressive lupus-like auto-immune disease. Importantly, apoptotic cell uptake by PPAR-δ or LXRα/β deficient macrophages was unable to mediate suppression of LPS-induced IL-12, TNF-α or IL-1β transcription and synthesis and failed to induce synthesis of TGF-β and IL-10, demonstrating that PPAR-δ and LXR contribute to the apoptotic cell induced suppression of inflammatory cytokine expression.

4 Innate Recognition of Apoptotic Cells During Microbial Infection Instructs T_H17 Cell Differentiation

T_H17 are a recently described subset of CD4 T cells that produce the cytokine interleukin-17 (IL-17). They differentiate as a distinct lineage from T_H1 or T_H2 subsets and play a crucial role in T cell-mediated adaptive immunity (Korn et al. 2009). T_H17 cells are generally thought to be pro-inflammatory, especially through the production of IL-17A and IL-17F (Korn et al. 2009). T_H17 cells also produce IL-21, IL-10 and IL-22 cytokines. T_H17 cells play an important role in host defense against infection by recruiting neutrophils and macrophages to infected tissues, promoting abscess formation, and inducing expression of antimicrobial peptides (Curtis and Way 2009). Induction of T_H17 cell differentiation in tissue culture paradoxically requires a unique combination of the pro-inflammatory cytokine IL-6 and the anti-inflammatory cytokine TGF-β (Mangan et al. 2006; Veldhoen et al. 2006). However, until recently, a single physiological stimulus that induced simultaneous synthesis of IL-6 and TGF-β from innate immune cells and subsequent differentiation of T_H17 cells was not yet known. Differentiation of CD4 T cells into immuno-regulatory T_{reg} cells, which suppress immune responses, also depends on TGF-β for differentiation, in the absence of IL-6. Not surprisingly, there is a reciprocal relationship between the development of these two T cell subsets depending on the presence or absence of IL-6 (Bettelli et al. 2006). We were very interested in gaining insight into the nature of the cytokines made in response to phagocytosis of infected apoptotic cells by DCs. In addition, we knew that antigens from infected apoptotic cells could be presented by MHC class II molecules to naïve CD4 T cells, so naturally we were curious what the consequences of this type of presentation would have on the differentiation fate of those activated CD4 T cells (Torchinsky et al. 2009). An infected apoptotic cell is expected to carry signatures of both self and non-self and present mixed signals to phagocytes. We hypothesized that phagocytosis of an infected apoptotic cell might constitute the type of signal that would not only engage pathways necessary for synthesis of TGF-β but also, because the phagocytic cargo is an infected cell, carry ligands for TLR engagement and contribute to the simultaneous synthesis of pro-inflammatory cytokines such as IL-6 (Torchinsky et al. 2009). This combination of cytokines would be conducive for T_H17 differentiation. As an extension of this hypothesis, phagocytosis of apoptotic cells in the absence of infection would induce synthesis of TGF-β, which has been shown to preferentially induce differentiation of regulatory T cells (Wan and Flavell 2007).

We tested our hypothesis in several ways. As a source of apoptotic cells, we used either B cells or peritoneal neutrophils. For infected apoptotic cells, we used two types of cells; either a model infected cell, taking advantage of the ability of live B cells to internalize and incorporate LPS into their membranes, or we injected mice intra-peritoneally with *Escherichia coli* and 12 h later collected neutrophils that were recruited to the peritoneal cavity and had phagocytosed the *E. coli*. To induce apoptosis. We used either antibody against CD95 (Fas), or we UV-irradiated the cells and confirmed that the cells underwent apoptosis and not necrosis. We then gave these bacteria- or TLR ligand-carrying 'infected' and uninfected apoptotic cells to bone marrow derived DC- an examined production of cytokines at both the protein and RNA transcript level. We found that apoptotic cells, 'infected' apoptotic cells and LPS all induced TGF-β synthesis. Interestingly though, only apoptotic cells and 'infected' apoptotic cells were able to induce synthesis of the biologically active TGF-β by DC. Phagocytosis of the 'infected' apoptotic cells induced synthesis of the T_H17 promoting cytokine IL-23, but so did free LPS. While both free LPS and apoptotic LPS blasts induced similar levels of IL-6, IL-12, was suppressed in response to apoptotic LPS blasts as compared to free LPS. We reasoned that this might also favor development of T_H17 over T_H1 cells.

Do these cytokines made in response to the phagocytosis of 'infected' apoptotic cells indeed favor T_H17 cell differentiation? To test this, we collected conditioned medium from DC that had phagocytosed 'infected' or uninfected apoptotic cells, activated naïve CD4 T cells with anti-CD3 and anti-CD28, without neutralizing IFN-γ and in the presence of this conditioned medium (CM), similar to a strategy originally used by Stockinger and Locksley (Veldhoen et al. 2006). We then examined subsequent T cell differentiation. We investigated the presence of IL-17A producing T cells by flow cytometry. As expected, activating T cells under the T_H17 skewing conditions TGF-β and IL-6 resulted in a large percentage of IL-17 producing T cells. When we activated T cells with CM from DC that had phagocytosed apoptotic cells, we did not observe any IL-17 expressing cells. However, when these apoptotic cells were 'infected' (i.e. contained TLR ligands) we found robust induction of T_H17 cells. Supplementing the CM derived from DC that had phagocytosed apoptotic cells with IL-6 restored IL-17 secretion. Because of the reciprocal relationship between T_H17 and regulatory T cells first described by Kuchroo and colleagues (Bettelli et al. 2006), we then examined FOXP3 expression. When we activated T cells in the presence of TGF-β, we confirmed the presence of a high percentage of FOXP3 expressing CD4 T cells. Notably, when we activated T cells in CM from DC that had taken up apoptotic cells, we also found induction of FOXP3 expressing CD4 T cells. Adding IL-6 to these cultures or using CM from DC that had phagocytosed 'infected' apoptotic cells suppressed the appearance of FOXP3 expressing cells. Collectively, our results showed a reciprocal relationship between the induction of IL-17 expressing cells and FOXP3 expressing cells depending on the presence or absence of TLR ligand in the apoptotic cells. When apoptotic cells carried TLR ligand, IL-17 expressing cells, and not FOXP3 expressing cells were induced. In the absence of TLR ligand, FOXP3 expressing cells were preferentially induced over IL-17 expressing cells.

We investigated the requirements for DC induction of T_H17 in response to uptake of 'infected' apoptotic cells (Torchinsky et al. 2009). T_H17 differentiation in response to 'infected' apoptotic cells required the presence of both TGF-β and IL-6 cytokines in the CM from DC, because IL-17 production by T cells was inhibited in a dose-dependent manner in the presence of neutralizing Abs to these cytokines. Neutralizing antibody to the p19 sub-unit of IL-23 showed only a partial effect on IL-17 production, consistent with the role of IL-23 in the maintenance, but not the initiation of a T_H17 response (Bettelli et al. 2006; Mangan et al. 2006; Veldhoen et al. 2006). Anti-TGF-β and anti-p19 together did not have an additive effect on inhibiting T_H17 cell differentiation. Neutralizing IL-6 with antibody to IL-6 also impaired T_H17 cell differentiation. Similarly, DC derived from mice deficient in IL-6 (*Il6*$^{-/-}$) showed severe impairment in their ability to induce differentiation of naive CD4 T cells into T_H17 cells.

TLR signaling within the DC was also required for T_H17 polarization in response to phagocytosis of infected apoptotic cells. CM collected from DC derived from mice deficient in the TLR adaptor proteins MyD88 and TRIF (*Myd88*$^{-/-}$*Trif*$^{-/-}$) was unable to induce T_H17 differentiation. On the other hand, IL-17 expression in response to curdlan (a fungal β-glucan that signals through Dectin-1 and has been shown by the Reis e Sousa group to induce DC to promote T_H17 differentiation in the presence of TGF-β) (LeibundGut-Landmann et al. 2007) was unaffected by MyD88/Trif deficiency. CM from *Myd88*$^{-/-}$*Trif*$^{-/-}$ DC that had phagocytosed 'infected' apoptotic cells not only did not support T_H17 differentiation, but instead supported differentiation of FOXP3-expressing T_{reg} cells. As expected, induction of FOXP3-expressing cells from naïve CD4 T cells upon phagocytosis of apoptotic cells was not affected by deficiency in MyD88 and TRIF. These results collectively showed that DC phagocytosis of apoptotic cells carrying TLR ligands induced differentiation of responding naïve T cells into T_H17 cells (Torchinsky et al. 2009). Given that apoptosis is an important component of the signals mediating T_H17 cell differentiation, the next section briefly reviews the current knowledge of the pathways that execute apoptosis.

5 Signaling Pathways Mediating Apoptosis

Central to apoptosis is the activation of members of the family of caspases, cysteinyl aspartate-directed proteases, which exert important functions in both inflammation and apoptosis. There are two pathways that result in the activation of caspases involved in executing the apoptosis program. The extrinsic pathway is mediated by cell surface receptors of the tumor necrosis factor (TNF) receptor family such as TNFR1, FAS, and TRAIL-R1 (Krammer 2000). The extrinsic pathway is characterized by the activation of initiator caspases such as caspase-8, leading to the activation and proteolytic cleavage of executioner caspases such as caspase-3, as well as cellular substrates such as the DNA-repair enzyme poly (ADP-ribose) polymerase (PARP). The intrinsic pathway is mediated by the mitochondria and is activated by various stress stimuli such as UV irradiation, DNA damage, and growth factor deprivation (Brenner and Mak 2009). The segregation of the extrinsic and intrinsic pathways, however, is not absolute. In some cell types such as liver cells, the intrinsic pathway can also be activated via surface receptors once BID, a BH3-only protein, is cleaved by caspase-8 to generate pro-apoptotic truncated BID (tBID) that translocates to the mitochondria (Yin et al. 1999). Because the intrinsic pathway is targeted by some microbial pathogens, this pathway is more relevant to our discussion here, and will be described in greater detail below.

5.1 Mitochondrial Pathways of Apoptosis

Central to the intrinsic or mitochondrial pathway of apoptosis is mitochondrial membrane permeabilization (MMP), which can occur as a result of two distinct processes, mitochondrial outer membrane permeabilization (MOMP) and permeability transition (PT) at the level of the inner mitochondrial membrane. The essential steps within the two processes are summarized below.

5.1.1 Mitochondrial Outer Membrane Permeabilization

A hallmark of the intrinsic or mitochondrial pathway of apoptosis is mitochondrial outer membrane permeabilization (MOMP) which results in leakage of molecules that are normally confined within the space between the inner and outer mitochondrial membranes (Green and Kroemer 2004; Kroemer et al. 2007). These molecules include cytochrome c, SMAC/DIABLO (second mitochondria-derived activator of caspase/direct IAP-bind protein with low pI), endonuclease G, and AIF (apoptosis inducing factor) among many others. Leakage of cytochrome c into the cytosol leads to its association with APAF1 (apoptosis protease activating factor 1), inducing its oligomerization and subsequent recognition of pro-caspase-9 to form a multiprotein complex called 'the apoptosome' which induces autocatalytic cleavage of pro-caspase-9. The mature caspase-9 then activates effector caspases, such as caspase-3.

On the other hand, the release of SMAC/DIABLO into the cytosol allows its interaction with the inhibitor of apoptosis (IAP) family of proteins, XIAP, c-IAP1, c-IAP2, and survivin, eliminating their effects by promoting the activation and enzymatic activity of mature caspase-3 (Chai et al. 2000). After its release from the mitochondrion, AIF enters the nucleus and initiates fragmentation of chromatin DNA into 50 kb fragments (Susin et al. 1999). The net result of these events are the phenotypic characteristic of apoptosis such as DNA laddering, nuclear condensation, blebbing, and exposure of PS on the plasma membrane (Brenner and Mak 2009; Chipuk and Green 2008).

MOMP is determined by the balance between pro-apoptotic and anti-apoptotic members of the B-cell CLL/Lymphoma-2 (BCL-2) family of proteins, all of which contain a varying number of BCL-2 homology (BH domains 1–4) (Brenner and Mak 2009; Chipuk and Green 2008). The anti-apoptotic Bcl-2 family members are BCL-2, BCL-xL, BCL-w, A1, and MCL-1 (myeloid cell leukemia 1). These proteins contain four BH domains (1–4), and function to directly bind and inhibit the pro-apoptotic family members. There are two groups of pro-apoptotic BCL-2 family members; BAX (BCL-1 associated x protein) and BAK (BCL-1 antagonist killer 1) are the effector proteins and contain three BH domains (1–3), while BAD (BCL-2 antagonist of cell death), BIM (BCL-2 interacting mediator of cell death), BIK (BCL-2 interacting killer), BID (BCL-2 interacting domain death agonist), bNIP3 (BCL-2/adenovirus E1B 19-KD protein-interacting protein 3), HRK (Harakiri), Noxa and PUMA (p53 upregulated modulator of apoptosis) are in a second group termed ‘BH-3 only’ because they contain one BH3 domain. These proteins function as both inhibitors of the anti-apoptotic BCL-2 family members as well as activators of BAX and BAK. Activation of BAX and/or BAK directly precedes MOMP and is defined as their stable insertion into the outer mitochondrial membrane and oligomerization to form proteolipid pores followed by MOMP (Chipuk and Green 2008).

5.1.2 Permeability Transition

Permeability transition (PT) involves a multi-protein pore called the permeability transition pore complex (PTPC) (Kroemer et al. 2007; Zhivotovsky et al. 2009). Although the composition of the PTPC is an intensely debated area (Zhivotovsky et al. 2009), the consensus is that two mitochondrial proteins, the voltage dependent anion channel (VDAC) in the outer membrane and the adenine nucleotide transporter (ANT) in the inner membrane in addition to a third protein present within the mitochondrial matrix, cyclophillin D, bring together the inner and outer mitochondrial membranes to form the PTPC at their junction (Kroemer et al. 2007). The PTPC is already assembled during physiological conditions where it functions in exchanging small metabolites between the cytosol and the mitochondrial matrix. However, in response to various forms of cell stress, the PTPC loses its regulated conductance leading to the unregulated entry of solutes and water, dissipation of the mitochondrial transmembrane potential ($\Delta\Psi_m$), and eventual rupture of the outer mitochondrial membrane. The VDAC1 and VDAC2 isoforms can also serve as receptors on the mitochondrial membrane for BAX and BAK in order to facilitate MMP (Green and Kroemer 2004; Kroemer et al. 2007). In the next several sections, we consider induction of the mitochondrial pathway of apoptosis by microbial pathogens.

6 Induction of Host Cell Apoptosis by Bacterial Infections

The mucosal epithelium protects deeper sterile tissues by performing two major functions; acting as a physical barrier against invasion by microbial pathogens and by secreting various anti-microbial peptides that serve as the first line of host defense against the possibility of infection. Pathogenic bacteria possess virulence factors that allow them to circumvent these essential functions of the epithelium. These virulence factors might encode adhesins, pore forming toxins, or type III secretion systems, which compromise epithelial barrier functions. Some of these virulence factors mediate death of the epithelial cells by targeting mitochondrial function and inducing apoptosis as a means of gaining access into sterile host tissues or of evading host defense.

6.1 *Apoptosis Induced by Enteropathogenic Escherichia coli*

In developing countries, the enteropathogenic *Escherichia coli* (EPEC) remains one of the leading causes of diarrhea in infants (Chen and Frankel 2005). EPEC attach to the intestinal epithelium, lead to loss (effacement) of microvilli, and the formation of actin-rich pedestals underneath the adherent bacteria, a process leading to attaching and effacing (A/E) lesions. EPEC disrupt intestinal epithelial tight junctions (Weflen et al. 2009) and induce apoptosis as demonstrated by the exposure of phosphatidyl serine on the outer leaflet of the plasma membrane, subsequent DNA fragmentation, and release of ATP (Abul-Milh et al. 2001; Baldwin et al. 1993; Barnett Foster et al. 2000; Crane et al. 1999, 2002; Malish et al. 2003).

A 35,624-bp genetic element known as the locus of enterocyte effacement (LEE) is necessary for the process of A/E (McDaniel et al. 1995; McDaniel and Kaper 1997). The LEE is a conserved pathogenicity island that is restricted to EPEC and entero-hemorrhagic strains of *E. coli* (EHEC), as well as the rodent pathogen *Citrobacter rodentium*. Indeed, *C. rodentium* has served as a faithful model for the study of the human infections with EPEC and EHEC unveiling the functions of the LEE in the pathogenesis of the infection. The LEE encodes a type III secretion system (T3SS), a translocation apparatus, regulator proteins, an outer membrane protein called intimin, the translocated intimin receptor (Tir), and various effector proteins including EspF, EspG and Map (Pallen et al. 2005). Intimin and Tir interactions are critical for A/E. Intimin is essential for firm adhesion of EPEC to the intestinal epithelial cell, while translocation of Tir into the host cell cytoplasm and its subsequent phosphorylation, initiates actin assembly leading to the formation of cup like structures or pedestals upon which the bacteria lay. More proteins involved in cytoskeletal remodeling are then recruited to the pedestal where they bind to the N-terminus of Tir, further anchoring the EPEC to the host cell cytoskeleton. EPEC can also induce its own internalization by non-phagocytic intestinal epithelial cells via a process termed as 'invasion' (Andrade et al. 1989; Donnenberg et al. 1989). Additionally, using the translocation apparatus, which consists of a needle-like structure extending from the bacterium to the plasma membrane of the intestinal epithelial cell, EPEC injects several effector proteins into the cytoplasm (Chen and Frankel 2005; Cossart and Sansonetti 2004; Dean et al. 2005). Among these effectors, Map and EspF possess an N-terminal mitochondrial targeting sequence (MTS) (Kenny and Jepson 2000; Nagai et al. 2005; Nougayrede and Donnenberg 2004). While both Map and EspF disrupt mitochondrial transmembrane potential ($\Delta\Psi_m$), EspF is reported to predominate and trigger apoptosis (Kenny and Jepson 2000; Nagai et al. 2005; Nougayrede and Donnenberg 2004). Although Map alters the morphology of the mitochondria likely leading to alterations in organelle function that are presently undefined (Ma et al. 2006; Papatheodorou et al. 2006), the main cellular functions attributed to Map rely on its location within the cytoplasm. For example, Map has been reported to mimic the active form of the Rho-family GTPase Cdc-42, resulting in the formation of stress fibers and filopodia (Alto et al. 2006). In the absence of a specific role for Map in its mitochondrial location, it is believed that mitochondrial targeting of Map serves to control its cytosolic functions (Papatheodorou et al. 2006). It remains possible that Map exerts an anti-apoptotic role counteracting the dominant pro-apoptotic roles of EspF.

The effector protein EspF induces the characteristic features of the mitochondrial pathway of apoptosis in host cells (Nagai et al. 2005; Nougayrede and Donnenberg 2004). This was demonstrated by generating a mutant strain lacking EspF (Crane et al. 2001; McNamara and Donnenberg 1998) which importantly behaved similarly to the wild type (WT) strain in all aspects of the A/E process including adherence to and invasion of colonic cell lines (Crane et al. 2001; McNamara and Donnenberg 1998), and ability to mediate cytoskeletal changes and tyrosine phosphorylation of Tir (McNamara and Donnenberg 1998). However, the EspF mutant EPEC strain lost its ability to induce dissipation of $\Delta\Psi_m$ and release of cytochrome c with concomitant inability to induce pro-caspase-9 and pro-caspase-3 cleavage (Nougayrede and Donnenberg 2004). Using site-directed mutagenesis of the 16th Leucine of EspF (L16E) in *C. rodentium*, the critical function of this amino acid residue in executing apoptosis was shown such that L16E impaired the mitochondrial localization of EspF and abrogated the ability of Citrobacter to result in dissipation of $\Delta\Psi_m$, release of cytochrome c and induction of host cell apoptosis (Nagai et al. 2005). EspF is an approximately 21 kDa protein containing three proline rich domains, the presence of which in proteins usually indicates potential interactions with other proteins containing SH3 domains or WW domains. Therefore, it is likely that EspF could interact with host cell proteins, but so far the search for host cell binding proteins has yielded a protein member of the ABC family of transporters, Abcf2 (McNamara et al. 2001). Infection of cells with EPEC reduced the levels of Abcf2, and knock down of this protein by RNA interference resulted in enhanced EspF-mediated caspase-9 and caspase-3 cleavage. These results point to a protective role for Abcf2 against EPEC-induced apoptosis.

In addition to inducing host cell apoptosis, EspF also disrupts tight junction formation both in vitro and in vivo resulting in compromised integrity of the intestinal epithelial barrier (Guttman et al. 2006a; Ma et al. 2006; McNamara et al. 2001; Shifflett et al. 2005; Weflen et al. 2009). This has led to the question of whether loss of tight junctions was a direct result of apoptosis of the intestinal epithelial cells. It was shown that apoptosis and tight junction disruption were independent processes. While EspF (L16E) was unable to promote apoptosis, tight junction disruption and barrier disruption were not impaired (Viswanathan et al. 2008). Furthermore, the pan-caspase inhibitor Q-VD-OPH had no effect on infection-induced loss of barrier function, despite its reduction of EPEC-induced host cell apoptosis (Viswanathan et al. 2008). Tight junction disruption was also not a result of the host inflammatory response to the infection. Colonic tight junctions were intact 21 days post infection of C57BL/6J mice with *C. rodentium,* a time point when bacteria had been cleared from the colon but a cellular inflammatory infiltrate persisted (Guttman et al. 2006b). In tissues where both bacteria are attached and inflammation persisted, tight junctions were altered only where bacteria were intimately attached, further demonstrating that tight junction disruption was not due to inflammatory cell infiltration (Guttman et al. 2006b).

6.2 *IL-17 and T_H17 Responses to Citrobacter rodentium*

Although *Citrobacter rodentium* infection was known to induce T_H1 responses (Higgins et al. 1999), the laboratory of Casey Weaver was the first to report that infection of mice with *C. rodentium* induced a robust T_H17 response within the intestinal lamina propria (Mangan et al. 2006). Mice infected with *C. rodentium* develop colitis that resolved within 2–3 weeks due to clearance of bacteria and development of a protective humoral and CD4 T cell mediated immune response (Bry and Brenner 2004; Maaser et al. 2004; Simmons et al. 2003). Deficiency in the p19 sub-unit of IL-23 led to significantly decreased inflammation within the colon, but resulted in impaired bacterial clearance and the mice succumbed to infection within 10–14 days (Mangan et al. 2006). T_H17 responses were comparable between WT and *p19*$^{-/-}$ mice, demonstrating that IL-23 was not critical for development of IL-17 secreting CD4 T cells in this model (Mangan et al. 2006). The difference appeared to be in the protection mediated by this CD4-T_H17 response, and while this response was protective in WT mice, it was not protective in *p19*$^{-/-}$ mice. This suggested that IL-23 mediated the induction of a factor that has a significant contribution towards protection against infection.

Subsequent studies by the laboratory of W. Ouyang showed that while IL-17 transcripts were similar in WT and *p19*$^{-/-}$ mice, IL-22 transcripts were decreased in *p19*$^{-/-}$ mice (Zheng et al. 2008). Thus, IL-23 regulated the production of IL-22 during infections with *C. rodentium*. To address whether IL-22 mediated protection, the authors inoculated *Il22*$^{-/-}$ mice and found 80–100% mortality in the second week post infection. These mice exhibited mucosal hyperplasia, sub-mucosal inflammation, multiple foci of colonic ulceration, and increased numbers of bacteria penetrating deeper into the colonic crypts and present within the mesenteric lymph nodes, spleen and liver (Zheng et al. 2008). This data supported the role for IL-22 in maintaining the integrity of the intestinal epithelial barrier. The protective role for IL-22 was mediated through the production of anti-microbial peptides of the Reg III family, Reg IIIγ and Reg IIIβ, as well as S100A8 and S100A9. Unlike IL-17A, the levels of IL-22 were similar in WT and *Rag2*$^{-/-}$ mice, indicating that the main source of IL-22 was innate immune cells. However, unlike WT mice, *Rag2*$^{-/-}$ mice succumbed to infection between days 24 and 30 post infection, indicating that the adaptive immune response is nonetheless important for mounting a protective immune response to Citrobacter infection.

In addition to IL-22, IL-17A, and IL-17F are also important for host defense against *C. rodentium* infection. Ishigame et al. compared the susceptibilities of *Il17a*$^{-/-}$, *Il17f*$^{-/-}$, and *Il17a*$^{-/-}$*Il17f*$^{-/-}$ mice to infection with *C. rodentium* (Ishigame et al. 2009). Deficiency in IL-17A or IL-17F resulted in greater bacterial burdens in the colons compared to WT mice, and these burdens were similar regardless if either IL-17A or IL-17F were absent. More severe splenomegaly, colonic hypertrophy, and inflammation were noted in the *Il17f*$^{-/-}$ and *Il17a*$^{-/-}$ *Il17f*$^{-/-}$ than *Il17a*$^{-/-}$ mice suggesting a more prominent role for IL-17F than IL-17A in this model. While the levels of neutrophil chemo-attractants were similar, a defect in β-defensin production was noted in the colons of *Il17f*$^{-/-}$ and *Il17a*$^{-/-}$ mice, likely explaining in part the increased susceptibilities of these mice to Citrobacter, especially since the levels of Citrobacter-specific IgG were not affected in the mutant mice. Of note, while IL-17F was produced mainly by colonic epithelial cells and innate immune cells, IL-17A was lost in *Rag2*$^{-/-}$ mice indicating that the main source of IL-17A was likely to be T_H17 cells. Collectively, the data showed a synergistic role for IL-17A and IL-17F in protection against Citrobacter infection.

6.3 Apoptosis and TLR Signaling: Critical Components of the T_H17 Response to Citrobacter Infection

As an extension to our experiments in vitro showing that phagocytosis of infected apoptotic cells by DCs instructed T_H17 differentiation of naïve CD4 T cells, we asked whether blockade of apoptosis impaired T_H17 development in vivo during a bacterial infection known to trigger T_H17 responses. We chose *Citrobacter rodentium* because the laboratories of C. Weaver and W. Ouyang had shown that Citrobacter infection induces a robust T_H17 response in the intestinal lamina propria of infected animals (Mangan et al. 2006; Zheng et al. 2008). When we examined the biology of Citrobacter infections, we found that the laboratories of B. Finlay, M. Donnenberg and C. Sasakawa had all reported that *C. rodentium* induced apoptosis both in vitro in cell culture as well as in vivo in the intestinal epithelium of infected animals (Crane et al. 2001; Nagai et al. 2005; Nougayrede and Donnenberg 2004; Vallance et al. 2003). Thus, infection with *C. rodentium* was perhaps the perfect in vivo model to test whether apoptosis of infected cells was an important component of the innate immune signals that instruct T_H17 cell differentiation.

We blocked apoptosis by treating Citrobacter infected mice with the pan-caspase inhibitor Q-VD-OPH (Torchinsky et al. 2009). Infection with Citrobacter-induced massive apoptosis demonstrated by the presence of a significant number of TUNEL positive cells in colonic sections of infected animals. When we infected mice with Citrobacter while administering Q-VD-OPH during the course of the infection, we found a statistically significant reduction in the number of TUNEL positive cells. We then investigated the consequences of blocking apoptosis within the intestinal tract on the Citrobacter specific T cell response. We infected two strains of mice, C57BL/6J and the more susceptible C3H/HeOuJ (Vallance et al. 2003). Citrobacter infected mice had slightly higher percentages of CD4 T cells in the lamina propria lymphocytes than uninfected mice, and this was not affected by Q-VD-OPH treatment (Torchinsky et al. 2009). As expected, we were able to reproduce the characteristic IL-17-producing cell population in the colonic LPL of B6 mice, and a very robust T_H17 response in the C3H/HeOuJ mice. Treatment with Q-VD-OPH decreased the population of IL-17 producing cells in both strains of mice to levels seen in uninfected animals.

To ensure that the caspase inhibitor Q-VD-OPH did not have an impact on T cell proliferation, we isolated T cells from the mesenteric lymph nodes (MLN) of uninfected mice or mice infected with Citrobacter, and treated or did not treat with Q-VD-OPH (Torchinsky et al. 2009). We then incubated these MLN cells with DC that had been exposed to heat-killed *C. rodentium* in culture. Proliferation was assessed by incorporation of ^{3}[H]thymidine. Only the MLN T cells when isolated from mice that were infected with Citrobacter proliferated specifically to Citrobacter antigens presented by the DC. Importantly, administration of Q-VD-OPH did not decrease proliferation of MLN T cells from infected mice. We also investigated whether Q-VD-OPH had an impact on T cell differentiation per se (Torchinsky et al. 2009). When we isolated CD4 T cells and then activated them in T_H1 or T_H17-skewing conditions, we found no differences in the induction of IFN-γ, or of IL-17, indicating that T cell differentiation was unaffected by Q-VD-OPH.

Although we were careful to exclude the effects of the pan-caspase inhibitor on immune function, it is nonetheless a broad-spectrum drug that could have various effects on other processes. We decided to use a fundamentally different approach to block apoptosis of the intestinal epithelium. We took advantage of a *C. rodentium* mutant that is unable to induce host cell apoptosis. We focused on two effector proteins, EspF (EPEC-secreted protein F) and Map (mitochondrial-associated protein), both of which are effectors encoded by the locus of enterocyte effacement present within attaching and effacing pathogenic *E. coli* (see Sect. 6.1 above). EspF has been shown to be important in inducing apoptosis of host cells (Nagai et al. 2005; Nougayrede and Donnenberg 2004). Therefore the EspF mutant should be unable to induce host cell apoptosis while still infecting the intestinal epithelium. As a control we turned to Map, which is not involved in inducing apoptosis of host cells. Therefore, the Map mutant should still be able to induce apoptosis. The laboratories of B. Finlay and G. Frankel had worked extensively with these mutants and had shown that despite similar colonization, shedding in the stool and colonic hyperplasia, only WT and Map mutants induce apoptosis and tight junction disruption in vitro and in vivo, whereas EspF mutants failed to do so (Nagai et al. 2005; Nougayrede and Donnenberg 2004; Vallance et al. 2003; Viswanathan et al. 2008). We confirmed this and found a significant increase the number of TUNEL positive cells in the distal colon of WT- and Map-infected, but not EspF-infected mice (Torchinsky et al. 2009). We also established similar intestinal colonization by these mutants using an antibody to *C. rodentium* LPS O antigen.

When we investigated the presence of IL-17 producing cells in these mice, we found that the mice infected with EspF, which does not induce apoptosis, had a reduction in IL-17 producing cells as compared to mice infected with WT *C. rodentium*, and closer to the levels in uninfected mice (Torchinsky et al. 2009). Infection with the Map mutant, which retains the ability to induce host cell apoptosis, also retained the IL-17 producing cells. The trend was similar in the small intestinal lamina propria, although the increases in IL-17 producing cells were more modest, as in both strains of mice. However, we noted that in C57BL/6J mice, all three strains of *C. rodentium* induced similar increases in the percentages of IFN-γ^+ CD4 T cells, compared to uninfected controls, and this was consistent with intact expression of the bacterial outer membrane protein intimin, which has been reported to drive T_H1 responses to *C. rodentium* (Kaper et al. 2004; Mangan et al. 2006). All three strains of *C. rodentium* induced similar increases in cellular infiltration in the percentages of CD4 T cells in the lamina propria as assessed by histological analyses (our unpublished observations) and flow cytometry (Torchinsky et al. 2009), respectively.

In all these mice, the numbers of Foxp3 positive cells were similar on day 9 of infection where peak T_H17 responses were observed, which suggested that the increases in IL-17 producing cells on day 9 did not come at the expense of Foxp3 expressing T_{reg} cells (Torchinsky et al. 2009). In other words, these results did not support conversion of T_{reg} cells to T_H17 cells when inflammatory cytokines were present. As a positive control for FOXP3 expression, mice were injected intraperitoneally with anti-CD3, and consistent with previous reports, an increase in Foxp3 expressing cells was observed under these conditions (Kamanaka et al. 2006). Confirming our results in vitro showed the importance of TLR signaling for the development of T_H17 cells in response to infected apoptotic cell recognition and we found that T_H17 responses in the intestinal lamina propria were impaired in *Myd88*$^{-/-}$*Trif*$^{-/-}$ mice (our unpublished observations), indicating the importance of TLRs in mounting T_H17 immunity against *C. rodentium* infection.

We have thus highlighted a novel role for host cell apoptosis during infection in determining the differentiation of T_H17 cells (Fig. 1). In addition to *C. rodentium*,

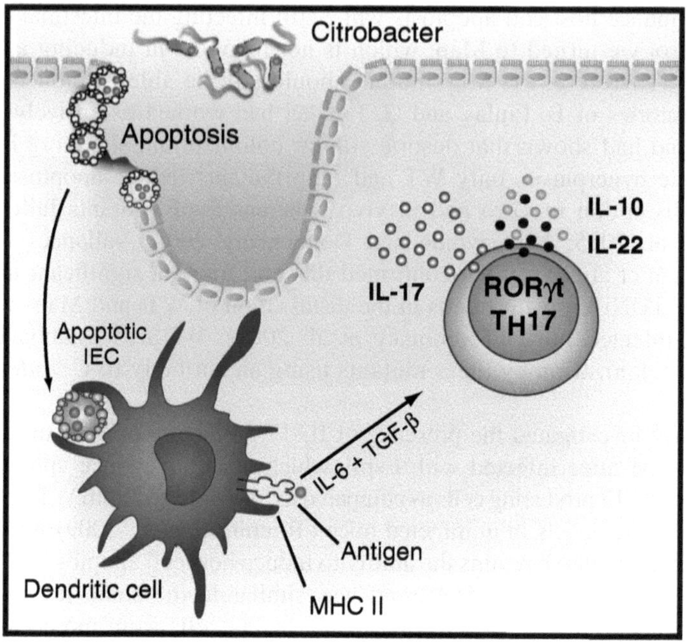

Fig. 1 Innate immune recognition of infected apoptotic intestinal epithelial cells triggers T_H17 responses in the lamina propria of infected animals. Infection with the enteric rodent pathogen *Citrobacter rodentium* induces apoptosis of intestinal epithelial cells. Our experiments in vitro have shown that phagocytosis of TLR ligand-carrying apoptotic cells by bone marrow derived dendritic cells (DC) leads to the simultaneous synthesis and production of IL-6 and TGF-β, a cytokine combination that instructs differentiation of CD4 T cells into IL-17, IL-22, and IL-10 producing RORγt⁺ T_H17 cells (Torchinsky et al. 2009). Based on this data in vitro, we propose that DC within the intestinal lamina propria can phagocytose infected intestinal epithelial cells and present Citrobacter-derived antigens within MHC class II molecules in context of the T_H17 skewing cytokines IL-6 and TGF-β. This leads to the activation of Citrobacter-specific naïve CD4 T cells and their differentiation into RORγt⁺ T_H17 cells. Indeed, we have shown that blocking apoptosis of intestinal epithelial cells during Citrobacter infection impairs the characteristic T_H17 response (Torchinsky et al. 2009). Production of IL-17, IL-10, and IL-22 by effector T_H17 cell populations within the intestinal lamina propria may be ideally suited for clearance of the infecting pathogen as well as repair of the intestinal epithelium damaged by the infection

a number of other bacteria have been shown to drive IL-17 associated immune responses (Khader et al. 2009). It will be important to investigate whether these bacteria also induce apoptosis of the host cells that they infect. In the next several sections, the pathogenesis of various bacteria that have been reported to induce IL-17 responses is considered. Particular emphasis is given to virulence factors that encode pore forming toxins, type III secretion systems (T3SS), and various modulators of host cell apoptosis.

7 Apoptosis and Other Bacterial Infections Known to Trigger IL-17 or T_H17 Responses

Many pathogens other than *Citrobacter rodentium* have been shown to elicit IL-17 responses, including *Helicobacter pylori, Staphylococcus aureus, Klebsiella pneumoniae, Pseudomonas aeruginosa, and Bordetella pertussis* (Khader et al. 2009). The relationship between the ability of these pathogens to induce apoptosis and to trigger IL-17-associated responses is not yet clear and certainly warrants investigation. Within the next several sections, studies showing the association of these infections with IL-17 responses are presented. Also, the pathogenesis of these bacteria is considered with special emphasis on studies that have examined the potential for these pathogens to induce apoptosis in the host cells that they infect. It is noteworthy, that in many of the infections associated with these pathogens, measurements have been made for levels of IL-17 cytokines as well as other cytokines involved in T_H17 responses, such as IL-23, IL-22, and IL-21. However, not all of these studies have shown a definitive involvement of CD4 T cells that have differentiated into the T_H17 lineage. The sources of IL-17 measured could presumably be derived from other cell types such as γδ T cells (Girardi 2006; Lockhart et al. 2006; Martin et al. 2009; Roark et al. 2008; Shibata et al. 2007; Umemura et al. 2007), NKT cells (Coquet et al. 2008; Doisne et al. 2009; Michel et al. 2007, 2008; Rachitskaya et al. 2008), NK cells (Colonna 2009; Cupedo et al. 2009; Passos et al. 2010), neutrophils (Ferretti et al. 2003; Li et al. 2010b), and eosinophils (Molet et al. 2001). Although we have shown a crucial role for apoptosis during infection in instructing T_H17 cell differentiation, whether apoptosis has any role in inducing the synthesis of IL-17 by these other cell types remains to be determined.

7.1 *Helicobacter pylori*

7.1.1 Pathogenesis of *Helicobacter pylori*

Helicobacter pylori, a Gram negative bacterium, constitutes the predominant microbiota of the stomach in about 50% of the human population (Cover and Blaser 2009). Its colonization can extend into the proximal duodenum or distal esophagus. *H. pylori* colonization is asymptomatic, but has been associated with increased risk for peptic ulcers, gastritis, gastric carcinomas, and mucosa-associated lymphomas (Cover and Blaser 2009). Most *H. pylori* reside in the mucus layer overlying the gastric epithelium, but some strains are considered more interactive with their host due to their expression of a *cag* pathogenicity island (PAI), which encodes a type IV secretion system, the effector protein CagA as well as other virulence factors. 'Host interactive' *H. pylori* can secrete a vacuolating cytotoxin (VacA) (also secreted by non-adherent *H. pylori*) encoded by the chromosomal gene *vacA*, or express outer membrane proteins including BabA and SabA, that allow their adhesion to gastric epithelial cells (Cover and Blaser 2009; Maeda and Mentis 2007). The expression of BabA and SabA differs among different strains of *H. pylori*, resulting in strain dependent differences in binding to gastric epithelial cells. Using these 'interaction factors', *H. pylori* leads to multiple alterations in host cells including dephosphorylation of cellular proteins, loss of epithelial cell polarity and tight junctions, and synthesis of the neutrophil chemo-attractant interleukin-8 (reviewed in (Cover and Blaser 2009)).

7.1.2 Induction of Apoptosis by *Helicobacter pylori*

Among the virulence factors of *H. pylori*, the focus here will be VacA because it has been shown to induce apoptosis of gastric epithelial cells (Cover et al. 2003; Galmiche et al. 2000; Kuck et al. 2001; Willhite and Blanke 2004; Willhite et al. 2003). Although VacA can remain surface bound and function as an adhesin, it is also secreted and mediates additional effects on gastric epithelial cells (reviewed in reference (Cover and Blanke 2005)). VacA appears to be important for persistent colonization of the human stomach in as much as all *H. pylori* strains isolated from human stomach contain the *vacA* gene. VacA also contributes to the ability of *H. pylori* to colonize the stomachs of non-primates such as gerbils and mice, because co-infection with $VacA^+$ and $VacA^-$ strains of *H. pylori* favors colonization by the $VacA^+$ strain. VacA inserts into the host cell plasma membrane, oligomerizes to form anion selective membrane channels, and becomes internalized into late endocytic compartments that are $Rab7^+$ and $LAMP\text{-}1^+$. Activity of the channels is promoted by the acidic pH within these compartments, leading to their osmotic swelling and cell vacuolation. VacA additionally reduces trans-epithelial electric resistance of polarized epithelial cells in vitro through an unknown mechanism thought to lead to selective permeabilization and release of molecules such as Fe^{3+}, Ni^{2+}, sugars, and amino acids which may support the growth of *H. pylori* in the gastric micro-environment.

Addition of VacA to cells and intracellular expression of VacA causes dissipation of mitochondrial transmembrane potential (Willhite and Blanke 2004) and release of cytochrome c (Galmiche et al. 2000; Willhite and Blanke 2004) consistent with MOMP (see Sect. 5 above). Transient transfection of VacA resulted in activation of pro-caspase-3 as determined by cleavage of PARP, which was inhibited by co-transfection of BCL-2 (Galmiche et al. 2000). Because VacA-induced apoptosis occurred at doses higher than those required for VacA-induced vacuolation, it was thought that MOMP was an indirect result of VacA cytotoxicity. However, studies showed that VacA translocates to the mitochondrion (Galmiche et al. 2000) suggesting a direct role in MOMP. Furthermore, inhibiting the vacuolating function of VacA with Bafilomycin A1, which blocks endosomal acidification, does not block MOMP – indicating independent roles of VacA in vacuolation and MOMP (Willhite et al. 2003).

Importantly, all *H. pylori* strains are $VacA^+$ and should theoretically induce apoptosis. However, successful induction of apoptosis is also determined by the expression of other virulence factors by the bacterium, especially expression of CagA. The importance of CagA expression was illustrated in a study by C. Sasakawa and colleagues, and showed that unlike the studies in vitro, *H. pylori* inhibits apoptosis of gastric epithelial cells in a Mongolian gerbil model of *H. pylori* infection (Mimuro et al. 2007). The authors induced apoptosis of the gastric superficial pit epithelium by oral administration of etoposide, which triggers the intrinsic pathway of apoptosis. Surprisingly, concomitant infection with *H. pylori* prevented apoptosis, and thereby enhanced the ability of *H. pylori* to colonize the stomach. This activity was mediated by CagA and the type IV secretion system (T4SS), as shown by the inability of Δ*cag* (Cag-A deficient) and Δ*virB7* (T4SS deficient) strains of *H. pylori* to prevent the etoposide-driven apoptosis of gastric epithelial cells. In the absence of etoposide treatment, the authors observed slightly increased numbers of apoptotic cells in the gastric pits 3 weeks post-infection with either the WT, Δ*Cag*, or Δ*virB7* strains, but no significant apoptosis was observed at 8 weeks post-infection; prompting the conclusion that the infection did not significantly affect gastric epithelial cell apoptosis long term in vivo. However, an earlier study also in Mongolian gerbils, by M. Blaser and colleagues had similarly reported a significantly increased pyloric antral apoptosis 2–4 weeks post-infection with *H. pylori*, followed by a decrease to baseline levels observed in uninfected animals (Peek et al. 2000). Given the known interactive nature of *H. pylori* with host epithelial cells, it is likely that $CagA^+$ strains of *H. pylori* successfully colonized the host epithelium by circumventing the normal renewal cycle of epithelial cell proliferation and apoptosis. Indeed, the study by C. Sasakawa showed that *H. pylori* prolongs epithelial cell survival by inducing CagA mediated upregulation of the pro-survival factor phospho-ERK and the anti-apoptotic protein MCL-1 (Mimuro et al. 2007).

7.1.3 IL-17 and T_H17 Responses to *Helicobacter pylori*

As mentioned above, *H. pylori* infection induces synthesis of IL-8 by gastric epithelial cells. The reported ability of IL-17 to induce synthesis of IL-8 by macrophages and epithelial cells prompted investigations into whether increased IL-8, which is associated with *H. pylori* colonization, could be due to the involvement of IL-17 in the local inflammatory response to *H. pylori* (Luzza et al. 2000). A consistently higher level of IL-17 RNA transcripts was detected in gastric mucosa and lamina propria mononuclear cells isolated from gastric biopsies from *H. pylori* positive compared to *H. pylori* negative patients presenting with gastritis (Luzza et al. 2000; Mizuno et al. 2005). A successful *H. pylori* eradication treatment correlated with a decrease in the levels of IL-17 protein within the mucosal tissue, while neutralization of IL-17 resulted in a significant reduction of IL-8 production by gastric lamina propria mononuclear cell cultures (Luzza et al. 2000). Conversely, the addition of recombinant IL-17 to a gastric epithelial cell line increased IL-8 secretion, indicating that IL-17 contributes to IL-8 production in *H. pylori* colonized mucosa. Of note, the study by Mizuno et al. found that all 36 gastric ulcer patients were *H. pylori*-positive, while 14 out of 29 non-ulcer patients were *H. pylori*-negative. All 51 *H. pylori* strains from both gastric ulcer and non-ulcer patients were $CagA^-$ and $VacA^+$, making it tempting to speculate that the IL-17 responses correlated with the lack of CagA expression and based on the studies discussed above (Mimuro et al. 2007; Peek et al. 2000), an intact ability to induce apoptosis. Comparing IL-17 responses in patients with $CagA^+$ versus $CagA^-$ strains will be important before making definitive conclusions regarding the preferential induction of IL-17 responses by $CagA^-$ strains of *H. pylori*.

The major source of IL-17 in *H. pylori* positive biopsies appeared to be both CD4 and CD8 T cells. Discrete populations of CD4 T cells that produced IFN-γ alone were also present at higher percentages in *H. pylori* positive biopsies (Caruso et al. 2008). Measurements of IL-23 by real time PCR and ELISA showed that it was constitutively present in all gastric biopsies, but its expression was significantly increased in biopsies of *H. pylori* positive patients compared to *H. pylori* negative patients (suggesting that it could sustain IL-17 production) (Caruso et al. 2008). Investigations of other T_H17-associated cytokines showed that while IL-21 is constitutively expressed in gastric mucosa, significantly increased levels of IL-21 protein were detectable in gastric biopsies from *H. pylori* positive patients compared to *H. pylori* negative patients (Caruso et al. 2007a). Epithelial cell extracts from all patients, regardless of *H. pylori* colonization, were shown to express similar levels of the IL-21 receptor (IL-21R) (Caruso et al. 2007a). Interestingly, treatment of primary gastric epithelial cells with IL-21 led to the synthesis of matrix metalloproteinases MMP2 and MMP9 (Caruso et al. 2007a), suggesting that the high levels of IL-21 detected within the gastric mucosa of *H. pylori* positive patients may contribute to both amplify T_H17 responses as well as trigger pathways that cause mucosal degradation and remodeling. In both studies (Caruso et al. 2007a, 2008), the authors mention that the IL-23 and IL-21 protein levels did not differ between patients carrying the $CagA^+$ versus the $CagA^-$ strains, but it is not clear whether their assessment of CagA status by Western blot analyses of total protein extracted from patient samples is as sensitive as the reverse hybridization line probe assay used by Mizuno et al. (2005). So although this issue remains to be clarified, the studies collectively point to the likely involvement of the IL-17/IL-23 axis in diseases associated with *H. pylori* colonization (Caruso et al. 2007b).

7.2 *Klebsiella pneumoniae*

7.2.1 IL-17 Associated Aesponses to *Klebsiella pneumoniae*

Jay Koll's group has shown that deficiency in either IL-12 or IL-23 led to reduced survival of mice during a pulmonary challenge with *K. pneumoniae* (Happel et al. 2005). Following infection, the mRNA levels of the p19 sub-unit, unique to IL-23, were detected sooner than those for IL-12 p40 or p35 sub-units within the bronchial alveolar lavage of infected mice. The source of IL-23 was alveolar macrophages and DCs. Within the lung of infected mice, IL-23 expression was required for the expression of IL-17A and IL-17F, while IL-12 was required for the production of IFN-γ. IL-23 regulated the expression of all the IL-17 dependent inflammatory mediators such as G-CSF, IL-6, MIP-1α, MIP-2, LIX, and KC within the lungs. IL-23 was additionally important for the induction of IL-22 that was detectable early after infection. The source of IL-22 did not appear to be innate immune cells, as it was no longer detectable at 24 h in *Rag2*$^{-/-}$*Il2rγ*$^{-/-}$ mice (deficient in natural killer, T and B cells) (Aujla et al. 2008). Instead, the authors show that CD90 (Thy-1)$^+$ T cells isolated from the lungs of infected mice produced significant levels of IL-22 (Aujla et al. 2008). Neutralization of IL-22 in vivo led to increased mortality of mice as a result of the infection as compared to either wild type or *Il17a*$^{-/-}$ mice, indicating a more critical role for IL-22 than IL-17 in host defense against *K. pneumoniae* infection. While *Il17a*$^{-/-}$ mice had higher bacterial burdens within the spleens and lungs than wild type mice, neutralization of IL-22 led to even greater numbers of bacteria within these organs. Furthermore, it appeared that IL-22 and IL-17 had differing roles. While *Il17a*$^{-/-}$ mice had lower levels of G-CSF and CXCL1 in the BAL compared to WT mice, these levels were not affected by neutralization of IL-22. In contrast, IL-6 and CCL3 were more dependent on IL-22 than IL-17. When human bronchial epithelial cells were treated with both IL-22 and IL-17, a synergistic induction of a number of host defense genes was observed. One function that was uniquely associated with IL-22 was the ability to increase the clonogenic frequency of human bronchial epithelial cells and increase repair of trans-epithelial resistance, consistent with a role of IL-22 in maintaining epithelial barrier function. Collectively, these studies showed that IL-23, IL-17, and IL-22 were critical players in the early phases of the host response to *K. pneumoniae* infection, at a time when the IL-12 and IFN-γ responses are not yet fully developed.

7.2.2 Apoptosis Induced by *Klebsiella pneumoniae*

Klebsiella pneumoniae has been reported to induce cytotoxicity of respiratory epithelial cells (Cano et al. 2009). Extensive characterization of the mode of cell death was not performed beyond a demonstration that DNA fragmentation accompanied this cytotoxicity, making it difficult to assess whether this observed cytotoxicity could be classified as apoptosis. In a different study, it was shown that Microcin E492, a 7.9 kDa channel-forming bacteriocin produced by *K. pneumoniae* (Lagos et al. 2009), induced cell shrinkage, dissipation of mitochondrial transmembrane potential ($\Delta\Psi_m$), release of cytochrome c, and activation of caspases 1 and 3 (Hetz et al. 2002). These events were blocked by pan-caspase inhibitor z-VAD-fmk. While high doses of microcin E492 induced necrosis, low doses induced apoptosis (Hetz et al. 2002), suggesting that caspase-1 activation, associated with an inflammatory type of cell death termed pyroptosis, may not be reflective of the levels of microcin E492 produced by bacteria at the infected site. Additional studies are needed to confirm whether *K. pneumoniae* can induce apoptosis of respiratory epithelial cells and whether this apoptosis is mediated through microcin E492 and/or dependent on the mitochondrial pathway.

7.3 *Pseudomonas aeruginosa*

Pseudomonas aeruginosa is an opportunistic pathogen. It is a Gram-negative bacterium that causes serious disease in immuno-compromised patients and patients with cystic fibrosis (Hauser 2009). This opportunistic pathogen colonizes and injures lung epithelial surfaces, resulting in pneumonia. It is also an important pathogen in infections of skin wounds and burns, urinary tract, and bloodstream. *P. aeruginosa* possesses a T3SS that encodes genes important for the assembly and regulation of the T3SS apparatus, as well as several protein effectors that are injected into host cells (Hauser 2009). These effectors will be discussed in more detail below owing to their roles in the induction of host cell death.

7.3.1 IL-17 Responses to Rnfection with *Pseudomonas aeruginosa*

Cystic fibrosis is characterized by chronic neutrophilic infiltrates (Chmiel et al. 2002), prompting J. Kolls and colleagues to investigate whether IL-17, known to induce neutrophil recruitment to sites of bacterial infection, was elevated in the lungs of patients with cystic fibrosis (McAllister et al. 2005). The authors analyzed IL-23, IL-17A, and IL-17F in the sputum of cystic fibrosis patients colonized with *P. aeruginosa* that were undergoing pulmonary exacerbation, and found elevated levels of these cytokines upon hospital admission that subsequently declined with antibiotic treatment. To understand the role of IL-17 associated responses in *P. aeruginosa* infection, a follow up study was conducted in mice where WT and $p19^{-/-}$ (IL-23 deficient) mice were challenged with agarose beads containing a clinical, mucoid isolate of *P. aeruginosa* to model airway infections seen in cystic fibrosis (Dubin and Kolls 2007). IL-17, keratinocyte-derived chemokine (KC), the metalloproteinase MMP-9, and IL-6 were significantly decreased in the bronchial alveolar lavage and lung homogenates of infected $p19^{-/-}$ mice compared to WT controls. $p19^{-/-}$ mice had significantly less airway inflammation and neutrophillic infiltrates. However, the levels of IFN-γ and TNF-α were similar between WT and $p19^{-/-}$ mice, and so were the bacterial loads in the lung indicating that the IL-23 was not important for bacterial clearance in this murine model. Nonetheless, the similarities observed in the increased IL-17 and IL-23 levels between the murine model and the human samples, suggested that these cytokines may be important for neutrophil recruitment, and that IL-23 targeting could be exploited as one form of immuno-therapy for cystic fibrosis patients.

An important role for IL-17-associated responses were demonstrated in a study by G. B. Pier and colleagues who were studying vaccine-induced protective responses in a murine model of acute fatal pneumonia caused by intranasal inoculation with *P. aeruginosa* (Priebe et al. 2008). Previous studies had suggested an important role for serum opsonic antibody against the *P. aeruginosa* lipopolysaccharide (LPS) O antigen, but recent clinical trials based on passively administered hyper-immune IgG specific to *P. aeruginosa* LPS O antigen proved unsuccessful, and did not reduce the incidence of infection (Donta et al. 1996). The authors compared vaccinations with two live attenuated strains of *P. aeruginosa*, PAO1ΔaroA and PA14ΔAroA with the prediction that the PA14 strain possesses more virulence factors and might elicit better protection (Priebe et al. 2008). To the authors' surprise, successful immunization with PA14 ΔAroA was not dependent on antibody responses, but rather depended on T cell secreted IL-17, and coincided with a reduction in the bacterial numbers within the bronchoalveolar lavage fluid. Curiously, although both PAO1ΔaroA and PA14ΔAroA mice showed neutrophilia and increased IL-17 production by splenic CD4 T cells, mice immunized with the PA14ΔAroA strain showed significantly increased neutrophil recruitment to airways and significantly higher levels of IL-17 produced by lung leukocytes and within the bronchoalveolar lavage compared to mice immunized with the PAO1ΔaroA strain. Neutralization of IL-17 or immunization of IL-17R deficient mice diminished vaccine-induced protection demonstrating that IL-17 played a crucial role.

7.3.2 Apoptosis Induced by *Pseudomonas aeruginosa*

Four effector proteins that are injected into host cells through the T3SS of *P. aeruginosa* have been identified. These are ExoS, ExoT, ExoU, and ExoY (Hauser 2009). Of these, ExoS, ExoT, and ExoU have been linked to the induction of host cell death. ExoS is a toxin that is expressed by invasive strains of *P. aeruginosa* and exhibits both GTPase-activating protein (GAP) activity and ADP ribosyl transferase (ADPRT) activity (Hauser 2009). The GAP activity of the protein allows it to interfere with proper activation of the small GTPase Rac-1, leading to disruption of the actin cytoskeleton and loss of cell–cell adherence, a property thought to impart on *P. aeruginosa* the ability to disrupt tight junctions within the epithelial barrier. Following binding to the host cell plasma membrane and subsequent transport into early endocytic compartments en route to the endoplasmic reticulum (ER) and golgi, the ADPRT activity of the protein is activated upon binding to the host cell protein 14-3-3. ADPRT activity further results in disruption of the actin cytoskeleton leading to cell rounding, inhibition of DNA synthesis, vesicular traffic, and eventual cell death. ExoS-induced cell death is associated with membrane blebbing, cell shrinkage, presence of apoptotic bodies, caspase-3 cleavage, DNA fragmentation, and chromatin condensation (Kaufman et al. 2000). Furthermore, complementation of a mutant *P. aeruginosa* lacking ExoS with a mutant form of ExoS incapable of ADP-ribosylation failed to induce apoptosis, demonstrating that the ADPRT activity of ExoS was essential for apoptosis induction (Kaufman et al. 2000). It is thought that inducing apoptosis in host cells allows Pseudomonas to penetrate the epithelial barrier, allowing it access to deeper sterile tissue. However, apoptosis relies on expression of the T3SS and is independent of the invasive property of the bacteria because mutant *P. aeruginosa* lacking ExsA, a global regulator of the T3SS machinery, were defective in their ability to induce apoptosis despite retaining their invasiveness (Kaufman et al. 2000).

Similar to ExoS, ExoT is also a bi-functional toxin with GAP and ADPRT activities. The GAP activity of ExoT disrupts the actin cytoskeleton and the ADPRT activity requires binding to the host cell protein 14-3-3 for activation in order to mediate synergistic disruption of the actin cytoskeleton, inhibition of cell migration, and cell adhesion. ExoT and ADP-ribosylates are host proteins that are distinct from those targeted by ExoS. ExoT has also been reported to delay wound healing, a property that may allow it to interfere with prompt repair of the epithelium, giving *P. aeruginosa* the advantage in breaching epithelial barriers. The ADPRT activity of ExoT has also been shown to cause a form of cell death that resembles apoptosis. ExoU possesses phospholipase A_2 activity that causes a rapid form of cell death mediated by the C-terminus of the protein (Finck-Barbancon and Frank 2001; Rabin and Hauser 2005; Sato et al. 2003; Stirling et al. 2006) and is characterized by loss of plasma membrane integrity similar to that seen in necrosis (Finck-Barbancon et al. 1997; Hauser et al. 1998). Similar to ExoS and ExoT, ExoU also requires a host cell protein for activation, with superoxide dismutase (SOD1) fulfilling this role for ExoU (Sato et al. 2006).

In addition to the induction of host cell death by *P. aeruginosa* via the activity of these T3SS effector proteins, Grassmé H. et al. showed a different mode of inducing cell death by *P. aeruginosa* mediated by CD95/CD95 ligand (Fas/FasL) interactions (Grassme et al. 2000). The authors showed that *P. aeruginosa* induced apoptosis of human Chang conjunctival cells and murine *ex-vivo* cultured lung fibroblasts and this apoptosis was blocked by a CD95-Fc fusion protein (Grassme et al. 2000). Similarly, genetic deficiency in CD95 or CD95 ligand using murine fibroblasts derived from *lpr* (lymphoproliferation) or *gld* (generalized lymphoproliferative disease) mice resulted in complete protection from *P. aeruginosa* mediated apoptosis (Grassme et al. 2000). Notably, intranasal infection with *P. aeruginosa* showed induction of apoptosis in lung epithelial cells as determined by TUNEL staining 6 h after infection, which was absent in *lpr* or *gld* mice (Grassme et al. 2000). The absence of apoptosis in infected *lpr* or *gld* mice correlated with sepsis, increased bacterial burdens within the spleen and lung, and increased mortality as compared to WT mice where only 10% of the mice died over 60 days after infection. In experiments with bone marrow (BM), chimeric mice demonstrated a crucial role for apoptosis of lung epithelial cells, and not hematopoietic cells, in the primary response to *P. aeruginosa* infection (Grassme et al. 2000). Transplantation of WT BM into irradiated recipient *lpr* or *gld* mice did not rescue their increased susceptibility to infection. On the other hand, WT mice reconstituted with BM cells from *lpr* or *gld* mice were just as resistant as un-transplanted WT mice. Likewise, WT mice transplanted with *lpr* or *gld* BM cells were just as susceptible as un-transplanted *lpr* or *gld* mice, and died of sepsis. These studies demonstrated an important role for apoptosis in host defense against *P. aeruginosa* infection, although the *P. aeruginosa* virulence factors mediating this apoptosis were not identified.

Consistent with an important role for lung epithelial cells in host defense against an infection with *P. aeruginosa*, Barlow P.G. et al. recently showed a surprising function for the human cathelicidin LL-37, which is produced by epithelial cells (Barlow et al. 2010). Production of this cationic, amphipathic peptide is upregulated in response to infection and inflammation, and mediates important anti-microbial functions. The authors showed that treatment of *P. aeruginosa*-infected human bronchial epithelial cells with LL-37 resulted in epithelial cell apoptosis (Barlow et al. 2010). Apoptosis was determined by nuclear DNA fragmentation visualized by TUNEL assay, cleavage of caspase-3 and 9, loss of mitochondrial transmembrane potential ($\Delta\Psi_m$), and release of cytochrome c into the cytosol. These effects were blocked with the pan caspase inhibitor z-VAD-fmk, and were not observed upon treatment of the epithelial cells with *P. aeruginosa* alone or LL-37 alone, indicating an essential synergistic activity of LL-37 and *P. aeruginosa* (Barlow et al. 2010). The concentration of LL-37 used approximated physiological levels (10–30 μg/ml) that would be present during lung inflammation. Curiously, none of the *P. aeruginosa* virulence factors suspected of causing apoptosis were necessary, since mutant strains of *P. aeruginosa* lacking exotoxin A (ExsA) and with a defective T3SS, or lacking pilus expression were still capable of inducing apoptosis synergistically with LL-37. Importantly, the synergistic LL-37 mediated apoptosis was lost when a deletion mutant of *P. aeruginosa*, MexAb-OprM that was incapable of invading the epithelial cells was used. Neither dead bacteria nor soluble factors produced by live bacteria were able to act synergistically with LL-37 in apoptosis induction. This data demonstrated that invasion of *P. aeruginosa* into epithelial cells in combination with LL-37 resulted in the induction of epithelial cell apoptosis. The mechanism by which LL-37 induces changes in the mitochondrion and subsequent apoptosis of the host epithelial cell is not known.

These studies with *P. aeruginosa* reveal a new mode of apoptosis of epithelial cells independent of bacterial virulence factors. Small antimicrobial peptides such as LL-37 can have the potential to preferentially induce apoptosis of infected epithelial cells as a means of ridding the host of cells that have been compromised by infection. Whether this mode of cell death initiated by the infected epithelial cell itself can also lead to IL-17 or T_H17 responses remains to be investigated.

8 Pyroptosis, a Different Form of Cell Death Triggered by Microbial Pathogens

Many intracellular bacterial pathogens such as *Legionella*, *Listeria*, *Shigella*, *Salmonella* and *Francisella* trigger a type of cell death that has morphological features different than those of apoptosis (Case et al. 2009; Cervantes et al. 2008; Fink et al. 2008; Hueffer and Galan 2004; Kim et al. 2010; Rathinam et al. 2010; Suzuki et al. 2007). This type of cell death is termed pyroptosis (reviewed in (Bergsbaken et al. 2009)). DNA damage is observed as marked nuclear condensation without intra-nucleosomal fragmentation of DNA or resultant DNA laddering morphology. There is no loss of MOMP, no activation of caspases 3 and 6, and no release of cytochrome c. Unique to this type of cell death is the activation of the inflammatory caspase, caspase-1 (Bergsbaken et al. 2009). Rupture of the plasma membrane is observed as a result of caspase-1 mediated formation of pores, which allow efflux of ions and influx of water resulting in osmotic lysis and release of inflammatory intracellular contents. Furthermore, caspase-1 activation results in the cleavage of inflammatory cytokines such as IL-1β, IL-18, and IL-33. An infected host cell undergoing pyroptosis releases these cytokines which results in a type of cell death that is inflammatory in nature (Green et al. 2009). Activation of caspase-1 that precedes pyroptosis occurs within a larger protein complex termed the inflammasome, the main components of which are the Nod-like receptors (NLRs), the adaptor protein ASC or Pycard, and caspase-1 itself (Bryant and Fitzgerald 2009). Inflammasomes assemble in response to the activation of the NLRs by various types of stimuli, which include the presence of microbial ligands within the cytosolic space and damage to the plasma membrane or endogenous triggers that form as a result of pathological inflammation (Bryant and Fitzgerald 2009). Thus, the ability of certain bacterial pathogens to trigger inflammasome activation becomes evident. For example, *Salmonella typhimurium* are equipped with a T3SS that assembles a needle complex on the plasma membrane of the host cell and injects their bacterial effectors into the cytosol (Galan and Wolf-Watz 2006). These effectors are recognized by NLRs, which serve as the pattern recognition receptors of the cytosol and recruit ASC in order to mediate the activation of caspase-1 (Lara-Tejero et al. 2006; Sutterwala et al. 2006). Unlike apoptosis, the consequences of pyroptosis on the adaptive immune system are currently not defined (Green et al. 2009). In particular, whether innate immune recognition of pyroptotic cells can lead to the differentiation of T_H17 cells has not been revealed. Given that pyroptosis and apoptosis are mediated by different cellular pathways and result in different immunological outcomes, the prediction would be that pyroptosis would not favor development of a T_H17 response. Indeed, infections with *Salmonella* and *Listeria* tend to induce T_H1 rather than T_H17 cell responses (Mittrucker and Kaufmann 2000; Neighbors et al. 2001).

9 Benefits of T_H17 Responses Against Infection and Tissue Injury

The effector functions of a T cell subset are determined by the nature of the cytokines it produces. For T_H17 cells, these cytokines include IL-17A, IL-17F, IL-21, IL-22, and IL-10 (Korn et al. 2009). It is likely that T_H17 immunity is the best tailored response against bacterial infections that cause significant apoptosis and tissue injury given that T_H17 cells, in addition to their anti-microbial functions, have also been associated with tissue repair through their production of the cytokine IL-22. In the next two sections, we will discuss the evidence for the roles of IL-22 and IL-17, two major effector cytokines produced by T_H17 cells. These cytokines allow T_H17 cells to communicate with a variety of cell types that express the corresponding cytokine receptors (Korn et al. 2009). For example, the IL-22 receptor (IL-22R), which consists of a heterodimer of IL-22R1 and IL-10R2, is not expressed on the surface of immune cells (Wolk et al. 2002, 2004). IL-22 acts to regulate tissue inflammation via induction of signal transduction in non-immune cells – primarily cells of the digestive tract (pancreas, liver, small and large intestine), lungs, kidney, and skin (Wolk and Sabat 2006; Wolk et al. 2010). Of note, these tissues form outer body barriers and IL-22R1 expression is particularly high on keratinocytes and epithelial cells with lower levels of expression on fibroblasts (Wolk et al. 2009, 2010). IL-17RA, the largest member of the family of IL-17 receptors, binds IL-17A in a complex with IL-17RC, and is expressed ubiquitously and at especially high levels in hematopoietic tissues (Gaffen 2009).

9.1 Effector Functions Associated with Protection against Infection

One major role that has been ascribed to IL-17 is the recruitment of neutrophils to sites of infection. Neutrophils are one of the first lines of host defense against infection, and their recruitment is mediated by cytokines and chemokines produced by professional phagocytes such as macrophages upon encountering microbial pathogens. These inflammatory mediators include CXCL8 (IL-8), which recruits neutrophils and mediates their activation and degranulation. Among other inflammatory cytokines such as IL-6 and IL-12, TNF-α and IL-1β are also produced, which activate the vascular endothelium and increase permeability and subsequent recruitment of neutrophils to the infection site. The relationship between IL-17 and neutrophils is shown in several ways. First, IL-17 induces G-CSF synthesis (Fossiez et al. 1996), and G-CSF is of key importance for neutrophil progenitor proliferation and neutrophil survival (Schwarzenberger et al. 1998; von Vietinghoff and Ley 2008). This has also been illustrated by the severe neutropenia observed in mice and humans that lack G-CSF or G-CSF receptors (Dong et al. 1994; Lieschke et al. 1994; Liu et al. 1996) and by the efficacy of G-CSF as a therapeutic for restoring normal neutrophil numbers in patients undergoing chemotherapy (Bhana 2007). Second, elevated levels of IL-17 and G-CSF were found in adhesion molecule deficient mice that exhibited severe neutrophilia (Forlow et al. 2001). Neutralization of IL-17 with a soluble IL-17 receptor or blockade of G-CSF restored the neutrophil counts in these mice back to normal. Third, *Il17ra*$^{-/-}$ mice have been reported to have decreased neutrophil counts (Kelly et al. 2005; Smith et al. 2008). Fourth, IL-17 protects from neutropenia induced by neutralization of G-CSF (Schwarzenberger et al. 2000). IL-17A has also been reported to induce expression of TNF-α, IL-1β, IL-6, IL-8, and in addition to CXCL8, other chemokines such as CCL2 (MCP-1), CXCL1 (GROα), and CXCL10 (IP-10) (Jovanovic et al. 1998; Laan et al. 1999; Witowski et al. 2000). In addition to G-CSF, IL-17 also induces expression of GM-CSF, which mobilizes more myeloid cells from the bone marrow (Fossiez et al. 1996).

The various effects of IL-17 described above are expected to precede development of the T_H17 adaptive immune response and are likely mediated by IL-17 derived from cells of the innate immune system. Indeed, γδ T cells, NK cells, NK T cells, basophils, and neutrophils have all been reported to produce IL-17 (Ferretti et al. 2003; Ley et al. 2006; Molet et al. 2001). Once at the site of infection, neutrophils clear the bacteria and subsequently undergo apoptosis. Phagocytosis and clearance of such 'bacteria loaded' apoptotic neutrophils would be expected to induce the production of IL-6, TGF-β, and IL-23 by macrophages and dendritic cells, forming the ideal cytokine milieu for T_H17 differentiation of CD4 T cells also recruited to the infection site (Torchinsky et al. 2009). The presence of T_H17 cells provides additional sources of IL-17, as well as other T_H17 related cytokines such as IL-22. The concerted action of these the two T_H17 derived cytokines, IL-17 and IL-22, plays a key role in conferring protection from infection. In addition to the induction of inflammatory cytokines, a plethora of studies both in vitro and in vivo indicate that IL-22 and IL-17 synergistically induce the expression of high levels of anti-microbial peptides such as the S100 family of calcium-binding proteins S100A7, S100A8, and S100A9, as well as cationic peptides such as the human β-defensins 2 and 3 (reviewed in reference (Kolls et al. 2008)). The mechanisms of synergy downstream of the IL-22 and IL-17 receptors are an active area of investigation at the moment.

In lung infections with *Klebsiella pneumoniae*, both IL-22 and IL-17 contributed to the production of cytokines and anti-microbial products from lung epithelial cells, but IL-22 was ultimately more important than IL-17 for defense against the bacteria (Aujla et al. 2008). In *Citrobacter rodentium* infection of the intestinal mucosa, blocking IL-22 resulted in increased susceptibility to the enteric pathogen (Zheng et al. 2008) supporting the notion that IL-22 is an important cytokine for bacterial defense at mucosal surfaces. IL-22 participated in protection by inducing the expression of β-defensins and the Reg family of anti-microbial proteins (RegIIIβ and RegIIIγ) by epithelial cells (Zheng et al. 2008). Other than a protective role in infections, IL-22 has also been shown to be protective in a mouse model of ulcerative colitis (Sugimoto et al. 2008). Although the downstream effects of IL-22 signaling can be protective as noted above, they can also be pro-inflammatory depending on the micro-environment and target tissue. In a mouse model of psoriasis for example, skin inflammation is dependent on IL-22 (Ma et al. 2008), indicating that IL-22 also contributes to pathologic inflammatory processes. This dual property of IL-22 reflects its ability to mediate both protection of host tissue against infection, and repair of the tissue damaged from infection. The reparative properties of IL-22 are discussed next.

9.2 Effector Functions Associated with Protection Against Tissue Injury

The one cytokine secreted by T_H17 cells which appears to play an important role in tissue repair and protection against infection and injury is IL-22. The evidence comes from a large number of studies looking at the effects of IL-22 in vivo as well as keratinocytes and hepatocytes in vitro. Over-expression of IL-22 in a human hepatoma cell line HepG2 promoted proliferation and protection from serum starvation-induced cell death (Radaeva et al. 2004). Treatment of HepG2 cells with IL-22 was shown to induce transcription of acute phase proteins such as serum amyloid A (SAA), haptoglobin and α1-antichymotrypsin, an effect that was recapitulated when IL-22 was injected into mice and induced increased hepatic SAA mRNA expression (Wolk et al. 2004). Transient overexpression of IL-22 in mice by hydrodynamic gene delivery protects against liver damage induced by a combination of carbon tetrachloride and Fas activation (Pan et al. 2004). In a Con-A induced T cell mediated liver injury model, the levels of IL-22 and IL-22R were elevated in the liver and neutralization of IL-22 showed that it was important in conferring protection from liver damage (Radaeva et al. 2004). Similarly, despite equal induction of inflammatory cytokine expression and T cell infiltration, *Il22*$^{-/-}$ mice were highly susceptible to Con-A induced hepatitis (Zenewicz et al. 2007). Compared to WT mice, *Il22*$^{-/-}$ mice exhibited increased hepatic lesions and high levels of the liver damage marker enzymes, alanine aminotransferase and aspartate aminotransferase, events that were dependent on increased susceptibility of hepatocytes to immune mediated damage and not greater susceptibility to Fas-mediated apoptosis (Zenewicz et al. 2007).

IL-22 was also protective against liver necrosis during *Salmonella enterica* serovar *enteritidis* infection (Schulz et al. 2008). In an effort to delineate the role of IL-23/IL-22 axis in the absence of protective IL-12-dependent T_H1 responses, G. Alber and colleagues found that IL-23 was essential for survival such that p35$^{-/-}$p19$^{-/-}$ doubly deficient mice succumbed to infection and showed diffuse liver inflammation, fibrin thrombi, and areas of necrosis while p35$^{-/-}$ mice were surprisingly protected and resistant. The livers of p35$^{-/-}$ mice showed granuloma formation, but no necrosis, whereas neutralization of IL-22 in these mice, while not affecting bacterial burdens, led to hepatocyte necrosis (Schulz et al. 2008).

Studies examining the effects of IL-22 on human keratinocytes revealed that IL-22 plays an important role in wound healing. IL-22 induced significant migration of keratinocytes upon in vitro wounding of mitomycin-treated confluent monolayers of normal human epidermal keratinocytes (Boniface et al. 2005). IL-22 triggered hyperplasia of keratinocyte layers in an in vitro reconstituted human epidermis, leading to an increase in the thickness of the epidermis (acanthosis) that was not due to increased proliferation of keratinocytes, but rather downregulation of genes involved in keratinocyte differentiation such as involucrin, loricrin, filaggrin, 27-kDa heat shock protein (Hsp), calmodulin-related proteins, and heme oxygenase 1 (Boniface et al. 2005). Other studies had also shown the IL-22 induced reduction in the transcriptional activation of genes involved in the terminal differentiation of keratinocytes (Boniface et al. 2005; Wolk et al. 2004, 2006). These genes also included profillagrin, keratins 1 and 10, calmodulin-like 5, keratinocyte differentiation associated protein, kallikrein 7, and late cornified envelop protein 1B, which mediate the process of forming the outermost layer of the epidermis, the stratum corneum (Wolk et al. 2004, 2006). It was also shown that IL-22 induced the expression of IL-20 by keratinocytes, which similarly to IL-22 also inhibited keratinocyte differentiation, suggesting that it may act as a positive regulatory factor in mediating

IL-22's effects (Sa et al. 2007; Wolk et al. 2009). IL-22 also increased the expression of proteins involved in cellular motility such as the matrix metalloproteinases (MMPs) 1 and 3, which degrade the extracellular matrix during tissue reabsorption and remodeling (Boniface et al. 2005; Wolk et al. 2004, 2006).

The observed effects of IL-22 in vitro were also observed in patients with psoriasis (Griffiths and Barker 2007; Sabat et al. 2007; Schon and Boehncke 2005) and confirmed in mouse models. Daily intradermal injection of mouse ears with IL-23 induced acanthosis and inflammation with increased levels of IL-17 and IL-22, and these effects were significantly decreased in *Il22*$^{-/-}$ mice (Zheng et al. 2007). In a psoriasis-like disease model in mice induced by the adoptive transfer of $CD4^{+}CD45RB^{hi}CD25^{-}$ T cells into pathogen-free *scid/scid* mice, neutralization of IL-22 led to amelioration of disease manifested by reduced acanthosis and decreased inflammatory infiltrates as well as IL-17A and IL-17F levels (Ma et al. 2008). Furthermore, IL-22 transgenic mice showed neonatal mortality with psoriasis-like skin lesions characterized by acanthosis (Wolk et al. 2009).

10 Concluding Remarks

The inflammatory trigger induced by an infected apoptotic cell comprises elements of both infection and tissue injury. We have shown that this combination of triggers results in T_H17 immunity (Torchinsky et al. 2009). T_H17 immunity is likely the best-tailored response against bacterial infections that cause significant apoptosis and tissue injury because T_H17 cells are particularly well suited not only for inducing the expression of protective inflammatory and anti-microbial peptides, but also for repairing the tissue damage associated with the extensive apoptosis that accompanies these infections. Thus, T_H17 immunity would not only be important for host defense against infection, but also for inducing repair of tissues damaged from the infection.

References

Abul-Milh, M., Wu, Y., Lau, B., Lingwood, C.A., and Barnett Foster, D. (2001). Induction of epithelial cell death including apoptosis by enteropathogenic Escherichia coli expressing bundle-forming pili. Infect Immun *69*, 7356–7364.

Aderem, A., and Underhill, D.M. (1999). Mechanisms of phagocytosis in macrophages. Annu Rev Immunol *17*, 593–623.

Alto, N.M., Shao, F., Lazar, C.S., Brost, R.L., Chua, G., Mattoo, S., McMahon, S.A., Ghosh, P., Hughes, T.R., Boone, C., *et al.* (2006). Identification of a bacterial type III effector family with G protein mimicry functions. Cell *124*, 133–145.

Amsen, D., Antov, A., and Flavell, R.A. (2009). The different faces of Notch in T-helper-cell differentiation. Nat Rev Immunol *9*, 116–124.

Andrade, J.R., Da Veiga, V.F., De Santa Rosa, M.R., and Suassuna, I. (1989). An endocytic process in HEp-2 cells induced by enteropathogenic Escherichia coli. J Med Microbiol *28*, 49–57.

Aujla, S.J., Chan, Y.R., Zheng, M., Fei, M., Askew, D.J., Pociask, D.A., Reinhart, T.A., McAllister, F., Edeal, J., Gaus, K., *et al.* (2008). IL-22 mediates mucosal host defense against Gram-negative bacterial pneumonia. Nat Med *14*, 275–281.

Baldwin, T.J., Lee-Delaunay, M.B., Knutton, S., and Williams, P.H. (1993). Calcium-calmodulin dependence of actin accretion and lethality in cultured HEp-2 cells infected with enteropathogenic Escherichia coli. Infect Immun *61*, 760–763.

Barlow, P.G., Beaumont, P.E., Cosseau, C., Mackellar, A., Wilkinson, T.S., Hancock, R.E., Haslett, C., Govan, J.R., Simpson, A.J., and Davidson, D.J. (2010). The Human Cathelicidin LL-37 Preferentially Promotes Apoptosis of Infected Airway Epithelium. Am J Respir Cell Mol Biol *43*, 692–702.

Barnett Foster, D., Abul-Milh, M., Huesca, M., and Lingwood, C.A. (2000). Enterohemorrhagic Escherichia coli induces apoptosis which augments bacterial binding and phosphatidylethanolamine exposure on the plasma membrane outer leaflet. Infect Immun *68*, 3108–3115.

Barton, G.M., and Kagan, J.C. (2009). A cell biological view of Toll-like receptor function: regulation through compartmentalization. Nat Rev Immunol *9*, 535–542.

Bergsbaken, T., Fink, S.L., and Cookson, B.T. (2009). Pyroptosis: host cell death and inflammation. Nat Rev Microbiol *7*, 99–109.

Bettelli, E., Carrier, Y., Gao, W., Korn, T., Strom, T.B., Oukka, M., Weiner, H.L., and Kuchroo, V.K. (2006). Reciprocal developmental pathways for the generation of pathogenic effector TH17 and regulatory T cells. Nature *441*, 235–238.

Bhana, N. (2007). Granulocyte colony-stimulating factors in the management of chemotherapy-induced neutropenia: evidence based review. Curr Opin Oncol *19*, 328–335.

Blander, J.M. (2007a). Coupling Toll-like receptor signaling with phagocytosis: potentiation of antigen presentation. Trends Immunol *28*, 19–25.

Blander, J.M. (2007b). Signalling and phagocytosis in the orchestration of host defence. Cell Microbiol *9*, 290–299.

Blander, J.M., and Medzhitov, R. (2004). Regulation of phagosome maturation by signals from toll-like receptors. Science *304*, 1014–1018.

Blander, J.M., and Medzhitov, R. (2006a). On regulation of phagosome maturation and antigen presentation. Nat Immunol *7*, 1029–1035.

Blander, J.M., and Medzhitov, R. (2006b). Toll-dependent selection of microbial antigens for presentation by dendritic cells. Nature *440*, 808–812.

Boniface, K., Bernard, F.X., Garcia, M., Gurney, A.L., Lecron, J.C., and Morel, F. (2005). IL-22 inhibits epidermal differentiation and induces proinflammatory gene expression and migration of human keratinocytes. J Immunol *174*, 3695–3702.

Bose, J., Gruber, A.D., Helming, L., Schiebe, S., Wegener, I., Hafner, M., Beales, M., Kontgen, F., and Lengeling, A. (2004). The phosphatidylserine receptor has essential functions during embryogenesis but not in apoptotic cell removal. J Biol *3*, 15.

Brenner, D., and Mak, T.W. (2009). Mitochondrial cell death effectors. Curr Opin Cell Biol *21*, 871–877.

Bry, L., and Brenner, M.B. (2004). Critical role of T cell-dependent serum antibody, but not the gut-associated lymphoid tissue, for surviving acute mucosal infection with Citrobacter rodentium, an attaching and effacing pathogen. J Immunol *172*, 433–441.

Bryant, C., and Fitzgerald, K.A. (2009). Molecular mechanisms involved in inflammasome activation. Trends Cell Biol *19*, 455–464.

Cano, V., Moranta, D., Llobet-Brossa, E., Bengoechea, J.A., and Garmendia, J. (2009). Klebsiella pneumoniae triggers a cytotoxic effect on airway epithelial cells. BMC Microbiol *9*, 156.

Caruso, R., Fina, D., Paoluzi, O.A., Del Vecchio Blanco, G., Stolfi, C., Rizzo, A., Caprioli, F., Sarra, M., Andrei, F., Fantini, M.C., *et al.* (2008). IL-23-mediated regulation of IL-17 production in Helicobacter pylori-infected gastric mucosa. Eur J Immunol *38*, 470–478.

Caruso, R., Fina, D., Peluso, I., Fantini, M.C., Tosti, C., Del Vecchio Blanco, G., Paoluzi, O.A., Caprioli, F., Andrei, F., Stolfi, C., *et al.* (2007a). IL-21 is highly produced in Helicobacter pylori-infected gastric mucosa and promotes gelatinases synthesis. J Immunol *178*, 5957–5965.

Caruso, R., Pallone, F., and Monteleone, G. (2007b). Emerging role of IL-23/IL-17 axis in H pylori-associated pathology. World J Gastroenterol *13*, 5547–5551.

Case, C.L., Shin, S., and Roy, C.R. (2009). Asc and Ipaf Inflammasomes direct distinct pathways for caspase-1 activation in response to Legionella pneumophila. Infect Immun *77*, 1981–1991.

Castrillo, A., and Tontonoz, P. (2004). Nuclear receptors in macrophage biology: at the crossroads of lipid metabolism and inflammation. Annu Rev Cell Dev Biol *20*, 455–480.

Cervantes, J., Nagata, T., Uchijima, M., Shibata, K., and Koide, Y. (2008). Intracytosolic Listeria monocytogenes induces cell death through caspase-1 activation in murine macrophages. Cell Microbiol *10*, 41–52.

Chai, J., Du, C., Wu, J.W., Kyin, S., Wang, X., and Shi, Y. (2000). Structural and biochemical basis of apoptotic activation by Smac/DIABLO. Nature *406*, 855–862.

Chang, M.K., Bergmark, C., Laurila, A., Horkko, S., Han, K.H., Friedman, P., Dennis, E.A., and Witztum, J.L. (1999). Monoclonal antibodies against oxidized low-density lipoprotein bind to apoptotic cells and inhibit their phagocytosis by elicited macrophages: evidence that oxidation-specific epitopes mediate macrophage recognition. Proc Natl Acad Sci USA *96*, 6353–6358.

Chen, H.D., and Frankel, G. (2005). Enteropathogenic Escherichia coli: unravelling pathogenesis. FEMS Microbiol Rev *29*, 83–98.

Chipuk, J.E., and Green, D.R. (2008). How do BCL-2 proteins induce mitochondrial outer membrane permeabilization? Trends Cell Biol *18*, 157–164.

Chmiel, J.F., Berger, M., and Konstan, M.W. (2002). The role of inflammation in the pathophysiology of CF lung disease. Clin Rev Allergy Immunol *23*, 5–27.

Colonna, M. (2009). Interleukin-22-producing natural killer cells and lymphoid tissue inducer-like cells in mucosal immunity. Immunity *31*, 15–23.

Coquet, J.M., Chakravarti, S., Kyparissoudis, K., McNab, F.W., Pitt, L.A., McKenzie, B.S., Berzins, S.P., Smyth, M.J., and Godfrey, D.I. (2008). Diverse cytokine production by NKT cell subsets and identification of an IL-17-producing CD4-NK1.1- NKT cell population. Proc Natl Acad Sci USA *105*, 11287–11292.

Cossart, P., and Sansonetti, P.J. (2004). Bacterial invasion: the paradigms of enteroinvasive pathogens. Science *304*, 242–248.

Cover, T.L., and Blanke, S.R. (2005). Helicobacter pylori VacA, a paradigm for toxin multifunctionality. Nat Rev Microbiol *3*, 320–332.

Cover, T.L., and Blaser, M.J. (2009). Helicobacter pylori in health and disease. Gastroenterology *136*, 1863–1873.

Cover, T.L., Krishna, U.S., Israel, D.A., and Peek, R.M., Jr. (2003). Induction of gastric epithelial cell apoptosis by Helicobacter pylori vacuolating cytotoxin. Cancer Res *63*, 951–957.

Crane, J.K., Majumdar, S., and Pickhardt, D.F., 3rd (1999). Host cell death due to enteropathogenic Escherichia coli has features of apoptosis. Infect Immun *67*, 2575–2584.

Crane, J.K., McNamara, B.P., and Donnenberg, M.S. (2001). Role of EspF in host cell death induced by enteropathogenic Escherichia coli. Cell Microbiol *3*, 197–211.

Crane, J.K., Olson, R.A., Jones, H.M., and Duffey, M.E. (2002). Release of ATP during host cell killing by enteropathogenic E. coli and its role as a secretory mediator. Am J Physiol Gastrointest Liver Physiol *283*, G74-86.

Cupedo, T., Crellin, N.K., Papazian, N., Rombouts, E.J., Weijer, K., Grogan, J.L., Fibbe, W.E., Cornelissen, J.J., and Spits, H. (2009). Human fetal lymphoid tissue-inducer cells are interleukin 17-producing precursors to RORC+ CD127+ natural killer-like cells. Nat Immunol *10*, 66–74.

Curtis, M.M., and Way, S.S. (2009). Interleukin-17 in host defence against bacterial, mycobacterial and fungal pathogens. Immunology *126*, 177–185.

Cvetanovic, M., and Ucker, D.S. (2004). Innate immune discrimination of apoptotic cells: repression of proinflammatory macrophage transcription is coupled directly to specific recognition. J Immunol *172*, 880–889.

Dardalhon, V., Awasthi, A., Kwon, H., Galileos, G., Gao, W., Sobel, R.A., Mitsdoerffer, M., Strom, T.B., Elyaman, W., Ho, I.C., *et al.* (2008). IL-4 inhibits TGF-beta-induced Foxp3+ T cells and, together with TGF-beta, generates IL-9+ IL-10+ Foxp3(−) effector T cells. Nat Immunol *9*, 1347–1355.

Dean, P., Maresca, M., and Kenny, B. (2005). EPEC's weapons of mass subversion. Curr Opin Microbiol *8*, 28–34.

Doisne, J.M., Becourt, C., Amniai, L., Duarte, N., Le Luduec, J.B., Eberl, G., and Benlagha, K. (2009). Skin and peripheral lymph node invariant NKT cells are mainly retinoic acid receptor-related orphan receptor (gamma)t+ and respond preferentially under inflammatory conditions. J Immunol *183*, 2142–2149.

Dong, F., Hoefsloot, L.H., Schelen, A.M., Broeders, C.A., Meijer, Y., Veerman, A.J., Touw, I.P., and Lowenberg, B. (1994). Identification of a nonsense mutation in the granulocyte-colony-stimulating factor receptor in severe congenital neutropenia. Proc Natl Acad Sci USA *91*, 4480–4484.

Donnenberg, M.S., Donohue-Rolfe, A., and Keusch, G.T. (1989). Epithelial cell invasion: an overlooked property of enteropathogenic Escherichia coli (EPEC) associated with the EPEC adherence factor. J Infect Dis *160*, 452–459.

Donta, S.T., Peduzzi, P., Cross, A.S., Sadoff, J., Haakenson, C., Cryz, S.J., Jr., Kauffman, C., Bradley, S., Gafford, G., Elliston, D., *et al.* (1996). Immunoprophylaxis against klebsiella and pseudomonas aeruginosa infections. The Federal Hyperimmune Immunoglobulin Trial Study Group. J Infect Dis *174*, 537–543.

Dubin, P.J., and Kolls, J.K. (2007). IL-23 mediates inflammatory responses to mucoid Pseudomonas aeruginosa lung infection in mice. Am J Physiol Lung Cell Mol Physiol *292*, L519–528.

Duhen, T., Geiger, R., Jarrossay, D., Lanzavecchia, A., and Sallusto, F. (2009). Production of interleukin 22 but not interleukin 17 by a subset of human skin-homing memory T cells. Nat Immunol *10*, 857–863.

Elliott, M.R., Chekeni, F.B., Trampont, P.C., Lazarowski, E.R., Kadl, A., Walk, S.F., Park, D., Woodson, R.I., Ostankovich, M., Sharma, P., *et al.* (2009). Nucleotides released by apoptotic cells act as a find-me signal to promote phagocytic clearance. Nature *461*, 282–286.

Erwig, L.P., and Henson, P.M. (2008). Clearance of apoptotic cells by phagocytes. Cell Death Differ *15*, 243–250.

Fadok, V.A., Bratton, D.L., Konowal, A., Freed, P.W., Westcott, J.Y., and Henson, P.M. (1998). Macrophages that have ingested apoptotic cells in vitro inhibit proinflammatory cytokine production through autocrine/paracrine mechanisms involving TGF-beta, PGE2, and PAF. J Clin Invest *101*, 890–898.

Fadok, V.A., Bratton, D.L., Rose, D.M., Pearson, A., Ezekewitz, R.A., and Henson, P.M. (2000). A receptor for phosphatidylserine-specific clearance of apoptotic cells. Nature *405*, 85–90.

Fadok, V.A., Voelker, D.R., Campbell, P.A., Cohen, J.J., Bratton, D.L., and Henson, P.M. (1992). Exposure of phosphatidylserine on the surface of apoptotic lymphocytes triggers specific recognition and removal by macrophages. J Immunol *148*, 2207–2216.

Ferretti, S., Bonneau, O., Dubois, G.R., Jones, C.E., and Trifilieff, A. (2003). IL-17, produced by lymphocytes and neutrophils, is necessary for lipopolysaccharide-induced airway neutrophilia: IL-15 as a possible trigger. J Immunol *170*, 2106–2112.

Finck-Barbancon, V., and Frank, D.W. (2001). Multiple domains are required for the toxic activity of Pseudomonas aeruginosa ExoU. J Bacteriol *183*, 4330–4344.

Finck-Barbancon, V., Goranson, J., Zhu, L., Sawa, T., Wiener-Kronish, J.P., Fleiszig, S.M., Wu, C., Mende-Mueller, L., and Frank, D.W. (1997). ExoU expression by Pseudomonas aeruginosa correlates with acute cytotoxicity and epithelial injury. Mol Microbiol *25*, 547–557.

Fink, S.L., Bergsbaken, T., and Cookson, B.T. (2008). Anthrax lethal toxin and Salmonella elicit the common cell death pathway of caspase-1-dependent pyroptosis via distinct mechanisms. Proc Natl Acad Sci USA *105*, 4312–4317.

Forlow, S.B., Schurr, J.R., Kolls, J.K., Bagby, G.J., Schwarzenberger, P.O., and Ley, K. (2001). Increased granulopoiesis through interleukin-17 and granulocyte colony-stimulating factor in leukocyte adhesion molecule-deficient mice. Blood *98*, 3309–3314.

Fossiez, F., Djossou, O., Chomarat, P., Flores-Romo, L., Ait-Yahia, S., Maat, C., Pin, J.J., Garrone, P., Garcia, E., Saeland, S., *et al.* (1996). T cell interleukin-17 induces stromal cells to produce proinflammatory and hematopoietic cytokines. J Exp Med *183*, 2593–2603.

Gaffen, S.L. (2009). Structure and signalling in the IL-17 receptor family. Nat Rev Immunol *9*, 556–567.

Galan, J.E., and Wolf-Watz, H. (2006). Protein delivery into eukaryotic cells by type III secretion machines. Nature *444*, 567–573.

Galmiche, A., Rassow, J., Doye, A., Cagnol, S., Chambard, J.C., Contamin, S., de Thillot, V., Just, I., Ricci, V., Solcia, E., *et al.* (2000). The N-terminal 34 kDa fragment of Helicobacter pylori vacuolating cytotoxin targets mitochondria and induces cytochrome c release. Embo J *19*, 6361–6370.

Girardi, M. (2006). Immuno-surveillance and immuno-regulation by gammadelta T cells. J Invest Dermatol *126*, 25–31.

Grassme, H., Kirschnek, S., Riethmueller, J., Riehle, A., von Kurthy, G., Lang, F., Weller, M., and Gulbins, E. (2000). CD95/CD95 ligand interactions on epithelial cells in host defense to Pseudomonas aeruginosa. Science *290*, 527–530.

Green, D.R., Ferguson, T., Zitvogel, L., and Kroemer, G. (2009). Immunogenic and tolerogenic cell death. Nat Rev Immunol *9*, 353–363.

Green, D.R., and Kroemer, G. (2004). The pathophysiology of mitochondrial cell death. Science *305*, 626–629.

Gregory, C.D., Devitt, A., and Moffatt, O. (1998). Roles of ICAM-3 and CD14 in the recognition and phagocytosis of apoptotic cells by macrophages. Biochem Soc Trans *26*, 644–649.

Griffiths, C.E., and Barker, J.N. (2007). Pathogenesis and clinical features of psoriasis. Lancet *370*, 263–271.

Gronski, M.A., Kinchen, J.M., Juncadella, I.J., Franc, N.C., and Ravichandran, K.S. (2009). An essential role for calcium flux in phagocytes for apoptotic cell engulfment and the anti-inflammatory response. Cell Death Differ *16*, 1323–1331.

Guttman, J.A., Li, Y., Wickham, M.E., Deng, W., Vogl, A.W., and Finlay, B.B. (2006a). Attaching and effacing pathogen-induced tight junction disruption in vivo. Cell Microbiol *8*, 634–645.

Guttman, J.A., Samji, F.N., Li, Y., Vogl, A.W., and Finlay, B.B. (2006b). Evidence that tight junctions are disrupted due to intimate bacterial contact and not inflammation during attaching and effacing pathogen infection in vivo. Infect Immun *74*, 6075–6084.

Hanayama, R., Tanaka, M., Miwa, K., Shinohara, A., Iwamatsu, A., and Nagata, S. (2002). Identification of a factor that links apoptotic cells to phagocytes. Nature *417*, 182–187.

Happel, K.I., Dubin, P.J., Zheng, M., Ghilardi, N., Lockhart, C., Quinton, L.J., Odden, A.R., Shellito, J.E., Bagby, G.J., Nelson, S., *et al.* (2005). Divergent roles of IL-23 and IL-12 in host defense against Klebsiella pneumoniae. J Exp Med *202*, 761–769.

Hauser, A.R. (2009). The type III secretion system of Pseudomonas aeruginosa: infection by injection. Nat Rev Microbiol *7*, 654–665.

Hauser, A.R., Kang, P.J., and Engel, J.N. (1998). PepA, a secreted protein of Pseudomonas aeruginosa, is necessary for cytotoxicity and virulence. Mol Microbiol *27*, 807–818.

Hetz, C., Bono, M.R., Barros, L.F., and Lagos, R. (2002). Microcin E492, a channel-forming bacteriocin from Klebsiella pneumoniae, induces apoptosis in some human cell lines. Proc Natl Acad Sci USA *99*, 2696–2701.

Higgins, L.M., Frankel, G., Connerton, I., Goncalves, N.S., Dougan, G., and MacDonald, T.T. (1999). Role of bacterial intimin in colonic hyperplasia and inflammation. Science *285*, 588–591.

Hueffer, K., and Galan, J.E. (2004). Salmonella-induced macrophage death: multiple mechanisms, different outcomes. Cell Microbiol *6*, 1019–1025.

Ishigame, H., Kakuta, S., Nagai, T., Kadoki, M., Nambu, A., Komiyama, Y., Fujikado, N., Tanahashi, Y., Akitsu, A., Kotaki, H., *et al.* (2009). Differential roles of interleukin-17A and -17F in host defense against mucoepithelial bacterial infection and allergic responses. Immunity *30*, 108–119.

Iwasaki, A., and Medzhitov, R. (2010). Regulation of adaptive immunity by the innate immune system. Science *327*, 291–295.

Janeway, C.A., Jr., and Medzhitov, R. (2002). Innate immune recognition. Annu Rev Immunol *20*, 197–216.

Johann, A.M., von Knethen, A., Lindemann, D., and Brune, B. (2006). Recognition of apoptotic cells by macrophages activates the peroxisome proliferator-activated receptor-gamma and attenuates the oxidative burst. Cell Death Differ *13*, 1533–1540.

Jovanovic, D.V., Di Battista, J.A., Martel-Pelletier, J., Jolicoeur, F.C., He, Y., Zhang, M., Mineau, F., and Pelletier, J.P. (1998). IL-17 stimulates the production and expression of proinflammatory cytokines, IL-beta and TNF-alpha, by human macrophages. J Immunol *160*, 3513–3521.
Kamanaka, M., Kim, S.T., Wan, Y.Y., Sutterwala, F.S., Lara-Tejero, M., Galan, J.E., Harhaj, E., and Flavell, R.A. (2006). Expression of interleukin-10 in intestinal lymphocytes detected by an interleukin-10 reporter knockin tiger mouse. Immunity *25*, 941–952.
Kaper, J.B., Nataro, J.P., and Mobley, H.L. (2004). Pathogenic Escherichia coli. Nat Rev Microbiol *2*, 123–140.
Kaufman, M.R., Jia, J., Zeng, L., Ha, U., Chow, M., and Jin, S. (2000). Pseudomonas aeruginosa mediated apoptosis requires the ADP-ribosylating activity of exoS. Microbiology *146 (Pt 10)*, 2531–2541.
Kelly, M.N., Kolls, J.K., Happel, K., Schwartzman, J.D., Schwarzenberger, P., Combe, C., Moretto, M., and Khan, I.A. (2005). Interleukin-17/interleukin-17 receptor-mediated signaling is important for generation of an optimal polymorphonuclear response against Toxoplasma gondii infection. Infect Immun *73*, 617–621.
Kenny, B., and Jepson, M. (2000). Targeting of an enteropathogenic Escherichia coli (EPEC) effector protein to host mitochondria. Cell Microbiol *2*, 579–590.
Khader, S.A., Gaffen, S.L., and Kolls, J.K. (2009). Th17 cells at the crossroads of innate and adaptive immunity against infectious diseases at the mucosa. Mucosal Immunol *2*, 403–411.
Kim, S., Bauernfeind, F., Ablasser, A., Hartmann, G., Fitzgerald, K.A., Latz, E., and Hornung, V. (2010). Listeria monocytogenes is sensed by the NLRP3 and AIM2 Inflammasome. Eur J Immunol *40*, 1545–1551.
Kinchen, J.M., and Ravichandran, K.S. (2008). Phagosome maturation: going through the acid test. Nat Rev Mol Cell Biol *9*, 781–795.
King, C. (2009). New insights into the differentiation and function of T follicular helper cells. Nat Rev Immunol *9*, 757–766.
King, C., Tangye, S.G., and Mackay, C.R. (2008). T follicular helper (TFH) cells in normal and dysregulated immune responses. Annu Rev Immunol *26*, 741–766.
Kobayashi, N., Karisola, P., Pena-Cruz, V., Dorfman, D.M., Jinushi, M., Umetsu, S.E., Butte, M.J., Nagumo, H., Chernova, I., Zhu, B., *et al.* (2007). TIM-1 and TIM-4 glycoproteins bind phosphatidylserine and mediate uptake of apoptotic cells. Immunity *27*, 927–940.
Kolls, J.K., McCray, P.B., Jr., and Chan, Y.R. (2008). Cytokine-mediated regulation of antimicrobial proteins. Nat Rev Immunol *8*, 829–835.
Korn, T., Bettelli, E., Oukka, M., and Kuchroo, V.K. (2009). IL-17 and Th17 Cells. Annu Rev Immunol *27*, 485–517.
Krammer, P.H. (2000). CD95's deadly mission in the immune system. Nature *407*, 789–795.
Kroemer, G., Galluzzi, L., and Brenner, C. (2007). Mitochondrial membrane permeabilization in cell death. Physiol Rev *87*, 99–163.
Kuck, D., Kolmerer, B., Iking-Konert, C., Krammer, P.H., Stremmel, W., and Rudi, J. (2001). Vacuolating cytotoxin of Helicobacter pylori induces apoptosis in the human gastric epithelial cell line AGS. Infect Immun *69*, 5080–5087.
Kumar, H., Kawai, T., and Akira, S. (2009). Toll-like receptors and innate immunity. Biochem Biophys Res Commun *388*, 621–625.
Kunisaki, Y., Masuko, S., Noda, M., Inayoshi, A., Sanui, T., Harada, M., Sasazuki, T., and Fukui, Y. (2004). Defective fetal liver erythropoiesis and T lymphopoiesis in mice lacking the phosphatidylserine receptor. Blood *103*, 3362–3364.
Laan, M., Cui, Z.H., Hoshino, H., Lotvall, J., Sjostrand, M., Gruenert, D.C., Skoogh, B.E., and Linden, A. (1999). Neutrophil recruitment by human IL-17 via C-X-C chemokine release in the airways. J Immunol *162*, 2347–2352.
Lagos, R., Tello, M., Mercado, G., Garcia, V., and Monasterio, O. (2009). Antibacterial and antitumorigenic properties of microcin E492, a pore-forming bacteriocin. Curr Pharm Biotechnol *10*, 74–85.
Lara-Tejero, M., Sutterwala, F.S., Ogura, Y., Grant, E.P., Bertin, J., Coyle, A.J., Flavell, R.A., and Galan, J.E. (2006). Role of the caspase-1 inflammasome in Salmonella typhimurium pathogenesis. J Exp Med *203*, 1407–1412.

Lauber, K., Bohn, E., Krober, S.M., Xiao, Y.J., Blumenthal, S.G., Lindemann, R.K., Marini, P., Wiedig, C., Zobywalski, A., Baksh, S., *et al.* (2003). Apoptotic cells induce migration of phagocytes via caspase-3-mediated release of a lipid attraction signal. Cell *113*, 717–730.
LeibundGut-Landmann, S., Gross, O., Robinson, M.J., Osorio, F., Slack, E.C., Tsoni, S.V., Schweighoffer, E., Tybulewicz, V., Brown, G.D., Ruland, J., *et al.* (2007). Syk- and CARD9-dependent coupling of innate immunity to the induction of T helper cells that produce interleukin 17. Nat Immunol *8*, 630–638.
Ley, K., Smith, E., and Stark, M.A. (2006). IL-17A-producing neutrophil-regulatory Tn lymphocytes. Immunol Res *34*, 229–242.
Li, F., Huang, Q., Chen, J., Peng, Y., Roop, D.R., Bedford, J.S., and Li, C.Y. (2010a). Apoptotic cells activate the "phoenix rising" pathway to promote wound healing and tissue regeneration. Sci Signal *3*, ra13.
Li, L., Huang, L., Vergis, A.L., Ye, H., Bajwa, A., Narayan, V., Strieter, R.M., Rosin, D.L., and Okusa, M.D. (2010b). IL-17 produced by neutrophils regulates IFN-gamma-mediated neutrophil migration in mouse kidney ischemia-reperfusion injury. J Clin Invest *120*, 331–342.
Li, M.O., Sarkisian, M.R., Mehal, W.Z., Rakic, P., and Flavell, R.A. (2003). Phosphatidylserine receptor is required for clearance of apoptotic cells. Science *302*, 1560–1563.
Lieschke, G.J., Grail, D., Hodgson, G., Metcalf, D., Stanley, E., Cheers, C., Fowler, K.J., Basu, S., Zhan, Y.F., and Dunn, A.R. (1994). Mice lacking granulocyte colony-stimulating factor have chronic neutropenia, granulocyte and macrophage progenitor cell deficiency, and impaired neutrophil mobilization. Blood *84*, 1737–1746.
Liu, F., Wu, H.Y., Wesselschmidt, R., Kornaga, T., and Link, D.C. (1996). Impaired production and increased apoptosis of neutrophils in granulocyte colony-stimulating factor receptor-deficient mice. Immunity *5*, 491–501.
Lockhart, E., Green, A.M., and Flynn, J.L. (2006). IL-17 production is dominated by gammadelta T cells rather than CD4 T cells during Mycobacterium tuberculosis infection. J Immunol *177*, 4662–4669.
Luzza, F., Parrello, T., Monteleone, G., Sebkova, L., Romano, M., Zarrilli, R., Imeneo, M., and Pallone, F. (2000). Up-regulation of IL-17 is associated with bioactive IL-8 expression in Helicobacter pylori-infected human gastric mucosa. J Immunol *165*, 5332–5337.
Ma, C., Wickham, M.E., Guttman, J.A., Deng, W., Walker, J., Madsen, K.L., Jacobson, K., Vogl, W.A., Finlay, B.B., and Vallance, B.A. (2006). Citrobacter rodentium infection causes both mitochondrial dysfunction and intestinal epithelial barrier disruption in vivo: role of mitochondrial associated protein (Map). Cell Microbiol *8*, 1669–1686.
Ma, H.L., Liang, S., Li, J., Napierata, L., Brown, T., Benoit, S., Senices, M., Gill, D., Dunussi-Joannopoulos, K., Collins, M., *et al.* (2008). IL-22 is required for Th17 cell-mediated pathology in a mouse model of psoriasis-like skin inflammation. J Clin Invest *118*, 597–607.
Maaser, C., Housley, M.P., Iimura, M., Smith, J.R., Vallance, B.A., Finlay, B.B., Schreiber, J.R., Varki, N.M., Kagnoff, M.F., and Eckmann, L. (2004). Clearance of Citrobacter rodentium requires B cells but not secretory immunoglobulin A (IgA) or IgM antibodies. Infect Immun *72*, 3315–3324.
Maeda, S., and Mentis, A.F. (2007). Pathogenesis of Helicobacter pylori infection. Helicobacter *12 Suppl 1*, 10–14.
Majai, G., Sarang, Z., Csomos, K., Zahuczky, G., and Fesus, L. (2007). PPARgamma-dependent regulation of human macrophages in phagocytosis of apoptotic cells. Eur J Immunol *37*, 1343–1354.
Malish, H.R., Freeman, N.L., Zurawski, D.V., Chowrashi, P., Ayoob, J.C., Sanger, J.W., and Sanger, J.M. (2003). Potential role of the EPEC translocated intimin receptor (Tir) in host apoptotic events. Apoptosis *8*, 179–190.
Mangan, P.R., Harrington, L.E., O'Quinn, D.B., Helms, W.S., Bullard, D.C., Elson, C.O., Hatton, R.D., Wahl, S.M., Schoeb, T.R., and Weaver, C.T. (2006). Transforming growth factor-beta induces development of the T(H)17 lineage. Nature *441*, 231–234.
Manicassamy, S., and Pulendran, B. (2009). Modulation of adaptive immunity with Toll-like receptors. Semin Immunol *21*, 185–193.

Martin, B., Hirota, K., Cua, D.J., Stockinger, B., and Veldhoen, M. (2009). Interleukin-17-producing gammadelta T cells selectively expand in response to pathogen products and environmental signals. Immunity *31*, 321–330.

McAllister, F., Henry, A., Kreindler, J.L., Dubin, P.J., Ulrich, L., Steele, C., Finder, J.D., Pilewski, J.M., Carreno, B.M., Goldman, S.J., *et al.* (2005). Role of IL-17A, IL-17F, and the IL-17 receptor in regulating growth-related oncogene-alpha and granulocyte colony-stimulating factor in bronchial epithelium: implications for airway inflammation in cystic fibrosis. J Immunol *175*, 404–412.

McDaniel, T.K., Jarvis, K.G., Donnenberg, M.S., and Kaper, J.B. (1995). A genetic locus of enterocyte effacement conserved among diverse enterobacterial pathogens. Proc Natl Acad Sci USA *92*, 1664–1668.

McDaniel, T.K., and Kaper, J.B. (1997). A cloned pathogenicity island from enteropathogenic Escherichia coli confers the attaching and effacing phenotype on E. coli K-12. Mol Microbiol *23*, 399–407.

McNamara, B.P., and Donnenberg, M.S. (1998). A novel proline-rich protein, EspF, is secreted from enteropathogenic Escherichia coli via the type III export pathway. FEMS Microbiol Lett *166*, 71–78.

McNamara, B.P., Koutsouris, A., O'Connell, C.B., Nougayrede, J.P., Donnenberg, M.S., and Hecht, G. (2001). Translocated EspF protein from enteropathogenic Escherichia coli disrupts host intestinal barrier function. J Clin Invest *107*, 621–629.

Medzhitov, R., and Janeway, C.A., Jr. (2002). Decoding the patterns of self and nonself by the innate immune system. Science *296*, 298–300.

Michel, M.L., Keller, A.C., Paget, C., Fujio, M., Trottein, F., Savage, P.B., Wong, C.H., Schneider, E., Dy, M., and Leite-de-Moraes, M.C. (2007). Identification of an IL-17-producing NK1.1(neg) iNKT cell population involved in airway neutrophilia. J Exp Med *204*, 995–1001.

Michel, M.L., Mendes-da-Cruz, D., Keller, A.C., Lochner, M., Schneider, E., Dy, M., Eberl, G., and Leite-de-Moraes, M.C. (2008). Critical role of ROR-gammat in a new thymic pathway leading to IL-17-producing invariant NKT cell differentiation. Proc Natl Acad Sci USA *105*, 19845–19850.

Mimuro, H., Suzuki, T., Nagai, S., Rieder, G., Suzuki, M., Nagai, T., Fujita, Y., Nagamatsu, K., Ishijima, N., Koyasu, S., *et al.* (2007). Helicobacter pylori dampens gut epithelial self-renewal by inhibiting apoptosis, a bacterial strategy to enhance colonization of the stomach. Cell Host Microbe *2*, 250–263.

Mittrucker, H.W., and Kaufmann, S.H. (2000). Immune response to infection with Salmonella typhimurium in mice. J Leukoc Biol *67*, 457–463.

Miyanishi, M., Tada, K., Koike, M., Uchiyama, Y., Kitamura, T., and Nagata, S. (2007). Identification of Tim4 as a phosphatidylserine receptor. Nature *450*, 435–439.

Mizuno, T., Ando, T., Nobata, K., Tsuzuki, T., Maeda, O., Watanabe, O., Minami, M., Ina, K., Kusugami, K., Peek, R.M., *et al.* (2005). Interleukin-17 levels in Helicobacter pylori-infected gastric mucosa and pathologic sequelae of colonization. World J Gastroenterol *11*, 6305–6311.

Moffatt, O.D., Devitt, A., Bell, E.D., Simmons, D.L., and Gregory, C.D. (1999). Macrophage recognition of ICAM-3 on apoptotic leukocytes. J Immunol *162*, 6800–6810.

Molet, S., Hamid, Q., Davoine, F., Nutku, E., Taha, R., Page, N., Olivenstein, R., Elias, J., and Chakir, J. (2001). IL-17 is increased in asthmatic airways and induces human bronchial fibroblasts to produce cytokines. J Allergy Clin Immunol *108*, 430–438.

Mukundan, L., Odegaard, J.I., Morel, C.R., Heredia, J.E., Mwangi, J.W., Ricardo-Gonzalez, R.R., Goh, Y.P., Eagle, A.R., Dunn, S.E., Awakuni, J.U., *et al.* (2009). PPAR-delta senses and orchestrates clearance of apoptotic cells to promote tolerance. Nat Med *15*, 1266–1272.

N, A.G., Bensinger, S.J., Hong, C., Beceiro, S., Bradley, M.N., Zelcer, N., Deniz, J., Ramirez, C., Diaz, M., Gallardo, G., *et al.* (2009). Apoptotic cells promote their own clearance and immune tolerance through activation of the nuclear receptor LXR. Immunity *31*, 245–258.

Nagai, T., Abe, A., and Sasakawa, C. (2005). Targeting of enteropathogenic Escherichia coli EspF to host mitochondria is essential for bacterial pathogenesis: critical role of the 16th leucine residue in EspF. J Biol Chem *280*, 2998–3011.

Nagata, K., Ohashi, K., Nakano, T., Arita, H., Zong, C., Hanafusa, H., and Mizuno, K. (1996). Identification of the product of growth arrest-specific gene 6 as a common ligand for Axl, Sky, and Mer receptor tyrosine kinases. J Biol Chem *271*, 30022–30027.

Neighbors, M., Xu, X., Barrat, F.J., Ruuls, S.R., Churakova, T., Debets, R., Bazan, J.F., Kastelein, R.A., Abrams, J.S., and O'Garra, A. (2001). A critical role for interleukin 18 in primary and memory effector responses to Listeria monocytogenes that extends beyond its effects on Interferon gamma production. J Exp Med *194*, 343–354.

Nougayrede, J.P., and Donnenberg, M.S. (2004). Enteropathogenic Escherichia coli EspF is targeted to mitochondria and is required to initiate the mitochondrial death pathway. Cell Microbiol *6*, 1097–1111.

Nurieva, R.I., Chung, Y., Hwang, D., Yang, X.O., Kang, H.S., Ma, L., Wang, Y.H., Watowich, S.S., Jetten, A.M., Tian, Q., *et al.* (2008). Generation of T follicular helper cells is mediated by interleukin-21 but independent of T helper 1, 2, or 17 cell lineages. Immunity *29*, 138–149.

O'Neill, L.A., and Bowie, A.G. (2007). The family of five: TIR-domain-containing adaptors in Toll-like receptor signalling. Nat Rev Immunol *7*, 353–364.

Oka, K., Sawamura, T., Kikuta, K., Itokawa, S., Kume, N., Kita, T., and Masaki, T. (1998). Lectin-like oxidized low-density lipoprotein receptor 1 mediates phagocytosis of aged/apoptotic cells in endothelial cells. Proc Natl Acad Sci USA *95*, 9535–9540.

Pallen, M.J., Beatson, S.A., and Bailey, C.M. (2005). Bioinformatics analysis of the locus for enterocyte effacement provides novel insights into type-III secretion. BMC Microbiol *5*, 9.

Pan, H., Hong, F., Radaeva, S., and Gao, B. (2004). Hydrodynamic gene delivery of interleukin-22 protects the mouse liver from concanavalin A-, carbon tetrachloride-, and Fas ligand-induced injury via activation of STAT3. Cell Mol Immunol *1*, 43–49.

Papatheodorou, P., Domanska, G., Oxle, M., Mathieu, J., Selchow, O., Kenny, B., and Rassow, J. (2006). The enteropathogenic Escherichia coli (EPEC) Map effector is imported into the mitochondrial matrix by the TOM/Hsp70 system and alters organelle morphology. Cell Microbiol *8*, 677–689.

Park, D., Hochreiter-Hufford, A., and Ravichandran, K.S. (2009). The phosphatidylserine receptor TIM-4 does not mediate direct signaling. Curr Biol *19*, 346–351.

Park, D., Tosello-Trampont, A.C., Elliott, M.R., Lu, M., Haney, L.B., Ma, Z., Klibanov, A.L., Mandell, J.W., and Ravichandran, K.S. (2007). BAI1 is an engulfment receptor for apoptotic cells upstream of the ELMO/Dock180/Rac module. Nature *450*, 430–434.

Park, S.Y., Jung, M.Y., Kim, H.J., Lee, S.J., Kim, S.Y., Lee, B.H., Kwon, T.H., Park, R.W., and Kim, I.S. (2008a). Rapid cell corpse clearance by stabilin-2, a membrane phosphatidylserine receptor. Cell Death Differ *15*, 192–201.

Park, S.Y., Kang, K.B., Thapa, N., Kim, S.Y., Lee, S.J., and Kim, I.S. (2008b). Requirement of adaptor protein GULP during stabilin-2-mediated cell corpse engulfment. J Biol Chem *283*, 10593–10600.

Passos, S.T., Silver, J.S., O'Hara, A.C., Sehy, D., Stumhofer, J.S., and Hunter, C.A. (2010). IL-6 Promotes NK Cell Production of IL-17 during Toxoplasmosis. J Immunol *184*, 1776–1783.

Peek, R.M., Jr., Wirth, H.P., Moss, S.F., Yang, M., Abdalla, A.M., Tham, K.T., Zhang, T., Tang, L.H., Modlin, I.M., and Blaser, M.J. (2000). Helicobacter pylori alters gastric epithelial cell cycle events and gastrin secretion in Mongolian gerbils. Gastroenterology *118*, 48–59.

Priebe, G.P., Walsh, R.L., Cederroth, T.A., Kamei, A., Coutinho-Sledge, Y.S., Goldberg, J.B., and Pier, G.B. (2008). IL-17 is a critical component of vaccine-induced protection against lung infection by lipopolysaccharide-heterologous strains of Pseudomonas aeruginosa. J Immunol *181*, 4965–4975.

Rabin, S.D., and Hauser, A.R. (2005). Functional regions of the Pseudomonas aeruginosa cytotoxin ExoU. Infect Immun *73*, 573–582.

Rachitskaya, A.V., Hansen, A.M., Horai, R., Li, Z., Villasmil, R., Luger, D., Nussenblatt, R.B., and Caspi, R.R. (2008). Cutting edge: NKT cells constitutively express IL-23 receptor and RORgammat and rapidly produce IL-17 upon receptor ligation in an IL-6-independent fashion. J Immunol *180*, 5167–5171.

Radaeva, S., Sun, R., Pan, H.N., Hong, F., and Gao, B. (2004). Interleukin 22 (IL-22) plays a protective role in T cell-mediated murine hepatitis: IL-22 is a survival factor for hepatocytes via STAT3 activation. Hepatology *39*, 1332–1342.

Rathinam, V.A., Jiang, Z., Waggoner, S.N., Sharma, S., Cole, L.E., Waggoner, L., Vanaja, S.K., Monks, B.G., Ganesan, S., Latz, E., *et al.* (2010). The AIM2 inflammasome is essential for host defense against cytosolic bacteria and DNA viruses. Nat Immunol *11*, 395–402.

Ravichandran, K.S., and Lorenz, U. (2007). Engulfment of apoptotic cells: signals for a good meal. Nat Rev Immunol *7*, 964–974.

Roark, C.L., Simonian, P.L., Fontenot, A.P., Born, W.K., and O'Brien, R.L. (2008). gammadelta T cells: an important source of IL-17. Curr Opin Immunol *20*, 353–357.

Sa, S.M., Valdez, P.A., Wu, J., Jung, K., Zhong, F., Hall, L., Kasman, I., Winer, J., Modrusan, Z., Danilenko, D.M., *et al.* (2007). The effects of IL-20 subfamily cytokines on reconstituted human epidermis suggest potential roles in cutaneous innate defense and pathogenic adaptive immunity in psoriasis. J Immunol *178*, 2229–2240.

Sabat, R., Philipp, S., Hoflich, C., Kreutzer, S., Wallace, E., Asadullah, K., Volk, H.D., Sterry, W., and Wolk, K. (2007). Immunopathogenesis of psoriasis. Exp Dermatol *16*, 779–798.

Sambrano, G.R., and Steinberg, D. (1995). Recognition of oxidatively damaged and apoptotic cells by an oxidized low density lipoprotein receptor on mouse peritoneal macrophages: role of membrane phosphatidylserine. Proc Natl Acad Sci USA *92*, 1396–1400.

Sato, H., Feix, J.B., and Frank, D.W. (2006). Identification of superoxide dismutase as a cofactor for the pseudomonas type III toxin, ExoU. Biochemistry *45*, 10368–10375.

Sato, H., Frank, D.W., Hillard, C.J., Feix, J.B., Pankhaniya, R.R., Moriyama, K., Finck-Barbancon, V., Buchaklian, A., Lei, M., Long, R.M., *et al.* (2003). The mechanism of action of the Pseudomonas aeruginosa-encoded type III cytotoxin, ExoU. Embo J *22*, 2959–2969.

Savill, J., Dransfield, I., Gregory, C., and Haslett, C. (2002). A blast from the past: clearance of apoptotic cells regulates immune responses. Nat Rev Immunol *2*, 965–975.

Savill, J., Dransfield, I., Hogg, N., and Haslett, C. (1990). Vitronectin receptor-mediated phagocytosis of cells undergoing apoptosis. Nature *343*, 170–173.

Savina, A., and Amigorena, S. (2007). Phagocytosis and antigen presentation in dendritic cells. Immunol Rev *219*, 143–156.

Schon, M.P., and Boehncke, W.H. (2005). Psoriasis. N Engl J Med *352*, 1899–1912.

Schulz, S.M., Kohler, G., Schutze, N., Knauer, J., Straubinger, R.K., Chackerian, A.A., Witte, E., Wolk, K., Sabat, R., Iwakura, Y., *et al.* (2008). Protective immunity to systemic infection with attenuated Salmonella enterica serovar enteritidis in the absence of IL-12 is associated with IL-23-dependent IL-22, but not IL-17. J Immunol *181*, 7891–7901.

Schwarzenberger, P., Huang, W., Ye, P., Oliver, P., Manuel, M., Zhang, Z., Bagby, G., Nelson, S., and Kolls, J.K. (2000). Requirement of endogenous stem cell factor and granulocyte-colony-stimulating factor for IL-17-mediated granulopoiesis. J Immunol *164*, 4783–4789.

Schwarzenberger, P., La Russa, V., Miller, A., Ye, P., Huang, W., Zieske, A., Nelson, S., Bagby, G.J., Stoltz, D., Mynatt, R.L., *et al.* (1998). IL-17 stimulates granulopoiesis in mice: use of an alternate, novel gene therapy-derived method for in vivo evaluation of cytokines. J Immunol *161*, 6383–6389.

Shibata, K., Yamada, H., Hara, H., Kishihara, K., and Yoshikai, Y. (2007). Resident Vdelta1+ gammadelta T cells control early infiltration of neutrophils after Escherichia coli infection via IL-17 production. J Immunol *178*, 4466–4472.

Shifflett, D.E., Clayburgh, D.R., Koutsouris, A., Turner, J.R., and Hecht, G.A. (2005). Enteropathogenic E. coli disrupts tight junction barrier function and structure in vivo. Lab Invest *85*, 1308–1324.

Simmons, C.P., Clare, S., Ghaem-Maghami, M., Uren, T.K., Rankin, J., Huett, A., Goldin, R., Lewis, D.J., MacDonald, T.T., Strugnell, R.A., *et al.* (2003). Central role for B lymphocytes and CD4+ T cells in immunity to infection by the attaching and effacing pathogen Citrobacter rodentium. Infect Immun *71*, 5077–5086.

Smith, E., Stark, M.A., Zarbock, A., Burcin, T.L., Bruce, A.C., Vaswani, D., Foley, P., and Ley, K. (2008). IL-17A inhibits the expansion of IL-17A-producing T cells in mice through "short-loop" inhibition via IL-17 receptor. J Immunol *181*, 1357–1364.

Stewart, C.R., Stuart, L.M., Wilkinson, K., van Gils, J.M., Deng, J., Halle, A., Rayner, K.J., Boyer, L., Zhong, R., Frazier, W.A., *et al.* (2010). CD36 ligands promote sterile inflammation through assembly of a Toll-like receptor 4 and 6 heterodimer. Nat Immunol *11*, 155–161.

Stirling, F.R., Cuzick, A., Kelly, S.M., Oxley, D., and Evans, T.J. (2006). Eukaryotic localization, activation and ubiquitinylation of a bacterial type III secreted toxin. Cell Microbiol *8*, 1294–1309.

Stuart, L.M., and Ezekowitz, R.A. (2005). Phagocytosis: elegant complexity. Immunity *22*, 539–550.

Sugimoto, K., Ogawa, A., Mizoguchi, E., Shimomura, Y., Andoh, A., Bhan, A.K., Blumberg, R.S., Xavier, R.J., and Mizoguchi, A. (2008). IL-22 ameliorates intestinal inflammation in a mouse model of ulcerative colitis. J Clin Invest *118*, 534–544.

Susin, S.A., Lorenzo, H.K., Zamzami, N., Marzo, I., Snow, B.E., Brothers, G.M., Mangion, J., Jacotot, E., Costantini, P., Loeffler, M., *et al.* (1999). Molecular characterization of mitochondrial apoptosis-inducing factor. Nature *397*, 441–446.

Sutterwala, F.S., Ogura, Y., Szczepanik, M., Lara-Tejero, M., Lichtenberger, G.S., Grant, E.P., Bertin, J., Coyle, A.J., Galan, J.E., Askenase, P.W., *et al.* (2006). Critical role for NALP3/CIAS1/Cryopyrin in innate and adaptive immunity through its regulation of caspase-1. Immunity *24*, 317–327.

Suzuki, T., Franchi, L., Toma, C., Ashida, H., Ogawa, M., Yoshikawa, Y., Mimuro, H., Inohara, N., Sasakawa, C., and Nunez, G. (2007). Differential regulation of caspase-1 activation, pyroptosis, and autophagy via Ipaf and ASC in Shigella-infected macrophages. PLoS Pathog *3*, e111.

Takeuchi, O., and Akira, S. (2010). Pattern recognition receptors and inflammation. Cell *140*, 805–820.

Torchinsky, M.B., Garaude, J., Martin, A.P., and Blander, J.M. (2009). Innate immune recognition of infected apoptotic cells directs T(H)17 cell differentiation. Nature *458*, 78–82.

Trifari, S., Kaplan, C.D., Tran, E.H., Crellin, N.K., and Spits, H. (2009). Identification of a human helper T cell population that has abundant production of interleukin 22 and is distinct from T(H)-17, T(H)1 and T(H)2 cells. Nat Immunol *10*, 864–871.

Trombetta, E.S., and Mellman, I. (2005). Cell biology of antigen processing in vitro and in vivo. Annu Rev Immunol *23*, 975–1028.

Umemura, M., Yahagi, A., Hamada, S., Begum, M.D., Watanabe, H., Kawakami, K., Suda, T., Sudo, K., Nakae, S., Iwakura, Y., *et al.* (2007). IL-17-mediated regulation of innate and acquired immune response against pulmonary Mycobacterium bovis bacille Calmette-Guerin infection. J Immunol *178*, 3786–3796.

Underhill, D.M., Ozinsky, A., Hajjar, A.M., Stevens, A., Wilson, C.B., Bassetti, M., and Aderem, A. (1999). The Toll-like receptor 2 is recruited to macrophage phagosomes and discriminates between pathogens. Nature *401*, 811–815.

Vallance, B.A., Deng, W., Jacobson, K., and Finlay, B.B. (2003). Host susceptibility to the attaching and effacing bacterial pathogen Citrobacter rodentium. Infect Immun *71*, 3443–3453.

Veldhoen, M., Hocking, R.J., Atkins, C.J., Locksley, R.M., and Stockinger, B. (2006). TGFbeta in the context of an inflammatory cytokine milieu supports de novo differentiation of IL-17-producing T cells. Immunity *24*, 179–189.

Veldhoen, M., Uyttenhove, C., van Snick, J., Helmby, H., Westendorf, A., Buer, J., Martin, B., Wilhelm, C., and Stockinger, B. (2008). Transforming growth factor-beta 'reprograms' the differentiation of T helper 2 cells and promotes an interleukin 9-producing subset. Nat Immunol *9*, 1341–1346.

Viswanathan, V.K., Weflen, A., Koutsouris, A., Roxas, J.L., and Hecht, G. (2008). Enteropathogenic E. coli-induced barrier function alteration is not a consequence of host cell apoptosis. Am J Physiol Gastrointest Liver Physiol *294*, G1165–1170.

Vogelzang, A., McGuire, H.M., Yu, D., Sprent, J., Mackay, C.R., and King, C. (2008). A fundamental role for interleukin-21 in the generation of T follicular helper cells. Immunity *29*, 127–137.

Voll, R.E., Herrmann, M., Roth, E.A., Stach, C., Kalden, J.R., and Girkontaite, I. (1997). Immunosuppressive effects of apoptotic cells. Nature *390*, 350–351.

von Vietinghoff, S., and Ley, K. (2008). Homeostatic regulation of blood neutrophil counts. J Immunol *181*, 5183–5188.

Wan, Y.Y., and Flavell, R.A. (2007). 'Yin-Yang' functions of transforming growth factor-beta and T regulatory cells in immune regulation. Immunol Rev *220*, 199–213.

Weflen, A.W., Alto, N.M., and Hecht, G.A. (2009). Tight junctions and enteropathogenic E. coli. Ann N Y Acad Sci *1165*, 169–174.

Willhite, D.C., and Blanke, S.R. (2004). Helicobacter pylori vacuolating cytotoxin enters cells, localizes to the mitochondria, and induces mitochondrial membrane permeability changes correlated to toxin channel activity. Cell Microbiol *6*, 143–154.

Willhite, D.C., Cover, T.L., and Blanke, S.R. (2003). Cellular vacuolation and mitochondrial cytochrome c release are independent outcomes of Helicobacter pylori vacuolating cytotoxin activity that are each dependent on membrane channel formation. J Biol Chem *278*, 48204–48209.

Wilson, C.B., Rowell, E., and Sekimata, M. (2009). Epigenetic control of T-helper-cell differentiation. Nat Rev Immunol *9*, 91–105.

Witowski, J., Pawlaczyk, K., Breborowicz, A., Scheuren, A., Kuzlan-Pawlaczyk, M., Wisniewska, J., Polubinska, A., Friess, H., Gahl, G.M., Frei, U., *et al.* (2000). IL-17 stimulates intraperitoneal neutrophil infiltration through the release of GRO alpha chemokine from mesothelial cells. J Immunol *165*, 5814–5821.

Wolk, K., Haugen, H.S., Xu, W., Witte, E., Waggie, K., Anderson, M., Vom Baur, E., Witte, K., Warszawska, K., Philipp, S., *et al.* (2009). IL-22 and IL-20 are key mediators of the epidermal alterations in psoriasis while IL-17 and IFN-gamma are not. J Mol Med *87*, 523–536.

Wolk, K., Kunz, S., Asadullah, K., and Sabat, R. (2002). Cutting edge: immune cells as sources and targets of the IL-10 family members? J Immunol *168*, 5397–5402.

Wolk, K., Kunz, S., Witte, E., Friedrich, M., Asadullah, K., and Sabat, R. (2004). IL-22 increases the innate immunity of tissues. Immunity *21*, 241–254.

Wolk, K., and Sabat, R. (2006). Interleukin-22: a novel T- and NK-cell derived cytokine that regulates the biology of tissue cells. Cytokine Growth Factor Rev *17*, 367–380.

Wolk, K., Witte, E., Wallace, E., Docke, W.D., Kunz, S., Asadullah, K., Volk, H.D., Sterry, W., and Sabat, R. (2006). IL-22 regulates the expression of genes responsible for antimicrobial defense, cellular differentiation, and mobility in keratinocytes: a potential role in psoriasis. Eur J Immunol *36*, 1309–1323.

Wolk, K., Witte, E., Witte, K., Warszawska, K., and Sabat, R. (2010). Biology of interleukin-22. Semin Immunopathol *32*, 17–31.

Xiao, Y.Q., Freire-de-Lima, C.G., Schiemann, W.P., Bratton, D.L., Vandivier, R.W., and Henson, P.M. (2008). Transcriptional and translational regulation of TGF-beta production in response to apoptotic cells. J Immunol *181*, 3575–3585.

Yin, X.M., Wang, K., Gross, A., Zhao, Y., Zinkel, S., Klocke, B., Roth, K.A., and Korsmeyer, S.J. (1999). Bid-deficient mice are resistant to Fas-induced hepatocellular apoptosis. Nature *400*, 886–891.

Zenewicz, L.A., Yancopoulos, G.D., Valenzuela, D.M., Murphy, A.J., Karow, M., and Flavell, R.A. (2007). Interleukin-22 but not interleukin-17 provides protection to hepatocytes during acute liver inflammation. Immunity *27*, 647–659.

Zheng, Y., Danilenko, D.M., Valdez, P., Kasman, I., Eastham-Anderson, J., Wu, J., and Ouyang, W. (2007). Interleukin-22, a T(H)17 cytokine, mediates IL-23-induced dermal inflammation and acanthosis. Nature *445*, 648–651.

Zheng, Y., Valdez, P.A., Danilenko, D.M., Hu, Y., Sa, S.M., Gong, Q., Abbas, A.R., Modrusan, Z., Ghilardi, N., de Sauvage, F.J., *et al.* (2008). Interleukin-22 mediates early host defense against attaching and effacing bacterial pathogens. Nat Med *14*, 282–289.

Zhivotovsky, B., Galluzzi, L., Kepp, O., and Kroemer, G. (2009). Adenine nucleotide translocase: a component of the phylogenetically conserved cell death machinery. Cell Death Differ *16*, 1419–1425.

IL-1 and Innate Immunity in the Differentiation of T_H17 Cells

Stephen J. Lalor, Caroline E. Sutton, and Kingston H.G. Mills

Abstract $CD4^+$ T cells that secrete IL-17 (T_H17 cells) play a key role in the pathogenesis of many auto-immune diseases and function with T_H1 cells to mediate protective immunity to pathogens. The differentiation and expansion of T_H17 cells from naïve T cells appears to involve signals from IL-6, IL-1 IL-21, and IL-23. The role of TGF-β is more controversial with evidence that it can promote differentiation but also inhibit activation of T_H17 cells. Although much of the focus has been on $CD4^+$ T cells, and more recently $CD8^+$ T cells, there is increasing evidence that innate cells, especially γδ T cells, but also NKT cells, are important sources of IL-17 and other T_H17 associated cytokines. We have found that IL-1α or IL-1β can synergize with IL-23 to promote IL-17 secretion from memory T cells, and that IL-1β and IL-23, produced by dendritic cells in response to Toll-like receptor and NOD-like receptor activation, can also promote IL-17 production by γδ T cells without T cell receptor engagement. IL-17-producing γδ T cells are found at high frequency in the brain and spinal cord of mice with experimental auto-immune encephalomyelitis, where they function with T_H17 cells to mediate auto-immune inflammation and pathology. γδ T cells appear to act as an important source of innate IL-17 and IL-21, which act in an amplification loop early in the immune response to promote further IL-17 production by T_H17 cells. This chapter discusses the role of IL-1 and IL-23 in promoting IL-17 production by antigen-specific T_H17 cells and innate IL-17 from γδ T cells and the role of IL-17-producing γδ T cells in auto-immunity and infection.

K.H.G. Mills (✉)
School of Biochemistry and Immunology, Trinity College Dublin, Dublin 2, Ireland
e-mail: millsk@tcd.ie

S. Jiang (ed.), *T_H17 Cells in Health and Disease*,
DOI 10.1007/978-1-4419-9371-7_6,

1 Introduction

CD4$^+$ T cells that produce IL-17 (termed T_H17 cells) have clearly been identified as a distinct subset of Th cells that promote inflammation and are pathogenic in many auto-immune disorders. CD4$^+$ T cells, although a major source of IL-17, are not only source of this inflammatory cytokine. A number of studies have shown that innate immune cells, including NKT cells, lymphoid tissue inducer-like (LTi) cells and γδ T cells (Shibata et al. 2007; Sutton et al. 2009; Takatori et al. 2008), as well as neutrophils and microglia (Ferretti et al. 2003; Kawanokuchi et al. 2008), are important sources of IL-17. Furthermore, innate IL-17 production has been shown to play a critical role in protection against certain infections and can mediate pathology in auto-immune diseases. Despite the fact that the orphan nuclear receptor RORγt was discovered to be the critical transcription factor required for development of T_H17 cells (Ivanov et al. 2006), it was also found to be expressed constitutively by NKT cells and γδ T cells. Recent studies have demonstrated that each of these cells express the IL-23R, either constitutively, or following activation. NKT cells, γδ T cells or LTi-like cells produce IL-17 following stimulation with IL-23 alone or in combination with other cytokines (Michel et al. 2007; Rachitskaya et al. 2008; Sutton et al. 2006).

2 Differentiation of T_H17 Cells

A pathogenic role for IL-17 in auto-immunity was established from observations that IL-23p19$^{-/-}$ mice and IL-12p40$^{-/-}$ mice, which failed to generated T_H17 cells, were resistant to experimental auto-immune encephalomyelitis (EAE), whereas IFN-γ$^{-/-}$ or signal transducer and activator of transcription (STAT)-1$^{-/-}$ mice that lacked T_H1 cells, developed more severe EAE (Bettelli et al. 2004; Ferber et al. 1996). It had been shown earlier that IL-23, which shares the IL-12p40 sub-unit with IL-12 and has a distinct IL-23p19 sub-unit (Oppmann et al. 2000), promoted induction of a distinct population of memory CD4$^+$ T cells characterized by the production of IL-17 (Aggarwal et al. 2003; Langrish et al. 2005; Murphy et al. 2003). It was quickly established that these IL-23-driven CD4$^+$ T cells were a distinct lineage, separate from T_H1 and T_H2 cells, and expressed IL-17A, IL-17F, and TNF-α and were subsequently called T_H17 cells (Langrish et al. 2005; McKenzie et al. 2005; Weaver et al. 2006). Although initial studies suggested that T_H17 and not T_H1 cells were the key pathogenic T cells, more recent studies have suggested that both populations may play complementary roles in auto-immunity. Studies in the EAE model showed that T_H1 cells promote expression of monocyte attracting chemokines and macrophage rich infiltrates into the spinal cord, whereas T_H17 cells activate neutrophil-attracting chemokines and promote neutrophil recruitment and activation, especially in the brain (Kroenke et al. 2008; Stromnes et al. 2008).

Following the initial demonstration that IL-23 promoted induction of T_H17 cells, it was later shown that IL-6 and TGF-β, not IL-23, were required for differentiation of naïve murine T_H17 cells (Mangan et al. 2006). This was consistent with the observation that naïve T cells do not express the IL-23R and do not differentiate into T_H17 cells in response to stimulation with IL-23 (Langrish et al. 2005; van Beelen et al. 2007). Conversely, it was reported that IL-23 could expand a population of T_H17 cells in vitro (Aggarwal et al. 2003; Langrish et al. 2005). The importance of IL-6 in the differentiation of T_H17 cells was provided by the demonstration that IL-$6^{-/-}$ mice do not develop T_H17 responses and are resistant to EAE; their peripheral repertoire is dominated by $Foxp3^+$ Treg cells (Eugster et al. 1998; Korn et al. 2008). Furthermore, IL-6 negates the suppressive effect of Foxp3 on RORγt function and inhibits the reciprocal generation of $Foxp3^+$ Treg cells induced by TGF-β (Bettelli et al. 2006; Zhou et al. 2009). The role of TGF-β in differentiation of T_H17 cells is more controversial. It has recently been demonstrated that TGF-β may function to suppress T_H1 and T_H2 cells, the products of which can inhibit the differentiation of T_H17 cells (Das et al. 2009). TGF-β blocks expression of the transcription factor STAT-4 and GATA-3, thus preventing T_H1 and T_H2 cell differentiation. TGF-β had no effect on the expression of Rar-associated orphan receptor C (RORc), also referred to as RORγt, the transcription factor associated with IL-17 production. Furthermore, mice deficient in both STAT-6 and T-bet, which are unable to generate T_H1 and T_H2 cells, IL-6 alone was sufficient to induce differentiation of T_H17 cells, whereas TGF-β had no effect (Das et al. 2009). Thus, the role of TGF-β in T_H17 differentiation appears to be indirect by inhibiting IFN-γ and IL-4 and, in their absence, TGF-β is dispensable for development of T_H17 cells.

IL-1 has also been implicated in promoting the induction of IL-17 production by naïve and memory $CD4^+$ T cells and by unconventional T cells. Studies in our laboratory provided key evidence that this cytokine is a major driver of pathogenic T cells. We found that IL-1 receptor I deficient (IL-$1RI^{-/-}$) mice, which lack the functional effects of IL-1α and IL-1β, had significantly weaker T_H17 responses following immunization with MOG and CFA and were resistant to development of EAE (Sutton et al. 2006). Furthermore, IL-1 and IL-23 synergised to promote IL-17 production by $CD3^+$, $CD4^+$ and $CD8^+$ T cells from naïve mice. Purified $CD4^+$ and $CD8^+$ T cells required co-stimulation with anti-CD3 and anti-CD28 or APC, but unseparated $CD3^+$ T cells secreted low concentrations of IL-17 following stimulation with IL-1α or IL-1β in combination with IL-23, in the absence of TCR engagement (Sutton et al. 2006). It has also been reported that IL-1 and IL-23 induced IL-17 production by human $CD45RO^+$ T cells (Acosta-Rodriguez et al. 2007a; Wilson et al. 2007). In addition, it has been demonstrated that IL-1, IL-6, and IL-23 may promote T_H17 cell differentiation acting via STAT-3 to induce RORγt expression (Chen et al. 2006; Ivanov et al. 2006; Milner et al. 2008; Wei et al. 2007). It is thought that IL-6, by inducing IL-21, upregulates the IL-23R in a STAT-3-dependent manner upon ligation of the TCR and thus promotes T_H17 cell differentiation in the presence of IL-23 (Zhou et al. 2007).

3 IL-1 Family of Cytokines and Receptors

IL-1 was first called endogenous pyrogen, a soluble factor that induced fever (Atkins and Wood 1955). It was subsequently shown that IL-1 has a range of pro-inflammatory effects and plays an important role in host protection against microbial infection. Interestingly, early studies revealed that this cytokine was capable of promoting the induction or expansion of T cells and was also known as lymphocyte activating factor (LAF) (Katz et al. 1978; Mizel et al. 1978). Following the cloning of murine and humans, IL-1α and IL-1β (Auron et al. 1984; Lomedico et al. 1984) and a number of further members of the IL-1 family were discovered, including IL-18, IL-33, and the inhibitory IL-1 receptor antagonist (IL-1Ra) (Dinarello 2009). There are now 11 known members of this family, most of which are involved in local or systemic inflammation, but some have anti-inflammatory properties. IL-1F5 and IL-1F7 appear to exert their anti-inflammatory effects non-specifically (Dinarello 2009) or through interaction with the orphan IL-1 receptor, single immunoglobulin IL-1R-related molecule (SIGIRR) (Costelloe et al. 2008). The closest member of the IL-1 family to IL-1α and IL-1β is IL-18, which has been shown to promote differentiation of T_H1 cells (Okamura et al. 1995). Finally, the more distantly related IL-33 plays a role in mast cell functions and drives allergic and T_H2 responses (Schmitz et al. 2005).

IL-1α (IL-1F1) and IL-1β (IL-1F2) are encoded by two different genes and are synthesized as 31 kDa proteins (pro-IL-1α and pro-IL-1β). The active forms of IL-1α and IL-1β bind to and signal through the IL-1 type I receptor (IL-1RI). The activity of IL-1α and IL-1β is regulated by a decoy non-signaling type II receptor and IL-1Ra (IL-1F3), a specific inhibitor of IL-1α and IL-1β, which competitively binds IL-1RI with high affinity. IL-1RI is similar in structure to the toll like receptor (TLR) family (Fig. 1), which recognizes conserved pathogen associated molecular patterns (PAMPs). IL-1RI and TLR both possess an internal Toll/IL-1 receptor (TIR) domain (Akira et al. 2001; Medzhitov 2001), which mediates the internal

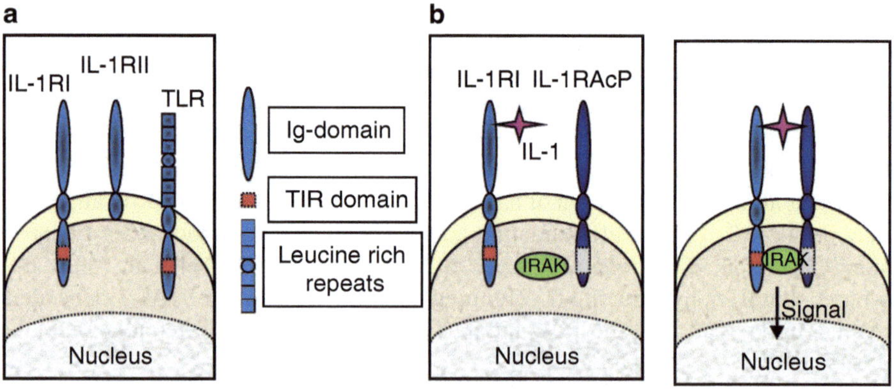

Fig. 1 (**a**) IL-1R and TLR structure. (**b**) IL-1RI activation and signaling

signaling for the receptor. On binding of IL-1 to IL-1RI, the accessory protein IL-1RAcP is recruited to form a high affinity heterodimeric receptor complex, initiating the IL-1 signaling cascade (Braddock and Quinn 2004).

IL-1α and IL-1β are produced by numerous cells of the immune system and are major mediators of inflammation and innate immune responses. They have pleiotropic effects, including induction of production of various growth factors, inflammatory mediators, adhesion molecules, chemokines and other cytokines, either directly or indirectly, as well as upregulating further IL-1 production in an autocrine manner. Thus, IL-1α and IL-1β play important roles in the initiation of many inflammatory responses. These inflammatory responses include induction of matrix metalloproteinases (MMPs), reactive oxygen species, peroxides, nitric oxide, and lipid mediators of inflammation, such as prostaglandins, leukotrienes and platelet-activating factor (Warabi et al. 2007). It has also been demonstrated that IL-1β can activate naïve and memory antigen-specific T cells (Badovinac et al. 1998; Nakae et al. 2003a).

4 Induction of IL-1α and IL-1β and the Role of the Inflammasome

IL-1β is primarily produced by cells of the monocytic lineage, including monocytes, macrophages and dendritic cells (DC). Activation of a number of signaling pathways can lead to IL-1β gene expression, including those induced by IL-1β itself, IL-17, TNF-α, as well as PAMPs, such as LPS, and damage associated molecular patterns (DAMPs) like ATP, which are recognized through membrane-bound TLR and nucleotide oligomerization domain (NOD) like receptors (NLR) (Braddock and Quinn 2004). IL-1β is not constitutively expressed by monocytes, macrophages, or DC. However, stimulation via TLR and NLR results in *Il1b* mRNA expression and its translation to pro-IL-1β protein (Eder 2009). Pro-IL-1β is subsequently cleaved by active caspase-1, also known as IL-1β-converting enzyme (ICE), to generate mature 17 kDa IL-1β protein. Caspase-1 is a cysteine protease that requires activation in a proteolytic cascade following assembly and activation of an inflammasome. The inflammasome is a large multi-protein complex, comprising the intracellular adaptor protein ASC (apoptosis-associated speck-like protein containing a CARD (caspase-activating recruitment domain)), which links caspase-1 with an NLR (Franchi et al. 2009) (Fig. 2). There are a number of different types of inflammasomes involving different NLRs. The NALP1 (NLRP1) inflammasome plays a crucial role in the innate immune response to the anthrax lethal toxin from *Bacillus anthracis* (Hsu et al. 2008). The NALP3 (NLRP3 or cryopyrin) inflammasome is involved in caspase-1 activation in response to a variety of microbial stimuli, including LPS, muramyl dipeptide (MDP), bacterial RNA, double stranded viral RNA (Hise et al. 2009; Kanneganti et al. 2006; Martinon et al. 2004), endogenous danger signals such as ATP and monosodium urate crystals (Mariathasan et al. 2006; Martinon and Tschopp 2004), and toxins such as maitotoxin,

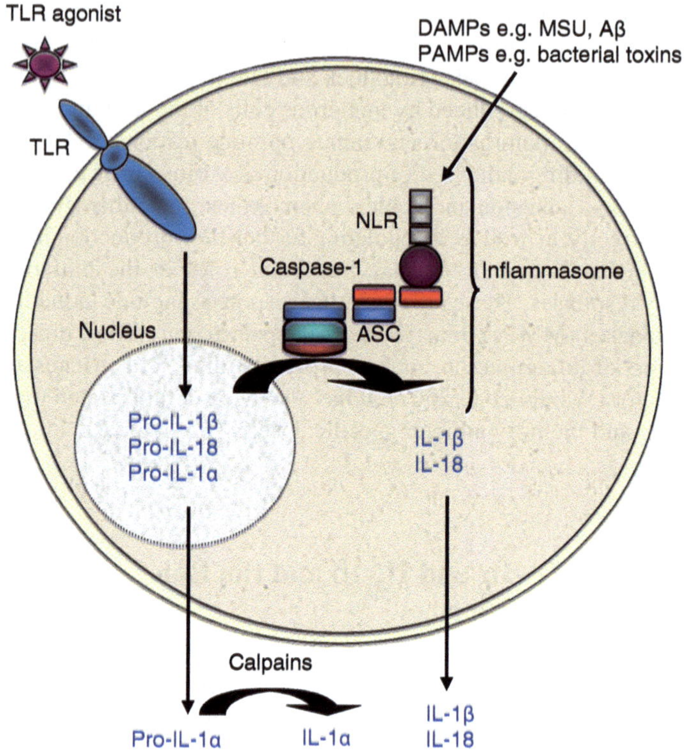

Fig. 2 Induction and activation of IL-1 and IL-18 in innate immune cells

nigericine (Mariathasan et al. 2006) and adenylate cyclise toxin from Bordetella pertussis (Dunne et al. 2010). The NLRC4 or Ipaf inflammasome regulates caspase-1 activation in response to cell infection with intracellular bacteria, including *Salmonella typhimurium, Legionella pneumophila* and *Pseudomonas aeruginosa* (Amer et al. 2006; Franchi et al. 2009; Sutterwala et al. 2007). The mechanism of activation of the inflammasomes is poorly understood, but K^+ efflux is thought to be an important initiating event, at least for NALP1 and NALP3 containing complexes, with downstream involvement of the pore-forming pannexin-1 protein in NALP3 inflammasomes (Eder 2009; Pelegrin and Surprenant 2006). Gain of function mutations in NALP3 that result in super-physiological levels of IL-1β have been associated with human auto-inflammatory diseases, such as Muckle-Wells syndrome (Agostini et al. 2004).

IL-1α is a cell-associated protein, believed to function in an autocrine manner. It is expressed by a variety of somatic cells, contributing to normal homeostasis. In severe illness, IL-1α is found at high levels in the circulation and thought to be released by dead and dying cells and cleaved by extracellular calpain proteases (Wakabayashi et al. 1991). Consequently, IL-1α is also thought to be DAMP or an alarmin.

5 The Role of IL-1 in Induction of IL-17 Production

It has been established for some time that IL-1β has the capacity to activate T cells (Badovinac et al. 1998; Nakae et al. 2003a). Recent studies from our laboratory have shown that it plays a critical role in promoting IL-17 production by T_H17 cells and by γδ T cells (Sutton et al. 2006, 2009). We found that both IL-1α and IL-1β synergize with IL-23 to promote IL-17 production by murine T cells in the presence or absence of T cell receptor engagement. Consistent with these findings, it was demonstrated that IL-1 synergises with IL-6 and IL-23 to induce expression of the transcription factors interferon regulatory factor (IRF)-4 and RORγt, thereby driving T_H17 polarization (Chung et al. 2009). Moreover, human memory CD45RO$^+$ T_H17 cells, but not naïve T cells, secrete IL-17 in vitro in response to stimulation with IL-1 and IL-23 (van Beelen et al. 2007).

IL-1RI$^{-/-}$ mice were found to be resistant to induction of EAE, and this was associated with deficient IL-17 production by T cells. However, adoptive transfer of wild type encephalitogenic T cells into IL-1RI$^{-/-}$ mice could induce disease, suggesting a role for IL-1 upstream of T cell activation (Sutton et al. 2006). Nakae et al. (2003b) have also demonstrated that IL-1Ra deficient mice that spontaneously develop auto-immune arthritis due to excessive IL-1 signaling, exhibited vastly elevated expression of IL-17. The spontaneous development of arthritis did not occur in IL-1Ra$^{-/-}$ mice also deficient in IL-17 (Nakae et al. 2003a). In addition, mutations in the *Nalp3* gene, which is required for cleavage of pro-IL-1β to its active form, lead to excess IL-1β production by innate immune cells, resulting in augmented T_H17 cell differentiation and consequent IL-17-dominant immunopathology (Meng et al. 2009). These studies suggest that IL-1 is required for the induction of T_H17 cells, possibly via a cascade of IL-1-induced IL-6 and IL-23, and is therefore critical during the development, but not at the effector stage of autoimmune diseases.

Unconventional T cells, such as γδ T cells, have been shown to express IL-1RI, as well as the IL-23R, and produce IL-17 in response to IL-1 and IL-23 in the absence of TCR engagement (McCandless et al. 2009; Sutton et al. 2009). In addition, these IL-17 producing γδ T cells have been shown to increase susceptibility to EAE. Thus, the DC-derived cytokines IL-1 and IL-23 play an important role in promoting IL-17 production in inflammation.

6 IL-1β in Auto-inflammatory and Auto-immune Diseases

IL-1β has been shown to be a critical component of disease pathogenesis in autoimmune disorders, such as multiple sclerosis (MS), rheumatoid arthritis (RA), and Crohn's disease and in the mouse models of these diseases. IL-1β and IL-1α induce synthesis of a range of pro-inflammatory mediators, including MMP, which are major mediators of pathology in joints of patients with RA (Burrage et al. 2006) and in the CNS of patients with MS. The concentration of IL-1β in the cerebro-spinal fluid (CSF) of MS patients has been shown to correlate with disease severity (Hauser et al. 1990) and an imbalance in the ratio of IL-1β to IL-1Ra predisposes to MS (de Jong et al. 2002). Furthermore, mutations in IL-1 associated genes, which result in excessive levels of IL-1 have been linked with the development of various auto-immune syndromes. Treatment with rhIL-1Ra (Kineret/Anakinra) is very effective against systemic-onset juvenile idiopathic arthritis, cryopyrin-associated periodic syndromes (CAPS) and other auto-inflammatory disorders, such as gout, and is moderately effective against RA (Dinarello 2009). IL-1 Trap (Rilonacept), an Fc fusion protein with the extra-cellular domains of IL-1RI and IL-1RAcP, as well as anti-IL-1β (Canakinumab), have been approved for use in CAPS. IL-1 targeted drugs, although very effective against some auto-inflammatory diseases are not that effective against all auto-immune disorders.

One explanation for the poor efficacy of IL-1 targeted drugs in auto-immunity may reflect the fact that a primary function of IL-1 is to promote induction or expansion rather than to mediate effector function of IL-17-producting T cells. The studies by Cua and colleagues showing that IL-23-driven autoantigen-specific T_H17 cells could induced EAE following transfer into naïve mice (Langrish et al. 2005), suggesting that IL-23 was not required for lineage commitment, but was essential for the terminal differentiation and full pathogenic activity of T_H17 cells (McGeachy et al. 2009). We and others extended these findings through the demonstration that IL-1α and IL-1β synergized with IL-23 to promote IL-17 production by memory T_H17 cells (Higgins et al. 2006; Sutton et al. 2006; van Beelen et al. 2007), but also promoted innate IL-17 production by γδ T cells (Sutton et al. 2009). Studies involving adoptive transfer of T cells into IL-23p19$^{-/-}$ mice suggested that IL-23 played a critical role in the development, but not in the effector function of encephalitogenic T cells (Thakker et al. 2007).

The most definitive evidence of IL-1 in the development of T_H17 cells that mediate auto-immunity came from studies showing that mice that lacked the signaling receptor (IL-1RI$^{-/-}$) were resistant to the development of MOG-induced EAE mice and had a corresponding reduction in MOG-specific T_H17 cells (Sutton et al. 2006). Conversely, IL-1RI$^{-/-}$ mice were capable of developing EAE after transfer of MOG-specific T cells from wild-type mice (Sutton et al. 2006). This suggested that IL-1 is essential upstream, but is dispensable downstream of T cell activation, perhaps due to redundancy in the actions of pro-inflammatory cytokines and chemokines in the induction of inflammatory pathology in EAE. These findings provide an alternative explanation for the poor efficacy of the IL-1 antagonist, IL-1Ra, as a therapy for auto-immune disease. It had previously been assumed that IL-1Ra blocked the inflammatory effect of IL-1 downstream of T cell activation. However, it now appears that IL-1Ra may be more effective at blocking the development of T_H17 cells. IL-1 and IL-23-induced IL-17 production from C57BL/6 spleen cells was successfully inhibited by treatment with IL-1Ra. In contrast, blocking IFN-γ, with a neutralizing antibody increased IL-1 and IL-23-induced IL-17 production, agreeing with reports suggesting that IFN-γ could inhibit the development of T_H17 cells (Murphy et al. 2003).

7 The Role of IL-18 in IL-17 Production

Like IL-1β, IL-18 also requires cleavage by caspase-1 for activation from its inactive precursor (24 kDa) to the mature 18 kDa cytokine. IL-18 functions with IL-12 and IL-15 to promote IFN-γ production by T_H1 cells (Okamura et al. 1995). In the absence of IL-12 however, IL-18 can also induce IL-4, driving T_H2 responses (Nakanishi et al. 2001). It has been demonstrated that production of both IFN-γ and the T_H2 related cytokines IL-4, IL-5, and IL-13 was significantly increased in transgenic mice expressing high levels of IL-18 (Hoshino et al. 2001). IL-18 can also act synergistically with IL-23 to induce T_H17 cytokine production independently of TCR stimulation (Weaver et al. 2006). However, IL-18 has been shown to have a less dominant role in T_H17 cell activation when compared with its role in T_H1/T_H2 differentiation in vivo. Van de Loo and colleagues have demonstrated that blockade of IL-18 signaling with a soluble form of the IL-18R accessory protein (sIL-18Rβ) leads to reduced IFN-γ and IL-4 production by splenic $CD3^+$ T cells and decreased circulating $CD4^+CD25^+Foxp3^+$ Treg cells, but enhanced T_H17 differentiation and IL-17 production in collagen-induced arthritis (CIA) (Veenbergen et al. 2010).

The role of IL-18 in auto-immunity is controversial. It has been reported that IL-$18^{-/-}$ mice are resistant to the development of EAE, as they have diminished IFN-γ production by NK cells and are unable to mount auto-reactive T_H1 responses (Shi et al. 2000). Conversely, Becher and colleagues (Gutcher et al. 2006) demonstrated that IL-$18^{-/-}$ mice are fully susceptible to MOG-induced EAE, with T cell responses and IFN-γ levels comparable to those found in WT mice. However, deletion of IL-18Rα conferred resistance to development of EAE, and the authors argue that an alternative, yet unidentified IL-18Rα ligand induces production of IL-23 and consequent expansion of T_H17 cells (Gutcher et al. 2006).

IL-18 plays a role in both innate and adaptive immunity by activating neutrophils, enhancing T cell and NK cell maturation, and inducing IFN-γ production by CD4 T cells from mice with EAE (Lalor et al. 2011). IL-18, in synergy with IL-23 also induces IFN-γ and IL-17 expression by unconventional $CD3^+$ T cells (Lalor et al. 2011). NKT cells express IL-23R (Rachitskaya et al. 2008) and secrete IFN-γ upon TCR ligation (Godfrey et al. 2000). γδ T cells also constitutively express IL-23R (Sutton et al. 2009) and IL-1RI (McCandless et al. 2009). Furthermore, γδ T cells produce IFN-γ in response to IL-23 signaling in the presence or absence of IL-1β (Lalor et al. 2011). IL-18 has also been shown to induce IFN-γ expression by γδ T cells (Haas et al. 2009). Cells of the innate immune system, including NK cells and γδ T cells, may be primed to respond rapidly to inflammatory cytokines, such as IL-23 and IL-1, or IL-18, and may play a role in driving T_H1 and T_H17 responses. IL-1 and IL-18 may have redundant roles in activating pathogenic T cells and this may explain why blocking either IL-1 or IL-18 are ineffective treatments for certain auto-immune diseases.

8 Innate IL-17 Production by γδ T Cells

While much of the focus has been on IL-17-producing CD4$^+$ T cells, γδ T cells have also been shown to express RORγt and IL-23R and are an important source of IL-17 (Martin et al. 2009; Sutton et al. 2009). It has been demonstrated that γδ T cells from the spleen or lymph nodes of naïve mice are capable of producing IL-17 *ex vivo*, without additional in vitro stimulation (Romani et al. 2008; Stark et al. 2005; Sutton et al. 2009). Expression of other T_H17 cell-associated cytokines, including IL-17F, IL-22, and TNF-α, were upregulated by purified γδ T cells in response to IL-23 signaling, and this was enhanced by addition of IL-1β (Lockhart et al. 2006; Shibata et al. 2007; Sutton et al. 2009). Purified naive γδ T cells secrete IL-17 in response to IL-1β and IL-23 in the absence of stimulation with anti-CD3 or anti-CD28 (Sutton et al. 2009). Interestingly, it has been shown that IL-17 production by γδ T cells is independent of IL-6, an essential cytokine for driving the development of T_H17 cells (Lochner et al. 2008). It seems possible that certain, perhaps tissue-specific, subsets of γδ T cells are already differentiated in the periphery, and are capable of rapidly producing IL-17 in response to increased local production of IL-23 (Roark et al. 2008) or IL-23 and IL-1 produced by DC or other innate immune cells in response to PAMPs or DAMPs (Sutton et al. 2009) (Fig. 3).

It is possible that γδ T cells may co-localize in the periphery with innate immune cells, such as DC and macrophages, subsequently promoting IL-17 production by αβ T cells. TLR-activated monocytes are required for optimum induction of human T_H17 cells (Evans et al. 2007), while murine DC derived IL-1 and IL-23 has been shown to drive T_H17 cells and EAE (Brereton et al. 2009; Shainheit et al. 2008). Furthermore, co-culture experiments have revealed that IL-1 and IL-23-activated γδ T cells promoted IL-17 production by CD4$^+$ and CD8$^+$ T cells through interaction with DC (Sutton et al. 2009). This study also found that DC produced IL-1 and IL-23 in response to stimulation with IL-17. The pro-inflammatory cytokines, IL-1, TNF-α, and IL-6 synergize with IL-23 to enhance the production of IL-17 from spleen cells. However, these cytokines were also capable of promoting the production of IL-23 and IL-1β from DC. Therefore, an amplification loop may exist through which IL-1 and IL-23 stimulate αβ and γδ T cells and consequently enhance IL-17 production by both of these T cell subsets.

Expression of the chemokine receptor CCR6 has been found on almost all IL-17A, IL-17F, and IL-22 producing CD4$^+$ T cells (Acosta-Rodriguez et al. 2007b; Singh et al. 2008). Recently, γδ T cells have also been shown to express CCR6, and expression of the chemokine receptor on γδ T cells is associated with IL-17 production, while CCR6$^-$ γδ T cells have been shown to produce IFN-γ (Haas et al. 2009). The sole ligand for CCR6, CCL20, has also been shown to be involved in the development of EAE and in T_H17 cell homing to inflamed joints in RA (Ambrosini et al. 2003; Hirota et al. 2007). γδ T cells produce CCL20 in response to IL-1 and IL-23 stimulation (Sutton et al. unpublished observation), indicating a feedback loop for the further attraction of IL-17 producing γδ T cells, as well as T_H17 cells.

Another surface receptor recently associated with IL-17 regulation is the TNF receptor family member CD27. This receptor has recently emerged as a thymic determinant of the cytokine profile of peripheral γδ T cells. Silva-Santos and colleagues demonstrated that the phenotypes of IFN-γ and IL-17-producing γδ T cell subsets are established during thymic development, when CD27 acts as a regulator of γδ T cell differentiation (Ribot et al. 2009). They showed that the majority of γδ T cells in the spleen and lymph nodes, as well as peripheral tissues including the lungs and gut, expressed CD27 and secreted IFN-γ. Between 10 and 30% of peripheral γδ T cells were $CD27^-$ and comprised essentially all IL-17-producing γδ cells. Consistent with these findings, we have found that γδ T cells infiltrating the brain at the onset of clinical symptoms of EAE were $CD27^-$, and a very high frequency of these produced IL-17 (Lalor et al. 2011).

It has been shown that *Il17a* and *Il17f* gene expression is rapidly upregulated in draining lymph nodes 4 h after injection of IL-1 and IL-23 into the footpad (Sutton et al. 2009). This observation is consistent with earlier reports in which intracellular staining revealed the frequency of IL-17-producing γδ T cells to be relatively high in the lymph nodes of naïve mice (Jensen et al. 2008). IL-17 expression was found to be enhanced in both $CD3^+$ T cells and γδ T cells following injection of IL-1β and IL-23(Brereton et al. 2009). These findings suggest that IL-1 and IL-23 promote rapid IL-17 gene expression and protein production in vitro and in vivo and that γδ T cells are a major source of innate IL-17.

Thus, γδ T cells may behave like IL-17-producing memory αβ T cells. Yoshikai and colleagues have demonstrated that resident $V\delta1^+$ γδ T cells in the peritoneal cavity behaved like memory CD4 T cells (Shibata et al. 2007). This study showed a rapid increase in IL-17 production following i.p. injection of *E. coli*, which coincided with upregulation of *Il23p19* mRNA expression. Furthermore, they confirmed that IL-23 production by another cell source was necessary for the activation of γδ T cells and their subsequent production of IL-17 (Shibata et al. 2007) (Fig. 3).

9 IL-17 Producing γδ T Cells in Infection and Auto-immunity

γδ T cells are an important source of inflammatory cytokines and chemokines that mediate immunity against infection, especially at mucosal surfaces (Goerlich et al. 1991; Kabelitz et al. 1990; van der Heyde et al. 1995). The control of infection with many intracellular microorganisms has been attributed to IFN-γ and TNF-α production by αβ T and γδ T cells. However, IL-17-producing $CD4^+$ T cell and/or γδ T cells have been detected in a number of bacterial and viral infections. In bacterial infections, γδ T cells can rapidly produce IL-17 (Lockhart et al. 2006; Shibata et al. 2007), and in some models they are the main source of IL-17 (Umemura et al. 2007). It has been shown that γδ T cells provide an early release of IL-17 in response to *E. coli* and *Listeria monocytogenes* infection, which is required for resolution of infection via neutrophil recruitment (Hamada et al. 2008; Shibata et al. 2007).

It has also been shown that γδ T cells have a pathogenic role in experimental models of auto-immunity and this was thought to be mediated through IFN-γ production. There is circumstantial evidence from studies in humans that γδ T cells may play a pathogenic role in MS and colitis (Nanno et al. 2008).

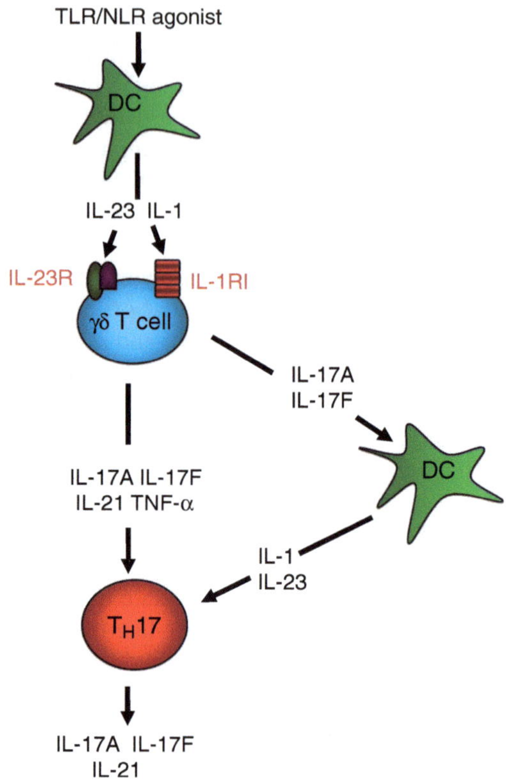

Fig. 3 Activation of γδ T cells and T_H17 cells

Oligoclonal expansion or activation of γδ T cells has been found in acute MS brain lesions and in the cerebrospinal fluid of MS patients with recent disease onset, suggesting that γδ T cells contribute to neuro-inflammation (Hvas et al. 1993; Poggi et al. 1999; Selmaj et al. 1991; Shimonkevitz et al. 1993; Wucherpfennig et al. 1992). In mouse models of auto-immune diseases, the role of γδ T cells is more controversial, with both protective and pathogenic functions ascribed to these cells. Nevertheless, evidence is emerging to suggest that γδ T cells are a major source of IL-17 in several murine models of auto-immunity (Roark et al. 2007; Shibata et al. 2007; Sutton et al. 2009). IL-17-producing γδ T cells have been shown to play an important role in the pathogenesis of CIA (Roark et al. 2007). Disease severity is markedly reduced in IL-$17^{-/-}$ mice (Nakae et al. 2003a), and mice depleted of the $V\gamma4^+$ T cells showed reduced incidence and milder clinical symptoms of disease (Roark et al. 2007). In these studies it appears that expansion of the IL-17-producing Vγ4 sub-population of γδ T cells was driven by self molecules that arose during inflammation, rather than the immunizing antigen.

In the EAE model, γδ T cells infiltrate the CNS prior to the onset of clinical symptoms. CNS infiltrating γδ T cells secrete IFN-γ and TNF-α (Smith and Barnum 2008) and a very high frequency produce IL-17 at the onset of clinical signs (Sutton et al. 2009). Deletion of γδ cells immediately prior to onset or during the chronic phase of disease significantly reduced clinical signs and pathological changes (Rajan et al. 1996). Disruption of the TCR δ chain gene led to reduced severity of disease in one study, although not in another (Clark and Lingenheld 1998; Spahn et al. 1999). Strangely, a further study failed to find a role for γδ T cells in the development of EAE (Clark and Lingenheld 1998). There are a number of factors that might explain some of the conflicting results in these studies. Firstly, the frequency of γδ subtypes varies between mouse strains, and different subtypes have different functions. Secondly, it has been suggested that antibody depletion of γδ T cells using either the GL3 or the UC7-13D5 antibodies, which have been shown to accelerate the onset of EAE, may activate rather than deplete γδ T cells by cross linking the γδ TCR (Koenecke et al. 2009). We have shown that $IL\text{-}1RI^{-/-}$ mice, which are resistant to induction of EAE by immunization with MOG and CFA, do develop EAE following transfer of total $CD3^+$ T cells, but not following transfer of $\gamma\delta^+$ T cells or $CD3^+$ T cells depleted of γδ T cells. This suggests that that both αβ and γδ T cells are required for optimum development of EAE (Sutton et al. 2009).

It is thought that γδ T cells may participate in the development of EAE by regulating leukocyte transfer across the blood–brain barrier and acting as a source of pro-inflammatory molecules. γδ T cell-derived IL-17 is involved in recruitment of neutrophils and other leukocytes to the site of inflammation (Carlson et al. 2009; Shibata et al. 2007). It has been suggested that neutrophils may be among the first leukocytes to infiltrate the CNS in some models of EAE (Brown et al. 1983), and have been implicated in blood–brain barrier breakdown (Veldhuis et al. 2003). Depletion of γδ T cells in mice with EAE resulted in a marked decrease in IFN-γ, IL-1, IL-6, and TNF-α, especially at disease onset (Rajan et al. 1998). Furthermore, we found that MOG-specific IL-17 production was significantly reduced in $Tcr\delta^{-/-}$ mice (Sutton et al. 2009). Thus, γδ T cells may play a role in facilitating the activation and migration of myelin-reactive T cells into the CNS by upregulating expression of appropriate pro-inflammatory cytokines and chemokines, but also by promoting IL-17 production by T_H17 cells. Such effects may not require large numbers of cells and could be facilitated by the numbers of γδ T cells routinely found in the spinal cord and brain of mice with EAE (Rajan et al. 1998; Sutton et al. 2009). This is in line with more recent data showing that IL-17 induces expression of IL-1, IL-6, and TNF-α in inflamed tissues (Langrish et al. 2005) and that IL-17 induces *Il17a* and *Il17f* mRNA expression in T cells (Sutton et al. 2009).

Studies by Antel and colleagues have demonstrated that γδ T cells are cytotoxic against brain-derived oligodendrocytes, the cells responsible for the ensheathment of neuronal axons with myelin, the target of pathogenic T cells in MS and EAE (Freedman et al. 1991, 1997). It is possible that the activation of γδ T cells during the effector phase of disease may result, not only in the recruitment of αβ T cells, neutrophils, and other leukocytes to the site of inflammation in the CNS, but also mediate irreversible damage of oligodendrocytes and ultimately lead to blocking of signal conduction and the development of neurological symptoms and signs of MS and EAE.

10 Role of γδ T Cells in Driving Differentiation of T_H17 Cells

γδ T cells appear to be an important source of innate IL-17 during the development of auto-immunity, preceding and influencing the induction of adaptive immune responses mediated by T_H17 cells. Our studies suggest that IL-1β and IL-23 produced by DC in response to TLR and NLR activation might be important stimuli for early production of innate IL-17 (Sutton et al. 2009). It has also been demonstrated that γδ T cells express TLR2 and TLR4 and direct stimulation with ligands for these PRR promoted IL-17 production by γδ T cells (Martin et al. 2009). Evidence of a role for IL-1 in driving pathogenic IL-17-secreting γδ T cells was provided by experiments which showed that transfer of $CD3^+$ T cells from naive WT mice into IL-$1RI^{-/-}$ mice (which are normally resistant to EAE) conferred susceptibility to EAE that was reduced by depletion of γδ T cells from the transferred T cells (Sutton et al. 2009). Furthermore, $CD4^+$ T cells express IL-17R and upon stimulation with supernatants from activated γδ T cells containing IL-17 and IL-21 could induce IL-17 production by $CD4^+$ T cells (Sutton et al. 2009). This is consistent with the demonstration that IL-21 plays a critical role in IL-17 production by T_H17 cells (Korn et al. 2007; Nurieva et al. 2007; Zhou et al. 2007). In addition, these findings reveal a positive feedback role of IL-17 production and indicate that early IL-17 production by γδ T cells may promote the induction or activation of T_H17 cells. DCs express IL-17R and secrete IL-23, IL-1β, and IL-6, in addition to a number of T_H17-related chemokines, in response to IL-17. This provides another potential mechanism whereby γδ T cells may indirectly promote development, expansion or recruitment of T_H17 cells (Sutton et al. 2009).

Since γδ T cells infiltrate the CNS prior to the onset of EAE and a large percentage of these secrete IL-17, it is possible that IL-17-producing γδ T cells have a pathogenic role early in the development of auto-immune disease, not only by substantially contributing to the pool of IL-17 in the target organ and thereby mediating further leukocyte recruitment, but also by controlling the generation or re-activation of αβ T_H17 cells. We feel that γδ T cells may represent an important source of innate IL-17 that may be pathogenic in auto-immune diseases by facilitating recruitment and activation of autoantigen-specific $CD4^+$ T cells in the target organ. γδ T cells represent a smaller and more discrete population, uniquely defined by their restricted TCR usage and tissue specific localization. Therefore, rather than targeting $CD4^+$ T cells with the possibility of global immuno-suppression, therapeutics that specifically target γδ T cells may be a more attractive approach for the treatment of auto-immune diseases.

References

Acosta-Rodriguez, E.V., Napolitani, G., Lanzavecchia, A., and Sallusto, F. (2007a). Interleukins 1beta and 6 but not transforming growth factor-beta are essential for the differentiation of interleukin 17-producing human T helper cells. Nat Immunol *8*, 942–949.

Acosta-Rodriguez, E.V., Rivino, L., Geginat, J., Jarrossay, D., Gattorno, M., Lanzavecchia, A., Sallusto, F., and Napolitani, G. (2007b). Surface phenotype and antigenic specificity of human interleukin 17-producing T helper memory cells. Nat Immunol *8*, 639–646.

Aggarwal, S., Ghilardi, N., Xie, M.H., de Sauvage, F.J., and Gurney, A.L. (2003). Interleukin-23 promotes a distinct CD4 T cell activation state characterized by the production of interleukin-17. J Biol Chem *278*, 1910–1914.

Agostini, L., Martinon, F., Burns, K., McDermott, M.F., Hawkins, P.N., and Tschopp, J. (2004). NALP3 forms an IL-1beta-processing inflammasome with increased activity in Muckle-Wells auto-inflammatory disorder. Immunity *20*, 319–325.

Akira, S., Takeda, K., and Kaisho, T. (2001). Toll-like receptors: critical proteins linking innate and acquired immunity. Nat Immunol *2*, 675–680.

Ambrosini, E., Columba-Cabezas, S., Serafini, B., Muscella, A., and Aloisi, F. (2003). Astrocytes are the major intracerebral source of macrophage inflammatory protein-3alpha/CCL20 in relapsing experimental auto-immune encephalomyelitis and in vitro. Glia *41*, 290–300.

Amer, A., Franchi, L., Kanneganti, T.D., Body-Malapel, M., Ozoren, N., Brady, G., Meshinchi, S., Jagirdar, R., Gewirtz, A., Akira, S., and Nunez, G. (2006). Regulation of Legionella phagosome maturation and infection through flagellin and host Ipaf. J Biol Chem *281*, 35217–35223.

Atkins, E., and Wood, W.B., Jr. (1955). Studies on the pathogenesis of fever. I. The presence of transferable pyrogen in the blood stream following the injection of typhoid vaccine. J Exp Med *101*, 519–528.

Auron, P.E., Webb, A.C., Rosenwasser, L.J., Mucci, S.F., Rich, A., Wolff, S.M., and Dinarello, C.A. (1984). Nucleotide sequence of human monocyte interleukin 1 precursor cDNA. Proc Natl Acad Sci USA *81*, 7907–7911.

Badovinac, V., Mostarica-Stojkovic, M., Dinarello, C.A., and Stosic-Grujicic, S. (1998). Interleukin-1 receptor antagonist suppresses experimental auto-immune encephalomyelitis (EAE) in rats by influencing the activation and proliferation of encephalitogenic cells. J Neuroimmunol *85*, 87–95.

Bettelli, E., Carrier, Y., Gao, W., Korn, T., Strom, T.B., Oukka, M., Weiner, H.L., and Kuchroo, V.K. (2006). Reciprocal developmental pathways for the generation of pathogenic effector TH17 and regulatory T cells. Nature *441*, 235–238.

Bettelli, E., Sullivan, B., Szabo, S.J., Sobel, R.A., Glimcher, L.H., and Kuchroo, V.K. (2004). Loss of T-bet, but not STAT1, prevents the development of experimental auto-immune encephalomyelitis. J Exp Med *200*, 79–87.

Braddock, M., and Quinn, A. (2004). Targeting IL-1 in inflammatory disease: new opportunities for therapeutic intervention. Nat Rev Drug Discov *3*, 330–339.

Brereton, C.F., Sutton, C.E., Lalor, S.J., Lavelle, E.C., and Mills, K.H. (2009). Inhibition of ERK MAPK Suppresses IL-23- and IL-1-Driven IL-17 Production and Attenuates Auto-immune Disease. J Immunol *183*, 1715–23.

Brown, R.D., Yuen, E., Kronenberg, H., and Rickard, K.A. (1983). In vitro clonogenic assays in selective neutropenia. Scand J Haematol *30*, 110–116.

Burrage, P.S., Mix, K.S., and Brinckerhoff, C.E. (2006). Matrix metalloproteinases: role in arthritis. Front Biosci *11*, 529–543.

Carlson, M.J., West, M.L., Coghill, J.M., Panoskaltsis-Mortari, A., Blazar, B.R., and Serody, J.S. (2009). In vitro-differentiated TH17 cells mediate lethal acute graft-versus-host disease with severe cutaneous and pulmonary pathologic manifestations. Blood *113*, 1365–1374.

Chen, Z., Laurence, A., Kanno, Y., Pacher-Zavisin, M., Zhu, B.M., Tato, C., Yoshimura, A., Hennighausen, L., and O'Shea, J.J. (2006). Selective regulatory function of Socs3 in the formation of IL-17-secreting T cells. Proc Natl Acad Sci USA *103*, 8137–8142.

Chung, Y., Chang, S.H., Martinez, G.J., Yang, X.O., Nurieva, R., Kang, H.S., Ma, L., Watowich, S.S., Jetten, A.M., Tian, Q., and Dong, C. (2009). Critical regulation of early Th17 cell differentiation by interleukin-1 signaling. Immunity *30*, 576–587.

Clark, R.B., and Lingenheld, E.G. (1998). Adoptively transferred EAE in gamma delta T cell-knockout mice. J Autoimmun *11*, 105–110.

Costelloe, C., Watson, M., Murphy, A., McQuillan, K., Loscher, C., Armstrong, M.E., Garlanda, C., Mantovani, A., O'Neill, L.A., Mills, K.H., and Lynch, M.A. (2008). IL-1F5 mediates anti-inflammatory activity in the brain through induction of IL-4 following interaction with SIGIRR/TIR8. J Neurochem *105*, 1960–1969.

Das, J., Ren, G., Zhang, L., Roberts, A.I., Zhao, X., Bothwell, A.L., Van Kaer, L., Shi, Y., and Das, G. (2009). Transforming growth factor beta is dispensable for the molecular orchestration of Th17 cell differentiation. J Exp Med *206*, 2407–2416.

de Jong, E.C., Vieira, P.L., Kalinski, P., Schuitemaker, J.H., Tanaka, Y., Wierenga, E.A., Yazdanbakhsh, M., and Kapsenberg, M.L. (2002). Microbial compounds selectively induce Th1 cell-promoting or Th2 cell-promoting dendritic cells in vitro with diverse th cell-polarizing signals. J Immunol *168*, 1704–1709.

Dinarello, C.A. (2009). Immunological and inflammatory functions of the interleukin-1 family. Annu Rev Immunol *27*, 519–550.

Dunne, A., Ross, P.J., Pospisilova, E., Masin, J., Meaney, A., Sutton, C.E., Iwakura, Y., Tschopp, J., Sebo, P., and Mills, K.H. (2010). Inflammasome activation by adenylate cyclase toxin directs Th17 responses and protection against Bordetella pertussis. J Immunol *185*, 1711–1719.

Eder, C. (2009). Mechanisms of interleukin-1beta release. Immunobiology *214*, 543–53.

Eugster, H.P., Frei, K., Kopf, M., Lassmann, H., and Fontana, A. (1998). IL-6-deficient mice resist myelin oligodendrocyte glycoprotein-induced auto-immune encephalomyelitis. Eur J Immunol *28*, 2178–2187.

Evans, H.G., Suddason, T., Jackson, I., Taams, L.S., and Lord, G.M. (2007). Optimal induction of T helper 17 cells in humans requires T cell receptor ligation in the context of Toll-like receptor-activated monocytes. Proc Natl Acad Sci USA *104*, 17034–17039.

Ferber, I.A., Brocke, S., Taylor-Edwards, C., Ridgway, W., Dinisco, C., Steinman, L., Dalton, D., and Fathman, C.G. (1996). Mice with a disrupted IFN-gamma gene are susceptible to the induction of experimental auto-immune encephalomyelitis (EAE). J Immunol *156*, 5–7.

Ferretti, S., Bonneau, O., Dubois, G.R., Jones, C.E., and Trifilieff, A. (2003). IL-17, produced by lymphocytes and neutrophils, is necessary for lipopolysaccharide-induced airway neutrophilia: IL-15 as a possible trigger. J Immunol *170*, 2106–2112.

Franchi, L., Warner, N., Viani, K., and Nunez, G. (2009). Function of Nod-like receptors in microbial recognition and host defense. Immunol Rev *227*, 106–128.

Freedman, M.S., Bitar, R., and Antel, J.P. (1997). gamma delta T-cell-human glial cell interactions. II. Relationship between heat shock protein expression and susceptibility to cytolysis. J Neuroimmunol *74*, 143–148.

Freedman, M.S., Ruijs, T.C., Selin, L.K., and Antel, J.P. (1991). Peripheral blood gamma-delta T cells lyse fresh human brain-derived oligodendrocytes. Ann Neurol *30*, 794–800.

Godfrey, D.I., Hammond, K.J., Poulton, L.D., Smyth, M.J., and Baxter, A.G. (2000). NKT cells: facts, functions and fallacies. Immunol Today *21*, 573–583.

Goerlich, R., Hacker, G., Pfeffer, K., Heeg, K., and Wagner, H. (1991). Plasmodium falciparum merozoites primarily stimulate the V gamma 9 subset of human gamma/delta T cells. Eur J Immunol *21*, 2613–2616.

Gutcher, I., Urich, E., Wolter, K., Prinz, M., and Becher, B. (2006). Interleukin 18-independent engagement of interleukin 18 receptor-alpha is required for auto-immune inflammation. Nat Immunol *7*, 946–953.

Haas, J.D., Gonzalez, F.H., Schmitz, S., Chennupati, V., Fohse, L., Kremmer, E., Forster, R., and Prinz, I. (2009). CCR6 and NK1.1 distinguish between IL-17A and IFN-gamma-producing gammadelta effector T cells. Eur J Immunol *39*, 3488–3497.

Hamada, S., Umemura, M., Shiono, T., Tanaka, K., Yahagi, A., Begum, M.D., Oshiro, K., Okamoto, Y., Watanabe, H., Kawakami, K., *et al.* (2008). IL-17A produced by gammadelta T cells plays a critical role in innate immunity against listeria monocytogenes infection in the liver. J Immunol *181*, 3456–3463.

Hauser, S.L., Doolittle, T.H., Lincoln, R., Brown, R.H., and Dinarello, C.A. (1990). Cytokine accumulations in CSF of multiple sclerosis patients: frequent detection of interleukin-1 and tumor necrosis factor but not interleukin-6. Neurology *40*, 1735–1739.

Higgins, S.C., Jarnicki, A.G., Lavelle, E.C., and Mills, K.H. (2006). TLR4 mediates vaccine-induced protective cellular immunity to Bordetella pertussis: role of IL-17-producing T cells. J Immunol *177*, 7980–7989.

Hirota, K., Yoshitomi, H., Hashimoto, M., Maeda, S., Teradaira, S., Sugimoto, N., Yamaguchi, T., Nomura, T., Ito, H., Nakamura, T., *et al.* (2007). Preferential recruitment of CCR6-expressing Th17 cells to inflamed joints via CCL20 in rheumatoid arthritis and its animal model. J Exp Med *204*, 2803–2812.

Hise, A.G., Tomalka, J., Ganesan, S., Patel, K., Hall, B.A., Brown, G.D., and Fitzgerald, K.A. (2009). An essential role for the NLRP3 inflammasome in host defense against the human fungal pathogen Candida albicans. Cell Host Microbe *5*, 487–497.

Hoshino, T., Kawase, Y., Okamoto, M., Yokota, K., Yoshino, K., Yamamura, K., Miyazaki, J., Young, H.A., and Oizumi, K. (2001). Cutting edge: IL-18-transgenic mice: in vivo evidence of a broad role for IL-18 in modulating immune function. J Immunol *166*, 7014–7018.

Hsu, L.C., Ali, S.R., McGillivray, S., Tseng, P.H., Mariathasan, S., Humke, E.W., Eckmann, L., Powell, J.J., Nizet, V., Dixit, V.M., and Karin, M. (2008). A NOD2-NALP1 complex mediates caspase-1-dependent IL-1beta secretion in response to Bacillus anthracis infection and muramyl dipeptide. Proc Natl Acad Sci USA *105*, 7803–7808.

Hvas, J., Oksenberg, J.R., Fernando, R., Steinman, L., and Bernard, C.C. (1993). Gamma delta T cell receptor repertoire in brain lesions of patients with multiple sclerosis. J Neuroimmunol *46*, 225–234.

Ivanov, II, McKenzie, B.S., Zhou, L., Tadokoro, C.E., Lepelley, A., Lafaille, J.J., Cua, D.J., and Littman, D.R. (2006). The orphan nuclear receptor RORgammat directs the differentiation program of proinflammatory IL-17+ T helper cells. Cell *126*, 1121–1133.

Jensen, K.D., Su, X., Shin, S., Li, L., Youssef, S., Yamasaki, S., Steinman, L., Saito, T., Locksley, R.M., Davis, M.M., *et al.* (2008). Thymic selection determines gammadelta T cell effector fate: antigen-naive cells make interleukin-17 and antigen-experienced cells make interferon gamma. Immunity *29*, 90–100.

Kabelitz, D., Bender, A., Schondelmaier, S., Schoel, B., and Kaufmann, S.H. (1990). A large fraction of human peripheral blood gamma/delta + T cells is activated by Mycobacterium tuberculosis but not by its 65-kD heat shock protein. J Exp Med *171*, 667–679.

Kanneganti, T.D., Ozoren, N., Body-Malapel, M., Amer, A., Park, J.H., Franchi, L., Whitfield, J., Barchet, W., Colonna, M., Vandenabeele, P., *et al.* (2006). Bacterial RNA and small antiviral compounds activate caspase-1 through cryopyrin/Nalp3. Nature *440*, 233–236.

Katz, S.P., Kierszenbaum, F., and Waksman, B.H. (1978). Mechanisms of action of "lymphocyte-activating factor" (LAF). III. Evidence that LAF acts on stimulated lymphocytes by raising cyclic GMP in G1. J Immunol *121*, 2386–2391.

Kawanokuchi, J., Shimizu, K., Nitta, A., Yamada, K., Mizuno, T., Takeuchi, H., and Suzumura, A. (2008). Production and functions of IL-17 in microglia. J Neuroimmunol *194*, 54–61.

Koenecke, C., Chennupati, V., Schmitz, S., Malissen, B., Forster, R., and Prinz, I. (2009). In vivo application of mAb directed against the gammadelta TCR does not deplete but generates "invisible" gammadelta T cells. Eur J Immunol *39*, 372–379.

Korn, T., Bettelli, E., Gao, W., Awasthi, A., Jager, A., Strom, T.B., Oukka, M., and Kuchroo, V.K. (2007). IL-21 initiates an alternative pathway to induce proinflammatory T(H)17 cells. Nature *448*, 484–487.

Korn, T., Mitsdoerffer, M., Croxford, A.L., Awasthi, A., Dardalhon, V.A., Galileos, G., Vollmar, P., Stritesky, G.L., Kaplan, M.H., Waisman, A., *et al.* (2008). IL-6 controls Th17 immunity in vivo by inhibiting the conversion of conventional T cells into Foxp3+ regulatory T cells. Proc Natl Acad Sci USA *105*, 18460–18465.

Kroenke, M.A., Carlson, T.J., Andjelkovic, A.V., and Segal, B.M. (2008). IL-12- and IL-23-modulated T cells induce distinct types of EAE based on histology, CNS chemokine profile, and response to cytokine inhibition. J Exp Med *205*, 1535–1541.

Lalor, S.J., Dungan, L.S., Sutton, C.E., Basdeo, S.A., Fletcher, J.M., and Mills, K.H. (2011). Caspase-1-processed cytokines IL-1β and IL-18 promote IL-17 production by γδ and CD4 T cells that mediate autoimmunity. J Immunol *186*, 5738–5748.

Langrish, C.L., Chen, Y., Blumenschein, W.M., Mattson, J., Basham, B., Sedgwick, J.D., McClanahan, T., Kastelein, R.A., and Cua, D.J. (2005). IL-23 drives a pathogenic T cell population that induces auto-immune inflammation. J Exp Med *201*, 233–240.

Lochner, M., Peduto, L., Cherrier, M., Sawa, S., Langa, F., Varona, R., Riethmacher, D., Si-Tahar, M., Di Santo, J.P., and Eberl, G. (2008). In vivo equilibrium of pro-inflammatory IL-17+ and regulatory IL-10+ Foxp3+ RORgamma t+ T cells. J Exp Med *205*, 1381–1393.

Lockhart, E., Green, A.M., and Flynn, J.L. (2006). IL-17 production is dominated by gammadelta T cells rather than CD4 T cells during Mycobacterium tuberculosis infection. J Immunol *177*, 4662–4669.

Lomedico, P.T., Gubler, U., Hellmann, C.P., Dukovich, M., Giri, J.G., Pan, Y.C., Collier, K., Semionow, R., Chua, A.O., and Mizel, S.B. (1984). Cloning and expression of murine interleukin-1 cDNA in Escherichia coli. Nature *312*, 458–462.

Mangan, P.R., Harrington, L.E., O'Quinn, D.B., Helms, W.S., Bullard, D.C., Elson, C.O., Hatton, R.D., Wahl, S.M., Schoeb, T.R., and Weaver, C.T. (2006). Transforming growth factor-beta induces development of the T(H)17 lineage. Nature *441*, 231–234.

Mariathasan, S., Weiss, D.S., Newton, K., McBride, J., O'Rourke, K., Roose-Girma, M., Lee, W.P., Weinrauch, Y., Monack, D.M., and Dixit, V.M. (2006). Cryopyrin activates the inflammasome in response to toxins and ATP. Nature *440*, 228–232.

Martin, B., Hirota, K., Cua, D.J., Stockinger, B., and Veldhoen, M. (2009). Interleukin-17-producing gammadelta T cells selectively expand in response to pathogen products and environmental signals. Immunity *31*, 321–330.

Martinon, F., Agostini, L., Meylan, E., and Tschopp, J. (2004). Identification of bacterial muramyl dipeptide as activator of the NALP3/cryopyrin inflammasome. Curr Biol *14*, 1929–1934.

Martinon, F., and Tschopp, J. (2004). Inflammatory caspases: linking an intracellular innate immune system to auto-inflammatory diseases. Cell *117*, 561–574.

McCandless, E.E., Budde, M., Lees, J.R., Dorsey, D., Lyng, E., and Klein, R.S. (2009). IL-1R signaling within the central nervous system regulates CXCL12 expression at the blood-brain barrier and disease severity during experimental auto-immune encephalomyelitis. J Immunol *183*, 613–620.

McGeachy, M.J., Chen, Y., Tato, C.M., Laurence, A., Joyce-Shaikh, B., Blumenschein, W.M., McClanahan, T.K., O'Shea, J.J., and Cua, D.J. (2009). The interleukin 23 receptor is essential for the terminal differentiation of interleukin 17-producing effector T helper cells in vivo. Nat Immunol *10*, 314–324.

McKenzie, B.S., Kastelein, R.A., and Cua, D.J. (2005). Understanding the IL-23-IL-17 immune pathway. Trends Immunol *27*, 17–23.

Medzhitov, R. (2001). Toll-like receptors and innate immunity. Nat Rev Immunol *1*, 135–145.

Meng, G., Zhang, F., Fuss, I., Kitani, A., and Strober, W. (2009). A mutation in the Nlrp3 gene causing inflammasome hyperactivation potentiates Th17 cell-dominant immune responses. Immunity *30*, 860–874.

Michel, M.L., Keller, A.C., Paget, C., Fujio, M., Trottein, F., Savage, P.B., Wong, C.H., Schneider, E., Dy, M., and Leite-de-Moraes, M.C. (2007). Identification of an IL-17-producing NK1.1(neg) iNKT cell population involved in airway neutrophilia. J Exp Med *204*, 995–1001.

Milner, J.D., Brenchley, J.M., Laurence, A., Freeman, A.F., Hill, B.J., Elias, K.M., Kanno, Y., Spalding, C., Elloumi, H.Z., Paulson, M.L., *et al.* (2008). Impaired T(H)17 cell differentiation in subjects with autosomal dominant hyper-IgE syndrome. Nature *452*, 773–776.

Mizel, S.B., Oppenheim, J.J., and Rosenstreich, D.L. (1978). Characterization of lymphocyte-activating factor (LAF) produced by the macrophage cell line, P388D1. I. Enhancement of LAF production by activated T lymphocytes. J Immunol *120*, 1497–1503.

Murphy, C.A., Langrish, C.L., Chen, Y., Blumenschein, W., McClanahan, T., Kastelein, R.A., Sedgwick, J.D., and Cua, D.J. (2003). Divergent pro- and antiinflammatory roles for IL-23 and IL-12 in joint auto-immune inflammation. J Exp Med *198*, 1951–1957.

Nakae, S., Nambu, A., Sudo, K., and Iwakura, Y. (2003a). Suppression of immune induction of collagen-induced arthritis in IL-17-deficient mice. J Immunol *171*, 6173–6177.

Nakae, S., Saijo, S., Horai, R., Sudo, K., Mori, S., and Iwakura, Y. (2003b). IL-17 production from activated T cells is required for the spontaneous development of destructive arthritis in mice deficient in IL-1 receptor antagonist. Proc Natl Acad Sci USA *100*, 5986–5990.

Nakanishi, H., Zhang, J., Koike, M., Nishioku, T., Okamoto, Y., Kominami, E., von Figura, K., Peters, C., Yamamoto, K., Saftig, P., and Uchiyama, Y. (2001). Involvement of nitric oxide released from microglia-macrophages in pathological changes of cathepsin D-deficient mice. J Neurosci *21*, 7526–7533.

Nanno, M., Kanari, Y., Naito, T., Inoue, N., Hisamatsu, T., Chinen, H., Sugimoto, K., Shimomura, Y., Yamagishi, H., Shiohara, T., *et al.* (2008). Exacerbating role of gammadelta T cells in chronic colitis of T-cell receptor alpha mutant mice. Gastroenterology *134*, 481–490.

Nurieva, R., Yang, X.O., Martinez, G., Zhang, Y., Panopoulos, A.D., Ma, L., Schluns, K., Tian, Q., Watowich, S.S., Jetten, A.M., and Dong, C. (2007). Essential autocrine regulation by IL-21 in the generation of inflammatory T cells. Nature *448*, 480–483.

Okamura, H., Tsutsi, H., Komatsu, T., Yutsudo, M., Hakura, A., Tanimoto, T., Torigoe, K., Okura, T., Nukada, Y., Hattori, K., and et al. (1995). Cloning of a new cytokine that induces IFN-gamma production by T cells. Nature *378*, 88–91.

Oppmann, B., Lesley, R., Blom, B., Timans, J.C., Xu, Y., Hunte, B., Vega, F., Yu, N., Wang, J., Singh, K., *et al.* (2000). Novel p19 protein engages IL-12p40 to form a cytokine, IL-23, with biological activities similar as well as distinct from IL-12. Immunity *13*, 715–725.

Pelegrin, P., and Surprenant, A. (2006). Pannexin-1 mediates large pore formation and interleukin-1beta release by the ATP-gated P2X7 receptor. EMBO J *25*, 5071–5082.

Poggi, A., Zocchi, M.R., Costa, P., Ferrero, E., Borsellino, G., Placido, R., Galgani, S., Salvetti, M., Gasperini, C., Ristori, G., *et al.* (1999). IL-12-mediated NKRP1A up-regulation and consequent enhancement of endothelial transmigration of V delta 2+ TCR gamma delta+ T lymphocytes from healthy donors and multiple sclerosis patients. J Immunol *162*, 4349–4354.

Rachitskaya, A.V., Hansen, A.M., Horai, R., Li, Z., Villasmil, R., Luger, D., Nussenblatt, R.B., and Caspi, R.R. (2008). Cutting Edge: NKT Cells Constitutively Express IL-23 Receptor and ROR{gamma}t and Rapidly Produce IL-17 upon Receptor Ligation in an IL-6-Independent Fashion. J Immunol *180*, 5167–5171.

Rajan, A.J., Gao, Y.L., Raine, C.S., and Brosnan, C.F. (1996). A pathogenic role for gamma delta T cells in relapsing-remitting experimental allergic encephalomyelitis in the SJL mouse. J Immunol *157*, 941–949.

Rajan, A.J., Klein, J.D., and Brosnan, C.F. (1998). The effect of gammadelta T cell depletion on cytokine gene expression in experimental allergic encephalomyelitis. J Immunol *160*, 5955–5962.

Ribot, J.C., deBarros, A., Pang, D.J., Neves, J.F., Peperzak, V., Roberts, S.J., Girardi, M., Borst, J., Hayday, A.C., Pennington, D.J., and Silva-Santos, B. (2009). CD27 is a thymic determinant of the balance between interferon-gamma- and interleukin 17-producing gammadelta T cell subsets. Nat Immunol *10*, 427–436.

Roark, C.L., French, J.D., Taylor, M.A., Bendele, A.M., Born, W.K., and O'Brien, R.L. (2007). Exacerbation of collagen-induced arthritis by oligoclonal, IL-17-producing gamma delta T cells. J Immunol *179*, 5576–5583.

Roark, C.L., Simonian, P.L., Fontenot, A.P., Born, W.K., and O'Brien, R.L. (2008). gammadelta T cells: an important source of IL-17. Curr Opin Immunol *20*, 353–357.

Romani, L., Zelante, T., De Luca, A., Fallarino, F., and Puccetti, P. (2008). IL-17 and therapeutic kynurenines in pathogenic inflammation to fungi. J Immunol *180*, 5157–5162.

Schmitz, J., Owyang, A., Oldham, E., Song, Y., Murphy, E., McClanahan, T.K., Zurawski, G., Moshrefi, M., Qin, J., Li, X., *et al.* (2005). IL-33, an interleukin-1-like cytokine that signals via the IL-1 receptor-related protein ST2 and induces T helper type 2-associated cytokines. Immunity *23*, 479–490.

Selmaj, K., Brosnan, C.F., and Raine, C.S. (1991). Colocalization of lymphocytes bearing gamma delta T-cell receptor and heat shock protein hsp65+ oligodendrocytes in multiple sclerosis. Proc Natl Acad Sci USA *88*, 6452–6456.

Shainheit, M.G., Smith, P.M., Bazzone, L.E., Wang, A.C., Rutitzky, L.I., and Stadecker, M.J. (2008). Dendritic cell IL-23 and IL-1 production in response to schistosome eggs induces Th17 cells in a mouse strain prone to severe immunopathology. J Immunol *181*, 8559–8567.
Shi, F.D., Takeda, K., Akira, S., Sarvetnick, N., and Ljunggren, H.G. (2000). IL-18 directs autoreactive T cells and promotes autodestruction in the central nervous system via induction of IFN-gamma by NK cells. J Immunol *165*, 3099–3104.
Shibata, K., Yamada, H., Hara, H., Kishihara, K., and Yoshikai, Y. (2007). Resident Vdelta1+ gammadelta T cells control early infiltration of neutrophils after Escherichia coli infection via IL-17 production. J Immunol *178*, 4466–4472.
Shimonkevitz, R., Colburn, C., Burnham, J.A., Murray, R.S., and Kotzin, B.L. (1993). Clonal expansions of activated gamma/delta T cells in recent-onset multiple sclerosis. Proc Natl Acad Sci USA *90*, 923–927.
Singh, S.P., Zhang, H.H., Foley, J.F., Hedrick, M.N., and Farber, J.M. (2008). Human T cells that are able to produce IL-17 express the chemokine receptor CCR6. J Immunol *180*, 214–221.
Smith, S.S., and Barnum, S.R. (2008). Differential expression of beta 2-integrins and cytokine production between gammadelta and alphabeta T cells in experimental auto-immune encephalomyelitis. J Leukoc Biol *83*, 71–79.
Spahn, T.W., Issazadah, S., Salvin, A.J., and Weiner, H.L. (1999). Decreased severity of myelin oligodendrocyte glycoprotein peptide 33 - 35-induced experimental auto-immune encephalomyelitis in mice with a disrupted TCR delta chain gene. Eur J Immunol *29*, 4060–4071.
Stark, M.A., Huo, Y., Burcin, T.L., Morris, M.A., Olson, T.S., and Ley, K. (2005). Phagocytosis of apoptotic neutrophils regulates granulopoiesis via IL-23 and IL-17. Immunity *22*, 285–294.
Stromnes, I.M., Cerretti, L.M., Liggitt, D., Harris, R.A., and Goverman, J.M. (2008). Differential regulation of central nervous system auto-immunity by T(H)1 and T(H)17 cells. Nat Med *14*, 337–342.
Sutterwala, F.S., Mijares, L.A., Li, L., Ogura, Y., Kazmierczak, B.I., and Flavell, R.A. (2007). Immune recognition of Pseudomonas aeruginosa mediated by the IPAF/NLRC4 inflammasome. J Exp Med *204*, 3235–3245.
Sutton, C., Brereton, C., Keogh, B., Mills, K.H., and Lavelle, E.C. (2006). A crucial role for interleukin (IL)-1 in the induction of IL-17-producing T cells that mediate auto-immune encephalomyelitis. J Exp Med *203*, 1685–1691.
Sutton, C.E., Lalor, S.J., Sweeney, C.M., Brereton, C.F., Lavelle, E.C., and Mills, K.H. (2009). Interleukin-1 and IL-23 Induce Innate IL-17 Production from gammadelta T Cells, Amplifying Th17 Responses and Auto-immunity. Immunity *31*, 331–341.
Takatori, H., Kanno, Y., Chen, Z., and O'Shea, J.J. (2008). New complexities in helper T cell fate determination and the implications for auto-immune diseases. Mod Rheumatol *18*, 533–541.
Thakker, P., Leach, M.W., Kuang, W., Benoit, S.E., Leonard, J.P., and Marusic, S. (2007). IL-23 is critical in the induction but not in the effector phase of experimental auto-immune encephalomyelitis. J Immunol *178*, 2589–2598.
Umemura, M., Yahagi, A., Hamada, S., Begum, M.D., Watanabe, H., Kawakami, K., Suda, T., Sudo, K., Nakae, S., Iwakura, Y., and Matsuzaki, G. (2007). IL-17-mediated regulation of innate and acquired immune response against pulmonary Mycobacterium bovis bacille Calmette-Guerin infection. J Immunol *178*, 3786–3796.
van Beelen, A.J., Zelinkova, Z., Taanman-Kueter, E.W., Muller, F.J., Hommes, D.W., Zaat, S.A., Kapsenberg, M.L., and de Jong, E.C. (2007). Stimulation of the Intracellular Bacterial Sensor NOD2 Programs Dendritic Cells to Promote Interleukin-17 Production in Human Memory T Cells. Immunity *27*, 660–669.
van der Heyde, H.C., Elloso, M.M., Chang, W.L., Kaplan, M., Manning, D.D., and Weidanz, W.P. (1995). Gamma delta T cells function in cell-mediated immunity to acute blood-stage Plasmodium chabaudi adami malaria. J Immunol *154*, 3985–3990.
Veenbergen, S., Smeets, R.L., Bennink, M.B., Arntz, O.J., Joosten, L.A., van den Berg, W.B., and van de Loo, F.A. (2010). The natural soluble form of IL-18 receptor beta exacerbates collagen-induced arthritis via modulation of T-cell immune responses. Ann Rheum Dis *69*, 276–283.

Veldhuis, W.B., Floris, S., van der Meide, P.H., Vos, I.M., de Vries, H.E., Dijkstra, C.D., Bar, P.R., and Nicolay, K. (2003). Interferon-beta prevents cytokine-induced neutrophil infiltration and attenuates blood-brain barrier disruption. J Cereb Blood Flow Metab *23*, 1060–1069.

Wakabayashi, G., Gelfand, J.A., Burke, J.F., Thompson, R.C., and Dinarello, C.A. (1991). A specific receptor antagonist for interleukin 1 prevents Escherichia coli-induced shock in rabbits. FASEB J *5*, 338–343.

Warabi, Y., Matsumoto, Y., and Hayashi, H. (2007). Interferon beta-1b exacerbates multiple sclerosis with severe optic nerve and spinal cord demyelination. J Neurol Sci *252*, 57–61.

Weaver, C.T., Harrington, L.E., Mangan, P.R., Gavrieli, M., and Murphy, K.M. (2006). Th17: an effector CD4 T cell lineage with regulatory T cell ties. Immunity *24*, 677–688.

Wei, L., Laurence, A., Elias, K.M., and O'Shea, J.J. (2007). IL-21 is produced by Th17 cells and drives IL-17 production in a STAT3-dependent manner. J Biol Chem *282*, 34605–34610.

Wilson, N.J., Boniface, K., Chan, J.R., McKenzie, B.S., Blumenschein, W.M., Mattson, J.D., Basham, B., Smith, K., Chen, T., Morel, F., *et al.* (2007). Development, cytokine profile and function of human interleukin 17-producing helper T cells. Nat Immunol *8*, 950–957.

Wucherpfennig, K.W., Newcombe, J., Li, H., Keddy, C., Cuzner, M.L., and Hafler, D.A. (1992). Gamma delta T-cell receptor repertoire in acute multiple sclerosis lesions. Proc Natl Acad Sci USA *89*, 4588–4592.

Zhou, L., Chong, M.M., and Littman, D.R. (2009). Plasticity of CD4+ T cell lineage differentiation. Immunity *30*, 646–655.

Zhou, L., Ivanov, II, Spolski, R., Min, R., Shenderov, K., Egawa, T., Levy, D.E., Leonard, W.J., and Littman, D.R. (2007). IL-6 programs T(H)-17 cell differentiation by promoting sequential engagement of the IL-21 and IL-23 pathways. Nat Immunol *8*, 967–974.

Negative Regulation of T_H17 Differentiation

Mark S. Sundrud and Sergei Koralov

Abstract T_H17 cells and their associated cytokines act on resident parenchymal cells within tissues, thereby setting the stage for chronic inflammation. This realization, together with the finding that T_H17 cell development is reciprocally linked to that of T regulatory (Treg) cells, has revolutionized the way T cell-mediated immune pathology is viewed, and challenged the long-standing binary view of T cell differentiation (i.e., T_H1/T_H2), thereby opening exciting new opportunities to treat autoimmune inflammation. Much effort is now placed on understanding how T_H17 cells are restrained through endogenous mechanisms; the goal being to negatively regulate T_H17 development or function in clinical disease settings. The T_H1 and T_H2 cytokines, IFNγ and IL-4, as well as IL-27 and IL-10, all repress T_H17 cell differentiation. TGFβ signaling, which supports T_H17 differentiation in some contexts, can also strongly induce expression of the signature regulatory T cell transcription factor, Foxp3, which in turn cripples T_H17 differentiation through direct antagonism of the T_H17-specific orphan nuclear receptor RORγt. Emerging evidence also suggests that T_H17 cells are both inherently unstable and uniquely sensitive to metabolic stress. Here, we discuss some of the key molecular features of T_H17 cell development and highlight examples of cell-intrinsic and cell-extrinsic pathways that negatively influence T_H17 differentiation, the latter of which could be exploited for therapeutic application.

M.S. Sundrud (✉)
Department of Pathology, Harvard Medical School, Immune Disease Institute
and the Program in Cellular and Molecular Medicine at Children's Hospital
Boston, Boston, MA, USA
and
Tempero Pharmaceuticals, 200 Technology Square, Suite 602, Cambridge, MA 02139, USA
e-mail: msundrud@temperopharma.com

S. Jiang (ed.), *T_H17 Cells in Health and Disease*,
DOI 10.1007/978-1-4419-9371-7_7, © Springer Science+Business Media, LLC 2011

1 Introduction

T_H17 cells are uniquely equipped among lymphocytes to establish chronic tissue inflammation. As with other helper T cell lineages, their function is derived from the repertoire of pro-inflammatory cytokines they produce. A key feature of T_H17 cell biology is that, unlike T_H1 and T_H2 cells, T_H17 cells produce cytokines that lack classic immuno-regulatory (i.e., helper) function. IL-17A, IL-17F, and IL-22 made by T_H17 cells act primarily on tissue parenchyma, particularly at mucosal surfaces (Awane et al. 1999; Wolk et al. 2004; Zrioual et al. 2008). These cytokines have been implicated in wound healing and the maintenance of barrier immunity, but they also elicit antimicrobial peptide expression and production of pro-inflammatory cytokines and chemokines (Crome et al. 2010; Eyerich et al. 2009; Wolk et al. 2010). In contrast, IFNγ (produced by T_H1 cells) or IL-4, IL-5, and IL-13 (made by T_H2 cells), activate phagocytic and cytolytic immunity, or mobilize antibody responses, respectively (Amsen et al. 2009; Zhu and Paul 2008). The result is that T_H17-infiltrated tissues become a staging ground for chronic and progressive immune-mediated tissue damage, precipitating further recruitment of B and T cells, and well as innate immune cells, neutrophils in particular. T_H17 cells and their associated cytokines are found elevated in many common human auto-immune disorders, though their pioneering role in immune pathology is perhaps best exemplified in experimental auto-immune encephalomyelitis (EAE), a mouse model of multiple sclerosis. In EAE, myelin-reactive T_H17 cells are among the first cells to reach lesions within the central nervous system (Reboldi et al. 2009). Subsequent production of T_H17 cytokines activates resident parenchymal cells and damages integrity of the blood/brain barrier, fostering neutrophil and further lymphocyte recruitment (Reboldi et al. 2009). Thus, the propensity of T_H17 cells to drive propagate tissue inflammation is a distinguishing feature among lymphocytes and has been implicated in the pathogenesis of auto-immunity, fibrosis, and even cancer.

By virtue of their effects on resident epithelial cells, T_H17 cells are also important for mucosal immunity, most notably against fungal and bacterial insults. This role has been elucidated using numerous mouse models of infection (Chung et al. 2003; Huang et al. 2004; Mangan et al. 2006; Zelante et al. 2007; Zheng et al. 2008) and it is further illustrated when considering hyper IgE syndrome (HIES) in humans. HIES patients suffer from a primary immuno-deficiency due to somatic loss-of-function mutations within *Stat3* (Holland et al. 2007; Minegishi et al. 2007). HIES patients have a paucity of circulating T_H17 cells and afflicted individuals present clinically with recurrent fungal and bacterial infections (Holland et al. 2007; Milner et al. 2008; Minegishi et al. 2007). In mice, T_H17 cells have been shown to be key in warding off certain species of bacteria that colonize in the gut, such as *Citrobacter rodentium* (Chung et al. 2003; Dubin and Kolls 2008; Mangan et al. 2006). These results support previous work demonstrating that the vast majority of T_H17 cells in mice at steady-state are present within mucosal layers of the gut, such as the lamina propria of the small intestine and colon (Ivanov et al. 2006, 2009). Development of T_H17 cells in the gut, however, does not require pathogenic infection, but rather appears to be driven by commensal flora (Ivanov et al. 2008, 2009; Wu et al. 2010). Accordingly, gut-resident T_H17 cells can be abrogated if mice are housed in a germ free environment or treated with antibiotics, such as vancomycin, which target filamentous bacteria (Ivanov et al. 2008).

While there is still some debate as to whether T_H17 cells are inherently "good" or "bad", there is little doubt that the identification of T_H17 cells dramatically expands our understanding of potentially pathogenic T cell subsets. The T_H17 paradigm also opens up new therapeutic opportunities to more specifically treat chronic and auto-immune inflammation whilst leaving intact other aspects of protective immunity (vis-à-vis general immuno-suppressants). To fully exploit such opportunities, we must first understand the endogenous mechanisms that normally keep T_H17 responses in check, and also appreciate how these checkpoints are subverted in context of immune pathology.

Like all T cell responses generated in the periphery, T_H17 cell differentiation is a multi-step process consisting of lineage-commitment, amplification, and stabilization. Underlying each of these steps is a highly coordinated and tightly regulated program of gene expression (Miller and Weinmann 2009; Wilson et al. 2009). T_H17 differentiation begins with cognate interactions between naïve T cells and antigen-presenting cells (APC). Depending on the type of pathogen encountered (and therefore the combination of pathogen-associated molecular pattern receptors engaged), APC can express a variety of co-stimulatory or inhibitory cell surface receptors and secrete inflammatory or regulatory cytokines (Adema 2009; Vance et al. 2009). The engagement of peptide-MHC complexes with cognate T cell antigen receptors (TCR) triggers a number of outside-in signal transduction pathways that mirror antigen dose. This information is integrated, together with the type and quality of co-stimulation and cytokine gradients, in the form of post-translational activation or *de novo* expression of lineage-defining transcription factors (Miller and Weinmann 2009; Sharpe 2009; Sundrud and Nolan 2010). As a general rule, lineage-defining transcription factors such as T-bet (for T_H1 cells), GATA-3 (for T_H2 cells), Foxp3 (for inducible Treg cells), and RORγt (for T_H17 cells) interact with ubiquitously expressed transcription factors to form a dense regulatory network that specifies T cell lineage commitment (Miller and Weinmann 2009; Sundrud and Nolan 2010). Still other soluble factors, including cytokines, hormones, and growth factors, are produced by local bystander cells; they are sensed by activated T cells via an equally dense network of metabolic signaling pathways and can play a profound role in tuning peripheral T cell responses (discussed below).

2 Control of T_H17 Responses by Cytokines and STAT Proteins

Although many factors can influence the outcome of naïve T cell differentiation, cytokines are arguably the most dominant force. Hematopoietic cytokines bind to multimeric receptors comprised of both unique and shared sub-units. Cytokine binding induces rapid clustering and tyrosine phosphorylation of receptors mediated by receptor-associated Janus kinases (JAKs). Phosphorylated receptors then serve as docking surfaces for the SH2 domains of signal transducer activator of transcription (STAT) proteins, which exist in the cytoplasm of resting cells as latent monomers. Newly recruited STAT proteins are promptly phosphorylated by JAKs, leading to their dimerization and subsequent nuclear translocation. Activated STAT dimers subsequently direct gene transcription through interactions with cognate promoter elements (Fig. 1a). A number of excellent reviews are available discussing the modes and mechanics of JAK/STAT signal transduction (Adamson et al. 2009; Hu and Ivashkiv 2009; O'Shea and Murray 2008). In addition to their initial activation, several mechanisms exist to control the amplitude and duration of STAT signaling following cytokine stimulation. These mechanisms include receptor downregulation, active dephosphorylation or ubiquitination by Protein Inhibitor of Activated Stat (PIAS) proteins, and feedback inhibition by Suppressor of Cytokine Signaling (SOCS) proteins (Fig. 1a). Relatively little is known about the roles of PIAS proteins in immune regulation compared to SOCS proteins, which have been the focus of extensive investigation [reviewed in (Alexander and Hilton 2004; Shuai and Liu 2005)] (discussed below).

That both the initial and sustained activation of STAT proteins are tightly regulated allows cells to respond rapidly to fluctuations in extra-cellular cytokine concentrations. Mammals have 7 STAT proteins (STAT1-4, STAT5a, STAT5b, STAT6), and these proteins and their associated regulatory networks ultimately determine the fate of T cell differentiation (Adamson et al. 2009; Alexander and Hilton 2004; O'Shea and Murray 2008). STAT proteins play unique and determinant roles in T cell fate determinism. STAT1 and STAT4 are critical for T_H1 responses and the upregulation of T-bet, whereas STAT6 is integral to IL-4-mediated T_H2 differentiation and induction of GATA-3 expression (Adamson et al. 2009; Amsen et al. 2009; Elo et al. 2010; Schulz et al. 2009). Moreover, STAT5 activated downstream of IL-2 is essential for Foxp3 expression in Treg cells, and STAT3 is a key transcriptional regulator of T_H17 differentiation (Adamson et al. 2009; Burchill et al. 2007; Laurence et al. 2007; O'Shea and Murray 2008; Passerini et al. 2008; Wei et al. 2008).

Though many cytokines can promote its activation, IL-6 is the principle activator of STAT3 in naïve T cells (Chen et al. 2006; Kishimoto 2005; Nishihara et al. 2007) (Fig. 1a). The functional IL-6 receptor is expressed by a variety of cell types in most tissues, and is comprised of two sub-units; a unique cytokine-binding chain (IL-6Rα), and the IL-6 signal transducer (IL6ST; a.k.a. gp130), which is shared between multiple cytokine receptors, and (as its name implies) mediates IL-6 dependent signal transduction (Kishimoto 2005; Nishihara et al. 2007; Wang et al. 2009). IL-6 is a principle regulator of T_H17 differentiation in vitro, and T_H17 responses in vivo. Genetic ablation of IL-6 or IL-6R in mice, or anti-IL-6R monoclonal antibody treatment impairs T_H17 cell development and is protective in a variety of T_H17-driven disease models (Serada et al. 2008; Jones et al. 2010; Okuda et al. 1998).

Activation of naïve T cells in T_H17-polarizing conditions (i.e., IL-6 plus TGFβ) leads to STAT3-dependent upregulation of both *Il21* and *Il23r* gene expression (Yang et al. 2007; Zhou et al. 2007). Subsequent IL-21 and IL-23 production by activated T_H17 cells and APC sustains STAT3 phosphorylation (Kwon et al. 2009;

McGeachy et al. 2009), thereby maintaining its activity in developing T_H17 cells. Persistent STAT3 signaling may be critical to allow for cooperation with the T_H17-specific orphan nuclear receptors RORγt and RORα, which are STAT3 target genes themselves, in remodeling of the *Il17a/Il17f* locus and driving high-level IL-17 expression (Yang et al. 2007; Zhou et al. 2007).

TGFβ, the other classical T_H17-inducing cytokine, has been suggested to exert its effects, at least in part, by enhancing tyrosine phosphorylation of STAT3 (Qin et al. 2009) (Fig. 1a). The precise mechanism remains unclear, but TGFβ-dependent regulation of STAT3 activation is likely indirect, as the type 1 TGFβ receptor (TGFβR1) is a serine/threonine kinase. TGFβ also promotes T_H17 responses via negative regulation of IFNγ and IL-4 signaling (Das et al. 2009; Li et al. 2006), both of which inhibit T_H17 differentiation (Park et al. 2005) (Fig. 1a). TGFβ likely utilizes multiple downstream signaling pathways to control both T_H17 differentiation and Foxp3 expression (Derynck and Zhang 2003; Martinez et al. 2009; Yang et al. 2008b). Further understanding of how these distinct signaling modules control T cell differentiation may shed new light on the diverse biological activities of TGFβ in vivo.

STAT3 directly *trans*-activates most of the T_H17 signature genes, including *Il17a*, *Il17f*, *Rorc* (RORγt), and *Rora* (RORα) (Durant et al. 2010; Nishihara et al. 2007; Yang et al. 2007, 2008c; Zhou et al. 2007). As noted above, synergy between STAT3 and RORγt may be critical for T_H17 differentiation; whereas STAT3 activation does not induce high-level IL-17 expression in the absence of RORγt, and ectopic expression of RORγt is likewise insufficient to drive T_H17 differentiation in Stat3$^{-/-}$ T cells (Yang et al. 2007, 2008c; Zhou et al. 2007).

In addition to positively regulating T_H17 differentiation, STAT3 prevents developing T_H17 cells from diverging into the inducible T regulatory (iTreg) lineage. IL-6, IL-21, and IL-23 each inhibit TGFβ-induced expression of Foxp3, and do so in a STAT3-dependent manner (Bettelli et al. 2006; Korn et al. 2008; Yang et al. 2007; Zhou et al. 2007, 2008). STAT3 may also modulate the function of Foxp3 protein. Whereas Foxp3 has been shown to directly bind and inhibit the transcriptional activity of both RORγt and RORα, RORγt function is restored in the context of Foxp3 expression by stimulation of cells with STAT3-activating cytokines (Ichiyama et al. 2008; Martinez et al. 2009; Samanta et al. 2008; Zhang et al. 2008; Zhou et al. 2008). Thus, mechanisms underlying STAT3-mediated interference with Foxp3 expression or function may lead to novel therapeutic approaches to inhibit T_H17 differentiation, while enhancing iTreg cell development or function.

STAT3 also maintains the pro-inflammatory function of differentiated T_H17 cells. Effector/memory T_H17 cells have been shown to be inherently unstable both in vitro and in vivo (Janke et al. 2010; Martin-Orozco et al. 2009; McGeachy et al. 2007); likely requiring continued inflammatory cytokine signaling, particularly through IL-23 and IL-1β, to maintain IL-17 expression. In vitro, T_H17 cells differentiated with TGFβ and IL-6 to produce both IL-17 and the anti-inflammatory cytokine IL-10 (McGeachy et al. 2007). Transfer of in vitro-differentiated T_H17 cells generates inflammatory lesions, but disease is often associated with a phenotypic switch of the transferred cells from a T_H17 to a T_H1 phenotype (Martin-Orozco et al. 2009). Restimulation of previously differentiated T_H17 cells in vitro with TGFβ and IL-6

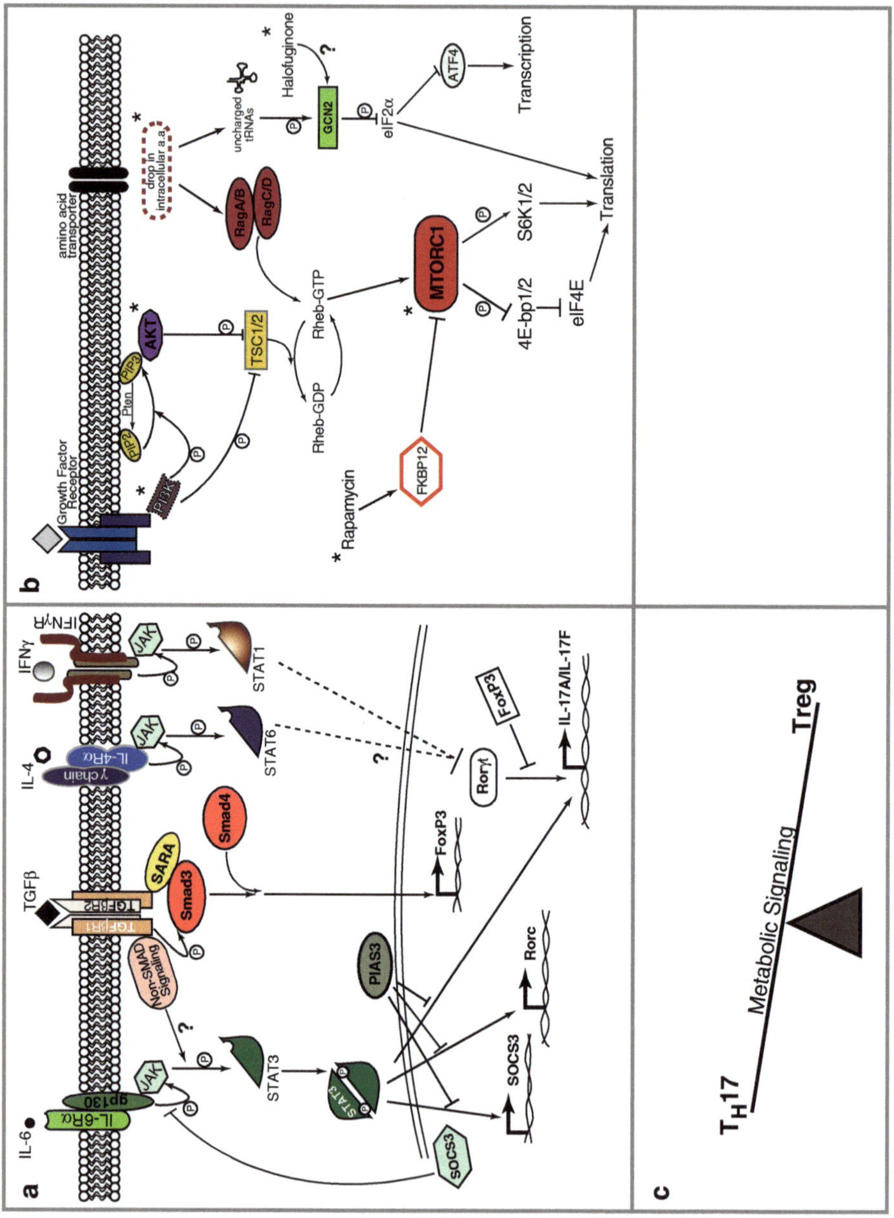
a
IL-6
IL-6Rα
gp130
JAK
STAT3
SOCS3
Non-SMAD Signaling
TGFβ
TGFβR1
TGFβR2
SARA
Smad3
Smad4
PIAS3
Rorc
FoxP3
IL-4
γ chain
IL-4Rα
STAT6
IFNγ
IFNγR
STAT1
Rorγt
IL-17A/IL-17F
b
Growth Factor Receptor
PI3K
PIP2
PIP3
Pten
AKT
TSC1/2
Rheb-GDP
Rheb-GTP
amino acid transporter
drop in intracellular a.a.
RagA/B
RagC/D
MTORC1
Rapamycin
FKBP12
4E-bp1/2
eIF4E
S6K1/2
Translation
uncharged tRNAs
Halofuginone
GCN2
eIF2α
ATF4
Transcription
c
T_H17
Metabolic Signaling
Treg

Fig. 1 Regulation of Th17 differentiation by cytokine and metabolic signaling pathways. (**a**) Cross-regulation of cytokine signaling pathways during T_H17 differentiation. Stimulation of activated T cells with IL-6 induces JAK activation, and subsequent recruitment, phosphorylation, and dimerization of STAT3. Active STAT3 dimers translocate to the nucleus where they *trans*-activate the expression of T_H17-signature genes (Rorc, *Il17a*, *Il17f*) as well as the feedback inhibitor of STAT3 signaling, SOCS3. SOCS3 binds the intracellular domain of gp130 and promotes its degradation. PIAS3 physically interacts with STAT3 and blocks its function by promoting tyrosinedephosphorylation and nuclear export. Co-stimulation of T cells with TGFβ leads to transcription of *FoxP3* in a Smad3/4-dependent pathway, which in turn interacts with and blocks Rorγt function. Smad3/4-independent TGFβ signaling may enhance IL-6-dependent phosphorylation of STAT3, possibly through repression of *Socs3* gene expression. IL-4 and IFNγ signaling inhibits *Rorc* upregulation and T_H17 differentiation. (**b**) Metabolic pathways and signal transduction. Activation of PI-3K downstream of growth factor stimulation promotes generation of the lipid second messenger PIP_3, which is regulated by the phosphatase Pten. Both PI-3K and Akt phosphorylate and inhibit the TSC1/2 complex, which in turn inhibits the small GTPase Rheb through its intrinsic GAP activity. Active (GTP-bound) Rheb, is required for the kinase activity of mTORC1 that drives cell metabolism in nutrient-rich conditions via phosphorylation of 4E-bp1/2 and S6K1. Amino acid deprivation, in contrast to nutrient-responsive pathways, inactivates the Rag family of small GTPases, which are necessary for Rheb-mediated activation of mTORC1. Amino acid limitation also leads to unaminoacylated (i.e., uncharged) tRNAs, which are non-discriminately recognized by the eIF2α kinase GCN2. Following tRNA binding, GCN2 undergoes autophosphorylation, subsequently activating the amino acid starvation response (AAR) pathway. Biochemical hallmarks of AAR pathway activation include phosphorylation of eIF2α and induction of ATF4 protein expression. Whereas inhibition of eIF2α reduces global protein synthesis, ATF4 activates a compensatory program of gene expression that confers relative stress resistance to cells. Furthermore, the macrolide rapamycin binds to and destabilizes mTORC1 in complex with its cellular ligand FKBP12. The small molecule halofuginone stimulates GCN2-dependent activation of the AAR. Asterisks indicate molecules or proteins shown to regulate the reciprocal differentiation of T_H17 and iTreg cells. (**c**) Metabolic stress promotes T cell tolerance by inhibiting T_H17 differentiation and enhancing iTreg cell development

for a second time results in cells that produce IL-10 but not IL-17 (McGeachy et al. 2007). This may represent the fact that IL-6R expression is down-regulated on T cells following activation (Betz and Muller 1998). In contrast, secondary stimulation of T_H17 cells with IL-23 confers stable expression of IL-17 and down-regulation of IL-10 (McGeachy et al. 2007). Upon transfer into mice, IL-23-stimulated T_H17 cells promote tissue inflammation, whereas T_H17 cells repeatedly stimulated with TGFβ and IL-6 actually protect against auto-immune tissue damage (McGeachy et al. 2007). Given that IL-23 is potent inducer of STAT3 activation, these findings collectively indicate that STAT3 initiates, amplifies, and stabilizes the pro-inflammatory function of T_H17 cells (Ahern et al. 2010; Korn et al. 2009).

Given these key roles of STAT3 in dictating T_H17 differentiation and function, a growing body of literature has been dedicated to investigating the regulation of STAT3 in T cells. As with all STAT proteins, STAT3 is controlled by the biology of the cytokine receptors that induce its activity. In the case of IL-6, both IL-6Rα and gp130 can be expressed either on the cell surface, or secreted as truncated proteins lacking their trans-membrane domains (Kishimoto 2005; Rose-John et al. 2006). Soluble IL-6Rα (sIL-6Rα) binds free extracellular IL-6; this complex in turn interacts with cell surface-expressed gp130 to initiate signaling. This mode of IL-6 signal transduction is termed trans-IL-6 signaling and has been implicated in diverse pathophysiologies such as auto-immunity, cancer, and tissue fibrosis (Igaz et al. 2000; Kishimoto 2005; McLoughlin et al. 2005; Nowell et al. 2003). Trans-IL-6 signaling generally amplifies IL-6 signaling by rendering otherwise non-responsive cells (i.e., cells that express gp130 but not IL-6Rα), responsive to IL-6:sIL6Rα complexes. Potentially, as a counter-balance to the generally pro-inflammatory actions of trans-IL-6 signaling, gp130 can also be expressed in a soluble form (sgp130), either via proteolytic cleavage or alternative splicing (Diamant et al. 1997; Graf et al. 2008). In contrast to soluble IL-6Rα, however, soluble gp130 is antagonistic in nature, binding to extracellular IL-6:sIL-6Rα complexes and preventing subsequent signaling through cell surface-expressed gp130 (Rose-John et al. 2006). Because it cannot bind to soluble IL-6 in the absence of sIL-6Rα, sgp130 only blocks trans-IL-6 signaling, not IL-6 signaling instigated by the binding of free IL-6 to cell surface receptors. Both sIL-6Rα and sgp130 are commonly elevated in the serum of auto-immune patients, as well as in affected tissue (Dayer and Choy 2010; Simon et al. 2008). In mice, recombinant sgp130 inhibits T_H17 differentiation and associated tissue inflammation via inhibition of STAT3 activation (Nowell et al. 2009). However, because gp130 is a shared component of all IL-6 family cytokine receptors, sgp130 is predicted to have broad effects on in vivo (Muller-Newen 2003).

The other STAT3-activating cytokines relevant for T_H17 cells, namely IL-21 and IL-23, signal through unique cell surface receptors (IL-21R and IL-23R). The IL-21R is comprised of IL-21Rα and the shared IL-2RG (a.k.a., gamma-common (γ_c)) (Rochman et al. 2009). IL-23 receptors consist of IL-23R and IL-12Rβ1, the latter of which is shared with the IL-12 receptor complex (Kastelein et al. 2007). In contrast to IL-6 receptors, the IL-21 and IL-23 receptors are expressed exclusively as transmembrane proteins and are regulated via gene expression. The IL-21 and IL-23 receptors are highly expressed on developing T_H17 cells (Yang et al. 2008a; Zhou et al. 2007). IL-23R transcripts are abundant in mouse and human IL-17-producing memory T cells as well as some myeloid cells that express IL-17 in response to IL-23 stimulation (Awasthi et al. 2009). Temporal regulation of the IL-6, IL-21, and IL-23 receptors, rather than utilization of distinct signaling pathways, may explain why these cytokines seem to play non-redundant roles in initiating, amplifying, and stabilizing T_H17 cell development; each cytokine receptor may be used by T_H17 cells to maintain STAT3 activation at distinct stages in their maturation.

In addition to cytokine receptors, physiological STAT3 activation is under the cell-intrinsic control of SOCS proteins (Alexander and Hilton 2004; O'Shea and Murray 2008) (Fig. 1a). Eight SOCS proteins exist in mammals (SOCS1-7, and CIS) and their gene expression is directly *trans*-activated by STAT proteins themselves; they can utilize distinct biochemical mechanisms to inhibit STAT signaling. Inhibitory mechanisms of SOCS proteins include: (1) competing with STAT proteins for cytokine receptor docking, (2) serving as pseuosubstrates for JAK phosphorylation, and (3) degrading cytokine receptors or JAKs directly via recruitment of the E3 ubiquitin ligase complex (Alexander and Hilton 2004; O'Shea and Murray 2008). Even though SOCS proteins are generally conserved, each family member displays remarkably specialized functions in T cell biology, with SOCS3 specifically regulating STAT3-driven responses (Chen et al. 2006).

Germline deletion of SOCS3 in mice, or transgenic overexpression at an embryonic stage, is lethal (Marine et al. 1999). Conditional deletion of SOCS3 in T cells, however, is viable, and these mice have grossly normal T cell compartments (Chen et al. 2006). T cells lacking SOCS3 display longer and more exaggerated STAT3 phosphorylation kinetics in response to cytokine stimulation (Chen et al. 2006). The increased activation kinetics of STAT3 in the context of SOCS3 deficiency is accompanied by increases in STAT3 promoter occupancy (Chen et al. 2006). In contrast, cytokine-mediated phosphorylation of other STAT family members (e.g., STAT1, STAT4, STAT5, STAT6) is largely unaffected by SOCS3 ablation (Chen et al. 2006). SOCS3 has been suggested to regulate STAT3 activation via binding to gp130 and inducing its degradation via an ubiquitin- and proteosome-dependent pathway (Alexander and Hilton 2004; O'Shea and Murray 2008). As such, transgenic expression of a mutant gp130 protein (Y759F) in mice that cannot interact with SOCS3 exacerbates T_H17 cell development in vitro, similar to mice lacking SOCS3 (Atsumi et al. 2002; Nishihara et al. 2007). Naïve T cells isolated from gp130 ^{Y759F} transgenic mice activated in T_H17-polarizing conditions also express less Foxp3, and the mice themselves tend to develop spontaneous arthritis (Nishihara et al. 2007). The interplay between SOCS3, STAT3, and STAT3-activating receptors thus represents a seminal pathway in the development of T_H17 cells (Chen et al. 2006; McLoughlin et al. 2005; O'Shea and Murray 2008; Qin et al. 2009).

In addition to the SOCS, the PIAS protein family contains intrinsic phosphatase and E3-ubiquitin ligase activity and has also been shown to negatively regulate STAT signaling (Shuai and Liu 2005). PIAS1 acutely regulates innate inflammatory responses through repression of STAT1 activity, whereas PIAS3 targets STAT3 for dephosphorylation (Liu et al. 2004; Dabir et al. 2009) (Fig. 1a). PIAS3, like all PIAS family members, is phosphorylated in response to inflammation; it interacts and translocates with tyrosine-phosporylated STAT3 to the nucleus before inducing dephosphorylation and nuclear export of STAT3 (Dabir et al. 2009). The STAT3/PIAS3 complex can be further regulated, by TRIM8 for example, which relieves PIAS3-mediated repression of STAT3 signaling when overexpressed in cells (Okumura et al. 2010). In addition to promoting STAT dephosphorylation, some PIAS proteins function primarily by inducing ubiquitin-dependent proteosomal degradation of interacting molecules (Liu et al. 2004; Shuai and Liu 2005). PIAS proteins can also regulate non-STAT transcription factors, such as p53 and CBP/p300, and can bind the small ubiquitin-like protein SUMO (Tan et al. 2010). Whether PIAS3 regulates STAT3-driven T_H17 differentiation directly remains to be determined.

Each mechanism of STAT3 regulation discussed above, with the exception of PIAS3, has lent key insight into the biology of T_H17 cells. However, an important concept to remember is that the STAT3 signaling pathway does not exist in a vacuum. T cells activated in vivo are simultaneously bombarded with an array of cytokines, each at gradient concentrations. T cells exposed to IL-6, therefore, are simultaneously confronted by a broader milieu of cytokines in the local micro-environment; any or all of which may augment or repress IL-6R signaling. As touched on earlier, TGFβ signaling can both synergize with IL-6 to enhance STAT3 phosphorylation (Qin et al. 2009), and inhibit RORγt activity via induction of Foxp3 (Martinez et al. 2009; Zhang et al. 2008; Zhou et al. 2008). Still other examples of cytokines that negatively regulate T_H17 differentiation are the T_H1- and T_H2-associated cytokines, IFNγ and IL-4, respectively (Park et al. 2005) (Fig. 1a). IL-2 in the mouse also inhibits T_H17 differentiation (Laurence et al. 2007), though it remains unclear if this is through enhancing Foxp3 expression, or through more direct regulation of STAT3 signaling. Regardless, the inhibitory role of IL-2 during T_H17 differentiation is confounded by the fact that IL-2 is paradoxically *required* for the differentiation of human T_H17 cells (Manel et al. 2008). Lastly, IL-27, which is produced by activated APC, is another potent inhibitor of T_H17 differentiation (Diveu et al. 2009; Murugaiyan et al. 2009). However, IL-27 does not affect the pro-inflammatory function of established T_H17 cells (El-behi et al. 2009). In general, little is known as to how these cytokines repress T_H17 differentiation, although it is interesting to note that IL-27, like IL-6, IL-21, and IL-23, also induces the phosphorylation of STAT3 (Xu et al. 2009). However, STAT3 activation downstream of IL-27 fails to induce IL-17 expression, and instead leads to the differentiation of IL-10-producing type 1 regulatory (Tr1) cells (Apetoh et al. 2010; Xu et al. 2009). How STAT3 functions in a context-dependent manner to prompt either pro- or anti-inflammatory gene expression in T cells remains an open question.

3 Dynamic Transcription Factor Interactions Regulating T_H17 Cell Development

T_H17 differentiation should be viewed as a continuum, with STAT3 representing only the first of many nodal points that ultimately gives rise to a pro-inflammatory T_H17 cell. Chronologically, the second step in the, 'how to become a T_H17 cell' manual is the induction of a transcription factor network conducive for T_H17

differentiation that also prevents diversion into alternative effector or regulatory lineages. At the center of the T_H17 transcriptional regulatory web is the lineage-defining orphan nuclear receptor RORγt (Chen et al. 2007; Zhou and Littman 2009). The notion of RORγt as a lineage-defining transcription factor for T_H17 cells was first illustrated by gene expression profiling showing its selective expressed in T_H17 cells vis-à-vis T_H1, T_H2, or iTreg cells (Ivanov et al. 2006). RORγt was also demonstrated to be necessary for the in vitro differentiation of mouse and human naïve T cells into T_H17 cells (Ivanov et al. 2006; Manel et al. 2008; Zhou et al. 2007). In vivo, nearly all $CD4^+$ T_H17 cells present in the intestinal lamina propria express RORγt (Ivanov et al. 2006). Furthermore, T cells from *Rorc*$^{-/-}$ mice fail to generate T_H17 cells upon antigen challenge, and these mice are largely resistant to EAE (Ivanov et al. 2006; Yang et al. 2008c). The related orphan nuclear receptor RORα (specifically Rora4) is also expressed in T_H17 cells; it is responsible for residual T_H17 differentiation observed in the absence of RORγt (Yang et al. 2008c). Collectively, this data demonstrates a pivotal role for RORγt in the developmental program of both mouse and human T_H17 cells.

RORγt, like other lineage-defining transcription factors in T cells (e.g., T-bet, Foxp3), is not expressed in conventional (i.e., non-Treg) naïve T cells, but is rapidly upregulated under T_H17-priming conditions (Ivanov et al. 2006; Yang et al. 2008c; Zhou et al. 2007) (Fig. 1a). As with STAT3, both the initial induction and subsequent function of RORγt are subject to strict regulation. Whereas STAT3 *trans*-activates *Rorc* gene expression, PPARγ, a broad-acting anti-inflammatory nuclear receptor, inhibits *Rorc* expression (Klotz et al. 2009). PPARγ does not bind directly to the *Rorc* promoter, but instead prevents clearance of silencing mediator for retinoid and thyroid hormone receptors (SMRT) from the *Rorc* promoter upon T cell activation (Klotz et al. 2009). Accordingly, mice lacking PPARγ display enhanced *Rorc* expression and increased T_H17 differentiation, whereas pharmacologic activation of PPARγ represses *Rorc* gene expression in wild-type, but not PPAR$^{-/-}$ T cells (Klotz et al. 2009).

MicroRNAs have also been implicated in controlling RORγt expression (Wei and Pei 2010). miRNA-326 is highly expressed in human and mouse T_H17 cells, and is elevated in inflamed CNS tissue from multiple sclerosis patients (Du et al. 2009). Regulation of RORγt expression by mir-326 is indirect, presumably functioning through its bio-informatically-predicted target, the transcription factor Ets-1 (Du et al. 2009). The mechanism by which Ets-1 regulates *Rorc* expression and T_H17 differentiation has yet to be clarified. In a broader sense, the regulation of T_H17 differentiation by miRNAs remains open for exploration (Wei and Pei 2010). Indeed, most of the signaling molecules discussed herein have conserved miRNA binding sites in their 3′ UTRs; identification of T_H17-regulating miRNAs, and the function of their target molecules in this context could have broad therapeutic implications.

Once expressed, RORγt interacts with a host of transcription factors (physically or functionally) to control T_H17 lineage commitment and specify patterns of cytokine expression within T_H17 cells. Most notable of the RORγt-interacting partners is Foxp3 (Zhou et al. 2008). This transcriptional complex is particularly intriguing because both RORγt and Foxp3 are lineage-defining transcription factors that are co-regulated by TGFβ signaling. Other examples of transcription factor interactions critical for T cell fate determinism include the interaction between T-bet and GATA-3, which regulates the T_H1/T_H2 balance, and the Foxp3/NFAT interaction that guides Treg cell development (Hwang et al. 2005; Wu et al. 2006). Several groups have now confirmed that Foxp3 directly interacts with RORγt (also RORα) via co-immuno-precipitation experiments (Du et al. 2008; Ichiyama et al. 2008; Yang et al. 2008b; Zhang et al. 2008). Such steady-state interactions between transcription factors are relatively uncommon, as many ternary complexes are only stabilized upon DNA binding (e.g., Foxp3:NFAT, NFAT:AP-1) (Jain et al. 1993; Wu et al. 2006). An N-terminal portion of Foxp3 encoded by exon 2, and the ligand-binding domain of RORγt mediate the Foxp3/RORγt interaction (Du et al. 2008; Ichiyama et al. 2008; Yang et al. 2008b; Zhou et al. 2008).

Functionally, the Foxp3/RORγt prevents RORγt transcriptional activity on target genes, either in primary T cells or in transiently-transfected cell lines (Du et al. 2008; Ichiyama et al. 2008; Yang et al. 2008b; Zhou et al. 2008) (Fig. 1a). While some details remain unclear, exon 2 of Foxp3 contains a conserved LQALL motif, which matches the LxxLL core motif used by transcriptional co-activators (NCoA) or co-repressors (NCoR) to bind to nuclear receptors such as RORγt and RORα (Du et al. 2008; Yang et al. 2008b). Thus, it is conceivable that Foxp3 passively interferes with RORγt transcriptional activity by competing with necessary co-factors for RORγt binding. However, this model of Foxp3-mediated RORγt antagonism is not without caveats. One perplexing factor is that Foxp3-mediated inhibition of RORγt function is relieved by pro-inflammatory cytokine stimulation, notably IL-6 or IL-23 (Yang et al. 2008b; Zhou et al. 2008). As detailed above, these cytokines are indispensible for STAT3-mediated induction of RORγt expression. How Foxp3 antagonizes RORγt function in vivo if these factors are only co-expressed in the context of inflammation remains to be elucidated.

TGFβ signaling may also utilize additional mechanisms to antagonize RORγt function. One specific example is the transcription factor Smad3, which is recruited to and directly phosphorylated by active TGFβ receptors. Smad3, particularly phosphorylated Smad3, has been shown capable of directly interacting with RORγt (Martinez et al. 2009). Similar to the Foxp3/RORγt interaction, Smad3/RORγt complexes antagonize RORγt transcriptional activity on transfected reporter constructs. In line with this model, conditional deletion of Smad3 in T cells leads to enhanced T_H17 differentiation and impaired Foxp3 upregulation (Martinez et al. 2009). Whether the T_H17-enhancing effects of Smad3 deficiency is due to increased RORγt activity or is secondary to impaired Foxp3 upregulation is presently unclear. Nonetheless, the interactions between RORγt, Foxp3, and Smad3 represent novel putative mechanisms by which TGFβ maintains peripheral T cell tolerance.

RORγt also interacts with a broader network of more ubiquitous transcription factors to coordinate T_H17 differentiation and to specify patterns of cytokine gene expression [reviewed in (Sundrud and Nolan 2010)]. A number of transcription factors have been shown to play key synergistic roles downstream of RORγt to control T_H17 differentiation. Notable among these 'secondary' T_H17 transcription factors are IRF4, the AP-1 transcription factor BATF, and STAT3 itself (Brustle et al. 2007;

Huber et al. 2008; Schraml et al. 2009; Yang et al. 2007; Zhou et al. 2007). All of these factors share two key functional features within the T_H17 framework: (1) unlike RORγt, they are broadly expressed in multiple T cell lineages, and (2) ectopic expression of RORγt does not compensate for their absence. With regard to broad expression in T cell lineages, each of these factors also regulates non-T_H17 T cell functions. For example, STAT3 regulates development of $CXCR5^+$ T follicular helper (Tfh) cells, as well as IL-10-producing Tr1 cells (Apetoh et al. 2010; Nurieva et al. 2008; Vogelzang et al. 2008). Likewise, IRF4 is important for the development of both T_H2, and IL-9-secreting 'T_H9' cells (Rengarajan et al. 2002; Staudt et al. 2010). BATF is more generally involved in adaptive immunity, regulating both T and B cell function (Betz et al. 2010). Thus, IRF4, BATF, and STAT3 constitute an integral network of secondary transcription factors usurped by RORγt to specify T_H17 differentiation. These findings indicate that transcriptional regulation of T_H17 differentiation is fundamentally synergistic, and that interactions between transcription factors (be they functional or physical) are key for achieving the diversity seen in adaptive immune responses.

All of the T_H17 regulatory mechanisms described above are defined by their influence on IL-17 expression. Yet still other transcription factors form a more peripheral regulatory network during T_H17 differentiation, specifying the pattern of cytokine gene expression in T_H17 cells without regulating IL-17 expression per se. The aryl hydrocarbon receptor (AHR) and c-maf are examples of transcription factors that regulate auxiliary cytokine expression in T_H17 cells. AHR is a cellular receptor for a variety of synthetic and naturally-occurring aromatic hydrocarbons (Esser et al. 2009). More recently, AHR has been found to be an important regulator of IL-22 expression in T_H17 cells (Alam et al. 2010; Ramirez et al. 2010; Veldhoen et al. 2008). Given the broad role of IL-22 in cutaneous biology (Wolk et al. 2004, 2010), AHR may play a more specialized role in tuning the functions of skin-homing T_H17 cells. AHR-deficient T cells can still generate IL-17A-producing T_H17 cells, albeit with slightly reduced efficiency compared with wild-type cells. However, IL-22 expression in $AHR^{-/-}$ T cells is profoundly impaired (Alam et al. 2010; Ramirez et al. 2010; Veldhoen et al. 2008). Further illustration is provided by experiments in which either RORγt or AHR are retrovirally overexpressed in T cells; RORγt strongly induces IL-17 expression without much affect on IL-22, whereas AHR expression in T cells has the opposite effect, inducing IL-22 but not IL-17 expression (Alam et al. 2010; Ramirez et al. 2010; Veldhoen et al. 2008).

C-maf also regulates auxiliary cytokine expression in T_H17 cells. C-maf was originally characterized as a T_H2-promoting transcription factor due to its direct regulation of the *Il4* locus (Kim et al. 1999). More recently, however, c-maf has been shown to be highly expressed in T_H17 cells, again in a STAT3-dependent manner, and to activate the expression of IL-10 (Xu et al. 2009). IL-10 displays potent anti-inflammatory properties in vivo, even in instances where cells are co-expressing IL-17 (Apetoh et al. 2010; McGeachy et al. 2007). Thus, AHR and c-maf have little impact on classical T_H17 differentiation (as defined by expression of IL-17), but rather serve to further specify whether or not T_H17 cells will make additional pro- or anti-inflammatory cytokines. From the parenchymal perspective, factors that dictate the balance of pro- and anti-inflammatory cytokine expression in

T_H17 cells are equally as important as those that regulate a single 'signature' cytokine. Indeed, a more holistic view of T cell responses could greatly impact the way we think about inflammation and immunity.

4 Tuning T Cell Responses by Metabolic Signaling Pathways

Although cytokines get the most publicity, they are not the only means of immunoregulation. Nutrients (i.e., vitamins, hormones, growth factors) and amino acids pervade all tissues, and their local concentrations are constantly read-out by cells via conserved metabolic signaling pathways (Fig. 1b). Metabolic signaling has historically been considered a means of ensuring that surrounding tissues have the nutrients necessary to support cellular function; inducing a state of cellular hyporesponsiveness to conserve nutrients and maintain homeostasis when nutrient starvation (i.e., metabolic stress) occurs (Glick et al. 2010; Lin et al. 2008; Reiling and Sabatini 2006). More nuanced effects of metabolic pathways on cell physiology, however, have begun to emerge. It is now clear that more modest fluctuations in nutrient and amino acid levels (as opposed to complete starvation) can shape, rather than abort, cellular responses to growth and differentiation cues (Delgoffe and Powell 2009; Maciver et al. 2008). Moreover, numerous plant and microbe-derived natural products target evolutionarily-conserved metabolic signaling pathways and these molecules have proven to be invaluable tools for dissecting cell physiology, while also providing a framework for rational drug design (Haustedt et al. 2006; Li and Vederas 2009).

A historical focus of cancer research, metabolic signaling within the immune system has emerged as a broad and relatively unexplored signal transduction network that has profound impacts on the differentiation and effector function of T cells. During the last 15 years, investigations into the effects of calorie restriction (CR) and small molecules such as the fungal macrolide rapamycin have led to the general principle that metabolic stress promotes T cell tolerance (Barshes et al. 2004; Delgoffe and Powell 2009; Piccio et al. 2008; Wu and Mohan 2009). More recent studies, particularly those involving gene targeting in mice, have taken this concept a step further, showing that metabolic signaling pathways, particularly those involved in amino acid and energy homeostasis, influence both T_H17 and Treg cells in a cell-intrinsic manner (Delgoffe and Powell 2009; Mellor and Munn 2008; Webb et al. 2008; Wu and Mohan 2009) (Fig. 1c).

5 mTORC1 in Control of T Cell Differentiation

One of the best-characterized metabolic signaling pathways that exert broad tolerogenic effects on T cell function is the mammalian target of rapamycin (mTOR) pathway [reviewed in (Huang and Manning 2008; Ma and Blenis 2009; Reiling and

Sabatini 2006)] (Fig. 1b). A serine/threonine kinase belonging to the phosphatidylinositol 3-kinase (PI-3K) family, mTOR can be found in two distinct macromolecular complexes at steady-state; mTOR complex 1 (mTORC1), which is comprised of mTOR bound to the regulatory proteins Raptor, mLST8/GβL, PRAS40, and DEPTOR, and mTORC2, in which mTOR is bound to Rictor, GβL, and mSIN1 (Huang and Manning 2008; Laplante and Sabatini 2009; Ma and Blenis 2009; Sarbassov et al. 2004). Whereas mTORC1 is a key cellular nutrient sensor that is directly regulated by intracellular amino acids, ATP, and growth factor signaling, mTORC2 is not directly regulated by nutrients. The two complexes also differ in their substrate specificity and their sensitivity to rapamycin (Laplante and Sabatini 2009; Reiling and Sabatini 2006).

When nutrients and amino acid levels are at a surplus, mTORC1 is maintained as an active kinase, phosphorylating substrates in a raptor-dependent manner (Kim et al. 2002). Substrates of mTORC1 include eIF4E-bp1/2 and S6K1/2 (Proud 2004; Schalm and Blenis 2002). These substrates control metabolic processes such as ribosome biogenesis, cap-dependent translation, and energy production through the regulation of glycolytic enzymes (Ma and Blenis 2009; Reiling and Sabatini 2006). Nutrient starvation, limited amino acid availability, or exposure to rapamycin, all inactivate mTORC1 by causing its disassociation with Raptor, and all have been demonstrated to reduce T_H17 differentiation and enhance Treg differentiation or function (Kopf et al. 2007; Locke et al. 2009; Sundrud et al. 2009) (Fig. 1b).

Importantly, distinct types of metabolic stress negatively regulate mTORC1 via unique upstream mechanisms (Huang and Manning 2008; Laplante and Sabatini 2009). Nutrients and amino acids control mTORC1 via the small GTPase Rheb, which activates mTORC1 when in the GTP-bound state (Huang and Manning 2008; Ma and Blenis 2009; Reiling and Sabatini 2006). Rapamycin bypasses Rheb by binding to the immunophilin FKBP12, which directly destabilizes mTORC1 (Yip et al. 2010). Furthermore, nutrients regulate Rheb activity via tuberous sclerosis complex (TSC)1/2, which are also called hamartin and tuberin. TSC1/2 inhibit Rheb function through its intrinsic GTPase activating protein (GAP) activity (Kwiatkowski 2003; Manning et al. 2005). Phosphorylation of the TSC1/2 complex by nutrient and energy-responsive kinases (e.g., PI-3K, Akt/PKB, Erk1/2, AMPK) inhibits TSC1/2 to maintain mTORC1 activity in nutrient-rich environments (Huang and Manning 2008; Kwiatkowski 2003). In contrast to nutrients, amino acid levels control Rheb-dependent mTORC1 activity through the Rag family of GTPases, not TCS1/2 (Sancak et al. 2008, 2010).

T cell-specific deletion of mTOR results in defective naïve T cell differentiation into T_H1, T_H2, or T_H17 effector lineages, while promoting Foxp3 upregulation and Treg suppressive function (Delgoffe et al. 2009). The diversion of mTOR-deficient T cells from effector to regulatory lineages is associated with dysregulated STAT activation downstream of multiple cytokine receptors, including IL-12, IL-4 and IL-6. $mTOR^{-/-}$ T cells also display increased TGFβ responsiveness, as seen by enhanced Smad3 phosphorylation (Delgoffe et al. 2009). In line with these genetic experiments, transient inhibition of mTORC1 signaling by either amino acid limitation or treatment with sub-immuno-suppressive doses of rapamycin also inhibits

T_H17 differentiation in vitro (Kopf et al. 2007; Sundrud et al. 2009). At least for amino acid depletion, the inhibition of T_H17 differentiation is selective, as it does not influence T_H1 or T_H2 differentiation (Sundrud et al. 2009).

Further implicating mTORC1 as a key regulator of the T_H17/Treg balance are studies showing that both PI-3K and Akt activation downstream of the TCR strongly reduces Foxp3 expression in recently activated T cells, even in the presence of TGFβ (Sauer et al. 2008). Reducing the strength or duration of TCR-activation in vitro strongly enhances Foxp3 expression while inhibiting T_H17 differentiation (Sauer et al. 2008). Furthermore, spontaneous antigen-specific iTreg differentiation is observed in vivo following low-dose soluble antigen challenge (Apostolou and von Boehmer 2004; Kretschmer et al. 2005). Collectively, this data suggest that the PI-3K/Akt/mTORC1 signaling axis modifies peripheral T cell responses to TGFβ signaling in a way that dominantly inhibits Foxp3 induction when active, allowing for synergy with IL-6 to induce T_H17 differentiation. In addition to driving mTORC1 activity, active Akt phosphorylates and inhibits Foxo family transcription factors, which have recently been shown to activate *Foxp3* gene expression (Harada et al. 2010; Ouyang et al. 2010). As a potential compensatory mechanism, natural thymic-derived Treg (nTreg) cells fail to activate Akt in response to TCR-agonists and this feature appears key for maintaining Treg cell suppressive function despite strong TCR engagement (Zeiser et al. 2008). Indeed, ectopic expression of a constitutively active Akt mutant in developing thymocytes or mature nTregs represses Foxp3 expression and extinguishes suppressive function (Haxhinasto et al. 2008).

6 Metabolic Stress Negatively Regulates T_H17 Cell Development

Amino acids are cellular multi-taskers. In addition to regulating mTORC1 activity, free intracellular amino acids are covalently linked onto cognate tRNA molecules for use in protein synthesis. Amino acid starvation thus blocks mTORC1 signaling and concomitantly prompts the accumulation of un-aminoacylated (i.e., uncharged) tRNAs (Deval et al. 2009; Staschke et al. 2010). Uncharged tRNAs are collectively recognized and bound by the protein kinase general control nonrepressed 2 (GCN2), which in turn inhibits cap-dependent protein synthesis through serine phosphorylation of the alpha sub-unit of the translation initiation factor eIF2 (eIF2α) (Harding et al. 2003; Wek et al. 2006). Whereas protein synthesis is globally reduced upon eIF2 inhibition in order to preserve amino acids, translation of a select set of stress-responsive mRNAs is actually induced. Among these stress-activated proteins is the transcription factor ATF4, which directly *trans*-activates expression of genes involved in amino acid transport and biosynthesis. This stress response system, termed the amino acid starvation response (AAR; also called general amino acid control (GAAC) in yeast), allows cells to compensate for amino acid-limiting environments (Harding et al. 2000, 2003) (Fig. 1b). The AAR pathway is biochemically linked to a broader network of stress response pathways, including those responsible for responses to hypoxia and ER stress, which is collectively termed integrated stress response (ISR) (Harding et al. 2003; Ron and Walter 2007; Wek et al. 2006). ISR pathways, including the AAR, share the feature of inhibiting eIF2-dependent translation as a means to maintain cell homeostasis, but can be distinguished by the activation of unique upstream eIF2 kinases as well as distinct transcriptional responses. Recent work indicates that the AAR pathway selectively inhibits T_H17 differentiation in vitro and reduces T_H17-mediated inflammation in vivo (Sundrud et al. 2009).

Activation of the AAR pathway is actively regulated in vivo in order to stem T cell-mediated inflammation. Indoleamine 2,3-dioxygenase (IDO) and arginase 1 (Arg1) are enzymes secreted by APC subsets in response to inflammation; they mediate the catabolic depletion of tryptophan and arginine, resulting in local amino acid starvation (Huang et al. 2010; Nicholson et al. 2009). Several studies have reported that IDO-expressing DC subsets dominantly induce activation of the AAR in bystander T cells, promote Treg-mediated tolerance, and suppress the development of T_H17 cells in vivo (Baban et al. 2009; Munn et al. 2005). In a more recent study of HIV-infected individuals, elevated IDO expression has been associated with fewer circulating T_H17 cells, increased frequency of Treg cells, and disease progression (Favre et al. 2010). By virtue of its catabolic activity on tryptophan, IDO also generates small molecules known collectively as kynurenines (Constantino 2009). Interestingly, kynurenines have also recently been shown to suppress T_H17 differentiation and enhance Foxp3 expression in the absence of tryptophan deprivation (Desvignes and Ernst 2009; Mezrich et al. 2010). Whether IDO-mediated effects on T_H17 and Treg cells in vivo are mediated through tryptophan depletion, kynurenine production, or some combination thereof, remains to be determined.

Amino acid starvation and subsequent inhibition of T_H17 differentiation can also be pharmacologically induced, by treatment with the small molecule halofuginone (HF). HF is a derivative of the naturally-occurring plant alkaloid, febrifugine [reviewed in (Pines and Nagler 1998)]. Treatment of T cells with HF rapidly induces GCN2 activation, eIF2α phosphorylation, ATF expression, and a transcriptional response similar to that of amino acid starvation (Sundrud et al. 2009). Functionally, HF treatment impairs T_H17 differentiation selectively, without affecting the development of T_H1 or T_H2 cells. These selective effects of HF on T_H17 differentiation are mirrored by amino acid deprivation, but differ from the more broad effects of mTORC1 deficiency on all effector lineages. HF also modulates T_H17 cell differentiation selectively in vivo; C57/B6 mice treated with HF are protected against T_H17-, but not T_H1-driven EAE (Sundrud et al. 2009). Accordingly, HF reduces the proportion of T_H17 cells in the periphery and CNS of diseased mice, but has no effect on the development of T_H1 cells. HF-treated T cells cultured in T_H17 polarizing cytokine conditions display elevated levels of Foxp3 concomitant with repression of IL-17. However, the upregulation of Foxp3 expression by HF is likely a bystander effect of inhibiting T_H17 lineage commitment, as HF treatment impairs IL-6-dependent STAT3 activation, and also strongly represses T_H17 differentiation in T cells lacking Foxp3 (Sundrud et al. 2009). Importantly, although amino acid starvation induced by HF is both necessary and sufficient for repressing T_H17 differentiation, the downstream pathways that link HF-induced stress to T cell regulation are unclear.

Collectively, this data highlights that T_H17 cells are uniquely sensitive to metabolic stress and nutrient availability. The obvious question is why? Although more work is needed to address this question, it is interesting to speculate that T_H17 cells are subject to a "metabolic checkpoint". Because T_H17 cells drive chronic and progressive tissue inflammation which involves activation and proliferation of resident cells as well as massive infiltration of metabolically active mononuclear cells, T_H17-targeted tissues are likely to require significant stores of nutrients and amino acids to maintain homeostasis. Thus, much in the way cell cycle checkpoints prevent unwarranted or malignant proliferation, a metabolic checkpoint for T_H17 differentiation would reduce the chance of a local inflammatory response spiraling into pathologic tissue damage. It would also stand to reason that such a checkpoint would involve mechanisms to induce or activate local regulatory cells, and leave intact other host-protective aspects of adaptive immunity (i.e., T_H1, T_H2 cells).

Interestingly, it seems not all forms of cellular stress are specialized to regulate T_H17 responses. For example, the unfolded protein response (UPR), which is a distinct stress response induced by increased ER load, has no effect on the T_H17/Treg balance (Sundrud et al. 2009). It will be interesting to determine whether additional forms of cellular stress (e.g., genotoxic stress, hypoxia) regulate T_H17 differentiation specifically, or T cell responses more broadly. Certainly future investigation into the link between metabolic stress and T_H17 cells is needed, but if harnessed, this concept and the underlying biology offers vast therapeutic potential for the treatment of T_H17-driven immune pathologies.

7 Concluding Remarks

Through all the data and discussion presented in this chapter, the central theme is a simple one: T cell responses to growth and differentiation signals are inherently shaped by the context in which they are seen. That context is not static or one-dimensional. Rather, the physiologic context of T cells in vivo is dynamic and multi-dimensional. Of course, T_H17 cell differentiation involves the establishment of a core, cell-intrinsic, and well-defined developmental program involving STAT3, RORγt, and a series of interacting transcription factors. However, as laid out herein, this core program is subject to synergy or antagonism by auxiliary cytokines, miRNAs, and even the metabolic state of cells. By example, IL-2, all-trans retinoic acid, and metabolic stress all create a cellular context in which TGFβ signaling drives Foxp3 expression and development of regulatory characteristics. Yet in other contexts, such as inflammation and active metabolic signaling, the response of cells to TGFβ signaling is molded from inducing tolerance to promoting T_H17 differentiation and inflammation. Indeed, great strides have been made in our understanding of T_H17 cell biology. Yet much more work is needed to translate these and future discoveries into medicines. The challenge now is to define these cellular contexts, determine how they are established or may be manipulated, and interrogate their molecular interactions with the core T_H17 differentiation program. Answers to these questions will no doubt provide new insight into T_H17 cell biology and may lead to novel strategies for the regulation of T_H17 cells in clinical disease settings.

References

Adamson, A. S., Collins, K., Laurence, A., and O'Shea, J. J. (2009). The Current STATus of lymphocyte signaling: new roles for old players. Curr Opin Immunol *21*, 161–166.

Adema, G. J. (2009). Dendritic cells from bench to bedside and back. Immunol Lett *122*, 128–130.

Ahern, P. P., Schiering, C., Buonocore, S., McGeachy, M. J., Cua, D. J., Maloy, K. J., and Powrie, F. (2010). Interleukin-23 Drives Intestinal Inflammation through Direct Activity on T Cells. Immunity *33*, 279–288.

Alam, M. S., Maekawa, Y., Kitamura, A., Tanigaki, K., Yoshimoto, T., Kishihara, K., and Yasutomo, K. (2010) Notch signaling drives IL-22 secretion in CD4+ T cells by stimulating the aryl hydrocarbon receptor. Proc Natl Acad Sci USA *107*, 5943–5948.

Alexander, W. S., and Hilton, D. J. (2004). The role of suppressors of cytokine signaling (SOCS) proteins in regulation of the immune response. Annu Rev Immunol 22, 503–529.

Amsen, D., Spilianakis, C. G., and Flavell, R. A. (2009). How are T(H)1 and T(H)2 effector cells made? Curr Opin Immunol *21*, 153–160.

Apetoh, L., Quintana, F. J., Pot, C., Joller, N., Xiao, S., Kumar, D., Burns, E. J., Sherr, D. H., Weiner, H. L., and Kuchroo, V. K. (2010). The aryl hydrocarbon receptor interacts with c-Maf to promote the differentiation of type 1 regulatory T cells induced by IL-27. Nat Immunol *11*, 854–861.

Apostolou, I., and von Boehmer, H. (2004). In vivo instruction of suppressor commitment in naive T cells. J Exp Med *199*, 1401–1408.

Atsumi, T., Ishihara, K., Kamimura, D., Ikushima, H., Ohtani, T., Hirota, S., Kobayashi, H., Park, S. J., Saeki, Y., Kitamura, Y., and Hirano, T. (2002). A point mutation of Tyr-759 in interleukin 6 family cytokine receptor subunit gp130 causes autoimmune arthritis. J Exp Med *196*, 979–990.

Awane, M., Andres, P. G., Li, D. J., and Reinecker, H. C. (1999). NF-kappa B-inducing kinase is a common mediator of IL-17-, TNF-alpha-, and IL-1 beta-induced chemokine promoter activation in intestinal epithelial cells. J Immunol *162*, 5337–5344.

Awasthi, A., Riol-Blanco, L., Jager, A., Korn, T., Pot, C., Galileos, G., Bettelli, E., Kuchroo, V. K., and Oukka, M. (2009). Cutting edge: IL-23 receptor gfp reporter mice reveal distinct populations of IL-17-producing cells. J Immunol *182*, 5904–5908.

Baban, B., Chandler, P. R., Sharma, M. D., Pihkala, J., Koni, P. A., Munn, D. H., and Mellor, A. L. (2009). IDO activates regulatory T cells and blocks their conversion into Th17-like T cells. J Immunol *183*, 2475–2483.

Barshes, N. R., Goodpastor, S. E., and Goss, J. A. (2004). Pharmacologic immuno-suppression. Front Biosci *9*, 411–420.

Bettelli, E., Carrier, Y., Gao, W., Korn, T., Strom, T. B., Oukka, M., Weiner, H. L., and Kuchroo, V. K. (2006). Reciprocal developmental pathways for the generation of pathogenic effector TH17 and regulatory T cells. Nature *441*, 235–238.

Betz, B. C., Jordan-Williams, K. L., Wang, C., Kang, S. G., Liao, J., Logan, M. R., Kim, C. H., and Taparowsky, E. J. (2010). Batf coordinates multiple aspects of B and T cell function required for normal antibody responses. J Exp Med *207*, 933–942.

Betz, U. A., and Muller, W. (1998). Regulated expression of gp130 and IL-6 receptor alpha chain in T cell maturation and activation. Int Immunol *10*, 1175–1184.

Brustle, A., Heink, S., Huber, M., Rosenplanter, C., Stadelmann, C., Yu, P., Arpaia, E., Mak, T. W., Kamradt, T., and Lohoff, M. (2007). The development of inflammatory T(H)-17 cells requires interferon-regulatory factor 4. Nat Immunol *8*, 958–966.

Burchill, M. A., Yang, J., Vogtenhuber, C., Blazar, B. R., and Farrar, M. A. (2007). IL-2 receptor beta-dependent STAT5 activation is required for the development of Foxp3+ regulatory T cells. J Immunol *178*, 280–290.

Chen, Z., Laurence, A., Kanno, Y., Pacher-Zavisin, M., Zhu, B. M., Tato, C., Yoshimura, A., Hennighausen, L., and O'Shea, J. J. (2006). Selective regulatory function of Socs3 in the formation of IL-17-secreting T cells. Proc Natl Acad Sci USA *103*, 8137–8142.

Chen, Z., Laurence, A., and O'Shea, J. J. (2007). Signal transduction pathways and transcriptional regulation in the control of Th17 differentiation. Semin Immunol *19*, 400–408.

Chung, D. R., Kasper, D. L., Panzo, R. J., Chitnis, T., Grusby, M. J., Sayegh, M. H., and Tzianabos, A. O. (2003). CD4+ T cells mediate abscess formation in intra-abdominal sepsis by an IL-17-dependent mechanism. J Immunol *170*, 1958–1963.

Constantino, G. (2009). New promises for manipulation of kynurenine pathway in cancer and neurological diseases. Expert Opin Ther Targets *13*, 247–258.

Crome, S. Q., Clive, B., Wang, A. Y., Kang, C. Y., Chow, V., Yu, J., Lai, A., Ghahary, A., Broady, R., and Levings, M. K. (2010). Inflammatory Effects of Ex Vivo Human Th17 Cells Are Suppressed by Regulatory T Cells. J Immunol *185*, 3199–3208.

Dabir, S., Kluge, A., and Dowlati, A. (2009). The association and nuclear translocation of the PIAS3-STAT3 complex is ligand and time dependent. Mol Cancer Res *7*, 1854–1860.

Das, J., Ren, G., Zhang, L., Roberts, A. I., Zhao, X., Bothwell, A. L., Van Kaer, L., Shi, Y., and Das, G. (2009). Transforming growth factor beta is dispensable for the molecular orchestration of Th17 cell differentiation. J Exp Med *206*, 2407–2416.

Dayer, J. M., and Choy, E. (2010). Therapeutic targets in rheumatoid arthritis: the interleukin-6 receptor. Rheumatology (Oxford) *49*, 15–24.

Delgoffe, G. M., Kole, T. P., Zheng, Y., Zarek, P. E., Matthews, K. L., Xiao, B., Worley, P. F., Kozma, S. C., and Powell, J. D. (2009). The mTOR kinase differentially regulates effector and regulatory T cell lineage commitment. Immunity *30*, 832–844.

Delgoffe, G. M., and Powell, J. D. (2009). mTOR: taking cues from the immune microenvironment. Immunology *127*, 459–465.

Derynck, R., and Zhang, Y. E. (2003). Smad-dependent and Smad-independent pathways in TGF-beta family signalling. Nature *425*, 577–584.

Desvignes, L., and Ernst, J. D. (2009). Interferon-gamma-responsive non-hematopoietic cells regulate the immune response to Mycobacterium tuberculosis. Immunity *31*, 974–985.

Deval, C., Chaveroux, C., Maurin, A. C., Cherasse, Y., Parry, L., Carraro, V., Milenkovic, D., Ferrara, M., Bruhat, A., Jousse, C., and Fafournoux, P. (2009). Amino acid limitation regulates the expression of genes involved in several specific biological processes through GCN2-dependent and GCN2-independent pathways. Febs J *276*, 707–718.

Diamant, M., Rieneck, K., Mechti, N., Zhang, X. G., Svenson, M., Bendtzen, K., and Klein, B. (1997). Cloning and expression of an alternatively spliced mRNA encoding a soluble form of the human interleukin-6 signal transducer gp130. FEBS Lett *412*, 379–384.

Diveu, C., McGeachy, M. J., Boniface, K., Stumhofer, J. S., Sathe, M., Joyce-Shaikh, B., Chen, Y., Tato, C. M., McClanahan, T. K., de Waal Malefyt, R., *et al.* (2009). IL-27 blocks RORc expression to inhibit lineage commitment of Th17 cells. J Immunol *182*, 5748–5756.

Du, C., Liu, C., Kang, J., Zhao, G., Ye, Z., Huang, S., Li, Z., Wu, Z., and Pei, G. (2009). MicroRNA miR-326 regulates TH-17 differentiation and is associated with the pathogenesis of multiple sclerosis. Nat Immunol *10*, 1252–1259.

Du, J., Huang, C., Zhou, B., and Ziegler, S. F. (2008). Isoform-specific inhibition of ROR alpha-mediated transcriptional activation by human FOXP3. J Immunol *180*, 4785–4792.

Dubin, P. J., and Kolls, J. K. (2008). Th17 cytokines and mucosal immunity. Immunol Rev *226*, 160–171.

Durant, L., Watford, W. T., Ramos, H. L., Laurence, A., Vahedi, G., Wei, L., Takahashi, H., Sun, H. W., Kanno, Y., Powrie, F., and O'Shea, J. J. (2010). Diverse targets of the transcription factor STAT3 contribute to T cell pathogenicity and homeostasis. Immunity *32*, 605–615.

El-behi, M., Ciric, B., Yu, S., Zhang, G. X., Fitzgerald, D. C., and Rostami, A. (2009). Differential effect of IL-27 on developing versus committed Th17 cells. J Immunol *183*, 4957–4967.

Elo, L. L., Jarvenpaa, H., Tuomela, S., Raghav, S., Ahlfors, H., Laurila, K., Gupta, B., Lund, R. J., Tahvanainen, J., Hawkins, R. D., *et al.* (2010). Genome-wide profiling of interleukin-4 and STAT6 transcription factor regulation of human Th2 cell programming. Immunity *32*, 852–862.

Esser, C., Rannug, A., and Stockinger, B. (2009). The aryl hydrocarbon receptor in immunity. Trends Immunol *30*, 447–454.

Eyerich, S., Eyerich, K., Pennino, D., Carbone, T., Nasorri, F., Pallotta, S., Cianfarani, F., Odorisio, T., Traidl-Hoffmann, C., Behrendt, H., *et al.* (2009). Th22 cells represent a distinct human T cell subset involved in epidermal immunity and remodeling. J Clin Invest *119*, 3573–3585.

Favre, D., Mold, J., Hunt, P. W., Kanwar, B., Loke, P., Seu, L., Barbour, J. D., Lowe, M. M., Jayawardene, A., Aweeka, F., *et al.* (2010). Tryptophan catabolism by indoleamine 2,3-dioxygenase 1 alters the balance of TH17 to regulatory T cells in HIV disease. Sci Transl Med *2*, 32ra36.

Glick, D., Barth, S., and Macleod, K. F. (2010). Autophagy: cellular and molecular mechanisms. J Pathol *221*, 3–12.

Graf, D., Haselow, K., Munks, I., Bode, J. G., and Haussinger, D. (2008). Caspase-mediated cleavage of the signal-transducing IL-6 receptor subunit gp130. Arch Biochem Biophys *477*, 330–338.

Harada, Y., Harada, Y., Elly, C., Ying, G., Paik, J. H., DePinho, R. A., and Liu, Y. C. (2010). Transcription factors Foxo3a and Foxo1 couple the E3 ligase Cbl-b to the induction of Foxp3 expression in induced regulatory T cells. J Exp Med *207*, 1381–1391.

Harding, H. P., Novoa, I., Zhang, Y., Zeng, H., Wek, R., Schapira, M., and Ron, D. (2000). Regulated translation initiation controls stress-induced gene expression in mammalian cells. Mol Cell *6*, 1099–1108.

Harding, H. P., Zhang, Y., Zeng, H., Novoa, I., Lu, P. D., Calfon, M., Sadri, N., Yun, C., Popko, B., Paules, R., *et al.* (2003). An integrated stress response regulates amino acid metabolism and resistance to oxidative stress. Mol Cell *11*, 619–633.

Haustedt, L. O., Mang, C., Siems, K., and Schiewe, H. (2006). Rational approaches to natural-product-based drug design. Curr Opin Drug Discov Devel *9*, 445–462.

Haxhinasto, S., Mathis, D., and Benoist, C. (2008). The AKT-mTOR axis regulates de novo differentiation of CD4+Foxp3+ cells. J Exp Med *205*, 565–574.

Holland, S. M., DeLeo, F. R., Elloumi, H. Z., Hsu, A. P., Uzel, G., Brodsky, N., Freeman, A. F., Demidowich, A., Davis, J., Turner, M. L., *et al.* (2007). STAT3 mutations in the hyper-IgE syndrome. N Engl J Med *357*, 1608–1619.

Hu, X., and Ivashkiv, L. B. (2009). Cross-regulation of signaling pathways by interferon-gamma: implications for immune responses and auto-immune diseases. Immunity *31*, 539–550.

Huang, J., and Manning, B. D. (2008). The TSC1-TSC2 complex: a molecular switchboard controlling cell growth. Biochem J *412*, 179–190.

Huang, L., Baban, B., Johnson, B. A., 3rd, and Mellor, A. L. (2010). Dendritic cells, indoleamine 2,3 dioxygenase and acquired immune privilege. Int Rev Immunol *29*, 133–155.

Huang, W., Na, L., Fidel, P. L., and Schwarzenberger, P. (2004). Requirement of interleukin-17A for systemic anti-Candida albicans host defense in mice. J Infect Dis *190*, 624–631.

Huber, M., Brustle, A., Reinhard, K., Guralnik, A., Walter, G., Mahiny, A., von Low, E., and Lohoff, M. (2008). IRF4 is essential for IL-21-mediated induction, amplification, and stabilization of the Th17 phenotype. Proc Natl Acad Sci USA *105*, 20846–20851.

Hwang, E. S., Szabo, S. J., Schwartzberg, P. L., and Glimcher, L. H. (2005). T helper cell fate specified by kinase-mediated interaction of T-bet with GATA-3. Science *307*, 430–433.

Ichiyama, K., Yoshida, H., Wakabayashi, Y., Chinen, T., Saeki, K., Nakaya, M., Takaesu, G., Hori, S., Yoshimura, A., and Kobayashi, T. (2008). Foxp3 inhibits RORgammat-mediated IL-17A mRNA transcription through direct interaction with RORgammat. J Biol Chem *283*, 17003–17008.

Igaz, P., Horvath, A., Horvath, B., Szalai, C., Pallinger, E., Rajnavolgyi, E., Toth, S., Rose-John, S., and Falus, A. (2000). Soluble interleukin-6 receptor (sIL-6R) makes IL-6R negative T cell line respond to IL-6; it inhibits TNF production. Immunol Lett *71*, 143–148.

Ivanov, II, Atarashi, K., Manel, N., Brodie, E. L., Shima, T., Karaoz, U., Wei, D., Goldfarb, K. C., Santee, C. A., Lynch, S. V., *et al.* (2009). Induction of intestinal Th17 cells by segmented filamentous bacteria. Cell *139*, 485–498.

Ivanov, II, Frutos Rde, L., Manel, N., Yoshinaga, K., Rifkin, D. B., Sartor, R. B., Finlay, B. B., and Littman, D. R. (2008). Specific microbiota direct the differentiation of IL-17-producing T-helper cells in the mucosa of the small intestine. Cell Host Microbe *4*, 337–349.

Ivanov, II, McKenzie, B. S., Zhou, L., Tadokoro, C. E., Lepelley, A., Lafaille, J. J., Cua, D. J., and Littman, D. R. (2006). The orphan nuclear receptor RORgammat directs the differentiation program of pro-inflammatory IL-17+ T helper cells. Cell *126*, 1121–1133.

Jain, J., McCaffrey, P. G., Miner, Z., Kerppola, T. K., Lambert, J. N., Verdine, G. L., Curran, T., and Rao, A. (1993). The T-cell transcription factor NFATp is a substrate for calcineurin and interacts with Fos and Jun. Nature *365*, 352–355.

Janke, M., Peine, M., Nass, A., Morawietz, L., Hamann, A., and Scheffold, A. (2010). In-vitro-induced Th17 cells fail to induce inflammation in vivo and show an impaired migration into inflamed sites. Eur J Immunol *40*, 1089–1098.

Jones, G. W., McLoughlin, R. M., Hammond, V. J., Parker, C. R., Williams, J. D., Malhotra, R., Scheller, J., Williams, A. S., Rose-John, S., Topley, N., and Jones, S. A. (2010). Loss of CD4+ T cell IL-6R expression during inflammation underlines a role for IL-6 trans signaling in the local maintenance of Th17 cells. J Immunol *184*, 2130–2139.

Kastelein, R. A., Hunter, C. A., and Cua, D. J. (2007). Discovery and biology of IL-23 and IL-27: related but functionally distinct regulators of inflammation. Annu Rev Immunol *25*, 221–242.

Kim, D. H., Sarbassov, D. D., Ali, S. M., King, J. E., Latek, R. R., Erdjument-Bromage, H., Tempst, P., and Sabatini, D. M. (2002). mTOR interacts with raptor to form a nutrient-sensitive complex that signals to the cell growth machinery. Cell *110*, 163–175.

Kim, J. I., Ho, I. C., Grusby, M. J., and Glimcher, L. H. (1999). The transcription factor c-Maf controls the production of interleukin-4 but not other Th2 cytokines. Immunity *10*, 745–751.

Kishimoto, T. (2005). Interleukin-6: from basic science to medicine--40 years in immunology. Annu Rev Immunol *23*, 1–21.

Klotz, L., Burgdorf, S., Dani, I., Saijo, K., Flossdorf, J., Hucke, S., Alferink, J., Nowak, N., Beyer, M., Mayer, G., *et al.* (2009). The nuclear receptor PPAR gamma selectively inhibits Th17 differentiation in a T cell-intrinsic fashion and suppresses CNS auto-immunity. J Exp Med *206*, 2079–2089.

Kopf, H., de la Rosa, G. M., Howard, O. M., and Chen, X. (2007). Rapamycin inhibits differentiation of Th17 cells and promotes generation of FoxP3+ T regulatory cells. Int Immunopharmacol *7*, 1819–1824.

Korn, T., Bettelli, E., Oukka, M., and Kuchroo, V. K. (2009). IL-17 and Th17 Cells. Annu Rev Immunol *27*, 485–517.

Korn, T., Mitsdoerffer, M., Croxford, A. L., Awasthi, A., Dardalhon, V. A., Galileos, G., Vollmar, P., Stritesky, G. L., Kaplan, M. H., Waisman, A., *et al.* (2008). IL-6 controls Th17 immunity in vivo by inhibiting the conversion of conventional T cells into Foxp3+ regulatory T cells. Proc Natl Acad Sci USA *105*, 18460–18465.

Kretschmer, K., Apostolou, I., Hawiger, D., Khazaie, K., Nussenzweig, M. C., and von Boehmer, H. (2005). Inducing and expanding regulatory T cell populations by foreign antigen. Nat Immunol *6*, 1219–1227.

Kwiatkowski, D. J. (2003). Rhebbing up mTOR: new insights on TSC1 and TSC2, and the pathogenesis of tuberous sclerosis. Cancer Biol Ther *2*, 471–476.

Kwon, H., Thierry-Mieg, D., Thierry-Mieg, J., Kim, H. P., Oh, J., Tunyaplin, C., Carotta, S., Donovan, C. E., Goldman, M. L., Tailor, P., *et al.* (2009). Analysis of interleukin-21-induced Prdm1 gene regulation reveals functional cooperation of STAT3 and IRF4 transcription factors. Immunity *31*, 941–952.

Laplante, M., and Sabatini, D. M. (2009). mTOR signaling at a glance. J Cell Sci *122*, 3589–3594.

Laurence, A., Tato, C. M., Davidson, T. S., Kanno, Y., Chen, Z., Yao, Z., Blank, R. B., Meylan, F., Siegel, R., Hennighausen, L., *et al.* (2007). Interleukin-2 signaling via STAT5 constrains T helper 17 cell generation. Immunity *26*, 371–381.

Li, J. W., and Vederas, J. C. (2009). Drug discovery and natural products: end of an era or an endless frontier? Science *325*, 161–165.

Li, M. O., Wan, Y. Y., Sanjabi, S., Robertson, A. K., and Flavell, R. A. (2006). Transforming growth factor-beta regulation of immune responses. Annu Rev Immunol *24*, 99–146.

Lin, J. H., Walter, P., and Yen, T. S. (2008). Endoplasmic reticulum stress in disease pathogenesis. Annu Rev Pathol *3*, 399–425.

Liu, B., Mink, S., Wong, K. A., Stein, N., Getman, C., Dempsey, P. W., Wu, H., and Shuai, K. (2004). PIAS1 selectively inhibits interferon-inducible genes and is important in innate immunity. Nat Immunol *5*, 891–898.

Locke, N. R., Patterson, S. J., Hamilton, M. J., Sly, L. M., Krystal, G., and Levings, M. K. (2009). SHIP regulates the reciprocal development of T regulatory and Th17 cells. J Immunol *183*, 975–983.

Ma, X. M., and Blenis, J. (2009). Molecular mechanisms of mTOR-mediated translational control. Nat Rev Mol Cell Biol *10*, 307–318.

Maciver, N. J., Jacobs, S. R., Wieman, H. L., Wofford, J. A., Coloff, J. L., and Rathmell, J. C. (2008). Glucose metabolism in lymphocytes is a regulated process with significant effects on immune cell function and survival. J Leukoc Biol *84*, 949–957.

Manel, N., Unutmaz, D., and Littman, D. R. (2008). The differentiation of human T(H)-17 cells requires transforming growth factor-beta and induction of the nuclear receptor RORgammat. Nat Immunol *9*, 641–649.

Mangan, P. R., Harrington, L. E., O'Quinn, D. B., Helms, W. S., Bullard, D. C., Elson, C. O., Hatton, R. D., Wahl, S. M., Schoeb, T. R., and Weaver, C. T. (2006). Transforming growth factor-beta induces development of the T(H)17 lineage. Nature *441*, 231–234.

Manning, B. D., Logsdon, M. N., Lipovsky, A. I., Abbott, D., Kwiatkowski, D. J., and Cantley, L. C. (2005). Feedback inhibition of Akt signaling limits the growth of tumors lacking Tsc2. Genes Dev *19*, 1773–1778.

Marine, J. C., McKay, C., Wang, D., Topham, D. J., Parganas, E., Nakajima, H., Pendeville, H., Yasukawa, H., Sasaki, A., Yoshimura, A., and Ihle, J. N. (1999). SOCS3 is essential in the regulation of fetal liver erythropoiesis. Cell *98*, 617–627.

Martin-Orozco, N., Chung, Y., Chang, S. H., Wang, Y. H., and Dong, C. (2009). Th17 cells promote pancreatic inflammation but only induce diabetes efficiently in lymphopenic hosts after conversion into Th1 cells. Eur J Immunol *39*, 216–224.

Martinez, G. J., Zhang, Z., Chung, Y., Reynolds, J. M., Lin, X., Jetten, A. M., Feng, X. H., and Dong, C. (2009). Smad3 differentially regulates the induction of regulatory and inflammatory T cell differentiation. J Biol Chem *284*, 35283–35286.

McGeachy, M. J., Bak-Jensen, K. S., Chen, Y., Tato, C. M., Blumenschein, W., McClanahan, T., and Cua, D. J. (2007). TGF-beta and IL-6 drive the production of IL-17 and IL-10 by T cells and restrain T(H)-17 cell-mediated pathology. Nat Immunol *8*, 1390–1397.

McGeachy, M. J., Chen, Y., Tato, C. M., Laurence, A., Joyce-Shaikh, B., Blumenschein, W. M., McClanahan, T. K., O'Shea, J. J., and Cua, D. J. (2009). The interleukin 23 receptor is essential for the terminal differentiation of interleukin 17-producing effector T helper cells in vivo. Nat Immunol *10*, 314–324.

McLoughlin, R. M., Jenkins, B. J., Grail, D., Williams, A. S., Fielding, C. A., Parker, C. R., Ernst, M., Topley, N., and Jones, S. A. (2005). IL-6 trans-signaling via STAT3 directs T cell infiltration in acute inflammation. Proc Natl Acad Sci USA *102*, 9589–9594.

Mellor, A. L., and Munn, D. H. (2008). Creating immune privilege: active local suppression that benefits friends, but protects foes. Nat Rev Immunol *8*, 74–80.

Mezrich, J. D., Fechner, J. H., Zhang, X., Johnson, B. P., Burlingham, W. J., and Bradfield, C. A. (2010). An Interaction between Kynurenine and the Aryl Hydrocarbon Receptor Can Generate Regulatory T Cells. J Immunol *185*, 3190–3198.

Miller, S. A., and Weinmann, A. S. (2009). Common themes emerge in the transcriptional control of T helper and developmental cell fate decisions regulated by the T-box, GATA and ROR families. Immunology *126*, 306–315.

Milner, J. D., Brenchley, J. M., Laurence, A., Freeman, A. F., Hill, B. J., Elias, K. M., Kanno, Y., Spalding, C., Elloumi, H. Z., Paulson, M. L., *et al.* (2008). Impaired T(H)17 cell differentiation in subjects with autosomal dominant hyper-IgE syndrome. Nature *452*, 773–776.

Minegishi, Y., Saito, M., Tsuchiya, S., Tsuge, I., Takada, H., Hara, T., Kawamura, N., Ariga, T., Pasic, S., Stojkovic, O., *et al.* (2007). Dominant-negative mutations in the DNA-binding domain of STAT3 cause hyper-IgE syndrome. Nature *448*, 1058–1062.

Muller-Newen, G. (2003). The cytokine receptor gp130: faithfully promiscuous. Sci STKE *2003*, PE40.

Munn, D. H., Sharma, M. D., Baban, B., Harding, H. P., Zhang, Y., Ron, D., and Mellor, A. L. (2005). GCN2 kinase in T cells mediates proliferative arrest and anergy induction in response to indoleamine 2,3-dioxygenase. Immunity *22*, 633–642.

Murugaiyan, G., Mittal, A., Lopez-Diego, R., Maier, L. M., Anderson, D. E., and Weiner, H. L. (2009). IL-27 is a key regulator of IL-10 and IL-17 production by human CD4+ T cells. J Immunol *183*, 2435–2443.

Nicholson, L. B., Raveney, B. J., and Munder, M. (2009). Monocyte dependent regulation of autoimmune inflammation. Curr Mol Med *9*, 23–29.

Nishihara, M., Ogura, H., Ueda, N., Tsuruoka, M., Kitabayashi, C., Tsuji, F., Aono, H., Ishihara, K., Huseby, E., Betz, U. A., *et al.* (2007). IL-6-gp130-STAT3 in T cells directs the development of IL-17+ Th with a minimum effect on that of Treg in the steady state. Int Immunol *19*, 695–702.

Nowell, M. A., Richards, P. J., Horiuchi, S., Yamamoto, N., Rose-John, S., Topley, N., Williams, A. S., and Jones, S. A. (2003). Soluble IL-6 receptor governs IL-6 activity in experimental arthritis: blockade of arthritis severity by soluble glycoprotein 130. J Immunol *171*, 3202–3209.

Nowell, M. A., Williams, A. S., Carty, S. A., Scheller, J., Hayes, A. J., Jones, G. W., Richards, P. J., Slinn, S., Ernst, M., Jenkins, B. J., *et al.* (2009). Therapeutic targeting of IL-6 trans-signaling counteracts STAT3 control of experimental inflammatory arthritis. J Immunol *182*, 613–622.

Nurieva, R. I., Chung, Y., Hwang, D., Yang, X. O., Kang, H. S., Ma, L., Wang, Y. H., Watowich, S. S., Jetten, A. M., Tian, Q., and Dong, C. (2008). Generation of T follicular helper cells is mediated by interleukin-21 but independent of T helper 1, 2, or 17 cell lineages. Immunity *29*, 138–149.

O'Shea, J. J., and Murray, P. J. (2008). Cytokine signaling modules in inflammatory responses. Immunity *28*, 477–487.

Okuda, Y., Sakoda, S., Bernard, C. C., Fujimura, H., Saeki, Y., Kishimoto, T., and Yanagihara, T. (1998). IL-6-deficient mice are resistant to the induction of experimental auto-immune encephalomyelitis provoked by myelin oligodendrocyte glycoprotein. Int Immunol *10*, 703–708.

Okumura, F., Matsunaga, Y., Katayama, Y., Nakayama, K.I., and Hatakeyama, S. (2010). TRIM8 modulates STAT3 activity through negative regulation of PIAS3. J Cell Sci *123*, 2238–2245.

Ouyang, W., Beckett, O., Ma, Q., Paik, J. H., DePinho, R. A., and Li, M. O. (2010). Foxo proteins cooperatively control the differentiation of Foxp3+ regulatory T cells. Nat Immunol *11*, 618–627.

Park, H., Li, Z., Yang, X. O., Chang, S. H., Nurieva, R., Wang, Y. H., Wang, Y., Hood, L., Zhu, Z., Tian, Q., and Dong, C. (2005). A distinct lineage of CD4 T cells regulates tissue inflammation by producing interleukin 17. Nat Immunol *6*, 1133–1141.

Passerini, L., Allan, S. E., Battaglia, M., Di Nunzio, S., Alstad, A. N., Levings, M. K., Roncarolo, M. G., and Bacchetta, R. (2008). STAT5-signaling cytokines regulate the expression of FOXP3 in CD4+CD25+ regulatory T cells and CD4+CD25- effector T cells. Int Immunol *20*, 421–431.

Piccio, L., Stark, J. L., and Cross, A. H. (2008). Chronic calorie restriction attenuates experimental auto-immune encephalomyelitis. J Leukoc Biol *84*, 940–948.

Pines, M., and Nagler, A. (1998). Halofuginone: a novel antifibrotic therapy. Gen Pharmacol *30*, 445–450.

Proud, C. G. (2004). mTOR-mediated regulation of translation factors by amino acids. Biochem Biophys Res Commun *313*, 429–436.

Qin, H., Wang, L., Feng, T., Elson, C. O., Niyongere, S. A., Lee, S. J., Reynolds, S. L., Weaver, C. T., Roarty, K., Serra, R., *et al.* (2009). TGF-beta promotes Th17 cell development through inhibition of SOCS3. J Immunol *183*, 97–105.

Ramirez, J. M., Brembilla, N. C., Sorg, O., Chicheportiche, R., Matthes, T., Dayer, J. M., Saurat, J. H., Roosnek, E., and Chizzolini, C. (2010). Activation of the aryl hydrocarbon receptor reveals distinct requirements for IL-22 and IL-17 production by human T helper cells. Eur J Immunol *40*, 2450–2459.

Reboldi, A., Coisne, C., Baumjohann, D., Benvenuto, F., Bottinelli, D., Lira, S., Uccelli, A., Lanzavecchia, A., Engelhardt, B., and Sallusto, F. (2009). C-C chemokine receptor 6-regulated entry of TH-17 cells into the CNS through the choroid plexus is required for the initiation of EAE. Nat Immunol *10*, 514–523.

Reiling, J. H., and Sabatini, D. M. (2006). Stress and mTORture signaling. Oncogene *25*, 6373–6383.

Rengarajan, J., Mowen, K. A., McBride, K. D., Smith, E. D., Singh, H., and Glimcher, L. H. (2002). Interferon regulatory factor 4 (IRF4) interacts with NFATc2 to modulate interleukin 4 gene expression. J Exp Med *195*, 1003–1012.

Rochman, Y., Spolski, R., and Leonard, W. J. (2009). New insights into the regulation of T cells by gamma(c) family cytokines. Nat Rev Immunol *9*, 480–490.

Ron, D., and Walter, P. (2007). Signal integration in the endoplasmic reticulum unfolded protein response. Nat Rev Mol Cell Biol *8*, 519–529.

Rose-John, S., Scheller, J., Elson, G., and Jones, S. A. (2006). Interleukin-6 biology is coordinated by membrane-bound and soluble receptors: role in inflammation and cancer. J Leukoc Biol *80*, 227–236.

Samanta, A., Li, B., Song, X., Bembas, K., Zhang, G., Katsumata, M., Saouaf, S. J., Wang, Q., Hancock, W. W., Shen, Y., and Greene, M. I. (2008). TGF-beta and IL-6 signals modulate chromatin binding and promoter occupancy by acetylated FOXP3. Proc Natl Acad Sci USA *105*, 14023–14027.

Sancak, Y., Bar-Peled, L., Zoncu, R., Markhard, A. L., Nada, S., and Sabatini, D. M. (2010). Ragulator-Rag complex targets mTORC1 to the lysosomal surface and is necessary for its activation by amino acids. Cell *141*, 290–303.

Sancak, Y., Peterson, T. R., Shaul, Y. D., Lindquist, R. A., Thoreen, C. C., Bar-Peled, L., and Sabatini, D. M. (2008). The Rag GTPases bind raptor and mediate amino acid signaling to mTORC1. Science *320*, 1496–1501.

Sarbassov, D. D., Ali, S. M., Kim, D. H., Guertin, D. A., Latek, R. R., Erdjument-Bromage, H., Tempst, P., and Sabatini, D. M. (2004). Rictor, a novel binding partner of mTOR, defines a rapamycin-insensitive and raptor-independent pathway that regulates the cytoskeleton. Curr Biol *14*, 1296–1302.

Sauer, S., Bruno, L., Hertweck, A., Finlay, D., Leleu, M., Spivakov, M., Knight, Z. A., Cobb, B. S., Cantrell, D., O'Connor, E., *et al.* (2008). T cell receptor signaling controls Foxp3 expression via PI3K, Akt, and mTOR. Proc Natl Acad Sci USA *105*, 7797–7802.

Schalm, S. S., and Blenis, J. (2002). Identification of a conserved motif required for mTOR signaling. Curr Biol *12*, 632–639.

Schraml, B. U., Hildner, K., Ise, W., Lee, W. L., Smith, W. A., Solomon, B., Sahota, G., Sim, J., Mukasa, R., Cemerski, S., *et al.* (2009). The AP-1 transcription factor Batf controls T(H)17 differentiation. Nature *460*, 405–409.

Schulz, E. G., Mariani, L., Radbruch, A., and Hofer, T. (2009). Sequential polarization and imprinting of type 1 T helper lymphocytes by interferon-gamma and interleukin-12. Immunity *30*, 673–683.

Serada, S., Fujimoto, M., Mihara, M., Koike, N., Ohsugi, Y., Nomura, S., Yoshida, H., Nishikawa, T., Terabe, F., Ohkawara, T., *et al.* (2008). IL-6 blockade inhibits the induction of myelin antigen-specific Th17 cells and Th1 cells in experimental auto-immune encephalomyelitis. Proc Natl Acad Sci USA *105*, 9041–9046.

Sharpe, A. H. (2009). Mechanisms of costimulation. Immunol Rev *229*, 5–11.

Shuai, K., and Liu, B. (2005). Regulation of gene-activation pathways by PIAS proteins in the immune system. Nat Rev Immunol *5*, 593–605.

Simon, D., Denniston, A. K., Tomlins, P. J., Wallace, G. R., Rauz, S., Salmon, M., Murray, P. I., and Curnow, S. J. (2008). Soluble gp130, an antagonist of IL-6 transsignaling, is elevated in uveitis aqueous humor. Invest Ophthalmol Vis Sci *49*, 3988–3991.

Staschke, K. A., Dey, S., Zaborske, J. M., Palam, L. R., McClintick, J. N., Pan, T., Edenberg, H. J., and Wek, R. C. (2010). Integration of general amino acid control and target of rapamycin

(TOR) regulatory pathways in nitrogen assimilation in yeast. J Biol Chem *285*, 16893–16911.

Staudt, V., Bothur, E., Klein, M., Lingnau, K., Reuter, S., Grebe, N., Gerlitzki, B., Hoffmann, M., Ulges, A., Taube, C., *et al.* (2010). Interferon-Regulatory Factor 4 Is Essential for the Developmental Program of T Helper 9 Cells. Immunity *33*, 192–202.

Sundrud, M. S., Koralov, S. B., Feuerer, M., Calado, D. P., Kozhaya, A. E., Rhule-Smith, A., Lefebvre, R. E., Unutmaz, D., Mazitschek, R., Waldner, H., *et al.* (2009). Halofuginone inhibits TH17 cell differentiation by activating the amino acid starvation response. Science *324*, 1334–1338.

Sundrud, M. S., and Nolan, M. A. (2010). Synergistic and combinatorial control of T cell activation and differentiation by transcription factors. Curr Opin Immunol *22*, 286–292.

Tan, J. A., Song, J., Chen, Y., and Durrin, L. K. (2010). Phosphorylation-dependent interaction of SATB1 and PIAS1 directs SUMO-regulated caspase cleavage of SATB1. Mol Cell Biol *30*, 2823–2836.

Vance, R. E., Isberg, R. R., and Portnoy, D. A. (2009). Patterns of pathogenesis: discrimination of pathogenic and non-pathogenic microbes by the innate immune system. Cell Host Microbe *6*, 10–21.

Veldhoen, M., Hirota, K., Westendorf, A. M., Buer, J., Dumoutier, L., Renauld, J. C., and Stockinger, B. (2008). The aryl hydrocarbon receptor links TH17-cell-mediated auto-immunity to environmental toxins. Nature *453*, 106–109.

Vogelzang, A., McGuire, H. M., Yu, D., Sprent, J., Mackay, C. R., and King, C. (2008). A fundamental role for interleukin-21 in the generation of T follicular helper cells. Immunity *29*, 127–137.

Wang, X., Lupardus, P., Laporte, S. L., and Garcia, K. C. (2009). Structural biology of shared cytokine receptors. Annu Rev Immunol *27*, 29–60.

Webb, A., Johnson, A., Fortunato, M., Platt, A., Crabbe, T., Christie, M. I., Watt, G. F., Ward, S. G., and Jopling, L. A. (2008). Evidence for PI-3K-dependent migration of Th17-polarized cells in response to CCR2 and CCR6 agonists. J Leukoc Biol *84*, 1202–1212.

Wei, B., and Pei, G. (2010). microRNAs: critical regulators in Th17 cells and players in diseases. Cell Mol Immunol *7*, 175–181.

Wei, L., Laurence, A., and O'Shea, J. J. (2008). New insights into the roles of Stat5a/b and Stat3 in T cell development and differentiation. Semin Cell Dev Biol *19*, 394–400.

Wek, R. C., Jiang, H. Y., and Anthony, T. G. (2006). Coping with stress: eIF2 kinases and translational control. Biochem Soc Trans *34*, 7–11.

Wilson, C. B., Rowell, E., and Sekimata, M. (2009). Epigenetic control of T-helper-cell differentiation. Nat Rev Immunol *9*, 91–105.

Wolk, K., Kunz, S., Witte, E., Friedrich, M., Asadullah, K., and Sabat, R. (2004). IL-22 increases the innate immunity of tissues. Immunity *21*, 241–254.

Wolk, K., Witte, E., Witte, K., Warszawska, K., and Sabat, R. (2010). Biology of interleukin-22. Semin Immunopathol *32*, 17–31.

Wu, H. J., Ivanov, II, Darce, J., Hattori, K., Shima, T., Umesaki, Y., Littman, D. R., Benoist, C., and Mathis, D. (2010). Gut-residing segmented filamentous bacteria drive auto-immune arthritis via T helper 17 cells. Immunity *32*, 815–827.

Wu, T., and Mohan, C. (2009). The AKT axis as a therapeutic target in auto-immune diseases. Endocr Metab Immune Disord Drug Targets *9*, 145–150.

Wu, Y., Borde, M., Heissmeyer, V., Feuerer, M., Lapan, A. D., Stroud, J. C., Bates, D. L., Guo, L., Han, A., Ziegler, S. F., *et al.* (2006). FOXP3 controls regulatory T cell function through cooperation with NFAT. Cell *126*, 375–387.

Xu, J., Yang, Y., Qiu, G., Lal, G., Wu, Z., Levy, D. E., Ochando, J. C., Bromberg, J. S., and Ding, Y. (2009). c-Maf regulates IL-10 expression during Th17 polarization. J Immunol *182*, 6226–6236.

Yang, L., Anderson, D. E., Baecher-Allan, C., Hastings, W. D., Bettelli, E., Oukka, M., Kuchroo, V. K., and Hafler, D. A. (2008a). IL-21 and TGF-beta are required for differentiation of human T(H)17 cells. Nature *454*, 350–352.

Yang, X. O., Nurieva, R., Martinez, G. J., Kang, H. S., Chung, Y., Pappu, B. P., Shah, B., Chang, S. H., Schluns, K. S., Watowich, S. S., *et al.* (2008b). Molecular antagonism and plasticity of regulatory and inflammatory T cell programs. Immunity *29*, 44–56.

Yang, X. O., Panopoulos, A. D., Nurieva, R., Chang, S. H., Wang, D., Watowich, S. S., and Dong, C. (2007). STAT3 regulates cytokine-mediated generation of inflammatory helper T cells. J Biol Chem *282*, 9358–9363.

Yang, X. O., Pappu, B. P., Nurieva, R., Akimzhanov, A., Kang, H. S., Chung, Y., Ma, L., Shah, B., Panopoulos, A. D., Schluns, K. S., *et al.* (2008c). T helper 17 lineage differentiation is programmed by orphan nuclear receptors ROR alpha and ROR gamma. Immunity *28*, 29–39.

Yip, C. K., Murata, K., Walz, T., Sabatini, D. M., and Kang, S. A. (2010). Structure of the human mTOR complex I and its implications for rapamycin inhibition. Mol Cell *38*, 768–774.

Zeiser, R., Leveson-Gower, D. B., Zambricki, E. A., Kambham, N., Beilhack, A., Loh, J., Hou, J. Z., and Negrin, R. S. (2008). Differential impact of mammalian target of rapamycin inhibition on CD4+CD25+Foxp3+ regulatory T cells compared with conventional CD4+ T cells. Blood *111*, 453–462.

Zelante, T., De Luca, A., Bonifazi, P., Montagnoli, C., Bozza, S., Moretti, S., Belladonna, M. L., Vacca, C., Conte, C., Mosci, P., *et al.* (2007). IL-23 and the Th17 pathway promote inflammation and impair antifungal immune resistance. Eur J Immunol *37*, 2695–2706.

Zhang, F., Meng, G., and Strober, W. (2008). Interactions among the transcription factors Runx1, RORgammat and Foxp3 regulate the differentiation of interleukin 17-producing T cells. Nat Immunol *9*, 1297–1306.

Zheng, Y., Valdez, P. A., Danilenko, D. M., Hu, Y., Sa, S. M., Gong, Q., Abbas, A. R., Modrusan, Z., Ghilardi, N., de Sauvage, F. J., and Ouyang, W. (2008). Interleukin-22 mediates early host defense against attaching and effacing bacterial pathogens. Nat Med *14*, 282–289.

Zhou, L., Ivanov, II, Spolski, R., Min, R., Shenderov, K., Egawa, T., Levy, D. E., Leonard, W. J., and Littman, D. R. (2007). IL-6 programs T(H)-17 cell differentiation by promoting sequential engagement of the IL-21 and IL-23 pathways. Nat Immunol *8*, 967–974.

Zhou, L., and Littman, D. R. (2009). Transcriptional regulatory networks in Th17 cell differentiation. Curr Opin Immunol *21*, 146–152.

Zhou, L., Lopes, J. E., Chong, M. M., Ivanov, II, Min, R., Victora, G. D., Shen, Y., Du, J., Rubtsov, Y. P., Rudensky, A. Y., *et al.* (2008). TGF-beta-induced Foxp3 inhibits T(H)17 cell differentiation by antagonizing RORgammat function. Nature *453*, 236–240.

Zhu, J., and Paul, W. E. (2008). CD4 T cells: fates, functions, and faults. Blood *112*, 1557–1569.

Zrioual, S., Toh, M. L., Tournadre, A., Zhou, Y., Cazalis, M. A., Pachot, A., Miossec, V., and Miossec, P. (2008). IL-17RA and IL-17RC receptors are essential for IL-17A-induced ELR+ CXC chemokine expression in synoviocytes and are overexpressed in rheumatoid blood. J Immunol *180*, 655–663.

Signal Transduction and T_H17 Cell Differentiation

Xiang-Ping Yang, John J. O'Shea, Kamran Ghoreschi, and Arian Laurence

Abstract The discovery of the interleukin (IL-)17-producing T cells (T_H17) has markedly changed our view of T cell differentiation and T cell-mediated immunity. Characterization of the signaling pathways involved in the T_H17 commitment has provided exciting new insights into the contributions of $CD4^+$ T cells to immunoregulation, host defense, and the pathogenesis of auto-immune diseases. Additionally, emerging data on conversion among polarized T helper cells have raised the question how we should view such concepts as lineage commitment, terminal differentiation, and plasticity of different T cell subtypes. The transcriptional regulatory events and epigenetic modifications that control T_H17 cell differentiation are diverse and complex, and despite the intensive efforts on this subject, many questions remain to be answered. In this chapter, we focus on our current understanding of the signaling pathways, molecular interactions, transcriptional events, and epigenetic modifications that result in T_H17 differentiation and effector functions.

1 Introduction

Following antigen engagement by the T cell receptor (TCR), naïve $CD4^+$ T cells differentiate into distinct cytokine-secreting effector and helper (Th) subsets primarily as a result of the local cytokine milieu. T_H1 cells are characterized by IFN-γ production that combats intracellular pathogens, and T_H2 cells produce IL-4, IL-5, and IL-13 to control helminthic infections. IL-12 drives T_H1 cell differentiation, whereas IL-4 induces T_H2 cell differentiation. The T_H1/T_H2 model of

X.-P. Yang (✉)
Lymphocyte Cell Biology Section, Molecular Immunology and Inflammation Branch, National Institutes of Arthritis and Musculoskeletal and Skin Diseases, National Institutes of Health, 10 Center Drive, Building 10, Room 13C103, Bethesda, MD 20892, USA
e -mail: yangx2@mail.nih.gov

S. Jiang (ed.), *T_H17 Cells in Health and Disease*,
DOI 10.1007/978-1-4419-9371-7_8, © Springer Science+Business Media, LLC 2011

differentiation, first proposed by Mosmann and Coffman over 2 decades ago, has helped explain many phenomena in adaptive immunity (Mosmann et al. 1986; Mosmann and Coffman 1989). However, in the past several years, the view of Th cell differentiation has been expanded and more T cell subsets have been described including regulatory T cells (Treg), IL-21-secreting T follicular cells, and interleukin 17-secreting T helper cells (T_H17 cells) (O'Shea and Paul 2010; Sakaguchi et al. 2008; King et al. 2008; Park et al. 2005; Harrington et al. 2005). T_H17 cells have a distinct cytokine-secretion profile from that of T_H1 or T_H2 cells, producing IL-17A, IL-17F, IL-21, and IL-22. Additionally, other lineage-specific cytokines (IFN-γ and IL-4) inhibit the production of IL-17 (Park et al. 2005). Thus, T_H17 cells have been regarded as a separate lineage of T cells from T_H1 and T_H2 cells, however, their stability as a distinct lineage remains uncertain (Shi et al. 2008; Lee et al. 2009). IL-17 acts on cells of the endothelium to induce the production of IL-6, G-CSF and chemokines, recruiting neutrophils to the sites of inflammation. T_H17 cells are important in protection from extracellular bacteria and fungi as mice that lack either IL-17A receptor (Huang et al. 2004) or the T_H17-secreted cytokine IL-22 are susceptible to infection with these organisms (Zheng et al. 2008).

In the present chapter, we review the receptors that influence T_H17 cell differentiation and describe how they lead to activation of signaling pathways and transcription factors that drive specification of this lineage. Then we will discuss the epigenetic modifications involved in T_H17 differentiation and their possible role in the plasticity of this T cell subset.

2 Regulation of T_H17 Development by T Cell Receptor Signaling

The first step in T cell activation is engagement of the T cell receptor. For a number of T cell subsets, TCR-dependent signals are sufficient to induce IL-17 production. This is the case for invariant NK T cells, γδT cells, and memory $CD4^+$ T cells (Michel et al. 2007; Stark et al. 2005; Chen et al. 2005). Binding of MHC/antigen complexes by the TCR induces multiple and diverse distinct signaling pathways (Smith-Garvin et al. 2009). The earliest events after TCR ligation are the activation of protein tyrosine kinases and recruitment of various adaptor molecules. One consequence is the release of Ca^{2+} from the ER, which leads to the dephosphorylation and activation of the transcription factor NFAT. The proximal promoter of the human and mouse *IL17A* genes contain multiple NFAT binding sites (Gomez-Rodriguez et al. 2009; Liu et al. 2004). However, there are four isoforms of NFAT, which exhibit incomplete functional redundancy (Macian 2005), and we know little about the specific roles of the individual NFAT isoforms in IL-17 production. Mice deficient in NFATc2 have exaggerated T_H2 responses (Erb et al. 2003). Accordingly, these mice have attenuated oxazolone-mediated colitis and impaired IL-6 and IL-17 production (Weigmann et al. 2008). Mice deficient in both NFATc2 and NFATc3 massively overproduce T_H2 cytokines (Rengarajan et al. 2002b; Klein et al. 2006). The expectation would be that IL-17 production would also be greatly impaired in this setting. However, this remains to be determined.

The Tec family tyrosine kinases including inducible T cell kinase (Itk) are activated by TCR signaling. Itk regulates TCR-dependent PLC-γ activation, actin polarization, and cell adhesion (Berg et al. 2005). *Itk*-deficient T cells showed reduced responses to TCR stimulation and impaired NFAT activation. Recently, Gomez-Rodriguez et al. found a selective role of Itk for *Il17a* induction but not *Il17f* induction (Gomez-Rodriguez et al. 2009), which correlated to impaired NFATc1 activation. They found that a conserved NF-ATc1 binding site is located in the region of 3 kb upstream of the first exon of *Il17a* but not in the region 20 kb upstream of *Il-17f* gene.

TCR signal strength is a well-known factor influencing T_H1 versus T_H2 differentiation (Zhu and Paul 2008). Strong TCR signaling upregulates the Th1-associated transcription factor T-bet, which is a major regulator of IFN-γ production, whereas weak signals favor IL-4 production. Precisely how signal strength influences IL-17 regulation is not completely understood. Lezzi et al. found that high dose antigen-loaded dendritic cells can induce T_H17 generation even in the absence of exogenous cytokines (Lezzi et al. 2009). High dose antigen activated T cells express high levels of CD40L, which interact with CD40 on the surface of dendritic cells to enhance the cytokine production (i.e., IL-6 and IL-12). Similarly, Raftlin, a protein located in lipid rafts, has been suggested to regulate the strength of TCR signaling (Saeki et al. 2009). Raftlin-deficient T cells produce less IL-17 and Raftlin-deficient mice show reduced severity of experimental auto-immune encephalomyelitis (EAE). Given that high signal strength favors both T_H1 and T_H17 differentiation, it will be important to define mechanisms that influence specificity.

In addition, signals from co-stimulatory molecules cooperate with TCR signaling to regulate full T cell activation (Acuto and Michel 2003). Initially, it has been proposed that both ICOS and CD28 are required for T_H17 differentiation (Park et al. 2005). However, this remains controversial. In fact, *Icos*$^{-/-}$ mice have exaggerated severity of a prototypical T_H17-mediated disease, EAE (Galicia et al. 2009). ICOS is also not required for the early stage of T_H17 differentiation, however, *Icos*-deficient mice have impaired IL-23R and c-Maf expression (Bauquet et al. 2009). A recent study argued that CD28, which is far from being required for T_H17 differentiation, inhibits it via the induction of two known inhibitors of IL-17 expression, IL-2 and IFN-γ (Bouguermouh et al. 2009).

3 Regulation of T_H17 Differentiation by Cytokines

3.1 Role of STAT3-Activating Cytokines in T_H17 Specification

The cytokines IL-6, IL-21, and IL-23 have all been identified as important regulators of IL-17 production (Aggarwal et al. 2003; Nurieva et al. 2007; Zhou et al. 2007; Korn et al. 2007). Importantly, these cytokines share the capacity to potently activate STAT3. Induction of IL-6 is a part of the acute phase response during severe infection, but it is now known to be produced by many immune and non-immune cells (Heinrich et al. 2003; Kishimoto 2005). In vitro, IL-6 is a very potent inducer of T_H17 cells, but only when combined with other cytokines (Veldhoen et al. 2006; Bettelli et al. 2006; Mangan et al. 2006). *Il-6*$^{-/-}$ mice have reduced, but not absent numbers of T_H17 cells (Bettelli et al. 2006), suggesting that other STAT3-activating cytokines can compensate for its actions in vivo.

IL-21 is another potent activator of STAT3 (Rochman et al. 2009) and is produced by T_H17 and follicular helper T cells (Tfh). In vitro, IL-21 is a very effective inducer of IL-17 and of itself (Korn et al. 2007; Nurieva et al. 2007; Wei et al. 2007; Zhou et al. 2007). IL-21 also induces the expression of IL-23R. However, T_H17 differentiation is not abrogated in *Il21*$^{-/-}$ and *Il21r*$^{-/-}$ mice (Sonderegger et al. 2008; Coquet et al. 2008). One study reported that the development of EAE was significantly reduced in IL-21-deficient mice (Nurieva et al. 2007), while in contrast, two other studies showed that the development of EAE was not impaired by the absence of IL-21 signals (Sonderegger et al. 2008; Coquet et al. 2008). Thus it appears that both IL-6 and IL-21 are sufficient to induce T_H17 differentiation, but neither appears to be absolutely necessary.

IL-23 is a heterodimeric cytokine (composed of p19 and p40) that shares the latter sub-unit with IL-12 (Oppmann et al. 2000) and has long been recognized to be an inducer of IL-17 (Aggarwal et al. 2003). Cua and colleagues first observed that p19-deficient mice were completely resistant to EAE and lacked IL-17-producing T cells in the CNS (Cua et al. 2003). Moreover, IL-23-expanded and adoptively-transferred T cells were capable of inducing auto-immune disease (Langrish et al. 2005; Murphy et al. 2003). It was recognized that IL-23 could not be the initial specification factor for T_H17 differentiation, in part because naïve T cells do not express the IL-23R (Ivanov et al. 2006; Zhou et al. 2007). This led to the in vitro finding that TGF-β1 in conjunction with IL-6 is sufficient for initial specification (Bettelli et al. 2006; Veldhoen et al. 2006). Naïve T cells, when activated in the presence of IL-6 or IL-21, express the IL-23R (Zhou et al. 2007). Mouse T cells that have been activated under T_H17 conditions and then subsequently had STAT3 deleted were unable to express IL-17 in response to IL-23 suggesting that STAT3 is important both for the expression of IL-23R and the downstream signaling of this cytokine receptor complex (McGeachy et al. 2009). McGeachy et al. found that although initial in vivo generation of T_H17 cells in the draining lymph nodes of IL-23R deficient mice was intact, the percentage of IL-17$^+$ T cells failed to increase at later time points (McGeachy et al. 2009). In addition, IL-23 may 'fine tune' the functions of T_H17 cells in vivo. T cells differentiated with TGF-β and IL-6 secrete the anti-inflammatory cytokine IL-10 and do not induce tissue inflammation unless further cultured with IL-23 (McGeachy et al. 2007). Zhou et al. has shown that T cells which retro-virally express IL-23R are able to express IL-17 in response to this cytokine alone. We hypothesized that in the presence of T cell activation and IL-6, enough IL-23R would be expressed to mimic this finding, and indeed this is the case (Ghoreschi et al. 2010), suggesting that IL-23 has a direct role in T_H17 development, again producing T_H17 cells that do not secrete IL-10.

Regulation of IL-17 production in human CD4$^+$ T cells shares several features with mouse T_H17 differentiation, although there are some inconsistencies among the findings from different investigators. Despite this, a common feature between groups is the requirement for a STAT3-activating cytokine including IL-6, IL-21, or IL-23 (Acosta-Rodriguez et al. 2007b; Chen et al. 2007; Wilson et al. 2007; Yang et al. 2008a).

3.2 *IL-1 and T_H17 Differentiation*

IL-1β synergizes with IL-6 and IL-23 to induce T_H17 differentiation in both mouse and human cells, but *Il1r*$^{-/-}$ CD4$^+$ T cells fail to differentiate towards T_H17 cells (Sutton et al. 2006; Acosta-Rodriguez et al. 2007a). IL-1 in combination with IL-23 is a very potent inducer of IL-17 in γδ T cells in the absence of TCR stimulation (Sutton et al. 2009; Martin et al. 2009).

Accordingly, *Il1r*$^{-/-}$ mice have reduced incidence of EAE (Sutton et al. 2006). Adoptive transfer of *Il1r*$^{-/-}$ CD4$^+$ cells into *Rag2*$^{-/-}$ mice showed that these cells were poor inducers of EAE arguing for a T cell intrinsic role for IL-1. IL-6 induces IL-1R1 in activated T cells and induces moderate RORγt expression. This stimulus does not induce IRF4 expression, but the combination of IL-1 and IL-6 greatly increases the expression of these two transcription factors (Chung et al. 2009).

The IL-1 receptor has homology to Toll-like receptors, and IL-1 engagement of its receptor results in the recruitment of the adapter MyD88 and IL-1 receptor-associated kinases (IRAK) (O'Neill 2008). This complex recruits another adapter molecule, tumor necrosis factor receptor-associated factor (TRAF-)6. This leads to the activation of other kinases including Jun kinase (JNK) and inhibitor of kappa-B kinases (IKKs). Inhibitors of NF-κB (IκB) are phosphorylated by activated IKKs and subsequently ubiquitinated, leading to the release of NF-κB dimers to the nucleus. JNK phosphorylates the transcription factor Jun, which dimerizes with Fos to form the transcription factor AP-1.

NF-κB-inducing kinase (NIK) has been reported to play a central role in mediating a non-canonical NF-κB activation, while an overexpression of NIK increases the transcription of NF-κB genes (Hacker and Karin 2006). *Nik*-deficient naive CD4 T cells have a deficiency in the differentiation to T_H17 cells, although they are competent in committing to the other effector lineages (Jin et al. 2009). *Nik*$^{-/-}$ mice are resistant to EAE and the deficiency in T_H17 commitment is due to a T cell intrinsic effect. Reconstitution of *Rag2*$^{-/-}$ mice with *Nik*$^{-/-}$ T cells also fails to evoke EAE, arguing for an intrinsic role of NIK.

3.3 *Complex Role of TGF-β1 Signaling in T_H17 Cells*

There are three isoforms of transforming growth factor β; TGF-β1 being the most immunologically relevant. Its receptor comprises two chains: TGF-β type 1 receptor (TGF-βRI) and TGF-β type 2 receptor (TGF-βRII), both of which are receptor serine–threonine kinases. Ligand binding leads to the phosphorylation and activation of receptor-associated latent cytosolic transcription factors known as Smads (Smad2 or Smad3), which dimerize with Smad4.

The essential immuno-suppressive functions of TGF-β1 are best illustrated in mice lacking this cytokine, its receptor sub-units, or key signaling molecules. Such animals develop a fatal, systemic auto-immune disease (Letterio and Roberts 1998; Shull et al. 1992; Yang et al. 1999). The absence of TGF-β-signaling in vivo results in massive T cell activation, characterized by a predominance of IFN-γ-producing T_H1 cells in aging mice. The identification of TGF-β as the primary inducer of Treg cells does much to explain its inhibitory role within the immune system (and explains the phenotype of the mice). Additionally though, TGF-β in the presence of IL-6 is sufficient to induce naive T cells to produce IL-17 (Bettelli et al. 2006; Mangan et al. 2006; Veldhoen et al. 2006). The generation of antigen-specific T_H17 cells by immunization with CFA and peptide under conditions of massive T cell activation and IFN-γ-production is limited. Transgenic overexpression of TGF-β in T cells was associated with increased susceptibility to the development of EAE in mice injected with MOG peptide (Bettelli et al. 2006). Conversely, mice with Th cells unable to secrete TGF-β are relatively protected from EAE and no T_H17 cells were found in the CNS (although the mice have overproduction of IFN-γ) (Li et al. 2007). This data has led to idea that the combination of TGF-β and IL-6 are required for initial specification to T_H17 commitment.

The mechanism by which TGF-β induces T_H17 development remains elusive. Recently, the essential role of TGF-β in T_H17 differentiation has been challenged by research showing that in the absence of T-bet and STAT6, IL-6 alone was sufficient to generate T_H17 cells (Das et al. 2009). This suggests that TGF-β may play an indirect role in T_H17 differentiation by suppressing the expression of STAT4, T-bet, GATA3, and by limiting differentiation to alternate fates. We have found that T_H17 differentiation occurs in vitro and in vivo in the absence of TGF-β (Ghoreschi et al. 2010). TGF-β-independent T_H17 differentiation requires STAT3-dependent signals derived from IL-6 and IL-23 and more importantly, the expression of IL-23 receptor (IL-23R), which is absent on naïve T_H cells. Interestingly, IL-23R can be rapidly induced in naïve T_H cells by IL-6. Whereas IL-23 itself amplifies IL-23R expression, TGF-β suppresses IL-23R expression (Ghoreschi et al. 2010). The expression of T-bet and other T_H1-characteristic markers by T_H17 cells may explain their pathogenicity. T_H17 cells induced with TGF-β1 are not pathogenic, rather the cells produce IL-17 and IL-10 (Yang et al. 2009; McGeachy et al. 2007).

The role of TGF-β in the human T_H17 differentiation is similarly controversial (Annunziato et al. 2009). Studies using lower doses of TGF-β and naïve T cells derived from cord blood have argued that TGF-β is required for T_H17 development in human T_H cells in a manner analogous to mice (Manel et al. 2008). The authors noted that concentration of TGF-β is also an important factor in regulating the T_H17/Treg balance (Manel et al. 2008; Yang et al. 2008a). However, human T_H17 differentiation can be induced in the presence of IL-23 and the absence of TGF-β and in these circumstances several groups have found that TGF-β is an inhibitor of IL-17 production (Chen et al. 2007; Acosta-Rodriguez et al. 2007b; Wilson et al. 2007). In keeping with the possibility of a non-essential role of TGF-β in T_H17 specification, there is no data indicating that Smads directly bind to genes associated with T_H17 development.

4 Regulation of IL-17 Production by G-Protein Coupled Receptors

Prostaglandins are a class of lipids that signal via G-protein coupled receptors (GPCRs). Prostaglandin E2 (PGE2) is highly induced during inflammation and promotes the expansion of IL-17-producing cells (Chizzolini et al. 2008). Cultures of human peripheral blood mononuclear cells with PGE2 and IL-23 resulted in high levels of IL-17 and C–C chemokine ligand 20 (CCL20), with less IFN-γ and IL-22. PGE2 acts through both prostaglandin EP2- and EP4- mediated signaling and cyclic AMP pathways to upregulate IL-23 and IL-1 receptor expression resulting in synergistic induction of T_H17 cells with IL-23 and IL-1β (Boniface et al. 2009). Application of an EP4-selective antagonist in vivo reduces the accumulation of both T_H1 and T_H17 cells in local lymph nodes and significantly suppresses the disease severity of EAE or contact hypersensitivity (Yao et al. 2009). In vivo administration of PGE2 induced IL-23-dependent IL-17 production and exacerbated collagen-induced arthritis (CIA) in mice (Sheibanie et al. 2007). PGE2 acts through EP4, not only amplifying IL-23-mediated T_H17 cell expansion and facilitating T_H1 cell differentiation, but pointing to overlaps between T_H1 and T_H17 cells. In addition to its direct effect on T cells, PGE2 has been reported to induce dendritic cell IL-23 production (Sheibanie et al. 2007).

Sphingosine 1-phosphate (S1P) is a lipid molecule which signals via a GPCR and regulates IL-17 production. S1P is an abundant lysophospholipid present in blood and lymph that regulates T cell trafficking and egress from lymphoid structures via its receptor S1PR1 (Rosen and Goetzl 2005). Addition of S1P to cultures of splenic $CD4^+$ T cells has been reported to increase IL-17 and to reduce IFN-γ and IL-4 production (Liao et al. 2007). Accordingly, transgenic overexpression of an S1P receptor results in enhanced IL-17 production (Huang et al. 2007). The immunosuppressive compound FTY720 (fingolimod) has efficacy in multiple sclerosis and rheumatoid arthritis. Its major effect is to alter lymphoid trafficking, but whether S1P or fingolimiod are relevant regulators of T_H17 differentiation remain to be determined. Nonetheless, the findings raise the intriguing possibility that ligands for multiple G-protein coupled receptors might also influence T cell differentiation. The fact that these receptors are so "druggable" provides potential opportunities for new therapeutic approaches.

5 Regulation of T_H17 Differentiation by Transcription Factors

5.1 Central Role of STAT3 in T_H17 Specification

The observation that the T_H17- associated cytokine, IL-23 was an inducer of STAT3 initially led us to investigate the requirement for this transcription factor in T_H17 differentiation (Chen et al. 2006). We found that STAT3 was able to directly bind to the *Il-17a* promoter. The critical role of STAT3 was confirmed in T cells deficient for STAT3 by us and others (Yang et al. 2007; Laurence et al. 2007). Conversely, retroviral overexpression of constitutively active STAT3 was sufficient to induce IL-17 production (Mathur et al. 2007). Importantly, conditional deficiency of *Stat3* in $CD4^+$ T cells impairs IL-17 production in vivo and limits IL-17-associated pathology (Liu et al. 2008; Harris et al. 2007; Durant et al. 2010).

The requirement for STAT3 in human IL-17 production was unraveled in patients with Hyper IgE syndrome (HIES). HIES is a primary immuno-deficiency disorder due to dominant negative mutations of STAT3 (Holland et al. 2007; Minegishi et al. 2007). We and others found that patients with HIES have severely impaired ability to produce T_H17 cells (Holland et al. 2007; Milner et al. 2008; Minegishi et al. 2007; Ma et al. 2008). The impairment of STAT3 signaling and consequently attenuated T_H17 cell generation may explain the patients' inability to clear bacterial and fungal infections.

The demonstration of the importance of STAT3 in mouse and man raised the issue of exactly how STAT3 contributes to T_H17 cell development. Chromatin immuno-precipitation and massive parallel sequencing is a new technology that enables genome-wide determination of transcription factor-bound genes (Mardis 2007; Park 2009). The functional relevance of the binding can be ascertained by genome-wide transcriptional profiling and by assessing the transcription factor-dependent changes in epigenetic modifications (see below). We have done this for STAT3 in T_H17 cells. Remarkably, we found that STAT3 binds promoters and enhancers of many genes involved in T_H17 development (Durant et al. 2010). STAT3 binds throughout the *Il17a/f* locus, binding not just within the promoter regions, but also in intergenic region. STAT3 also binds and regulates the *Il21*, *Il21r,* and *Il23r* genes (Fig. 1). STAT3 directly controls expression of many of the other transcription factors that participate in T_H17 differentiation including Rorγt, IRF4, and Batf.

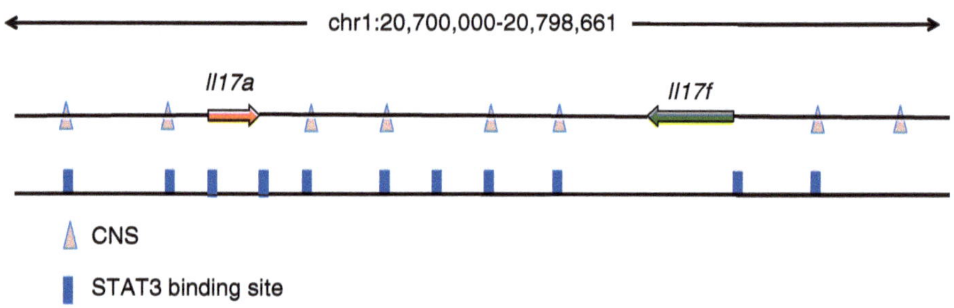

Fig. 1 STAT3 binding sites in the Il17a/Il17f locus. In addition to the previous identified STAT3 binding sites in the promoters of *Il17a* and *Il17f* in T_H17 conditions (Yang et al. 2007), STAT3 have been found to bind strongly with multiple conserved non-coding sequences located in the intergenic region between *Il17a* and *Il17f* (Durant et al. 2010). The significance of those individual binding sites is of interest for further investigations

This research revealed a greatly expanded role for STAT3 in T_H17 differentiation. The study also indicated an important role for STAT3 in proliferation and survival of T cells, which was not apparent in the original analysis of mice in which STAT3 was conditionally deleted in T cells (Harris et al. 2007).

5.2 *Retinoid Receptors and T_H17 Cells*

Retinoid receptors comprise a family of nuclear receptors, only some members of which are regulated by metabolites of vitamin A. They are part of a larger family of steroid nuclear receptors and, have direct transcription factor properties. They have pleiotropic actions in a wide variety of cell types and their importance in T_H cell differentiation is only recently becoming elucidated. Retinoid receptors can be divided into three families: classical retinoic acid receptors (RARα-γ), retinoid X receptors (RXRα-γ), and the retinoic acid orphan receptors (RORα-γ). The latter do not directly bind, nor are directly regulated by retinoids, and their ligands are currently unknown. RORγt is a splice variant of the *Rorc* gene that results from initiation by a distinct promoter within the full length *Rorc* gene, resulting in products that differ in their amino terminus (Jetten 2009). How RORγ and its splice variants are regulated is not clear. RORγ regulates gene expression by binding to specific DNA response elements consisting of the RGGTCA core motif preceded by an A/T rich sequence. RORγt has been found to be associated with co-activators and co-repressors as well, suggesting RORγt can function as both activator and repressor of gene expression.

Whereas RORγ is ubiquitously expressed, RORγt is exclusively expressed in lymphoid cells. RORγ is critical for T cell development, as evidenced by the fact that *Rorc*$^{-/-}$ mice exhibit reduced numbers of double positive and CD4^{+} single-positive thymocytes. *Rorc*$^{-/-}$ mice also show a lack of lymph nodes, Peyer's patches, lymphoid tissue-inducer cells (Sun et al. 2000), and are also critical for lymphoid organogenesis (Kurebayashi et al. 2000; Eberl and Littman 2003). *Rorc*$^{-/-}$ mice have reduced T_H17 differentiation and have reduced disease severity in the EAE model (Ivanov et al. 2006). Conversely, overexpression of RORγt promotes IL-17 expression, and in this regard, RORγt acts similarly to transcription factors such as T-bet and GATA3. In T_H1 and T_H2 differentiation it has been proposed to be a "master regulator" for T_H17 differentiation. Multiple RORγt binding sites are present in the *Il17a* promoter, and by the use of chromatin immuno-precipitation, RORγt has been found to bind the *Il17a* gene (Ichiyama et al. 2008).

Stimulation of naïve CD4 T cells with TGF-β and IL-6 induces RORγt mRNA expression (Ivanov et al. 2006). However, RORγt can be induced in human and mouse T_H17 cells by IL-23 (Ivanov et al. 2006; Wilson et al. 2007) and can be induced in naïve CD4+ T cells in the absence of TGF-β1 by IL-1β, IL-6, and IL-23 (Ghoreschi et al. 2010). As indicated, STAT3-deficient T cells fail to induce RORγt expression (Yang et al. 2007; Laurence et al. 2007) and STAT3 binds the *Rorc* gene (Durant et al. 2010). However, overexpression of active STAT3 in *Rorc*-deficient cells resulted in poor IL-17 induction (Zhou et al. 2007), suggesting that STAT3 is necessary, but not sufficient for IL-17 expression. It is highly likely that STAT3 and RORγt act cooperatively on the *Il17* locus to drive production of this cytokine.

Another related retinoic acid nuclear receptor, RORα, is also preferentially expressed in T_H17 cells (Yang et al. 2008b). In contrast to RORγt though, *Rora* deletion in mice had a minimal effect on IL-17 production. However, deficiency of both *Rora* and *Rorc* completely abolished IL-17 production and protected against EAE (Yang et al. 2008b). Co-expression of the two factors enhanced the number of IL-17-producing cells. Collectively, the findings suggest that there is some redundancy between the functions of RORα and RORγt.

5.3 Runx Transcription Factors

Runx transcription factors represent another family of proteins that regulate $CD4^+$ T helper cell differentiation. There are three mammalian Runt domain transcription factors, Runx1, Runx2, and Runx3. These factors form heterodimers with an obligatory, non-DNA-binding partner (Cbfβ protein) which is required to stabilize the interaction of Runx proteins with DNA. Runx1 is crucial for normal hematopoiesis including thymic T cell development (Collins et al. 2009). Additionally, both Runx1 and Runx3 are reported to act as repressors of CD4 and Zbtb7b (the latter being encodes Th-POK, a transcription factor that regulates $CD4^+$ differentiation), but at different stages of thymocyte development affect the CD4/CD8 lineage choice (Taniuchi et al. 2002; Setoguchi et al. 2008). Runx3 also functions as a CD4 silencer in $CD8^+$ T cells and is involved in a feed-forward regulatory circuit in which T-bet induces Runx3 and then 'partners' with Runx3 to direct activation of *Ifng* and silencing of *Il4* (Djuretic et al. 2007; Naoe et al. 2007).

Runx1 has been reported to bind the *Il2* promoter at a site upstream of NFAT, AP-1, and NF-κB binding sites and to enhance the TCR-stimulated IL-2 expression. In Treg cells, Runx1 binds the N-terminus of Foxp3. This interaction is required for Foxp3-mediated suppression of IL-2 and suppression activity of Treg cells (Hu et al. 2007). Recent work has indicated that Runx1 also plays a role in T_H17 differentiation (Zhang et al. 2008). Both Runx1 and RORγt bind to the *Il17* promoters and by interaction with distinct functional partners (RORγt versus Foxp3), Runx1 may contribute to both Treg and T_H17 differentiation.

5.4 Aryl Hydrocarbon Receptor

The aryl hydrocarbon receptor (AhR) belongs to a bHLH-PAS family of transcription factors and is highly conserved in evolution (Esser et al. 2009). AhR senses a wide range of small synthetic compounds and natural ligands. Ligands of AhR have diverse structure and only need to meet minimal requirements of planar shape and size.

The natural ligands include metabolites of tryptophan and arachidonic acids, as well as dietary flavonols and kaempferols. Synthetic ligands include both planar and aryl hydrocarbons, found in environmental toxins. 2,3,7,8-tetrachlorodibenzo-*p*-dioxin (TCDD) are the most toxic members of a large group of structurally similar halogenated aromatic hydrocarbons polychlorinated dibenzodioxin congeners (PCDDs), as well as polychlorinated biphenyls (PCBs) and polychlorinated dibenzofurans (PCDFs). Dioxin is widely used as a surrogate ligand for AhR signaling. Without ligand binding, AhR is inactive in the cytosol and associated with the chaperone Hsp90, p23 and AhR-interacting protein. Upon ligand binding, AhR undergoes conformational change, dissociates with its chaperone Hsp90 to expose the nuclear translocation sites, and migrates into the nucleus. Subsequently, AhR associates with the aryl receptor nuclear translocator (ARNT) forming a heterodimeric complex which binds to the Ahr responsible element 5′-TNGCGTG-3′ to regulate target gene expression.

Various reports indicate that TCDD is profoundly immuno-suppressive (Kerkvliet 2002) and activation of AhR by TCDD expands $CD25^+$ $Foxp3^+$ Treg cells which suppress EAE (Quintana et al. 2008). Paradoxically, it has been found that T_H17 cells preferentially express AhR. Indeed, another ligand of AhR, tryptophan-derived photoproduct 6-formylindolo [3,2-b] carbazole (FICZ), induces high expression of IL-17A, IL-17F, and IL-22 (Veldhoen et al. 2008). In line with this, medium containing AhR ligands potentiates the in vitro T_H17 differentiation (Veldhoen et al. 2009). AhR is required for IL-22 production, but not IL-17 (Veldhoen et al. 2008). However, mice lacking AhR have reduced severity in EAE. How two different AhR agonists, TCDD and FICZ, could have such distinct effects remains a puzzle, and what the function of AhR in IL-22 induction is, remain to be fully delineated. Regardless of the existing controversies, AhR expression in T_H17 cells provides an intriguing link as to how environmental toxins might contribute to the pathogenesis of auto-immune diseases.

It has been reported that AhR selectively interacts with STAT1 and STAT5, but not with STAT3 or STAT6 (Kimura et al. 2008). AhR deficiency was associated with prolonged STAT1 phosphorylation, so it is intriguing that induction of AhR provides a counter-regulatory mechanism that would attenuate the effects of cytokines like IFN-γ and IL-27 that activate STAT1.

5.5 IRF4 and T_H17 Differentiation

IRF4, a member of the interferon regulatory factor (IRF) family of transcription factors, was originally implicated as a key inducer of GATA3 expression in T_H2 lineage differentiation (Lohoff et al. 2002). However, T cells from $IRF4^{-/-}$ mice have impaired IL-17 production in response to TGF-β and IL-6, and $IRF4^{-/-}$ mice are resistant to EAE (Brustle et al. 2007). However, $IRF4^{-/-}$ mice show a degree of resistance to EAE in excess of *Il17*, *Il23*, *Rorc* or *Stat3*-null animals, implying that there are additional effects of IRF deficiency. As indicated above, IL-1β is a major inducer of IRF4 expression (Chung et al. 2009). It has been suggested that IRF4 might co-operate with STAT3 to induce RORγt expression, but IRF4 also interacts with NFATc1 and c2 (Hu et al. 2002; Rengarajan et al. 2002a). In this manner, it may affect other Th cell lineages. This is a rare example of a factor required for both T_H17 and T_H2 development, which is in contrast with the previously accepted links with T_H17 and T_H1 development via IL-23 (Harrington et al. 2005) and T_H17 and Treg development via TGF-β (Bettelli et al. 2006).

5.6 *Batf and T_H17 Differentiation*

Batf is a member of the AP-1 transcription factor family that contains a basic region and leucine zipper and lacks a transcriptional activation domain, so has been considered an inhibitor of AP-1 transcriptional activity (Wagner and Eferl 2005). Recently, Murphy et al. demonstrated that Batf is essential for T_H17 cell differentiation (Schraml et al. 2009). *Batf*$^{-/-}$ mice show a complete loss of T_H17 differentiation and are resistant to the induction of EAE. Batf mRNA is expressed in activated T cells and in all major subsets including T_H1, T_H2, and T_H17 cells. Nonetheless, in vitro T_H1 and T_H2 differentiations are not affected the absence of Batf. Instead of acting as a transcriptional repressor, Batf heterodimerizes with JunB in T_H17 cells and positively regulated T_H17-associated genes. Batf binds the promoters of *Il-17a*, *Il-17f*, *Il-21*, and *Il-22* and the intergenic regions between *IL-17a* and *IL-17f*. In addition to STAT3, it is likely that TCR signaling drives Batf expression, as it is induced in multiple lineages.

6 Negative Regulation

6.1 *Socs3*

Suppressor of cytokine signaling (Socs3) is a cytokine-inducible negative regulator of STAT3; Socs proteins are among the most prominent Stat-regulated genes (Yoshimura et al. 2007) providing a negative feedback to Stat signaling. Chen et al. first showed that T cell-specific deletion of Socs3 is associated with enhanced STAT3 phosphorylation and increased IL-17 production (Chen et al. 2006). Deletion of Socs3 in the hematopoietic and endothelial cell compartment is associated with widespread auto-immune disease and increased susceptibility to both CIA and EAE (Wong et al. 2006).

6.2 *Inhibition of T_H17 Differentiation by STAT Proteins*

In addition to positive roles in regulating T_H17 differentiation, cytokines that bind Type I/II cytokine receptors can negatively regulate T_H17 differentiation. Consistent with work indicating that IFN-γ and IL-4 also antagonize T_H2 and T_H1 differentiation,

IFN-γ and IL-4 also inhibit IL-17 production (Park et al. 2005). IL-27 is an IL-12-related cytokine that consists of two sub-units, p28 and EBI3 (Pflanz et al. 2002). Like IL-6, its receptor complex contains gp130, but it also has a ligand-specific sub-unit termed WSX-1. Despite its similarities to IL-6, IL-27 has the opposite effect in regulating IL-17 expression and it is a critical negative regulator of T_H17 differentiation (Batten et al. 2006; Stumhofer et al. 2006). Both IFN-γ and IL-27 activate STAT1 and the inhibitory effect of IL-27 is abrogated in STAT1-deficient mice (Stumhofer et al. 2006). Additionally, mice lacking STAT1 develop airway disease following viral infections that is associated with increased IL-17 expression (Hashimoto et al. 2005). Further investigation is needed to understand the exact mechanisms by STAT1 inhibit T_H17 generation.

Activated by IL-12, STAT4 is well known as a critical positive regulator of T_H1 differentiation and IFN-γ production. As both IL-12 and IFN-γ suppress IL-17 production, in an analogy to STAT1, the expectation might be that STAT4 would negatively regulate IL-17 expression. However, two studies have reported that IL-17 production is decreased in STAT4-deficient T cells, arguing for a positive role of STAT4 in IL-17 production (Hildner et al. 2007; Mathur et al. 2007). The decreased IL-17 production in STAT4-deficient T cells might be related to the impaired IL-23 signaling, as IL-23 also activates STAT4 (Watford et al. 2004). It will be of interest to further examine the role of STAT4 in T_H17 differentiation.

IL-2 is a well-known T cell growth factor in vitro, but at the same time, the deficiency of IL-2 results in severe multi-organ auto-immune disease in vivo (Malek 2008). This is in part due to its role in promoting the differentiation of Tregs by STAT5, but recent work has shown that IL-2 also suppresses T_H17 differentiation in a STAT5-dependent manner (Laurence et al. 2007; Kryczek et al. 2007). Like *Il2*$^{-/-}$ mice, STAT5-deficient mice suffer from inflammatory auto-immune disease that is associated with a loss of Treg cells and the simultaneous expansion of T_H17 cells. STAT5a/b appear to be essential for Foxp3 expression and in constraining T_H17 cells (Laurence et al. 2007; Yao et al. 2007). However, the mechanisms underlying the blockade of T_H17 differentiation by IL-2 via STAT5 are not entirely clear. It is possible that this is mediated indirectly via induction of Foxp3, a factor known to bind and inhibit RORγt. Alternatively, STAT5a/b might act as a direct repressor to inhibit *Il17a* expression (Laurence et al. 2007) (Fig. 2)

6.3 Single Ig IL1 Receptor-Related Molecule

Single Ig IL1 receptor-related molecule (SIGGR), contains an extracellular immunoglobulin (Ig) domain and a TIR (Toll-IL1R) intracellular domain; SIGGR inhibits IL-1 and LPS signaling via its interaction with IL-1R and TLR complexes. Recently Gulen et al. 2010 showed that SIGGR is induced over the course of T_H17 differentiation and T cells with deficiency of SIGGR are more pathogenic in EAE (Gulen et al. 2010). In the absence of SIGGR, IL-1 induces elevated mTOR signaling with hyperphosphorylation of 4E-BP1 (an mTOR substrate) and S6 (a downstream

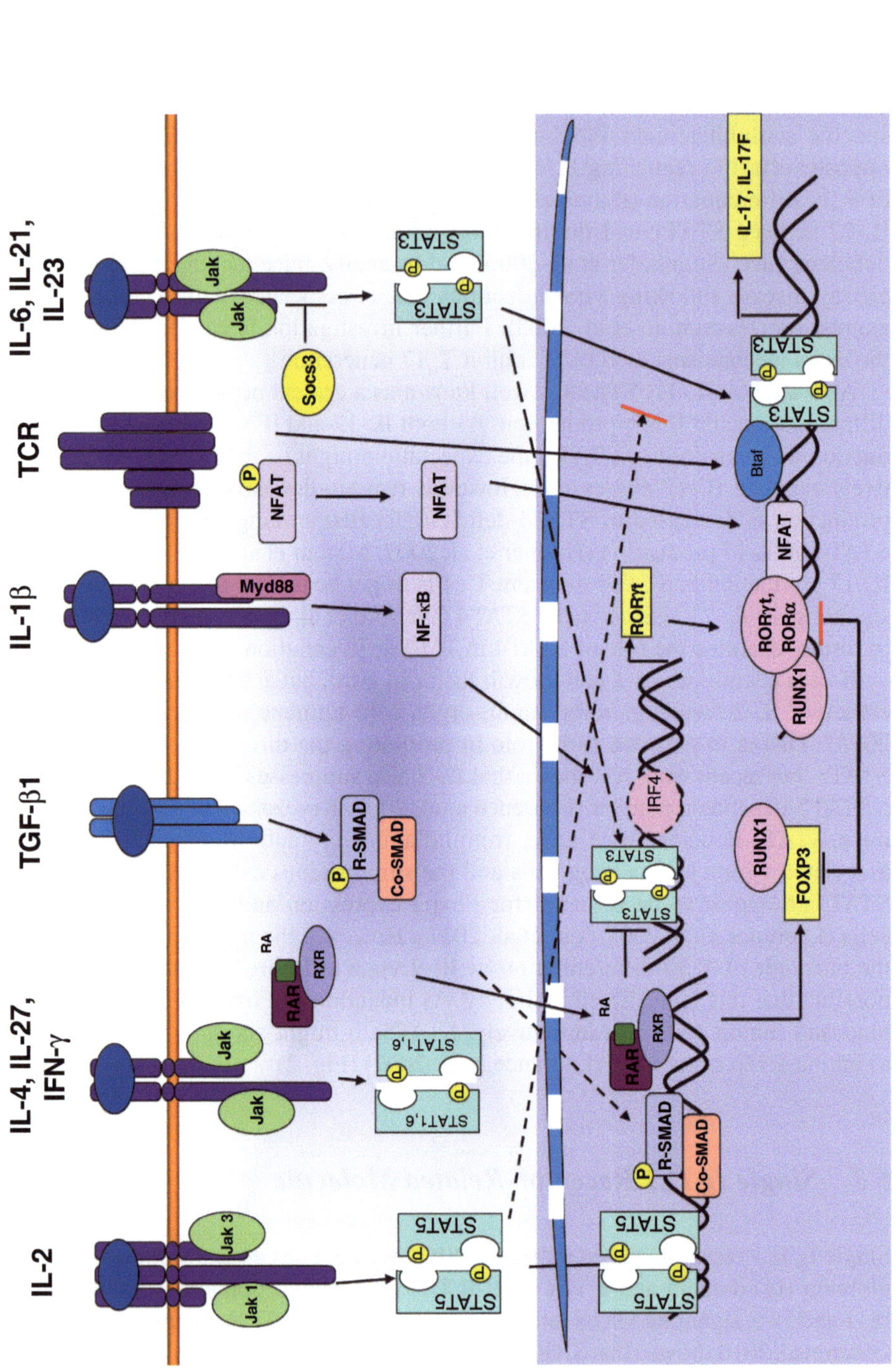

IL-2
IL-4, IL-27, IFN-γ
TGF-β1
IL-1β
TCR
IL-6, IL-21, IL-23
Jak 1
Jak 3
Jak
Socs3
Myd88
NFAT
NF-κB
P
STAT5
STAT1,6
STAT3
R-SMAD
Co-SMAD
RA
RAR
RXR
IRF4
RORγt
RORγt, RORα
RUNX1
FOXP3
Btaf
IL-17, IL-17F

Fig. 2 Signaling pathways and transcription factors that regulate T_H17 differentiation. TCR stimulation activates the transcription factor NFAT, which regulates IL-17A and IL-17F differentially. IL-6, IL-21, and IL-23 induce STAT3 activation, which in turn binds the *Il17a, Il17f and Il21* genes. STAT3 binds to *Rorc* and Irf4 genes and IRF4 cooperates with STAT3 to induce RORγt expression. TGF-β1 signaling involves the activation of SMAD proteins, although the mechanism by which it promotes both iTreg and T_H17 differentiation remains largely unknown. One possibility is that it does not provide inductive signals, but only attenuates expression of T-bet, GATA3, and other factors. IL-1 signals to potentiate IL-6-induced IRF4 expression and thus promotes T_H17 differentiation. Both RORγt and RORα bind the *Il17a* gene. In contrast, IL-2, IL-4, IL-27, and IFN-γ inhibit T_H17 differentiation through STAT5, STAT6, and STAT1 activation respectively, although the underlying mechanism(s) are not completely understood. Reciprocally, STAT5 upregulates Foxp3 expression. Runx1 associates with both RORγt and Foxp3 and possibly regulates differentiation towards either the iTreg or T_H17 lineage. Cytokines also upregulate Socs3, which attenuates STAT3 activation

component of mTOR signaling). The mTOR signaling pathway is required for the IL-1-mediated proliferation of T_H17 cells. Blockade of the mTOR signaling induced by IL-1 with rapamycin restricts the IL-1 effect on T_H17 differentiation.

6.4 Negative Regulation of T_H17 by Nuclear Receptors

In contrast to RORγt and RORα, other retinoic acid nuclear receptors have been suggested to inhibit T_H17 differentiation. Several groups have shown that retinoid acid inhibits T_H1, T_H2, and T_H17 differentiation in vitro. Retinoic acid reciprocally downregulates expression of RORγt and enhances the expression of Foxp3 (Mucida et al. 2007; Coombes et al. 2007; Sun et al. 2007; Elias et al. 2008). Conversely, a RAR antagonist inhibited Foxp3 expression. Foxp3 binds to the *Il17a* promoter (Zhou et al. 2008), thus Foxp3 and RORγt, directly interact and modify each other's function (Zhou et al. 2008). T cells cultured in retinoid acid and TGF-β are functionally immuno-suppressive, even in the absence of IL-2.

Peroxisome proliferator-activated receptor γ (PPARγ), a member of the nuclear receptor superfamily of transcription factors, plays important roles in adipogenesis, development and inflammation. Multiple natural and synthetic PPARγ ligands have been described to exert anti-inflammatory effects on various levels (Szanto and Nagy 2008). Upon ligand binding, PPARγ translocates into the nucleus and forms a heterodimer with the retinoid X receptor and then binds to PPARγ response elements located in the promoter region of target genes. Similar to the ligands of Ahr, the ligands of PPARγ can be divided into two groups: natural ligands, i.e. the linoleic acid derivative 13s-hydroxyoctadecadienoic acid (HODE) and several synthetic agonists i.e. the anti-diabetic thiazolidinediones, pioglitazone (PIO). Recently, the relevance of PPARγ in T_H17 differentiation has been demonstrated with the use of the synthetic ligand PIO (Klotz et al. 2009). T_H17 differentiation and expression of T_H17-associated genes (*Il23r*, *Il21*, *Il17a* and *Il17f*) were dramatically inhibited by treatment of the $CD4^+$ T cells with PIO, but this had no effect on T_H1 differentiation. PIO ameliorated EAE clinical scores and mice had fewer IL-17^+ cells in the CNS. Consistent with this, the mice lacking PPARγ in $CD4^+$ T cells showed earlier onset and aggravated disease in the initial T cell-dependent phase of EAE. However, there was no difference in the later effector phase of disease. PPARγ activation suppresses the induction of RORγt by inhibiting the cytokine-induced clearance of SMRT, leading to persistent repression of RORγt expression rather than binding to the promoter of RORγt. It remains unclear if the activation of PPARγ can alter the binding of STAT3, Batf, or IRF4 to the *Il17* locus. Interestingly, epidermal fatty acid-binding protein, E-FABP, a lipid chaperone, promotes regulation of the expression of PPARγ and thereby enhances T_H17 differentiation and inhibits the development of $FoxP3^+$ Tregs in vivo and in vitro (Li et al. 2009). Treatment of FABP-deficient $CD4^+$ T cells with the PPARγ antagonist restored expression of RORγt, RORα, IL-21 and IL-17. PPARγ has been reported to suppress IL-6-induced STAT3 activity in this study.

NR2F6, another nuclear receptor, has been reported to negatively regulate T_H17 differentiation both in vitro and in vivo. NR2F6-deficient mice exhibit both faster onset and more severe pathology in the EAE model (Hermann-Kleiter et al. 2008). NR2F6 is phosphorylated by protein kinase C following TCR stimulation and is reported to inhibit NFAT DNA binding activity. Curiously, NR2F6 suppresses both IL-17 and IFN-γ, but not IL-4 production.

6.5 MicroRNA and Negative Regulation of IL-17

Ets-1 is a transcription factor that associates with T-bet and binds the *Ifng* promoter. Ets-1 deficiency results in enhanced T_H17 differentiation, along with increased levels of mRNA for IL-22 and IL-23R (Moisan et al. 2007). Ets-1's actions are likely to be indirect, as no binding of Ets-1 to the *Il17a* promoter was observed. Rather, Ets-1 appears to negatively regulate T_H17 differentiation through effects on IL-2. Specifically, Ets-1 deficient T cells were found to secrete less IL-2 and have impaired responsiveness to this cytokine. Ets-1 has been found to be a target of microRNA (mir-326) and mir-326 has been reported to be upregulated in patients with MS and in murine T_H17 cells generated in vitro (Du et al. 2009). Overexpression of mir-326 enhances T_H17 generation and disease severity in EAE, whereas blocking mir-326 reduces T_H17 generation and ameliorates disease severity of the EAE. This has been explained by regulating expression of Ets-1.

7 Plasticity, Lineage Commitment and Epigenetic Regulation of T_H17 Cells

Cellular differentiation is associated with heritable changes through chromatin in daughter cells that preserve gene expression. Epigenetic modifications such as DNA methylation, chromatin remodeling, histone modifications, and incorporation of histone variants can all contribute to regulate gene expression (Kouzarides 2007). Cytokine genes in helper T cell subsets are similarly regulated (Lee et al. 2006; Wilson et al. 2009).

The *Il17a* gene is linked to the *Il17f* gene on chromosome 1 (human chromosome 6) in a tail-to-tail configuration. Expression of these two cytokines is linked, although there are some circumstances in which there is selective expression of one or the other (Chang and Dong 2007; Liang et al. 2007). The promoter regions of both the *Il17a* and *Il17f* genes undergo histone H3 acetylation and K4 tri-methylation upon T_H17 differentiation, implying increased accessibility of the locus (Yang et al. 2007). In addition, there are eight conserved noncoding sequences in this locus, four of which reside in the intergenic region, which also undergo preferential histone acetylation (Yang et al. 2007). In addition to the promoters, STAT3 binds in this intergenic region and the accessible histone marks are STAT3-dependent (Durant et al. 2010).

Although the standard model of T_H1 and T_H2 differentiation implies that these subsets behave like terminally differentiated cells, recent findings indicate more flexibility than envisioned and provide mechanisms for flexibility in expression of key transcription factors (Hegazy et al. 2010; Wei et al. 2009; O'Shea and Paul 2010). While the histone methylation at the proximal promoters of cytokines genes showed reciprocal permissive (H3K4me3) and repressive (H3K27me3) marks in cell subsets that express the signature cytokines, the histone methylation patterns of the key transcriptional factors for lineage specification exhibit bivalent modifications of those genes not expressed (Wei et al. 2009). For example, the promoters of *Tbx21* and *Gata3* are modified by both H3K4me3 and H3K27me3 in T_H17 cells, suggesting that these genes are poised for expression. This is consistent with the reported transition of T_H2 to produce IFN-γ following viral infection (Hegazy et al. 2010). In this context, it is perhaps not surprising that T_H17 cells generated in vitro are intrinsically unstable (Shi et al. 2008) (Bending et al. 2009). In contrast, isolated memory T_H17 cells appear to be more stable (Lexberg et al. 2008).

The *Rorc* locus acquires repressive histone methylation under T_H1 and T_H2 differentiation conditions, consistent with the refractoriness of conversion of Th1 and T_H2 to T_H17 cells (Wei et al. 2009). Despite the progress made in the field of epigenetic regulation of T_H17 generation, there are still many more questions that remain to be answered. The mechanisms through which either TGF-β or IL-23, in combination with IL-6, promotes accessibility to this locus are unclear, although STAT3 is an important driver. How RORγt and STAT3 work together to influences chromatin modifications in the *Il17a–Il17f* locus is less clear. It is also not known how the inhibitory cytokines of T_H17 differentiation (IL-2, IL-4 and IL-27) regulate the epigenetics of the *Il-17A* and related genes involved in T_H17 differentiation.

8 Conclusions

Since the discovery of T_H17 cells in 2005, remarkable advances have been made about this new T cell subset: the major cytokines required for T_H17 differentiation have been identified, key transcription factors involved in their generation are continuingly to be identified, their function in animal models of auto-immune diseases and in models of infectious diseases have been established, and the epigenetic regulation of their commitment is emerging. Although TGF-β in combination with IL-6 is undoubted to be an excellent inducer of T_H17 cells in vitro, it is hard to delineate whether it is required for T_H17 generation in vivo or modulates expression of factors such as T-bet and GATA3. Interestingly, T_H17 cells generated with different cocktails of cytokines seem to have distinct pathogenicity. Whether there are different subtypes of T_H17 cells under different pathogenic events in vivo will be of interest. The finding of conversion of T_H17 cells to other T_H subsets begs the question of what roles T_H17 cells play in tissue inflammation. If so, it would be very important to know what effector cytokine or what combination of T_H17-secreted cytokines mediate the effector function of T_H17 cells in inflammatory disease.

Acknowledgements We apologize to the colleagues whose work was not cited owing to space limitations. Work of the authors was supported by NIH/NIAMS Intramural Research Program (IRP).

References

Acosta-Rodriguez, E.V., Napolitani, G., Lanzavecchia, A., and Sallusto, F. (2007a). Interleukins 1beta and 6 but not transforming growth factor-beta are essential for the differentiation of interleukin 17-producing human T helper cells. Nat Immunol *8*, 942–949.

Acosta-Rodriguez, E.V., Rivino, L., Geginat, J., Jarrossay, D., Gattorno, M., Lanzavecchia, A., Sallusto, F., and Napolitani, G. (2007b). Surface phenotype and antigenic specificity of human interleukin 17-producing T helper memory cells. Nat Immunol *8*, 639–646.

Acuto, O., and Michel, F. (2003). CD28-mediated co-stimulation: a quantitative support for TCR signalling. Nat Rev Immunol *3*, 939–951.

Aggarwal, S., Ghilardi, N., Xie, M.H., de Sauvage, F.J., and Gurney, A.L. (2003). Interleukin-23 promotes a distinct CD4 T cell activation state characterized by the production of interleukin-17. J Biol Chem *278*, 1910–1914.

Annunziato, F., Cosmi, L., Liotta, F., Maggi, E., and Romagnani, S. (2009). Human Th17 cells: are they different from murine Th17 cells? Eur J Immunol *39*, 637–640.

Batten, M., Li, J., Yi, S., Kljavin, N.M., Danilenko, D.M., Lucas, S., Lee, J., de Sauvage, F.J., and Ghilardi, N. (2006). Interleukin 27 limits auto-immune encephalomyelitis by suppressing the development of interleukin 17-producing T cells. Nat Immunol *7*, 929–936.

Bauquet, A.T., Jin, H., Paterson, A.M., Mitsdoerffer, M., Ho, I.C., Sharpe, A.H., and Kuchroo, V.K. (2009). The costimulatory molecule ICOS regulates the expression of c-Maf and IL-21 in the development of follicular T helper cells and TH-17 cells. Nat Immunol *10*, 167–175.

Bending, D., De La Pena, H., Veldhoen, M., Phillips, J.M., Uyttenhove, C., Stockinger, B., and Cooke, A. (2009). Highly purified Th17 cells from BDC2.5NOD mice convert into Th1-like cells in NOD/SCID recipient mice. J Clin Invest *119*, 565–572

Berg, L.J., Finkelstein, L.D., Lucas, J.A., and Schwartzberg, P.L. (2005). Tec family kinases in T lymphocyte development and function. Annu Rev Immunol *23*, 549–600.

Bettelli, E., Carrier, Y., Gao, W., Korn, T., Strom, T.B., Oukka, M., Weiner, H.L., and Kuchroo, V.K. (2006). Reciprocal developmental pathways for the generation of pathogenic effector TH17 and regulatory T cells. Nature *441*, 235–238.

Boniface, K., Bak-Jensen, K.S., Li, Y., Blumenschein, W.M., McGeachy, M.J., McClanahan, T.K., McKenzie, B.S., Kastelein, R.A., Cua, D.J., and de Waal Malefyt, R. (2009). Prostaglandin E2 regulates Th17 cell differentiation and function through cyclic AMP and EP2/EP4 receptor signaling. J Exp Med *206*, 535–548.

Bouguermouh, S., Fortin, G., Baba, N., Rubio, M., and Sarfati, M. (2009). CD28 co-stimulation down regulates Th17 development. PLoS One *4*, e5087.

Brustle, A., Heink, S., Huber, M., Rosenplanter, C., Stadelmann, C., Yu, P., Arpaia, E., Mak, T.W., Kamradt, T., and Lohoff, M. (2007). The development of inflammatory T(H)-17 cells requires interferon-regulatory factor 4. Nat Immunol *8*, 958–966.

Chang, S.H., and Dong, C. (2007). A novel heterodimeric cytokine consisting of IL-17 and IL-17F regulates inflammatory responses. Cell Res *17*, 435–440.

Chen, C.W., Lin, T.Y., Chen, T.C., and Juang, J.L. (2005). Distinct translation regulation by two alternative 5'UTRs of a stress-responsive protein--dPrx I. J Biomed Sci *12*, 729–739.

Chen, Z., Laurence, A., Kanno, Y., Pacher-Zavisin, M., Zhu, B.M., Tato, C., Yoshimura, A., Hennighausen, L., and O'Shea, J.J. (2006). Selective regulatory function of Socs3 in the formation of IL-17-secreting T cells. Proc Natl Acad Sci USA *103*, 8137–8142.

Chen, Z., Tato, C.M., Muul, L., Laurence, A., and O'Shea, J.J. (2007). Distinct regulation of interleukin-17 in human T helper lymphocytes. Arthritis Rheum *56*, 2936–2946.

Chizzolini, C., Chicheportiche, R., Alvarez, M., de Rham, C., Roux-Lombard, P., Ferrari-Lacraz, S., and Dayer, J.M. (2008). Prostaglandin E2 synergistically with interleukin-23 favors human Th17 expansion. Blood *112*, 3696–3703.

Chung, Y., Chang, S.H., Martinez, G.J., Yang, X.O., Nurieva, R., Kang, H.S., Ma, L., Watowich, S.S., Jetten, A.M., Tian, Q., and Dong, C. (2009). Critical regulation of early Th17 cell differentiation by interleukin-1 signaling. Immunity *30*, 576–587.

Collins, A., Littman, D.R., and Taniuchi, I. (2009). RUNX proteins in transcription factor networks that regulate T-cell lineage choice. Nat Rev Immunol *9*, 106–115.

Coombes, J.L., Siddiqui, K.R., Arancibia-Carcamo, C.V., Hall, J., Sun, C.M., Belkaid, Y., and Powrie, F. (2007). A functionally specialized population of mucosal CD103+ DCs induces Foxp3+ regulatory T cells via a TGF-beta and retinoic acid-dependent mechanism. J Exp Med *204*, 1757–1764.

Coquet, J.M., Chakravarti, S., Smyth, M.J., and Godfrey, D.I. (2008). Cutting edge: IL-21 is not essential for Th17 differentiation or experimental auto-immune encephalomyelitis. J Immunol *180*, 7097–7101.

Cua, D.J., Sherlock, J., Chen, Y., Murphy, C.A., Joyce, B., Seymour, B., Lucian, L., To, W., Kwan, S., Churakova, T., *et al.* (2003). Interleukin-23 rather than interleukin-12 is the critical cytokine for auto-immune inflammation of the brain. Nature *421*, 744–748.

Das, J., Ren, G., Zhang, L., Roberts, A.I., Zhao, X., Bothwell, A.L., Van Kaer, L., Shi, Y., and Das, G. (2009). Transforming growth factor beta is dispensable for the molecular orchestration of Th17 cell differentiation. J Exp Med *206*, 2407–2416.

Djuretic, I.M., Levanon, D., Negreanu, V., Groner, Y., Rao, A., and Ansel, K.M. (2007). Transcription factors T-bet and Runx3 cooperate to activate Ifng and silence Il4 in T helper type 1 cells. Nat Immunol *8*, 145–153.

Durant, L., Watford, W.T., Ramos H.L., Laurence, A., Vahedi, G., Wei, L., Takahashi, H., Sun, H.W., Kanno, Y., Powrie, F., O'Shea, J.J. (2010). Diverse targets of the transcription factor STAT3 contribute to T cell pathogenicity and homeostasis. Immunity *32*, 605–15.

Du, C., Liu, C., Kang, J., Zhao, G., Ye, Z., Huang, S., Li, Z., Wu, Z., and Pei, G. (2009). MicroRNA miR-326 regulates TH-17 differentiation and is associated with the pathogenesis of multiple sclerosis. Nat Immunol *10*, 1252–1259.

Eberl, G., and Littman, D.R. (2003). The role of the nuclear hormone receptor RORgammat in the development of lymph nodes and Peyer's patches. Immunol Rev *195*, 81–90.

Elias, K.M., Laurence, A., Davidson, T.S., Stephens, G., Kanno, Y., Shevach, E.M., and O'Shea, J.J. (2008). Retinoic acid inhibits Th17 polarization and enhances FoxP3 expression through a Stat-3/Stat-5 independent signaling pathway. Blood *111*, 1013–1020.

Erb, K.J., Twardzik, T., Palmetshofer, A., Wohlleben, G., Tatsch, U., and Serfling, E. (2003). Mice deficient in nuclear factor of activated T-cell transcription factor c2 mount increased Th2 responses after infection with Nippostrongylus brasiliensis and decreased Th1 responses after mycobacterial infection. Infect Immun *71*, 6641–6647.

Esser, C., Rannug, A., and Stockinger, B. (2009). The aryl hydrocarbon receptor in immunity. Trends Immunol *30*, 447–454.

Galicia, G., Kasran, A., Uyttenhove, C., De Swert, K., Van Snick, J., and Ceuppens, J.L. (2009). ICOS deficiency results in exacerbated IL-17 mediated experimental auto-immune encephalomyelitis. J Clin Immunol *29*, 426–433.

Ghoreschi, K., Laurence, A., Yang, X.P., Tato, C.M., McGeachy, M.J., Konkel, J.E., Ramos, H.L., Wei, L., Davidson, T.S., Bouladoux, N. *et al.* (2010) Generation of pathogenic T(H)17 cells in the absence of TGF-β signalling. Nature *467*, 967–971.

Gomez-Rodriguez, J., Sahu, N., Handon, R., Davidson, T.S., Anderson, S.M., Kirby, M.R., August, A., and Schwartzberg, P.L. (2009). Differential expression of interleukin-17A and -17F is coupled to T cell receptor signaling via inducible T cell kinase. Immunity *31*, 587–597.

Gulen, M.F., Kang, Z., Bulek, K., Youzhong, W., Kim, T.W., Chen, Y., Altuntas, C.Z., Sass Bak-Jensen, K., McGeachy, M.J., Do, J.S., *et al.* (2010). The receptor SIGIRR suppresses Th17 cell proliferation via inhibition of the interleukin-1 receptor pathway and mTOR kinase activation. Immunity *32*, 54–66.

Hacker, H., and Karin, M. (2006). Regulation and function of IKK and IKK-related kinases. Sci STKE *2006*, re13.

Harrington, L.E., Hatton, R.D., Mangan, P.R., Turner, H., Murphy, T.L., Murphy, K.M., and Weaver, C.T. (2005). Interleukin 17-producing CD4+ effector T cells develop via a lineage distinct from the T helper type 1 and 2 lineages. Nat Immunol *6*, 1123–1132.

Harris, T.J., Grosso, J.F., Yen, H.R., Xin, H., Kortylewski, M., Albesiano, E., Hipkiss, E.L., Getnet, D., Goldberg, M.V., Maris, C.H., *et al.* (2007). Cutting edge: An in vivo requirement for STAT3 signaling in TH17 development and TH17-dependent autoimmunity. J Immunol *179*, 4313–4317.

Hashimoto, K., Durbin, J.E., Zhou, W., Collins, R.D., Ho, S.B., Kolls, J.K., Dubin, P.J., Sheller, J.R., Goleniewska, K., O'Neal, J.F., *et al.* (2005). Respiratory syncytial virus infection in the absence of STAT 1 results in airway dysfunction, airway mucus, and augmented IL-17 levels. J Allergy Clin Immunol *116*, 550–557.

Hegazy, A.N., Peine, M., Helmstetter, C., Panse, I., Frohlich, A., Bergthaler, A., Flatz, L., Pinschewer, D.D., Radbruch, A., and Lohning, M. (2010). Interferons direct Th2 cell reprogramming to generate a stable GATA-3(+)T-bet(+) cell subset with combined Th2 and Th1 cell functions. Immunity *32*, 116–128.

Heinrich, P.C., Behrmann, I., Haan, S., Hermanns, H.M., Muller-Newen, G., and Schaper, F. (2003). Principles of interleukin (IL)-6-type cytokine signalling and its regulation. Biochem J *374*, 1–20.

Hermann-Kleiter, N., Gruber, T., Lutz-Nicoladoni, C., Thuille, N., Fresser, F., Labi, V., Schiefermeier, N., Warnecke, M., Huber, L., Villunger, A., *et al.* (2008). The nuclear orphan receptor NR2F6 suppresses lymphocyte activation and T helper 17-dependent auto-immunity. Immunity *29*, 205–216.

Hildner, K.M., Schirmacher, P., Atreya, I., Dittmayer, M., Bartsch, B., Galle, P.R., Wirtz, S., and Neurath, M.F. (2007). Targeting of the transcription factor STAT4 by antisense phosphorothioate oligonucleotides suppresses collagen-induced arthritis. J Immunol *178*, 3427–3436.

Holland, S.M., DeLeo, F.R., Elloumi, H.Z., Hsu, A.P., Uzel, G., Brodsky, N., Freeman, A.F., Demidowich, A., Davis, J., Turner, M.L., *et al.* (2007). STAT3 mutations in the hyper-IgE syndrome. N Engl J Med *357*, 1608–1619.

Hu, C.M., Jang, S.Y., Fanzo, J.C., and Pernis, A.B. (2002). Modulation of T cell cytokine production by interferon regulatory factor-4. J Biol Chem *277*, 49238–49246.

Hu, H., Djuretic, I., Sundrud, M.S., and Rao, A. (2007). Transcriptional partners in regulatory T cells: Foxp3, Runx and NFAT. Trends Immunol *28*, 329–332.

Huang, M.C., Watson, S.R., Liao, J.J., and Goetzl, E.J. (2007). Th17 augmentation in OTII TCR plus T cell-selective type 1 sphingosine 1-phosphate receptor double transgenic mice. J Immunol *178*, 6806–6813.

Huang, W., Na, L., Fidel, P.L., and Schwarzenberger, P. (2004). Requirement of interleukin-17A for systemic anti-Candida albicans host defense in mice. J Infect Dis *190*, 624–631.

Ichiyama, K., Yoshida, H., Wakabayashi, Y., Chinen, T., Saeki, K., Nakaya, M., Takaesu, G., Hori, S., Yoshimura, A., and Kobayashi, T. (2008). Foxp3 inhibits RORgammat-mediated IL-17A mRNA transcription through direct interaction with RORgammat. J Biol Chem *283*, 17003–17008.

Ivanov, II, McKenzie, B.S., Zhou, L., Tadokoro, C.E., Lepelley, A., Lafaille, J.J., Cua, D.J., and Littman, D.R. (2006). The orphan nuclear receptor RORgammat directs the differentiation program of proinflammatory IL-17+ T helper cells. Cell *126*, 1121–1133.

Jetten, A.M. (2009). Retinoid-related orphan receptors (RORs): critical roles in development, immunity, circadian rhythm, and cellular metabolism. Nucl Recept Signal *7*, e003.

Jin, W., Zhou, X.F., Yu, J., Cheng, X., and Sun, S.C. (2009). Regulation of Th17 cell differentiation and EAE induction by MAP3K NIK. Blood *113*, 6603–6610.

Kerkvliet, N.I. (2002). Recent advances in understanding the mechanisms of TCDD immunotoxicity. Int Immunopharmacol *2*, 277–291.

Kimura, A., Naka, T., Nohara, K., Fujii-Kuriyama, Y., and Kishimoto, T. (2008). Aryl hydrocarbon receptor regulates Stat1 activation and participates in the development of Th17 cells. Proc Natl Acad Sci USA *105*, 9721–9726.

King, C., Tangye, S.G., and Mackay, C.R. (2008). T follicular helper (TFH) cells in normal and dysregulated immune responses. Annu Rev Immunol *26*, 741–766.

Kishimoto, T. (2005). Interleukin-6: from basic science to medicine--40 years in immunology. Annu Rev Immunol *23*, 1–21.

Klein, M., Klein-Hessling, S., Palmetshofer, A., Serfling, E., Tertilt, C., Bopp, T., Heib, V., Becker, M., Taube, C., Schild, H., *et al.* (2006). Specific and redundant roles for NFAT transcription factors in the expression of mast cell-derived cytokines. J Immunol *177*, 6667–6674.

Klotz, L., Burgdorf, S., Dani, I., Saijo, K., Flossdorf, J., Hucke, S., Alferink, J., Novak, N., Beyer, M., Mayer, G., *et al.* (2009). The nuclear receptor PPAR{gamma} selectively inhibits Th17 differentiation in a T cell-intrinsic fashion and suppresses CNS autoimmunity. J Exp Med.

Korn, T., Bettelli, E., Gao, W., Awasthi, A., Jager, A., Strom, T.B., Oukka, M., and Kuchroo, V.K. (2007). IL-21 initiates an alternative pathway to induce proinflammatory T(H)17 cells. Nature *448*, 484–487.

Kouzarides, T. (2007). Chromatin modifications and their function. Cell *128*, 693–705.

Kryczek, I., Wei, S., Zou, L., Altuwaijri, S., Szeliga, W., Kolls, J., Chang, A., and Zou, W. (2007). Cutting edge: Th17 and regulatory T cell dynamics and the regulation by IL-2 in the tumor microenvironment. J Immunol *178*, 6730–6733.

Kurebayashi, S., Ueda, E., Sakaue, M., Patel, D.D., Medvedev, A., Zhang, F., and Jetten, A.M. (2000). Retinoid-related orphan receptor gamma (RORgamma) is essential for lymphoid organogenesis and controls apoptosis during thymopoiesis. Proc Natl Acad Sci USA *97*, 10132–10137.

Langrish, C.L., Chen, Y., Blumenschein, W.M., Mattson, J., Basham, B., Sedgwick, J.D., McClanahan, T., Kastelein, R.A., and Cua, D.J. (2005). IL-23 drives a pathogenic T cell population that induces auto-immune inflammation. J Exp Med *201*, 233–240.

Laurence, A., Tato, C.M., Davidson, T.S., Kanno, Y., Chen, Z., Yao, Z., Blank, R.B., Meylan, F., Siegel, R., Hennighausen, L., *et al.* (2007). Interleukin-2 signaling via STAT5 constrains T helper 17 cell generation. Immunity *26*, 371–381.

Lee, G.R., Kim, S.T., Spilianakis, C.G., Fields, P.E., and Flavell, R.A. (2006). T helper cell differentiation: regulation by cis elements and epigenetics. Immunity *24*, 369–379.

Lee, Y.K., Turner, H., Maynard, C.L., Oliver, J.R., Chen, D., Elson, C.O., and Weaver, C.T. (2009). Late developmental plasticity in the T helper 17 lineage. Immunity *30*, 92–107.

Letterio, J.J., and Roberts, A.B. (1998). Regulation of immune responses by TGF-beta. Annu Rev Immunol *16*, 137–161.

Lexberg, M.H., Taubner, A., Forster, A., Albrecht, I., Richter, A., Kamradt, T., Radbruch, A., and Chang, H.D. (2008). Th memory for interleukin-17 expression is stable in vivo. Eur J Immunol *38*, 2654–2664.

Lezzi, G., Sonderegger, I., Ampenberger, F., Schmitz, N., Marsland, B.J., and Kopf, M. (2009). CD40-CD40L cross-talk integrates strong antigenic signals and microbial stimuli to induce development of IL-17-producing CD4+ T cells. Proc Natl Acad Sci USA *106*, 876–881.

Li, B., Reynolds, J.M., Stout, R.D., Bernlohr, D.A., and Suttles, J. (2009). Regulation of Th17 differentiation by epidermal fatty acid-binding protein. J Immunol *182*, 7625–7633.

Li, M.O., Wan, Y.Y., and Flavell, R.A. (2007). T cell-produced transforming growth factor-beta1 controls T cell tolerance and regulates Th1- and Th17-cell differentiation. Immunity *26*, 579–591.

Liang, S.C., Long, A.J., Bennett, F., Whitters, M.J., Karim, R., Collins, M., Goldman, S.J., Dunussi-Joannopoulos, K., Williams, C.M., Wright, J.F., and Fouser, L.A. (2007). An IL-17F/A heterodimer protein is produced by mouse Th17 cells and induces airway neutrophil recruitment. J Immunol *179*, 7791–7799.

Liao, J.J., Huang, M.C., and Goetzl, E.J. (2007). Cutting edge: Alternative signaling of Th17 cell development by sphingosine 1-phosphate. J Immunol *178*, 5425–5428.

Liu, X., Lee, Y.S., Yu, C.R., and Egwuagu, C.E. (2008). Loss of STAT3 in CD4+ T cells prevents development of experimental auto-immune diseases. J Immunol *180*, 6070–6076.

Liu, X.K., Lin, X., and Gaffen, S.L. (2004). Crucial role for nuclear factor of activated T cells in T cell receptor-mediated regulation of human interleukin-17. J Biol Chem *279*, 52762–52771.

Lohoff, M., Mittrucker, H.W., Prechtl, S., Bischof, S., Sommer, F., Kock, S., Ferrick, D.A., Duncan, G.S., Gessner, A., and Mak, T.W. (2002). Dysregulated T helper cell differentiation in the absence of interferon regulatory factor 4. Proc Natl Acad Sci USA *99*, 11808–11812.

Ma, C.S., Chew, G.Y., Simpson, N., Priyadarshi, A., Wong, M., Grimbacher, B., Fulcher, D.A., Tangye, S.G., and Cook, M.C. (2008). Deficiency of Th17 cells in hyper IgE syndrome due to mutations in STAT3. J Exp Med *205*, 1551–1557.

Macian, F. (2005). NFAT proteins: key regulators of T-cell development and function. Nat Rev Immunol *5*, 472–484.

Malek, T.R. (2008). The biology of interleukin-2. Annu Rev Immunol *26*, 453–479.

Manel, N., Unutmaz, D., and Littman, D.R. (2008). The differentiation of human T(H)-17 cells requires transforming growth factor-beta and induction of the nuclear receptor RORgammat. Nat Immunol *9*, 641–649.

Mangan, P.R., Harrington, L.E., O'Quinn, D.B., Helms, W.S., Bullard, D.C., Elson, C.O., Hatton, R.D., Wahl, S.M., Schoeb, T.R., and Weaver, C.T. (2006). Transforming growth factor-beta induces development of the T(H)17 lineage. Nature *441*, 231–234.

Mardis, E.R. (2007). ChIP-seq: welcome to the new frontier. Nat Methods *4*, 613–614.

Martin, B., Hirota, K., Cua, D.J., Stockinger, B., and Veldhoen, M. (2009). Interleukin-17-producing gammadelta T cells selectively expand in response to pathogen products and environmental signals. Immunity *31*, 321–330.

Mathur, A.N., Chang, H.C., Zisoulis, D.G., Stritesky, G.L., Yu, Q., O'Malley, J.T., Kapur, R., Levy, D.E., Kansas, G.S., and Kaplan, M.H. (2007). Stat3 and Stat4 direct development of IL-17-secreting Th cells. J Immunol *178*, 4901–4907.

McGeachy, M.J., Bak-Jensen, K.S., Chen, Y., Tato, C.M., Blumenschein, W., McClanahan, T., and Cua, D.J. (2007). TGF-beta and IL-6 drive the production of IL-17 and IL-10 by T cells and restrain T(H)-17 cell-mediated pathology. Nat Immunol *8*, 1390–1397.

McGeachy, M.J., Chen, Y., Tato, C.M., Laurence, A., Joyce-Shaikh, B., Blumenschein, W.M., McClanahan, T.K., O'Shea, J.J., and Cua, D.J. (2009). The interleukin 23 receptor is essential for the terminal differentiation of interleukin 17-producing effector T helper cells in vivo. Nat Immunol *10*, 314–324.

Michel, M.L., Keller, A.C., Paget, C., Fujio, M., Trottein, F., Savage, P.B., Wong, C.H., Schneider, E., Dy, M., and Leite-de-Moraes, M.C. (2007). Identification of an IL-17-producing NK1.1(neg) iNKT cell population involved in airway neutrophilia. J Exp Med *204*, 995–1001.

Milner, J.D., Brenchley, J.M., Laurence, A., Freeman, A.F., Hill, B.J., Elias, K.M., Kanno, Y., Spalding, C., Elloumi, H.Z., Paulson, M.L., *et al.* (2008). Impaired T(H)17 cell differentiation in subjects with autosomal dominant hyper-IgE syndrome. Nature *452*, 773–776.

Minegishi, Y., Saito, M., Tsuchiya, S., Tsuge, I., Takada, H., Hara, T., Kawamura, N., Ariga, T., Pasic, S., Stojkovic, O., *et al.* (2007). Dominant-negative mutations in the DNA-binding domain of STAT3 cause hyper-IgE syndrome. Nature *448*, 1058–1062.

Moisan, J., Grenningloh, R., Bettelli, E., Oukka, M., and Ho, I.C. (2007). Ets-1 is a negative regulator of Th17 differentiation. J Exp Med *204*, 2825–2835.

Mosmann, T.R., Cherwinski, H., Bond, M.W., Giedlin, M.A., and Coffman, R.L. (1986). Two types of murine helper T cell clone. I. Definition according to profiles of lymphokine activities and secreted proteins. J Immunol *136*, 2348–2357.

Mosmann, T.R., and Coffman, R.L. (1989). TH1 and TH2 cells: different patterns of lymphokine secretion lead to different functional properties. Annu Rev Immunol *7*, 145–173.

Mucida, D., Park, Y., Kim, G., Turovskaya, O., Scott, I., Kronenberg, M., and Cheroutre, H. (2007). Reciprocal TH17 and regulatory T cell differentiation mediated by retinoic acid. Science *317*, 256–260.

Murphy, C.A., Langrish, C.L., Chen, Y., Blumenschein, W., McClanahan, T., Kastelein, R.A., Sedgwick, J.D., and Cua, D.J. (2003). Divergent pro- and anti-inflammatory roles for IL-23 and IL-12 in joint auto-immune inflammation. J Exp Med *198*, 1951–1957.

Naoe, Y., Setoguchi, R., Akiyama, K., Muroi, S., Kuroda, M., Hatam, F., Littman, D.R., and Taniuchi, I. (2007). Repression of interleukin-4 in T helper type 1 cells by Runx/Cbf beta binding to the Il4 silencer. J Exp Med *204*, 1749–1755.

Nurieva, R., Yang, X.O., Martinez, G., Zhang, Y., Panopoulos, A.D., Ma, L., Schluns, K., Tian, Q., Watowich, S.S., Jetten, A.M., and Dong, C. (2007). Essential autocrine regulation by IL-21 in the generation of inflammatory T cells. Nature *448*, 480–483.

O'Neill, L.A. (2008). The interleukin-1 receptor/Toll-like receptor superfamily: 10 years of progress. Immunol Rev *226*, 10–18.

O'Shea, J.J., and Paul, W.E (2010). Mechanisms underlying lineage commitment and plasticity of helper CD4+ T cells. Science *327*, 1098–1102.

Oppmann, B., Lesley, R., Blom, B., Timans, J.C., Xu, Y., Hunte, B., Vega, F., Yu, N., Wang, J., Singh, K., *et al.* (2000). Novel p19 protein engages IL-12p40 to form a cytokine, IL-23, with biological activities similar as well as distinct from IL-12. Immunity *13*, 715–725.

Park, H., Li, Z., Yang, X.O., Chang, S.H., Nurieva, R., Wang, Y.H., Wang, Y., Hood, L., Zhu, Z., Tian, Q., and Dong, C. (2005). A distinct lineage of CD4 T cells regulates tissue inflammation by producing interleukin 17. Nat Immunol *6*, 1133–1141.

Park, P.J. (2009). ChIP-seq: advantages and challenges of a maturing technology. Nat Rev Genet *10*, 669–680.

Pflanz, S., Timans, J.C., Cheung, J., Rosales, R., Kanzler, H., Gilbert, J., Hibbert, L., Churakova, T., Travis, M., Vaisberg, E., *et al.* (2002). IL-27, a heterodimeric cytokine composed of EBI3 and p28 protein, induces proliferation of naive CD4(+) T cells. Immunity *16*, 779–790.

Quintana, F.J., Basso, A.S., Iglesias, A.H., Korn, T., Farez, M.F., Bettelli, E., Caccamo, M., Oukka, M., and Weiner, H.L. (2008). Control of T(reg) and T(H)17 cell differentiation by the aryl hydrocarbon receptor. Nature *453*, 65–71.

Rengarajan, J., Mowen, K.A., McBride, K.D., Smith, E.D., Singh, H., and Glimcher, L.H. (2002a). Interferon regulatory factor 4 (IRF4) interacts with NFATc2 to modulate interleukin 4 gene expression. J Exp Med *195*, 1003–1012.

Rengarajan, J., Tang, B., and Glimcher, L.H. (2002b). NFATc2 and NFATc3 regulate T(H)2 differentiation and modulate TCR-responsiveness of naive T(H)cells. Nat Immunol *3*, 48–54.

Rochman, Y., Spolski, R., and Leonard, W.J. (2009). New insights into the regulation of T cells by gamma(c) family cytokines. Nat Rev Immunol *9*, 480–490.

Rosen, H., and Goetzl, E.J. (2005). Sphingosine 1-phosphate and its receptors: an autocrine and paracrine network. Nat Rev Immunol *5*, 560–570.

Saeki, K., Fukuyama, S., Ayada, T., Nakaya, M., Aki, D., Takaesu, G., Hanada, T., Matsumura, Y., Kobayashi, T., Nakagawa, R., and Yoshimura, A. (2009). A major lipid raft protein raftlin modulates T cell receptor signaling and enhances th17-mediated auto-immune responses. J Immunol *182*, 5929–5937.

Sakaguchi, S., Yamaguchi, T., Nomura, T., and Ono, M. (2008). Regulatory T cells and immune tolerance. Cell *133*, 775–787.

Schraml, B.U., Hildner, K., Ise, W., Lee, W.L., Smith, W.A., Solomon, B., Sahota, G., Sim, J., Mukasa, R., Cemerski, S., *et al.* (2009). The AP-1 transcription factor Batf controls T(H)17 differentiation. Nature *460*, 405–409.

Setoguchi, R., Tachibana, M., Naoe, Y., Muroi, S., Akiyama, K., Tezuka, C., Okuda, T., and Taniuchi, I. (2008). Repression of the transcription factor Th-POK by Runx complexes in cytotoxic T cell development. Science *319*, 822–825.

Sheibanie, A.F., Khayrullina, T., Safadi, F.F., and Ganea, D. (2007). Prostaglandin E2 exacerbates collagen-induced arthritis in mice through the inflammatory interleukin-23/interleukin-17 axis. Arthritis Rheum *56*, 2608–2619.

Shi, G., Cox, C.A., Vistica, B.P., Tan, C., Wawrousek, E.F., and Gery, I. (2008). Phenotype switching by inflammation-inducing polarized Th17 cells, but not by Th1 cells. J Immunol *181*, 7205–7213.

Shull, M.M., Ormsby, I., Kier, A.B., Pawlowski, S., Diebold, R.J., Yin, M., Allen, R., Sidman, C., Proetzel, G., Calvin, D., and et al. (1992). Targeted disruption of the mouse transforming growth factor-beta 1 gene results in multi-focal inflammatory disease. Nature *359*, 693–699.

Smith-Garvin, J.E., Koretzky, G.A., and Jordan, M.S. (2009). T cell activation. Annu Rev Immunol *27*, 591–619.

Sonderegger, I., Kisielow, J., Meier, R., King, C., and Kopf, M. (2008). IL-21 and IL-21R are not required for development of Th17 cells and autoimmunity in vivo. Eur J Immunol *38*, 1833–1838.

Stark, M.A., Huo, Y., Burcin, T.L., Morris, M.A., Olson, T.S., and Ley, K. (2005). Phagocytosis of apoptotic neutrophils regulates granulopoiesis via IL-23 and IL-17. Immunity *22*, 285–294.

Stumhofer, J.S., Laurence, A., Wilson, E.H., Huang, E., Tato, C.M., Johnson, L.M., Villarino, A.V., Huang, Q., Yoshimura, A., Sehy, D., *et al.* (2006). Interleukin 27 negatively regulates the development of interleukin 17-producing T helper cells during chronic inflammation of the central nervous system. Nat Immunol *7*, 937–945.

Sun, C.M., Hall, J.A., Blank, R.B., Bouladoux, N., Oukka, M., Mora, J.R., and Belkaid, Y. (2007). Small intestine lamina propria dendritic cells promote de novo generation of Foxp3 T reg cells via retinoic acid. J Exp Med *204*, 1775–1785.

Sun, Z., Unutmaz, D., Zou, Y.R., Sunshine, M.J., Pierani, A., Brenner-Morton, S., Mebius, R.E., and Littman, D.R. (2000). Requirement for RORgamma in thymocyte survival and lymphoid organ development. Science *288*, 2369–2373.

Sutton, C., Brereton, C., Keogh, B., Mills, K.H., and Lavelle, E.C. (2006). A crucial role for interleukin (IL)-1 in the induction of IL-17-producing T cells that mediate auto-immune encephalomyelitis. J Exp Med *203*, 1685–1691.

Sutton, C.E., Lalor, S.J., Sweeney, C.M., Brereton, C.F., Lavelle, E.C., and Mills, K.H. (2009). Interleukin-1 and IL-23 induce innate IL-17 production from gammadelta T cells, amplifying Th17 responses and auto-immunity. Immunity *31*, 331–341.

Szanto, A., and Nagy, L. (2008). The many faces of PPARgamma: anti-inflammatory by any means? Immunobiology *213*, 789–803.

Taniuchi, I., Osato, M., Egawa, T., Sunshine, M.J., Bae, S.C., Komori, T., Ito, Y., and Littman, D.R. (2002). Differential requirements for Runx proteins in CD4 repression and epigenetic silencing during T lymphocyte development. Cell *111*, 621–633.

Veldhoen, M., Hirota, K., Christensen, J., O'Garra, A., and Stockinger, B. (2009). Natural agonists for aryl hydrocarbon receptor in culture medium are essential for optimal differentiation of Th17 T cells. J Exp Med *206*, 43–49.

Veldhoen, M., Hirota, K., Westendorf, A.M., Buer, J., Dumoutier, L., Renauld, J.C., and Stockinger, B. (2008). The aryl hydrocarbon receptor links TH17-cell-mediated auto-immunity to environmental toxins. Nature *453*, 106–109.

Veldhoen, M., Hocking, R.J., Atkins, C.J., Locksley, R.M., and Stockinger, B. (2006). TGFbeta in the context of an inflammatory cytokine milieu supports de novo differentiation of IL-17-producing T cells. Immunity *24*, 179–189.

Wagner, E.F., and Eferl, R. (2005). Fos/AP-1 proteins in bone and the immune system. Immunol Rev *208*, 126–140.

Watford, W.T., Hissong, B.D., Bream, J.H., Kanno, Y., Muul, L., and O'Shea, J.J. (2004). Signaling by IL-12 and IL-23 and the immuno-regulatory roles of STAT4. Immunol Rev *202*, 139–156.

Wei, G., Wei, L., Zhu, J., Zang, C., Hu-Li, J., Yao, Z., Cui, K., Kanno, Y., Roh, T.Y., Watford, W.T., *et al.* (2009). Global mapping of H3K4me3 and H3K27me3 reveals specificity and plasticity in lineage fate determination of differentiating CD4+ T cells. Immunity *30*, 155–167.

Wei, L., Laurence, A., Elias, K.M., and O'Shea, J.J. (2007). IL-21 is produced by Th17 cells and drives IL-17 production in a STAT3-dependent manner. J Biol Chem *282*, 34605–34610.

Weigmann, B., Lehr, H.A., Yancopoulos, G., Valenzuela, D., Murphy, A., Stevens, S., Schmidt, J., Galle, P.R., Rose-John, S., and Neurath, M.F. (2008). The transcription factor NFATc2 controls IL-6-dependent T cell activation in experimental colitis. J Exp Med *205*, 2099–2110.

Wilson, C.B., Rowell, E., and Sekimata, M. (2009). Epigenetic control of T-helper-cell differentiation. Nat Rev Immunol *9*, 91–105.

Wilson, N.J., Boniface, K., Chan, J.R., McKenzie, B.S., Blumenschein, W.M., Mattson, J.D., Basham, B., Smith, K., Chen, T., Morel, F., *et al.* (2007). Development, cytokine profile and function of human interleukin 17-producing helper T cells. Nat Immunol *8*, 950–957.

Wong, P.K., Egan, P.J., Croker, B.A., O'Donnell, K., Sims, N.A., Drake, S., Kiu, H., McManus, E.J., Alexander, W.S., Roberts, A.W., and Wicks, I.P. (2006). SOCS-3 negatively regulates

innate and adaptive immune mechanisms in acute IL-1-dependent inflammatory arthritis. J Clin Invest *116*, 1571–1581.
Yang, L., Anderson, D.E., Baecher-Allan, C., Hastings, W.D., Bettelli, E., Oukka, M., Kuchroo, V.K., and Hafler, D.A. (2008a). IL-21 and TGF-beta are required for differentiation of human T(H)17 cells. Nature *454*, 350–352.
Yang, X., Letterio, J.J., Lechleider, R.J., Chen, L., Hayman, R., Gu, H., Roberts, A.B., and Deng, C. (1999). Targeted disruption of SMAD3 results in impaired mucosal immunity and diminished T cell responsiveness to TGF-beta. EMBO J *18*, 1280–1291.
Yang, X.O., Panopoulos, A.D., Nurieva, R., Chang, S.H., Wang, D., Watowich, S.S., and Dong, C. (2007). STAT3 regulates cytokine-mediated generation of inflammatory helper T cells. J Biol Chem *282*, 9358–9363.
Yang, X.O., Pappu, B.P., Nurieva, R., Akimzhanov, A., Kang, H.S., Chung, Y., Ma, L., Shah, B., Panopoulos, A.D., Schluns, K.S., *et al.* (2008b). T helper 17 lineage differentiation is programmed by orphan nuclear receptors ROR alpha and ROR gamma. Immunity *28*, 29–39.
Yang, Y., Weiner, J., Liu, Y., Smith, A.J., Huss, D.J., Winger, R., Peng, H., Cravens, P.D., Racke, M.K., and Lovett-Racke, A.E. (2009). T-bet is essential for encephalitogenicity of both Th1 and Th17 cells. J Exp Med *206*, 1549–1564.
Yao, C., Sakata, D., Esaki, Y., Li, Y., Matsuoka, T., Kuroiwa, K., Sugimoto, Y., and Narumiya, S. (2009). Prostaglandin E2-EP4 signaling promotes immune inflammation through Th1 cell differentiation and Th17 cell expansion. Nat Med *15*, 633–640.
Yao, Z., Kanno, Y., Kerenyi, M., Stephens, G., Durant, L., Watford, W.T., Laurence, A., Robinson, G.W., Shevach, E.M., Moriggl, R., *et al.* (2007). Nonredundant roles for Stat5a/b in directly regulating Foxp3. Blood *109*, 4368–4375.
Yoshimura, A., Naka, T., and Kubo, M. (2007). SOCS proteins, cytokine signalling and immune regulation. Nat Rev Immunol *7*, 454–465.
Zhang, F., Meng, G., and Strober, W. (2008). Interactions among the transcription factors Runx1, RORgammat and Foxp3 regulate the differentiation of interleukin 17-producing T cells. Nat Immunol *9*, 1297–1306.
Zheng, Y., Valdez, P.A., Danilenko, D.M., Hu, Y., Sa, S.M., Gong, Q., Abbas, A.R., Modrusan, Z., Ghilardi, N., de Sauvage, F.J., and Ouyang, W. (2008). Interleukin-22 mediates early host defense against attaching and effacing bacterial pathogens. Nat Med *14*, 282–289.
Zhou, L., Ivanov, II, Spolski, R., Min, R., Shenderov, K., Egawa, T., Levy, D.E., Leonard, W.J., and Littman, D.R. (2007). IL-6 programs T(H)-17 cell differentiation by promoting sequential engagement of the IL-21 and IL-23 pathways. Nat Immunol *8*, 967–974.
Zhou, L., Lopes, J.E., Chong, M.M., Ivanov, II, Min, R., Victora, G.D., Shen, Y., Du, J., Rubtsov, Y.P., Rudensky, A.Y., *et al.* (2008). TGF-beta-induced Foxp3 inhibits T(H)17 cell differentiation by antagonizing RORgammat function. Nature *453*, 236–240.
Zhu, J., and Paul, W.E. (2008). CD4 T cells: fates, functions, and faults. Blood *112*, 1557–1569.

Structure and IL-17 Receptor Signaling

Michael Tsang and Sarah L. Gaffen

Abstract The IL-17/IL-17 receptor family is the newest family of cytokine systems to be defined. Although considerable progress has been made in recent years to decipher the structure-function relationships and downstream signaling events mediated by this family, there are still many gaps in our understanding of this area. This chapter will provide an overview of the molecular biology of the IL-17R family, with a major focus on the IL-17A/F and IL-17RA/IL-17RC families. Some discussion of the IL-17E/IL-25 and IL-17RB system and the SEF/IL-17RD system will also be reviewed.

1 Introduction to IL-17 and the IL-17R Families

As noted in earlier sections of this book, the IL-17 cytokine family came into prominence in 2005 with the seminal discovery of the T_H17 lineage (Korn et al. 2009; Steinman 2007). However, IL-17 (IL-17A) was originally cloned in 1993, termed CTLA-8 (Rouvier et al. 1993) (Table 1). Interestingly, CTLA-8 was homologous to a viral open reading frame located in the genome of *Herpesvirus saimiri*, a T cell tropic γ-herpesvirus. Such homology presumably indicates that this IL-17-like viral protein is beneficial for some aspect of viral pathogenesis, but the biological significance of this homology remains unclear.

The gene encoding IL-17A was noteworthy even at the start due to its distinct sequence structure and lack of homology with known cytokines. Similarly, the first IL-17A receptor was notably different from all cytokine receptors characterized

S.L. Gaffen (✉)
Division of Rheumatology & Clinical Immunology, Department of Medicine,
University of Pittsburgh, BST S708, 3500 Terrace Street, Pittsburgh, PA 15261, USA
e-mail: sig65@pitt.edu

S. Jiang (ed.), *T_H17 Cells in Health and Disease*,
DOI 10.1007/978-1-4419-9371-7_9, © Springer Science+Business Media, LLC 2011

Table 1 Timeline of major discoveries in defining IL-17A/IL-17RA signaling pathways

Year	Discovery	References
1993	CTLA-8 (later termed IL-17 or IL-17A) cloned from rodent cDNA library, homology to *Herpesvirus saimiri* gene noted	Rouvier et al. (1993)
1995	IL-17R (later termed IL-17RA) cloned, recognized as a unique type of cytokine receptor, shown to activate NF-κB	Yao et al. (1995a)
2000	IL-17RA shown to signal through TRAF6 to induce NF-κB	Schwandner et al. (2000)
2001	Crystal structure of IL-17F solved, found to be member of cysteine knot family cytokines	Hymowitz et al. (2001)
2003	Bioinformatics study identified the SEFIR domain, a subdomain the IL-17R family cytoplasmic tails with homology to the Toll-IL-1 receptor family	Novatchkova et al. (2003)
2004	IL-17 first shown to signal through C/EBPβ and C/EBPδ transcription factors	Ruddy et al. (2004)
2006	IL-17RC shown to be obligate co-receptor for IL-17A and IL-17F signaling	Toy et al. (2006)
2006, 2007	IL-17RA shown to interact with Act1, which is upstream of NF-κB and TRAF6 activation	Chang et al. (2006), Qian et al. (2007)
2007	First structure-function analysis of IL-17RA cytoplasmic tail reported	Maitra et al. (2007)
2008	IL-17RA shown to partner with IL-17RB to mediate signaling through IL-17E/IL-25	Rickel et al. (2008)
2009	IL-17 shown to synergize with BAFF in B cells via competition for Act1 and activation of Twist1	Doreau et al. (2009)
2009	Crystal structure of IL-17RA with IL-17F solved	Ely et al. (2009)

to date, and could not be classified within any of the prototypical receptor subfamilies (Yao et al. 1995a). As cytokines with significant homology to IL-17A were identified in the ensuing years, it became clear that this represented a bona fide sub-family of cytokines and cognate receptors (Aggarwal and Gurney 2002). Moreover, this unique sequence structure also suggested that there might be new modalities of signal transduction mediated by this family, which is now indeed recognized to be the case.

There are now six known mammalian IL-17 ligands (IL-17A-F) and five receptors (IL-17RA-RE) (Table 2) (Gaffen 2009). Considerable progress in defining the ligand-receptor relationships within this family has been made, but there are still remain orphan receptors and ligands. In addition, it is still not clear in all cases whether all co-receptors have been identified. This chapter will focus primarily on the IL-17A/F-IL-17RA/IL-17RC system, since that is by far the best characterized in terms of molecular structure and signal transduction. Some recent insights into IL-17E (IL-25)-IL-17RA/IL-17RB signaling and SEF (IL-17RD) will also be discussed.

Table 2 IL-17 ligand and receptor relationships

Ligand	Receptor(s)	Major references
IL-17A	IL-17RA/IL-17RC	Kuestner et al. (2007), Rouvier et al. (1993), Toy et al. (2006), Yao et al. (1995a)
IL-17A/F	IL-17RA/IL-17RC	Chang and Dong 2007, Wright et al. (2007)
IL-17F	IL-17RA/IL-17RC	Kuestner et al. (2007), Starnes et al. (2001)
IL-17B	IL-17RB	Li et al. (2000), Shi et al. (2000)
IL-17C	IL-17RE	Li et al. (2000)
IL-17D	Unknown	Starnes et al. (2002)
IL-17E/IL-25	IL-17RA/IL-17RB	Lee et al. (2001), Rickel et al. (2008)
Unknown	IL-17RD (SEF)/FGFR	Tsang et al. (2002)
vIL-17 (*H. saimiri*)	IL-17RA/IL-17RC?	Yao et al. (1995b)

2 Structure and Receptor Dynamics of the IL-17 Receptor Complex

Upon its discovery, murine IL-17RA was found to bind to the human, mouse and viral forms of IL-17A (Yao et al. 1995a). This receptor was striking with respect to its lack of homology with other known receptors, and raised the possibility that this molecule might engage novel modes of signaling. The extracellular domain was later recognized to have homology to fibronectin III-like (FNIII) domains (Kramer et al. 2007; Novatchkova et al. 2003), which are commonly found in Type I and Type II hematopoietic receptors and (Murphy and Young 2006). The cytoplasmic tail is unusually large (525 amino acids) and at the time had no detectable similarities with other known receptors (Yao et al. 1997). IL-17RA is ubiquitously expressed, with particularly high expression in hematopoietic tissue. However, the main responder cells appeared from early studies to be non-hematopoietic tissue cells, including epithelial, endothelial and mesenchymal cell types [reviewed in (Onishi and Gaffen 2010)].

Data from early studies suggested that IL-17RA might pair with an undefined co-receptor, since the concentration of IL-17A required for signaling was less than that needed to effectively bind based on affinity calculations (Yao et al. 1995a). Indeed, in 2006 Toy et al. (Toy et al. 2006) showed that fibroblasts from IL-17RA$^{-/-}$ mice (Ye et al. 2001) could not restore IL-17A-dependent signaling when reconstituted with human IL-17RA, whereas co-expression with human IL-17RC could rescue this defect. They also showed co-immuno-precipitation of IL-17RA with IL-17RC, supporting a physical association of these receptors. These findings further suggested that IL-17RA pairs with IL-17RC to mediate signal transduction, and that human IL-17RA cannot effectively pair with the mouse IL-17RC that is endogenous to fibroblasts (Toy et al. 2006).

Although a member of the IL-17R sub-family of receptors, IL-17RC [originally termed IL-17RL (Haudenschild et al. 2002)] has many unique features, and is still not well characterized [reviewed in (Ho and Gaffen 2010; Moseley et al. 2003)]. Unlike IL-17RA, IL-17RC exists in multiple splice forms, only some of which bind to IL-17A and IL-17F (Haudenschild et al. 2006; Kuestner et al. 2007). The expression patterns of IL-17RC are surprisingly distinct from IL-17RA, perhaps implying that this receptor may bind a different ligand. The cytoplasmic tail is much shorter than that of IL-17RA, but is still necessary for mediating signal transduction (discussed in more depth below) (Toy et al. 2006). In humans, the full length isoform of IL-17RC binds to both IL-17A and IL-17F, whereas in mice it directly contacts only IL-17F (Kuestner et al. 2007). However, the limited studies on the IL-17RC$^{-/-}$ mice reported to date indicate that there is no detectable IL-17-dependent signaling (Zheng et al. 2008) (and SLG, unpublished data).

IL-17F, which has the closest sequence similarity to IL-17A (55%), also requires IL-17RA to mediate signal transduction, although its effects are quantitatively much weaker. Studies in knockout mice have revealed qualitative differences as well (Ishigame et al. 2009; Yang et al. 2008), and no doubt further distinctions will be made in the future. The crystal structure of IL-17F revealed this family to be a member of the cystine-knot family of growth factors, akin to PDGF and NGF (Hymowitz et al. 2001). The salient features of the other IL-17 family members such as cysteine placement are conserved, indicating this is true for the entire subgroup. IL-17A and IL-17F exist as homodimers (Fossiez et al. 1996), and so it was not unexpected to find that they can also form heterodimers. The IL-17A/F homodimers also use both IL-17RA and IL-17RC to mediate signaling, and generally have a potency that is intermediate between IL-17A and IL-17F (Chang and Dong 2007; Wright et al. 2008; Wright et al. 2007).

IL-17RA appears to participate in multiple receptor complexes, suggesting it may function in a manner similar to shared receptor sub-units in other cytokine subfamilies, such as the "common β" (βc), "common γ" (γc) and gp130 sub-units (Ozaki and Leonard 2002) (Fig. 1). For example, it was unexpectedly revealed that IL-17E (better known as IL-25, Table 2) uses IL-17RA in combination with IL-17RB (IL-25R) to signal, as both IL-17RB$^{-/-}$ and IL-17RA$^{-/-}$ mice are defective in responses to IL-25 (Rickel et al. 2008). An overexpression study suggested that IL-17RA co-immuno-precipitates with IL-17RD (also known as SEF, see subsequent sections) and may help to mediate signaling through SEF (Rong et al. 2009). However, this has not been independently verified, nor is it consistent with data in IL-17RD$^{-/-}$ mice (Gaffen 2009).

The stoichiometry and receptor sub-unit dynamics within the various complexes with which IL-17RA participates have not been fully defined. The classic view of cytokine receptors is that the constituent sub-units are monomeric on the plasma membrane surface, and are induced to form a multimeric complex only in the presence of ligand. However, studies using fluorescence resonance energy transfer (FRET) to analyze fluorescently-tagged receptor sub-units have often challenged this model. In the TNFR and TLR systems, for example, FRET was used to demonstrate pre-assembly of constituent sub-units, and revealed that addition of ligand strengthens this association, in some cases leading to a re-organization of the sub-units permitting efficient signal transduction (Chan 2007; Chan et al. 2000; Latz et al. 2007).

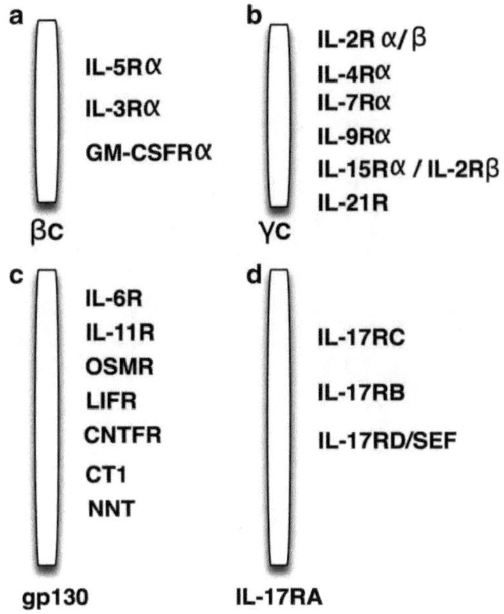

Fig. 1 Shared cytokine receptor families. Many Type I hematopoietin receptors use common signaling receptors. (**a**) The βc family. (**b**) The γc/IL-2 family. (**c**) The gp130 receptor family. (**d**) The IL-17RA family

A FRET analysis of murine IL-17RA (performed before IL-17RC was discovered) indicated that at least some molecules of this receptor exist at the membrane as a pre-assembled multimer (Kramer et al. 2006). Following ligand binding, however, FRET signals between IL-17RA sub-units were reduced, indicating that the receptor undergoes some sort of major conformational alteration. This could be due to replacement of one of the IL-17RA sub-units with another sub-unit (such as IL-17RC), rotation of the receptor sub-units with respect to one another, physical distancing of the receptors to accommodate a signaling molecule (such as Act1, see below), or a combination of the above.

In 2009, a crystal structure of IL-17RA with the IL-17F homodimer was reported (Ely et al. 2009) (Fig. 2). This work provided the first direct view of the receptor, and confirmed the unusual nature of FNIII domains in IL-17RA. The dominant complex was IL-17RA with IL-17F in a 1:2 ratio (Fig. 2a), although at high concentrations a 2:2 ratio was observed. The ECD was confirmed to contain 2 FNIII domains joined by an 18-residue helical linker. The overall structure was in some ways similar to the Type I hematopoietin cytokine receptors in the sense that it contained tandem β-sandwich domains, but the manner of engaging ligand was distinct. Five of the seven predicted N-linked glycans were observed. In terms of binding IL-17F, the FNIII domains bind in a side-on orientation and interact with ligand in a pocket formed at the dimeric interface. Three main interaction sites were identified.

To assess issues of stoichiometry, Ely et al. performed elegant surface plasmon resonance (SPR) studies with soluble receptors and ligands to measure affinities of various combinations of receptors. By immobilizing receptor at low density in order to favor monomers on the chip surface and then adding ligand, they were able

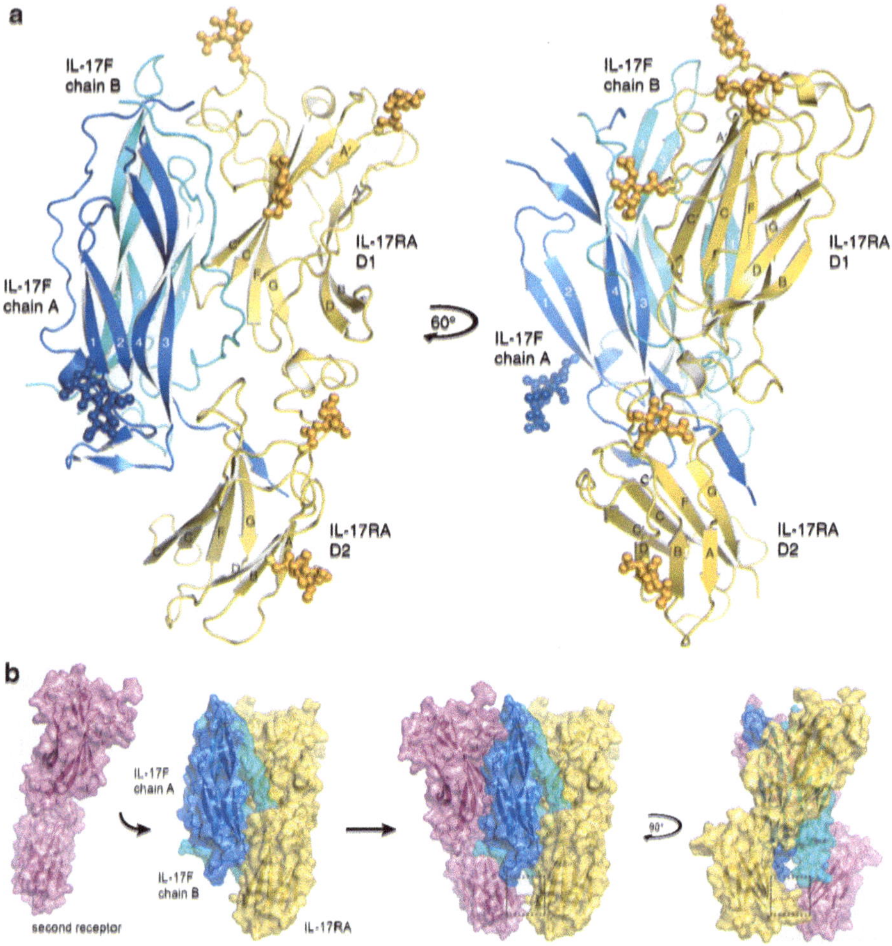

Fig. 2 IL-17R structure. (**a**) Crystal structure of IL-17RA with an IL-17F homodimer. (**b**) Model of IL-17R heterodimerization. Figures from Ely et al. (2009) with permission

to evaluate the "preferences" of pre-formed ligand-receptor complexes for a second sub-unit. In so doing, they demonstrated that IL-17A bound to both IL-17RA and IL-17RC with high affinity, consistent with previous studies (Kuestner et al. 2007). The subsequent affinity for a heterodimeric sub-unit (i.e., IL-17RA:IL-17A for IL-17RC) was high, but the affinity for a homodimer (i.e., IL-17RA:IL-17A for IL-17RA) was low. This approach thus argues against an IL-17RA-homodimer, at least in the liganded state. Further modelling studies supported this model, due to predicted steric interference between IL-17RA sub-units (Fig. 2b). Thus, these studies indicate that most likely stoichiometry of the human IL-17A or IL-17F for an IL-17RA/RC complex is a heterodimer, and further argue that IL-17RC is likely recruited to the complex following ligand binding. However, it is still unclear whether

IL-17RA (or IL-17RC) may form homodimers in the unliganded state, as suggested for murine IL-17RA in the FRET studies discussed above (Kramer et al. 2006). In addition, there are major differences between mouse and human IL-17R complexes in terms of ligand binding (Kuestner et al. 2007). For example, both human IL-17A and IL-17F bind with high affinity to IL-17RC, whereas murine IL-17A exhibits no detectable binding to IL-17RC (Ho and Gaffen 2010; Kuestner et al. 2007). Accordingly, there may be significant structural differences in how murine versus human IL-17R complexes assemble, which will await further analysis.

3 IL-17 Downstream Signal Transduction Pathways

As noted above, the initial cloning of the IL-17RA cytoplasmic tail did not yield any clues about likely downstream signaling pathways, due to the striking lack of homology with other receptor types. However, IL-17A was shown to induce inflammatory genes such as IL-6, IL-8 and other CXC chemokines, and to activate pro-inflammatory signals such as NF-κB and MAPK [reviewed in (Gaffen 2009; Shen and Gaffen 2008)]. Thus, despite being produced largely by T lymphocytes, IL-17A exhibits properties in common with innate inflammatory cytokines like IL-1 and TNFα (Yu and Gaffen 2008). Indeed, IL-17A was shown to cooperate potently with other inflammatory signals, particularly TNFα (Miossec 2003), placing it in the camp of pro-inflammatory cytokines. Related to this, the first proximal signaling molecule identified in the IL-17R pathway was TRAF6, which is also used by the IL-1 and TLR families (Schwandner et al. 2000).

An important insight regarding signal transduction mechanisms came from a bioinformatic study published in 2003, which suggested that there is in fact homology between a region within IL-17 receptor family members' cytoplasmic tails and the Toll/IL-1 receptor family "TIR" signaling motif (Novatchkova et al. 2003). The authors dubbed this a "SEFIR" domain, and further noted that a similar motif is found in Act1, an adaptor and E3 ubiquitin ligase linked to negative regulation of NF-κB activation in the BAFF and CD40L systems (Qian et al. 2004). However, this study provided no empirical evidence that the SEFIR domain was required for IL-17A-mediated signal transduction, or that Act1 participates in IL-17R signaling. Proof came from studies using Act1$^{-/-}$ mice, which are resistant to IL-17A-mediated inflammatory conditions such as EAE and DSS-colitis (Qian et al. 2007). Moreover, Act1$^{-/-}$ fibroblasts were shown to be deficient in most IL-17-dependent signaling events, including activation of NF-κB (Chang et al. 2006; Qian et al. 2007). IL-17A stimulates rapid recruitment of Act1 to IL-17RA, promoting subsequent recruitment of TRAF6 and TAK1. Act1 also has E3 ubiquitin ligase activity, and targets TRAF6, an event necessary for NF-κB mobilization (Liu et al. 2009). Consistently, deletion of the SEFIR domain within either Act1 or IL-17RA abrogates IL-17A-induced signaling. Thus, homotypic SEFIR–SEFIR interactions mediate signaling, analogous to TIR domains in the TLR/IL-1R system (Maitra et al. 2007; Qian et al. 2007). Accordingly, a model has now emerged whereby Act1 is the most proximal adaptor molecule

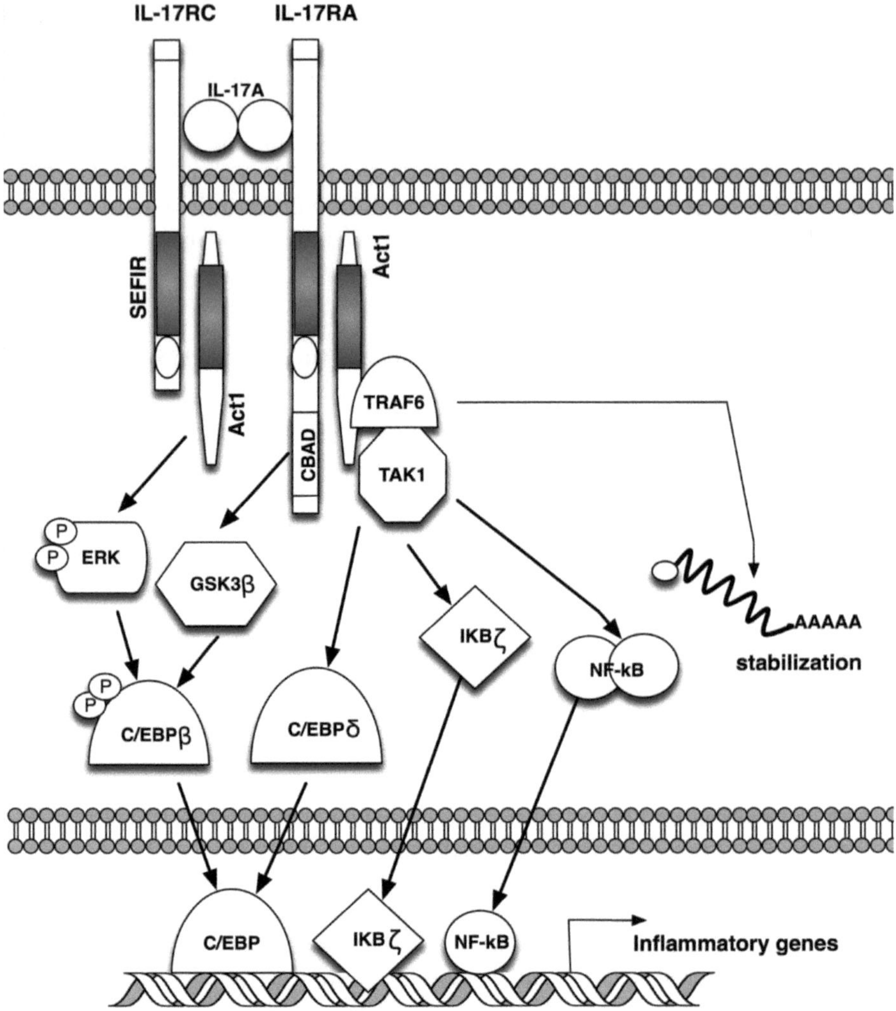

Fig. 3 IL-17RA signaling pathway. The primary signaling events triggered by IL-17 signaling are indicated

recruited to IL-17RA, which consequently permits assembly of the TRAF6-TAK1 complex that leads to activation of the classic NF-κB pathway (Fig. 3). Interestingly, Act1 is a mechanism by which IL-17A and BAFF mediate cooperative signaling in B cells. Specifically, Act1 is a negative regulator IL-17RA preferentially recruits Act1 away from BAFF, thus enhancing the classical NF-κB pathway and promoting target genes that contribute to B cell survival and differentiation (Doreau et al. 2009). Furthermore, Act1 was shown to bind to IL-17RB in response to IL-25 signaling (Claudio et al. 2009; Swaidani et al. 2009) as well as to IL-17RC in response to IL-17A (Ho et al. 2010). Thus, Act1 seems to be a common signaling molecule used by the IL-17R family to mediate signal transduction.

Many IL-17A target genes are chemokines and cytokines such as CXCL1 (KC, Groα) and IL-6 that are regulated strongly at the level of mRNA stability by means of AU-rich elements (AREs) located in the 3′ untranslated region. For the CXCL1 gene, IL-17A regulation of the 3′UTR has been analyzed in considerable detail. As expected, Act1 is necessary for IL-17A to promote message stability, but surprisingly TRAF6 is not (Hartupee et al. 2007; Hartupee et al. 2009). Although the p38 MAPK pathway is important for mediating cytokine gene mRNA stability in many systems such as IL-1 signaling, it also appears not to be essential for IL-17A-mediated signaling to this pathway. Consistent with this, TTP-deficient cells still promote IL-17A-mediated stability of target chemokine genes such as CXCL1 (Datta et al. 2010).

As indicated, the SEFIR domain is a conserved motif in the IL-17R family that mediates binding to Act1. However, detailed mapping of the functional subdomains within IL-17RA revealed that an extended, non-conserved region downstream of the SEFIR is also needed for activation of NF-κB and other pathways (Maitra et al. 2007). This motif bears some similarity with a sub-structure of TIR domains called the BB-loop, and hence we dubbed it a "TIR-like loop" or TILL. A deletion of or point mutation within the TILL domain renders IL-17RA completely non-functional (Maitra et al. 2007), and illustrates that the functional signaling domain of IL-17RA is surprisingly complicated. The TILL domain is not found in other IL-17R family members or Act1. Further dissection of the IL-17RA cytoplasmic tail indicates that the functional domain required for Act1 activation and subsequent signaling extends well beyond the SEFIR/TILL motif (Onishi et al. 2010) (Fig. 3). Similarly, IL-17RC also requires sequences well beyond its SEFIR domain to mediate signaling (Ho et al. 2010). Presumably the SEFIR plus the extended domains comprise a single composite domain, with unique features among IL-17R family members.

IL-17A activates the classical NF-κB pathway via Act1 and TRAF6, but also activates other transcription factors. A bioinformatics comparison of various IL-17A target gene promoters indicated enrichment not only of NF-κB promoter elements but also CCAAT/Enhancer Binding Protein (C/EBP) binding sites (Shen et al. 2006). Empirical analyses of the IL-6 and lipocalin-2/24p3 promoters confirmed that both the NF-κB and C/EBP binding elements are necessary for mediating IL-17A responsiveness (Ruddy et al. 2004; Shen et al. 2006). Similar observations were made for C-reactive protein (Patel et al. 2007). IL-17A activation of C/EBP is quite complex. Both C/EBPβ and C/EBPδ are induced transcriptionally by IL-17 in cooperation with TNFα. The SEFIR/TILL domain of IL-17RA and Act1 are required for induction of both proteins (Maitra et al. 2007).

While C/EBPδ expression appears to be regulated primarily at the transcriptional level, C/EBPβ is regulated by alternative translation and inducible phosphorylation. The major isoform of C/EBPβ is translated from the second methionine (M20) and is termed LAP [liver activated protein, reviewed in (Ramji and Foka 2002)]. The full length variant, known as LAP*, is induced by IL-17 signaling in a manner that is not well defined (Shen et al. 2006). C/EBPβ also encodes two "regulatory domains" (RD1 and RD2), the latter of which is the target of phosphorylation by a variety of kinases. IL-17A signaling leads to inducible phosphorylation of RD2 on two sites, T188 and T179 (Shen et al. 2009). The ERK and GSK3β kinases appear to be the mediators of this event. Although inducible phosphorylation of RD2 is a positive regulator of C/EBPβ transcriptional activity in the context of adipogenesis (Tang et al. 2005), in the IL-17A system the phosphorylation is surprisingly inhibitory. Nonetheless, C/EBPβ, both in the LAP and LAP* isoforms, can promote induction of IL-17A target genes (Shen et al. 2009). Thus the dynamics of C/EBPβ regulation and the coordinated activities of C/EBPδ still remain to be elucidated.

Interestingly, the distal tail of IL-17RA is required for regulation of both C/EBPβ alternative translation from LAP to LAP* as well as the inducible phosphorylation of RD2 (Fig. 3) (Maitra et al. 2007; Shen et al. 2009). Hence, this domain has been termed a "C/EBPβ activation domain" (C-BAD). Deletion of this C-terminal subdomain within IL-17RA led to a marked reduction in both events, perhaps suggesting that they are in some way linked. However, there is no obvious homology of the C-BAD domain with other proteins, and a detailed structure-function analysis has not been performed to date.

In addition to C/EBP, IL-17A signaling activates expression of IκBζ/MAIL (Shen et al. 2005). Despite its homology to the inhibitory family of NF-κB proteins, IκBζ is a positive activator of IL-6 transcription, as well as other genes. IL-17A induces cooperative upregulation of I-κBζ, but whether this is critical for downstream gene expression is not well defined.

4 Other Regulation of IL-17R

IL-17RA is regulated by a variety of other mechanisms, though none have been deeply explored. Although IL-17R expression is fairly ubiquitous, its expression varies considerably among tissue types. High IL-17RA expression is seen in hematopoietic cells, with much reduced expression in other cell types. The basis for these differences is unknown. Although IL-17A-dependent responses were recently identified in B cells, to date there is little data supporting a positive role for effects in T cells, perhaps because the latter lack significant IL-17RC expression (Ishigame et al. 2009; Lindemann et al. 2008). However, IL-21 and IL-15 enhance IL-17RA expression in T cells (Lindemann et al. 2008), whereas IL-2 suppresses it. This is at least partly regulated by PI3K pathways. What is the function of IL-17RA in T cells? It is possible that IL-17RA is acting as a co-receptor for another of the IL-17 ligand family. Alternatively, IL-17A treatment of cells bearing IL-17RA leads to rapid downregulation of the receptor (Lindemann et al. 2008), and it is plausible that T cell expression of IL-17RA serves as a "sink" to remove IL-17A from the system and hence dampen inflammatory signals.

Most IL-17R family members apart from IL-17RA are highly spliced in the extracellular domain (Moseley et al. 2003). In most cases the significance of this is unknown, but for IL-17RC it has been demonstrated that this regulates binding to specific ligands. IL-17RC has been predicted to encode as many as 90 different splice variants (Haudenschild et al. 2002). At least four variants are found frequently in cells that have been evaluated. In terms of binding ligand, splice variant function is different between species. Whereas unspliced human IL-17RC binds both IL-17A and IL-17F, murine IL-17RA only binds detectably to IL-17F. Human IL-17RCΔ7 (lacking the seventh exon) also binds both ligands, while murine IL-17RCΔ7 does not (Ho and Gaffen 2010; Kuestner et al. 2007). A similar theme is likely to emerge for other receptors, but has not been reported. The mechanisms by which splicing is controlled remain unclear, but this area provides another avenue for controlling this intriguing receptor family.

Similar to other cytokine receptors, IL-17R family members are glycosylated (Ely et al. 2009; O'Connor et al. 2009). It is not known the extent to which this affects signaling or stability of expression, and represents another avenue of potential research in this field.

5 Synergy in Signaling

A hallmark feature of IL-17A activity is the potent synergy it exhibits with other cytokines. The most dramatic is with TNFα, but cooperation with IL-22, BAFF, lymphotoxin and IFNγ have been described [reviewed in (Gaffen 2009)]. This synergy probably occurs at multiple levels. There is a clear contribution of IL-17A signaling to regulation transcript stability of many IL-17 target genes, particularly chemokines such as CXCL1 (Hartupee et al. 2007). As noted above, this is independent of TRAF6 (Hartupee et al. 2009) and TTP (Datta et al. 2010). IL-6 is also regulated by message stabilization, which shows cooperation between IL-17A and TNFα. However, not all IL-17-target genes are regulated by mRNA stability, such as lipocalin 2/24p3, where synergy at the level of the proximal promoter is observed (Shen et al. 2006). In addition, the IL-6 promoter is also targeted at least additively by IL-17 and TNFα, which is mediated at least partially through C/EBP transcription factors (Ruddy et al. 2004). In total, it is likely that the combined effects of these cytokines on several facets of gene/protein expression may cumulatively lead to dramatic synergistic production.

5.1 Other IL-17R Family Members: SEF

The original discovery of IL-17RD/SEF was by two independent groups employing identical gene expression screens to identify novel genes in zebra fish by random whole mount in situ hybridization (Furthauer et al. 2002; Tsang et al. 2002). Although amino acid alignment showed homology to IL-17Rs, it was the expression pattern during zebra fish development that provided clues to its function. This IL-17R-like gene exhibited expression that was almost identical to fibroblast growth factor (FGF) ligands, *fgf8* and *fgf3*, and was thus named SEF (*s*imilar *e*xpression to *f*gf genes) (Thisse and Thisse 2005; Tsang and Dawid 2004). Because FGFs activate the RAS/MAPK pathway that led to gene transcription, it was hypothesized that SEF expression was under the control of FGF activity (Fig. 4). This was confirmed as activation of the FGF pathway induced and expanded SEF. In addition, blockade

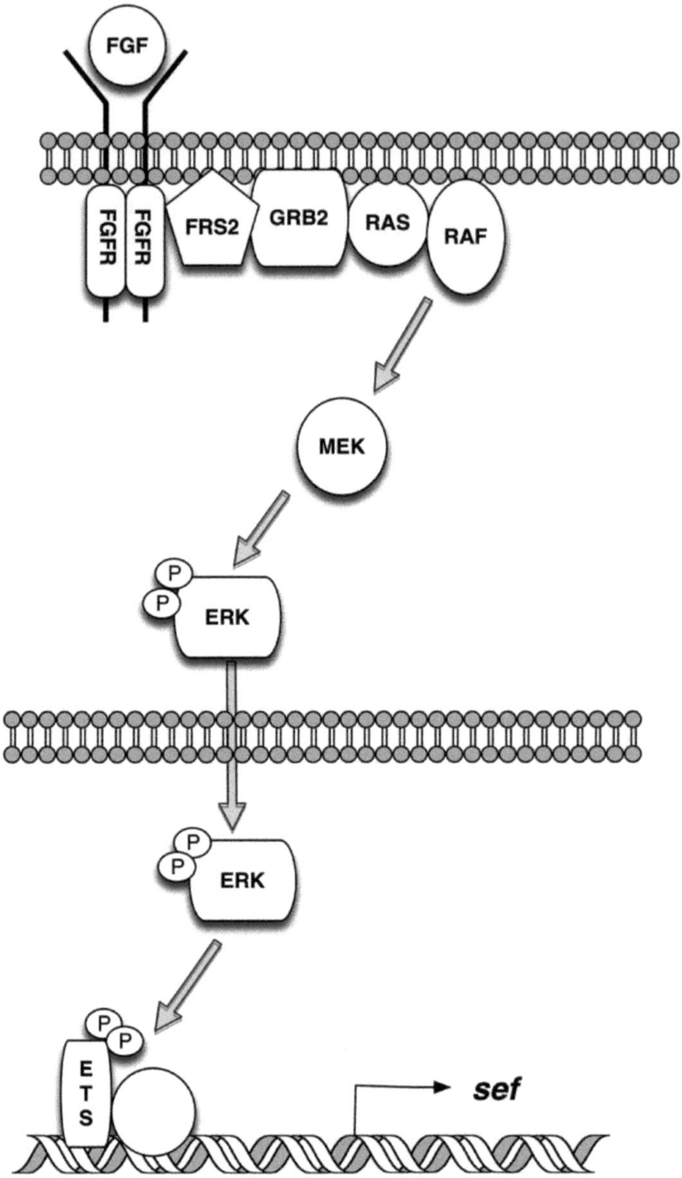

Fig. 4 SEF signal transduction pathway. Main signaling events triggered by SEF in conjunction with the FGFR are indicated

of FGF activity suppressed SEF expression. These experiments suggested that SEF expression is under the direct transcriptional control of FGF signaling in zebra fish. Subsequent studies with mouse and chick orthologs confirmed that SEF is controlled by FGF activity, revealing an evolutionary conservation in gene regulation (Furthauer et al. 2002; Harduf et al. 2005; Lin et al. 2002).

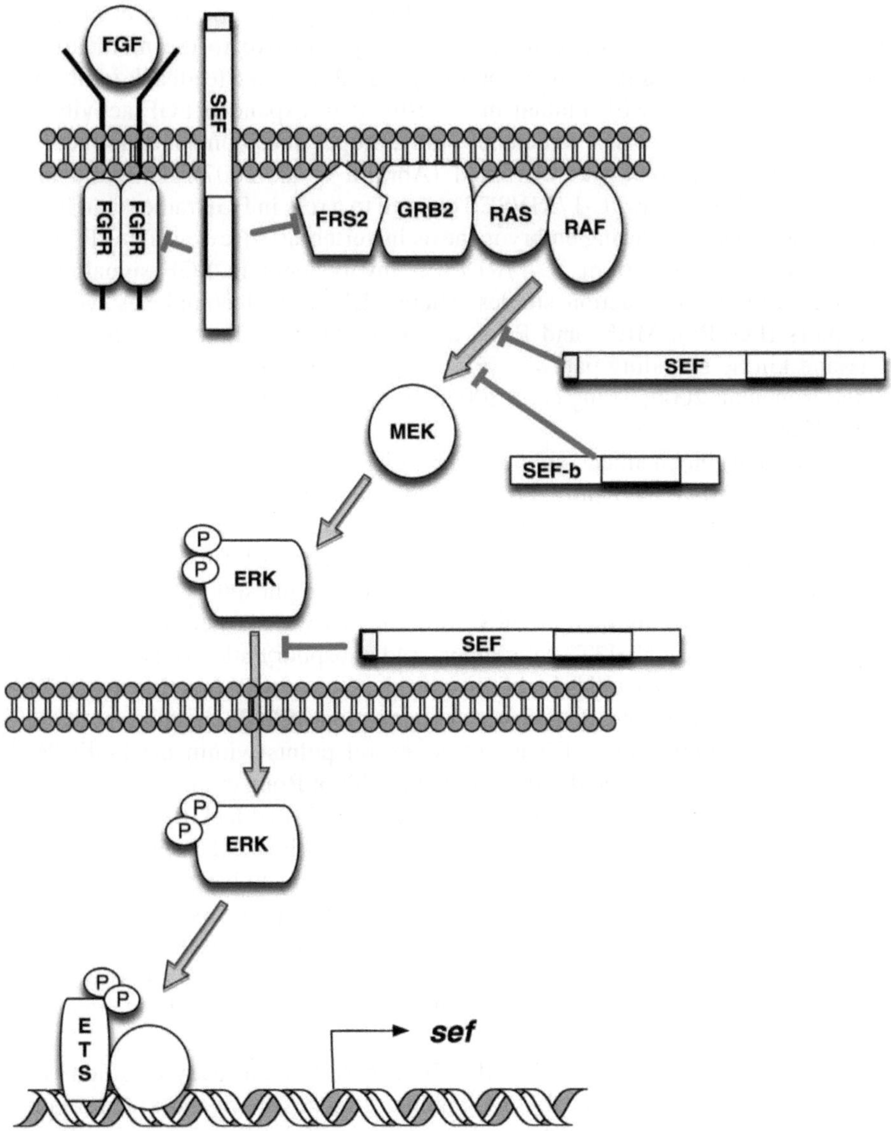

Fig. 4 (continued)

SEF function was explored initially in zebra fish by gain-of-function studies, where ectopic SEF blocked FGF signaling and evoked phenotypes that were reminiscent of FGF deficiencies in zebra fish and Xenopus embryos (Furthauer et al. 2002; Tsang et al. 2002). These studies suggested that SEF could function as a feedback inhibitor of FGF signaling given FGFs control of SEF's transcription. In contrast, loss-of-function studies produced phenotypes that were similar to overexpression of the Fgf8 ligand in early development. In mice, knock-out of

SEF did not result in any discernable embryonic phenotype (Lin et al. 2002). However, SEF-deficient mice were particularly sensitive to the modulation of FGF signaling, as ectopic expression of Sprouty2, another feedback inhibitor of the FGF pathway in SEF mutant mice resulted in expanded FGF activity (Lin et al. 2002). In another SEF deficiency study, proper development and function of the auditory brainstem was disrupted (Abraira et al. 2007). Thus the initial discovery and studies of IL-17RD/SEF pointed to a role in FGF rather than IL-17R signaling, especially during embryogenesis in vertebrates [reviewed in (Ron et al. 2008)]. Further support of IL-17RD/SEF involvement in FGF signaling was revealed through interaction studies where SEF formed complexes with FGF receptors (FGFRs), MEK, and RAS, all are cellular mediators of the receptor tyrosine kinase signaling pathway (Fig. 4) (Kovalenko et al. 2003; Preger et al. 2004; Ren et al. 2006; Tsang et al. 2002; Xiong et al. 2003; Yang et al. 2003; Ziv et al. 2006).

Subsequent biochemical studies were directed at the mechanism of how SEF suppresses FGF signaling. From amino acid alignment SEF was predicted to be a single pass trans-membrane protein as both signal sequence and putative trans-membrane domains were identified. The discovery that SEF forms a complex with FGFRs implied that SEF could block FGFR tyrosine phosphorylation and downstream adaptor proteins such as FRS2 (Fig. 4b) (Kovalenko et al. 2003). Indeed expression of SEF in NIH3T3 cells suppressed phosphorylation of both FGFR and FRS2. Although these studies highlight a mechanism of SEF function at the FGFR level, several studies have demonstrated attenuation of FGF signaling at the level of MEK, indicating that SEF can act at several points within the FGFR/RAS/MAPK pathway (Preger et al. 2004; Ren et al. 2006; Rong et al. 2007; Torii et al. 2004; Ziv et al. 2006). Shorter splice variants of SEF that lack the signal peptide have been isolated from human and mice that clearly block MEK activity (Fig. 4b) (Preger et al. 2004; Rong et al. 2007). Another mechanism has proposed that SEF inhibits the nuclear localization of activated ERK, and prevents it from activating transcription factors such as Elk1 (Torii et al. 2004). Taken together, it seems that SEF can act at multiple points within the FGFR/RAS/MAPK pathway adding another level of fine-tuning for proper FGF signaling levels in development and homeostasis.

Given the homology of SEF to IL-17Rs, there remain many questions as to whether SEF can function in IL-17 signaling. To date, the ligand(s) that binds to SEF is not known. More recently, evidence for IL-17D/SEF responding to IL-17A and activation of IL-17A-responsive promoters has been demonstrated (Rong et al. 2009). In these studies, heterodimers with other IL-17Rs and co-localization of these IL-17Rs to the membrane was reported. Furthermore, co-immune complexes of IL-17D/SEF and TRAF6, an essential intracellular signaling component of IL-17Rs, were observed. Whether these studies in cells are representative of in vivo biology remains to be demonstrated. However, this provides the first evidence of a potential role for IL-17RD/SEF in IL-17 signaling, and future studies will certainly uncover how and whether SEF can function in both FGF and IL-17 signaling.

6 Conclusions and Perspectives

In summary, the IL-17 receptor family has provided a fascinating new window for exploring cytokine receptor biology. With unique structures and unusual modes of signaling, this family has revealed new insights into this area. There are still orphan ligands and receptors whose functions remain obscure (Table 2), and no doubt there are still surprises ahead.

Acknowledgments SLG is supported by the NIH (AR054389) and Amgen. MT is funded by NIH (HL088016).

References

Abraira, V.E., Hyun, N., Tucker, A.F., Coling, D.E., Brown, M.C., Lu, C., Hoffman, G.R., and Goodrich, L.V. (2007). Changes in Sef levels influence auditory brainstem development and function. J Neurosci *27*, 4273–4282.

Aggarwal, S., and Gurney, A.L. (2002). IL-17: A prototype member of an emerging family. J. Leukoc. Biol *71*, 1–8.

Chan, F.K. (2007). Three is better than one: pre-ligand receptor assembly in the regulation of TNF receptor signaling. Cytokine *37*, 101–107.

Chan, F.K., Chun, H.J., Zheng, L., Siegel, R.M., Bui, K.L., and Lenardo, M.J. (2000). A domain in TNF receptors that mediates ligand-independent receptor assembly and signaling. Science *288*, 2351–2354.

Chang, S.H., and Dong, C. (2007). A novel heterodimeric cytokine consisting of IL-17 and IL-17F regulates inflammatory responses. Cell Res *17*, 435–440.

Chang, S.H., Park, H., and Dong, C. (2006). Act1 adaptor protein is an immediate and essential signaling component of interleukin-17 receptor. The Journal of biological chemistry *281*, 35603–35607.

Claudio, E., Sonder, S.U., Saret, S., Carvalho, G., Ramalingam, T.R., Wynn, T.A., Chariot, A., Garcia-Perganeda, A., Leonardi, A., Paun, A., *et al.* (2009). The adaptor protein CIKS/Act1 is essential for IL-25-mediated allergic airway inflammation. J Immunol *182*, 1617–1630.

Datta, S., Novotny, M., Pavicic, P.G., Jr., Zhao, C., Herjan, T., Hartupee, J., and Hamilton, T. (2010). IL-17 regulates CXCL1 mRNA stability via an AUUUA/tristetraprolin-independent sequence. J Immunol *184*, 1484–1491.

Doreau, A., Belot, A., Bastid, J., Riche, B., Trescol-Biemont, M.C., Ranchin, B., Fabien, N., Cochat, P., Pouteil-Noble, C., Trolliet, P., *et al.* (2009). Interleukin 17 acts in synergy with B cell-activating factor to influence B cell biology and the pathophysiology of systemic lupus erythematosus. Nature immunology *10*, 778–785.

Ely, L.K., Fischer, S., and Garcia, K.C. (2009). Structural basis of receptor sharing by interleukin 17 cytokines. Nat Immunol *10*, 1245–1251.

Fossiez, F., Djossou, O., Chomarat, P., Flores-Romo, L., Ait-Yahia, S., Maat, C., Pin, J.-J., Garrone, P., Garcia, E., Saeland, S., *et al.* (1996). T cell interleukin-17 induces stromal cells to produce proinflammatory and hematopoietic cytokines. J Exp Med *183*, 2593–2603.

Furthauer, M., Lin, W., Ang, S.L., Thisse, B., and Thisse, C. (2002). Sef is a feedback-induced antagonist of Ras/MAPK-mediated FGF signaling. Nat Cell Biol *4*, 170–174.

Gaffen, S.L. (2009). Structure and signaling in the IL-17 receptor family. Nat Rev Immunol *9*, 556–567.

Harduf, H., Halperin, E., Reshef, R., and Ron, D. (2005). Sef is syn-expressed with FGFs during chick embryogenesis and its expression is differentially regulated by FGFs in the developing limb. Dev Dyn *233*, 301–312.

Hartupee, J., Liu, C., Novotny, M., Li, X., and Hamilton, T. (2007). IL-17 enhances chemokine gene expression through mRNA stabilization. J Immunol *179*, 4135–4141.

Hartupee, J., Liu, C., Novotny, M., Sun, D., Li, X., and Hamilton, T.A. (2009). IL-17 signaling for mRNA stabilization does not require TNF receptor-associated factor 6. J Immunol *182*, 1660–1666.

Haudenschild, D., Moseley, T., Rose, L., and Reddi, A.H. (2002). Soluble and trans-membrane isoforms of novel interleukin-17 receptor-like protein by RNA splicing and expression in prostate cancer. The Journal of biological chemistry *277*, 4309–4316.

Haudenschild, D.R., Curtiss, S.B., Moseley, T.A., and Reddi, A.H. (2006). Generation of interleukin-17 receptor-like protein (IL-17RL) in prostate by alternative splicing of RNA. Prostate *66*, 1268–1274.

Ho, A., and Gaffen, S.L. (2010). IL-17RC: A partner in IL-17 signaling and beyond. Semin Immunopathol *32*, 33–42.

Ho, A.W., Shen, F., Conti, H.R., Patel, N., Childs, E.E., Peterson, A.C., Hernández-Santos, N., Kolls, J.K., Kane, L.P., Ouyang, W., *et al.* (2010). IL-17RC is required for immune signaling via an extended SEF/IL-17R signaling domain in the cytoplasmic tail. J Immunol *185*, 1063–1070.

Hymowitz, S.G., Filvaroff, E.H., Yin, J.P., Lee, J., Cai, L., Risser, P., Maruoka, M., Mao, W., Foster, J., Kelley, R.F., *et al.* (2001). IL-17s adopt a cystine knot fold: structure and activity of a novel cytokine, IL-17F, and implications for receptor binding. Embo J *20*, 5332–5341.

Ishigame, H., Kakuta, S., Nagai, T., Kadoki, M., Nambu, A., Komiyama, Y., Fujikado, N., Tanahashi, Y., Akitsu, A., Kotaki, H., *et al.* (2009). Differential roles of interleukin-17A and -17F in host defense against muco-epithelial bacterial infection and allergic responses. Immunity *30*, 108–119.

Korn, T., Bettelli, E., Oukka, M., and Kuchroo, V.K. (2009). IL-17 and Th17 Cells. Annual review of immunology *27*, 485–518.

Kovalenko, D., Yang, X., Nadeau, R.J., Harkins, L.K., and Friesel, R. (2003). Sef inhibits fibroblast growth factor signaling by inhibiting FGFR1 tyrosine phosphorylation and subsequent ERK activation. The Journal of biological chemistry *278*, 14087–14091.

Kramer, J., Hanel, W., Shen, F., Isik, N., Malone, J., Maitra, A., Sigurdson, W., Swart, D., Tocker, J., Jin, T., and Gaffen, S.L. (2007). Cutting Edge: Identification of the pre-ligand assembly domain (PLAD) and ligand binding site in the IL-17 receptor. J Immunol *179*, 6379–6383.

Kramer, J., Yi, L., Shen, F., Maitra, A., Jiao, X., Jin, T., and Gaffen, S. (2006). Cutting Edge: Evidence for ligand-independent multimerization of the IL-17 receptor. J Immunol *176*, 711–715.

Kuestner, R., Taft, D., Haran, A., Brandt, C., Brender, T., Lum, K., Harder, B., Okada, S., Osatrander, C., Kreindler, J., *èt al.* (2007). Identification of the IL-17 receptor related molecule, IL-17RC, as the receptor for IL-17F. J Immunol *179*, 5462–5473.

Latz, E., Verma, A., Visintin, A., Gong, M., Sirois, C.M., Klein, D.C., Monks, B.G., McKnight, C.J., Lamphier, M.S., Duprex, W.P., *et al.* (2007). Ligand-induced conformational changes allosterically activate Toll-like receptor 9. Nature immunology *8*, 772–779.

Lee, J., Ho, W.H., Maruoka, M., Corpuz, R.T., Baldwin, D.T., Foster, J.S., Goddard, A.D., Yansura, D.G., Vandlen, R.L., Wood, W.I., and Gurney, A.L. (2001). IL-17E, a novel proinflammatory ligand for the IL-17 receptor homolog IL-17Rh1. The Journal of biological chemistry *276*, 1660–1664.

Li, H., Chen, J., Huang, A., Stinson, J., Heldens, S., Foster, J., Dowd, P., Gurney, A.L., and Wood, W.I. (2000). Cloning and characterization of IL-17B and IL-17C, two new members of the IL-17 family. Proc Natl Acad Sci USA *97*, 773–778.

Lin, W., Furthauer, M., Thisse, B., Thisse, C., Jing, N., and Ang, S.L. (2002). Cloning of the mouse Sef gene and comparative analysis of its expression with Fgf8 and Spry2 during embryogenesis. Mech Dev *113*, 163–168.

Lindemann, M.J., Hu, Z., Benczik, M., Liu, K.D., and Gaffen, S.L. (2008). Differential Regulation of the IL-17 Receptor by gamma-c Cytokines: INHIBITORY SIGNALING BY THE PHOSPHATIDYLINOSITOL 3-KINASE PATHWAY. The Journal of biological chemistry *283*, 14100–14108.

Liu, C., Qian, W., Qian, Y., Giltiay, N.V., Lu, Y., Swaidani, S., Misra, S., Deng, L., Chen, Z.J., and Li, X. (2009). Act1, a U-box E3 ubiquitin ligase for IL-17 signaling. Sci Signal *2*, ra63.

Maitra, A., Shen, F., Hanel, W., Mossman, K., Tocker, J., Swart, D., and Gaffen, S.L. (2007). Distinct functional motifs within the IL-17 receptor regulate signal transduction and target gene expression. Proc Natl Acad Sci USA *104*, 7506–7511.

Miossec, P. (2003). Interleukin-17 in rheumatoid arthritis: if T cells were to contribute to inflammation and destruction through synergy. Arthritis and rheumatism *48*, 594–601.

Moseley, T.A., Haudenschild, D.R., Rose, L., and Reddi, A.H. (2003). Interleukin-17 family and IL-17 receptors. Cytokine Growth Factor Rev *14*, 155–174.

Murphy, J.M., and Young, I.G. (2006). IL-3, IL-5, and GM-CSF signaling: crystal structure of the human beta-common receptor. Vitam Horm *74*, 1–30.

Novatchkova, M., Leibbrandt, A., Werzowa, J., Neubuser, A., and Eisenhaber, F. (2003). The STIR-domain superfamily in signal transduction, development and immunity. Trends Biochem Sci *28*, 226–229.

O'Connor, W., Jr., Kamanaka, M., Booth, C.J., Town, T., Nakae, S., Iwakura, Y., Kolls, J.K., and Flavell, R.A. (2009). A protective function for interleukin 17A in T cell-mediated intestinal inflammation. Nature immunology *10*, 603–609.

Onishi, R., and Gaffen, S.L. (2010). IL-17 and its Target Genes: Mechanisms of IL-17 Function in Disease. Immunology *129*, 311–321.

Onishi, R.M., Park, S.J., Hanel, W., Ho, A.W., Maitra, A., Gaffen, S.L. (2010). SEF/IL-17R (SEFIR) is not enough: an extended SEFIR domain is required for IL-17RA-mediated signal transduction. J Biol Chem *285*, 32751–32759.

Ozaki, K., and Leonard, W.J. (2002). Cytokine and cytokine receptor pleiotropy and redundancy. The Journal of biological chemistry *277*, 29355–29358.

Patel, D.N., King, C.A., Bailey, S.R., Holt, J.W., Venkatachalam, K., Agrawal, A., Valente, A.J., and Chandrasekar, B. (2007). Interleukin-17 stimulates C-reactive protein expression in hepatocytes and smooth muscle cells via p38 MAPK and ERK1/2-dependent NF-kappaB and C/EBPbeta activation. The Journal of biological chemistry *282*, 27229–27238.

Preger, E., Ziv, I., Shabtay, A., Sher, I., Tsang, M., Dawid, I.B., Altuvia, Y., and Ron, D. (2004). Alternative splicing generates an isoform of the human Sef gene with altered subcellular localization and specificity. Proc Natl Acad Sci USA *101*, 1229–1234.

Qian, Y., Liu, C., Hartupee, J., Altuntas, C.Z., Gulen, M.F., Jane-Wit, D., Xiao, J., Lu, Y., Giltiay, N., Liu, J., *et al.* (2007). The adaptor Act1 is required for interleukin 17-dependent signaling associated with autoimmune and inflammatory disease. Nature immunology *8*, 247–256.

Qian, Y., Qin, J., Cui, G., Naramura, M., Snow, E.C., Ware, C.F., Fairchild, R.L., Omori, S.A., Rickert, R.C., Scott, M., *et al.* (2004). Act1, a negative regulator in CD40- and BAFF-mediated B cell survival. Immunity *21*, 575–587.

Ramji, D.P., and Foka, P. (2002). CCAAT/enhancer-binding proteins: structure, function and regulation. Biochem J *365*, 561–575.

Ren, Y., Cheng, L., Rong, Z., Li, Z., Li, Y., Li, H., Wang, Z., and Chang, Z. (2006). hSef co-localizes and interacts with Ras in the inhibition of Ras/MAPK signaling pathway. Biochem Biophys Res Commun *347*, 988–993.

Rickel, E.A., Siegel, L.A., Yoon, B.R., Rottman, J.B., Kugler, D.G., Swart, D.A., Anders, P.M., Tocker, J.E., Comeau, M.R., and Budelsky, A.L. (2008). Identification of functional roles for both IL-17RB and IL-17RA in mediating IL-25-induced activities. J Immunol *181*, 4299–4310.

Ron, D., Fuchs, Y., and Chorev, D.S. (2008). Know thy Sef: a novel class of feedback antagonists of receptor tyrosine kinase signaling. Int J Biochem Cell Biol *40*, 2040–2052.

Rong, Z., Cheng, L., Ren, Y., Li, Z., Li, Y., Li, X., Li, H., Fu, X.Y., and Chang, Z. (2007). Interleukin-17F signaling requires ubiquitination of interleukin-17 receptor via TRAF6. Cell Signal *19*, 1514–1520.

Rong, Z., Wang, A., Li, Z., Ren, Y., Cheng, L., Li, Y., Wang, Y., Ren, F., Zhang, X., Hu, J., and Chang, Z. (2009). IL-17RD (Sef or IL-17RLM) interacts with IL-17 receptor and mediates IL-17 signaling. Cell Res *19*, 208–215.

Rouvier, E., Luciani, M.-F., Mattei, M.-G., Denizot, F., and Golstein, P. (1993). CTLA-8, cloned from an activated T cell, bearing AU-rich messenger RNA instability sequences, and homologous to a Herpesvirus Saimiri gene. J Immunol *150*, 5445–5456.

Ruddy, M.J., Wong, G.C., Liu, X.K., Yamamoto, H., Kasayama, S., Kirkwood, K.L., and Gaffen, S.L. (2004). Functional cooperation between interleukin-17 and tumor necrosis factor-α is mediated by CCAAT/enhancer binding protein family members. The Journal of biological chemistry *279*, 2559–2567.

Schwandner, R., Yamaguchi, K., and Cao, Z. (2000). Requirement of tumor necrosis factor-associated factor (TRAF)6 in interleukin 17 signal transduction. J Exp Med *191*, 1233–1239.

Shen, F., and Gaffen, S.L. (2008). Structure-function relationships in the IL-17 receptor: Implications for signal transduction and therapy. Cytokine *41*, 92–104.

Shen, F., Hu, Z., Goswami, J., and Gaffen, S.L. (2006). Identification of common transcriptional regulatory elements in interleukin-17 target genes. The Journal of biological chemistry *281*, 24138–24148.

Shen, F., Li, N., Gade, P., Kalvakolanu, D.V., Weibley, T., Doble, B., Woodgett, J.R., Wood, T.D., and Gaffen, S.L. (2009). IL-17 Receptor Signaling Inhibits C/EBP{beta} by Sequential Phosphorylation of the Regulatory 2 Domain. Sci Signal *2*, ra8.

Shen, F., Ruddy, M.J., Plamondon, P., and Gaffen, S.L. (2005). Cytokines link osteoblasts and inflammation: microarray analysis of interleukin-17- and TNF-alpha-induced genes in bone cells. J Leukoc Biol *77*, 388–399.

Shi, Y., Ullrich, S.J., Zhang, J., Connolly, K., Grzegorzewski, K.J., Barber, M.C., Wang, W., Wathen, K., Hodge, V., Fisher, C.L., *et al.* (2000). A novel cytokine receptor-ligand pair. Identification, molecular characterization, and in vivo immuno-modulatory activity. The Journal of biological chemistry *275*, 19167–19176.

Starnes, T., Broxmeyer, H.E., Robertson, M.J., and Hromas, R. (2002). Cutting edge: IL-17D, a novel member of the IL-17 family, stimulates cytokine production and inhibits hemopoiesis. J Immunol *169*, 642–646.

Starnes, T., Robertson, M.J., Sledge, G., Kelich, S., Nakshatri, H., Broxmeyer, H.E., and Hromas, R. (2001). Cutting edge: IL-17F, a novel cytokine selectively expressed in activated T cells and monocytes, regulates angiogenesis and endothelial cell cytokine production. J Immunol *167*, 4137–4140.

Steinman, L. (2007). A brief history of T(H)17, the first major revision in the T(H)1/T(H)2 hypothesis of T cell-mediated tissue damage. Nature Med *13*, 139–145.

Swaidani, S., Bulek, K., Kang, Z., Liu, C., Lu, Y., Yin, W., Aronica, M., and Li, X. (2009). The critical role of epithelial-derived Act1 in IL-17- and IL-25-mediated pulmonary inflammation. J Immunol *182*, 1631–1640.

Tang, Q.Q., Gronborg, M., Huang, H., Kim, J.W., Otto, T.C., Pandey, A., and Lane, M.D. (2005). Sequential phosphorylation of CCAAT enhancer-binding protein beta by MAPK and glycogen synthase kinase 3beta is required for adipogenesis. Proc Natl Acad Sci USA *102*, 9766–9771.

Thisse, B., and Thisse, C. (2005). Functions and regulations of fibroblast growth factor signaling during embryonic development. Dev Biol *287*, 390–402.

Torii, S., Kusakabe, M., Yamamoto, T., Maekawa, M., and Nishida, E. (2004). Sef is a spatial regulator for Ras/MAP kinase signaling. Dev Cell *7*, 33–44.

Toy, D., Kugler, D., Wolfson, M., Vanden Bos, T., Gurgel, J., Derry, J., Tocker, J., and Peschon, J.J. (2006). Cutting Edge: Interleukin-17 signals through a heteromeric receptor complex. J Immunol *177*, 36–39.

Tsang, M., and Dawid, I. (2004). Promotion and Attenuation of FGF Signaling Through the Ras-MAPK Pathway. Sci STKE *228*, PE17.

Tsang, M., Friesel, R., Kudoh, T., and Dawid, I. (2002). Identification of Sef, a novel modulator of FGF signaling. Nature Cell Biol *4*, 165–169.

Wright, J.F., Bennett, F., Li, B., Brooks, J., Luxenberg, D.P., Whitters, M.J., Tomkinson, K.N., Fitz, L.J., Wolfman, N.M., Collins, M., *et al.* (2008). The human IL-17F/IL-17A heterodimeric cytokine signals through the IL-17RA/IL-17RC receptor complex. J Immunol *181*, 2799–2805.

Wright, J.F., Guo, Y., Quazi, A., Luxenberg, D.P., Bennett, F., Ross, J.F., Qiu, Y., Whitters, M.J., Tomkinson, K.N., Dunussi-Joannopoulos, K., *et al.* (2007). Identification of an Interleukin 17F/17A Heterodimer in Activated Human CD4+ T Cells. The Journal of biological chemistry *282*, 13447–13455.

Xiong, S., Zhao, Q., Rong, Z., Huang, G., Huang, Y., Chen, P., Zhang, S., Liu, L., and Chang, Z. (2003). hSef inhibits PC-12 cell differentiation by interfering with Ras-mitogen-activated protein kinase MAPK signaling. The Journal of biological chemistry *278*, 50273–50282.

Yang, R.B., Ng, C.K., Wasserman, S.M., Komuves, L.G., Gerritsen, M.E., and Topper, J.N. (2003). A novel interleukin-17 receptor-like protein identified in human umbilical vein endothelial cells antagonizes basic fibroblast growth factor-induced signaling. The Journal of biological chemistry *278*, 33232–33238.

Yang, X.O., Chang, S.H., Park, H., Nurieva, R., Shah, B., Acero, L., Wang, Y.H., Schluns, K.S., Broaddus, R.R., Zhu, Z., and Dong, C. (2008). Regulation of inflammatory responses by IL-17F. The Journal of experimental medicine *205*, 1063–1075.

Yao, Z., Fanslow, W.C., Seldin, M.F., Rousseau, A.-M., Painter, S.L., Comeau, M.R., Cohen, J.I., and Spriggs, M.K. (1995a). Herpesvirus Saimiri encodes a new cytokine, IL-17, which binds to a novel cytokine receptor. Immunity *3*, 811–821.

Yao, Z., Painter, S.L., Fanslow, W.C., Ulrich, D., Macduff, B.M., Spriggs, M.K., and Armitage, R.J. (1995b). Cutting Edge: Human IL-17: A novel cytokine derived from T cells. J Immunol *155*, 5483–5486.

Yao, Z., Spriggs, M.K., Derry, J.M.J., Strockbine, L., Park, L.S., VandenBos, T., Zappone, J., Painter, S.L., and Armitage, R.J. (1997). Molecular characterization of the human interleukin-17 receptor. Cytokine *9*, 794–800.

Ye, P., Rodriguez, F.H., Kanaly, S., Stocking, K.L., Schurr, J., Schwarzenberger, P., Oliver, P., Huang, W., Zhang, P., Zhang, J., *et al.* (2001). Requirement of interleukin 17 receptor signaling for lung CXC chemokine and granulocyte colony-stimulating factor expression, neutrophil recruitment, and host defense. The Journal of experimental medicine *194*, 519–527.

Yu, J., and Gaffen, S.L. (2008). Interleukin-17: A novel inflammatory cytokine that bridges innate and adaptive immunity. Front Biosci *13*, 170–177.

Zheng, Y., Valdez, P.A., Danilenko, D.M., Hu, Y., Sa, S.M., Gong, Q., Abbas, A.R., Modrusan, Z., Ghilardi, N., de Sauvage, F.J., and Ouyang, W. (2008). Interleukin-22 mediates early host defense against attaching and effacing bacterial pathogens. Nature medicine *14*, 282–289.

Ziv, I., Fuchs, Y., Preger, E., Shabtay, A., Harduf, H., Zilpa, T., Dym, N., and Ron, D. (2006). The human sef-a isoform utilizes different mechanisms to regulate receptor tyrosine kinase signaling pathways and subsequent cell fate. The Journal of biological chemistry *281*, 39225–39235.

Trafficking Receptors and Migration of T_H17 Cell Subsets

Chang H. Kim

Abstract T_H17 cells produce IL17 and other cytokines that regulate anti-microbial and auto-immune responses. It is well established that T_H1 and T_H2 effector T cells differ from each other in migratory behavior. $FoxP3^+$ regulatory T cells apparently have the migratory features of both naïve and effector T cells. It is a question of interest if T_H17 cells would display a unique trafficking behavior specific for this T cell lineage. T_H17 cells are highly enriched in the intestine and other tissues with certain types of inflammatory diseases. T_H17 cells are unique in that almost all of these T cells express the chemokine receptor CCR6. Otherwise, T_H17 cells are considered conventional, heterogeneously expressing a number of other memory/effector type trafficking receptors, which are required for migration of T_H17 cells into various tissues. I will review our current understanding of the trafficking receptors and migration behavior of T_H17 cell subsets.

1 Introduction

Naïve T helper cells become heterogeneous effector T cells following antigen priming. The effector function of T helper cells is generally defined by the cytokines that they produce. T cell cytokines (or cell surface effector molecules) activate downstream target cell types for diverse responses including cell-mediated and/or antibody responses. Naïve T cells can become IFN-γ-producing T_H1 cells, IL4-producing T_H2 cells, IL21/IL4-producing follicular helper (T-FH) cells, or induced $FoxP3^+$ suppressor T cells. One should note that there are many more types of effector T cells that are yet to be characterized. The presence of T_H17 cells defined by expression of IL17A and IL17F is now well established through the research performed in the last

C.H. Kim (✉)
Laboratory of Immunology and Hematopoiesis, Department of Comparative Pathobiology, Purdue Cancer Center, Purdue University, 725 Harrison Street, West Lafayette, IN 47907, USA
e-mail: chkim@purdue.edu

S. Jiang (ed.), *T_H17 Cells in Health and Disease*,
DOI 10.1007/978-1-4419-9371-7_10, © Springer Science+Business Media, LLC 2011

5 years (Dong 2009; Fouser et al. 2008; Iwakura et al. 2008; Lee et al. 2009; Zhou et al. 2009). The best characterized function of T_H17 cells is to induce certain types of auto-immune diseases such as experimental allergic encephalomyelitis (EAE) and arthritis (Annunziato et al. 2009; Deenick and Tangye 2007; Iwakura et al. 2008; Tesmer et al. 2008; van den Berg and Miossec 2009). Most T_H17 cells, however, are not reactive to self antigens and do not cause auto-immune diseases. Indeed, the primary function of T_H17 cells is to mount effective immune responses upon infection by fungi or bacteria (Curtis and Way 2009; Dubin and Kolls 2008; Iwakura et al. 2008; Khader and Cooper 2008; Liu et al. 2009; O'Quinn et al. 2008). The list of pathogens that T_H17 cells respond to has been expanded to include some viruses and even protozoa (O'Quinn et al. 2008). In this regard, induction of T_H17 cells is positively regulated by bacterial products such as Toll-like receptor (TLR) ligands (Abdollahi-Roodsaz et al. 2008; Kattah et al. 2008; Tanaka et al. 2009). Human T_H17 cells can be induced by fungal products (Acosta-Rodriguez et al. 2007; Zhou et al. 2008). Certain commensal bacteria provide a major inducing signal for intestinal T_H17 cells (Ivanov et al. 2008, 2009; Niess et al. 2008; Zaph et al. 2008). T_H17 cells are induced in peculiar cytokine milieu composed of both the immuno-suppressive cytokine TGF-β1 and certain inflammatory cytokines such as IL-23, IL-6, IL-1β, and TNF-α. These inflammatory cytokines are induced upon infection by microbial agents and so microbial products are thought to play a major role in generation of T_H17 cells. A similar cytokine response can occur in chronic inflammation and autoimmune T_H17 cells are induced as the consequence. In a manner similar to other effector T cells, the migration ability is important for the function of T_H17 cells because they need to migrate and stay in certain tissues for effective effector functions. I will focus this review on migration of T_H17 cells with relevant information on tissue tropism, expression of trafficking receptors, and factors that regulate the migration of T_H17 cells.

2 A General Overview of T Cell Migration

Most T cells are made in the thymus as naïve T cells. These naïve T cells penetrate the thymus and join the blood circulation for controlled homeostatic recirculation through the body. Controlled recirculation refers to continuous migration of lymphocytes from blood to secondary lymphoid tissues and then back to blood (Stekel et al. 1997). T cells stop recirculating upon antigen priming in secondary lymphoid tissues. Homing of naïve T cells into secondary lymphoid tissues requires the expression of the chemokine receptor CCR7 by T cells and the presence of CCR7 ligands such as CCL19 and CCL21 on high endothelial cell venules (Forster et al. 1999; Gunn et al. 1999; Warnock et al. 2000). After homing, T cell co-localization with dendritic cells occurs also in response to the CCR7 ligand chemokines expressed in the T cell area of the secondary lymphoid tissues because both naïve T cells and mature dendritic cells express CCR7 (Gunn et al. 1999; Kellermann et al. 1999; Saeki et al. 1999). Thus, CCR7 ligands are important at a specific stage of recirculation. Also important is the expression of certain adhesion molecules that mediate rolling and firm adhesion of T cells on endothelial cells (Tedder et al. 1995; von Andrian and M'Rini 1998). Another vital regulator of T cell recirculation is a lipid factor called sphingosin-1-phosphate (S1P) (Lo et al. 2005; Sanna et al. 2004). S1P is a chemo-attractant for cells expressing S1P receptors (Goetzl et al. 2008). There are 5 receptors from S1P1 through S1P5 for S1P. S1P is present at the highest concentration in the blood and lowest within organs such as SLT. The concentration in lymph is lower than that of blood but higher than that of lymphoid tissues (Schwab et al. 2005). This forms an increasing chemotactic gradient of S1P from lymphoid tissues to lymph and then to blood (Schwab et al. 2005). S1P could function through their action on endothelial cells in promoting transendothelial migration of lymphocytes (Schwab and Cyster 2007). S1P1 expression on T cells quickly goes down upon T cell activation (Marelli-Berg et al. 2008). This down-regulation makes the T cells insensitive to the S1P gradient so that activated T cells become unresponsive to the S1P gradient. The consequence is that T cells stop the migration and instead become sessile in lymphoid tissues until they finish the activation process and become resting memory T cells. Where they down-regulate S1P1, activated T cells upregulate adhesion molecules for T cell interaction with antigen presenting cells or stromal cells (Marelli-Berg et al. 2008). T cells regain the expression of S1P1 and this is important for their infiltration following the S1P gradient.

In addition to CCR7, naïve T cells express CXCR4 and CD62L. CXCR4 can partly replace CCR7 for homing of T cells into secondary lymphoid tissues (Kim 2006; Lim et al. 2004, 2006). CD62L is a selectin molecule and mediates rolling of T cells on endothelial cells to initiate the margination process to enter the tissues. Expression of $\alpha4\beta7$ would help these T cells to migrate into gut and gut-associated lymphoid tissues (Berlin et al. 1993; Hamann et al. 1994; Wagner et al. 1996). When T cells undergo activation, expression of CCR7 and CD62L is greatly reduced, and the T cells instead express many other chemokine receptors and adhesion molecules to confer the T cells new and heterogeneous migratory capacities (Kim 2005; Lee et al. 2007). These receptors that are heterogeneously expressed by memory and effector T cells include CCR2, CCR4, CCR5, CCR6, CXCR3, CXCR5, CXCR6, P-selectin ligand, and E-selectin ligand. Effector T cell subsets are heterogeneous in expression of the trafficking receptors. For example, T_H1 cells typically express CCR5, CXCR3 and CXCR6, while T_H2 cells express CRTH2 (human but not mouse) and CCR4 (Austrup et al. 1997; Kim 2005).

Another T cell subset called follicular T cells, which are specialized in helping B cells, express CXCR5 for migration into follicles (Hardtke et al. 2005; Haynes et al. 2007). Apart from the polarization status, T cells that are present in distinctive tissues express different trafficking receptors. T cells that are present in the intestinal lamina propria express $\alpha4\beta7$ at high levels (Bradley et al. 1998; Rott et al. 1997; Williams and Butcher 1997). The T cells that are present in the small intestine express CCR9 (Kunkel et al. 2000; Papadakis et al. 2000; Stenstad et al. 2006; Svensson et al. 2002; Wurbel et al. 2001). Skin homing T cells express E/P-selectin ligands, CCR4, and CCR8 (Hwang 2001; Kim 2004; Marelli-Berg et al. 2008; Schaerli et al. 2004). The benefit of having such distinct requirements for migration is perhaps to recruit only certain type of T cells appropriate for each type of immune response. If T_H1 and T_H2 cells would migrate to the same tissue site, the two T cell subsets would cancel each other out in function and would not mount effective immune responses to clear pathogens. Another benefit is to limit the area for T cells to patrol to promote effective regional immune responses.

3 Generation and Tissue Distribution of T_H17 Cells

T_H17 cells are distributed widely in most lymphoid and non-lymphoid tissues (Wang et al. 2009). However, the frequencies of T_H17 cells in most tissues are low (<0.5% of $CD4^+$ cells) in the absence of any ongoing immune responses. The intestine is the tissue where T_H17 cells are most highly enriched (Ivanov et al. 2008; Wang et al. 2009). 5–10% of $CD4^+$ T cells are T_H17 cells in the intestine. It appears that the frequencies of T_H17 cells in the intestine are determined by the housing environment and diet. The frequencies of T_H17 cells in the intestine are regulated by certain types of commensals in the gut (Ivanov et al. 2008; Niess et al. 2008; Zaph et al. 2008). In this regard, colonization of the mice with segmented filamentous bacteria was sufficient to induce T_H17 cells in the lamina propria of the small intestine (Ivanov et al. 2009). The low frequencies in other tissue sites reflect the fact that other tissues would lack these T_H17 cell-inducing factors, including commensals. Antibiotics such as vancomycin effectively suppress the population of gut T_H17 cells, presumably eliminating the T_H17 cell-inducing commensals including the segmented filamentous bacteria (Ivanov et al. 2008).

The thymus is the place where uncommitted naïve T cells are made. However, there is an indication that certain T cells that are made in the thymus are already committed into T_H17 cells (Marks et al. 2009). Marks et al. demonstrated this by using a transgenic TCR system that T_H17 cells can differentiates in the thymus in a manner influenced by recognition of self antigen and by the cytokines IL-6 and TGF-β1. These T_H17 cells have all the features of the peripheral T_H17 cells. A caveat with this study is the artificial nature intrinsic with the antigen-TCR transgenic mouse model, where T_H17 cells were abnormally increased in frequency in the thymus. The fact that even the T_H1 cell frequency was greatly increased in the thymus suggests this model is certainly abnormal. It is possible that the T_H17 cells in the thymus are the result of colonization from the periphery rather than induction. Despite the limitations, the study suggests the possibility that small numbers of T_H17 cells could be made in the thymus. Some suggested that there is a T_H17 cell precursor in the human thymus and periphery (Cosmi et al. 2008). These cells express CD4 and CD161. CD161 is associated with NKT cells but these cells are not NKT cells and can efficiently differentiate into T_H17 cells in response to IL-1β and IL-23 in vitro.

Outside the intestine, the major site of T_H17 cell production is probably the secondary lymphoid tissues. T_H17 cells migrate out of the secondary lymphoid tissues; and they settle into sites of inflammation and undergo expansion there.

As discussed, the intestinal mucosa is a major place where T_H17 cells are constitutively induced and expanded. Induction of T_H17 cells requires both TGF-β1 and certain types of cytokines that activate the transcription factor STAT3 (i.e. IL-6, IL-23, TNF-α, and IL-1β) (Annunziato et al. 2009; Dong 2009; Lee et al. 2009). In addition to STAT3, several transcription factors can promote the generation of T_H17 cells. These are ROR-γt, aryl hydrocarbon receptor (AHR), and ROR-α (Ivanov et al. 2006; Kimura et al. 2008; Veldhoen et al. 2008; Yang et al. 2008). The T_H17 cell-inducing cytokines activate these transcription factors to promote T cell polarization into T_H17 cells. Many cytokines that are associated with $FoxP3^+$ cells, T_H1, or T_H2 cells have suppressive effects on generation of T_H17 cells. These T_H17 cell-inhibiting cytokines include IL-2 (FoxP3, T_H1, and T_H2), IFN-γ, IL-12, IL-27 (T_H1), and IL-4 (T_H2) (Annunziato et al. 2009; Dong 2009; Lee et al. 2009). Thus, these cytokines promote T cell polarization into a particular T cell lineage at the expense of T_H17 cells for an apparently selfish reason. This suggests that T_H17 cells do not co-exist well with T_H1 and T_H2 cells during ongoing immune responses. It is common during inflammation to observe that when T_H1 cells are dominant there are few T_H17 cells found in the same tissue, and vice versa. Thus, T_H17 cells should have a trafficking behavior different from that of T_H1 cells to effectively mount a T_H17 cell response.

4 Trafficking Receptors of T_H17 Cells

Among all of the trafficking receptors examined so far by a number of scientists, CCR6 is the most T_H17-specific trafficking receptor (Acosta-Rodriguez et al. 2007; Annunziato et al. 2007; Brucklacher-Waldert et al. 2009; Kim 2009; Lim et al. 2008; Romagnani et al. 2009; Singh et al. 2008; Yamazaki et al. 2008). This is true for both mouse and human T_H17 cells. Virtually all of the T_H17 cells, regardless of their origin of tissues, express CCR6. CCR6 can be expressed by other T cells (Kleinewietfeld et al. 2005). Particularly, some $FoxP3^+$ T cells can express CCR6 (Kleinewietfeld et al. 2005; Lim et al. 2008; Lugering et al. 2005; Yamazaki et al. 2008). The reason why T_H17 cells and $FoxP3^+$ T cells express CCR6 is because CCR6 is induced by TGF-β1 (Wang et al. 2009), which is the cytokine that induces the generation of both T_H17 cells and $FoxP3^+$ T cells (Bettelli et al. 2006; Mangan et al. 2006). While most T_H17 cells express CCR6, some $FoxP3^+$ T cells do not express CCR6 (Wang et al. 2009). This is because IL-2 that helps to induce and maintain $FoxP3^+$ T cells can suppress the expression of CCR6 (Wang et al. 2009). Therefore, when there are both TGF-β1 and IL-2 at significant concentrations, $CCR6^-$ $FoxP3^+$ T cells are induced. When TGFβ1 and the T_H17-cell inducing inflammatory cytokines are present at high levels, $CCR6^+$ T_H17 cells are made (Fig. 1). When TGF-β1 is present with low levels of IL-2, $CCR6^+$ $FoxP3^+$ T cells would be generated (Wang et al. 2009).

In addition to CCR6, T_H17 cells express other trafficking receptors. T_H17 cells additionally express CXCR3, CCR5, CCR7, CD62L, α4β7, CCR4, and CXCR5

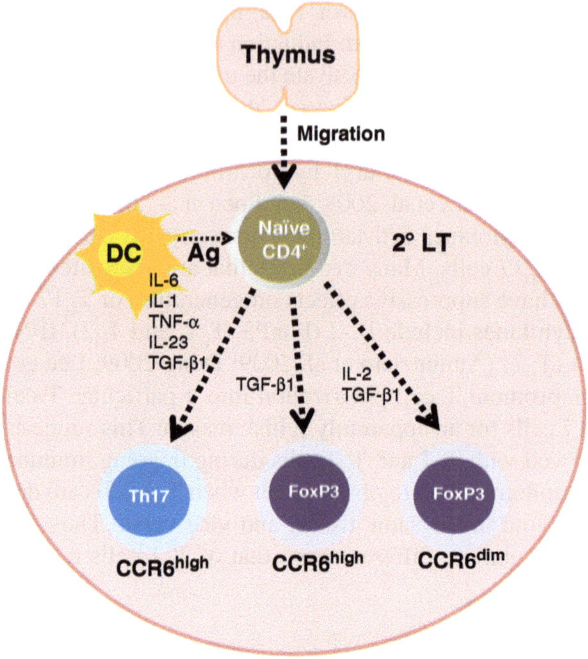

Fig. 1 Signals required for emergence of CCR6⁺ T_H17 cells. Naïve T cells, progenitors for T_H17 cells, are made in the thymus and migrate into secondary lymphoid tissues. Upon antigen priming, naïve T cells become effector T cells such as T_H17 cells. Normally, the induction rate of T_H17 cells from naïve T cells is very low in most tissues. This is because T_H17 cells are generated when the tissue environment is rich with IL-6, TGF-β1, IL-21, and IL-23, but these cytokines are not abundantly produced during general immune responses. However, naïve T cells become T_H17 cells in the intestine and associated tissues, and this appears to be due to filamentous bacterial commensals in the intestine. TGF-β1 plays a major role in both induction of T_H17 cells and their expression of CCR6. TGF-β1 also induces FoxP3⁺ T cells, but many FoxP3⁺ T cells do not express CCR6 because a Treg-inducing cytokine (IL-2) suppresses the expression of CCR6. CCR6 plays a role in migration of T_H17 cells into the gut and, potentially, other tissues as well

(Lim et al. 2008; Wang et al. 2009). Among these receptors, CXCR3 and CCR5 are frequently expressed by T_H1 cells (Kim et al. 2001). One should note that many more human peripheral T_H17 cells are found in the T cell population that do not express CXCR3 than in those that do express CXCR3 (Acosta-Rodriguez et al. 2007; Lim et al. 2008). However, this is not true in human tonsils where both T_H1 and T_H17 cells are expressed equallyCXCR3 (Lim et al. 2008).

CCR7 and CD62L are expressed by secondary lymphoid tissue-homing memory T cells. α4β7 and CCR9 are expressed by gut-homing T cells. CXCR5 is expressed by B cell follicle-homing T cells. CCR4 is expressed by many T_H2 cells (Andrew et al. 2001; Kim et al. 2001; Syrbe et al. 1999). One single T_H17 cell does not express all of the receptors. Rather, T_H17 cells are heterogeneous in that they express

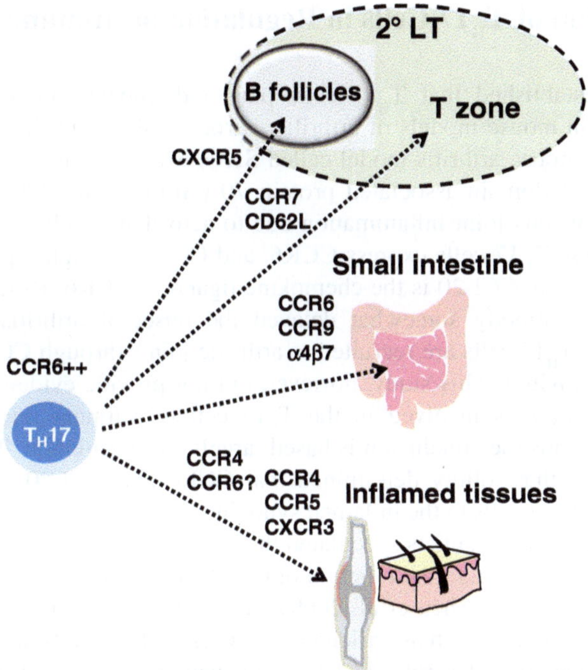

Fig. 2 Trafficking receptors and migration potential of T_H17 cells. While CCR6 is a chemokine receptor most widely expressed by T_H17 cells, T_H17 cells express diverse receptors that are shared with other effector T cell lineages. For example, many T_H17 cells express CXCR5 in the spleen and Peyer's patches and can migrate toward B cell follicles. It is unclear if T_H17 cells can actively help B cells in a manner similar to follicular T cells, which are specialized in helping B cells. T_H17 cells express CCR7 and CD62L to migrate into the T cell zone of secondary lymphoid tissues. T_H17 cells additionally express α4β7 and CCR9 to migrate to the intestine. More specifically, CCR9 is involved in localization of T_H17 cells to the small intestine. CCR6 is a homing receptor for T_H17 cells to Peyer's patches and small intestine. α4β7 is a general trafficking receptor required for T_H17 cell migration to the whole intestine and associated lymphoid tissues. T_H17 cells express CCR4, CXCR3, and/or CCR5 to migrate to various tissues sites of inflammation or infection

different combinations of the trafficking receptors (Lim et al. 2008; Wang et al. 2009). This suggests that T_H17 cells appear to be heterogeneous in migration behavior and perhaps in function as well (Fig. 2). The additional trafficking receptors would modify the trafficking behavior of T_H17 cells by working together with or even overriding the function of CCR6. In general, T cells express diverse chemokine receptors to navigate sequentially multiple chemo-attractant domains. T_H17 cells have a similar pattern of migration. In our study, the small intestine-residing T_H17 cells highly express CCR9 and α4β7 (Wang et al. 2009). The Peyer's patch and spleen-residing T_H17 cells highly expressed CXCR5. Secondary lymphoid tissue-residing T_H17 cells express CD62L and CCR7. Thus, T_H17 cells faithfully follow the general trafficking receptor expression rule of T cells (Kim et al. 2001, 2005).

5 Migration of T_H17 Cells in Regulation of Immune Responses

It has been established that T_H17 cells play a dominant role in induction of inflammation in mouse models of arthritis. Hirota et al. found that T_H17 cells are increased in a mouse arthritis model called SKG. Mice that are deficient with the Src homology 2 domain-associated protein 70 (Hirota et al. 2007). These mice develop spontaneous joint inflammation due to activated T cells. Similar to other T_H17 cells, these T_H17 cells express CCR6, and CCL20 is highly produced in the inflamed synovium. CCL20 is the chemokine ligand of CCR6. Blocking of CCR6 by anti-CCR6 antibody somewhat delayed the onset of arthritis. The authors concluded that T_H17 cells are recruited to arthritic joints through CCL20 produced by arthritic synovium. This study, however, did not provide evidence that CCR6-mediated migration is involved in the T_H17 cell recruitment into the inflamed synovium, and thus the conclusion is based largely on speculation. There has been no single study that solidly determined that CCR6 and CCL20 are involved in recruitment of T_H17 cells to the inflamed synovium. In a manner similar to the study by Hirota et al. many scientists speculated that T_H17 cells would be recruited to the inflamed synovium through the function of CCR6 (Jandus et al. 2008; Nistala et al. 2008; Pene et al. 2008; Shen et al. 2009). Again, their conclusions are based on circumstantial evidence such as enrichment of $CCR6^+$ T_H17 cells and expression of CCL20 in the synovium. The role of CCR6 in regulation of the migration of T_H17 cells in arthritis is still unclear. Similarly, while CCR6 is required for optimal induction of IL-23-injection-mediated psoriasis, the role of CCR6 in migration of T_H17 cells to the inflamed skin has not been determined (Hedrick et al. 2009).

Experimental allergic encephalomyelitis (EAE) is perhaps the most well-known animal model where T_H17 cells are involved in the pathogenesis. Yamazaki et al. reported that T_H17 cells regulate CCR6-dependent migration of T_H17 cells and Treg cells via production of CCL20. Therefore, CCR6-CCL20 axis regulates the function of T_H17 cells in EAE (Yamazaki et al. 2008). Again, this group did not provide evidence that CCR6 and CCL20 are involved in migration of T_H17 cells to inflamed tissues in a mouse model of EAE. They provided data that CCR6-deficiency led to somewhat delayed induction of EAE. If CCR6 would have a dominant role in migration of T_H17 cells to inflamed intestine, why could CCR6 deficiency not be able to prevent the onset and progression of the disease? Yamazaki et al. provided the data that CCR6-deficient T_H17 cell, when transferred, were not able to effectively induce EAE. Another group reported that the CCR6 ligand, CCL20, is constitutively expressed in epithelial cells of choroid plexus in mice and humans, and T_H17 cells migrate into the inflamed central nervous system via this area (Reboldi et al. 2009). However, other groups reported contradicting results. Elhofy et al. reported that CCR6-deficient mice developed significantly more severe inflammation as compared to wild type mice (Elhofy et al. 2009). This study noted decreased numbers of the PD-L1+ dendritic cell subset in the spleen and proposed that CCR6 may be involved in functions of certain regulatory dendritic cells. Overall, the role of CCR6-CCL20 in migration of T_H17 cells to the inflamed central nervous system in EAE has not been clearly established.

The role of CCR6 in migration of T_H17 cells to the intestine has been established (Wang et al. 2009). CCL20 is expressed in Peyer's patches and small intestine, while CCR6-deficient T_H17 cells fail to migrate to these tissues. Interestingly, CCR6-defcient T_H17 cells induced more vigorous inflammation in the intestine of Rag1-deficient mice compared to CCR6 (+/+) T_H17 cells. While the mechanism remains to be determined, T_H17 cells fail to migrate to the tissue area close to the epithelial cell layer and more readily trans-differentiate to inflammatory T_H1 cells in the absence of CCR6 (Wang et al. 2009). Thus, CCR6 appears to have a beneficial role in pathogenesis in the intestine.

6 Concluding Remarks

In this chapter, the trafficking receptors and migration behavior of T_H17 cells are discussed. Population of T_H17 cells in a certain tissue of the body involves not only migration but also retention and other non-migratory events. Non-migratory events include survival, proliferation, and maintenance of effector T cell identity, which are regulated by antigens, antigen presenting cells and the tissue cytokine milieu. In this regard, trafficking of T_H17 cells is highly coordinated with the non-migratory events to mount a sustaining T_H17 cell immune response. In the absence of such coordination, T_H17 cells transferred into certain tissues die quickly or trans-differentiate into other types of T cells such as T_H1 cells or FoxP3$^+$ T cells. As discussed in detail, the role of CCR6 in migration of T_H17 cells to various sites of inflammation has not been established. What is clear is that CCR6 is important for migration and retention of T_H17 cells within certain tissue sites in the intestine. More studies should be conducted to determine the specific roles of various trafficking receptors in not only T_H17 cell migration and but also T_H17-cell mediated immunity and pathogenesis.

References

Abdollahi-Roodsaz, S., Joosten, L.A., Koenders, M.I., Devesa, I., Roelofs, M.F., Radstake, T.R., Heuvelmans-Jacobs, M., Akira, S., Nicklin, M.J., Ribeiro-Dias, F., and van den Berg, W.B. (2008). Stimulation of TLR2 and TLR4 differentially skews the balance of T cells in a mouse model of arthritis. The Journal of clinical investigation *118*, 205–216.

Acosta-Rodriguez, E.V., Rivino, L., Geginat, J., Jarrossay, D., Gattorno, M., Lanzavecchia, A., Sallusto, F., and Napolitani, G. (2007). Surface phenotype and antigenic specificity of human interleukin 17-producing T helper memory cells. Nature immunology *8*, 639–646.

Andrew, D.P., Ruffing, N., Kim, C.H., Miao, W., Heath, H., Li, Y., Murphy, K., Campbell, J.J., Butcher, E.C., and Wu, L. (2001). C-C chemokine receptor 4 expression defines a major subset of circulating non-intestinal memory T cells of both Th1 and Th2 potential. J Immunol *166*, 103–111.

Annunziato, F., Cosmi, L., Liotta, F., Maggi, E., and Romagnani, S. (2009). Type 17 T helper cells-origins, features and possible roles in rheumatic disease. Nat Rev Rheumatol *5*, 325–331.

Annunziato, F., Cosmi, L., Santarlasci, V., Maggi, L., Liotta, F., Mazzinghi, B., Parente, E., Fili, L., Ferri, S., Frosali, F., *et al.* (2007). Phenotypic and functional features of human Th17 cells. The Journal of experimental medicine *204*, 1849–1861.

Austrup, F., Vestweber, D., Borges, E., Lohning, M., Brauer, R., Herz, U., Renz, H., Hallmann, R., Scheffold, A., Radbruch, A., and Hamann, A. (1997). P- and E-selectin mediate recruitment of T-helper-1 but not T-helper-2 cells into inflamed tissues. Nature *385*, 81–83.

Berlin, C., Berg, E.L., Briskin, M.J., Andrew, D.P., Kilshaw, P.J., Holzmann, B., Weissman, I.L., Hamann, A., and Butcher, E.C. (1993). Alpha 4 beta 7 integrin mediates lymphocyte binding to the mucosal vascular address in MAdCAM-1. Cell *74*, 185–195.

Bettelli, E., Carrier, Y., Gao, W., Korn, T., Strom, T.B., Oukka, M., Weiner, H.L., and Kuchroo, V.K. (2006). Reciprocal developmental pathways for the generation of pathogenic effector TH17 and regulatory T cells. Nature *441*, 235–238.

Bradley, L.M., Malo, M.E., Fong, S., Tonkonogy, S.L., and Watson, S.R. (1998). Blockade of both L-selectin and alpha4 integrins abrogates naive CD4 cell trafficking and responses in gut-associated lymphoid organs. International immunology *10*, 961–968.

Brucklacher-Waldert, V., Steinbach, K., Lioznov, M., Kolster, M., Holscher, C., and Tolosa, E. (2009). Phenotypical characterization of human Th17 cells unambiguously identified by surface IL-17A expression. J Immunol *183*, 5494–5501.

Cosmi, L., De Palma, R., Santarlasci, V., Maggi, L., Capone, M., Frosali, F., Rodolico, G., Querci, V., Abbate, G., Angeli, R., *et al.* (2008). Human interleukin 17-producing cells originate from a CD161+CD4+ T cell precursor. The Journal of experimental medicine *205*, 1903–1916.

Curtis, M.M., and Way, S.S. (2009). Interleukin-17 in host defense against bacterial, mycobacterial and fungal pathogens. Immunology *126*, 177–185.

Deenick, E.K., and Tangye, S.G. (2007). Autoimmunity: IL-21: a new player in Th17-cell differentiation. Immunology and cell biology *85*, 503–505.

Dong, C. (2009). Differentiation and function of pro-inflammatory Th17 cells. Microbes and infection/Institut Pasteur *11*, 584–588.

Dubin, P.J., and Kolls, J.K. (2008). Th17 cytokines and mucosal immunity. Immunological reviews *226*, 160–171.

Elhofy, A., Depaolo, R.W., Lira, S.A., Lukacs, N.W., and Karpus, W.J. (2009). Mice deficient for CCR6 fail to control chronic experimental auto-immune encephalomyelitis. Journal of neuroimmunology *213*, 91–99.

Forster, R., Schubel, A., Breitfeld, D., Kremmer, E., Renner-Muller, I., Wolf, E., and Lipp, M. (1999). CCR7 coordinates the primary immune response by establishing functional microenvironments in secondary lymphoid organs. Cell *99*, 23–33.

Fouser, L.A., Wright, J.F., Dunussi-Joannopoulos, K., and Collins, M. (2008). Th17 cytokines and their emerging roles in inflammation and autoi-mmunity. Immunological reviews *226*, 87–102.

Goetzl, E.J., Liao, J.J., and Huang, M.C. (2008). Regulation of the roles of sphingosine 1-phosphate and its type 1 G protein-coupled receptor in T cell immunity and auto-immunity. Biochim Biophys Acta *1781*, 503–507.

Gunn, M.D., Kyuwa, S., Tam, C., Kakiuchi, T., Matsuzawa, A., Williams, L.T., and Nakano, H. (1999). Mice lacking expression of secondary lymphoid organ chemokine have defects in lymphocyte homing and dendritic cell localization. The Journal of experimental medicine *189*, 451–460.

Hamann, A., Andrew, D.P., Jablonski-Westrich, D., Holzmann, B., and Butcher, E.C. (1994). Role of alpha 4-integrins in lymphocyte homing to mucosal tissues in vivo. J Immunol *152*, 3282–3293.

Hardtke, S., Ohl, L., and Forster, R. (2005). Balanced expression of CXCR5 and CCR7 on follicular T helper cells determines their transient positioning to lymph node follicles and is essential for efficient B-cell help. Blood *106*, 1924–1931.

Haynes, N.M., Allen, C.D., Lesley, R., Ansel, K.M., Killeen, N., and Cyster, J.G. (2007). Role of CXCR5 and CCR7 in follicular Th cell positioning and appearance of a programed cell death gene-1high germinal center-associated subpopulation. J Immunol *179*, 5099–5108.

Hedrick, M.N., Lonsdorf, A.S., Shirakawa, A.K., Richard Lee, C.C., Liao, F., Singh, S.P., Zhang, H.H., Grinberg, A., Love, P.E., Hwang, S.T., and Farber, J.M. (2009). CCR6 is required for IL-23-induced psoriasis-like inflammation in mice. The Journal of clinical investigation *119*, 2317–2329.

Hirota, K., Hashimoto, M., Yoshitomi, H., Tanaka, S., Nomura, T., Yamaguchi, T., Iwakura, Y., Sakaguchi, N., and Sakaguchi, S. (2007). T cell self-reactivity forms a cytokine milieu for spontaneous development of IL-17+ Th cells that cause auto-immune arthritis. The Journal of experimental medicine *204*, 41–47.

Hwang, S.T. (2001). Mechanisms of T-cell homing to skin. Adv Dermatol *17*, 211–241.

Ivanov, II, Atarashi, K., Manel, N., Brodie, E.L., Shima, T., Karaoz, U., Wei, D., Goldfarb, K.C., Santee, C.A., Lynch, S.V., *et al.* (2009). Induction of Intestinal Th17 Cells by Segmented Filamentous Bacteria. Cell 139, 485–98

Ivanov, II, Frutos Rde, L., Manel, N., Yoshinaga, K., Rifkin, D.B., Sartor, R.B., Finlay, B.B., and Littman, D.R. (2008). Specific microbiota direct the differentiation of IL-17-producing T-helper cells in the mucosa of the small intestine. Cell host & microbe *4*, 337–349.

Ivanov, II, McKenzie, B.S., Zhou, L., Tadokoro, C.E., Lepelley, A., Lafaille, J.J., Cua, D.J., and Littman, D.R. (2006). The orphan nuclear receptor RORgammat directs the differentiation program of proi-nflammatory IL-17+ T helper cells. Cell *126*, 1121–1133.

Iwakura, Y., Nakae, S., Saijo, S., and Ishigame, H. (2008). The roles of IL-17A in inflammatory immune responses and host defense against pathogens. Immunological reviews *226*, 57–79.

Jandus, C., Bioley, G., Rivals, J.P., Dudler, J., Speiser, D., and Romero, P. (2008). Increased numbers of circulating polyfunctional Th17 memory cells in patients with seronegative spondylarthritides. Arthritis and rheumatism *58*, 2307–2317.

Kattah, M.G., Wong, M.T., Yocum, M.D., and Utz, P.J. (2008). Cytokines secreted in response to Toll-like receptor ligand stimulation modulate differentiation of human Th17 cells. Arthritis and rheumatism *58*, 1619–1629.

Kellermann, S.A., Hudak, S., Oldham, E.R., Liu, Y.J., and McEvoy, L.M. (1999). The CC chemokine receptor-7 ligands 6Ckine and macrophage inflammatory protein-3 beta are potent chemo-attractants for in vitro- and in vivo-derived dendritic cells. J Immunol *162*, 3859–3864.

Khader, S.A., and Cooper, A.M. (2008). IL-23 and IL-17 in tuberculosis. Cytokine *41*, 79–83.

Kim, C.H. (2004). Chemokine-chemokine receptor network in immune cell trafficking. Curr Drug Targets Immune Endocr Metabol Disord *4*, 343–361.

Kim, C.H. (2005). The greater chemotactic network for lymphocyte trafficking: chemokines and beyond. Curr Opin Hematol *12*, 298–304.

Kim, C.H. (2006). Migration and function of FoxP3+ regulatory T cells in the hematolymphoid system. Exp Hematol *34*, 1033–1040.

Kim, C.H. (2009). Migration and function of Th17 cells. Inflammation & allergy drug targets *8*, 221–228.

Kim, C.H., Rott, L., Kunkel, E.J., Genovese, M.C., Andrew, D.P., Wu, L., and Butcher, E.C. (2001). Rules of chemokine receptor association with T cell polarization in vivo. The Journal of clinical investigation *108*, 1331–1339.

Kimura, A., Naka, T., Nohara, K., Fujii-Kuriyama, Y., and Kishimoto, T. (2008). Aryl hydrocarbon receptor regulates Stat1 activation and participates in the development of Th17 cells. Proceedings of the National Academy of Sciences of the United States of America *105*, 9721–9726.

Kleinewietfeld, M., Puentes, F., Borsellino, G., Battistini, L., Rotzschke, O., and Falk, K. (2005). CCR6 expression defines regulatory effector/memory-like cells within the CD25(+)CD4+ T-cell subset. Blood *105*, 2877–2886.

Kunkel, E.J., Campbell, J.J., Haraldsen, G., Pan, J., Boisvert, J., Roberts, A.I., Ebert, E.C., Vierra, M.A., Goodman, S.B., Genovese, M.C., *et al.* (2000). Lymphocyte CC chemokine receptor 9 and epithelial thymus-expressed chemokine (TECK) expression distinguish the small intestinal immune compartment: Epithelial expression of tissue-specific chemokines as an organizing principle in regional immunity. The Journal of experimental medicine *192*, 761–768.

Lee, J.H., Kang, S.G., and Kim, C.H. (2007). FoxP3+ T Cells Undergo Conventional First Switch to Lymphoid Tissue Homing Receptors in Thymus but Accelerated Second Switch to Non-lymphoid Tissue Homing Receptors in Secondary Lymphoid Tissues. J Immunol *178*, 301–311.

Lee, Y.K., Mukasa, R., Hatton, R.D., and Weaver, C.T. (2009). Developmental plasticity of Th17 and Treg cells. Current opinion in immunology *21*, 274–280.

Lim, H.W., Broxmeyer, H.E., and Kim, C.H. (2006). Regulation of Trafficking Receptor Expression in Human Forkhead Box P3+ Regulatory T Cells. J Immunol *177*, 840–851.

Lim, H.W., Hillsamer, P., and Kim, C.H. (2004). Regulatory T cells can migrate to follicles upon T cell activation and suppress GC-Th cells and GC-Th cell-driven B cell responses. The Journal of clinical investigation *114*, 1640–1649.

Lim, H.W., Lee, J., Hillsamer, P., and Kim, C.H. (2008). Human Th17 cells share major trafficking receptors with both polarized effector T cells and FOXP3+ regulatory T cells. J Immunol *180*, 122–129.

Liu, J.Z., Pezeshki, M., and Raffatellu, M. (2009). Th17 cytokines and host-pathogen interactions at the mucosa: dichotomies of help and harm. Cytokine *48*, 156–160.

Lo, C.G., Xu, Y., Proia, R.L., and Cyster, J.G. (2005). Cyclical modulation of sphingosine-1-phosphate receptor 1 surface expression during lymphocyte recirculation and relationship to lymphoid organ transit. The Journal of experimental medicine *201*, 291–301.

Lugering, A., Floer, M., Westphal, S., Maaser, C., Spahn, T.W., Schmidt, M.A., Domschke, W., Williams, I.R., and Kucharzik, T. (2005). Absence of CCR6 inhibits CD4+ regulatory T-cell development and M-cell formation inside Peyer's patches. Am J Pathol *166*, 1647–1654.

Mangan, P.R., Harrington, L.E., O'Quinn, D.B., Helms, W.S., Bullard, D.C., Elson, C.O., Hatton, R.D., Wahl, S.M., Schoeb, T.R., and Weaver, C.T. (2006). Transforming growth factor-beta induces development of the T(H)17 lineage. Nature *441*, 231–234.

Marelli-Berg, F.M., Cannella, L., Dazzi, F., and Mirenda, V. (2008). The highway code of T cell trafficking. J Pathol *214*, 179–189.

Marks, B.R., Nowyhed, H.N., Choi, J.Y., Poholek, A.C., Odegard, J.M., Flavell, R.A., and Craft, J. (2009). Thymic self-reactivity selects natural interleukin 17-producing T cells that can regulate peripheral inflammation. Nature immunology *10*, 1125–1132.

Niess, J.H., Leithauser, F., Adler, G., and Reimann, J. (2008). Commensal gut flora drives the expansion of pro-inflammatory CD4 T cells in the colonic lamina propria under normal and inflammatory conditions. J Immunol *180*, 559–568.

Nistala, K., Moncrieffe, H., Newton, K.R., Varsani, H., Hunter, P., and Wedderburn, L.R. (2008). Interleukin-17-producing T cells are enriched in the joints of children with arthritis, but have a reciprocal relationship to regulatory T cell numbers. Arthritis and rheumatism *58*, 875–887.

O'Quinn, D.B., Palmer, M.T., Lee, Y.K., and Weaver, C.T. (2008). Emergence of the Th17 pathway and its role in host defense. Advances in immunology *99*, 115–163.

Papadakis, K.A., Prehn, J., Nelson, V., Cheng, L., Binder, S.W., Ponath, P.D., Andrew, D.P., and Targan, S.R. (2000). The role of thymus-expressed chemokine and its receptor CCR9 on lymphocytes in the regional specialization of the mucosal immune system. J Immunol *165*, 5069–5076.

Pene, J., Chevalier, S., Preisser, L., Venereau, E., Guilleux, M.H., Ghannam, S., Moles, J.P., Danger, Y., Ravon, E., Lesaux, S., *et al.* (2008). Chronically inflamed human tissues are infiltrated by highly differentiated Th17 lymphocytes. J Immunol *180*, 7423–7430.

Reboldi, A., Coisne, C., Baumjohann, D., Benvenuto, F., Bottinelli, D., Lira, S., Uccelli, A., Lanzavecchia, A., Engelhardt, B., and Sallusto, F. (2009). C-C chemokine receptor 6-regulated entry of TH-17 cells into the CNS through the choroid plexus is required for the initiation of EAE. Nature immunology *10*, 514–523.

Romagnani, S., Maggi, E., Liotta, F., Cosmi, L., and Annunziato, F. (2009). Properties and origin of human Th17 cells. Mol Immunol *47*, 3–7.

Rott, L.S., Rose, J.R., Bass, D., Williams, M.B., Greenberg, H.B., and Butcher, E.C. (1997). Expression of mucosal homing receptor alpha4beta7 by circulating CD4+ cells with memory for intestinal rotavirus. The Journal of clinical investigation *100*, 1204–1208.

Saeki, H., Moore, A.M., Brown, M.J., and Hwang, S.T. (1999). Cutting edge: secondary lymphoid-tissue chemokine (SLC) and CC chemokine receptor 7 (CCR7) participate in the emigration pathway of mature dendritic cells from the skin to regional lymph nodes. J Immunol *162*, 2472–2475.

Sanna, M.G., Liao, J., Jo, E., Alfonso, C., Ahn, M.Y., Peterson, M.S., Webb, B., Lefebvre, S., Chun, J., Gray, N., and Rosen, H. (2004). Sphingosine 1-phosphate (S1P) receptor subtypes S1P1 and S1P3, respectively, regulate lymphocyte recirculation and heart rate. The Journal of biological chemistry *279*, 13839–13848.

Schaerli, P., Ebert, L., Willimann, K., Blaser, A., Roos, R.S., Loetscher, P., and Moser, B. (2004). A Skin-selective Homing Mechanism for Human Immune Surveillance T Cells. The Journal of experimental medicine *199*, 1265–1275.

Schwab, S.R., and Cyster, J.G. (2007). Finding a way out: lymphocyte egress from lymphoid organs. Nature immunology *8*, 1295–1301.

Schwab, S.R., Pereira, J.P., Matloubian, M., Xu, Y., Huang, Y., and Cyster, J.G. (2005). Lymphocyte sequestration through S1P lyase inhibition and disruption of S1P gradients. Science (New York, N.Y) *309*, 1735–1739.

Shen, H., Goodall, J.C., and Hill Gaston, J.S. (2009). Frequency and phenotype of peripheral blood Th17 cells in ankylosing spondylitis and rheumatoid arthritis. Arthritis and rheumatism *60*, 1647–1656.

Singh, S.P., Zhang, H.H., Foley, J.F., Hedrick, M.N., and Farber, J.M. (2008). Human T cells that are able to produce IL-17 express the chemokine receptor CCR6. J Immunol *180*, 214–221.

Stekel, D.J., Parker, C.E., and Nowak, M.A. (1997). A model of lymphocyte recirculation. Immunology today *18*, 216–221.

Stenstad, H., Ericsson, A., Johansson-Lindbom, B., Svensson, M., Marsal, J., Mack, M., Picarella, D., Soler, D., Marquez, G., Briskin, M., and Agace, W.W. (2006). Gut-associated lymphoid tissue-primed CD4+ T cells display CCR9-dependent and -independent homing to the small intestine. Blood *107*, 3447–3454.

Svensson, M., Marsal, J., Ericsson, A., Carramolino, L., Broden, T., Marquez, G., and Agace, W.W. (2002). CCL25 mediates the localization of recently activated CD8alphabeta(+) lymphocytes to the small-intestinal mucosa. The Journal of clinical investigation *110*, 1113–1121.

Syrbe, U., Siveke, J., and Hamann, A. (1999). Th1/Th2 subsets: distinct differences in homing and chemokine receptor expression. Springer Semin Immunopathol *21*, 263–285.

Tanaka, J., Watanabe, N., Kido, M., Saga, K., Akamatsu, T., Nishio, A., and Chiba, T. (2009). Human TSLP and TLR3 ligands promote differentiation of Th17 cells with a central memory phenotype under Th2-polarizing conditions. Clin Exp Allergy *39*, 89–100.

Tedder, T.F., Steeber, D.A., Chen, A., and Engel, P. (1995). The selectins: vascular adhesion molecules. Faseb J *9*, 866–873.

Tesmer, L.A., Lundy, S.K., Sarkar, S., and Fox, D.A. (2008). Th17 cells in human disease. Immunological reviews *223*, 87–113.

van den Berg, W.B., and Miossec, P. (2009). IL-17 as a future therapeutic target for rheumatoid arthritis. Nat Rev Rheumatol *5*, 549–553.

Veldhoen, M., Hirota, K., Westendorf, A.M., Buer, J., Dumoutier, L., Renauld, J.C., and Stockinger, B. (2008). The aryl hydrocarbon receptor links TH17-cell-mediated auto-immunity to environmental toxins. Nature *453*, 106–109.

von Andrian, U.H., and M'Rini, C. (1998). In situ analysis of lymphocyte migration to lymph nodes. Cell adhesion and communication *6*, 85–96.

Wagner, N., Lohler, J., Kunkel, E.J., Ley, K., Leung, E., Krissansen, G., Rajewsky, K., and Muller, W. (1996). Critical role for beta7 integrins in formation of the gut-associated lymphoid tissue. Nature *382*, 366–370.

Wang, C., Kang, S.G., Lee, J., Sun, Z., and Kim, C.H. (2009). The roles of CCR6 in migration of Th17 cells and regulation of effector T-cell balance in the gut. Mucosal immunology *2*, 173–183.

Warnock, R.A., Campbell, J.J., Dorf, M.E., Matsuzawa, A., McEvoy, L.M., and Butcher, E.C. (2000). The role of chemokines in the micro-environmental control of T versus B cell arrest in Peyer's patch high endothelial venules. The Journal of experimental medicine *191*, 77–88.

Williams, M.B., and Butcher, E.C. (1997). Homing of naive and memory T lymphocyte subsets to Peyer's patches, lymph nodes, and spleen. J Immunol *159*, 1746–1752.

Wurbel, M.A., Malissen, M., Guy-Grand, D., Meffre, E., Nussenzweig, M.C., Richelme, M., Carrier, A., and Malissen, B. (2001). Mice lacking the CCR9 CC-chemokine receptor show a mild impairment of early T- and B-cell development and a reduction in T-cell receptor gammadelta(+) gut intraepithelial lymphocytes. Blood *98*, 2626–2632.

Yamazaki, T., Yang, X.O., Chung, Y., Fukunaga, A., Nurieva, R., Pappu, B., Martin-Orozco, N., Kang, H.S., Ma, L., Panopoulos, A.D., *et al.* (2008). CCR6 regulates the migration of inflammatory and regulatory T cells. J Immunol *181*, 8391–8401.

Yang, X.O., Pappu, B.P., Nurieva, R., Akimzhanov, A., Kang, H.S., Chung, Y., Ma, L., Shah, B., Panopoulos, A.D., Schluns, K.S., *et al.* (2008). T Helper 17 Lineage Differentiation Is Programmed by Orphan Nuclear Receptors RORalpha and RORgamma. Immunity *28*, 29–39.

Zaph, C., Du, Y., Saenz, S.A., Nair, M.G., Perrigoue, J.G., Taylor, B.C., Troy, A.E., Kobuley, D.E., Kastelein, R.A., Cua, D.J., *et al.* (2008). Commensal-dependent expression of IL-25 regulates the IL-23-IL-17 axis in the intestine. The Journal of experimental medicine *205*, 2191–2198.

Zhou, L., Chong, M.M., and Littman, D.R. (2009). Plasticity of CD4+ T cell lineage differentiation. Immunity *30*, 646–655.

Zhou, M., Yang, B., Ma, R., and Wu, C. (2008). Memory Th-17 cells specific for C. albicans are persistent in human peripheral blood. Immunology letters *118*, 72–81.

Plastic T Cells: Recycling Effector Functions?

Eleftherios I. Agorogiannis, Stephen P. Cobbold, and Herman Waldmann

Abstract Polarized helper T cells were once thought to represent stable lineages with limited or no potential to ac predictable behavior for in vivo adoptive transfer studies should allow proper assessment of the role of individual T cell subsets in immune-mediated tissue damage.

1 Introduction

The T_H1/T_H2 paradigm (Mosmann et al. 1986) dominated immunological thought for almost two decades, as it was considered sufficient to explain the ability of the immune system to mount tailored responses to diverse classes of invading pathogens and describe the pathogenesis of immune-mediated tissue damage in auto-immune diseases, allergy and transplant rejection. The definitive identification of $CD4^+FoxP3^+$ regulatory T (T_{reg}) cells (Fontenot et al. 2003; Hori et al. 2003) ended a long-standing dispute over the existence of committed immune-suppressing cells and satisfied the need for a fail-safe mechanism able to prevent collateral damage due to excessive T_H1 and T_H2 activity. The discovery of IL-17A (Yao et al. 1995), IL-23 (Oppmann et al. 2000), and the observation that the latter is implicated in several auto-immune disorders (Cua et al. 2003; Langrish et al. 2005; Murphy et al. 2003) laid the foundations for the identification of T_H17 cells as a third independent effector $CD4^+$ T cell subset (Harrington et al. 2005; Park et al. 2005). Consequently, the immunological community became more receptive to additional helper T cell subsets, even though earlier studies of murine and human $CD4^+$ T cell clones had already argued that the T_H1/T_H2 model could not efficiently accommo-

H. Waldmann (✉)
Sir William Dunn School of Pathology, University of Oxford, South Parks Road, Oxford, OX1 3RE, UK
e-mail: herman.waldmann@path.ox.ac.uk

S. Jiang (ed.), *T_H17 Cells in Health and Disease*,
DOI 10.1007/978-1-4419-9371-7_11,

date the range of cytokines produced in the immune response (Kelso 1995; Kelso and Gough 1988). Accordingly, the recent description of T_H9 cells as yet another separate CD4+ T cell subset (Dardalhon et al. 2008; Veldhoen et al. 2008) met with little scepticism, although the combined effect of TGFβ and IL-4 to induce IL-9 production had been known for 14 years (Schmitt et al. 1994). This was accompanied by the identification of follicular helper T (T_{FH}) cells (Yu et al. 2009) and T_H22 cells (Duhen et al. 2009; Trifari et al. 2009).

The initial observations that culminated in the description of the T_H1/T_H2 model were based on the study of long-term T cell clones (Mosmann et al. 1986), leading to the prediction that polarized CD4+ T cells should retain a stable profile of cytokine expression (Mosmann and Coffman 1989). It was later shown that individual cells within cloned populations exhibit variable patterns of cytokine production (Kelso and Gough 1988; Openshaw et al. 1995) and that short-term activated T_H1 and T_H2 cells may still express cytokines of the opposing lineage when cultured in the presence of IL-4 or IL-12 (Murphy et al. 1996). However, such reversibility was not observed when established CD4+ T cell clones or T cells activated for longer periods were subjected to changes in culture conditions (Murphy et al. 1996). IL-4 was still able to decrease the secretion of IFNγ by stable T_H1 clones and IL-12 was able to induce transient expression of IFNγ by fully polarized T_H2 cells (Manetti et al. 1994; Murphy et al. 1996), implying that even T cell clones retain a limited ability to adapt to the cytokine micro-environment. These findings indicated that long-term exposure to polarizing conditions, coupled with antigenic activation, could bestow a stable phenotype on helper T cells, but there remained a narrow time window after initial activation during which cells maintained the capacity to modify their phenotype.

Studies on T_H1 and T_H2 cells generated in vitro and subsequently transferred in vivo showed that 8 weeks after transfer, recovered cells maintained their memory of previous cytokine expression, and secreted significant amounts of their signature cytokines following antigen-specific re-stimulation (Swain 1994). Other observations suggested that protection against lymphocytic choriomeningitis virus (LCMV) infection, following adoptive transfer of in vitro polarized LCMV-specific T_H2 cells, relied on interferon-dependent upregulation of T-bet and IFNγ, leading to the development of T-bet+GATA-3+ cells producing both IL-4 and IFNγ (Hegazy et al. 2010). In addition, in vivo generated memory T_H1 and T_H2 cells that can produce IL-4 and IFNγ under opposing polarizing conditions (Krawczyk et al. 2007). Collectively, these findings indicated that both T_H1 and T_H2 cells display significant phenotypic flexibility contingent upon the context within which secondary stimulation takes place.

The capacity of polarized T_H1 and T_H2 cells to express both IFNγ and IL-4 under appropriate circumstances suggested that the stable T_H1 and T_H2 clones described by Mosmann and Coffman most probably represented the two extremes of a wide spectrum of potential CD4+ T cell differentiation states. These "hybrid" phenotypes would be explained with the assumption of a third effector subset, termed T_H0, stably combining features of both T_H1 and T_H2 cells (Firestein et al. 1989). However, such interpretation eventually left little room to speculate that phenotypic plasticity could be a major determinant of helper T cell function.

2 Studies on Additional CD4$^+$ T Cell Subsets Reveal Phenotypic Flexibility as a Common Feature of Helper T Cell Responses

With the demonstration of the requirement of TGFβ for the induction of both T_{reg} and T_H17 cells and the ability of IL-6 to promote the differentiation of the latter (Bettelli et al. 2006; Mangan et al. 2006; Veldhoen et al. 2006), it became evident that combinatorial cytokine signaling on T cells can induce phenotypes distinct from those emerging from the influence of each individual cytokine. This dependence of different subsets on the same cytokine for differentiation implies that there may be significant similarities in the phenotype and behavior of certain subsets and raises the question what the relationship among the different effector cell populations is.

After realizing that an inflammatory cytokine like IL-6 was able to divert the immune regulatory potential of TGFβ to pro-inflammatory activity and the induction of T_H17 cells, it was soon reported that IL-6 and/or other inflammatory cytokines could induce both murine and human isolated T_{reg} cells to produce IL-17A in vitro (Beriou et al. 2009; Koenen et al. 2008; Xu et al. 2007). An inflammatory environment seemed to mediate in vivo loss of the T_{reg}-specific transcription factor FoxP3 allowing the expression of IL-17A and other effector cytokines (Yang et al. 2008; Zhou et al. 2009). Downregulation of FoxP3 was accompanied by the loss of other T_{reg}-associated markers like CD25, GITR, and CTLA-4 (Zhou et al. 2009), and could be observed even after in vivo transfer into both lymphopenic and wild-type hosts in the absence of antigen (Komatsu et al. 2009). Importantly, high expression of CD25 seemed to define a population of T_{reg} cells that resisted such conversion implying considerable heterogeneity within the natural T_{reg} cell pool (Komatsu et al. 2009).

In addition, natural T_{reg} cells have been shown to differentiate into T_{FH} cells in the Peyer's patches of T cell-deficient mice in a process that requires downregulation of FoxP3 expression and interaction with B cells (Tsuji et al. 2009). T_{FH} cells help B cells by providing essential signals for isotype switching and plasma cell differentiation (Yu et al. 2009) and recent evidence suggests that in helminth infections these T_{FH} cells may be derived from T_H2 cells (Zaretsky et al. 2009). On the other hand, T_H2 cells can be induced to produce IL-9 in the presence of TGFβ (Veldhoen et al. 2008), further emphasizing the capacity of the tissue milieu to modify the pattern of cytokine expression of preformed effectors. This dynamic relationship among different subsets implies that the functional diversity of the adaptive immune system extends beyond the anticipation of the T_H1/T_H2 model for cells that assumed a limited adaptability of their alternative subsets to their highly specialized roles.

3 Auto-immune Disease Models Suggest Phenotypic Plasticity of T_H17 Cells

A large body of evidence derived from both experimental and clinical studies suggests that T_H17 cells hold a prominent position in the pathogenesis of auto-immune diseases, like multiple sclerosis and rheumatoid arthritis (Korn et al. 2009; Miossec et al. 2009). These had been traditionally linked to aberrant T_H1 responses mainly due to the efficacy of anti-p40 antibodies in reducing disease severity (Constantinescu et al. 1998) and the absence of disease in $p40^{-/-}$ mice (Segal et al. 1998), as it was unknown that the p40 sub-unit is shared between IL-12 and IL-23 (Hunter 2005; Oppmann et al. 2000). The role of IL-17A had been suspected before the recognition of T_H17 cells as a distinct T cell subset (Nakae et al. 2003a, b). The identification of T_H17-inducing factor (Bettelli et al. 2006; Korn et al. 2007; Mangan et al. 2006; Nurieva et al. 2007; Veldhoen et al. 2006; Zhou et al. 2007) RORγt (Ivanov et al. 2006) as a master transcriptional regulator and the observation that these cells produce a panoply of pro-inflammatory cytokines (Korn et al. 2009; Miossec et al. 2009), motivated more thorough understanding of the involvement of T_H17 cells in auto-immunity. T_H17 cells and their associated cytokines came to provide new targets for future therapeutic interventions. T_H17 cells were then suggested to be the sole player in certain auto-immune diseases, whereas a protective role was assumed for IFNγ-producing T_H1 cells. It was later shown, however, that both T_H1 and T_H17 cells are able to induce tissue damage (Kroenke et al. 2008; Luger et al. 2008), which is compatible with earlier findings that *t-bet*$^{-/-}$ mice are resistant to experimental auto-immune encephalomyelitis (Bettelli et al. 2004) and *rorγ*$^{-/-}$ mice show only mild disease manifestations (Ivanov et al. 2006).

Many of these observations relied on in vivo adoptive transfer of T_H1 or T_H17 cells, but these revealed a capacity of transferred T_H17 cells to produce both IL-17A and IFNγ in vivo, unlike transferred T_H1 cells which did not express IL-17A (Lee et al. 2009; O'Connor et al. 2008). Additional studies suggested that conversion into IFNγ-producing cells after in vivo transfer was a prerequisite for the diabetogenic potential of T_H17 cells that could be abrogated by antibodies against IFNγ, but not IL-17A (Bending et al. 2009; Martin-Orozco et al. 2009). T_H17 cells used in these adoptive transfer studies were generated by short-term in vitro activation in the presence of polarizing conditions, or after re-stimulation of T cells primed in vivo with IL-23 in vitro. In some cases cells were sorted on the basis of IL-17F (Croxford et al. 2009; Lee et al. 2009; Nurieva et al. 2009) or IL-17A expression and absence of IFNγ secretion (Bending et al. 2009), providing unequivocal evidence that the observed acquisition of IFNγ expression by transferred cells was not due to contaminating non-polarized or T_H1 cells.

What could the biological significance of this capacity of seemingly differentiated $CD4^+$ T cells to acquire phenotypic characteristics of other subsets be? It is certainly tempting to imagine the immune system as a flexible organization of plastic elements that can adjust their phenotype depending on the tissue microenvironment. Perhaps such flexibility has evolved to accelerate secondary immune responses against the same antigen by exploiting established memory cells through fine tuning of their cytokine profile in order to provide appropriate tissue-tailored immunity. This may therefore be a favorable property allowing rapid establishment of protective immunity.

Regarding the demonstrated propensity of T_H17 cells to lose IL-17A expression and produce IFNγ, it is worth considering again the common requirement of T_H17 and regulatory T cells for TGFβ. If pro-inflammatory cytokines act as a binary switch controlling the development of T_H17 cells at the expense of regulatory T cell induction, then T_H17 cells could have emerged in evolution as a mechanism that enforces protective immunity by preventing the establishment of tolerance. Once the immune system has decided to respond against a potential threat, T_H17 cells can acquire alternative phenotypes able to confer long-lasting protection. In such cases, T_H17 cells would be short-lived but might still hold an important role by preparing the inflammatory scene to accept the influx of T_H1 or other immune effector cells, For example, in recall responses against *Mycobacterium tuberculosis* (Khader et al. 2007).

4 Do Current Murine Auto-immunity Models Faithfully Represent Human Disease?

How can we reconcile the findings in experimental models demonstrating the instability of transferred T_H17 cells with their presumed role in the pathogenesis of chronic inflammatory disorders? In principle, it seems counter-intuitive that we can detect memory T_H17 cells in both humans (Acosta-Rodriguez et al. 2007; Wilson et al. 2007) and mice (Lexberg et al. 2008) when animal studies suggest that these cells are prone to loss of IL-17A expression and upregulation of IFNγ. Apparently, T_H17 cells should have dedicated effector functions beyond the early stages of immune activation, even though it is becoming increasingly accepted that, at least in certain auto-immune disease models, T_H17-derived cytokines may not hold a dominant pathogenic role (Haak et al. 2009; Luger et al. 2008). One explanation is that in vitro activated T_H17 cells may differ from in vivo primed cells. Ex vivo isolated memory T_H17 cells maintain IL-17A expression and show only minor upregulation of IFNγ or IL-4 upon re-stimulation with IL-12 or IL-4 (Lexberg et al. 2008), whereas in vitro generated T_H17 cells rapidly downregulate IL-17A under these conditions (Bending et al. 2009; Lee et al. 2009; Lexberg et al. 2008).

Large scale profiling of histone modifications accompanying $CD4^+$ T cell differentiation suggested that helper T cells retain the capacity to express genes associated with different subsets (Wei et al. 2009) in accordance with the phenotypic flexibility observed. Short-term cultured cells were assayed in these experiments which were similar to the populations used for the induction of tissue-specific auto-immunity in most adoptive transfer studies. It is uncertain though, whether these faithfully represent $CD4^+$ T cells found in vivo. It would be informative to know whether ex vivo isolated memory cells share similar epigenetic profiles. As this may not be technically feasible, cells activated in vitro for longer periods or "stable" T cell clones may allow for better testing of the epigenetic modifications underlying helper T cell development and function by minimizing the heterogeneity of tested populations. Although there is considerable scepticism whether prolonged growth in culture can reliably depict physiological T cell states (Bluestone et al. 2009; O'Shea and Paul 2010), predictions on the functional adaptability of helper T cells have for many years been made successfully by using T cell clones, long before the use of elegant reporter systems allowed efficient and highly specific T cell isolation.

Short-term $CD4^+$ and $CD8^+$ T cell clones display great variability of cytokine expression, implying that the observed plasticity lies at the single-cell level (Kelso and Gough 1988; Kelso and Groves 1997). It is then probable that early after activation cells may express combinations of genes that do not necessarily comply with a specific mode of adaptive response. As the effector phenotype depends primarily on environmental cues, a naive T cell recognizing cognate antigen has to integrate a wide variety of sometimes opposing cytokine stimuli. Therefore, it should be expected that cells might express features of seemingly opposing effector fates. Transient expression of FoxP3 is common in recently activated human $CD4^+$ T cells (Wang et al. 2007). Perhaps we are also misplaced in thinking of the differentiation process as acquisition of a specific phenotype rather than loss of the potential for functional plasticity. In such case, cells that have not yet reached their terminal differentiation state are the ones being examined.

In addition, repeated exposure of T cells to antigenic stimuli in the context of a uniform cytokine micro-environment, as in cases of chronic tissue-specific inflammation or secondary infectious challenges in the same anatomical compartment, may induce stable epigenetic modifications of critical loci allowing cells to maintain a dedicated cytokine expression program. Long-term T cell lines or clones may better depict such "stable" memory cells. Our unpublished observations show that antigen-specific T_H17 cells generated after long-term culture (at least 12 weeks) maintain IL-17A expression without upregulation of IFNγ, even after withdrawal of polarizing conditions in vitro or after adoptive transfer into lymphopenic hosts. The molecular program of memory cells may still be challenged when they enter inflamed tissues in which the cytokine milieu dictates a different effector phenotype. They might produce diverse cytokines without necessarily losing their memory of previous effector states. In fact, the current trend is to analyze the effects of subjecting helper T cells to opposing conditions without consideration of whether they can maintain their original phenotype upon withdrawing these or reinstating the initial conditions.

It is proper to conclude then, that helper T cells are in practice neither plastic nor subject to stochastic production of random cytokine combinations; they just retain a plastic potential that can be unleashed under appropriate circumstances. Moreover, an unambiguous way of defining helper T cell subsets is currently lacking, so that immunologists may describe these as expressing distinct combinations of transcription factors or cytokines or both. Expression of two opposing transcription factors should not be direct evidence of a hybrid phenotype unless it becomes apparent that both are competent to gain access to cognate transcription sites and functionally important for cellular activity. Effectors cells may occasionally need to utilize specific cytokines or chemokine receptors in order to fulfil their highly specialized roles. In a similar fashion, T_{reg} cells need to express T-bet, IRF4, or STAT3 in order to efficiently downregulate T_H1, T_H2, or T_H17 responses, suggesting that dedicated adaptation of a cell's molecular program to the characteristics of the habitat is important for proper function (Chaudhry et al. 2009; Koch et al. 2009; Zheng et al. 2009).

5 Conclusion

The functional multiplicity and heterogeneity of the immune response is unambiguous. It is possible that T cells participating in a given immune response consist of sub-populations that stably express a specific effector phenotype together with sub-populations that retain the potential to acquire additional or alternative effector properties, depending on the cytokine micro-environment. The incipient immune event in the pathogenesis of chronic auto-immune diseases usually takes place several years or even decades before clinical manifestations occur, at which point inflamed tissues may contain cells with different antigen recognition capacities and variable exposure to polarizing cytokine stimuli. Accumulating evidence also suggests that different helper T cell subsets are able to induce tissue-specific auto-immunity (Kroenke et al. 2008; Luger et al. 2008).

In graft rejection, both memory and naive T cells contribute to tissue damage. In fact, the transplantation picture is even more confusing. The immune response during transplant rejection seems to be dominated by opportunism, as diverse cell types like $CD8^+$ T cells, B cells, and natural killer cells can all participate (Alegre et al. 2007). The ability of T_H1 and T_H2 cells to induce graft rejection on their own is well established (Zelenika et al. 1998). Experimental and clinical evidence suggests an early role for IL-17A in mediating graft rejection (Tang et al. 2001; Vanaudenaerde et al. 2006), which would fit with the potential of IL-17A to attract neutrophils to the site of inflammation and the well established relationship between neutrophil infiltration and acute graft rejection (Alegre et al. 2007). Nevertheless, we have failed to observe any evidence of IL-17A upregulation in rejecting skin grafts after both major histocompatibility complex-mismatched and single minor antigen-mismatched transplantation (unpublished data). Published literature suggests a major role for $CD4^+$ and $CD8^+$ IL-17A-producing cells in *t-bet*$^{-/-}$ hosts that are unable to mount effective T_H1 responses (Burrell et al. 2008; Yuan et al. 2008, 2009). This means that T_H1 responses may normally dominate over T_H17 cells in graft rejection, but that involvement of T_H17 cells may become clinically relevant in cases where immuno-suppressive treatments preferentially inhibit development of T_H1 and T_H2 cells, or when graft antigens are being recognized by pre-existent memory T_H17 cells (heterologous immunity) (Adams et al. 2003).

It is crucial then, that adoptive transfer of isolated populations with predictable phenotype and behavior in vivo should be utilized to gain knowledge of function. The use of long-term T cell lines may allow more careful dissection of the mechanisms underlying auto-immune pathogenesis and graft rejection. Existing technology now allows a comprehensive analysis of helper T cell subsets and thorough description of their developmental and functional relationship. Proper assessment of the real diversity of the adaptive immune response will require examination of the complete variety of T cell differentiation states.

References

Acosta-Rodriguez, E.V., Rivino, L., Geginat, J., Jarrossay, D., Gattorno, M., Lanzavecchia, A., Sallusto, F., and Napolitani, G. (2007). Surface phenotype and antigenic specificity of human interleukin 17-producing T helper memory cells. Nat Immunol *8*, 639–646.

Adams, A.B., Pearson, T.C., and Larsen, C.P. (2003). Heterologous immunity: an overlooked barrier to tolerance. Immunol Rev *196*, 147–160.

Alegre, M.L., Florquin, S., and Goldman, M. (2007). Cellular mechanisms underlying acute graft rejection: time for reassessment. Curr Opin Immunol *19*, 563–568.

Bending, D., De La Pena, H., Veldhoen, M., Phillips, J.M., Uyttenhove, C., Stockinger, B., and Cooke, A. (2009). Highly purified Th17 cells from BDC2.5NOD mice convert into Th1-like cells in NOD/SCID recipient mice. J Clin Invest *119*, 565–572.

Beriou, G., Costantino, C.M., Ashley, C.W., Yang, L., Kuchroo, V.K., Baecher-Allan, C., and Hafler, D.A. (2009). IL-17-producing human peripheral regulatory T cells retain suppressive function. Blood *113*, 4240–4249.

Bettelli, E., Carrier, Y., Gao, W., Korn, T., Strom, T.B., Oukka, M., Weiner, H.L., and Kuchroo, V.K. (2006). Reciprocal developmental pathways for the generation of pathogenic effector TH17 and regulatory T cells. Nature *441*, 235–238.

Bettelli, E., Sullivan, B., Szabo, S.J., Sobel, R.A., Glimcher, L.H., and Kuchroo, V.K. (2004). Loss of T-bet, but not STAT1, prevents the development of experimental auto-immune encephalomyelitis. J Exp Med *200*, 79–87.

Bluestone, J.A., Mackay, C.R., O'Shea, J.J., and Stockinger, B. (2009). The functional plasticity of T cell subsets. Nat Rev Immunol *9*, 811–816.

Burrell, B.E., Csencsits, K., Lu, G., Grabauskiene, S., and Bishop, D.K. (2008). CD8+ Th17 mediate costimulation blockade-resistant allograft rejection in T-bet-deficient mice. J Immunol *181*, 3906–3914.

Chaudhry, A., Rudra, D., Treuting, P., Samstein, R.M., Liang, Y., Kas, A., and Rudensky, A.Y. (2009). CD4+ regulatory T cells control TH17 responses in a Stat3-dependent manner. Science *326*, 986–991.

Constantinescu, C.S., Wysocka, M., Hilliard, B., Ventura, E.S., Lavi, E., Trinchieri, G., and Rostami, A. (1998). Antibodies against IL-12 prevent superantigen-induced and spontaneous relapses of experimental auto-immune encephalomyelitis. J Immunol *161*, 5097–5104.

Croxford, A.L., Kurschus, F.C., and Waisman, A. (2009). Cutting edge: an IL-17F-CreEYFP reporter mouse allows fate mapping of Th17 cells. J Immunol *182*, 1237–1241.

Cua, D.J., Sherlock, J., Chen, Y., Murphy, C.A., Joyce, B., Seymour, B., Lucian, L., To, W., Kwan, S., Churakova, T., *et al.* (2003). Interleukin-23 rather than interleukin-12 is the critical cytokine for auto-immune inflammation of the brain. Nature *421*, 744–748.

Dardalhon, V., Awasthi, A., Kwon, H., Galileos, G., Gao, W., Sobel, R.A., Mitsdoerffer, M., Strom, T.B., Elyaman, W., Ho, I.C., *et al.* (2008). IL-4 inhibits TGF-beta-induced Foxp3+ T cells and, together with TGF-beta, generates IL-9+ IL-10+ Foxp3(−) effector T cells. Nat Immunol *9*, 1347–1355.

Duhen, T., Geiger, R., Jarrossay, D., Lanzavecchia, A., and Sallusto, F. (2009). Production of interleukin 22 but not interleukin 17 by a subset of human skin-homing memory T cells. Nat Immunol *10*, 857–863.

Firestein, G.S., Roeder, W.D., Laxer, J.A., Townsend, K.S., Weaver, C.T., Hom, J.T., Linton, J., Torbett, B.E., and Glasebrook, A.L. (1989). A new murine CD4+ T cell subset with an unrestricted cytokine profile. J Immunol *143*, 518–525.

Fontenot, J.D., Gavin, M.A., and Rudensky, A.Y. (2003). Foxp3 programs the development and function of CD4+CD25+ regulatory T cells. Nat Immunol *4*, 330–336.

Haak, S., Croxford, A.L., Kreymborg, K., Heppner, F.L., Pouly, S., Becher, B., and Waisman, A. (2009). IL-17A and IL-17F do not contribute vitally to auto-immune neuro-inflammation in mice. J Clin Invest *119*, 61–69.

Harrington, L.E., Hatton, R.D., Mangan, P.R., Turner, H., Murphy, T.L., Murphy, K.M., and Weaver, C.T. (2005). Interleukin 17-producing CD4+ effector T cells develop via a lineage distinct from the T helper type 1 and 2 lineages. Nat Immunol *6*, 1123–1132.

Hegazy, A.N., Peine, M., Helmstetter, C., Panse, I., Frohlich, A., Bergthaler, A., Flatz, L., Pinschewer, D.D., Radbruch, A., and Lohning, M. (2010). Interferons direct Th2 cell reprogramming to generate a stable GATA-3(+)T-bet(+) cell subset with combined Th2 and Th1 cell functions. Immunity *32*, 116–128.

Hori, S., Nomura, T., and Sakaguchi, S. (2003). Control of regulatory T cell development by the transcription factor Foxp3. Science *299*, 1057–1061.

Hunter, C.A. (2005). New IL-12-family members: IL-23 and IL-27, cytokines with divergent functions. Nat Rev Immunol *5*, 521–531.

Ivanov, II, McKenzie, B.S., Zhou, L., Tadokoro, C.E., Lepelley, A., Lafaille, J.J., Cua, D.J., and Littman, D.R. (2006). The orphan nuclear receptor RORgammat directs the differentiation program of pro-inflammatory IL-17+ T helper cells. Cell *126*, 1121–1133.

Kelso, A. (1995). Th1 and Th2 subsets: paradigms lost? Immunol Today *16*, 374–379.

Kelso, A., and Gough, N.M. (1988). Coexpression of granulocyte-macrophage colony-stimulating factor, gamma interferon, and interleukins 3 and 4 is random in murine alloreactive T-lymphocyte clones. Proc Natl Acad Sci USA *85*, 9189–9193.

Kelso, A., and Groves, P. (1997). A single peripheral CD8+ T cell can give rise to progeny expressing type 1 and/or type 2 cytokine genes and can retain its multipotentiality through many cell divisions. Proc Natl Acad Sci USA *94*, 8070–8075.

Khader, S.A., Bell, G.K., Pearl, J.E., Fountain, J.J., Rangel-Moreno, J., Cilley, G.E., Shen, F., Eaton, S.M., Gaffen, S.L., Swain, S.L., *et al.* (2007). IL-23 and IL-17 in the establishment of protective pulmonary CD4+ T cell responses after vaccination and during Mycobacterium tuberculosis challenge. Nat Immunol *8*, 369–377.

Koch, M.A., Tucker-Heard, G., Perdue, N.R., Killebrew, J.R., Urdahl, K.B., and Campbell, D.J. (2009). The transcription factor T-bet controls regulatory T cell homeostasis and function during type 1 inflammation. Nat Immunol *10*, 595–602.

Koenen, H.J., Smeets, R.L., Vink, P.M., van Rijssen, E., Boots, A.M., and Joosten, I. (2008). Human CD25highFoxp3pos regulatory T cells differentiate into IL-17-producing cells. Blood *112*, 2340–2352.

Komatsu, N., Mariotti-Ferrandiz, M.E., Wang, Y., Malissen, B., Waldmann, H., and Hori, S. (2009). Heterogeneity of natural Foxp3+ T cells: a committed regulatory T-cell lineage and an uncommitted minor population retaining plasticity. Proc Natl Acad Sci USA *106*, 1903–1908.

Korn, T., Bettelli, E., Gao, W., Awasthi, A., Jager, A., Strom, T.B., Oukka, M., and Kuchroo, V.K. (2007). IL-21 initiates an alternative pathway to induce pro-inflammatory T(H)17 cells. Nature *448*, 484–487.

Korn, T., Bettelli, E., Oukka, M., and Kuchroo, V.K. (2009). IL-17 and Th17 Cells. Annu Rev Immunol *27*, 485–517.

Krawczyk, C.M., Shen, H., and Pearce, E.J. (2007). Functional plasticity in memory T helper cell responses. J Immunol *178*, 4080–4088.

Kroenke, M.A., Carlson, T.J., Andjelkovic, A.V., and Segal, B.M. (2008). IL-12- and IL-23-modulated T cells induce distinct types of EAE based on histology, CNS chemokine profile, and response to cytokine inhibition. J Exp Med *205*, 1535–1541.

Langrish, C.L., Chen, Y., Blumenschein, W.M., Mattson, J., Basham, B., Sedgwick, J.D., McClanahan, T., Kastelein, R.A., and Cua, D.J. (2005). IL-23 drives a pathogenic T cell population that induces auto-immune inflammation. J Exp Med *201*, 233–240.

Lee, Y.K., Turner, H., Maynard, C.L., Oliver, J.R., Chen, D., Elson, C.O., and Weaver, C.T. (2009). Late developmental plasticity in the T helper 17 lineage. Immunity *30*, 92–107.

Lexberg, M.H., Taubner, A., Forster, A., Albrecht, I., Richter, A., Kamradt, T., Radbruch, A., and Chang, H.D. (2008). Th memory for interleukin-17 expression is stable in vivo. Eur J Immunol *38*, 2654–2664.

Luger, D., Silver, P.B., Tang, J., Cua, D., Chen, Z., Iwakura, Y., Bowman, E.P., Sgambellone, N.M., Chan, C.C., and Caspi, R.R. (2008). Either a Th17 or a Th1 effector response can drive autoimmunity: conditions of disease induction affect dominant effector category. J Exp Med *205*, 799–810.

Manetti, R., Gerosa, F., Giudizi, M.G., Biagiotti, R., Parronchi, P., Piccinni, M.P., Sampognaro, S., Maggi, E., Romagnani, S., Trinchieri, G., and et al. (1994). Interleukin 12 induces stable priming for interferon gamma (IFN-gamma) production during differentiation of human T helper (Th) cells and transient IFN-gamma production in established Th2 cell clones. J Exp Med *179*, 1273–1283.

Mangan, P.R., Harrington, L.E., O'Quinn, D.B., Helms, W.S., Bullard, D.C., Elson, C.O., Hatton, R.D., Wahl, S.M., Schoeb, T.R., and Weaver, C.T. (2006). Transforming growth factor-beta induces development of the T(H)17 lineage. Nature *441*, 231–234.

Martin-Orozco, N., Chung, Y., Chang, S.H., Wang, Y.H., and Dong, C. (2009). Th17 cells promote pancreatic inflammation but only induce diabetes efficiently in lymphopenic hosts after conversion into Th1 cells. Eur J Immunol *39*, 216–224.

Miossec, P., Korn, T., and Kuchroo, V.K. (2009). Interleukin-17 and type 17 helper T cells. N Engl J Med *361*, 888–898.

Mosmann, T.R., Cherwinski, H., Bond, M.W., Giedlin, M.A., and Coffman, R.L. (1986). Two types of murine helper T cell clone. I. Definition according to profiles of lymphokine activities and secreted proteins. J Immunol *136*, 2348–2357.

Mosmann, T.R., and Coffman, R.L. (1989). TH1 and TH2 cells: different patterns of lymphokine secretion lead to different functional properties. Annu Rev Immunol *7*, 145–173.

Murphy, C.A., Langrish, C.L., Chen, Y., Blumenschein, W., McClanahan, T., Kastelein, R.A., Sedgwick, J.D., and Cua, D.J. (2003). Divergent pro- and anti-inflammatory roles for IL-23 and IL-12 in joint auto-immune inflammation. J Exp Med *198*, 1951–1957.

Murphy, E., Shibuya, K., Hosken, N., Openshaw, P., Maino, V., Davis, K., Murphy, K., and O'Garra, A. (1996). Reversibility of T helper 1 and 2 populations is lost after long-term stimulation. J Exp Med *183*, 901–913.

Nakae, S., Nambu, A., Sudo, K., and Iwakura, Y. (2003a). Suppression of immune induction of collagen-induced arthritis in IL-17-deficient mice. J Immunol *171*, 6173–6177.

Nakae, S., Saijo, S., Horai, R., Sudo, K., Mori, S., and Iwakura, Y. (2003b). IL-17 production from activated T cells is required for the spontaneous development of destructive arthritis in mice deficient in IL-1 receptor antagonist. Proc Natl Acad Sci USA *100*, 5986–5990.

Nurieva, R., Yang, X.O., Chung, Y., and Dong, C. (2009). Cutting edge: in vitro generated Th17 cells maintain their cytokine expression program in normal but not lymphopenic hosts. J Immunol *182*, 2565–2568.

Nurieva, R., Yang, X.O., Martinez, G., Zhang, Y., Panopoulos, A.D., Ma, L., Schluns, K., Tian, Q., Watowich, S.S., Jetten, A.M., and Dong, C. (2007). Essential autocrine regulation by IL-21 in the generation of inflammatory T cells. Nature *448*, 480–483.

O'Connor, R.A., Prendergast, C.T., Sabatos, C.A., Lau, C.W., Leech, M.D., Wraith, D.C., and Anderton, S.M. (2008). Cutting edge: Th1 cells facilitate the entry of Th17 cells to the central nervous system during experimental auto-immune encephalomyelitis. J Immunol *181*, 3750–3754.

O'Shea, J.J., and Paul, W.E. (2010). Mechanisms underlying lineage commitment and plasticity of helper CD4+ T cells. Science *327*, 1098–1102.

Openshaw, P., Murphy, E.E., Hosken, N.A., Maino, V., Davis, K., Murphy, K., and O'Garra, A. (1995). Heterogeneity of intracellular cytokine synthesis at the single-cell level in polarized T helper 1 and T helper 2 populations. J Exp Med *182*, 1357–1367.

Oppmann, B., Lesley, R., Blom, B., Timans, J.C., Xu, Y., Hunte, B., Vega, F., Yu, N., Wang, J., Singh, K., *et al.* (2000). Novel p19 protein engages IL-12p40 to form a cytokine, IL-23, with biological activities similar as well as distinct from IL-12. Immunity *13*, 715–725.

Park, H., Li, Z., Yang, X.O., Chang, S.H., Nurieva, R., Wang, Y.H., Wang, Y., Hood, L., Zhu, Z., Tian, Q., and Dong, C. (2005). A distinct lineage of CD4 T cells regulates tissue inflammation by producing interleukin 17. Nat Immunol *6*, 1133–1141.

Schmitt, E., Germann, T., Goedert, S., Hoehn, P., Huels, C., Koelsch, S., Kuhn, R., Muller, W., Palm, N., and Rude, E. (1994). IL-9 production of naive CD4+ T cells depends on IL-2, is synergistically enhanced by a combination of TGF-beta and IL-4, and is inhibited by IFN-gamma. J Immunol *153*, 3989–3996.

Segal, B.M., Dwyer, B.K., and Shevach, E.M. (1998). An interleukin (IL)-10/IL-12 immunoregulatory circuit controls susceptibility to auto-immune disease. J Exp Med *187*, 537–546.

Swain, S.L. (1994). Generation and in vivo persistence of polarized Th1 and Th2 memory cells. Immunity *1*, 543–552.

Tang, J.L., Subbotin, V.M., Antonysamy, M.A., Troutt, A.B., Rao, A.S., and Thomson, A.W. (2001). Interleukin-17 antagonism inhibits acute but not chronic vascular rejection. Transplantation *72*, 348–350.

Trifari, S., Kaplan, C.D., Tran, E.H., Crellin, N.K., and Spits, H. (2009). Identification of a human helper T cell population that has abundant production of interleukin 22 and is distinct from T(H)-17, T(H)1 and T(H)2 cells. Nat Immunol *10*, 864–871.

Tsuji, M., Komatsu, N., Kawamoto, S., Suzuki, K., Kanagawa, O., Honjo, T., Hori, S., and Fagarasan, S. (2009). Preferential generation of follicular B helper T cells from Foxp3+ T cells in gut Peyer's patches. Science *323*, 1488–1492.

Vanaudenaerde, B.M., Dupont, L.J., Wuyts, W.A., Verbeken, E.K., Meyts, I., Bullens, D.M., Dilissen, E., Luyts, L., Van Raemdonck, D.E., and Verleden, G.M. (2006). The role of interleukin-17 during acute rejection after lung transplantation. Eur Respir J *27*, 779–787.

Veldhoen, M., Hocking, R.J., Atkins, C.J., Locksley, R.M., and Stockinger, B. (2006). TGFbeta in the context of an inflammatory cytokine milieu supports de novo differentiation of IL-17-producing T cells. Immunity *24*, 179–189.

Veldhoen, M., Uyttenhove, C., van Snick, J., Helmby, H., Westendorf, A., Buer, J., Martin, B., Wilhelm, C., and Stockinger, B. (2008). Transforming growth factor-beta 'reprograms' the differentiation of T helper 2 cells and promotes an interleukin 9-producing subset. Nat Immunol *9*, 1341–1346.

Wang, J., Ioan-Facsinay, A., van der Voort, E.I., Huizinga, T.W., and Toes, R.E. (2007). Transient expression of FOXP3 in human activated nonregulatory CD4+ T cells. Eur J Immunol *37*, 129–138.

Wei, G., Wei, L., Zhu, J., Zang, C., Hu-Li, J., Yao, Z., Cui, K., Kanno, Y., Roh, T.Y., Watford, W.T., *et al.* (2009). Global mapping of H3K4me3 and H3K27me3 reveals specificity and plasticity in lineage fate determination of differentiating CD4+ T cells. Immunity *30*, 155–167.

Wilson, N.J., Boniface, K., Chan, J.R., McKenzie, B.S., Blumenschein, W.M., Mattson, J.D., Basham, B., Smith, K., Chen, T., Morel, F., *et al.* (2007). Development, cytokine profile and function of human interleukin 17-producing helper T cells. Nat Immunol *8*, 950–957.

Xu, L., Kitani, A., Fuss, I., and Strober, W. (2007). Cutting edge: regulatory T cells induce CD4+CD25-Foxp3- T cells or are self-induced to become Th17 cells in the absence of exogenous TGF-beta. J Immunol *178*, 6725–6729.

Yang, X.O., Nurieva, R., Martinez, G.J., Kang, H.S., Chung, Y., Pappu, B.P., Shah, B., Chang, S.H., Schluns, K.S., Watowich, S.S., *et al.* (2008). Molecular antagonism and plasticity of regulatory and inflammatory T cell programs. Immunity *29*, 44–56.

Yao, Z., Fanslow, W.C., Seldin, M.F., Rousseau, A.M., Painter, S.L., Comeau, M.R., Cohen, J.I., and Spriggs, M.K. (1995). Herpesvirus Saimiri encodes a new cytokine, IL-17, which binds to a novel cytokine receptor. Immunity *3*, 811–821.

Yu, D., Batten, M., Mackay, C.R., and King, C. (2009). Lineage specification and heterogeneity of T follicular helper cells. Curr Opin Immunol *21*, 619–625.

Yuan, X., Ansari, M.J., D'Addio, F., Paez-Cortez, J., Schmitt, I., Donnarumma, M., Boenisch, O., Zhao, X., Popoola, J., Clarkson, M.R., *et al.* (2009). Targeting Tim-1 to overcome resistance to transplantation tolerance mediated by CD8 T17 cells. Proc Natl Acad Sci USA *106*, 10734–10739.

Yuan, X., Paez-Cortez, J., Schmitt-Knosalla, I., D'Addio, F., Mfarrej, B., Donnarumma, M., Habicht, A., Clarkson, M.R., Iacomini, J., Glimcher, L.H., *et al.* (2008). A novel role of CD4 Th17 cells in mediating cardiac allograft rejection and vasculopathy. J Exp Med *205*, 3133–3144.

Zaretsky, A.G., Taylor, J.J., King, I.L., Marshall, F.A., Mohrs, M., and Pearce, E.J. (2009). T follicular helper cells differentiate from Th2 cells in response to helminth antigens. J Exp Med *206*, 991–999.

Zelenika, D., Adams, E., Mellor, A., Simpson, E., Chandler, P., Stockinger, B., Waldmann, H., and Cobbold, S.P. (1998). Rejection of H-Y disparate skin grafts by monospecific CD4+ Th1 and Th2 cells: no requirement for CD8+ T cells or B cells. J Immunol *161*, 1868–1874.

Zheng, Y., Chaudhry, A., Kas, A., deRoos, P., Kim, J.M., Chu, T.T., Corcoran, L., Treuting, P., Klein, U., and Rudensky, A.Y. (2009). Regulatory T-cell suppressor program co-opts transcription factor IRF4 to control T(H)2 responses. Nature *458*, 351–356.

Zhou, L., Ivanov, II, Spolski, R., Min, R., Shenderov, K., Egawa, T., Levy, D.E., Leonard, W.J., and Littman, D.R. (2007). IL-6 programs T(H)-17 cell differentiation by promoting sequential engagement of the IL-21 and IL-23 pathways. Nat Immunol *8*, 967–974.

Zhou, X., Bailey-Bucktrout, S.L., Jeker, L.T., Penaranda, C., Martinez-Llordella, M., Ashby, M., Nakayama, M., Rosenthal, W., and Bluestone, J.A. (2009). Instability of the transcription factor Foxp3 leads to the generation of pathogenic memory T cells in vivo. Nat Immunol *10*, 1000–1007.

Part II
T_H17 Cells in Health and Disease

Human T_H17 Cells

Francesco Annunziato, Lorenzo Cosmi, Francesco Liotta, Enrico Maggi, and Sergio Romagnani

Abstract CD4+ T helper (T_H) lymphocytes represent a heterogeneous population of cells that play an essential role in adaptive immunity. In addition to type 1 (T_H1) and type 2 (T_H2) cells, a third subset of CD4+ T_H effector cells have recently been discovered and named as type 17 (T_H17), because of its unique ability to produce interleukin (IL)-17. Studies in humans have demonstrated the plasticity of T_H17 cells and their possibility to shift to cells producing IL-17A and IFN-γ or IL-17A and IL-4. The plasticity of T_H17 to T_H1 cells has recently been confirmed in mice, where it has been found that T_H17 cells are pathogenic only after their shifting to T_H1 cells. Human T_H17 are also different from murine T_H17 cells because all of them express CD161. They exclusively originate from CD161+ precursors present in umbilical cord blood and newborn thymus after their culturing in presence of IL-1β plus IL-23, and whereas TGF-β is not critical, it plays an indirect role in fostering their development.

1 T Helper Subsets

1.1 The T_H1-T_H2 Paradigm

CD4+ T helper (T_H) lymphocytes represent a heterogeneous population of cells playing an essential role in adaptive immunity. These cells include effector cells, devoted to the protection against pathogens, and regulatory T cells (Treg), which defend the body from effector responses to auto-antigens and exogenous antigens when they become dangerous for the host. Effector CD4+ T_H cells are heterogeneous with regard to their protective function, enabling a type of response which is

S. Romagnani (✉)
Department of Internal Medicine, University of Florence,
Viale Morgagni 85, Firenze 50134, Italy
e-mail: s.romagnani@dmi.unifi.it

S. Jiang (ed.), *T_H17 Cells in Health and Disease*,
DOI 10.1007/978-1-4419-9371-7_12, © Springer Science+Business Media, LLC 2011

different according to the nature of the invading microorganism. Over 20 years ago, two main subsets of CD4+ T_H cells with different functions and patterns of cytokine secretion were identified in both mice and humans, which were named as type 1 T_H (T_H1) and type 2 T_H (T_H2) lymphocytes, respectively (Mosmann et al. 1986; Romagnani 1991). T_H1 cells produce high levels of interferon (IFN)-γ and are responsible for both phagocyte activation and the production of opsonizing and complement-fixing antibodies, thus playing an important role in the protection against intracellular pathogens. T_H2 cells produce interleukin (IL)-4, IL-5, IL-9, and IL-13 and are mainly involved in the protection against parasitic helminths (Romagnani 1997; Gately et al. 1998). In addition to their protective functions against invading pathogens, T_H1 and T_H2 cells responses can contribute to the development of human disorders: T_H1 cells have been thought to be involved in the pathogenesis of organ-specific auto-immune diseases, as well as other chronic inflammatory disorders, such as Crohn's disease (CD), sarcoidosis, and atherosclerosis (Gately et al. 1998). T_H2 cells certainly play a central role in the development of allergic disorders (Romagnani 1997). The T_H1–T_H2 paradigm was maintained until some years ago when a third subset of CD4+ effector T_H cells, named T_H17 cells, was identified (Oppman et al. 2000; Cua et al. 2003).

1.2 The Third Actor: T_H17 Lymphocytes

Although the existence of IL-17 as a product of activated CD4+ T cells has been known for more than 10 years, only recently has the existence of T_H17 as a distinct subset of T lymphocytes been recognized (Oppman et al. 2000; Cua et al. 2003). The main features of this population in human beings have been described by some studies performed in the last 3 years (Acosta-Rodriguez et al. 2007; Annunziato et al. 2007; Wilson et al. 2007; Cosmi et al. 2008).

T_H17 lymphocytes produce the distinctive cytokines IL-17A and IL-17F. The IL-17 family of cytokines comprehends five members, designated IL-17A–F. IL-17A is disulfide-linked homodimeric glycoprotein, consisting of 155 amino acids (Yao et al. 1995) and sharing great homology with IL-17F (55%). IL-17A and IL-17F can exist as IL-17A homodimers and IL-17F homodimers or as IL-17A-IL-17F heterodimers. The other IL-17 family members, IL-17B, IL-17C, and IL-17D, are produced by a non-T cell source. T_H17 cells play a critical role in the recruitment, activation and migration of neutrophil granulocytes, both directly, through IL-8 production (Pelletier et al. 2010) and indirectly, by inducing, via IL-17, the production of colony stimulatory factors (CSF) and CXCL8 (Ouyang et al. 2008) in tissue resident cells. T_H17 lymphocytes are also able to stimulate CXCL chemokines production and to increase mRNA and protein for the mucins, MUC5AC and MUC5B, in primary human bronchial epithelial cells in vitro (Chen et al. 2003), as well as the expression of human beta defensin-2 (Kao et al. 2004) and CCL20 in lung epithelial cells (Kao et al. 2005). T_H17 lymphocytes also produce cytokines different from IL-17, such as IL-21 and IL-22 that contribute to the activation of mononuclear and/or resident cells and therefore may induce and/or maintain a chronic inflammatory process. However, because of their unique ability to recruit neutrophils, the main protective function of T_H17 cells appears to be the clearance of extracellular pathogens, including fungi (Dubin and Kolls 2008), as confirmed by the demonstration in humans of *Candida albicans*-specific IL-22 producing lymphocytes (Liu et al. 2009). Moreover, it has recently been shown that patients with autoimmune polyendocrine syndrome type I (APS-I) displaying chronic mucocutaneous candidiasis (CMC) exhibit high titers of auto-antibodies against IL-17A, IL-17F, and/or IL-22 (Kisand 2010; Puel et al. 2010). Finally, the association between STAT3 mutations, which have been associated to impaired T_H17 differentiation (Harris et al. 2007; de Beaucoudrey et al. 2008), and the increased susceptibility to fungal infections in patients with the hyper-immunoglobulin E syndrome (Ma et al. 2008), indirectly confirms the crucial role of T_H17 cells in the protection against extracellular pathogens, where they probably bridge innate and adaptive immunity to produce a robust antimicrobial inflammatory response (Peck and Mellins 2010).

2 Human T_H17 lymphocytes

Mice and human T_H17 cells are phenotypically and functionally similar, but not identical. Human T_H17 cells exhibit expression of RORC, which is the human horologe of mouse RORγt, and of IL-23 receptor (IL-23R). Human T_H17 cells express CCR6 and CCR4 in the absence of CXCR3, which is alternatively expressed by T_H1 cells (Acosta-Rodriguez et al. 2007; Annunziato et al. 2007), a finding that had not previously reported in mice. Interestingly, *Candida albicans*-specific T cells were CCR6+CCR4+ and produced IL-17A but no IFN-γ, whereas *Mycobacterium tuberculosis* PPD-specific T cells, which were CCR6+CXCR3+, produced IFN-γ but not IL-17A (Acosta-Rodriguez et al. 2007). Different studies have also demonstrated the existence of substantial numbers of human CD4+ T cells able to produce both IL-17A and IFN-γ (Acosta-Rodriguez et al. 2007; Annunziato et al. 2007; Chen et al. 2007; van Beelen et al. 2007). Based on the demonstration that several human T-cell clones show the production of both these cytokines, the term of T_H17/T_H1 cells was proposed (Annunziato et al. 2007; Cosmi et al. 2008). Of note, the CCR6+CXCR3– circulating cell fraction contained mainly T_H17 and some T_H17/T_H1 cells, and virtually all the CCR6-CXCR3+ cells were T_H1 cells; the CCR6+CXCR3+ circulating cell fraction can contain all the three types of cells (T_H17, T_H17/T_H1 and T_H1) (Annunziato et al. 2007). A rare population of human CD4+ T cells is able to produce both IL-17A and IL-4 has been described (Cosmi et al. 2010). The increased numbers of these T_H17/T_H2 cells in the circulation of patients with bronchial asthma, suggests that they may play some role in the pathogenesis of this disease.

Based on the results of a microarray analysis performed on human T_H17 in comparison with T_H1 and T_H2 clones derived from the same donor, another unexpected phenotypic property of human T_H17 cells was discovered. Virtually all human T_H17 clones expressed CD161 (Cosmi et al. 2008), which is human homologue of murine

NK1.1 (Lanier et al. 1994). NK1.1 has been never described to be expressed on murine T_H17 cells. Virtually all IL-17A-producing CD4+ T cells were contained within the CD161+ circulating cell fraction or within the CD161+ cell fraction infiltrating the inflamed gut of patients with Crohn's disease or the inflamed skin of patients with psoriasis (Cosmi et al. 2008; Kleinschek et al. 2009).

3 The Origin of T_H17 Lymphocytes

The origin of human T_H17 cells, as well as the cytokines that induce their development have been a matter of intense debate for the last few years, probably because of some substantial differences existing between mice and men. In mice, several studies indicate that transforming growth factor (TGF)-β is required for initiation and that IL-6 is a critical co-factor for T_H17 cell development, whereas IL-23 is prevalently involved in the maintenance and expansion of this process. IL-1β and tumour necrosis factor (TNF)-α were found to amplify this process, but they could not substitute for either TGF-β or IL-6 (Bettelli et al. 2006; Mangan and Harrington 2006; Veldhoen et al. 2006). IL-21, a cytokine produced even by T_H17 cells themselves, was reported to provide an additional autocrine amplificatory signal (Korn et al. 2007; Nurieva et al. 2007).

The origin of human T_H17 cells, and in particular the role of TGF-β, in their differentiation, has also been the object of much examination. Acosta-Rodriguez et al. (2007) showed that human T_H17 originated from naïve CD4+ T cells following their activation in the presence of a mixture of IL-1β and IL-6, whereas the addition of TGF-β was inhibitory. Wilson et al. (2007) found that naïve T cells could be induced to differentiate into T_H17 cells in response to either IL-1β or IL-23, but not to TGF-β. Chen et al. (2007) also concluded that under the same conditions, IL-23, but not TGF-β, was critical for the human T_H17 cell development, with particular regard to its ability to induce the production by these cells of IL-22. Finally, van Beelen et al. (2007a) demonstrated that human T_H17 cells could be derived from memory, but not from naïve, T cells. This effect was due to the nucleotide oligomerization domain-2 ligand muramyl peptide which enhanced IL-23 and IL-1 production by dendritic cells. All these studies were performed by using naïve (CD45RA+CD45RO-) CD4+ T cells sorted from adult peripheral blood and raised the question of the true 'naiveté' of these cells and the extent to which the cytokines were acting truly naïve or activated on memory-contaminating CD4+ T cells (Laurence and O'Shea 2007). Moreover, human naive CD4+ T cells were cultured in bovine or human serum, which can contain traces of TGF-β, suggesting a possible underestimation of its importance in the differentiation process. Subsequently, three independent studies (Manel et al. 2008; Volpe et al. 2008; Yang et al. 2008) showed that even the development of human T_H17 cells requires the activity of TGF-β, as it happens in mouse, even if the results of these studies were somehow contradictory. Manel et al. (2008) found that TGF-β, IL-1 β, IL-6, and IL-21 or TGF-β, IL-1β, and IL-23 were necessary and sufficient to induce IL-17 expression in naïve umbilical cord blood (UCB) human CD4+ T cells. TGF-β upregulated RORC expression but simultaneously inhibited its ability to induce IL-17 expression, an activity which was antagonized by the inflammatory cytokines. Volpe et al. (2008) found that TGF-β, IL-23, IL-1β, and IL-6 were all essential for human T_H17 development, but they differentially modulated the cytokines produced by T_H17 cells. More importantly, the absence of TGF-β induced a shift from a T_H17-like to a T_H1-like profile. By contrast, Yang et al. (2008) found that TGF-β and IL-21 uniquely promoted the differentiation of human naïve CD4+ T cells into T_H17 cells, accompanied by the expression of RORC. The results of these studies were enthusiastically welcome (O'Garra et al. 2008), because they seemed to exclude the unfortunate and unhelpful possibility for the biomedical research that mice are of limited use as models for the development of T_H17 cells in the human immune system. However, we found that IL-17A-producing cells originate in humans from CD161+ CD4+ T-cell precursors, obtained from both UCB and thymus when these cells were activated in presence of a combination of IL-1β plus IL-23. The presence of both these cytokines induced not only an increase in the expression of RORC and IL-23R, but also of T-bet and IL-12Rβ2 and allowed the development of higher numbers of T_H1 cells, suggesting a possible developmental relationship between human T_H17 and T_H1 cells. No other cytokine or cytokine combinations (including TGF-β, IL-6 and IL-21) were able to induce IL-17A mRNA expression and IL-17A production. CD161+ naive CD4+ T could also be induced to differentiate into T_H1 or T_H2 cells when cultured in the presence of exogenous IL-12 or IL-4, respectively. In contrast, CD161-, naïve CD4+ T cells from both UCB or thymus could be induced to differentiate only into T_H1 or T_H2 cells under appropriate polarizing conditions (IL1β plus IL-23, or IL-12 alone, for T_H1 cells and IL-4 for T_H2 cells), but they never differentiated into IL-17A-producing cells (Cosmi et al. 2008). To clarify the role of TGF-β in human T_H17 differentiation we performed further experiments in the presence of a neutralizing anti-TGF-β antibody in serum-containing cultures and also under serum-free conditions in order to exclude the possibility of undesirable effects exerted by TGF-β contained in bovine serum already present in culture (Santarlasci et al. 2009). Even under these conditions, IL-1β and IL-23 were found to be sufficient to induce the development of IL-17A-producing cells from purified UCB CD161+CD4+ T cells (Santarlasci et al. 2009). When TGF-β was added in serum-free cultures containing IL-1β and IL-23, it increased the relative proportions of CD161+ T cells that differentiated into T_H17 cells, but inhibited both T-bet expression and T_H1 development, suggesting that T_H17 and T_H1 cells may have a different susceptibility to the suppressive activity of TGF-β. In agreement with this hypothesis, when the proliferation of human T_H17 clones and also of circulating T_H17 cells in the presence or absence of TGF-β was examined. These cells showed a lower susceptibility to the anti-proliferative effect of TGF-β than T_H1 or T_H2 clones and circulating IFN-γ-producing (T_H1) cells. This difference appeared to be due to the reduced apoptotic cell death of T_H17 cells in the presence of TGF-β in comparison with T_H1 cells, a finding which appeared to be consistent with the demonstration that human T_H17 cells exhibit lower expression of clusterin (TGF-β signaling, pro-apoptotic), and higher Bcl-2 (anti-apoptotic) in comparison with

T_H1 or T_H2 clones (Santarlasci et al. 2009). Taken together, these findings support the concept that TGF-β does not have a direct effect on the development of human T_H17 cells, but can indirectly favor their development by selectively inhibiting both T-bet expression and the development of T_H1 cells. This is in keeping with the results of a recent study showing that patients with mutations of *TGFB1* or *TGFBR2* did not exhibit any difference in the numbers of IL-17A-producing T cells compared to healthy controls, whereas patients with *STAT3* and *IL-12Rβ1* mutations have impaired T_H17 development (de Beaucoudrey et al. 2008). The involvement of TGF-β in the induction of T_H17 phenotype has recently been questioned, even in mice (Das et al. 2009; Chung et al. 2009), and it has been reported that IL-1 plays a crucial role. Das et al. (2009) reported that TGF-beta does not directly promote T_H17 cell differentiation, but instead acts indirectly by blocking STAT4 and GATA-3expression, preventing T_H1 and T_H2 cell differentiation, as it has been reported by us in humans (Santarlasci et al. 2009; Annunziato and Romagnani 2009). Finally, T_H17 cells have also been observed in the thymus of wild-type mice and it has been suggested that they directly migrate to lymphoid tissues of some organs, such as gut and lungs (Marks et al. 2009).

4 T_H17 Involvement in the Pathogenesis of Auto-immune and Inflammatory Diseases

Beyond their protective role in the clearance of extracellular pathogens, a major role of T_H17 lymphocytes seems to be their involvement in the promotion and maintenance of chronic inflammatory processes. This has been observed in several models of auto-immunity, such as experimental auto-immune encephalomyelitis (EAE), collagen-induce arthritis (CIA) and inflammatory bowel disorders (IBD). EAE and CIA have historically been associated with unchecked T_H1 responses, largely based on studies in which disease development was ablated by neutralizing the IL-12p40 chain or targeting the *p40* or the *IL-12* genes (Gately et al. 1998). The link with IL-12 in these diseases was called into question by the discovery that a new IL-12 family member, IL-23, shared with IL-12 the p40 sub-unit, the heterodimer of IL-12 being composed of p40 and p35, and that of IL-23 being composed of p40 and p19. After this discovery, it was found that EAE and CIA did not develop in mice deficient in IL-23p19 sub-unit, whereas they could develop in those deficient in IL-12p35 sub-unit, suggesting that IL-23, but not IL-12, is critically linked to auto-immunity, at least in these models. This concept has been immediately and probably erroneously extrapolated to human disorders which are considered as equivalent to the above mentioned murine models, such as multiple sclerosis, rheumatoid arthritis and IBD (Fouser et al. 2008), psoriasis and contact dermatitis (van Beelen et al. 2007b). T_H17 cells were thought to be pathogenic in virtually all chronic inflammatory disorders, where the effect of T_H1 cells, that had been shown to be important in hundreds of previous studies, was underestimated or even seen as protective against the

T_H17-mediated inflammation (Chen et al. 2007). This sort of iconoclastic fury against the past dogma was well expressed in a commentary, which got the idea from the expression used in the age of monarchy "Le Roi est mort; vive le Roi", to mark the passing of eras ushering out one regime while introducing another (Tato et al. 2006). The new dogma that T_H17 are pathogenic, while T_H1 cells are protective in murine models of auto-immune disorders has recently been challenged. It has been shown that T_H17 cells can be pathogenic in certain forms of EAE (those characterized by granulocyte infiltration), but T_H1 cells are mainly responsible for the EAE models characterized by prevalent mononuclear cell infiltration and T_H17 cells do not seem to play any pathogenic role (Kroenke et al. 2009). Moreover, in the murine model of the auto-immune disorder known as proteoglycan-induced arthritis (PIA), T_H1, but not T_H17, cells are pathogenic (Doodes et al. 2008), and either T_H1 or T_H17 cells have been found to be pathogenic in experimental auto-immune uveitis (Luger et al. 2008). It should be taken into account that even if EAE and CIA develop in IL-12p35 –/– mice, they did not occur in mice deficient in IL-23p19 sub-unit. The deficiency of p35 gene did not only result in the neutralization of IL-12, but also of IL-35 (Collison and Vignali 2008), a recently discovered cytokine with immunosuppressive activity produced by Treg cells, whose deficiency could contribute to the development of auto-immune and inflammatory disorders. In addition, even if IL-17 gene knock out mice showed a milder form of EAE (Komiyama et al. 2006), therapeutic treatment with an anti-IL-17A antibody was able to attenuate, but not to abrogate EAE (Hofstetter et al. 2005) aggravated, DSS-induced colitis (Ogawa et al. 2004). This indicates that other cytokines are also involved in the pathogenesis of these diseases. To definitively confirm this hypothesis, it has recently been shown that T_H17 cells can promote pancreatic inflammation, but only induce type 1 insulin-dependent diabetes mellitus (IDDM) in lymphopenic mice after their conversion into T_H1 cells (Martin-Orozco et al. 2009). Accordingly, highly purified T_H17 cells from BDC2.5NOD mice shift into T_H1-like cells in NOD/SCD recipient mice. The transferred T_H17 cells, completely devoid of IFN-γ at the time of the transfer rapidly converted to secrete IFN-γ in the NOD/SCID recipients. The development of IDDM was prevented by treatment with anti-IFN-γ-, but not with anti-IL17A-, specific antibody (Bending et al. 2009). Thus, the plasticity of T_H17 cells into T_H1 cells initially observed in humans (Annunziato et al. 2007) is confirmed and provides the basis for supporting a major pathogenic role of T_H17-derived T_H1 cells in murine auto-immune or other chronic inflammatory disorders.

5 T_H17 Lymphocyte Plasticity

In order to reconcile contradictory results of the literature on the role exerted by T_H1 and T_H17 effector cells in the pathogenesis of auto-immune and inflammatory diseases, it is useful to remember that T_H17 cells have been shown to possess a plasticity that allow them to acquire both T_H1- and T_H2-like functional features.

We recently described the existence of a remarkable number of double positive (IFN-γ+IL-17+) cells or T-cell clones (named as 'T_H17/T_H1'), derived from peripheral blood of healthy donors (Annunziato et al. 2007). In the same study, we also showed that both T_H17 and T_H17/T_H1 clones expressed RORc and T-bet. Moreover, we found that T_H17/T_H1 cells exhibit the same marker observed on T_H17 lymphocytes, such as IL-23R, CCR6 (Annunziato et al. 2007), and CD161 (Cosmi et al. 2008). More importantly, the in vitro culture of human T_H17 clones in the presence of IL-12 allowed these cells to become producers of IFN-γ in addition to IL-17, and this effect was associated with reduced RORc and increased T-bet expression. This suggests a flexibility of T_H17 towards T_H1 cells and a possible developmental relationship between the two cell types (Annunziato et al. 2007). This data has been confirmed by a recent study performed on a murine model in vivo, which demonstrates that T_H17 cells have a plasticity that allows the transition of primed T_H17 effectors to the IFN-γ-producing progeny (Lee et al. 2009). Accordingly, it is the combination of IL-1β and IL-23 that induces human naïve CD4+ T lymphocytes to differentiate not only into T_H17, but also into T_H17/T_H1, and even into T_H1 cells, suggesting that these sub-populations can develop and coexist in the same micro-environment. In this context, it is not surprising that treatment with ustekinumab, a human immunoglobulin monoclonal antibody that binds the shared p40 sub-unit of human IL-12 and IL-23, blocks the activity of both cytokines and results in a rapid and significant improvement of symptoms in psoriatic patients. This data fully support the notion that both IL-12 and IL-23 have a key role in the immuno-pathophysiology of the disease (Leonardi et al. 2008; Papp et al. 2008) and reconcile contradictory results of the literature on the role exerted by T_H1 and T_H17 effector cells in the pathogenesis of auto-immune and inflammatory diseases. Very recently, we also showed the plasticity of human T_H17 toward the T_H2 phenotype, which may happen when human memory T_H17 cells are exposed to an IL-4-rich micro-environment. These T_H17/T_H2 cells produced both T_H17- and T_H2-related cytokines and were able to induce the synthesis of IgE by B lymphocytes (Cosmi et al. 2010). The demonstration of the existence in vitro and ex vivo of $CD4^+$ T cells able to produce both T_H17- and T_H2-related cytokines, together with their increase in the circulation of patients with asthma, raises the important question of the pathophysiological role of this novel subset in allergic disorders. It is reasonable to suggest that T_H17 lymphocytes exhibit a broad plasticity which allows them to shift towards both T_H1 and T_H2 phenotype. As a consequence, it is possible to hypothesize that a shift from T_H17 to T_H1 occurs during the development of some auto-immune and inflammatory diseases. Likewise, a shift from T_H17 cells to T_H17/T_H2 cells could happen during chronic bronchial asthma, contributing to the perpetuation of inflammation. The origin and the plasticity of human T_H17 cells are depicted in Fig. 1.

In conclusion, the field of T_H17 cells seems to represent a prototypic example of how studies in humans might better illustrate biological events than studies in mice, and therefore cannot be undervalued, but rather encouraged, either to correct wrong views derived from the results obtained in murine models or to open completely new avenues.

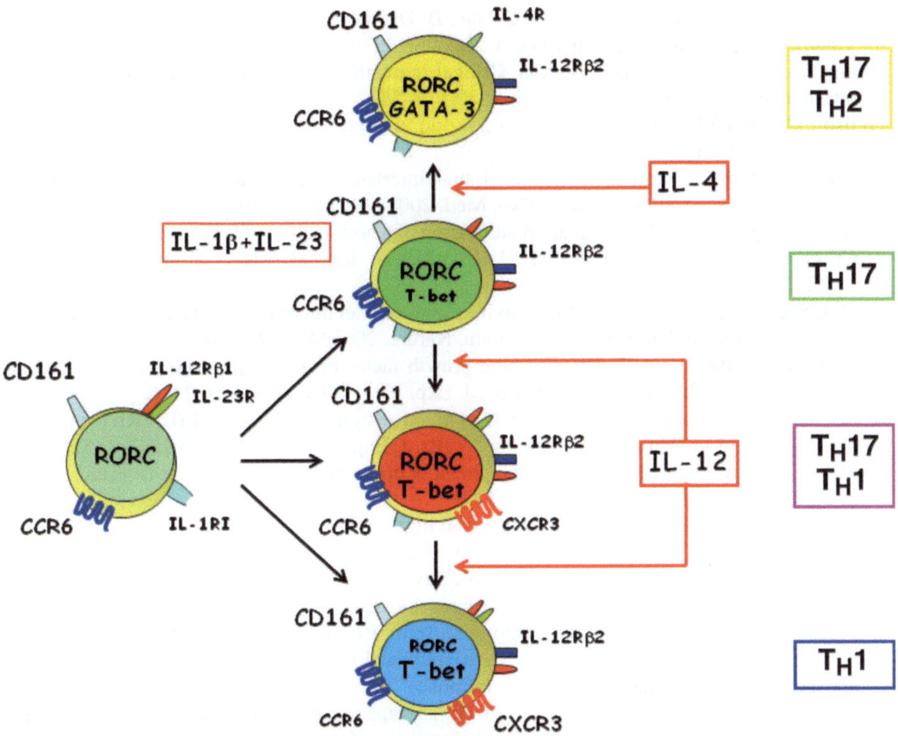

Fig. 1 Origin and plasticity of human T_H17 cells. Human T_H17 cells originate from a small subset of CD4+CD161+ T-cell precursors present in newborn thymus and UCB which already express RORC, IL-23R, IL-1RI, and CCR6. These cells differentiate in vitro in response to the combined activity of IL-1β and IL-23 into mature T_H17 cells which express RORC, T-bet, IL-12Rβ2, CD161, and CCR6. Under the same conditions, they can give rise to T_H17/T_H1 and T_H1 cells which express T-bet and CXCR3. In presence of IL-12, T-bet expression is up-regulated in T_H17 cells, which shifts to the production of IFN-γ. In presence of IL-4, GATA3 expression is up-regulated in T_H17 cells, which shifts to the production of IL-4

References

Acosta-Rodriguez EV, Rivino L, Geginat J, et al. Surface phenotype and antigenic specificity of human interleukin 17-producing T helper memory cells. Nat. Immunol. 2007; 8:639–646.

Annunziato F, Cosmi L, Santarlasci V, et al. Phenotypic and functional features of human Th17 cells. J. Exp. Med. 2007; 204:1849–1861.

Annunziato F, Romagnani S. Do studies in humans better depict Th17 cells? Blood. 2009; 114:2213–9.

Bending D, De La Peña H, Veldhoen M et al. Highly purified Th17 cells from BDC2.5NOD mice convert into Th1-like cells in NOD/SCID recipient mice. J. Clin. Invest. 2009; 119:565–572.

Bettelli E, Carrier Y, Gao W, et al. Reciprocal developmental pathways for the generation of pathogenic effector Th17 and regulatory T cells. Nature. 2006; 44:235–238.

Chen Y, Thai P, Zhao YH, et al. Stimulation of airway mucin gene expression by interleukin (IL)-17 through IL-6 paracrine/autocrine loop. J. Biol. Chem. 2003; 278:17036–17043.

Chen Z, Tato CM, Muul L, Laurence A, O'Shea JJ. Distinct regulation of interleukin-17 in human T helper lymphocytes. Arthritis Rheum. 2007; 56:2936–2946.

Chung Y, Chang SH, Martinez GJ, et al. Critical regulation of early Th17 cell differentiation by interleukin-1 signaling. Immunity. 2009; 30:576–587.

Collison LW, and Vignali DA. Interleukin-35: odd one out or part of the family? Immunol Rev. 2008; 226:248–262.

Cosmi L, De Palma R, Santarlasci V, et al. Human interleukin-17-producing cells originate from a CD161+ CD4+ T-cell precursor. J. Exp. Med. 2008; 205:1903–1916.

Cosmi L, Maggi L, Santarlasci V et al. Identification of a novel subset of human circulating memory CD4+ T cells that produce both IL-17A and IL-4. Journal Allergy Clin. Immunol. 2010; 125:222–30.

Cua DJ, Sherlock J, Chen Y et al. Interleukin-23 rather than interleukin-12 is the critical cytokine for auto-immune inflammation of the brain. Nature. 2003; 421:744–748.

Das J, Ren G, Zhang L, et al. Transforming growth factor beta is dispensable for the molecular orchestration of Th17 cell differentiation. J. Exp. Med. 2009; 206:2407–2416.

de Beaucoudrey L, Puel A, Filipe-Santos O et al. Mutations in STAT3 and IL12RB1 impair the development of human IL-17-producing T cells. J. Exp. Med. 2008 7; 205:1543–1550.

Doodes PD, Cao Y, Hamel KM et al. Development of proteoglycan-induced arthritis is independent of IL-17. J. Immunol. 2008; 181:329–337.

Dubin PJ, Kolls JK. Th17 cytokines and mucosal immunity. Immunol. Rev. 2008; 226:160–171.

Fouser LA, Wright JF, Dunussi-Joannopoulos K, et al. Th17 cytokines and their emerging roles in inflammation and auto-immunity. Immunol. Rev. 2008; 226:87–102.

Gately MK, Renzetti LM, Magram J et al. The interleukin-12/interleukin-12-receptor system: role in normal and pathologic immune responses. Annu. Rev. Immunol. 1998; 16:495–521.

Harris TJ, Grosso JF, Yen HR, et al. An in vivo requirement for STAT3 signaling in TH17 development and TH17-dependent auto-immunity. J. Immunol. 2007; 179:4313–4317.

Hofstetter HH, Ibrahim SM, Koczan D, et al. Therapeutic efficacy of IL-17 neutralization in murine experimental auto-immune encephalomyelitis. Cell. Immunol. 2005; 237:123–130.

Kao CY, Chen Y, Thai P, et al. IL-17 markedly up-regulates beta-defensin-2 expression in human airway epithelium via JAK and NF-kappaB signaling pathways. J. Immunol. 2004. 173:3482–3491.

Kao CY, Huang F, Chen Y, Thaiet al. Up-regulation of CC chemokine ligand 20 expression in human airway epithelium by IL-17 through a JAK-independent but MEK/NF-kappaB-dependent signaling pathway. J. Immunol. 2005; 175:6676–6685.

Kisand K, Bøe Wolff AS, Podkrajsek KT, et al. Chronic mucocutaneous candidiasis in APECED or thymoma patients correlates with auto-immunity to Th17-associated cytokines. J. Exp. Med. 2010; 207:299–308.

Kleinschek MA, Boniface K, Sadekova S, et al. Circulating and gut-resident human Th17 cells express CD161 and promote intestinal inflammation. J. Exp. Med. 2009; 206:525–534.

Komiyama Y, Nakae S, Matsuki T, et al. IL-17 plays an important role in the development of experimental auto-immune encephalomyelitis. J. Immunol. 2006; 177:566–573.

Korn T, Bettelli E, Gao W, et al. IL-21 initiates an alternative pathway to induce pro-inflammatory T(H)17 cells. Nature. 2007; 448:484–487.

Kroenke MA, Carlson TJ, Andjelkovic AV, et al. IL-12- and IL-23-modulated T cells induce distinct types of EAE based on histology, CNS chemokine profile, and response to cytokine inhibition. J. Exp. Med. 2009; 205:1535–1541.

Lanier LL, Chang C, Philips JH. Human NKR-P1A. A disulphide-linked homodimer of the C-type lectin superfamily expressed by a subset of NK and T lymphocytes. J. Immunol.1994; 153:2417–2428 .

Laurence A, O'Shea JJ. T(H)-17 differentiation: of mice and men. Nat Immunol. 2007;8:903–905.

Lee YK, Turner H, Maynard CL, et al. Late developmental plasticity in the T helper 17 lineage. Immunity. 2009; 30:92–107.

Leonardi CL, Kimball AB, Papp KA, et al. Efficacy and safety of ustekinumab, a human interleukin-12/23 monoclonal antibody in patients with psoriasis: 76-week results from a randomised, double-blind, placebo-controlled trial (PHOENIX 1). Lancet. 2008; 371:1665–1674.

Liu Y, Yang B, Zhou M, et al. Memory IL-22-producing CD4+ T cells specific for Candida albicans are present in humans. Eur. J. Immunol. 2009; 39:1472–1479.
Luger D, Silver PB, Tang J, et al. Either a Th17 or a Th1 effector response can drive auto-immunity: conditions of disease induction affect dominant effector category. J. Exp. Med. 2008; 205: 799–810.
Ma CS, Chew GY, Simpson N, et al. Deficiency of Th17 cells in hyper IgE syndrome due to mutations in STAT3. J. Exp. Med. 2008; 205:1551–1557.
Manel N, Unutmaz D, Littman DR. The differentiation of human Th17 cells requires transforming growth factor-β and induction of the nuclear receptor RORγt. Nat. Immunol. 2008; 9:641–649.
Mangan PR, Harrington LE, O'Quinn, et al. Transforming growth factor-beta induces development of the T(H)17 lineage. Nature. 2006; 441:231–234.
Marks BR, Nowyhed HN, Choi JY, et al. Thymic self-reactivity selects natural interleukin 17-producing T cells that can regulate peripheral inflammation. Nat. Immunol. 2009; 10: 1125–1132.
Martin-Orozco N, Chung Y, Chang SH, et al. Th17 cells promote pancreatic inflammation but only induce diabetes efficiently in lymphopenic hosts after conversion into Th1 cells. Eur. J. Immunol. 2009; 39:216–24.
Mosmann TR, Cherwinski C, Bond MW, et al. Two types of murine helper T cell clones. Definition according to profiles of lymphokine activities and secreted proteins. J. Immunol. 1986; 136: 2348–2357.
Nurieva R, Yang XO, Martinez G, et al. Essential autocrine regulation by IL-21 in the generation of inflammatory T cells. Nature. 2007; 448:480–483.
Ogawa A, Andoh A, Araki Y, et al. Neutralization of interleukin-17 aggravates dextran sulfate sodium-induced colitis in mice. Clin. Immunol. 2004; 110:55–62.
Ouyang W, Koli JK, and Zheng Y. The biological functions of T helper 17 cell effector cytokines in inflammation. Immunity. 2008; 28:454–467.
Papp KA, Langley RG, Lebwohl M, et al. Efficacy and safety of ustekinumab, a human interleukin-12/23 monoclonal antibody, in patients with psoriasis: 52-week results from a randomised, double-blind, placebo-controlled trial (PHOENIX 2). Lancet. 2008; 371:1675–1684.
Pelletier M., Maggi L, Micheletti A, et al., Evidence for a cross-talk between human neutrophils and Th17 cells. Blood. 2010; 115:335–343.
O'Garra A, Stockinger B, Veldhoen M. Differentiation of human T(H)-17 cells does require TGF-beta! Nat. Immunol. 2008; 9:588–90.
Oppman B, Lesley R, Blom B, et al. Novel p19 protein engages IL-12p40 to form a cytokine, IL-23, with biological activities similar as well as distinct from IL-12. Immunity. 2000; 13:715–725.
Peck A, Mellins ED. Precarious balance: Th17 cells in host defense. Infect. Immun. 2010; 78:32–38.
Puel A, Döffinger R, Natividad A, et al. Autoantibodies against IL-17A, IL-17F, and IL-22 in patients with chronic mucocutaneous candidiasis and auto-immune polyendocrine syndrome type I. J. Exp. Med. 2010; 207:291–297.
Romagnani S. Human TH1 and TH2 subsets: doubt no more. Immunol Today 1991;12:256–257.
Romagnani S. The Th1/Th2 paradigm. Immunol. Today. 1997; 18:263–266.
Santarlasci V, Maggi L, Capone M et al. TGF-beta indirectly favors the development of human Th17 cells by inhibiting Th1 cells. Eur. J. Immunol. 2009;39: 207–215.
Tato CM, Laurence A, O'Shea JJ. J Helper T cell differentiation enters a new era: le roi est mort; vive le roi! J. Exp. Med. 2006; 203:809–812.
van Beelen AJ, Zelinkova Z, Taanman-Kueter EW et al. Stimulation of the intracellular bacterial sensor NOD2 programs dendritic cells to promote interleukin-17 production in human memory T cells. Immunity. 2007a; 27:660–669.
van Beelen AJ, Teunissen MB, Kapsenberg ML, de Jong EC. Interleukin-17 in inflammatory skin disorders. Curr. Opin. Allergy Clin. Immunol. 2007b; 7:374–381
Veldhoen M, Hocking RJ, Atkins CJ, et al. TGF-beta in the context of an inflammatory cytokine milieu supports differentiation of IL-17-producing T cells. Immunity. 2006; 25:179–189.

Volpe E, Servant N, Zollinger R, et al. A critical function for transforming growth factor-β, interleukin-23 and pro-inflammatory cytokines in driving and modulating human Th17 responses. Nat. Immunol. 2008; 9: 650–657.
Wilson NJ, Boniface K, Chan JR, et al. Development, cytokine profile and function of human interleukin 17-producing helper T cells. Nat. Immunol. 2007; 8:950–957.
Yang L, Anderson DE, Baecher-Allan C, et al. IL-21 and TGF-beta are required for differentiation of human T(H)17 cells. Nature. 2008; 454:350–352.
Yao Z, et al. Herpesvirus Saimiri encodes a new cytokine, IL-17, which binds to a novel cytokine receptor. Immunity. 1995; 3:811–821.

T_H17 Cytokines in Primary Mucosal Immunity

Jay K. Kolls and Shabaana A. Khader

Abstract T helper type 17 (T_H17) cells are a recently discovered lineage of T cells that produce several effector molecules including IL-17A, IL-17F, IL-21, and IL-22. Scientific evidence to date strongly implicates that this arm of the immune system plays a critical role in mucosal immunity to many extracellular pathogens as well as coordinate adaptive immunity to some intracellular pathogens. In this paper, recent progress in our understanding of these cytokines, their cellular or sources and mechanism of their effector function in the mucosa in various infections is reviewed.

1 Introduction

CD4+ T helper cells are critical mediators of adaptive immune responses. Following interaction with antigen presenting cells, T cells receive signals by engagement of the T cell receptor (signal 1), co-stimulatory molecules (signal 2), and a complex network of cytokine signals (signal 3), then undergo activation and differentiation into effector CD4+ T cells. The critical role these cells play in mucosal immunity was clearly evidenced by the HIV epidemic that leads to the depletion of these cells from the mucosa and periphery (Kader et al. 2009; Mattapallil et al. 2005; Li et al. 2005), leading to the acquired immuno-deficiency syndrome (AIDS).

Five years after the initial clinical description of AIDS in 1981, Mossmann and Coffman described the first two CD4+ T-cells subsets based on discrete cytokine profiles (Mosmann et al. 1986). T_H1 effectors produce Interferon-gamma (IFN-γ) and regulate cellular immunity against intracellular infections, whereas T_H2 cells

J.K. Kolls (✉)
Department of Genetics, Louisiana State University Health Sciences Center,
CSRB 657, 533 Bolivar St., New Orleans, LA 70112, USA
e-mail: jkolls@lsuhsc.edu

S. Jiang (ed.), *T_H17 Cells in Health and Disease*,
DOI 10.1007/978-1-4419-9371-7_13,

produce Interleukin (IL)-4, IL-5, and IL-13 and mediate humoral immunity against parasite infections.

However, these two T-cell subsets do not completely account for the opportunistic infections seen in congenital or acquired absence of CD4+ T-cells such as mucosal candidiasis, *Pneumocystis carinii* pneumonia, or some bacterial pneumonias. Mice deficient in T_H1 or T_H2 responses (or both) are not permissive for *P. carinii* pneumonia (Garvy et al. 1997a, b), a hallmark infection in AIDS patients with low CD4+ T-cell counts. This data suggests that other CD4+ T-cell lineages are critical for host defenses against opportunistic infections. IL-17 was cloned from CD4+ memory cells in 1993 (Rouvier et al. 1993) and Infante-Duarte demonstrated that IL-17 cells could be produced in response to bacterial lipopeptides (Infante-Duarte et al. 2000). This early study showed that IL-17 producing cells were distinct from T_H1 cells and thus provided the first evidence that T-cell derived IL-17 may have unique effector functions in host resistance against pathogens (Infante-Duarte et al. 2000). Recent evidence has shown that a major subset of cells that produce IL-17 are distinct from T_H1 and T_H2 cells and consist of a third subset of T cells referred to as T_H17 cells (Langrish et al. 2005; Harrington et al. 2005; Park et al. 2005). T_H17 cells produce the cytokines IL-17A (IL-17) (Harrington et al. 2005; Park et al. 2005) and IL-17F (Langrish et al. 2005), as well as the cytokines IL-21 (Nurieva et al. 2007; Korn et al. 2007) and IL-22 (Chung et al. 2006; Liang et al. 2006) (Fig. 1). This new T_H17 cell lineage fills in some of the missing gaps in host immunity not fully explained by the T_H1/T_H2 paradigm.

IL-17 is the prototype of T_H17 cytokines and is the best studied of the T_H17 cytokines. The first identified receptor for IL-17, IL-17RA is a Type I transmembrane protein and is ubiquitously expressed on various organs including lung, kidney, and spleen (Yao et al. 1995). This receptor can bind three members of the IL-17 family including IL-17A, the closely related IL-17F, as well as the most distally related IL-17 family member, IL-17E (Gaffen 2009). The best studied examples of cells that express the receptor for IL-17 are leukocytes, epithelial cells, mesothelial cells, vascular endothelial cells, keratinocytes, and fibroblasts. They respond to IL-17R-mediated signaling by production of granulocyte colony-stimulating factor (G-CSF), IL-6, and IL-8 and mediate granulopoiesis, neutrophil recruitment, and inflammatory responses (Ouyang et al. 2008) (Fig. 1). Recently it has been shown that macrophages (Ishigame et al. 2009) and dendritic cells (Bai et al. 2009; Lin et al. 2009) also respond to IL-17 and produce cytokines and chemokines. Overexpression of IL-17 in mice results in massive extra-medullary hematopoiesis (Schwarzenberger et al. 1998) via the induction of G-CSF and stem cell factor (Schwarzenberger et al. 2000). IL-17RA KO mice can develop spontaneous infection with *S. aureus* infection around the eyes and nares (Schwarzenberger and Kolls 2002) and this phenotype is also observed in mice deficient in two of the three ligands for this receptor (Ishigame et al. 2009). Interestingly, this phenotype is not observed in IL-17A or IL-17F single KO mice (Ishigame et al. 2009), suggesting that these two cytokines can compensate for each other in this model. Both IL-17A and IL-17RA KO mice show susceptibility to pulmonary infection with the

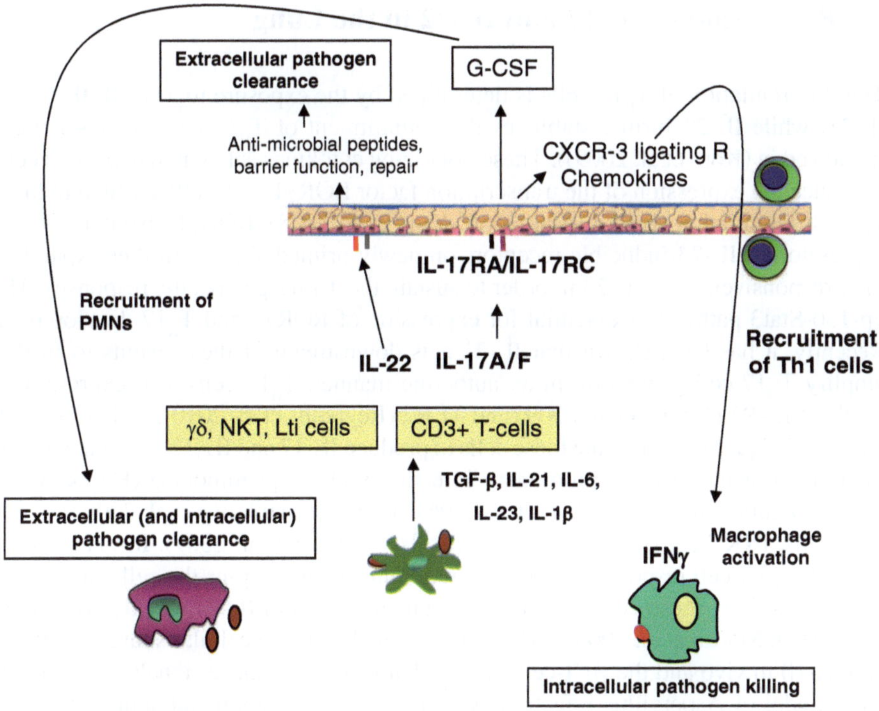

Fig. 1 Overview of Th17 cytokines and immunity at the mucosa. IL-17 and IL-22 can be produced by several immune cells found at mucosal sites in response to a variety of infectious stimuli. IL-17 and IL-22 and signal to the mucosal epithelium, where IL-17 induces G-CSF and CXC chemokine production. These two factors together result in neutrophil recruitment, which is required for bacterial and fungal clearance at mucosal sites. IL-22 and IL-17 can also augment the expression of antimicrobial peptides. IL-22, likely through the activation of STAT3 can also mediate epithelial repair, which is critical to control of extracellular bacterial pathogens. In the setting of vaccine-induced immunity, Th17 cells can induce the production of ligands for CXCR3 and augment the recruitment of IFNγ-producing Th1 cells to control intracellular pathogen growth

extracellular gram negative bacteria *K. pneumoniae,* and again with IL-17RA KO mice showing greater susceptibility than IL-17A KO mice (Aujla et al. 2008). In this model, IL-17RA KO mice fail to recruit neutrophils into the lung, in part due to reduced CXC chemokine production (Ye et al. 2001), but also reduced granulopoiesis through reduced G-CSF production (Ye et al. 2001). Through its effects on dendritic cells, IL-17 has been shown to regulate IL-12p70 production by dendritic cells, thereby regulating T_H1 response to the intracellular pathogen *Francisella tularensis* (Lin et al. 2009) and Chlamydia muridarum (Bai et al. 2009). In contrast, IL-17 appears to be dispensable for T_H1 immunity to *Mycobacterium tuberculosis and Listeria.* Thus, data strongly implicates T_H17 cytokines in host immunity to both extracellular pathogens and certain intracellular pathogens.

2 Regulation of IL-17 and IL-22 in the Lung

The differentiation of T_H17 cells is determined by the exposure to TGF-β, IL-6, and IL-21, while IL-23 further stabilizes the commitment of T_H17 cells to this lineage (reviewed in (Korn et al. 2009)). These polarizing cytokines act on newly primed cells to induce the expression of the transcription factor RORγT and RORα which induces T_H17 differentiation (Ivanov et al. 2006; Yang et al. 2008). RORγT also controls the expression of IL-23-inducible receptors on newly primed T cells, further expanding their responsiveness to IL-23 in order to sustain the T lineage-specific responses. The gp-130-Stat3 pathway is essential for expression of RORγT and T_H17 development. Recently, it has been shown that IL-21 acts downstream of these events to further amplify T_H17 cell generation in an autocrine manner. T_H17 cells also express high levels of IL-R1 (Chung et al. 2009) and TLR2 (Reynolds et al. 2010), and both IL-1β and TLR2 ligands can activate these cells to produce IL-17 and IL-22. Cellular sources of IL-17 in acute primary pulmonary infection with K. pneumoniae (Happel et al. 2003), M. tuberculosis (Lockhart et al. 2006), or Influenza (Crowe et al. 2009) include γδ T-cells, iNKT cells, and possibly LTi cells (Takatori et al. 2009). In many infections, the γδ T-cell response can be the predominant source, plus the cellular production of IL-17 by these cells is critically regulated by both IL-23 and IL-1β (Sutton et al. 2009; Martin et al. 2009). What remains unclear is the cellular sources of IL-23 and IL-1β in vivo and the γδ T-cell subsets that respond to these signals. In a pulmonary model of Aspergillus infection, Vγ1 T-cells are the dominant source of IL-17 (Romani et al. 2008). However, in a model of chronic *Bacillus subtilis* infection, the Vγ6 subset dominates (Simonian et al. 2009), while in a pulmonary model of M.bovis BCG, both Vγ6 and Vγ4 dominate (Umemura et al. 2007).

Since over-induction of T_H17 cells can impact tissue damage due to induction of inflammatory pathways, the generation of T_H17 cells is strictly regulated. For example, cytokines such as IL-27 (Batten et al. 2006; Stumhofer et al. 2006), those belonging to T_H1 (IFNγ) and T_H2 (IL-4) (Harrington et al. 2005; Park et al. 2005), and IL-2 (Laurence et al. 2007) tightly regulate the induction of T_H17 cells. Endogenous lipid mediators like prostaglandin E2 (PGE_2) released under inflammatory conditions promote T_H17 cells differentiation, suggesting that external infection-induced mediators can also influence the decision between a $T_H1/T_H2/T_H17$ and T regulatory cell responses.

Several of these T_H17 polarizing cytokines such as IL-23, TGF-β, IL-6, and IL-1β are induced in dendritic cells activated by components of microbes. Several bacteria and its products such as *Kleibsiella pneumoniae* (Happel et al. 2003, 2005), *Mycobacterium tuberculosis* (Gerosa et al. 2008; Khader et al. 2005), *Helicobacter pylori* (Mitchell et al. 2007), *Francisella tularensis* (Butchar et al. 2007), *Salmonella enterica* (Siegemund et al. 2007), and *Bordetella pertussis* (Fedele et al. 2005) induce these cytokines through TLR signaling. Further bacterial peptidoglycans can induce the generation of T_H17 cells through nucleotide oligomerization domain 2 (NOD2) receptor signaling in dendritic cells (van Beelen et al. 2007). Viruses such as Herpes simplex virus (Kim et al. 2008) and fungus and fungal components such as β-glucans (Gerosa et al. 2008; LeibundGut-Landmann et al. 2007), mannans (Saijo et al. 2010; van de Veerdonk et al. 2009), *Cryptococcus* (Siegemund and

Alber 2008), *Candida albicans* (LeibundGut-Landmann et al. 2007) and *Aspergillus fumigatus* (Zelante et al. 2007) can all induce some or all of these polarizing cytokines from dendritic cells and play a role in differentiation of T_H17 cells. These studies suggest that relative amounts of the polarizing cytokines induced by the pathogen will define the final outcome of differentiation of naïve T cells to T_H1, T_H2, T_H17, or T regulatory cells during infection.

3 T_H17 Effector Responses in the Mucosa

Human bronchial lung epithelial (HBE) cells express both IL-17RA, IL-17RC, IL-22R, and IL-10R2 and thus can respond to IL-17A, IL-17F, and IL-22 (McAllister et al. 2005; Kuestner et al. 2007; Aujla et al. 2008; Huang et al. 2007, Fig. 1). The expression of IL-17RA and IL-17RC also appears in a basolateral dominant fashion (McAllister et al. 2005; Huang et al. 2007) such that signaling only occurs in polarized epithelial cells which are exposed to ligands provided by the basolateral surface (Uesugi et al. 2001; Huang et al. 2007). Treatment of HBE cells with IL-17 induces CXC chemokines such as IL-8 (McAllister et al. 2005; Jones and Chan 2002), G-CSF (McAllister et al. 2005), and antimicrobial proteins such as human β-defensin 2 (Kao et al. 2004). IL-17 treatment also induces IL-19 (Huang et al. 2008), which may be important in regulating T_H2 responses in the mucosa. IL-17 also induced the expression of the polymeric immunoglobulin receptor (Aujla et al. 2008) and indeed T_H17 cytokines have been shown to be critical in generating mucosal IgA responses (Jaffar et al. 2009). IL-22 can activate STAT3 (Wolk et al. 2010) in both murine and human lung epithelial cells and synergizes with IL-17 to increases the expression of human antimicrobial genes such as HBD2, lipocalin (Aujla et al. 2008), and the calgranulins (Aujla et al. 2008). An activity that is unique to IL-22 and not shared by IL-17A or IL-17F in human lung epithelial cells is the fact that IL-22 can augment the clonogenic potential of these cells and accelerate wound repair (Aujla et al. 2008). In skin keratinocytes, IL-22 can also cause acanthosis and hyper-proliferation (Zheng et al. 2007). Thus, IL-22 may be important to epithelial repair in infection, as well as augmenting barrier function (Fig. 1). Neutralizing IL-22 in vivo leads to rapid dissemination of bacteria from the lung during *K. pneumoniae* infection (Aujla et al. 2008). Mucosal IL-22 administration can reduce local bacterial growth as well as prevent dissemination in this model (Aujla et al. 2008). The cellular source of IL-22 in the model is unclear, as IL-22 producing cells are present in Rag 2–/– mice but not Rag 2–/–, and γC double KO mice demonstrate that IL-22 producers are γC dependent.

4 T_H17 Cytokines and Bacterial Infections at the Mucosa

As mentioned above, early work in IL-17RA KO mice demonstrates increased sensitivity to cutaneous *S. aureus* (Schwarzenberger and Kolls 2002; Ishigame et al. 2009; Cho et al. 2010) and pulmonary *K. pneumoniae* (Ye et al. 2001) infection and established a critical role for IL-17 in protective immunity against extracellular bacteria. Further studies show that T_H17 cells play a protective role against extracellular bacterial infections in the gut mucosa. *Citrobacter rodentum*, a naturally occurring mouse pathogen, requires the generation of IL-23-dependent T_H17 cells in the lamina propria for protection (Mangan et al. 2006; Ishigame et al. 2009). Also, IL-22 contributes to the early host defense against *C.rodentium* through the direct induction of the Reg family of antimicrobial proteins in colonic epithelial cells (Zheng et al. 2008). IL-17 signaling also appears to be host-protective in the oral mucosa, as IL-17R-deficient mice are highly susceptible to infection by the gram-negative anaerobic periodontal pathogen, *Porphyromonas gingivalis,* due to reduced neutrophil mobilization and recruitment (Yu et al. 2008).

Although these studies strongly suggest that T_H17 cytokine responses are protective against most extracellular pathogens, in some cases, T_H17 responses contribute to tissue pathology. This was first shown in response to chronic biofilm infection with *Pseudomonas aeruginosa* where IL-23 deficient mice had markedly reduced IL-17 responses and less tissue pathology in response to chronic mucoid *P. aeruginosa* infection. Furthermore, in whooping cough, an infection caused by *B. pertussis*, a gram negative extracellular bacterial infection in the respiratory tract resulted in persistent cough as one of the hallmarks of the clinical disease. Accumulating evidence suggests that the *B. pertussis* infection may bias the host response towards the production of T_H17 cytokines (Siciliano et al. 2006; Fedele et al. 2008) by preferentially inhibiting IL-12 and inducing IL-23 (Fedele et al. 2008). *B. pertussis* causes severe respiratory pathology including bronchiectasis, while *B. parapertussis* causes less severe disease pathology (Carbonetti 2007). Interestingly, *B. pertussis* lipooligosaccharide induces potent production of IL-23, IL-1β, and IL-6 from DCs and drives a more robust differentiation of naïve CD4 T cells to T_H17 cells when compared to DCs activated with LPS from *B. parapertussis* (Fedele et al. 2008). The current emerging hypothesis is that host bias towards T_H17 following *B. pertussis* infection results in inflammation and destruction of the airways, leading to bronchiectasis and persistent cough. This hypothesis is further supported by another cause of bronchiectasis, cystic fibrosis, which is highly associated with bronchiectasis due to chronic biofilm infection with *P. aeruginosa* and elevated IL-17 and IL-22 responses in draining lung lymph node cells (Aujla et al. 2008). Following *M. tuberculosis* infection, the lung inflammation was caused by repeated vaccinations with BCG and resulted in pathological consequences that were IL-17 dependent (Cruz et al. 2010). These disease models may serve as an example for the role of IL-17 in mediating pathology while conferring protection against extracellular bacterial infections in the respiratory mucosa. These findings are not confined to the lung. In *H.pylori* infection in the gastrointestinal tract, the T_H17 response may culminate in a pathogenic inflammatory response in the gut mucosa (Luzza et al. 2000; Caruso et al. 2008). In contrast, certain intracellular pathogens that may require both T_H1 cells and neutrophils for protection at mucosal sites may be dependent on IL-23/IL-17 for pathogen control. For example, the induction of IL-17 and IL-17F production following acute *Mycoplasma pneumonia* pulmonary infection is IL-23-dependent and contributes to neutrophil recruitment and mediates protection against the infection (Wu et al. 2007). Also, infection with *Salmonella typhimurium,* which can exist as an intracellular pathogen, results in induction of IL-17 and IL-22 in the ileal mucosa. The absence of IL-17R signaling results in increased translocation of the bacteria to the mesenteric lymph nodes, reduced induction of chemokines, anti-microbials, and reduced neutrophilic recruitment to the ileal mucosa (Raffatellu et al. 2008). Using a macaque model of *Simian Immuno-deficiency Virus* (SIV) to study HIV human disease and related complications arising due to bacterial co-infections such as *S.typhimurium*, it was found that SIV co-infection selectively inhibits T_H17 responses elicited by *S. typhimurium,* probably due to depletion of $CD4^+$ T cells in the ileal mucosa (Raffatellu et al. 2008). This results in an inability of SIV-infected macaques to mount normal mucosal inflammatory response to *S. typhimurium* infection and allows dissemination of bacteria into the mesenteric lymph node. This data may provides a basis for the observation that people with HIV are at a increased risk of developing bacteremia due to dissemination of bacteria resulting from reduced CD4 T_H17 responses. HIV-infected individuals that receive anti-retroviral therapy undergo effective CD4 T-cell restoration and this is usually associated with enhanced CD4 T_H17-cell accumulation (Macal et al. 2008).

5 T_H17 Cytokines and Fungal Infections at the Mucosa

The role of T_H17 cytokines in fungal infection has been controversial, but emerging evidence in both animal and human models shows that these cytokines play a critical role in mucosal infections. Initial observations suggested that T_H17 cytokines, particularly IL-17, contribute to tissue pathology in invasive aspergillus infections in the lung, predominantly in the setting of NADPH oxidase deficiency (Romani et al. 2008). However, IL-17 and IL-17RA signaling are critical for host resistance to oro-pharyngeal candidiasis and for the expression of mouse β-defensin 3 that has anti-Candicidal activity (Conti et al. 2009). Patients with Hyper IgE syndrome (HIES) that frequently suffer from oro-pharyngeal candidiasis have mutations in STAT3 (Holland et al. 2007; Minegishi et al. 2007). They fail to generate Candida specific T_H17 cells, adding further evidence that this phenotype is tightly linked to T_H17 immunity (Milner et al. 2008; de et al. 2008). Treatment of skin keratinocytes with T_H17 cytokines markedly increases anti-Candicidal activity in vitro, while activated T-cells from HIES patients are unable to induce this anti-Candicidal activity. It has recently been shown that patients with chronic mucocutaneous candidiasis also have antibodies that can block IL-17 and/or IL-22 which may also explain the susceptibility of these patients to mucosal Candida infection (Puel et al. 2010). In respiratory tract models of fungal infections using *Pneumocytis carnii* (Rudner et al. 2007) and Aspergillus fumigatus (Werner et al. 2009), induction of IL-23 and IL-17 following pathogen challenge is protective, since IL-23KO mice or neutralization of the IL-23/IL-17 axis resulted in impaired clearance of the pathogen.

6 T_H17 Cytokines and Viral Infection

The role of T_H17 cytokines in viral infection is a rapidly emerging area of research. *Herpes simplex virus* (HSV-1) infection of the cornea results in early induction of both IL-23 (Kim et al. 2008) and IL-17 (Molesworth-Kenyon et al. 2008). IL-17RA KO mice have also reduced early infiltration of neutrophils and corneal opacity following HSV infection (Molesworth-Kenyon et al. 2008). In contrast, IL-23KO mice have exacerbated disease and pathology possibly due to HSV increased IL-12 responses and increased IFNγ producing cells (Kim et al. 2008). In pulmonary influenza infection, IL-17R KO mice had record tissue pathology and weight loss suggesting that in this model, IL-17 contributes to tissue pathology as well (Crowe et al. 2009). In a heterotypic influenza model, CD8+ cells that produce IL-17 can mediate protection against influenza challenge (Hamada et al. 2009). Thus, the role of the IL-17 response to primary influenza infection versus heterotypic infection appears to differ. Human *rhinovirus* (HRV) infections are associated with exacerbations of asthma and chronic obstructive pulmonary infiltration. IL-17 was shown to function synergistically with HRV to induce IL-8 from epithelial cells and may contribute to the recruitment of neutrophils, immature DCs, and memory T cells to the lung; contributing to severe inflammatory profiles seen during viral exacerbations of airway disease (Wiehler and Proud 2007).

7 Conclusions

We specifically did not address the role of T_H17 cytokines vaccine-induced immunity as this has been recently reviewed (Khader et al. 2009). Current data suggests that T_H17 cells have evolved to mediate protective immunity against a variety of pathogens at different mucosal sites. Moreover, deficient T_H17 responses explain in part, the increased susceptibility to certain infections such as mucocutaneous in HIES patients and the depletion of T_H17 cells by HIV. Taken together, strategies to augment these cells or recover these cell populations will be important for improved vaccine or therapeutic efficacy in these patients. It is important to remember that IL-17 is evolutionary conserved and that the gene exists in mollusks and *Ciona intestinalis.* This predates the development of adaptive T-cell immunity, and this cytokine likely bridged innate and adaptive immunity. What remains unclear is how evolutionary pressure forced this gene to be expressed in memory CD4+ T-cells in mammals. This will be an important area of future investigation.

References

Aujla, S.J., Chan, Y.R., Zheng, M., Fei, M., Askew, D.J., Pociask, D.A., Reinhart, T.A., McAllister, F., Edeal, J., Gaus, K., Husain, S., Kreindler, J.L., Dubin, P.J., Pilewski, J.M., Myerburg, M.M., Mason, C.A., Iwakura, Y., and Kolls, J.K. (2008). IL-22 mediates mucosal host defense against Gram-negative bacterial pneumonia. Nat. Med. 14, 275–281.

Bai, H., Cheng, J., Gao, X., Joyee, A.G., Fan, Y., Wang, S., Jiao, L., Yao, Z., and Yang, X. (2009). IL-17/Th17 promotes type 1 T cell immunity against pulmonary intracellular bacterial infection through modulating dendritic cell function. J. Immunol. 183, 5886–5895.

Batten, M., Li, J., Yi, S., Kljavin, N.M., Danilenko, D.M., Lucas, S., Lee, J., De Sauvage, F.J., and Ghilardi, N. (2006). Interleukin 27 limits auto-immune encephalomyelitis by suppressing the development of interleukin 17-producing T cells. Nat. Immunol. 7, 929–936.

Butchar, J.P., Rajaram, M.V., Ganesan, L.P., Parsa, K.V., Clay, C.D., Schlesinger, L.S., and Tridandapani, S. (2007). Francisella tularensis induces IL-23 production in human monocytes. J. Immunol. 178, 4445–4454.

Carbonetti, N.H. (2007). Immuno-modulation in the pathogenesis of Bordetella pertussis infection and disease. Curr. Opin. Pharmacol. 7, 272–278.

Caruso, R., Fina, D., Paoluzi, O.A., Del Vecchio, B.G., Stolfi, C., Rizzo, A., Caprioli, F., Sarra, M., Andrei, F., Fantini, M.C., Macdonald, T.T., Pallone, F., and Monteleone, G. (2008). IL-23-mediated regulation of IL-17 production in Helicobacter pylori-infected gastric mucosa. Eur. J. Immunol. 38, 470–478.

Cho, J.S., Pietras, E.M., Garcia, N.C., Ramos, R.I., Farzam, D.M., Monroe, H.R., Magorien, J.E., Blauvelt, A., Kolls, J.K., Cheung, A.L., Cheng, G., Modlin, R.L., and Miller, L.S. (2010). IL-17 is essential for host defense against cutaneous Staphylococcus aureus infection in mice. J. Clin. Invest 120, 1762–1773.

Chung, Y., Chang, S.H., Martinez, G.J., Yang, X.O., Nurieva, R., Kang, H.S., Ma, L., Watowich, S.S., Jetten, A.M., Tian, Q., and Dong, C. (2009). Critical regulation of early Th17 cell differentiation by interleukin-1 signaling. Immunity 30, 576–587.

Chung, Y., Yang, X., Chang, S.H., Ma, L., Tian, Q., and Dong, C. (2006). Expression and regulation of IL-22 in the IL-17-producing CD4+ T lymphocytes. Cell. Res. 16, 902–907.

Conti, H.R., Shen, F., Nayyar, N., Stocum, E., Sun, J.N., Lindemann, M.J., Ho, A.W., Hai, J.H., Yu, J.J., Jung, J.W., Filler, S.G., Masso-Welch, P., Edgerton, M., and Gaffen, S.L. (2009). Th17 cells and IL-17 receptor signaling are essential for mucosal host defense against oral candidiasis. J. Exp. Med. 206, 299–311.

Crowe, C.R., Chen, K., Pociask, D.A., Alcorn, J.F., Krivich, C., Enelow, R.I., Ross, T.M., Witztum, J.L., and Kolls, J.K. (2009). Critical role of IL-17RA in immuno-pathology of influenza infection. J. Immunol. 183, 5301–5310.

Cruz, A., Fraga, A.G., Fountain, J.J., Rangel-Moreno, J., Torrado, E., Saraiva, M., Pereira, D.R., Randall, T.D., Pedrosa, J., Cooper, A.M., and Castro, A.G. (2010). Pathological role of interleukin 17 in mice subjected to repeated BCG vaccination after infection with Mycobacterium tuberculosis. J. Exp. Med. 207, 1609–1616.

de, B.L., Puel, A., Filipe-Santos, O., Cobat, A., Ghandil, P., Chrabieh, M., Feinberg, J., von, B.H., Samarina, A., Janniere, L., Fieschi, C., Stephan, J.L., Boileau, C., Lyonnet, S., Jondeau, G., Cormier-Daire,V., Le,M.M., Hoarau,C., Lebranchu,Y., Lortholary,O., Chandesris,M.O., Tron, F., Gambineri, E., Bianchi, L., Rodriguez-Gallego, C., Zitnik, S.E., Vasconcelos, J., Guedes, M., Vitor, A.B., Marodi, L., Chapel, H., Reid, B., Roifman, C., Nadal, D., Reichenbach, J., Caragol, I., Garty, B.Z., Dogu, F., Camcioglu, Y., Gulle, S., Sanal, O., Fischer, A., Abel, L., Stockinger, B., Picard, C., and Casanova, J.L. (2008). Mutations in STAT3 and IL12RB1 impair the development of human IL-17-producing T cells. J. Exp. Med. 205, 1543–1550.

Fedele, G., Nasso, M., Spensieri, F., Palazzo, R., Frasca, L., Watanabe, M., and Ausiello, C.M. (2008). Lipopolysaccharides from Bordetella pertussis and Bordetella parapertussis differently modulate human dendritic cell functions resulting in divergent prevalence of Th17-polarized responses. J. Immunol. 181, 208–216.

Fedele, G., Stefanelli, P., Spensieri, F., Fazio, C., Mastrantonio, P., and Ausiello, C.M. (2005). Bordetella pertussis infected human monocyte-derived dendritic cells undergo maturation and induce Th1 polarization and interleukin-23 expression. Infect. Immun. 73, 1590–1597.

Gaffen, S.L. (2009). Structure and signaling in the IL-17 receptor family. Nat. Rev. Immunol. 9, 556–567.

Garvy, B.A., Ezekowitz, R.A., and Harmsen, A.G. (1997a). Role of gamma interferon in the host immune and inflammatory responses to Pneumocystis carinii infection. Infect. Immun. 65, 373–379.

Garvy, B.A., Wiley, J.A., Gigliotti, F., and Harmsen, A.G. (1997b). Protection against Pneumocystis carinii pneumonia by antibodies generated from either T helper 1 or T helper 2 responses. Infection & Immunity 65, 5052–5056.

Gerosa, F., Baldani-Guerra, B., Lyakh, L.A., Batoni, G., Esin, S., Winkler-Pickett, R.T., Consolaro, M.R., De, M.M., Giachino, D., Robbiano, A., Astegiano, M., Sambataro, A., Kastelein, R.A., Carra, G., and Trinchieri, G. (2008). Differential regulation of interleukin 12 and interleukin 23 production in human dendritic cells. J. Exp. Med. 205, 1447–1461.

Hamada, H., Garcia-Hernandez, M.L., Reome, J.B., Misra, S.K., Strutt, T.M., McKinstry, K.K., Cooper, A.M., Swain, S.L., and Dutton, R.W. (2009). Tc17, a unique subset of CD8 T cells that can protect against lethal influenza challenge. J. Immunol. 182, 3469–3481.

Happel, K.I., Dubin, P.J., Zheng, M., Ghilardi, N., Lockhart, C., Quinton, L.J., Odden, A.R., Shellito, J.E., Bagby, G.J., Nelson, S., and Kolls, J.K. (2005). Divergent roles of IL-23 and IL-12 in host defense against Klebsiella pneumoniae. J. Exp. Med. 202, 761–769.

Happel, K.I., Zheng, M., Young, E., Quinton, L.J., Lockhart, E., Ramsay, A.J., Shellito, J.E., Schurr, J.R., Bagby, G.J., Nelson, S., and Kolls, J.K. (2003). Cutting Edge: Roles of Toll-Like Receptor 4 and IL-23 in IL-17 Expression in Response to Klebsiella pneumoniae Infection. J. Immunol. 170, 4432–4436.

Harrington, L.E., Hatton, R.D., Mangan, P.R., Turner, H., Murphy, T.L., Murphy, K.M., and Weaver,C.T. (2005). Interleukin 17-producing CD4+ effector T cells develop via a lineage distinct from the T helper type 1 and 2 lineages. Nat. Immunol. 6, 1123–1132.

Holland, S.M., DeLeo, F.R., Elloumi, H.Z., Hsu, A.P., Uzel, G., Brodsky, N., Freeman, A.F., Demidowich, A., Davis, J., Turner, M.L., Anderson, V.L., Darnell, D.N., Welch, P.A., Kuhns, D.B., Frucht, D.M., Malech, H.L., Gallin, J.I., Kobayashi, S.D., Whitney, A.R., Voyich, J.M., Musser, J.M., Woellner, C., Schaffer, A.A., Puck, J.M., and Grimbacher, B. (2007). STAT3 mutations in the hyper-IgE syndrome. N. Engl. J. Med. 357, 1608–1619.

Huang, F., Kao, C.Y., Wachi, S., Thai, P., Ryu, J., and Wu, R. (2007). Requirement for both JAK-mediated PI3K signaling and ACT1/TRAF6/TAK1-dependent NF-kappaB activation by IL-17A in enhancing cytokine expression in human airway epithelial cells. J. Immunol. 179, 6504–6513.

Huang, F., Wachi, S., Thai, P., Loukoianov, A., Tan, K.H., Forteza, R.M., and Wu, R. (2008). Potentiation of IL-19 expression in airway epithelia by IL-17A and IL-4/IL-13: important implications in asthma. J. Allergy Clin. Immunol. 121, 1415–21, 1421.

Infante-Duarte, C., Horton, H.F., Byrne, M.C., and Kamradt, T. (2000). Microbial lipopeptides induce the production of IL-17 in Th cells. J. Immunol. 165, 6107–6115.

Ishigame, H., Kakuta, S., Nagai, T., Kadoki, M., Nambu, A., Komiyama, Y., Fujikado, N., Tanahashi, Y., Akitsu, A., Kotaki, H., Sudo, K., Nakae, S., Sasakawa, C., and Iwakura,Y. (2009). Differential roles of interleukin-17A and -17F in host defense against mucoepithelial bacterial infection and allergic responses. Immunity *30*, 108–119.

Ivanov, I.I., McKenzie, B.S., Zhou, L., Tadokoro, C.E., Lepelley, A., Lafaille, J.J., Cua, D.J., and Littman, D.R. (2006). The orphan nuclear receptor RORgammat directs the differentiation program of pro-inflammatory IL-17+ T helper cells. Cell. 126, 1121–1133.

Jaffar, Z., Ferrini, M.E., Herritt, L.A., and Roberts, K. (2009). Cutting edge: Lung mucosal Th17-mediated responses induce polymeric Ig receptor expression by the airway epithelium and elevate secretory IgA levels. J. Immunol. 182, 4507–4511.

Jones, C.E. and Chan, K. (2002). Interleukin-17 stimulates the expression of interleukin-8, growth-related oncogene-alpha, and granulocyte-colony-stimulating factor by human airway epithelial cells. Am. J. Respir. Cell Mol. Biol. 26, 748–753.

Kader, M., Wang, X., Piatak, M., Lifson, J., Roederer, M., Veazey, R., and Mattapallil, J.J. (2009). Alpha4(+)beta7(hi)CD4(+) memory T cells harbor most Th-17 cells and are preferentially infected during acute SIV infection. Mucosal. Immunol. 2, 439–449.

Kao, C.Y., Chen, Y., Thai, P., Wachi, S., Huang, F., Kim, C., Harper, R.W., and Wu,R. (2004). IL-17 markedly up-regulates beta-defensin-2 expression in human airway epithelium via JAK and NF-kappaB signaling pathways. J. Immunol. 173, 3482–3491.

Khader, S.A., Gaffen, S.L., and Kolls, J.K. (2009). Th17 cells at the crossroads of innate and adaptive immunity against infectious diseases at the mucosa. Mucosal. Immunol. 2, 403–411.

Khader, S.A., Pearl, J.E., Sakamoto, K., Gilmartin, L., Bell, G.K., Jelley-Gibbs, D.M., Ghilardi, N., deSauvage, F., and Cooper, A.M. (2005). IL-23 compensates for the absence of IL-12p70 and is essential for the IL-17 response during tuberculosis but is dispensable for protection and antigen-specific IFN-gamma responses if IL-12p70 is available. J. Immunol. 175, 788–795.

Kim, B., Sarangi, P.P., Azkur, A.K., Kaistha, S.D., and Rouse, B.T. (2008). Enhanced viral immunoinflammatory lesions in mice lacking IL-23 responses. Microbes. Infect. 10, 302–312.

Korn, T., Bettelli, E., Gao, W., Awasthi, A., Jager, A., Strom, T.B., Oukka, M., and Kuchroo, V.K. (2007). IL-21 initiates an alternative pathway to induce pro-inflammatory T(H)17 cells. Nature 448, 484–487.

Korn, T., Bettelli, E., Oukka, M., and Kuchroo, V.K. (2009). IL-17 and Th17 Cells. Annu. Rev. Immunol. 27, 485–517.

Kuestner, R.E., Taft, D.W., Haran, A., Brandt, C.S., Brender, T., Lum, K., Harder, B., Okada, S., Ostrander, C.D., Kreindler, J.L., Aujla, S.J., Reardon, B., Moore, M., Shea, P., Schreckhise, R., Bukowski, T.R., Presnell, S., Guerra-Lewis, P., Parrish-Novak, J., Ellsworth, J.L., Jaspers, S., Lewis, K.E., Appleby, M., Kolls, J.K., Rixon, M., West, J.W., Gao, Z., and Levin, S.D. (2007). Identification of the IL-17 receptor related molecule IL-17RC as the receptor for IL-17F. J. Immunol. 179, 5462–5473.

Langrish, C.L., Chen, Y., Blumenschein, W.M., Mattson, J., Basham, B., Sedgwick, J.D., McClanahan, T., Kastelein, R.A., and Cua, D.J. (2005). IL-23 drives a pathogenic T cell population that induces auto-immune inflammation. J. Exp. Med. 201, 233–240.

Laurence, A., Tato, C.M., Davidson, T.S., Kanno, Y., Chen, Z., Yao, Z., Blank, R.B., Meylan, F., Siegel, R., Hennighausen, L., Shevach, E.M., and O'Shea, J.J. (2007). Interleukin-2 signaling via STAT5 constrains T helper 17 cell generation. Immunity. 26, 371–381.

LeibundGut-Landmann, S., Gross, O., Robinson, M.J., Osorio, F., Slack, E.C., Tsoni, S.V., Schweighoffer, E., Tybulewicz, V., Brown, G.D., Ruland, J., and Reis E Sousa (2007). Syk- and CARD9-dependent coupling of innate immunity to the induction of T helper cells that produce interleukin 17. Nat. Immunol. 8, 630–638.

Li, Q., Duan, L., Estes, J.D., Ma, Z.M., Rourke, T., Wang, Y., Reilly, C., Carlis, J., Miller, C.J., and Haase, A.T. (2005). Peak SIV replication in resting memory CD4+ T cells depletes gut lamina propria CD4+ T cells. Nature 434, 1148–1152.

Liang, S.C., Tan, X.Y., Luxenberg, D.P., Karim, R., Dunussi-Joannopoulos, K., Collins, M., and Fouser, L.A. (2006). Interleukin (IL)-22 and IL-17 are co-expressed by Th17 cells and cooperatively enhance expression of antimicrobial peptides. J. Exp. Med. 203, 2271–2279.

Lin, Y., Ritchea, S., Logar, A., Slight, S., Messmer, M., Rangel-Moreno, J., Guglani, L., Alcorn, J.F., Strawbridge, H., Park, S.M., Onishi, R., Nyugen, N., Walter, M.J., Pociask, D., Randall, T.D., Gaffen, S.L., Iwakura, Y., Kolls, J.K., and Khader, S.A. (2009). Interleukin-17 is required for T helper 1 cell immunity and host resistance to the intracellular pathogen Francisella tularensis. Immunity 31, 799–810.

Lockhart, E., Green, A.M., and Flynn, J.L. (2006). IL-17 production is dominated by gammadelta T cells rather than CD4 T cells during Mycobacterium tuberculosis infection. J. Immunol. 177, 4662–4669.

Luzza, F., Parrello, T., Monteleone, G., Sebkova, L., Romano, M., Zarrilli, R., Imeneo, M., and Pallone, F. (2000). Up-regulation of IL-17 is associated with bioactive IL-8 expression in Helicobacter pylori-infected human gastric mucosa. J. Immunol. 165, 5332–5337.

Macal, M., Sankaran, S., Chun, T.W., Reay, E., Flamm, J., Prindiville, T.J., and Dandekar, S. (2008). Effective CD4+ T-cell restoration in gut-associated lymphoid tissue of HIV-infected patients is associated with enhanced Th17 cells and polyfunctional HIV-specific T-cell responses. Mucosal. Immunol. 1, 475–488.

Mangan, P.R., Harrington, L.E., O'Quinn, D.B., Helms, W.S., Bullard, D.C., Elson, C.O., Hatton, R.D., Wahl, S.M., Schoeb, T.R., and Weaver, C.T. (2006). Transforming growth factor-β induces development of the TH17 lineage. Nature 441, 231–234.

Martin, B., Hirota, K., Cua, D.J., Stockinger, B., and Veldhoen, M. (2009). Interleukin-17-producing gammadelta T cells selectively expand in response to pathogen products and environmental signals. Immunity 31, 321–330.

Mattapallil, J.J., Douek, D.C., Hill, B., Nishimura, Y., Martin, M., and Roederer,M. (2005). Massive infection and loss of memory CD4+ T cells in multiple tissues during acute SIV infection. Nature 434, 1093–1097.

McAllister, F., Henry, A., Kreindler, J.L., Dubin, P.J., Ulrich, L., Steele, C., Finder, J.D., Pilewski, J.M., Carreno, B.M., Goldman, S.J., Pirhonen, J., and Kolls, J.K. (2005). Role of IL-17A, IL-17F, and the IL-17 Receptor in Regulating Growth-Related Oncogene-{alpha} and Granulocyte Colony-Stimulating Factor in Bronchial Epithelium: Implications for Airway Inflammation in Cystic Fibrosis. J. Immunol. 175, 404–412.

Milner,J.D., Brenchley,J.M., Laurence,A., Freeman,A.F., Hill,B.J., Elias,K.M., Kanno,Y., Spalding,C., Elloumi,H.Z., Paulson,M.L., Davis,J., Hsu,A., Asher,A.I., O'Shea,J., Holland,S.M., Paul,W.E., and Douek,D.C. (2008). Impaired T(H)17 cell differentiation in subjects with autosomal dominant hyper-IgE syndrome. Nature 452, 773–776.

Minegishi,Y., Saito,M., Tsuchiya,S., Tsuge,I., Takada,H., Hara,T., Kawamura,N., Ariga,T., Pasic,S., Stojkovic,O., Metin,A., and Karasuyama,H. (2007). Dominant-negative mutations in the DNA-binding domain of STAT3 cause hyper-IgE syndrome. Nature 448, 1058–1062.

Mitchell, P., Germain, C., Fiori, P.L., Khamri, W., Foster, G.R., Ghosh, S., Lechler, R.I., Bamford, K.B., and Lombardi, G. (2007). Chronic exposure to Helicobacter pylori impairs dendritic cell function and inhibits Th1 development. Infect. Immun. 75, 810–819.

Molesworth-Kenyon, S.J., Yin, R., Oakes, J.E., and Lausch, R.N. (2008). IL-17 receptor signaling influences virus-induced corneal inflammation. J. Leukoc. Biol. 83, 401–408.

Mosmann, T.R., Cherwinski, H., Bond, M.W., Giedlin, M.A., and Coffman, R.L. (1986). Two types of murine helper T cell clone. I. Definition according to profiles of lymphokine activities and secreted proteins. J. Immunol. 136, 2348–2357.

Nurieva, R., Yang, X.O., Martinez, G., Zhang, Y., Panopoulos, A.D., Ma, L., Schluns, K., Tian, Q., Watowich, S.S., Jetten, A.M., and Dong, C. (2007). Essential autocrine regulation by IL-21 in the generation of inflammatory T cells. Nature 448, 480–483.

Ouyang, W., Kolls, J.K., and Zheng, Y. (2008). The biological functions of T helper 17 cell effector cytokines in inflammation. Immunity 28, 454–467.

Park, H., Li, Z., Yang, X.O., Chang, S.H., Nurieva, R., Wang, Y.H., Wang, Y., Hood, L., Zhu, Z., Tian, Q., and Dong, C. (2005). A distinct lineage of CD4 T cells regulates tissue inflammation by producing interleukin 17. Nat. Immunol. 6, 1133–1141.

Puel, A., Doffinger, R., Natividad, A., Chrabieh, M., Barcenas-Morales, G., Picard, C., Cobat, A., Ouachee-Chardin, M., Toulon, A., Bustamante, J., Al-Muhsen, S., Al-Owain, M., Arkwright, P.D., Costigan, C., McConnell, V., Cant, A.J., Abinun, M., Polak, M., Bougneres, P.F., Kumararatne, D., Marodi, L., Nahum, A., Roifman, C., Blanche, S., Fischer, A., Bodemer, C., Abel, L., Lilic, D., and Casanova, J.L. (2010). Auto-antibodies against IL-17A, IL-17F, and IL-22 in patients with chronic mucocutaneous candidiasis and auto-immune polyendocrine syndrome type I. J. Exp. Med. 207, 291–297.

Raffatellu, M., Santos, R.L., Verhoeven, D.E., George, M.D., Wilson, R.P., Winter, S.E., Godinez, I., Sankaran, S., Paixao, T.A., Gordon, M.A., Kolls, J.K., Dandekar, S., and Baumler, A.J. (2008). Simian immunodeficiency virus-induced mucosal interleukin-17 deficiency promotes Salmonella dissemination from the gut. Nat. Med. 14, 421–428.

Reynolds, J.M., Pappu, B.P., Peng, J., Martinez, G.J., Zhang, Y., Chung, Y., Ma, L., Yang, X.O., Nurieva, R.I., Tian, Q., and Dong, C. (2010). Toll-like receptor 2 signaling in CD4(+) T lymphocytes promotes T helper 17 responses and regulates the pathogenesis of auto-immune disease. Immunity 32, 692–702.

Romani, L., Fallarino, F., De Luca, A., Montagnoli, C., D'Angelo, C., Zelante, T., Vacca, C., Bistoni, F., Fioretti, M.C., Grohmann, U., Segal, B.H., and Puccetti, P. (2008). Defective tryptophan catabolism underlies inflammation in mouse chronic granulomatous disease. Nature 451, 211–215.

Rouvier, E., Luciani, M.F., Mattei, M.G., Denizot, F., and Golstein, P. (1993). CTLA-8, cloned from an activated T cell, bearing AU-rich messenger RNA instability sequences, and homologous to a herpes virus saimiri gene. J. Immunol. 150, 5445–5456.

Rudner, X.L., Happel, K.I., Young, E.A., and Shellito, J.E. (2007). Interleukin-23 (IL-23)-IL-17 cytokine axis in murine Pneumocystis carinii infection. Infect. Immun. 75, 3055–3061.

Saijo, S., Ikeda, S., Yamabe, K., Kakuta, S., Ishigame, H., Akitsu, A., Fujikado, N., Kusaka, T., Kubo, S., Chung, S.H., Komatsu, R., Miura, N., Adachi, Y., Ohno, N., Shibuya, K., Yamamoto, N., Kawakami, K., Yamasaki, S., Saito, T., Akira, S., and Iwakura, Y. (2010). Dectin-2 recognition of alpha-mannans and induction of Th17 cell differentiation is essential for host defense against Candida albicans. Immunity 32, 681–691.

Schwarzenberger, P., Huang, W., Ye, P., Oliver, P., Manuel, M., Zhang, Z., Bagby, G., Nelson, S., and Kolls, J.K. (2000). Requirement of Endogenous Stem Cell Factor and Granulocyte-Colony-Stimulating Factor for IL-17-Mediated Granulopoiesis. J. Immunol. 164, 4783–4789.

Schwarzenberger,P. and Kolls,J.K. (2002). Interleukin 17: an example for gene therapy as a tool to study cytokine mediated regulation of hematopoiesis. J. Cell Biochem. Suppl Suppl 38, 88–95.

Schwarzenberger, P., La, R., V, Miller, A., Ye, P., Huang, W., Zieske, A., Nelson, S., Bagby, G.J., Stoltz, D., Mynatt, R.L., Spriggs, M., and Kolls, J.K. (1998). IL-17 stimulates granulopoiesis in mice: use of an alternate, novel gene therapy-derived method for in vivo evaluation of cytokines. J. Immunol. 161, 6383–6389.

Siciliano, N.A., Skinner, J.A., and Yuk, M.H. (2006). Bordetella bronchiseptica modulates macrophage phenotype leading to the inhibition of CD4+ T cell proliferation and the initiation of a Th17 immune response. J. Immunol. 177, 7131–7138.

Siegemund, S. and Alber, G. (2008). Cryptococcus neoformans activates bone marrow-derived conventional dendritic cells rather than plasmacytoid dendritic cells and down-regulates macrophages. FEMS Immunol. Med. Microbiol. 52, 417–427.

Siegemund, S., Schutze, N., Freudenberg, M.A., Lutz, M.B., Straubinger, R.K., and Alber, G. (2007). Production of IL-12, IL-23 and IL-27p28 by bone marrow-derived conventional dendritic cells rather than macrophages after LPS/TLR4-dependent induction by Salmonella Enteritidis. Immunobiology 212, 739–750.

Simonian, P.L., Roark, C.L., Born, W.K., O'Brien, R.L., and Fontenot, A.P. (2009). Gammadelta T cells and Th17 cytokines in hypersensitivity pneumonitis and lung fibrosis. Transl. Res. 154, 222–227.

Stumhofer, J.S., Laurence, A., Wilson, E.H., Huang, E., Tato, C.M., Johnson, L.M., Villarino, A.V., Huang, Q., Yoshimura, A., Sehy, D., Saris, C.J., O'Shea, J.J., Hennighausen, L., Ernst, M., and Hunter, C.A. (2006). Interleukin 27 negatively regulates the development of interleukin 17-producing T helper cells during chronic inflammation of the central nervous system. Nat. Immunol. 7, 937–945.

Sutton, C.E., Lalor, S.J., Sweeney, C.M., Brereton, C.F., Lavelle, E.C., and Mills, K.H. (2009). Interleukin-1 and IL-23 induce innate IL-17 production from gammadelta T cells, amplifying Th17 responses and auto-immunity. Immunity 31, 331–341.

Takatori, H., Kanno, Y., Watford, W.T., Tato, C.M., Weiss, G., Ivanov, I.I., Littman, D.R., and O'Shea, J.J. (2009). Lymphoid tissue inducer-like cells are an innate source of IL-17 and IL-22. J. Exp. Med. 206, 35–41.

Uesugi, T., Froh, M., Arteel, G.E., Bradford, B.U., Gabele, E., Wheeler, M.D., and Thurman, R.G. (2001). Delivery of IkappaB superrepressor gene with adenovirus reduces early alcohol-induced liver injury in rats. Hepatology 34, 1149–1157.

Umemura, M., Yahagi, A., Hamada, S., Begum, M.D., Watanabe, H., Kawakami, K., Suda, T., Sudo, K., Nakae, S., Iwakura, Y., and Matsuzaki, G. (2007). IL-17-mediated regulation of innate and acquired immune response against pulmonary Mycobacterium bovis bacille Calmette-Guerin infection. J. Immunol. 178, 3786–3796.

van Beelen, A.J., Zelinkova, Z., Taanman-Kueter, E.W., Muller, F.J., Hommes, D.W., Zaat, S.A., Kapsenberg, M.L., and de Jong, E.C. (2007). Stimulation of the intracellular bacterial sensor NOD2 programs dendritic cells to promote interleukin-17 production in human memory T cells. Immunity 27, 660–669.

van de Veerdonk, F.L., Marijnissen, R.J., Kullberg, B.J., Koenen, H.J., Cheng, S.C., Joosten, I., Van den Berg, W.B., Williams, D.L., van der Meer, J.W., Joosten, L.A., and Netea, M.G. (2009). The macrophage mannose receptor induces IL-17 in response to Candida albicans. Cell Host. Microbe 5, 329–340.

Werner, J.L., Metz, A.E., Horn, D., Schoeb, T.R., Hewitt, M.M., Schwiebert, L.M., Faro-Trindade, I., Brown, G.D., and Steele,C. (2009). Requisite role for the dectin-1 beta-glucan receptor in pulmonary defense against Aspergillus fumigatus. J. Immunol. 182, 4938–4946.

Wiehler, S. and Proud, D. (2007). Interleukin-17A modulates human airway epithelial responses to human rhinovirus infection. Am. J. Physiol Lung Cell Mol. Physiol 293, L505–L515.

Wolk, K., Witte, E., Witte, K., Warszawska, K., and Sabat, R. (2010). Biology of interleukin-22. Semin. Immunopathol. 32, 17–31.

Wu, Q., Martin, R.J., Rino, J.G., Breed, R., Torres, R.M., and Chu, H.W. (2007). IL-23-dependent IL-17 production is essential in neutrophil recruitment and activity in mouse lung defense against respiratory Mycoplasma pneumoniae infection. Microbes. Infect. 9, 78–86.

Yang, X.O., Pappu, B.P., Nurieva, R., Akimzhanov, A., Kang, H.S., Chung, Y., Ma, L., Shah, B., Panopoulos, A.D., Schluns, K.S., Watowich, S.S., Tian, Q., Jetten, A.M., and Dong, C. (2008). T helper 17 lineage differentiation is programmed by orphan nuclear receptors ROR alpha and ROR gamma. Immunity 28, 29–39.

Yao, Z., Fanslow, W.C., Seldin, M.F., Rousseau, A.-M., Painter, S.L., Comeau, M.R., Cohen, J.I., and Spriggs, M.K. (1995). Herpes virus saimiri encodes a new cytokine, IL-17, which binds to novel cytokine receptor. Immunity 3, 811–821.

Ye, P., Rodriguez, F.H., Kanaly, S., Stocking, K.L., Schurr, J., Schwarzenberger, P., Oliver, P., Huang, W., Zhang, P., Zhang, J., Shellito, J.E., Bagby, G.J., Nelson, S., Charrier, K., Peschon, J.J., and Kolls, J.K. (2001). Requirement of interleukin 17 receptor signaling for lung CXC chemokine and granulocyte colony-stimulating factor expression, neutrophil recruitment, and host defense. J. Exp. Med. 194, 519–527.

Yu, J.J., Ruddy, M.J., Conti, H.R., Boonanantanasarn, K., and Gaffen, S.L. (2008). The interleukin-17 receptor plays a gender-dependent role in host protection against Porphyromonas gingivalis-induced periodontal bone loss. Infect. Immun. 76, 4206–4213.

Zelante, T., De Luca, A., Bonifazi, P., Montagnoli, C., Bozza, S., Moretti, S., Belladonna, M.L., Vacca, C., Conte, C., Mosci, P., Bistoni, F., Puccetti, P., Kastelein, R.A., Kopf, M., and Romani,L. (2007). IL-23 and the Th17 pathway promote inflammation and impair antifungal immune resistance. Eur. J. Immunol. 37, 2695–2706.

Zheng, Y., Danilenko, D.M., Valdez, P., Kasman, I., Eastham-Anderson, J., Wu, J., and Ouyang, W. (2007). Interleukin-22, a T(H)17 cytokine, mediates IL-23-induced dermal inflammation and acanthosis. Nature 445, 648–651.

Zheng, Y., Valdez, P.A., Danilenko, D.M., Hu, Y., Sa, S.M., Gong, Q., Abbas, A.R., Modrusan, Z., Ghilardi, N., De Sauvage, F.J., and Ouyang, W. (2008). Interleukin-22 mediates early host defense against attaching and effacing bacterial pathogens. Nat. Med. 14, 282–289.

T_H17 Cells and IL-23 in Gut Inflammation

Giovanni Monteleone, Massimiliano Sarra, Francesco Pallone, and Thomas T. MacDonald

Abstract Gut inflammation in patients with inflammatory bowel disease (IBD) has traditionally been considered to be mediated by an exaggerated T_H1 response (Crohn's disease) or an atypical T_H2 response (ulcerative colitis). Recent work however, has also shown that in both types of IBD there is enhanced synthesis of cytokines made by a distinct lineage of T helper cells, termed T_H17 cells. Targeting these T_H17 cells may be a new avenue for the development of therapeutic strategies in IBD. Interleukin (IL)-23, a heterodimeric cytokine produced by activated antigen presenting cells, is necessary for amplifying T_H17 cell responses. Paradoxically, recent studies have also shown that T_H17-related cytokines have protective effects in the gut. In this chapter, we review the available data regarding the role of T_H17 cells in chronic intestinal inflammation in an attempt to resolve this conundrum.

1 Introduction

Crohn's disease (CD) and ulcerative colitis (UC), the main inflammatory bowel dieases (IBD), are relapsing conditions where chronic tissue inflammation alters the integrity and function of the gut. The etiology of IBD is still unknown, but evidence suggests that both CD and UC result from the interaction of genetic and environmental factors that ultimately promote an excessive and poorly controlled mucosal immune response directed against components of the normal microflora of the gut (Strober et al. 2007). Experimental models of IBD show that CD4+ T cells play a major role in initiating and shaping this immuno-pathologic process

G. Monteleone (✉)
Department of Internal Medicine, University "Tor Vergata" of Rome, Via Montpellier, 100133 Rome, Italy
e-mail: Gi.Monteleone@Med.uniroma2.it

S. Jiang (ed.), *T_H17 Cells in Health and Disease*,
DOI 10.1007/978-1-4419-9371-7_14,

(Strober et al. 2002; Zenewicz et al. 2009). Consistently, T cell-directed therapies have been employed with clinical success in IBD patients (Baumgart and Sandborn 2007).

The ability of CD4+ T cells to promote/expand gut inflammation is largely dependent on the production of cytokines. In particular, the inflamed gut in CD is massively infiltrated with T helper type 1 (T_H1)-CD4+ cells that produce large quantities of interferon (IFN)-γ, tumor necrosis factor (TNF)-α, and express high levels of T_H1-driving transcription factors, such as T-bet and Stat4 (Fuss et al. 1996; Parrello et al. 2000; Neurath et al. 2002). In the same diseased mucosal tissues, there is also an exaggerated production of IL-12, the major T_H1-inducing factor in humans (Monteleone et al. 1997). By contrast, in UC, the inflammatory response is associated with high production of T_H2-cytokines, such as IL-5 and IL-13 (Fuss et al. 1996; Heller et al. 2005). Although simplified pathways are useful to understand disease processes, more complex networks of immune interactions are now being discovered in IBD. For instance, studies from several laboratories have shed light onto the role of another subset of Th cells, namely T_H17 cells, in the development of gut inflammation (Fujino et al. 2003; Fina et al. 2008; Leppkes et al. 2009; Rovedatti et al. 2009). T_H17 cells, as their name suggests, produce IL-17, also termed IL-17A, and express the transcription factors retinoic acid-related orphan receptor (ROR)-γt and RORα (Ivanov et al. 2006; Yang et al. 2008b). T_h17 cells can also synthesize additional cytokines, such as IL-17F, IL-22, and IL-21 (Dubin and Kolls 2008). The molecular requirements governing T_H17 cell development and functions are not yet fully understood yet. In mice, T_H17 cells are induced from virgin T cells by the dual actions of TGF-β1 and IL-6, while IL-23 is necessary for expanding and/or maintaining T_H17 cell responses (Bettelli et al. 2006; Langrish et al. 2005; Ouyang et al. 2008) (Fig. 1). The tissue damaging effects of T_H17 cells are strictly linked to their ability to enhance the recruitment and facilitate the activation of neutrophils, to stimulate the production of chemokines and other inflammatory cytokines, and to promote the synthesis of tissue-damaging proteases (Ye et al. 2001; Korn et al. 2007b; Wilson et al. 2007) (Fig. 1). Nonetheless, recent studies have demonstrated that T_H17 cytokines, particularly IL-17 and IL-22, may exert anti-inflammatory function in the gut (Sugimoto et al. 2008; Zenewicz et al. 2008; O'Connor et al. 2009). In this chapter we will review the available data regarding the involvement of T_H17 cell populations in the modulation of intestinal inflammatory processes.

2 Expression of T_H17-Related Cytokines in Human IBD

The first report on IL-17 in IBD came from a descriptive study showing that the inflamed gut of patients with CD or UC contains high levels of IL-17-producing cells as compared to normal colonic mucosa, or colonic mucosa of patients with ischemic colitis (Fujino et al. 2003). In active UC, IL-17-expressing cells were located mainly within the lamina propria, while in active CD, cells were scattered

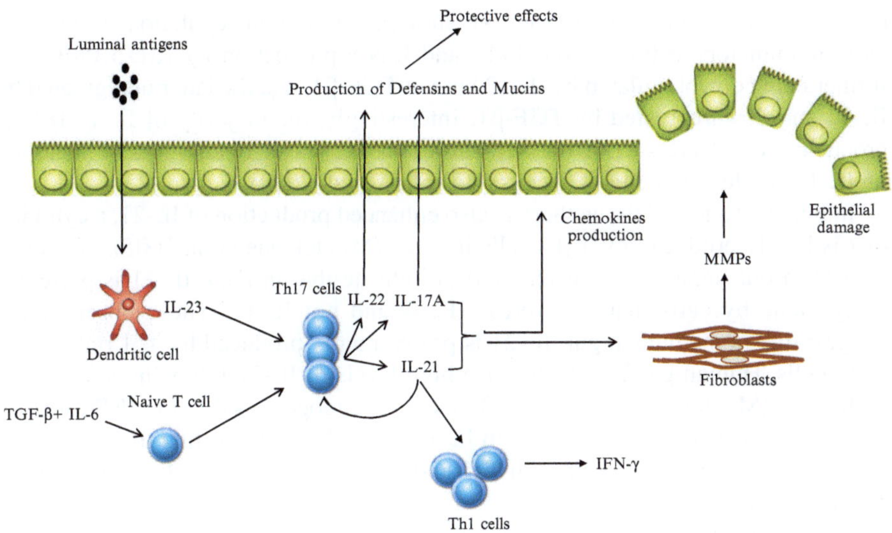

Fig. 1 Role of T_h17-related cytokines in gut inflammation. T_h17 cells differentiate from naïve T lymphocytes under the stimulus of TGF-β and IL-6, while IL-23, a cytokine produced by bacteria-stimulated antigen presenting cells, and IL-21, a Th17 cell-product, expand T_h17 cell populations. T_h17-derived cytokines, such as IL-17A, IL-21, and IL-22 stimulate epithelial cells to make chemokines, promoting the recruitment of inflammatory cells in the intestinal lamina propria. Moreover, IL-17A and IL-21 stimulate fibroblasts to make matrix metalloproteinases, a family of enzymes that play a major role in the mucosal degradation occurring in IBD. IL-21 is also able to positively regulate the production of cytokines (i.e. IFN-γ) by Th1 cells. Since IL-17A and IL-22 enhance the production of defensins and mucins by epithelial cells, these cytokines can exert protective effects in the gut. They are as follows: (1) antigen presenting cell (APC), (2) transforming growth factor-β (TGFβ), (3) matrix metalloproteinases (MMPs), (4) interleukin (IL), and interferon (IFN)

throughout the sub-mucosa and muscularis propria (Fujino et al. 2003). CD3+ T cells and CD68+ cells (macrophages) were the major sources of IL-17 (Fujino et al. 2003). High levels of IL-17 have been also found in serum of IBD patients (Fujino et al. 2003). Consistent with these initial studies, it was then shown that RNA transcripts for IL-17A and IL-17F were high in the inflamed mucosa of UC patients and CD patients (Fujino et al. 2003; Seiderer et al. 2008). More recently, Annuziato et al. (2007) has demonstrated that the number of IL-17-producing T cells is higher in CD than in normal gut mucosa, and that some of these cells co-express IFN-γ. The phenotype of such cells does not appear to be stable however, because in vitro stimulation with IL-12 enhances the expression of T-bet and IFN-γ, and down-regulates RORγt and IL-17 (Annunziato et al. 2007; Lee et al. 2009). Studies conducted by our group have confirmed and expanded on this data by showing that IL-17 production in gut biopsies grown ex vivo and lamina propria mononuclear cells (LPMC) cultured in vitro is higher in IBD patients than in control subjects. It is also associated with an increased percentage of IL-17A- and IL-17A/IFN-γ-producing cells

(Rovedatti et al. 2009). Moreover, we showed differential regulation by pro and anti-inflammatory cytokines on IL-17 and IFN-γ production by IBD LPMC, the former being downregulated by IL-12 but not TGF-β1, and the latter upregulated by IL-12 and downregulated by TGF-β1. Interestingly, the majority of IL-17/IFN-γ-producing cells express CD161, a well known marker of NK T cells recently identified on IL-17-producing memory T cells (Rovedatti et al. 2009).

In both CD and UC tissue there is also enhanced production of IL-21, a cytokine that is highly produced by T_H17 cells in mice (Monteleone et al. 2005). However, data from our unpublished recent studies indicate that in IBD, IL-21 is preferentially made by cells that co-express IFN-γ and not IL-17. These observations suggest that in the human gut, IL-21 is preferentially produced by T_H1 rather than T_H17 cells. Human gut T cells produce high levels of IL-21 following stimulation with IL-12 (Monteleone et al. 2005). Blockade of endogenous IL-21 in IBD mucosal samples reduces the synthesis of both IL-17 and IFN-γ , supporting the key role of this cytokine in the control of both T_H1 and T_H17 cell responses (Fina et al. 2008; Monteleone et al. 2005).

The inflamed gut of IBD patients also contains high levels of IL-22, another cytokine made by T_H17 cells (Brand et al. 2006; Kobayshi et al. 2008). Little is known about the cellular source of IL-22 in the human IBD, and it remains to be ascertained if IL-22 is really made by T_H17 cells. In fact, studies in other systems have shown that the IL-22 expression profile may differ from that of IL-17A, and that IL-22-producing T cells, termed T_H22 cells, could represent a T cell subset distinct from typical T_H17 cells (Eyerich et al. 2009).

3 Involvement of T_H17 Cells in the Pathogenesis of Experimental Colitis

The role of T_H17-related cytokines in the control of gut inflammation has been evaluated in experimental murine colitis. Studies on IL-17 receptor A (IL-17RA) knockout mice have demonstrated that IL-17 is necessary for development of acute gut inflammation induced by intra-rectal administration of trinitrobenzenesulfonic acid (TNBS), a T cell-mediated colitis with striking similarities to CD (Zhang et al. 2006). Blockade of IL-17 signaling by an IL-17RA IgG1 fusion protein significantly attenuated colonic inflammation and prevented weight loss after TNBS administration (Zhang et al. 2006). It is noteworthy that IL-17RA mediates the functional activities of both IL-17A and IL-17F (Moseley et al. 2003), making difficult to establish the exact contribution of each of these cytokines in the pathogenesis of TNBS-colitis. Studies in dextran sulfate sodium (DSS)-induced colitis also showed that IL-17F deficiency lead to reduced colitis, whereas IL-17A-knockout mice developed more severe disease (Yang et al. 2008a). Although more work is needed in this area, the available data would seem to suggest that IL-17F, rather than IL-17A is crucial in sustaining DSS-colitis.

T_H17 cells are also involved in the pathogenesis of colitis induced by transfer of the cecal bacterial, antigen-specific C3H/HeJBir (C3Bir) CD4 (+) T-cell line to C3H/HeSnJ SCID mice (Elson et al. 2007). In this model, gut inflammation is associated with enhanced production of IL-17, and adoptive transfer of IL-17-secreting T cells to SCID recipients resulted in a more severe colitis induced by transfer of T_H1 cells (Elson et al. 2007). Giving mice a monoclonal anti-IL-23p19 prevented, and in fact ameliorated, ongoing colitis. It also downregulated the synthesis of a broad array of inflammatory cytokines and chemokines in the colon and promoted apoptosis of colitogenic T_H17 cells (Elson et al. 2007). By using a model of CD8+ T cell-dependent colitis, Tajima et al. (2008) also showed that a single adoptive transfer of naïve CD8+ T cells into syngeneic RAG-deficient mice was followed by rapid spontaneous proliferation of these cells in the mesenteric lymph nodes and severe colitis. Analysis of cytokine-secreting CD8+ T cells in the mesenteric lymph nodes showed the existence of IL-17 and IFN-γ-double-positive cells. Notably, adoptive transfer of naïve CD8+ T cells derived from either IL-17- or IFN-γ-knockout mice was associated with less severe colitis, raising the intriguing possibility that IL-17 and IFN-γ can cooperate to cause pathology (Tajima et al. 2008). In line with these findings, we have shown that IL-21-deficient mice were largely protected against the development of DSS-colitis and TNBS-relapsing colitis (Fina et al. 2008). This protection was associated with a reduced colonic expression of several T_H17-related genes, including IL-17, IL-17F, and ROR-γt, consistent with the role of IL-21 in promoting T_H17 cell differentiation (Korn et al. 2007a; Nurieva et al. 2007). Additionally, blockade of IL-21 activity with a specific IL-21R-fusion protein reduced intestinal inflammation and T_H17 response during the course of DSS colitis (Fina et al. 2008).

Recently it has been demonstrated that enterotoxigenic Bacteroides fragilis (ETBF), a human colonic bacterium that chronically colonizes mice, promotes colitis and colonic tumors. ETBF-driven colitis is associated with a marked activation of STAT-3 and pathogenic T_H17 response in the colon. Antibody-mediated blockade of IL-17 and IL-23R inhibits ETBF-induced colitis and tumor formation (Wu et al. 2009).

4 T_H17-Related Cytokines Counter-Regulate Intestinal Inflammation

The aforementioned studies would seem to suggest that T_H17 cytokines are crucial factors for enhancing the effector phase of T cell responses that cause intestinal tissue inflammation and damage. Since T_H17 cells play an important role in the protection of the host against bacteria and fungi and are constitutively present in the gut, it may be expected that T_H17 cytokines can exert protective effects in the intestine. Indeed, O'Connor et al. has recently shown that adoptive transfer of CD45RBhiCD25-CD4+ T cells from IL-17A-deficient mice to recipient immuno-deficient mice initiates an overly aggressive colitis that manifests with rapid weight loss and severe colonic ulcerations (2009). Notably, IL-17R-deficient T cells induce the same aggressive disease, indicating that IL-17 exerts its effects directly on T cells (O'Connor et al. 2009). The greater severity of colitis induced by transfer of IL-17A-deficient T cells is not due to their enhanced migratory and infiltration capacity, but is instead related to enhanced T_H1 cell effector function (O'Connor et al. 2009). It is possible that the anti-inflammatory effect of IL-17A in the gut relies on the negative regulation of T_H1 cell responses. IL-22 mRNA was substantially up-regulated in the gut of recipients of IL-17A-deficient T cells, even though experiments conducted using T cells deficient in both IL-17 and IL-22 showed that IL-22 was not responsible for the accelerated colitis (O'Connor et al. 2009).

In contrast to this interesting data, Leppkes et al. (2009) has shown that transfer of IL-17A, IL-17F, or IL-22-deficient T lymphocytes into RAG1-null mice induces severe colitis that is indistinguishable from that caused by wild-type cells. Transfer of RORγ-null T cells fails to increase mucosal IL-17 cytokine levels and does not induce colitis. Treatment of RAG1 mice that received IL-17F-null T cells with a neutralizing anti-IL-17A antibody significantly suppresses disease, indicating redundant biological effects of IL-17A and IL-17F (Leppkes et al. 2009). Similarly, Noguchi et al. (2007) showed that colitis induced in RAG mice by transfer of naive CD4+ T cells prepared from IL-17-knockout mice does not differ in terms of severity from that induced by transfer of wild-type cells. The reason why these studies provided us with different results remains unknown. However, it is increasingly apparent that T_H17 subset is not a homogenous population of cells, and based on the cytokine milieu during differentiation, IL-17-expressing cells can be inflammatory or protective depending of the co-expression of other cytokines, particularly IL-10 (McGeachy et al. 2007).

Approaches similar to those described above have been used to explore the role of IL-22 in the control of gut inflammation. Adoptive transfer of IL-22-deficient T cells into RAG1 mice induces a colitis that is similar to that induced by wild-type T cells (Leppkes et al. 2009), clearly indicating that T cell-derived IL-22 is not sufficient to provide protection against colitis. IL-22 can also be made by other cell types, such as NK cells and CD11c+ cells, and there is evidence that IL-22 derived from these cell sources enhances the production of mucous and defensins by epithelial cells, thereby triggering an effective counter-regulatory mechanism in the gut (Sugimoto et al. 2008; Zenewicz et al. 2008; Zheng et al. 2008; Pickert et al. 2009).

5 IL-23 and Colitis

IL-23 is a heterodimeric cytokine made by a specific p19 sub-unit and a shared p40 sub-unit with IL-12 (Oppmann et al. 2000). IL-23 biological activity is mediated by a receptor complex consisting of IL-23R and IL-12Rβ1 (Oppmann et al. 2000). Murine models in which IL-23 expression/function has been experimentally regulated have contributed to defining the role of this cytokine in intestinal inflammation. By backcrossing IL-10-deficient mice with mice lacking IL-12p35 or IL-23 p19,

Yen et al. (2006) showed that IL-23 was essential for manifestation of chronic intestinal inflammation, whereas IL-12 was not. CD4+ T cells from IL-10/p19-knockout mice produced large amounts of IFN-γ, indicating that T_H1 responses developed normally in the absence of IL-23, but disease manifestations required the presence of IL-23 (Yen et al. 2006). Moreover, administration of exogenous IL-23 in RAG mice reconstituted with naïve CD4+ T cells causes a more severe colitis associated with enhanced production of IL-6 and IL-17 and is preventable by treatment with a blocking IL-6 or IL-17 antibody (Yen et al. 2006). Although in this model, IL-6 and IL-17 were made by memory T cells, there is no doubt that some of the pathogenic functions of IL-23 in the gut are T_H17 cell-independent. This was first shown by Powrie and co-workers who analyzed the effect of an agonistic anti-CD40 antibody in RAG mice lacking IL-23p19 or IL-12p35 (Uhlig et al. 2006). Administration of anti-CD40 caused a systemic and local inflammation characterized by wasting disease, splenomegaly, increases in serum pro-inflammatory cytokines, and colitis. The systemic inflammatory response and the elevated concentrations of pro-inflammatory cytokines in the serum were driven by IL-12, while the local intestinal inflammation and production of IL-17 in the intestine were controlled by IL-23 (Uhlig et al. 2006). However, the protection of inflammation seen in RAG mice lacking IL-23 is not associated with decrease in the levels of IL-17 (Izcue et al. 2008). Lack of IL-23 did not alter the expression of RORγt in the colon (Izcue et al. 2008), indicating that T_H17 cell responses are not specifically impaired in the intestine of IL-23-deficient mice and that IL-23-mediated colitis is not strictly dependent on IL-17.

Another model of T cell-independent intestinal inflammation is one caused by infection of 129SvEvRAG-deficient mice with the pathogenic bacterium Helicobacter (H) hepaticus. In this model, the development of chronic colitis is mediated through activation and accumulation of innate immune cells including granulocytes and monocytic cells, and is characterized by induction of IL-23/p19 and IL-12/p40, but not IL-12/p35. Treatment of mice with a neutralizing p19 antibody ameliorates colitis (Hue et al. 2006). Analysis of inflammatory cytokines in the colons of infected mice showed concomitant with the increased IL-23 expression, that there was a marked increase in IL-17 expression, indicating that IL-23 induces the secretion of IL-17 by non-T cells in an inflammatory environment (Hue et al. 2006).

H hepaticus has also been used to infect IL-10-knockout mice. In this model, the developing colitis is T cell-dependent and associates with enhanced production of IL-17A and a marked T_H1 response (Kullberg et al. 2006). However, the absence of IFN-γ, IL-17A is not sufficient to induce maximal colitis (Kullberg et al. 2006), suggesting that T_H1 and T_H17-cytokines likely synergize to elicit maximal intestinal pathology.

A more detailed analysis of the mechanisms underlying the IL-23-dependent pathologies revealed that IL-23 can facilitate colitis not only via direct effects on inflammatory mediators but also indirectly by counteracting regulatory mechanisms. Indeed, protection of colitis seen after transfer of naive T cells to RAG mice lacking IL-23 was associated with an increase in the frequency of Foxp3-expressing

T regulatory cells in the intestine (Izcue et al. 2008). Upon naive T cell transfer and administration of either an antagonistic IL-10R antibody or a blocking TGF-β antibody, increased colonic inflammation occurred when compared to untreated controls. Moreover, naive T cells isolated from transgenic mice expressing a dominant-negative form of TGF-β receptor II and unable to respond to TGF-β induced significant colitis when transferred to IL-23-deficient RAG mice (Izcue et al. 2008). High levels of IFN-γ, but not IL-17, were seen in colitic mice, thus suggesting that IFN-γ might drive the chronic intestinal inflammation in this setting. Notably, transfer of Foxp3-deficient T cells to IL-23-deficient RAG mice caused severe colitis, indicating that IL-23 is not essential to the pathogenesis of intestinal inflammation if counter-regulatory mechanisms are defective or absent (Izcue et al. 2008). These later findings fit with the concept that the requirement of IL-23 for the initiation and progress of gut inflammation varies, depending on the model. In fact, acute colitis induced by TNBS is driven by IL-12 and negatively regulated by IL-23 (Becker et al. 2006).

The synthesis of the p40 sub-unit and the functional heterodimeric IL-23 is greatly enhanced in patients with CD (Monteleone et al. 1997; Schmidt et al. 2005). Genome-wide association studies have led to the identification of multiple single nucleotide polymorphisms (SNPs) in the IL-23R gene region (chromosome 1p31) as having an association with both CD and UC (Duerr et al. 2006; Dubinski et al. 2007). One of these polymorphisms, Arg381Gln, confers a two- to threefold protection against developing CD (Tremelling and Parkes 2007). Nonetheless, the mechanism through which these SNPs confer either risk or protection from IBD remains unknown.

Taken together, these results suggest that inhibiting IL-23 activity may have therapeutic potential in IBD. In agreement with these observations, an anti-IL-12/IL-23p40 antibody demonstrated its clinical potential showing a high rate of remission and clinical responses compared to placebo. Ustekinumab, a similar antibody directed against the p40 sub-unit of IL-12 and IL-23, induced clinical response in patients with moderate-to-severe CD, especially in patients previously treated with anti-TNF-α (Sandborn et al. 2008). It was also shown that apilimod mesylate, an oral IL-12 and IL-23 inhibitor, has some clinical efficacy in patients with active CD (Burakoff et al. 2006).

6 Conclusions

The recent discovery that during gut inflammation many of the functions traditionally attributed to IL-12 are actually due to IL-23, and that IL-23 contributes to immunopathology by acting in part on T_H17 cells has provided a new picture of the way the local immune response can promote intestinal tissue damage. This new data suggests that the IL-23/T_H17 axis could be a promising target for suppressing gut inflammation. However, some important issues remain to be resolved. For example, it is unknown how T_H17 cells are induced in human IBD and whether these cells interact in vivo with the other regulatory and effector mucosal T cell subsets. Similarly, the role of each T_H17 cytokine in the initiation and progression of gut inflammation remains elusive. It also remains to be determined whether T_H17-related cytokines can enhance the production of T_H1-type cytokines in the gut and contribute to the tissue-damaging immune response (Lin et al. 2009).

T_H17-cytokines are constitutively produced in the human and mouse gut (Monteleone et al. 2009) where they seem to be involved in the maintenance of immunological homeostasis and/or in the control of specific inflammatory pathways. If this is the case, blocking T_H17-cytokines could have deleterious, rather than beneficial effects for the host.

Acknowledgments The authors received support for their work from the "Fondazione Umberto di Mario", Rome, the Broad Medical Research Program Foundation (Grant No. IBD-0154R), and Giuliani Spa, Milan, Italy.

References

Annunziato F, Cosmi L, Santarlasci V et al. (2007) Phenotypic and functional features of human Th17 cells. J Exp Med 204 (8):1849–1861.

Baumgart DC and Sandborn WJ (2007) Inflammatory bowel disease: clinical aspects and established and evolving therapies. Lancet 369 (9573):1641–1657.

Becker C, Dornhoff H, Neufert C et al. (2006) Cutting edge: IL-23 cross-regulates IL-12 production in T cell-dependent experimental colitis. J Immunol 177 (5):2760–2764.

Bettelli E, Carrier Y, Gao W et al. (2006) Reciprocal development pathways for the generation of pathogenic effector TH17 and regulatory T cells. Nature 441 (7090):235–238.

Brand S, Beigel F, Olszak T et al. (2006) IL-22 is increased in Crohn's disease and promotes proinflammatory gene expression and intestinal epithelial cell migration. Am J Physiol Gastrointest Liver Physiol 290 (4):G827–838.

Burakoff R, Barish CF, Riff D, et al. (2006) A phase 1/2A trial of STA 5326, an oral interleukin-12/23 inhibitor, in patients with active moderate to severe Crohn's disease. Inflamm Bowel Dis 12 (7):558–565.

Dubin PJ and Kolls JK (2008) Th17 cytokines and mucosal immunity. Immunol Rev 226: 160–171.

Dubinski MC, Wang D, Picomell Y et al. (2007) IL-23 receptor (IL-23R) gene protects against pediatric Crohn's disease. Inflamm Bowel Dis 13 (5):511–515.

Duerr RH, Taylor KD, Brandt SR et al. (2006) A genome-wide association study identifies IL23R as an inflammatory bowel disease gene. Science 314 (5804):1461–1463.

Elson CO, Cong Y, Weaver CT et al. (2007) Monoclonal anti-interleukin 23 reverses active colitis in a T cell-mediated model in mice. Gastroenterology 132 (7):2359–2370.

Eyerich S, Eyerich K, Pennino D et al. (2009) Th22 cells represent a distinct human T cell subset involved in epidermal immunity and remodeling. J Clin Invest 119 (12):3573–3585.

Fina D, Sarra M, Fantini MC et al. (2008) Regulation of gut inflammation and Th17 cell response by interleukin-21. Gastroenterology 134 (4):1038–1048.

Fujino S, Andoh A, Bamba S et al. (2003) Increased expression of interleukin 17 in inflammatory bowel disease. Gut 52 (1):65–70.

Fuss IJ, Neurath M, Boirivant M et al. (1996) Disparate CD4+ lamina Propria (LP) lymphokine secretion profiles in inflammatory bowel disease. Crohn's disease LP cells manifest increased secretion of IFN-gamma, whereas ulcerative colitis LP cells manifest increased secretion of IL-5. J Immunol 157 (3):1261–1270.

Heller F, Florian P, Bojarski C et al. (2005) Interleukin-13 is the key effector Th2 cytokine in ulcerative colitis that affects epithelial tight junction, apoptosis and cell restitution. Gastroenterology 129 (2):550–564.

Hue S, Ahem P, Buonocore S et al. (2006) Interleukin-23 drives innate and T cell-mediated intestinal inflammation. J Exp Med 203 (11):2473–2483.
Ivanov II, McKenzie BS, Zhou L et al. (2006) The orphan nuclear receptor RORgammat directs the differentiation program of pro-inflammatory IL-17+ T helper cells. Cell 126 (6):1121–1133.
Izcue A, Hue S, Buonocore S et al. (2008) Interleukin-23 restrains regulatory T cell activity to drive T cell-dependent colitis. Immunity 28 (4):559–570.
Kobayshi T, Okamoto S, Hisamatsu T et al. (2008) IL23 differentially regulates the Th1/Th17 balance in ulcerative colitis and Crohn's disease. Gut 57 (12):1682–1689.
Korn T, Bettelli E, Gao W et al. (2007a) IL-21 initiates an alternative pathway to induce proinflammatory T(H)17 cells. Nature 448 (7152):484–487.
Korn T, Oukka M, Kuchroo V et al. (2007b) Th17 cells: effector T cells with inflammatory properties. Semin Immunol 19 (6):362–371.
Kullberg MC, Jankovic D, Feng CG et al. (2006) IL-23 plays a key role in Helicobacter hepaticus-induced T cell- colitis. J Exp Med 203 (11):2485–2494.
Langrish CL, Chen Y, Blumenschein WM et al. (2005) IL-23 drives pathogenic T cell population that induces auto-immune inflammation. J Exp Med 201 (2):233–240.
Lee YK, Turner H, Maynard CL et al. (2009) Late development plasticity in the T helper 17 lineage. Immunity 30 (1):92–107.
Leppkes M, Becker C, Ivanov II et al. (2009) RORgamma-expressing Th17 cells induce murine chronic intestinal inflammation via redundant effects of IL-17A and IL-17F. Gastroenterology 136 (1):257–267.
Lin Y, Ritchea S, Logar A et al. (2009) Interleukin-17 is required for T helper 1 cell immunity and host resistance to the intracellular pathogen Francisella tularensis. Immunity 31 (5):799–810.
McGeachy MJ, Bak-Jensen KS, Chen Y, et al. (2007) TGF-beta and IL-6 drive the production of IL-17 and IL-10 by T cells and restrain T(H)-17 cell-mediated pathology. Nat Immunol 8(12):1390–1397.
Monteleone G, Biancone L, Marasco R et al. (1997) Interleukin 12 is expressed and actively released by Crohn's disease intestinal lamina propria mononuclear cells. Gastroenterology 112 (4):1169–1178.
Monteleone G, Monteleone I, Fina D et al. (2005) Interleukin-21 enhances T-helper cell type I signaling and interferon-gamma production in Crohn's disease. Gastroenterology 128 (3):687–694.
Monteleone I, Pallone F and Monteleone G (2009) Interleukin-23 and Th17 cells in the control of gut inflammation. Mediators Inflamm 2009:297645.
Moseley TA, Haudenschild DR, Rose L et al. (2003) Interleukin-17 family and IL-17 receptors. Cytokine Growth Factor Rev 14 (2):155–174.
Neurath MF, Weigmann B, Finotto S et al. (2002) The transcription factor T-bet regulates mucosal T cell activation in experimental colitis and Crohn's disease. J Exp Med 195 (9):1129–1143.
Noguchi D, Wakita D, Tajima M et al. (2007) Blocking of IL-6 signaling pathway prevents CD4+ T cell-mediated colitis in a T(h)17-independent manner. Int Immunol 19 (12):1431–1440.
Nurieva R, Yang XO, Martinez G et al. (2007) Essential autocrine regulation by IL-21 in the generation of inflammatory T cells. Nature 448 (7152):480–483.
O'Connor W Jr, Kamanaka M, Booth CJ et al. (2009) A protective function for interleukin 17A in T cell-mediated intestinal inflammation. Nat Immunol 10 (6):603–609.
Oppmann B, Lesley R, Blom B et al (2000) Novel p19 protein engages IL-12p40 to form a cytokine, IL-23, with biological activities similar as well as distinct from IL-12. Immunity 13 (5):715–725.
Ouyang W, Kolls JK and Zheng Y (2008) The biological functions of T helper 17 cell effector cytokines in inflammation. Immunity 28 (4):454–467.
Parrello T, Monteleone G, Cucchiara S et al. (2000) Up-regulation of the IL-12 receptor beta 2 chain in Crohn's disease. J Immunol 165 (12):7234–7239.
Pickert G, Neufert C, Leppkes M et al. (2009) STAT3 links IL-22 signaling in intestinal epithelial cell to mucosal wound healing. J Exp Med 206 (7):1465–1472.

Rovedatti L, Kudo T, Bianchieri P et al. (2009) Differential regulation of Interleukin 17 and Interferon gamma production in inflammatory bowel disease. Gut 58 (12):1629–1636.
Sandborn WJ, Feagan BG, Fedorak RN et al. (2008) A randomized trial of Ustekinumab, a human interleukin-12/23 monoclonal antibody, in patients with moderate-to-severe Crohn's disease. Gastroenterology 135 (4):1130–1141.
Schmidt C, Giese T, Ludwig B et al. (2005) Expression of Interleukin-12 related cytokine transcript in inflammatory bowel disease: elevated interleukin-23p19 and interleukin-27p28 in Crohn's disease but not in ulcerative colitis. Inflamm Bowel Dis 11 (1):16–23.
Seiderer J, Elben I, Diegelmann J et al. (2008) Role of the novel Th17 cytokine IL-17F in inflammatory bowel disease (IBD): upregulated colonic IL-17F expression in active Crohn's disease and analysis of the IL17Fp.His161arg polymorphism in IBD. Inflamm Bowel Dis 14 (4):437–445.
Strober W, Fuss IJ and Mannon P (2007) The fundamental basis of inflammatory bowel disease. J Clin Invest 117 (3):514–521.
Strober W, Fuss IJ and Blumberg RS (2002) The immunology of mucosal models of inflammation. Annu Rev Immunol 20 495–549.
Sugimoto K, Ogawa A, Mizoguchi E et al. (2008) IL-22 ameliorates intestinal inflammation in a mouse model of ulcerative colitis. J Clin Invest 118 (2):534–544.
Tajima M, Wakita D, Noguchi D et al. (2008) IL-6-dependent spontaneous proliferations required for the induction of colitogenic IL-17-producing CD8+ T cells. J Exp Med 205 (5):1019–1027.
Tremelling M and Parkes M (2007) Genome-wide association scans identify multiple confirmed susceptibility loci for Crohn's disease: lessons for study design. Inflamm Bowel Dis 13 (12): 1554–1560.
Uhlig HH, Mckenzie BS, Hue S et al. (2006) Differential activity of IL-12 and IL-23 in mucosal and systemic innate immune pathology. Immunity 25 (2):309–318.
Wilson NJ, Boniface K, Chan JR et al. (2007) Development, cytokine profile and function of human interleukin 17-producing helper T cells. Nat Immunol 8 (9):950–957.
Wu S, Rhee KJ, Albesiano E et al. (2009) A human colonic commensal promotes colon tumorigenesis via activation of T helper type 17 T cell responses. Nat Ned 15 (9):1016–1022.
Yang XO, Chang SH, Park H et al. (2008a) Regulation of inflammatory responses by IL-17F. J Exp Med 205 (5):1063–1075.
Yang XO, Pappu BP, Nurieva R et al. (2008b) T helper 17 lineage differentiation is programmed by orphan nuclear receptors ROR alpha and ROR gamma. Immunity 28 (1):29–39.
Ye P, Rodriguez FH, Kanaly S et al (2001) Requirement of interleukin 17 receptor signaling for lung CXC chemokines and granulocytes colony-stimulating factor expression, neutrophil recruitment, and host defense. J Exp Med 194 (4):519–527.
Yen D, Cheung J, Scheerns H et al. (2006) IL-23 is essential for T cell-mediated colitis and promotes inflammation via IL-17 and IL-16. J Clin Invest 116 (5):1310–1316.
Zenewicz LA, Antov A and Flavell RA (2009) CD4 T-cell differentiation and inflammatory bowel disease. Trends Mol Med 15 (5):199–207.
Zenewicz LA, Yancopoulos GD, Valenzuela DM et al. (2008) Innate and adaptive interleukin-22 protects mice from inflammatory bowel disease. Immunity 29 (6):947–957.
Zhang Z, Zheng M, Bindas J et al. (2006) Critical Role of IL-17 receptor signaling in acute TNBS-induced colitis. Inflamm Bowel Dis 12 (5):382–388.
Zheng Y, Valdez PA, Danilenko DM et al. (2008) Interleukin-22 mediates early host defense against attaching and effacing bacterial pathogens. Nat Med 14 (3):282–289.

The Roles of IL-17A and IL-17F in Mucosal Infection and Allergy

Harumichi Ishigame, Susumu Nakae, and Yoichiro Iwakura

Abstract T_H17 cells are a subset of CD4+ helper T cells that produce IL-17A, IL-17F, IL-9, IL-21, and IL-22. They play an important role in promoting allergic and auto-immune responses as well as in protecting hosts against pathogens. Because IL-17A and IL-17F have the highest homology among IL-17 family members and bind the same IL-17RA and IL-17RC receptor complex, it is suggested that these two cytokines have similar functions. However, accumulating evidence suggests that these cytokines have overlapping yet distinct roles in the immune system. In this review, we introduce how IL-17A and IL-17F are involved in inflammatory immune responses and host defense mechanisms and discuss their relationship with other cytokines in the development of inflammatory and infectious diseases.

1 IL-17A and IL-17F

The IL-17A gene, originally called CTLA-8 (cytotoxic T lymphocyte associated antigen-8) gene, was first cloned from a murine cytotoxic T lymphocyte (CTL) hybridoma cDNA library. Murine IL-17A is a 21-kDa glycoprotein with 147 amino acids and 63% amino acid homology with human IL-17A (155 amino acids). Recently, five additional related cytokines were identified (IL-17B, IL-17C, IL-17D, IL-17E also called IL-25, and IL-17F) with 16–50% amino acid identity with IL-17A (Aggarwal and Gurney 2002; Kolls and Linden 2004). Among these IL-17 family members, IL-17F has the highest amino acid sequence homology to IL-17A. The *Il17f* gene is located close to the *Il17a* gene in both humans and mice.

Y. Iwakura (✉)
Center for Experimental Medicine and Systems Biology, The Institute of Medical Science, The University of Tokyo, 4-6-1 Shirokanedai, Minato-ku, Tokyo 108-8639, Japan
and
Core Research for Evolutional Science and Technology (CREST), Japan Science and Technology Agency, Saitama 332-0012, Japan
e-mail: iwakura@ims.u-tokyo.ac.jp

S. Jiang (ed.), *T_H17 Cells in Health and Disease*,
DOI 10.1007/978-1-4419-9371-7_15,

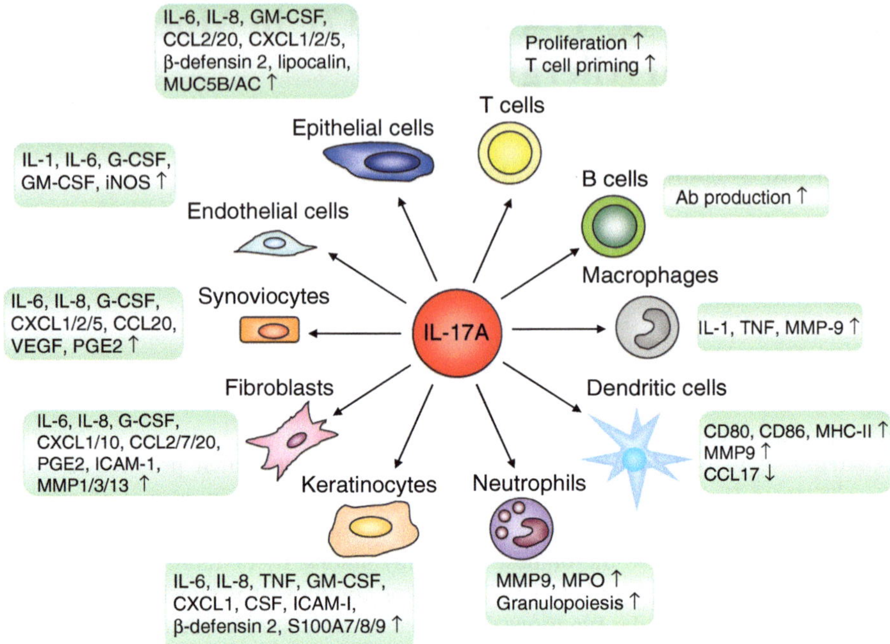

Fig. 1 Pleiotropic effects of IL-17A on multiple target cells

Both IL-17A and IL-17F induce the production of antimicrobial peptides (defensins and S100 proteins), cytokines (IL-6, G-CSF, and GM-CSF), chemokines (CXCL1, CXCL5, IL-8, CCL2, and CCL7), and matrix metalloproteinases (MMP1, MMP3, and MMP13) from fibroblasts, endothelial cells, and epithelial cells (Fig. 1). IL-17A induces intercellular cell adhesion molecule 1 (ICAM-1) in keratinocytes, IL-1 and TNF in macrophages, and iNOS and cyclooxygenase-2 (COX-2) in chondrocytes. IL-17A also promotes SCF and G-CSF-mediated granulopoiesis. Overexpression of IL-17A and IL-17F in the lungs of mice leads to increased pro-inflammatory cytokine and chemokine expression, causing inflammation associated with neutrophil infiltration (Oda et al. 2005; Park et al. 2005; Yang et al. 2008a). These observations suggest that these cytokines have similar biological functions. Furthermore, IL-17A and IL-17F are secreted as both homodimers and heterodimers. The IL-17A-IL-17F heterodimer is more potent than IL-17F, but less potent than IL-17A in inducing chemokine expression (Liang et al. 2007; Wright et al. 2007).

The IL-17 receptor family members (IL-17RA–IL-17RE) have also been identified (Gaffen 2009). Both IL-17A and IL-17F bind the same receptor complexes, IL-17RA and IL-17RC (Zheng et al. 2008), as both IL-17A and IL-17F failed to induce chemokine expression in either *Il17ra*$^{-/-}$ or *Il17rc*$^{-/-}$ cells (Yang et al. 2008a; Zheng et al. 2008). The expression of IL-17RA and IL-17RC is quite different; IL-17RA is highly expressed in lymphoid tissues, whereas IL-17RC is mainly expressed in non-hematopoietic tissues (Yao et al. 1995; Kuestner et al. 2007; Ishigame et al. 2009), suggesting that these receptors have different functions. Because the binding affinity of IL-17F to IL-17RA is much lower than that of IL-17A (Hymowitz et al. 2001; Wright et al. 2008), and only IL-17F binds to IL-17RC in the mouse (Kuestner et al. 2007), it seems likely that IL-17A and IL-17F differentially use these receptors. In fact, the effects of IL-17A and IL-17F are different among colonic epithelial cells, macrophages, and T cells; both IL-17A and IL-17F can induce neutrophil chemo-attractants and β-defensins in colonic epithelial cells, while only IL-17A can efficiently induce cytokines in macrophages and T cells (Ishigame et al. 2009; Lin et al. 2009). These results suggest that in addition to the different binding affinity of these cytokines to these receptors, the distribution of IL-17RA and IL-17RC may determine the biological activity of IL-17A and IL-17F. It is also possible that receptors with compositions other than IL-17RA–IL-17RC heterodimer complex may determine the cell-type specificity among different cell types. Indeed, IL-17RA and IL-17RC may also form homodimers (Kramer et al. 2006). Recent findings show that IL-17RA also forms combined with IL-17RB to transduce IL-25 signaling (Rickel et al. 2008), suggesting that IL-17RA serves as a common receptor for several IL-17 family members. Further studies to elucidate the ligand–receptor relationship in the IL-17 system are needed to address these issues.

The IL-17 family members use a unique signaling pathway (Gaffen 2009). IL-17A activates the MAP kinase, NF-κB, PI3-Akt, and C/EBPδ pathways. It has also been shown that IL-17RA signaling, like IL-1/Toll-like receptor (TLR) signaling, is dependent on TRAF6. Although IL-17A shows biological activities similar to IL-1 in several immune responses, the adaptor molecules MyD88, TRIF, and IRAK4 are not required for IL-17A signaling. Recent studies showed that Act1, which physically interacts with IL-17RA and mediates TRAF6 recruitment, is an essential adaptor protein for IL-17A and IL-17F signaling and function (Li 2008; Gaffen 2009).

2 Regulation of IL-17A and IL-17F Production

Upon antigenic stimulation, naive $CD4^+$ T cells differentiate into distinct functional T cell subsets including T_H1 and T_H2 cells that are characterized by different cytokine production profiles and effector functions (Fig. 2). T_H1 cells produce large quantities of IFN-γ and mediate cellular immunity, while T_H2 cells are involved in humoral immunity and mainly produce IL-4, IL-5, and IL-13. IL-12 induces the differentiation of naive $CD4^+$ T cells into IFN-γ-producing T_H1 cells through STAT4 activation. IFN-γ signals are transduced by STAT1, which activates the downstream transcription factor, T-bet, that enhances the expression of T_H1 cell-specific genes. In contrast, IL-4 induces STAT6 activation, followed by the expression of GATA-3, a transcription factor essential for both IL-4 production and T_H2 cell differentiation. Recently, a new $CD4^+$ T cell subset (T_H17) that preferentially produces IL-17A, IL-17F, IL-9, IL-21, and IL-22 was identified. T_H17 cells have been widely accepted as important effector cells in the development of auto-immune and allergic diseases and host defenses against a group of pathogens.

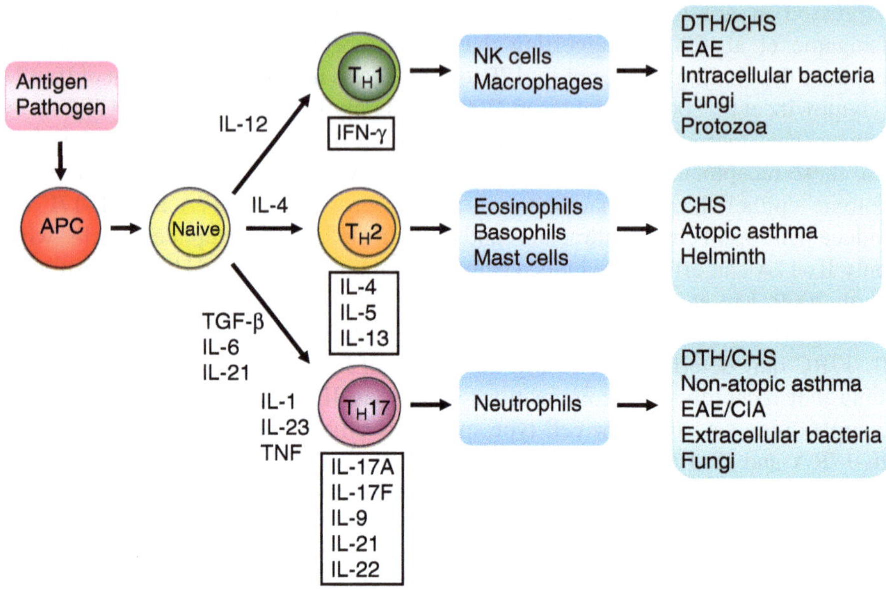

Fig. 2 Effector function of Th cell subsets

IL-17A was originally described as a product of memory CD4⁺ T cells. The discovery of the link between IL-17A-producing T cells and IL-23 effector function led to the concept that IL-17A-producing T_H17 cells belong to a CD4⁺ T cell subset that is distinct from the classical T_H1 and T_H2 cell subsets (Dong 2008; McGeachy and Cua 2008). Subsequent studies showed that T_H17 cell differentiation is induced by TGF-β plus IL-6 or IL-21 and accelerated by the coordinated activities of IL-1 and TNF (Fig. 2). Furthermore, T_H17 cell differentiation depends on transcription factors, including interferon-regulatory factor 4 (IRF4), aryl hydrocarbon receptor (AHR), STAT3, retinoic acid receptor-related orphan receptor α (RORα), and RORγ. It is now accepted that IL-23 is required for the growth, survival, and effector functions of T_H17 cells and promotes IL-17A, IL-17F, IL-9, IL-21, and IL-22 production by this T cell subset.

It is reported that T_H17 lineage is a heterogenous population. In addition to IL-17A and IL-17F double-positive cells, there is an IL-17A- or IL-17F single-positive population. Regulation of IL-17A and IL-17F production is also different, as IL-17F is expressed early in T_H17 development compared to IL-17A (Liang et al. 2007; Lee et al. 2009), suggesting that IL-17A and IL-17F production are regulated differently, depending on the stage of distinction. Although underlying molecular mechanisms are still not identified, it is likely that some factors, such as transcriptional factors or T cell receptor (TCR) signal strength, may distinctly regulate these productions. Indeed, deficiency of RORα only selectively reduces IL-17A, but not

IL-17F production (Yang et al. 2008b), and expression of IL-17A is more sensitive to the strength of TCR signaling compared to that of IL-17F (Gomez-Rodriguez et al. 2009). Furthermore, there are other sub-populations of T_H17 cells that selectively produce IL-9, IL-21, or IL-22, and the development of these subsets is differentially regulated (Annunziato and Romagnani 2009).

In addition to T_H17 cells, a wide variety of immune cells, including $CD8^+$ T cells, γδ T cells, and NKT cells, produce IL-17A and IL-17F under various conditions. It is likely that IL-17A production from $CD8^+$ T cells is dependent on TGF-β/IL-6 and IL-23 (He et al. 2006; Stumhofer et al. 2006). IL-23, but not IL-6, is required for IL-17A production from NKT and γδ T cells (Shibata et al. 2007; Rachitskaya et al. 2008). These cells constitutively express IL-23R and RORγt, unlike naive $CD4^+$ T cells (Lochner et al. 2008; Rachitskaya et al. 2008). Innate immune cells, such as neutrophils (Ferretti et al. 2003; Hoshino et al. 2008), monocytes (Starnes et al. 2001; Hue et al. 2006), NK cells (Satoh-Takayama et al. 2008), and lymphoid tissue inducer-like cells (LTi-like cells) (Luci et al. 2009; Sanos et al. 2009; Takatori et al. 2009), also produce IL-17A. It has now become evident that IL-17A production by these cells also contributes to various immune responses.

3 The Role of IL-17A and IL-17F in Allergic Responses

3.1 *Delayed-Type Hypersensitivity*

Delayed-type hypersensitivity (DTH) responses are elicited by immunization with exogenous antigens such as cells, protein antigens, and pathogens, and are believed to be mediated by $CD4^+$ T cells, especially T_H1 cells (Fig. 2). The contribution of IFN-γ in the induction of DTH is different among antigens. *Ifng*$^{-/-}$ mice have suppressed KLH-mediated DTH (Akahira-Azuma et al. 2004; Gao et al. 2006), but exacerbated OVA- and mBSA-induced DTH (Feuerer et al. 2006; Irmler et al. 2007). Thus, the pathogenic mechanism of DTH cannot be explained completely by the action of IFN-γ. Recent studies using *Il17a*$^{-/-}$and *Il17f*$^{-/-}$ showed that *Il17a*$^{-/-}$, but not *Il17f*$^{-/-}$ mice had attenuated mBSA-induced DTH (Ishigame et al. 2009). mBSA-specific T cell proliferation and mBSA-specific Ab production were also impaired in *Il17a*$^{-/-}$ mice, whereas *Il17f*$^{-/-}$ mice did not have this defect (Nakae et al. 2002; Ishigame et al. 2009). *Il17a*$^{-/-}$, but not *Il17f*$^{-/-}$, mice also showed significantly decreased KLH-specific Ab production (Yang et al. 2008a). In addition, similar to *Il17a*$^{-/-}$ mice, *Il23a*$^{-/-}$ mice show attenuated DTH responses against mBSA (Ghilardi et al. 2004). In these mice, antigen-specific T cell expansion and cytokine production (IL-2, IFN-γ, IL-4, IL-10, and GM-CSF) were normal, while IL-17A production was markedly impaired (Ghilardi et al. 2004). These observations suggest that IL-17A, rather than IL-17F, is important in DTH response. The relative contribution of IL-17A and IFN-γ in DTH response induced by different types of antigen has not yet been examined.

3.2 Contact Hypersensitivity

Contact hypersensitivity (CHS), which is induced by an epicutaneous exposure to chemicals, is considered to be a classic DTH response. Recently however, CHS is thought to be a different type of hypersensitive response, because the roles of $CD4^+$ T cells and $CD8^+$ T cells are opposite in these systems. $CD4^+$ T cells play effector function and $CD8^+$ T cells play regulatory role in DTH, while their functions are opposite in CHS (Grabbe and Schwarz 1998; Kimber and Dearman 2002). Involvement of T_H1- and IFN-γ-producing $CD8^+$ T cells (Tc1 cells) is suggested in CHS because a defect in IFN-γ signaling suppresses FITC-induced CHS response (Lu et al. 1998). Other reports demonstrate that IFN-γ is not pathogenic in oxazolone, TNCB, or DNFB-induced CHS (Saulnier et al. 1995; Lu et al. 1998; Reeve et al. 1999; Nakae et al. 2003a), suggesting involvement of other T cell subsets. In CHS induced by TNCB and DNFB, but not by oxazolone, the response is attenuated in *Il4*$^{-/-}$ mice, suggesting that T_H2 cells mediate the response (Berg et al. 1995; Weigmann et al. 1997; Dieli et al. 1999; Traidl et al. 1999). Consistently, CHS responses induced by TNCB, DNFB, FITC, and oxazolone are remarkably reduced in *Stat6*$^{-/-}$ mice (Yokozeki et al. 2000; Takeshita et al. 2004). IL-13, however, is not essential for the induction of CHS by DNFB (Herrick et al. 2003). These observations suggest that both IFN-γ and IL-4 differentially regulate CHS responses, depending on the mouse genetic background and chemicals used.

IL-17A has also been suggested in the pathogenesis of contact dermatitis because nickel-specific T cell clones established from contact dermatitis patients produce IL-17A (Albanesi et al. 1999). Both IL-17A and IL-17F strongly induce IL-6, IL-8, CXCL1, GM-CSF, SCF, and ICAM-1 expression/production in human keratinocytes (Fig. 1). CHS responses induced by TNCB, DNFB, or FITC are attenuated in *Il17a*$^{-/-}$ mice, while *Il17f*$^{-/-}$ mice display similar CHS response (Nakae et al. 2002; Oboki et al. 2008; Ishigame et al. 2009). Although IL-17A enhances T cell activation by promoting DC maturation (Antonysamy et al. 1999), an IL-17A deficiency does not affect dermal DC/Langerhans cell functions such as migration, maturation, and antigen presentation in the CHS response (Nakae et al. 2002). Instead, IL-17A is important for hapten-specific $CD4^+$, but not $CD8^+$, T cell expansion in the sensitization phase (Nakae et al. 2002). Wild-type mice engrafted with DNFB-sensitized *Il17a*$^{-/-}$ $CD4^+$ cells also exhibited reduced sensitivity to CHS (Nakae et al. 2002). It was suggested that IL-17A-producing $CD8^+$ T cells (Tc17 cells) are also important for the induction of CHS, because CHS induced by adaptive transfer of DNFB-sensitized $CD8^+$ T cell are suppressed in *Il17ra*$^{-/-}$ mice (He et al. 2006, 2009). These results indicate that IL-17A, but not IL17F, is responsible for the development of CHS in an IL-17RA-dependent manner, and IL-17A is involved in both sensitization and elicitation phases of CHS. However, relative contribution of T_H17 and Tc17 cells, and the molecular mechanisms by which IL-17A, IFN-γ and IL-4 orchestrate the development of CHS remain to be elucidated.

3.3 Allergic Airway Inflammation

Allergic asthma is mostly classified into two types: T_H2-type (atopic) and non-T_H2-type (non-atopic) (Oboki et al. 2008). Atopic asthma is characterized by an increase in serum IgE and the accumulation and activation of T_H2 cells, eosinophils, and mast cells, while non-atopic asthma is characterized by the accumulation of IL-8^+ cells, neutrophils, and mast cells, without elevated serum IgE (Amin et al. 2000). IL-17A and IL-17F mRNA and protein were increased in asthmatic patients compared to healthy subjects. Protein levels were profoundly elevated in the sputum of severe asthmatic patients with increased neutrophilia. IL-17A and IL-17F can activate bronchial fibroblasts, epithelial cells, and smooth muscle cells to produce various pro-inflammatory mediators such as IL-6, IL-8 and CXCL1, which are important for granulopoiesis and neutrophil recruitment (Fig. 1) (Oboki et al. 2008). IL-17A or IL-17F overexpression resulted in the induction of neutrophilia rather than eosinophilia in the lungs of rodents (Oda et al. 2005; Park et al. 2005; Yang et al. 2008a). Collectively, these observations suggest that both IL-17A and IL-17F contribute to the pathogenesis of non-atopic asthma rather than atopic asthma.

T_H2-dominant airway eosinophilia induced by immunization with OVA and antigens from fungi, cockroaches, or house dust mites is a well established rodent model for atopic asthma. Airway hypersensitivity responses (AHR) and inflammation in the lung induced by OVA with aluminum hydroxide (alum) immunization are normally induced in *Il17a*$^{-/-}$ mice and is associated with increased IL-4 and IL-5 levels in the bronchoalveolar lavages (BALs) (Nakae et al. 2002; Pichavant et al. 2008; Ishigame et al. 2009). By contrast, other studies reported that *Il17ra*$^{-/-}$ mice and *Il17a*$^{-/-}$ mice exhibit reduced pulmonary eosinophilia (Schnyder-Candrian et al. 2006; Yang et al. 2008a). IL-17F is not involved in this response because *Il17f*$^{-/-}$ mice have exacerbated or normal pulmonary eosinophilia during OVA/alum-induced airway inflammation (Yang et al. 2008a; Ishigame et al. 2009). These apparent discrepancies may be explained by different experimental conditions, such as sensitization protocols, immunization routes, antigens, and mouse backgrounds. In this regard, the molecular mechanism for the induction of airway inflammation and AHR induced by OVA changes depending on the adjuvant, such as alum (Oboki et al. 2008). He et al. showed that epicutaneous OVA sensitization potently induces T_H17 cells in the draining LNs, spleen, and lungs and recruits neutrophils in BALs, while intraperitoneal OVA immunization with alum induces weak T_H17 cell development and neutrophil recruitment (He et al. 2007). The eosinophil influx in the lungs was not affected by anti-IL-17A mAb treatment in mice epicutaneously immunized with OVA, whereas neutrophil recruitment was inhibited under these conditions. Intraperitoneal OVA sensitization with alum induced T_H17 cells in the spleen but not in the draining LNs (He et al. 2007). On the other hand, subcutaneous OVA immunization with alum did not induce T_H17 cells in the spleen, but did induce T_H17 cells in the draining LNs (Schnyder-Candrian et al. 2006). In this case, the eosinophil influx in the lungs was exacerbated by anti-IL-17A mAb treatment. Thus, T_H17 cell development in the spleen after OVA sensitization is influenced by the immunization route (epicutaneously >> intraperitoneally > subcutaneously), and this may explain the differences in IL-17A dependency among T_H2 cell-mediated and eosinophil-dominant murine asthma models.

In contrast to T_H2 cell-dominant eosinophilic asthma models, the importance of IL-17A in the pathogenesis of non-T_H2-type neutrophilic asthma models has been clearly shown in mice (Fig. 2). OVA-specific TCR-expressing DO11.10 and OTII mice exhibit AHR and airway inflammation after OVA inhalation without prior OVA sensitization (Knott et al. 2000; Wilder et al. 2001). Similar to non-atopic asthma (Amin et al. 2000), airway inflammation in OVA-inhaled DO11.10 and OTII mice is characterized by a predominant infiltration of neutrophils rather than eosinophils in the lungs without total and OVA-specific IgE elevation in sera (Knott et al. 2000; Wilder et al. 2001; Nakae et al. 2007). In addition, T_H1 cells and T_H17 cells, but not T_H2 cells, were increased in BALs from OVA-inhaled OTII mice (Nakae et al. 2007). OVA-induced airway neutrophilia in DO11.10 and OTII mice was profoundly suppressed by the deficiency of IL-17A, whereas *Il17f*$^{-/-}$ DO11.10 mice showed normal airway neutrophilia (Nakae et al. 2002, 2007; Ishigame et al. 2009). Airway inflammation was aggravated in *Ifng*$^{-/-}$ OTII mice (Nakae et al. 2007), suggesting that IL-17A, but not IL-17F, is an effector and IFN-γ is a negative regulator of this response. IL-17A induces TNF production by mast cells independent of IgE-mediated signals, leading to neutrophil influx in airways (Nakae et al. 2007). Thus, TNF functions downstream of IL-17A in the antigen-induced airway neutrophilia in the OTII model, while IL-6 is not required in this model (Tanaka et al. 2009).

Neutrophil-dominant airway inflammation also can be induced by OVA inhalation without prior OVA sensitization in mice adoptively transferred with DO11.10 T_H17 cells (Liang et al. 2007). In this setting, T_H17 cell-derived IL-17A, rather than IL-17F, is responsible for neutrophil recruitment to the airway since T_H17 cell-mediated airway neutrophilia is suppressed by an anti-IL-17A neutralizing mAb but not by an anti-IL-17F mAb (Liang et al. 2007). Likewise, *Rag2*$^{-/-}$ mice engrafted with *Tbx21*$^{-/-}$ DO11.10 $CD4^+$ T cells, which contain a larger amount of T_H2 cells and T_H17 cells (but fewer T_H1 cells) than wild-type DO11.10 $CD4^+$ T cells, show increased eosinophil and neutrophil counts in BALs compared to *Rag2*$^{-/-}$ mice engrafted with wild-type DO11.10 $CD4^+$ T cells (Fujiwara et al. 2007). Treatment with an anti-IL-17A neutralizing mAb suppressed neutrophil, but not eosinophil recruitment in *Tbx21*$^{-/-}$ DO11.10 $CD4^+$ T cell-transferred *Rag2*$^{-/-}$ mice after OVA inhalation (Fujiwara et al. 2007), indicating that IL-17A is responsible for the neutrophil accumulation.

Neutrophil-dominant allergic airway inflammation is also elicited by inhalation of fungal antigens (proteinase from *Aspergillus oryzae*) in mice independently of T and B cells (Kiss et al. 2007). In contrast to active and passive models using DO11.10 or OTII mice, fungal proteinase-induced airway neutrophilia was normal in *Il17a*$^{-/-}$ mice but was significantly attenuated in *Il17f*$^{-/-}$ or *Il17ra*$^{-/-}$ mice (Yang et al. 2008a), indicating that IL-17F, rather than IL-17A, is responsible for fungus-induced airway neutrophilia in an IL-17RA-dependent manner.

Taken together, IL-17A, and to a lesser extent IL-17F, plays more important role in the induction of non-T_H2-type neutrophilic airway inflammation than in T_H2-type eosinophilic airway inflammation.

4 The Role of IL-17A and IL-17F in Auto-immunity

4.1 Rheumatoid Arthritis

Rheumatoid arthritis (RA) is one of the most serious auto-immune diseases, mainly affecting multiple joints of the body. The development of RA was previously believed to be mediated by T_H1 cells because high levels of IL-12 and IFN-γ were detected in inflammatory sites (Feldmann et al. 1996; Gately et al. 1998). It is now clear, however, that T_H17 cells play crucial roles in this disease (Fig. 2) (Tesmer et al. 2008). Inhibition of either TNF, IL-1, or IL-6 activity in RA patients shows prominent beneficial effects on disease progression (Feldmann and Maini 2008). The roles of cytokines in the development of arthritis have been extensively examined using mouse models with different cytokine dependency.

Collagen-induced arthritis (CIA) is a typical induced arthritis model that is produced by immunizing animals with type II collagen (IIC). The development of CIA is largely dependent on IL-23, as *Il23a*$^{-/-}$ mice, but not *Il12a*$^{-/-}$ mice, are resistant to disease (Murphy et al. 2003). Both IL-17A and IL-17F are expressed in RA synovium and activates synoviocytes, fibroblasts, and endothelial cells to induce various inflammatory cytokines and chemokines, including IL-1 and TNF (Fig. 1). IL-17A also directly promotes osteoclast differentiation by inducing RANKL in osteoblasts (Sato et al. 2006). The development of this arthritis critically depends on IL-17A. *Il17a*$^{-/-}$ mice displayed significantly less severe arthritis development (Table 1) (Nakae et al. 2003b; Ishigame et al. 2009). The sensitization of T cells following immunization with IIC and IIC-specific antibody production were significantly reduced in *Il17a*$^{-/-}$ mice (Nakae et al. 2003b). These results suggest that IL-17A is involved in T cell sensitization and antibody production in addition to pro-inflammatory cytokine induction in the effector phase. As the deficiency of IL-17A could not completely suppress CIA, involvement of IL-17F was suggested. Although adoptive transfer of IL-17F gene-transduced $CD4^+$ T cells exacerbated arthritis (Yamaguchi et al. 2007), CIA was developed normally in *Il17f*$^{-/-}$ mice (Ishigame et al. 2009), indicating that IL-17A plays a more important role than IL-17F in this model. On the other hand, *Il22*$^{-/-}$ mice significantly suppressed the disease (Geboes et al. 2009) and blockade of IL-21 signaling by IL-21R-Fc fusion protein also attenuated the development of CIA (Young et al. 2007). The development of arthritis was also markedly suppressed in either $Il1\alpha^{-/-}$ $Il1\beta^{-/-}$ or *Il6*$^{-/-}$ mice (Alonzi et al. 1998; Saijo et al. 2002). These observations suggest that IL-17A, IL-22 and IL-21 corporately induce arthritis development downstream of IL-23, IL-1, and IL-6.

Various animal disease models other than CIA have been developed, including spontaneous, induced, and gene-manipulated animal models. The importance of IL-17A in the development of arthritis is also reported in several RA models. Transgenic mice carrying the HTLV-1 *Tax* gene with its own LTR promoter (HTLV-I Tg mice) developed chronic inflammatory polyarthropathy resembling RA in humans at a high incidence (Iwakura et al. 1991). The expression of pro-inflammatory cytokine genes, including IL-1α, IL-1β, IL-2, IL-6, TNF, IFN-γ, and IL-17A, is

Table 1 Knockout mouse phenotypes associated with T_H1/T_H17-related cytokines in auto-immune disease models

Auto-immune disease model					
Deficient mice	CIA	EAE	$CD4^+$ $CD45RB^{hi}$ T cell-induced colitis	DSS-induced colitis	Type I diabetes
IFN-γ/IFN-γR	↑ (Vermeire et al. 1997; Guedez et al. 2001)	↑ (Ferber et al. 1996; Krakowski and Owens 1996; Willenborg et al. 1996)	↓ (Ito and Fathman 1997)	→ (Suzuki et al. 2001)	→ (Hultgren et al. 1996; Kanagawa et al. 2000; Serreze et al. 2000)
IL-17A	↓ (Nakae et al. 2003b; Ishigame et al. 2009)	↓ (Komiyama et al. 2006; Yang et al. 2008a; Ishigame et al. 2009), → (Haak et al. 2009)	→ (Noguchi et al. 2007; Izcue et al. 2008; Leppkes et al. 2009), ↑ (O'Connor et al. 2009)	↑ (Yang et al. 2008a), ↓ (Ito et al. 2008)	ND
IL-17F	→ (Ishigame et al. 2009)	→ (Yang et al. 2008a; Haak et al. 2009; Ishigame et al. 2009)	→ (Leppkes et al. 2009)	↓ (Yang et al. 2008a)	ND
IL-9/IL-9R	ND	↓ (Nowak et al. 2009)	ND	ND	ND
IL-21/IL-21R	↓ (Young et al. 2007)	↓(Korn et al. 2007; Nurieva et al. 2007), ↑ (Coquet et al. 2008; Liu et al. 2008)	↓ (Fina et al. 2008)	↓ (Fina et al. 2008)	↓ (Spolski et al. 2008; Sutherland et al. 2009)
IL-22	↓ (Geboes et al. 2009)	→ (Kreymborg et al. 2007)	↑ (Zenewicz et al. 2008)	↑ (Zenewicz et al. 2008)	ND

↑: aggravated, →: no difference, ↓: suppressed, ND: not determined
EAE experimental auto-immune encephalomyelitis, *CIA* collagen-induced arthritis, *DSS* dextran sodium sulfate

Table 2 Cytokine dependency of spontaneous RA models

	IL-1	IL-6	TNF	IL-17A	IL-17F
HTLV-I Tg	↓ (Saijo et al. 2002)	↓ (Iwakura et al. 2008)	→ (Iwakura et al. 2008)	↓ (Iwakura et al. 2008)	ND
Il1rn$^{-/-}$	ND	→ (Iwakura et al. 2008)	↓↓ (Horai et al. 2004)	↓↓ (Nakae et al. 2003c)	↓ (Ishigame et al. 2009)
SKG	↓ (Hata et al. 2004)	↓↓ (Hata et al. 2004)	↓ (Hata et al. 2004)	↓↓ (Hirota et al. 2007a, b)	ND

→: independent, ↓: involved, ↓↓: dependent, ND: not determined
HTLV-I human T cell leukemia virus type I, *Il1rn* IL-1R antagonist

enhanced in transgenic joints (Iwakura et al. 1995). The development of arthritis was greatly suppressed in either $Il1\alpha^{-/-}Il1\beta^{-/-}$, $Il6^{-/-}$, or $Il17a^{-/-}$ HTLV-I Tg mice (Table 2) (Iwakura et al. 2008), indicating the importance of these cytokines in the development of arthritis in this model. Excess IL-6 signaling enhances the development of arthritis, as the development of arthritis is accelerated in HTLV-I Tg mice carrying a Y759F mutation in the IL-6R gp130 which is important for SOCS3-mediated negative feedback (Ishihara et al. 2004). In contrast, a TNF deficiency did not affect disease development at all (Iwakura et al. 2008). An IFN-γ and IL-4 deficiency also did not affect disease onset (Iwakura et al. 2008), suggesting that neither T_H1 nor T_H2 cells are involved in the pathogenesis of this arthritis.

Similar cytokine dependency is also observed in SKG mice, which develop autoimmune arthritis because of a mutation in ZAP 70 of the TCR complex. An IL-17A deficiency completely suppresses the development of arthritis (Table 2) (Hirota et al. 2007a, b). Consistent with this result, the development of arthritis in SKG mice depends on IL-6, IL-1, and TNF (Hata et al. 2004; Hirota et al. 2007a, b), indicating that these cytokines play important roles in the pathogenesis. The role of IL-17F, IL-21 and IL-22 in these models remains to be elucidated.

K/BxN mice carry the KRN transgene, which encodes a TCR reactive against a peptide from glucose-6-phosphate isomerase (GPI) presented by the A^{g7} MHC class II molecule (Matsumoto et al. 1999). KRN-transgenic mice on the NOD (A^{g7}) background spontaneously develop auto-immune arthritis. The development of arthritis in this model depends on both T cells and B cells, and serum from K/BxN mice can induce arthritis in recipient mice (Korganow et al. 1999). Auto-antibodies to GPI are responsible for the disease (Matsumoto et al. 1999) because immune complexes activate the C5a-containing complement activation pathway in mast cells through FcγRIII, resulting in the induction of inflammatory cytokines (Ji et al. 2002a; Nigrovic et al. 2007). It was shown that mast cell-derived IL-1 plays a crucial role in the development of arthritis in this serum transfer model (Nigrovic et al. 2007). In addition, TNF is involved in the development of arthritis in this model (Ji et al. 2002b). Neutralization of IL-17A does not affect K/BxN serum induced arthritis, indicating that IL-17A is not required in the effector phase. In this model however, autoreactive KRN T cells enhance K/BxN serum-transferred arthritis in a

IL-17 dependent manner, suggesting that IL-17 can enhance inflammation to some extent even in this setting (Jacobs et al. 2009).

It is widely accepted that IL-1 is a potent inducer of IL-17A and IL-17F production (Fig. 2). IL-1 receptor antagonist (Ra) is an endogenous negative regulator of IL-1 signaling and IL-1Ra deficient (*Il1rn*$^{-/-}$) mice spontaneously develop chronic inflammatory arthropathy (Horai et al. 2000). The IL-17A and IL-17F-producing T cell population is significantly expanded in the LNs of *Il1rn*$^{-/-}$ mice (Nakae et al. 2003c; Ishigame et al. 2009), suggesting that IL-1Ra deficiency may be sufficient to render T cells highly sensitive to IL-1, thereby leading to the activation of autoreactive IL-17A and IL-17F producing T cells by a physiological level of IL-1 in vivo. Interestingly, the development of arthritis in *Il1rn*$^{-/-}$ mice was almost completely suppressed in *Il17a*$^{-/-}$ *Il1rn*$^{-/-}$ mice (Nakae et al. 2003c), and slightly suppressed in *Il17f*$^{-/-}$ *Il1rn*$^{-/-}$ mice (Table 2) (Ishigame et al. 2009). In contrast, the deficiency of T-bet does not affect the development of arthritis (Wang et al. 2006), suggesting that T_H17 cells, not T_H1 cells, are involved in the pathogenesis of this model. A TNF deficiency also completely suppressed the development of arthritis in these mice (Horai et al. 2004). In contrast to HTLV-I Tg and SKG mice, *Il6*$^{-/-}$ *Il1rn*$^{-/-}$ mice did not suppress disease onset (Iwakura et al. 2008), signifying that IL-1 can bypass the requirement of IL-6 for T_H17 cell differentiation. Several studies have been reported that IL-1 directly acts on $CD4^+$ T cells to produce IL-17A (Ben-Sasson et al. 2009; Chung et al. 2009), although underlying mechanisms are unknown. Several reports have shown that IL-6 is not absolutely required for T_H17 differentiation. IL-21 can substitute for IL-6 to induce T_H17 cell development in an autocrine manner (Korn et al. 2007; Nurieva et al. 2007; Zhou et al. 2007). Furthermore, IL-6 is not required for the induction of IL-17A from NKT cells or γδ T cells (Shibata et al. 2007; Rachitskaya et al. 2008). Further study is necessary to address whether IL-1 alone can directly induce IL-17A production/T_H17 differentiation under certain conditions in vivo, or which T cell subset is the direct target for IL-1 to induce auto-immune arthritis in this model.

The role of TNF in RA is well established, as anti-TNF Ab treatment has been proved to be efficient for most RA patients (Feldmann and Maini 2008). However, a minor proportion of the patients are refractory against this treatment. It is possible that other cytokines such as IL-1, IL-6, and IL-17A are activated in these patients like HTLV-I Tg mouse model, which develop TNF-independent auto-immune arthritis. Because these cytokines act independently in the development of arthritis as revealed by these disease models, inhibitors for IL-17A, TNF, and IL-6 may be used in a complementary manner in the treatment of RA.

4.2 Experimental Auto-immune Encephalomyelitis

Experimental auto-immune encephalomyelitis (EAE) is a well established murine model of multiple sclerosis (MS) and has long been believed to be a T_H1 cytokine-mediated auto-immune disease (Kuchroo et al. 2002). However, the mechanisms of

EAE development are more complex than previously thought. EAE is much more severe in mice which are defective in IL-12/IFN-γ activity (Gran et al. 2002; Zhang et al. 2003), arguing against the importance of T_H1 cells in this disease. Several studies have demonstrated that IL-23, rather than IL-12 is essential for EAE development, as EAE was greatly attenuated in mice that lack an IL-23 signal (Cua et al. 2003; Zhang et al. 2003). Consistent with these observations, *Il17a*$^{-/-}$ mice were resistant to EAE (Table 1) (Komiyama et al. 2006; Yang et al. 2008a; Ishigame et al. 2009). However, one report claimed that IL-17A did not contribute to EAE development because $CD4^+$ T cell-specific IL-17A overexpression did not have a major impact on the development of EAE (Haak et al. 2009). It was not shown whether the IL-17A expression levels in these transgenic mice were enough for the development of EAE. In support of the importance of T_H17 cells, the development of EAE was also diminished in *Rorγ*$^{-/-}$ and *Rorα*$^{-/-}$ mice which lacked T_H17 cells (Ivanov et al. 2006; Yang et al. 2008b). Unlike the *Il23a*$^{-/-}$ mice, *Il17a*$^{-/-}$ mice still developed significant inflammation after MOG immunization, proposing that other IL-23-induced mediators, such as IL-17F and IL-22 may also contribute to the development of EAE. IL-17F, however, is not required for the pathogenesis of EAE (Yang et al. 2008a; Haak et al. 2009; Ishigame et al. 2009). Mice deficient in both IL-17A and IL-17F showed no additional suppression (Ishigame et al. 2009), showing that IL-17F is not only dispensable for the induction of these responses, but also does not have any substantial additive, synergistic, or compensatory effects to those of IL-17A in these disorders. T_H17 cell-derived IL-17A, as well as IL-22, disrupts tight junctions that form the blood-brain barrier, resulting in an infiltration of T_H17 cells into the central nervous system (Kebir et al. 2007) It is shown that IL-22 is also dispensable for EAE development (Kreymborg et al. 2007), while the contribution of IL-21 in the induction of EAE is still controversial. Although recombinant IL-21 can induce T_H17 cell differentiation from naive T cells in the presence of rhTGF-β in vitro, endogenous IL-21 is not essential for the T_H17 cell differentiation during EAE in vivo. EAE development is suppressed in some reports using *Il21*$^{-/-}$ and BALB/c-*Il21r*$^{-/-}$ mice (Korn et al. 2007; Nurieva et al. 2007), but is aggravated in other reports using the C57BL/6-*Il21*$^{-/-}$ and -*Il21r*$^{-/-}$ mice (Coquet et al. 2008; Liu et al. 2008). Recent studies also demonstrated that IL-9 produced by T_H17 cells is also involved in EAE development (Nowak et al. 2009). However, IL-9 blockade by antibody or IL-9R deficiency only partially ameliorates disease. Therefore, the downstream mechanism of IL-23 still remains to be elucidated.

In contrast to the results observed from mice lacking IL-12 or IL-23 signaling, EAE can be induced by transfer of either IL-12 or IL-23-stimulated $CD4^+$ T cells (Kroenke et al. 2008; Lees et al. 2008; Stromnes et al. 2008). Patterns of CNS infiltration and cytokine requirement in disease development are quite different between each T_H cell-transplanted mice. The majority of CNS infiltrating cells from recipient mice of IL-12-stimulated $CD4^+$ T cells are macrophages and lymphocytes, whereas significant neutrophil recruitment is observed in mice given IL-23-stimulated $CD4^+$ T cells (Kroenke et al. 2008; Lees et al. 2008; Stromnes et al. 2008). Neutralization of either IL-17A or GM-CSF delayed onset of disease induced by IL-23-stimulated $CD4^+$ T cells, whereas IL-12-stimulated $CD4^+$ T cell-mediated disease was not

(Kroenke et al. 2008). Induction of EAE mediated by ether cell types was not dependent on IFN-γ, because neither IL-12 or IL-23-mediated disease was suppressed by the treatment of anti-IFN-γ antibody (Kroenke et al. 2008). Several reports suggest that the T_H17/T_H1 ratio of infiltrating cells determines the sites of inflammation within the CNS (Kroenke et al. 2008; Lees et al. 2008; Stromnes et al. 2008). These observations suggest that T_H17 and T_H1 cell contribute to induction of EAE by using different pro-inflammatory pathways, and the balance between T_H1 and T_H17 cells is critical for the pathogenesis of EAE.

It is believed that T_H17 cell-derived IL-17A is required for the induction of EAE because $CD4^+$ T cells from *Il17a*$^{-/-}$ mice could not efficiently induce EAE (Komiyama et al. 2006). Several studies showed that IL-17A-producing γδ T cells are also important for the development of EAE. γδ T cell-deficient mice showed a delayed onset and reduced severity of EAE (Jensen et al. 2008). γδ T cells isolated from MOG-immunized mice could rapidly induce IL-17A production compared to that of T_H17 cells after in vitro re-stimulation, and this IL-17A induction is mediated by IL-1 and IL-23 (Sutton et al. 2009). These γδ T cells were able to promote IL-17A production by $CD4^+$ T cells as well as disease susceptibility, suggesting that IL-17A producing γδ T cells cooperate with $CD4^+$ T cells to induce EAE. IL-17A producing γδ T cells are also increased in the arthritic joints and deletion of $V\gamma4^+$ TCR, which is predominant source of IL-17A among γδ T cells during CIA and significantly reduces disease incidence (Roark et al. 2007).

4.3 *Inflammatory Bowel Disease*

Dysregulation of intestinal homeostasis causes inflammatory bowel diseases in which T cells play important roles. Recently, several studies have established that IL-23 is an essential cytokine for T cell-dependent intestinal inflammation. Transfer of $CD4^+CD45RB^{hi}$ T cells into lymphopenic mice is a well established model of IBD. In this model, adoptive transfer of $CD4^+CD45RB^{hi}$ T cells into IL-23 deficient *Rag*$^{-/-}$ mice could not induce colitis while wasting disease was observed in IL-12 deficient *Rag*$^{-/-}$ recipient mice (Uhlig et al. 2006). Similarly, deficiency of IL-23 suppressed the development of T cell-dependent spontaneous IBD in *Il10*$^{-/-}$ mice (Yen et al. 2006). These results indicate that IL-23, rather than IL-12, plays an important role in the intestinal inflammatory response. Consistent with these observations, naïve $CD4^+$ T cells isolated from *Rorgt*$^{-/-}$ mice induced less severe colitis (Leppkes et al. 2009), and adoptive transfer of $IL\text{-}17F^+$ $CD4^+$ T cells induced significantly rapid colitis (Lee et al. 2009). However, the role of IL-17A and IL-17F in colitis is still controversial. Neither IL-17A or IL-17F deficiency suppressed colitis in $CD4^+CD45RB^{hi}$ T cell adoptive transfer model (Table 1) (Noguchi et al. 2007; Izcue et al. 2008; Leppkes et al. 2009). Transfer of *Il17f*$^{-/-}$ $CD4^+CD45RB^{hi}$ T cells, in combination with anti-IL-17A Ab, significantly reduced colitis (Leppkes et al. 2009), suggesting that IL-17A and IL-17F have redundant function during colitis development in this model. Recent studies reported that mice transferred with

Il17a$^{-/-}$ $CD4^+$ T cells displayed an accelerated wasting disease, which is associated with increased T_H1-related cytokine production, and suggests a protective function of IL-17A (O'Connor et al. 2009). It should be noted that the protective function of IL-17A is only observed in recipient mouse weight loss and the extent of cellular infiltration does not correlate with the wasting aspect of the disease. Thus, the function of IL-17A in inflammation and in maintaining homeostasis of the gut may be different. On the other hand, many studies suggest that T_H1 cells are also a key mediator in the development of IBD. Colitis was suppressed by the adoptive transfer of $CD4^+CD45RB^{hi}$ T cells deficient in a T_H1 cell-associated gene, such as IFN-γ, T-bet, or STAT4. In addition, both T_H1 and T_H17 cells were required for the development of *Hericobactor hepaticus*-induced colitis downstream of IL-23 (Hue et al. 2006; Kullberg et al. 2006). It is interesting to investigate what kind of conditions determine the relative contribution of T_H1 and T_H17 cells and how other T_H17 cell-related cytokines, such as IL-17A, IL-17F, IL-22, and IL-21, are involved in the pathogenesis of T cell-mediated IBD.

The role of IL-17A in a dextran sodium sulfate (DSS)-induced acute colitis model is also still controversial. Although one study reported that *Il17a*$^{-/-}$ mice displayed significantly reduced clinical score (Ito et al. 2008), other studies demonstrated that mice deficient in IL-17A or given anti-IL-17A Ab showed severe weight loss and colonic epithelial damage (Table 1) (Ogawa et al. 2004; Yang et al. 2008a). On the other hand, *Il17f*$^{-/-}$ mice developed less severe colonic inflammation, which is associated with reduced chemokine expression (Yang et al. 2008a). IL-21 deficiency also suppressed body weight loss by inhibiting induction of T_H17 associated molecules such as IL-6, IL17A, and IL-17F (Fina et al. 2008). IL-22 has a protective role in both DSS- and $CD4^+CD45RB^{hi}$ T cell-induced colitis (Zenewicz et al. 2008). This protective function of IL-22 in these models appears to be mediated not only by $CD4^+$ T cell-derived IL-22, but also by NK cell-derived IL-22. Although T cell-produced IL-22 is dispensable for the protection of $CD4^+$ T cell-induced colitis, deficiency of both donor and recipient derived IL-22 production resulted in more severe colitis. It has been also demonstrated that IL-23 is also a key player in T cell-independent colitis induced by an agonistic anti-CD40 antibody (Uhlig et al. 2006). Because lymphopenic mice can also produce IL-17A, IL-17F, and IL-21 from innate immune cells such as LTi-like cells, NK cells, and monocytes, further studies are needed to investigate whether these cytokines are also involved in innate immune cell-mediated IBD models.

4.4 Type I Diabetes

Type I diabetes mellitus (T1D) is an auto-immune disease caused by the invasion of islets of Langerhans by mononuclear cells resulting in the destruction of β cells. Both $CD4^+$ and $CD8^+$ T cells are involved in the islet destruction. Deficiency of IFN-γ or IFN-γR on the NOD background did not suppress the development of diabetes (Hultgren et al. 1996; Kanagawa et al. 2000; Serreze et al. 2000) (Table 1), suggesting that other mediators contribute to the pathogenesis. Several groups reported that

IL-17A is involved in the pathogenesis of diabetes. Both IL-17A and IL-17F expression are increased in diabetic NOD pancreas (Martin-Orozco et al. 2009). Neutralization of IL-17A by anti-IL-17A Ab treatment or suppression of IL-17A production by inducing IFN-γ restored normoglycemia at the pre-diabetic stage of NOD mice is important (Jain et al. 2008; Emamaullee et al. 2009). IL-21 also plays a key role in the development of type 1 diabetes. *Il21r*$^{-/-}$ NOD mice are highly resistant to insulitis, auto-antibody production against insulin, and diabetes development (Spolski et al. 2008; Sutherland et al. 2009). Mice expressing IL-21 in pancreatic β cells produced elevated levels of pro-inflammatory cytokines including IL-17A, IL-17F, and IFN-γ, and spontaneously developed diabetes even on the diabetes-resistant background (Sutherland et al. 2009). It is still controversial whether or not IL-21-induced diabetes depends on T_H17 cells. Further studies are required to examine the role of IL-17A and IL-17F in the pathogenesis of diabetes by using IL-17A or IL-17F deficient NOD mice.

Transfer of the diabetogenic $CD4^+$ BDC2.5 TCR^+ T cell into NOD.scid mice can also cause diabetes (Katz et al. 1995). In this model, it is clear that T_H1 cells play an important role in the induction of diabetes, because in vitro differentiated IFN-γ-producing BDC2.5 TCR^+ T_H1 cells potently induce disease (Katz et al. 1995). Recent studies showed that diabetogenic T_H17 cells can also rapidly induce diabetes in NOD.scid mice (Bending et al. 2009; Martin-Orozco et al. 2009). Interestingly, several reports showed that adoptive transfer of in vitro generated T_H17 cells and maintained their differentiation program in normal recipients, whereas these cells were reprogrammed into T_H1 cells in lymphopenic recipients (Bending et al. 2009; Lee et al. 2009; Martin-Orozco et al. 2009). Accordingly, diabetes development induced by the transfer of in vitro differentiated T_H17 cells in NOD.scid mice was dependent on IFN-γ, but not IL-17A (Bending et al. 2009; Martin-Orozco et al. 2009). In $CD8^+$ T cell-induced diabetes model, adoptive transfer of IL-23-treated OTI Tc17 cells can induce diabetes in normal recipients that express OVA under the control of rat insulin promoter, and disease development is dependent on both IL-17A and IL-17F (Ciric et al. 2009). Thus, it is interesting to study what kinds of mediators regulate the plasticity of T_H17 and Tc17 developmental program, and the impact of these on the development of auto-immunity, such as type I diabetes.

5 The Role of IL-17A and IL-17F in Host Defense Against Infections

5.1 Microbe Stimulation and IL-17A and IL-17F Production

Recent studies suggest that IL-17A and IL-17F are also involved in host defense against infection. When stimulated by microbial products through pattern recognition receptors, APCs acquire the capacity to activate naive T cells to differentiate into effector T cells that mediate adaptive immune responses. APCs stimulated with pathogens such as *Bordetella pertusis*, *Klebsiella pneumoniae*, and *Mycobacterium tuberculosis* produce large amounts of IL-23, resulting in the development of T_H17 cells (Khader et al. 2009). Several Toll-like receptor (TLR) agonists including LPS (TLR4 ligand), CpG-containing oligonucleotides (TLR9 ligand), R848 (TLR7/8 ligand), and peptidoglycans (PGN) (TLR2 ligand) are thought to induce IL-12 production in DCs (Napolitani et al. 2005; Gerosa et al. 2008). These TLR agonists also induce IL-23 production that facilitates T_H17 differentiation (Napolitani et al. 2005; Gerosa et al. 2008) as CpG and PGN can substitute for complete Freund's adjuvant containing killed mycobacteria to induce EAE, in which T_H17 cells play a crucial role (Segal et al. 2000; Visser et al. 2005). Interestingly, in addition to TLRs, the intracellular receptor, NOD2, also plays a critical role in the generation of T_H17 cells (van Beelen et al. 2007). NOD2 recognizes muramyldipeptides (MDP), the minimal motif in PGN from bacterial cell walls. Although stimulation of DCs with MDP alone does not induce cytokine production, MDP in combination with other bacterial-derived TLR agonists enhances TLR-mediated IL-1 and IL-23 production and promotes T_H17 cell differentiation (van Beelen et al. 2007). These findings indicate that many pathogen-derived molecules can induce both IL-12 and IL-23. However, the precise conditions or mechanisms by which either IL-12 or IL-23 is preferentially induced and lead to the preferential expansion of T_H1 or T_H17 cells remain to be elucidated.

5.2 *Bacterial Infection*

The importance of IL-12 in host defense against various bacteria is widely accepted (Fig. 2). Mice deficient in IFN-γ, IFN-γR, or STAT1 are highly susceptible to many pathogens, including *Listeria monocytogenes*, *M. tuberculosis*, and *Salmonella enteritidis* (Shtrichman and Samuel 2001). The IL-12-IFN-γ axis is primarily involved in host defense against intracellular pathogens by activating cellular immunity to kill bacteria and infected cells. In contrast, the IL-23-IL-17A axis is thought to be critical for host defense against extracellular bacteria by inducing CXC chemokine and G-CSF production and antimicrobial peptides, such as β-defensins, lipocalin-2, and S100A family proteins in epithelial cells and keratinocytes (Fig. 1). Indeed, *Il17ra*$^{-/-}$, *IL17a*$^{-/-}$, and *Il23a*$^{-/-}$ mice are more susceptible to the extracellular bacterium *K. pneumoniae* in the lungs (Ye et al. 2001; Happel et al. 2005; Aujla et al. 2008) (Fig. 2). These mice show impaired neutrophil recruitment at the site of infection which is associated with defective G-CSF and CXC chemokine production. IL-12 and IFN-γ signaling have also been shown to be critical for host defense against *K. pneumoniae*. IL-22 is induced with kinetics that are similar to those of IL-17A and IL-17F from T_H17 cells and may also be involved in host defense mechanism. The protective role of IL-23 in colonic mucosal infection is also reported. IL-23, rather than IL-12, is required for host defense in the colon during the early phase of *Citrobacter rodentium* infection

(Mangan et al. 2006; Zheng et al. 2008). IL-17A and IL-17F are also involved in responses against *C. rodentium* by inducing β-defensin production. Notably, splenomegaly and colon hypertrophy, which are associated with severe colonic inflammation, were more pronounced in *Il17f*$^{-/-}$ mice than in *Il17a*$^{-/-}$ mice (Ishigame et al. 2009), suggesting that IL-17F is more important than IL-17A in protecting colonic epithelial cells from the pathogenic effects of this bacterium. However, a recent study using *Il22*$^{-/-}$ and *Il17rc*$^{-/-}$ mice, which do not respond to IL-17A and IL-17F, demonstrated that IL-22, but not IL-17A and IL-17F, expressed in response to IL-23 is essential for the early host response against *C. rodentium* (Zheng et al. 2008). IL-22 is produced by innate immune cells, including dendritic cells and NK cells during *C. rodentium* infection and induces the expression of Reg family antimicrobial proteins in colonic epithelial cells (Zheng et al. 2008). It is interesting to note that T_H17 cells differentiation in the gastrointestinal tract is largely dependent on commensal microbiota (Gaboriau-Routhiau et al. 2009; Ivanov et al. 2009). Germ-free mice have much fewer lamina propria T_H17 cells compared to specific-pathogen-free mice (Niess et al. 2008; Gaboriau-Routhiau et al. 2009; Ivanov et al. 2009), although one report showed increased T_H17 population in the colon due to decreased IL-25 production by colonic epithelial cells (Zaph et al. 2008). Thus, intestinal commensal bacteria may differentially influence gut immune cells, such as T_H17 cells, γδ T cells, and NK cells, causing various effects on the intestinal host immune responses.

These findings indicate that T_H17 cells and their related cytokines play critical roles in host defense against extracellular pathogens at epithelial and mucosal tissues such as the skin, lung, and intestine. T_H17 cells also appear to play important roles in humans. Subjects with mutations in STAT3, which is critical for T_H17 differentiation, often suffer from fungal and extracellular pathogen infections such as *Candida albicans* and *Staphylococcus aureus* in the skin and lung (Milner et al. 2008). T cells from these subjects fail to produce IL-17A, while IL-2, TNF, and IFN-γ production are normal (Milner et al. 2008). Similarly, *Il17ra*$^{-/-}$ or *Il17a*$^{-/-}$*Il17f*$^{-/-}$ mice are highly susceptible to opportunistic *S. aureus* infection (Schwarzenberger and Kolls 2002; Ishigame et al. 2009), whereas *Il17f*$^{-/-}$ and *Il17a*$^{-/-}$ mice show normal sensitivity to this pathogen (Ishigame et al. 2009). These results suggest that the increased susceptibility of these subjects to infection is at least partially due to impaired T_H17 cell differentiation and function, and that IL-17A and IL-17F complement each other in this setting.

Several lines of evidence have suggested that the IL-23-IL-17A axis is also required for host defense against intracellular pathogens in mice. Mice lacking both IL-12 and IL-23 are more susceptible to *M. tuberculosis* and *S. enteritidis* infections than mice lacking IL-12 alone (Holscher et al. 2001; Lehmann et al. 2001; Lieberman et al. 2004). Similar results are also reported in protozoa infection, such as *Toxoplasma gondii* (Kelly et al. 2005). *Il17ra*$^{-/-}$, *Il17a*$^{-/-}$, and *Il23a*$^{-/-}$ mice do not show increased susceptibility to *M. tuberculosis* infection (Khader et al. 2005; Umemura et al. 2007; Aujla et al. 2008) but IL-17RA signaling is required for the recruitment of T_H1 cells to the site of infection to induce a recall response against *M. tuberculosis* following vaccination (Khader et al. 2007). Thus, T_H1 cells, rather than T_H17 cells play a more important role in host protection against intracellular bacteria. In contrast to these findings, it has been reported that the IL-23/IL-17A pathway is required for host resistance to the intracellular pathogen *Francisella tularensis* by inducing T_H1-type immune responses (Lin et al. 2009). Impaired IFN-γ production is also observed when *Il17a*$^{-/-}$ mice are infected with *Mycobacterium bovis* bacilli Calmette-Guerin (BCG) (Umemura et al. 2007), indicating that under certain conditions, IL-23/IL-17A pathway-dependent induction of T_H1 immune responses is essential for effective clearance of intracellular bacteria. Furthermore, IL-17A and IL-17F produced by γδ T cells are also involved in innate immune response against *L. monocytegenes* in the liver. IL-17A and IL-17F are mainly produced by γδ T cells at early stages of infection and deficiency of IL-17A, IL-17RA, or IL-23 results in increased bacterial burden (Hamada et al. 2008; Meeks et al. 2009). Increased IL-17A production from γδT cells is also observed when mice are infected with *M. tuberculosis*, *M. bovis* BCG, or *S. enteritidis* (Lockhart et al. 2006; Umemura et al. 2007; Siegemund et al. 2009). Further studies are required to elucidate the relative contribution of $CD4^+$ T cells and γδ T cells in IL-17A/IL-17F-mediated host protective immunity against different bacteria.

5.3 *Fungal Infection*

Both IL-17A and IL-17F are also induced by β-glucans, the components of yeast, fungus, and mushroom cell walls, through a TLR-independent pathway. Dectin-1 is a C type lectin receptor that is widely expressed on myeloid cells such as DCs and macrophages. Dectin-1 recognizes β-glucans in zymosans and fungi and plays an important role in host defense against fungi (Saijo et al. 2007; Taylor et al. 2007). Dectin-1 activates the Syc-CARD9 pathway through the ITAM in the cytoplasmic domain causing induction of IL-23, TGF-β, and IL-6, but little IL-12, which preferentially promotes T_H17 cell differentiation (LeibundGut-Landmann et al. 2007). Dectin-2 is another C type lectin receptor for fungi and also contributes to fungus-induced IL-17A production (Robinson et al. 2009). Unlike dectin-1, interaction of Dectin-2 with FcRγ is required for the activation of Syk-CARD9 complex because Dectin-2 has no signaling motif in the cytoplasmic domain (Robinson et al. 2009). It is reported that candida mannan also induce T_H17 response via mannose receptor (van de Veerdonk et al. 2009), suggesting that these receptors induce T_H17 response in cooperation during fungal infections. Fungal zymosans in incomplete Freund's adjuvant (IFA) show potent adjuvant activity in EAE induction upon immunization with MOG peptide (Veldhoen et al. 2006). Furthermore, β-glucans derived from *C. albicans* act as an adjuvant in CIA (Hida et al. 2005) and fungal infection causes the development of arthritis in SKG mice in which T_H17 cells play an important role (Yoshitomi et al. 2005).

A protective role for the IL-23/IL-17A pathway in fungal infections is suggested by the finding that mice lacking IL-12/IL-23 are more susceptible to systemic *Cryptococcus neoformans* infection than mice lacking IL-12 (Decken et al. 1998). An IL-23 deficiency results in increased susceptibility to *C. neoformans* infection, although IL-12 plays a more important role (Kleinschek et al. 2006). The involvement of IL-17R signaling is also evident in systemic *C. albicans* infection (Huang et al. 2004). *Il23a*$^{-/-}$ and *Il17ra*$^{-/-}$, but not *Il12a*$^{-/-}$ mice, are also highly susceptible to oral candidiasis due to defective neutrophil recruitment and antimicrobial peptide induction (Conti et al. 2009). As *Il22*$^{-/-}$ mice are only mildly susceptible to oral candidiasis, it is suggested that IL-17A and IL-17F, rather than IL-22, is important for the host defense in this model.

Although IL-23/IL-17A is required for host defense against some fungi (Fig. 2), dysregulated production of these cytokines induces tissue damage in infected tissues. It has been reported that the IL-23/IL-17A pathway promotes inflammation and susceptibility in gastric *C. albicans* and pulmonary *A. fumigatus* infections. Although IL-17R signaling is critical for systemic and oral *C. albicans* infection, both IL-17A and IL-23 impair the antifungal activities of neutrophils by negatively regulating IFN-γ-mediated induction of indoleamine 2, 3-dioxygenase (IDO) (Zelante et al. 2007; Romani et al. 2008), which has potent regulatory effects on inflammatory and T cell responses. Similarly, in chronic *T. gondii* infection, a deficiency in IL-27, which negatively regulates T_H17 cell differentiation, caused severe neuro-inflammation associated with an increased number of T_H17 cells (Stumhofer et al. 2006). Thus, IL-23 and IL-17A function in both a protective and detrimental manner, depending on the pathogen and infection conditions.

6 Concluding Remarks

Clinical studies showed that blockade of IL-12/IL-23 (p40), IL-1, IL-6 or TNF activity is effective in the treatment of inflammatory diseases, such as RA, MS, inflammatory bowel diseases, and psoriasis. As these cytokines are involved in the development of T_H17 cells, neutralization of T_H17 cell-related cytokine activity may be an attractive strategy for the treatment of these diseases in humans. However, these treatments may increase the risk of opportunistic infections, because both IL-17A and IL-17F are involved in mucosal host defense. As most of available data suggests that IL-17A is more important mediator than IL-17F in allergic and auto-immune diseases, specific neutralization of IL-17A is an attractive treatment of inflammatory diseases without compromising host defense activity.

Accumulating evidence suggests that at least three independent effector T cell pathways are involved in inflammatory responses: IL-12/T_H1, IL-4/T_H2, and IL-23/T_H17. Identifying the major immune pathways responsible for the development of each disease is important for treatment because suppression of one pathway may accelerate the others. Understanding these cytokine networks will lead to the development of more effective treatment of allergic and auto-immune diseases.

Acknowledgement This work is supported by Grants in Aid from the Ministry of Education, Culture, Sports, Science and Technology of Japan and Core Research for Evolutional Science and Technology of the Japan.

References

Aggarwal S and Gurney AL (2002) IL-17: prototype member of an emerging cytokine family. J Leukoc Biol 71:1–8

Akahira-Azuma M, Szczepanik M, Tsuji RF et al. (2004) Early delayed-type hypersensitivity eosinophil infiltrates depend on T helper 2 cytokines and interferon-gamma via CXCR3 chemokines. Immunology 111:306–317

Albanesi C, Cavani A and Girolomoni G (1999) IL-17 is produced by nickel-specific T lymphocytes and regulates ICAM-1 expression and chemokine production in human keratinocytes: synergistic or antagonist effects with IFN-gamma and TNF-alpha. J Immunol 162:494–502

Alonzi T, Fattori E, Lazzaro D et al. (1998) Interleukin 6 is required for the development of collagen-induced arthritis. J Exp Med 187:461–468

Amin K, Ludviksdottir D, Janson C et al. (2000) Inflammation and structural changes in the airways of patients with atopic and non-atopic asthma. BHR Group. Am J Respir Crit Care Med 162:2295–2301

Annunziato F and Romagnani S (2009) Heterogeneity of human effector CD4$^+$ T cells. Arthritis Res Ther 11:257

Antonysamy MA, Fanslow WC, Fu F et al. (1999) Evidence for a role of IL-17 in organ allograft rejection: IL-17 promotes the functional differentiation of dendritic cell progenitors. J Immunol 162:577–584

Aujla SJ, Chan YR, Zheng M et al. (2008) IL-22 mediates mucosal host defense against Gram-negative bacterial pneumonia. Nat Med 14:275–281

Ben-Sasson SZ, Hu-Li J, Quiel J et al. (2009) IL-1 acts directly on CD4 T cells to enhance their antigen-driven expansion and differentiation. Proc Natl Acad Sci USA 106:7119–7124

Bending D, De La Pena H, Veldhoen M et al. (2009) Highly purified Th17 cells from BDC2.5NOD mice convert into Th1-like cells in NOD/SCID recipient mice. J Clin Invest 119:565–572

Berg DJ, Leach MW, Kuhn R et al. (1995) Interleukin 10 but not interleukin 4 is a natural suppressant of cutaneous inflammatory responses. J Exp Med 182:99–108

Chung Y, Chang SH, Martinez GJ et al. (2009) Critical regulation of early Th17 cell differentiation by interleukin-1 signaling. Immunity 30:576–587

Ciric B, El-behi M, Cabrera R et al. (2009) IL-23 drives pathogenic IL-17-producing CD8$^+$ T cells. J Immunol 182:5296–5305

Conti HR, Shen F, Nayyar N et al. (2009) Th17 cells and IL-17 receptor signaling are essential for mucosal host defense against oral candidiasis. J Exp Med 206:299–311

Coquet JM, Chakravarti S, Smyth MJ et al. (2008) Cutting edge: IL-21 is not essential for Th17 differentiation or experimental auto-immune encephalomyelitis. J Immunol 180:7097–7101

Cua DJ, Sherlock J, Chen Y et al. (2003) Interleukin-23 rather than interleukin-12 is the critical cytokine for auto-immune inflammation of the brain. Nature 421:744–748

Decken K, Kohler G, Palmer-Lehmann K et al. (1998) Interleukin-12 is essential for a protective Th1 response in mice infected with Cryptococcus neoformans. Infect Immun 66:4994–5000

Dieli F, Sireci G, Scire E et al. (1999) Impaired contact hypersensitivity to trinitrochlorobenzene in interleukin-4-deficient mice. Immunology 98:71–79

Dong C (2008) T_H17 cells in development: an updated view of their molecular identity and genetic programming. Nat Rev Immunol 8:337–348

Emamaullee JA, Davis J, Merani S et al. (2009) Inhibition of Th17 cells regulates auto-immune diabetes in NOD mice. Diabetes 58:1302–1311

Feldmann M, Brennan FM and Maini RN (1996) Role of cytokines in rheumatoid arthritis. Annu Rev Immunol 14:397–440

Feldmann M and Maini SR (2008) Role of cytokines in rheumatoid arthritis: an education in pathophysiology and therapeutics. Immunol Rev 223:7–19

Ferber IA, Brocke S, Taylor-Edwards C et al. (1996) Mice with a disrupted IFN-gamma gene are susceptible to the induction of experimental auto-immune encephalomyelitis (EAE). J Immunol 156:5–7

Ferretti S, Bonneau O, Dubois GR et al. (2003) IL-17, produced by lymphocytes and neutrophils, is necessary for lipopolysaccharide-induced airway neutrophilia: IL-15 as a possible trigger. J Immunol 170:2106–2112

Feuerer M, Eulenburg K, Loddenkemper C et al. (2006) Self-limitation of Th1-mediated inflammation by IFN-gamma. J Immunol 176:2857–2863

Fina D, Sarra M, Fantini MC et al. (2008) Regulation of gut inflammation and th17 cell response by interleukin-21. Gastroenterology 134:1038–1048

Fujiwara M, Hirose K, Kagami S et al. (2007) T-bet inhibits both TH2 cell-mediated eosinophil recruitment and TH17 cell-mediated neutrophil recruitment into the airways. J Allergy Clin Immunol 119:662–670

Gaboriau-Routhiau V, Rakotobe S, Lecuyer E et al. (2009) The key role of segmented filamentous bacteria in the coordinated maturation of gut helper T cell responses. Immunity 31:677–689

Gaffen SL (2009) Structure and signaling in the IL-17 receptor family. Nat Rev Immunol 9:556–567

Gao D, Kasten-Jolly J and Lawrence DA (2006) The paradoxical effects of lead in interferon-gamma knockout BALB/c mice. Toxicol Sci 89:444–453

Gately MK, Renzetti LM, Magram J et al. (1998) The interleukin-12/interleukin-12-receptor system: role in normal and pathologic immune responses. Annu Rev Immunol 16:495–521

Geboes L, Dumoutier L, Kelchtermans H et al. (2009) Pro-inflammatory role of the Th17 cytokine interleukin-22 in collagen-induced arthritis in C57BL/6 mice. Arthritis Rheum 60:390–395

Gerosa F, Baldani-Guerra B, Lyakh LA et al. (2008) Differential regulation of interleukin 12 and interleukin 23 production in human dendritic cells. J Exp Med 205:1447–1461

Ghilardi N, Kljavin N, Chen Q et al. (2004) Compromised humoral and delayed-type hypersensitivity responses in IL-23-deficient mice. J Immunol 172:2827–2833

Gomez-Rodriguez J, Sahu N, Handon R et al. (2009) Differential expression of interleukin-17A and -17F is coupled to T cell receptor signaling via inducible T cell kinase. Immunity 31:587–597

Grabbe S and Schwarz T (1998) Immuno-regulatory mechanisms involved in elicitation of allergic contact hypersensitivity. Immunol Today 19:37–44

Gran B, Zhang GX, Yu S et al. (2002) IL-12p35-deficient mice are susceptible to experimental auto-immune encephalomyelitis: evidence for redundancy in the IL-12 system in the induction of central nervous system auto-immune demyelination. J Immunol 169:7104–7110

Guedez YB, Whittington KB, Clayton JL et al. (2001) Genetic ablation of interferon-gamma up-regulates interleukin-1beta expression and enables the elicitation of collagen-induced arthritis in a nonsusceptible mouse strain. Arthritis Rheum 44:2413–2424

Haak S, Croxford AL, Kreymborg K et al. (2009) IL-17A and IL-17F do not contribute vitally to auto-immune neuro-inflammation in mice. J Clin Invest 119:61–69

Hamada S, Umemura M, Shiono T et al. (2008) IL-17A produced by gammadelta T cells plays a critical role in innate immunity against listeria monocytogenes infection in the liver. J Immunol 181:3456–3463

Happel KI, Dubin PJ, Zheng M et al. (2005) Divergent roles of IL-23 and IL-12 in host defense against Klebsiella pneumoniae. J Exp Med 202:761–769

Hata H, Sakaguchi N, Yoshitomi H et al. (2004) Distinct contribution of IL-6, TNF-alpha, IL-1, and IL-10 to T cell-mediated spontaneous auto-immune arthritis in mice. J Clin Invest 114:582–588

He D, Wu L, Kim HK et al. (2006) $CD8^+$ IL-17-producing T cells are important in effector functions for the elicitation of contact hypersensitivity responses. J Immunol 177:6852–6858

He D, Wu L, Kim HK et al. (2009) IL-17 and IFN-gamma mediate the elicitation of contact hypersensitivity responses by different mechanisms and both are required for optimal responses. J Immunol 183:1463–1470

He R, Oyoshi MK, Jin H et al. (2007) Epicutaneous antigen exposure induces a Th17 response that drives airway inflammation after inhalation challenge. Proc Natl Acad Sci USA 104: 15817–15822

Herrick CA, Xu L, McKenzie AN et al. (2003) IL-13 is necessary, not simply sufficient, for epicutaneously induced Th2 responses to soluble protein antigen. J Immunol 170:2488–2495

Hida S, Miura NN, Adachi Y et al. (2005) Effect of *Candida albicans* cell wall glucan as adjuvant for induction of auto-immune arthritis in mice. J Autoimmun 25:93–101

Hirota K, Hashimoto M, Yoshitomi H et al. (2007a) T cell self-reactivity forms a cytokine milieu for spontaneous development of IL-17$^+$ Th cells that cause auto-immune arthritis. J Exp Med 204:41–47

Hirota K, Yoshitomi H, Hashimoto M et al. (2007b) Preferential recruitment of CCR6-expressing Th17 cells to inflamed joints via CCL20 in rheumatoid arthritis and its animal model. J Exp Med 204:2803–2812

Holscher C, Atkinson RA, Arendse B et al. (2001) A protective and agonistic function of IL-12p40 in mycobacterial infection. J Immunol 167:6957–6966

Horai R, Nakajima A, Habiro K et al. (2004) TNF-alpha is crucial for the development of auto-immune arthritis in IL-1 receptor antagonist-deficient mice. J Clin Invest 114:1603–1611

Horai R, Saijo S, Tanioka H et al. (2000) Development of chronic inflammatory arthropathy resembling rheumatoid arthritis in interleukin 1 receptor antagonist-deficient mice. J Exp Med 191:313–320

Hoshino A, Nagao T, Nagi-Miura N et al. (2008) MPO-ANCA induces IL-17 production by activated neutrophils in vitro via its Fc region- and complement-dependent manner. J Autoimmune 31:79–89

Huang W, Na L, Fidel PL et al. (2004) Requirement of interleukin-17A for systemic anti-*Candida albicans* host defense in mice. J Infect Dis 190:624–631

Hue S, Ahern P, Buonocore S et al. (2006) Interleukin-23 drives innate and T cell-mediated intestinal inflammation. J Exp Med 203:2473–2483

Hultgren B, Huang X, Dybdal N et al. (1996) Genetic absence of gamma-interferon delays but does not prevent diabetes in NOD mice. Diabetes 45:812–817

Hymowitz SG, Filvaroff EH, Yin JP et al. (2001) IL-17s adopt a cystine knot fold: structure and activity of a novel cytokine, IL-17F, and implications for receptor binding. EMBO J 20:5332–5341

Irmler IM, Gajda M and Brauer R (2007) Exacerbation of antigen-induced arthritis in IFN-gamma-deficient mice as a result of unrestricted IL-17 response. J Immunol 179:6228–6236

Ishigame H, Kakuta S, Nagai T et al. (2009) Differential roles of interleukin-17A and -17F in host defense against mucoepithelial bacterial infection and allergic responses. Immunity 30:108–119

Ishihara K, Sawa S, Ikushima H et al. (2004) The point mutation of tyrosine 759 of the IL-6 family cytokine receptor gp130 synergizes with HTLV-1 pX in promoting rheumatoid arthritis-like arthritis. Int Immunol 16:455–465

Ito H and Fathman CG (1997) CD45RB^{high} CD4$^+$ T cells from IFN-gamma knockout mice do not induce wasting disease. J Auto-immune 10:455–459

Ito R, Kita M, Shin-Ya M et al. (2008) Involvement of IL-17A in the pathogenesis of DSS-induced colitis in mice. Biochem Biophys Res Commun 377:12–16

Ivanov, II, Atarashi K, Manel N et al. (2009) Induction of intestinal Th17 cells by segmented filamentous bacteria. Cell 139:485–498

Ivanov, II, McKenzie BS, Zhou L et al. (2006) The orphan nuclear receptor RORgammat directs the differentiation program of pro-inflammatory IL-17$^+$ T helper cells. Cell 126:1121–1133

Iwakura Y, Nakae S, Saijo S et al. (2008) The roles of IL-17A in inflammatory immune responses and host defense against pathogens. Immunol Rev 226:57–79

Iwakura Y, Saijo S, Kioka Y et al. (1995) Auto-immunity induction by human T cell leukemia virus type 1 in transgenic mice that develop chronic inflammatory arthropathy resembling rheumatoid arthritis in humans. J Immunol 155:1588–1598.

Iwakura Y, Tosu M, Yoshida E et al. (1991) Induction of inflammatory arthropathy resembling rheumatoid arthritis in mice transgenic for HTLV-I. Science 253:1026–1028

Izcue A, Hue S, Buonocore S et al. (2008) Interleukin-23 restrains regulatory T cell activity to drive T cell-dependent colitis. Immunity 28:559–570

Jacobs JP, Wu HJ, Benoist C et al. (2009) IL-17-producing T cells can augment auto-antibody induced arthritis. Proc Natl Acad Sci USA 106:21789–21794

Jain R, Tartar DM, Gregg RK et al. (2008) Innocuous IFNgamma induced by adjuvant-free antigen restores normoglycemia in NOD mice through inhibition of IL-17 production. J Exp Med 205:207–218

Jensen KD, Su X, Shin S et al. (2008) Thymic selection determines gammadelta T cell effector fate: antigen-naive cells make interleukin-17 and antigen-experienced cells make interferon gamma. Immunity 29:90–100

Ji H, Ohmura K, Mahmood U et al. (2002a) Arthritis critically dependent on innate immune system players. Immunity 16:157–168

Ji H, Pettit A, Ohmura K et al. (2002b) Critical roles for interleukin 1 and tumor necrosis factor alpha in antibody-induced arthritis. J Exp Med 196:77–85

Kanagawa O, Xu G, Tevaarwerk A et al. (2000) Protection of nonobese diabetic mice from diabetes by gene(s) closely linked to IFN-gamma receptor loci. J Immunol 164:3919–3923

Katz JD, Benoist C and Mathis D (1995) T helper cell subsets in insulin-dependent diabetes. Science 268:1185–1188

Kebir H, Kreymborg K, Ifergan I et al. (2007) Human T_H17 lymphocytes promote blood-brain barrier disruption and central nervous system inflammation. Nat Med 13:1173–1175

Kelly MN, Kolls JK, Happel K et al. (2005) Interleukin-17/interleukin-17 receptor-mediated signaling is important for generation of an optimal polymorphonuclear response against *Toxoplasma gondii* infection. Infect Immun 73:617–621

Khader SA, Bell GK, Pearl JE et al. (2007) IL-23 and IL-17 in the establishment of protective pulmonary $CD4^+$ T cell responses after vaccination and during *Mycobacterium tuberculosis* challenge. Nat Immunol 8:369–377

Khader SA, Gaffen SL and Kolls JK (2009) Th17 cells at the crossroads of innate and adaptive immunity against infectious diseases at the mucosa. Mucosal Immunol 2:403–411

Khader SA, Pearl JE, Sakamoto K et al. (2005) IL-23 compensates for the absence of IL-12p70 and is essential for the IL-17 response during tuberculosis but is dispensable for protection and antigen-specific IFN-gamma responses if IL-12p70 is available. J Immunol 175:788–795

Kimber I and Dearman RJ (2002) Allergic contact dermatitis: the cellular effectors. Contact Dermatitis 46:1–5

Kiss A, Montes M, Susarla S et al. (2007) A new mechanism regulating the initiation of allergic airway inflammation. J Allergy Clin Immunol 120:334–342

Kleinschek MA, Muller U, Brodie SJ et al. (2006) IL-23 enhances the inflammatory cell response in *Cryptococcus neoformans* infection and induces a cytokine pattern distinct from IL-12. J Immunol 176:1098–1106

Knott PG, Gater PR and Bertrand CP (2000) Airway inflammation driven by antigen-specific resident lung $CD4^+$ T cells in alphabeta-T cell receptor transgenic mice. Am J Respir Crit Care Med 161:1340–1348

Kolls JK and Linden A (2004) Interleukin-17 family members and inflammation. Immunity 21:467–476

Komiyama Y, Nakae S, Matsuki T et al. (2006) IL-17 plays an important role in the development of experimental auto-immune encephalomyelitis. J Immunol 177:566–573

Korganow AS, Ji H, Mangialaio S et al. (1999) From systemic T cell self-reactivity to organ specific auto-immune disease via immunoglobulins. Immunity 10:451–461

Korn T, Bettelli E, Gao W et al. (2007) IL-21 initiates an alternative pathway to induce proinflammatory T_H17 cells. Nature 448:484–487

Krakowski M and Owens T (1996) Interferon gamma confers resistance to experimental allergic encephalomyelitis. Eur J Immunol 26:1641–1646

Kramer JM, Yi L, Shen F et al. (2006) Evidence for ligand-independent multimerization of the IL-17 receptor. J Immunol 176:711–715

Kreymborg K, Etzensperger R, Dumoutier L et al. (2007) IL-22 is expressed by Th17 cells in an IL-23-dependent fashion, but not required for the development of auto-immune encephalomyelitis. J Immunol 179:8098–8104

Kroenke MA, Carlson TJ, Andjelkovic AV et al. (2008) IL-12- and IL-23-modulated T cells induce distinct types of EAE based on histology, CNS chemokine profile, and response to cytokine inhibition. J Exp Med 205:1535–1541

Kuchroo VK, Anderson AC, Waldner H et al. (2002) T cell response in experimental auto-immune encephalomyelitis (EAE): role of self and cross-reactive antigens in shaping, tuning, and regulating the auto-pathogenic T cell repertoire. Annu Rev Immunol 20:101–123

Kuestner RE, Taft DW, Haran A et al. (2007) Identification of the IL-17 receptor related molecule IL-17RC as the receptor for IL-17F. J Immunol 179:5462–5473

Kullberg MC, Jankovic D, Feng CG et al. (2006) IL-23 plays a key role in *Helicobacter hepaticus*-induced T cell-dependent colitis. J Exp Med 203:2485–2494

Lee YK, Turner H, Maynard CL et al. (2009) Late developmental plasticity in the T helper 17 lineage. Immunity 30:92–107

Lees JR, Golumbek PT, Sim J et al. (2008) Regional CNS responses to IFN-gamma determine lesion localization patterns during EAE pathogenesis. J Exp Med 205:2633–2642

Lehmann J, Bellmann S, Werner C et al. (2001) IL-12p40-dependent agonistic effects on the development of protective innate and adaptive immunity against Salmonella enteritidis. J Immunol 167:5304–5315

LeibundGut-Landmann S, Gross O, Robinson MJ et al. (2007) Syk- and CARD9-dependent coupling of innate immunity to the induction of T helper cells that produce interleukin 17. Nat Immunol 8:630–638

Leppkes M, Becker C, Ivanov, II et al. (2009) RORgamma-expressing Th17 cells induce murine chronic intestinal inflammation via redundant effects of IL-17A and IL-17F. Gastroenterology 136:257–267

Li X (2008) Act1 modulates auto-immunity through its dual functions in CD40L/BAFF and IL-17 signaling. Cytokine 41:105–113

Liang SC, Long AJ, Bennett F et al. (2007) An IL-17F/A heterodimer protein is produced by mouse Th17 cells and induces airway neutrophil recruitment. J Immunol 179:7791–7799

Lieberman LA, Cardillo F, Owyang AM et al. (2004) IL-23 provides a limited mechanism of resistance to acute toxoplasmosis in the absence of IL-12. J Immunol 173:1887–1893

Lin Y, Ritchea S, Logar A et al. (2009) Interleukin-17 is required for T helper 1 cell immunity and host resistance to the intracellular pathogen *Francisella tularensis*. Immunity 31:799–810

Liu R, Bai Y, Vollmer TL et al. (2008) IL-21 receptor expression determines the temporal phases of experimental auto-immune encephalomyelitis. Exp Neurol 211:14–24

Lochner M, Peduto L, Cherrier M et al. (2008) In vivo equilibrium of pro-inflammatory IL-17$^+$ and regulatory IL-10$^+$ Foxp3$^+$ RORgamma t$^+$ T cells. J Exp Med 205:1381–1393

Lockhart E, Green AM and Flynn JL (2006) IL-17 production is dominated by gammadelta T cells rather than CD4 T cells during *Mycobacterium tuberculosis* infection. J Immunol 177:4662–4669

Lu B, Ebensperger C, Dembic Z et al. (1998) Targeted disruption of the interferon gamma receptor 2 gene results in severe immune defects in mice. Proc Natl Acad Sci USA 95:8233–8238

Luci C, Reynders A, Ivanov, II et al. (2009) Influence of the transcription factor RORgammat on the development of NKp46$^+$ cell populations in gut and skin. Nat Immunol 10:75–82

Mangan PR, Harrington LE, O'Quinn DB et al. (2006) Transforming growth factor beta induces development of the T_H17 lineage. Nature 441:231–234

Martin-Orozco N, Chung Y, Chang SH et al. (2009) Th17 cells promote pancreatic inflammation but only induce diabetes efficiently in lymphopenic hosts after conversion into Th1 cells. Eur J Immunol 39:216–224

Matsumoto I, Staub A, Benoist C et al. (1999) Arthritis provoked by linked T and B cell recognition of a glycolytic enzyme. Science 286:1732–1735

McGeachy MJ and Cua DJ (2008) Th17 cell differentiation: the long and winding road. Immunity 28:445–453

Meeks KD, Sieve AN, Kolls JK et al. (2009) IL-23 is required for protection against systemic infection with Listeria monocytogenes. J Immunol 183:8026–8034

Milner JD, Brenchley JM, Laurence A et al. (2008) Impaired T_H17 cell differentiation in subjects with autosomal dominant hyper-IgE syndrome. Nature 452:773–776

Murphy CA, Langrish CL, Chen Y et al. (2003) Divergent pro and anti-inflammatory roles for IL-23 and IL-12 in joint auto-immune inflammation. J Exp Med 198:1951–1957

Nakae S, Komiyama Y, Nambu A et al. (2002) Antigen specific T cell sensitization is impaired in IL-17-deficient mice, causing suppression of allergic cellular and humoral responses. Immunity 17:375–387

Nakae S, Komiyama Y, Narumi S et al. (2003a) IL-1-induced tumor necrosis factor alpha elicits inflammatory cell infiltration in the skin by inducing IFN-gamma inducible protein 10 in the elicitation phase of the contact hypersensitivity response. Int Immunol 15:251–260

Nakae S, Nambu A, Sudo K et al. (2003b) Suppression of immune induction of collagen induced arthritis in IL-17-deficient mice. J Immunol 171:6173–6177

Nakae S, Saijo S, Horai R et al. (2003c) IL-17 production from activated T cells is required for the spontaneous development of destructive arthritis in mice deficient in IL-1 receptor antagonist. Proc Natl Acad Sci U S A 100:5986–5990

Nakae S, Suto H, Berry GJ et al. (2007) Mast cell-derived TNF can promote Th17 cell-dependent neutrophil recruitment in ovalbumin-challenged OTII mice. Blood 109:3640–3648

Napolitani G, Rinaldi A, Bertoni F et al. (2005) Selected Toll-like receptor agonist combinations synergistically trigger a T helper type 1 polarizing program in dendritic cells. Nat Immunol 6:769–776

Niess JH, Leithauser F, Adler G et al. (2008) Commensal gut flora drives the expansion of proinflammatory CD4 T cells in the colonic lamina propria under normal and inflammatory conditions. J Immunol 180:559–568

Nigrovic PA, Binstadt BA, Monach PA et al. (2007) Mast cells contribute to initiation of auto-antibody mediated arthritis via IL-1. Proc Natl Acad Sci USA 104:2325–2330

Noguchi D, Wakita D, Tajima M et al. (2007) Blocking of IL-6 signaling pathway prevents $CD4^+$ T cell-mediated colitis in a T_h17-independent manner. Int Immunol 19:1431–1440

Nowak EC, Weaver CT, Turner H et al. (2009) IL-9 as a mediator of Th17-driven inflammatory disease. J Exp Med 206:1653–1660

Nurieva R, Yang XO, Martinez G et al. (2007) Essential autocrine regulation by IL-21 in the generation of inflammatory T cells. Nature 448:480–483

O'Connor W, Jr., Kamanaka M, Booth CJ et al. (2009) A protective function for interleukin 17A in T cell-mediated intestinal inflammation. Nat Immunol 10:603–609

Oboki K, Ohno T, Saito H et al. (2008) Th17 and Allergy. Allergol Int 57:121–134

Oda N, Canelos PB, Essayan DM et al. (2005) Interleukin-17F induces pulmonary neutrophilia and amplifies antigen-induced allergic response. Am J Respir Crit Care Med 171:12–18

Ogawa A, Andoh A, Araki Y et al. (2004) Neutralization of interleukin-17 aggravates dextran sulfate sodium-induced colitis in mice. Clin Immunol 110:55–62

Park H, Li Z, Yang XO et al. (2005) A distinct lineage of CD4 T cells regulates tissue inflammation by producing interleukin 17. Nat Immunol 6:1133–1141

Pichavant M, Goya S, Meyer EH et al. (2008) Ozone exposure in a mouse model induces airway hyperreactivity that requires the presence of natural killer T cells and IL-17. J Exp Med 205:385–393

Rachitskaya AV, Hansen AM, Horai R et al. (2008) Cutting edge: NKT cells constitutively express IL-23 receptor and RORgammat and rapidly produce IL-17 upon receptor ligation in an IL-6-independent fashion. J Immunol 180:5167–5171

Reeve VE, Bosnic M and Nishimura N (1999) Interferon-gamma is involved in photo immunoprotection by UVA (320–400 nm) radiation in mice. J Invest Dermatol 112:945–950

Rickel EA, Siegel LA, Yoon BR et al. (2008) Identification of functional roles for both IL-17RB and IL-17RA in mediating IL-25-induced activities. J Immunol 181:4299–4310

Roark CL, French JD, Taylor MA et al. (2007) Exacerbation of collagen-induced arthritis by oligoclonal, IL-17-producing gamma delta T cells. J Immunol 179:5576–5583

Robinson MJ, Osorio F, Rosas M et al. (2009) Dectin-2 is a Syk coupled pattern recognition receptor crucial for Th17 responses to fungal infection. J Exp Med 206:2037–2051

Romani L, Fallarino F, De Luca A et al. (2008) Defective tryptophan catabolism underlies inflammation in mouse chronic granulomatous disease. Nature 451:211–215

Saijo S, Asano M, Horai R et al. (2002) Suppression of auto-immune arthritis in interleukin-1-deficient mice in which T cell activation is impaired due to low levels of CD40 ligand and OX40 expression on T cells. Arthritis Rheum 46:533–544

Saijo S, Fujikado N, Furuta T et al. (2007) Dectin-1 is required for host defense against *Pneumocystis carinii* but not against *Candida albicans*. Nat Immunol 8:39–46

Sanos SL, Bui VL, Mortha A et al. (2009) RORgammat and commensal microflora are required for the differentiation of mucosal interleukin 22-producing NKp46+ cells. Nat Immunol 10:83–91

Sato K, Suematsu A, Okamoto K et al. (2006) Th17 functions as an osteoclastogenic helper T cell subset that links T cell activation and bone destruction. J Exp Med 203:2673–2682

Satoh-Takayama N, Vosshenrich CA, Lesjean-Pottier S et al. (2008) Microbial flora drives interleukin 22 production in intestinal NKp46+ cells that provide innate mucosal immune defense. Immunity 29:958–970

Saulnier M, Huang S, Aguet M et al. (1995) Role of interferon-gamma in contact hypersensitivity assessed in interferon-gamma receptor-deficient mice. Toxicology 102:301–312

Schnyder-Candrian S, Togbe D, Couillin I et al. (2006) Interleukin-17 is a negative regulator of established allergic asthma. J Exp Med 203:2715–2725

Schwarzenberger P and Kolls JK (2002) Interleukin 17: an example for gene therapy as a tool to study cytokine mediated regulation of hematopoiesis. J Cell Biochem Suppl 38:88–95

Segal BM, Chang JT and Shevach EM (2000) CpG oligonucleotides are potent adjuvants for the activation of auto-reactive encephalitogenic T cells in vivo. J Immunol 164:5683–5688

Serreze DV, Post CM, Chapman HD et al. (2000) Interferon-gamma receptor signaling is dispensable in the development of auto-immune type 1 diabetes in NOD mice. Diabetes 49:2007–2011

Shibata K, Yamada H, Hara H et al. (2007) Resident Vdelta1+ gammadelta T cells control early infiltration of neutrophils after Escherichia coli infection via IL-17 production. J Immunol 178:4466–4472

Shtrichman R and Samuel CE (2001) The role of gamma interferon in antimicrobial immunity. Curr Opin Microbiol 4:251–259

Siegemund S, Schutze N, Schulz S et al. (2009) Differential IL-23 requirement for IL-22 and IL-17A production during innate immunity against *Salmonella enterica* serovar Enteritidis. Int Immunol 21:555–565

Spolski R, Kashyap M, Robinson C et al. (2008) IL-21 signaling is critical for the development of type I diabetes in the NOD mouse. Proc Natl Acad Sci USA 105:14028–14033

Starnes T, Robertson MJ, Sledge G et al. (2001) Cutting edge: IL-17F, a novel cytokine selectively expressed in activated T cells and monocytes, regulates angiogenesis and endothelial cell cytokine production. J Immunol 167:4137–4140

Stromnes IM, Cerretti LM, Liggitt D et al. (2008) Differential regulation of central nervous system auto-immunity by T_H1 and T_H17 cells. Nat Med 14:337–342

Stumhofer JS, Laurence A, Wilson EH et al. (2006) Interleukin 27 negatively regulates the development of interleukin 17-producing T helper cells during chronic inflammation of the central nervous system. Nat Immunol 7:937–945

Sutherland AP, Van Belle T, Wurster AL et al. (2009) Interleukin-21 is required for the development of type 1 diabetes in NOD mice. Diabetes 58:1144–1155

Sutton CE, Lalor SJ, Sweeney CM et al. (2009) Interleukin-1 and IL-23 induce innate IL-17 production from gammadelta T cells, amplifying Th17 responses and auto-immunity. Immunity 31:331–341

Suzuki A, Hanada T, Mitsuyama K et al. (2001) CIS3/SOCS3/SSI3 plays a negative regulatory role in STAT3 activation and intestinal inflammation. J Exp Med 193:471–481

Takatori H, Kanno Y, Watford WT et al. (2009) Lymphoid tissue inducer-like cells are an innate source of IL-17 and IL-22. J Exp Med 206:35–41

Takeshita K, Yamasaki T, Akira S et al. (2004) Essential role of MHC II-independent CD4+ T cells, IL-4 and STAT6 in contact hypersensitivity induced by fluorescein isothiocyanate in the mouse. Int Immunol 16:685–695

Tanaka S, Yoshimoto T, Naka T et al. (2009) Natural occurring IL-17 producing T cells regulate the initial phase of neutrophil mediated airway responses. J Immunol 183:7523–7530

Taylor PR, Tsoni SV, Willment JA et al. (2007) Dectin-1 is required for beta glucan recognition and control of fungal infection. Nat Immunol 8:31–38

Tesmer LA, Lundy SK, Sarkar S et al. (2008) Th17 cells in human disease. Immunol Rev 223:87–113

Traidl C, Jugert F, Krieg T et al. (1999) Inhibition of allergic contact dermatitis to DNCB but not to oxazolone in interleukin-4-deficient mice. J Invest Dermatol 112:476–482

Uhlig HH, McKenzie BS, Hue S et al. (2006) Differential activity of IL-12 and IL-23 in mucosal and systemic innate immune pathology. Immunity 25:309–318

Umemura M, Yahagi A, Hamada S et al. (2007) IL-17-mediated regulation of innate and acquired immune response against pulmonary *Mycobacterium bovis* bacille Calmette-Guerin infection. J Immunol 178:3786–3796

Van Beelen AJ, Zelinkova Z, Taanman-Kueter EW et al. (2007) Stimulation of the intracellular bacterial sensor NOD2 programs dendritic cells to promote interleukin-17 production in human memory T cells. Immunity 27:660–669

Van de Veerdonk FL, Marijnissen RJ, Kullberg BJ et al. (2009) The macrophage mannose receptor induces IL-17 in response to *Candida albicans*. Cell Host Microbe 5:329–340

Veldhoen M, Hocking RJ, Flavell RA et al. (2006) Signals mediated by transforming growth factor-beta initiate auto-immune encephalomyelitis, but chronic inflammation is needed to sustain disease. Nat Immunol 7:1151–1156

Vermeire K, Heremans H, Vandeputte M et al. (1997) Accelerated collagen-induced arthritis in IFN-gamma receptor-deficient mice. J Immunol 158:5507–5513

Visser L, Jan de Heer H, Boven LA et al. (2005) Pro-inflammatory bacterial peptidoglycan as a co-factor for the development of central nervous system auto-immune disease. J Immunol 174:808–816

Wang J, Fathman JW, Lugo-Villarino G et al. (2006) Transcription factor T-bet regulates inflammatory arthritis through its function in dendritic cells. J Clin Invest 116:414–421

Weigmann B, Schwing J, Huber H et al. (1997) Diminished contact hypersensitivity response in IL-4 deficient mice at a late phase of the elicitation reaction. Scand J Immunol 45:308–314

Wilder JA, Collie DD, Bice DE et al. (2001) Ovalbumin aerosols induce airway hyperreactivity in naive DO11.10 T cell receptor transgenic mice without pulmonary eosinophilia or OVA-specific antibody. J Leukoc Biol 69:538–547

Willenborg DO, Fordham S, Bernard CC et al. (1996) IFN-gamma plays a critical downregulatory role in the induction and effector phase of myelin oligodendrocyte glycoprotein induced auto-immune encephalomyelitis. J Immunol 157:3223–3227

Wright JF, Bennett F, Li B et al. (2008) The human IL-17F/IL-17A heterodimeric cytokine signals through the IL-17RA/IL-17RC receptor complex. J Immunol 181:2799–2805

Wright JF, Guo Y, Quazi A et al. (2007) Identification of an interleukin 17F/17A heterodimer in activated human $CD4^+$ T cells. J Biol Chem 282:13447–13455

Yamaguchi Y, Fujio K, Shoda H et al. (2007) IL-17B and IL-17C are associated with TNF-alpha production and contribute to the exacerbation of inflammatory arthritis. J Immunol 179:7128–7136

Yang XO, Chang SH, Park H et al. (2008a) Regulation of inflammatory responses by IL-17F. J Exp Med 205:1063–1075

Yang XO, Pappu BP, Nurieva R et al. (2008b) T helper 17 lineage differentiation is programmed by orphan nuclear receptors ROR alpha and ROR gamma. Immunity 28:29–39

Yao Z, Fanslow WC, Seldin MF et al. (1995) *Herpes virus Saimiri* encodes a new cytokine, IL-17, which binds to a novel cytokine receptor. Immunity 3:811–821

Ye P, Rodriguez FH, Kanaly S et al. (2001) Requirement of interleukin 17 receptor signaling for lung CXC chemokine and granulocyte colony stimulating factor expression, neutrophil recruitment, and host defense. J Exp Med 194:519–527

Yen D, Cheung J, Scheerens H et al. (2006) IL-23 is essential for T cell mediated colitis and promotes inflammation via IL-17 and IL-6. J Clin Invest 116:1310–1316

Yokozeki H, Ghoreishi M, Takagawa S et al. (2000) Signal transducer and activator of transcription 6 is essential in the induction of contact hypersensitivity. J Exp Med 191:995–1004

Yoshitomi H, Sakaguchi N, Kobayashi K et al. (2005) A role for fungal beta glucans and their receptor Dectin-1 in the induction of auto-immune arthritis in genetically susceptible mice. J Exp Med 201:949–960
Young DA, Hegen M, Ma HL et al. (2007) Blockade of the interleukin-21/interleukin-21 receptor pathway ameliorates disease in animal models of rheumatoid arthritis. Arthritis Rheum 56:1152–1163
Zaph C, Du Y, Saenz SA et al. (2008) Commensal-dependent expression of IL-25 regulates the IL-23-IL-17 axis in the intestine. J Exp Med 205:2191–2198
Zelante T, De Luca A, Bonifazi P et al. (2007) IL-23 and the Th17 pathway promote inflammation and impair antifungal immune resistance. Eur J Immunol 37:2695–2706
Zenewicz LA, Yancopoulos GD, Valenzuela DM et al. (2008) Innate and adaptive interleukin-22 protects mice from inflammatory bowel disease. Immunity 29:947–957
Zhang GX, Gran B, Yu S et al. (2003) Induction of experimental auto-immune encephalomyelitis in IL-12 receptor-beta 2-deficient mice: IL-12 responsiveness is not required in the pathogenesis of inflammatory demyelination in the central nervous system. J Immunol 170:2153–2160
Zheng Y, Valdez PA, Danilenko DM et al. (2008) Interleukin-22 mediates early host defense against attaching and effacing bacterial pathogens. Nat Med 14:282–289
Zhou L, Ivanov, II, Spolski R et al. (2007) IL-6 programs T_H-17 cell differentiation by promoting sequential engagement of the IL-21 and IL-23 pathways. Nat Immunol 8:967–974

T_H17 Cells in Fungal Infections

Teresa Zelante, Antonella De Luca, and Luigina Romani

Abstract Although T_H1 responses driven by the IL-12/IFN-gamma axis are central to protection against fungi, the paradigm has been revisited with two new T cell populations entering the scene: the T_H17 cells involved in inflammatory responses, and the T regulatory cells (Tregs), which minimize immune responses to avoid damage to the host. While many studies have focused on the pathological aspects of IL-17-producing T cells in many auto-immune diseases, their role in protective antimicrobial immunity has also been increasingly recognized. Some degree of inflammation is required for protection, particularly in mucosal tissues during the transitional response occurring between the rapid innate and slower adaptive response. However, progressive inflammation worsens disease, limits protective antifungal immune responses and ultimately prevents pathogen eradication. In this scenario, deregulated activity of T_H17 cells and Tregs in mediating and restraining inflammation may occur. The enzyme indoleamine 2, 3-dioxygenase and tryptophan metabolites crucially contribute to immune homeostasis by limiting T_H17 cell activation and inducing Tregs-taming heightened inflammatory responses. The new findings support a view in which the immune system tailors protective responses to suit infecting fungi while limiting host damage through distinct modules of immunity.

1 Introduction

The kingdom of fungi comprises over 1.5 million fungal species and yet only a handful are associated with a wide spectrum of diseases in humans and animals, ranging from allergy and auto-immunity to life-threatening infections (Fig. 1). Most fungi (such as *Aspergillus* spp.) are ubiquitous in the environment. Some, including *Malassezia* spp.

L. Romani (✉)
Microbiology Section, Department of Experimental Medicine and Biochemical Sciences, University of Perugia, Via del Giochetto, Perugia 06122, Italy
e-mail: lromani@unipg.it

S. Jiang (ed.), *T_H17 Cells in Health and Disease*,
DOI 10.1007/978-1-4419-9371-7_16, © Springer Science+Business Media, LLC 2011

Organism	Source	Morphologies	Disease Spectrum
Candida albicans	Commensal of the skin, GI tract, and vagina	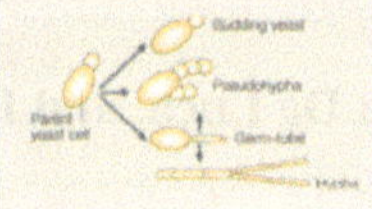	• Hematogenously disseminated infections • Mucocutaneous infections – Oropharyngeal infections – Skin/nail infections – Vaginitis
Aspergillus fumigatus	Found in the environment		• Pneumonia • Allergy
Malassezia spp.	Commensal of the skin		• Cutaneous infections – Pityriasis versicolor – Seborrheic dermatitis • Allergic atopic eczema
Cryptococcus neoformans	Found in the environment		• Pneumonia • Meningitis
Blastomyces dermatitidis	Found in the environment	25°C 37°C 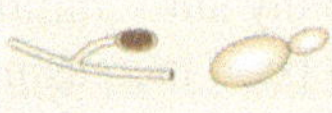	• Acute/chronic pulmonary infections • Skin lesions/subcutaneous nodules • Disseminated disease
Coccidioides immitis	Found in the environment	25°C 37°C 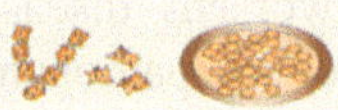	• Pneumonia • Disseminated disease
Histoplasma capsulatum	Found in the environment	25°C 37°C 	• Self-limited flu-like syndrome • Acute pneumonia • Chronic pulmonary infections • Disseminated disease
Paracoccidioides brasiliensis	Found in the environment	25°C 37°C 	• Pneumonia • Disseminated disease

Fig. 1 Major human pathogenic fungi and associated diseases. For each fungus, the source and disease spectrum are described along with an accompanying picture

and *Candida albicans*, establish lifelong commensalism on human skin and body cavities without necessarily causing disease. The fact that fungi are capable of colonizing in almost every niche within the human body suggests that they must possess particular adaptation mechanisms of co-existence to adapt to environmental abiotic stress conditions which could deviate into disease under conditions of either too low or too high immune responses (Romani 2001). Indeed, the occurrence of fungal diseases in primary immune deficiencies suggests the pivotal contribution of the underlying deregulated inflammatory immunity with susceptibility to fungal infections and diseases (Romani 2004; Shoham and Levitz 2005). The most serious of the human diseases caused by fungi are the opportunistic fungal infections that occur in patients with defective immunity, including AIDS and leukemia. *Candida* species remain one most important cause of hospital-acquired bloodstream infections. Invasive aspergillosis, mostly by *Aspergillus fumigatus* and *A. terreus*, and other mold infections are a leading cause of infection related death in hematopoietic stem cell transplant recipients. Other fungal diseases include mucosal infections (i.e., thrush and chronic mucocutaneous candidiasis CMC) and both IgE and eosinophilia-driven hypersensitivity diseases, including many cases of severe asthma, allergic bronchopulmonary mycoses, chronic sinusitis, hypersensitivity pneumonitis, atopic eczema/dermatitis syndrome and gut inflammation (Romani and Puccetti 2008). Sensitivity to molds has been reported in patients with asthma and allergic bronchopulmonary aspergillosis and is frequent in patients with asthma and cystic fibrosis. *Malassezia* yeasts are a part of the normal skin microbiota and have been associated with a number of diseases affecting the human skin, such as pityriasis versicolor, folliculitis, seborrheic dermatitis and dandruff, atopic eczema/dermatitis, psoriasis, and less commonly with other dermatologic disorders (Levin 2009). Although the burden of fungal diseases may rival those of many of the best known bacterial diseases, these are not often highlighted to the general public due to the relatively low mortalities and morbidities as compared to the severely invasive fungal infections.

This chapter attempts to position the new findings on the IL-17/T_H17 cell responses to fungi within the conceptual framework of a two component antifungal response that includes resistance (the ability to limit fungal burden) and tolerance (the ability to limit the host damage caused by either the immune response or other mechanisms). Evolutionary conserved from plants to vertebrates (Schneider and Ayres 2008), this new concept may help to define the best suitability in response to fungi and its integration into new medical practices.

1.1 Fungal Infections and Diseases as Examples of IRS-Like Diseases

We are currently entering an exciting transition period from studying the molecular and cellular bases of the virulence of pathogens to deciphering the mechanisms of immune activation and tolerance to them. The control of inflammation leading to tolerance, the molecular bases of regulation and rupture, and the way pathogens themselves regulate inflammation are essential issues (O'Garra and Vieira 2004). Infectious agents can induce auto-immune diseases in several experimental settings, some of which have clinical counterparts (Bach 2002). A variety of mechanisms have been invoked to explain these observations, including molecular mimicry and an increase in the immuno-genicity of auto-antigens caused by inflammation in the target organ. The inflammatory response, initiated by cells of the innate immune system, is followed by adaptive immunity which simultaneously responds and regulates signals emanating from the innate system. Unresolved infection and inflammation are major epigenetic and environmental factors that contribute to chronic diseases and auto-immunity, and in specific settings, to an increased risk of cancer. Paradoxically, infectious agents can also suppress allergic and auto-immune disorders. The central questions are whether immune dysregulation precedes and/or promotes infection and alternatively (or in addition), whether microbial exposure/colonization contributes to the burst and perpetuation of pathogenic inflammation and auto-immunity (Romani 2008a, b; Kivity et al. 2009).

Fungi provide a spectrum of examples of immune regulation and therapeutic regulation of the mucosal immune system. Although inflammation is an essential component of the protective response to fungi, its dysregulation may significantly worsen fungal diseases (Romani and Puccetti 2007; Romani 2008a, b). The balance of pro and anti-inflammatory signaling is a prerequisite for successful host/fungal interactions and requires the coordinate actions of both innate and adaptive immune systems. Fungi can exploit or subvert a host's inflammatory response, and thus affect carriage and pathogenicity. A hyper-inflammatory response does, in fact, enhance virulence in *Saccharomyces cerevisiae* and *Aspergillus nidulans* infections. In the normal skin, the fungus *Malassezia* downregulates inflammation via TGF-β1 and IL-10, and establishes itself as a commensal. In atopic dermatitis and psoriasis the skin barrier acts to enhance release of allergens and molecules involved in hyperproliferation, cell migration, and disease exacerbation. Thus, although the inflammatory response to fungi may serve to limit infection, an overzealous or heightened inflammatory response may contribute to pathogenicity and paradoxically promote infection (Romani and Puccetti 2007; Romani 2008a, b; Romani et al. 2008a, b). Bi-directional influences between infection and immune-related pathology have been known to exist in CMC in which the inability to clear *C. albicans* yeasts persists in recurring lesions of the skin, nails, and mucous membranes (Lilic 2002). Although occasionally associated with auto-immune polyendocrinopathy, candidiasis-ectodermal dystrophy (a condition of dysfunctional T-cell activity), CMC encompasses a variety of clinical entities, the pathogenesis of which is largely unclear. CMC patients often develop endocrine and inflammatory disorders, which suggest deregulation of the inflammatory and immune responses. These observations highlight a truly bipolar nature of the inflammatory process in infection, at least by specific fungi. Early inflammation prevents or limits infection, but an uncontrolled response may eventually oppose disease eradication. This condition is crucially exemplified by recent findings in mice with chronic granulomatous disease (CGD) in which an intrinsic, genetically determined failure to control inflammation to sterile fungal components determines the animals' inability to resolve an actual infection with *A. fumigatus* (Romani et al. 2008a, b). A main implication of these findings is that an exaggerated inflammatory response likely compromises the ability to cope with infecting/colonizing fungi and not an 'intrinsic' susceptibility to infection that determines a state of chronic or intractable disease. Thus, fungal diseases represent an important paradigm in immunology since they can result from either the lack of recognition or over-activation of the inflammatory response.

Clinically, severe fungal infections occur in patients with immune reconstitution syndrome (IRS), an entity characterized by local and systemic reactions that have both beneficial and deleterious effects on infection (Singh and Perfect 2007).

Intriguingly, IRS responses are also found in immuno-competent individuals and after rapid resolution of immuno-suppression, indicating that inflammatory responses can result in quiescent or latent infections manifesting as opportunistic mycoses. This likely reflects the association of the severity of disease with high levels of pro-inflammatory cytokines in patients with paracoccidioidomycosis (Corvino et al. 2007). Although host immunity is crucial in eradicating infection, immunological recovery can also be detrimental and may contribute towards worsening disease in opportunistic and non-opportunistic infections.

1.2 Resistance and Tolerance to Fungi

Resistance and tolerance are two types of host defense mechanisms that increase suitability in response to fungi (Zelante et al. 2009a, b). In experimental candidiasis and aspergillosis, the balance between T_H1 cells (that provide anti-fungal resistance) and regulatory T cells (Tregs) that limit the consequences of the associated inflammatory pathology may provide the basis for the occurrence of functionally-distinct modules of immunity for resistance and tolerance (Zelante et al. 2009a). The inflammatory response is initially mediated by cells of the innate immune system followed by a later adaptive immune response which responds to the signals originated by the innate immune system. The decision of how to respond will primarily be determined by interactions between pathogens and cells of the innate immune system, but the actions of T cells will feed back into this dynamic equilibrium to suppress overzealous innate responses.

Generation of a dominant T_H1 response driven by IL-12 is essentially required for the expression of protective immunity to fungi, while IL-4 acts as the most potent proximal signal for commitment to T_H2 reactivity that dampens protective T_H1 responses and favors fungal allergy. Through the production of the signature cytokine IFN-γ and help for opsonizing antibodies, the activation of T_H1 cells is instrumental in the optimal activation of phagocytes at sites of infection (Fig. 2). Therefore, the failure to deliver activating signals to effector phagocytes may predispose patients to overwhelming infections, limit the therapeutic efficacy of antifungals and antibodies, and favor persistency and/or commensalism (Romani 2004). Immunological studies in patients with polar forms of paracoccidioidomycosis demonstrate an association between T_H1-biased reactivity and the asymptomatic and mild forms of the infection, as opposed to the positive correlation of T_H2 responses with the severity of the disease and poor prognosis. Patients with inborn errors of the IL-12/IL-23/IFNγ-mediated immunity are indeed susceptible to disseminated paracoccidioidomycosis (Romani 2008a, b). However, patients with inborn deficits in the IL-12/IL-23/IFN-γ loop do not demonstrate increased susceptibility to other fungal infections (Romani 2008a, b), a finding showing that other cytokine pathways may also play a role.

In pathogens like fungi that have a complex pathogenesis, multiple types of regulatory cells could influence the outcome. Tregs have become an integral component of the immune response and through their capacity to inhibit aspects of innate and

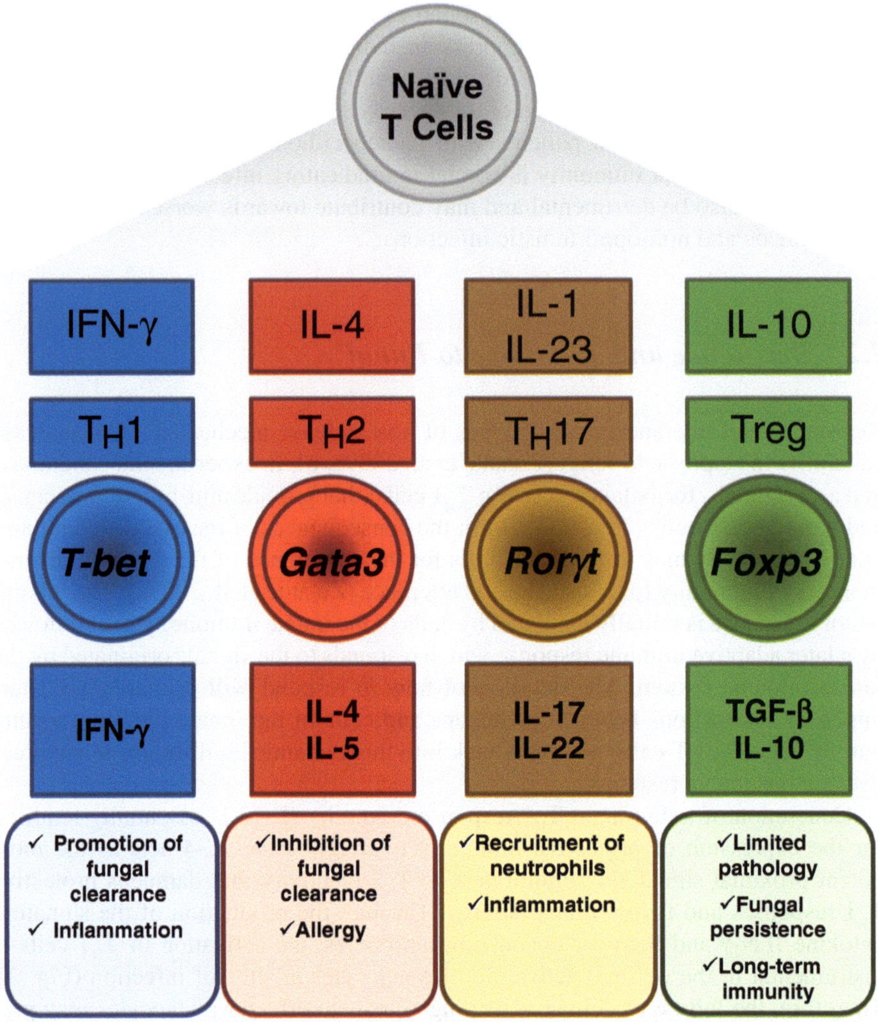

Fig. 2 Th cell subsets in response to fungi. The figure shows the Th/Treg cell subsets, their transcription factors, cytokine production and possible effector/regulatory functions in fungal infections. See text for details

adaptive anti-fungal immunity are strictly required for protective tolerance to fungi (Hori et al. 2002; Montagnoli et al. 2002, 2006; Cavassani et al. 2006; McKinley et al. 2006; Deepe and Gibbons 2008; Lazar-Molnar et al. 2008; Loures et al. 2010). The ability to control both class and magnitude of the immune response may confer an evolutionary advantage whereby the host would effectively fight infection but limit collateral immuno-pathology. Alternatively, fungus-induced immuno-suppression could be viewed as a powerful immuno-evasion strategy for the invading pathogen, while the consequence of Treg activity is less damage to the host, but also fungal persistence (Moreira et al. 2008). In experimental fungal infections, fungal growth, inflammatory immunity, and tolerance in the respiratory or the gastrointesftinal mucosa were all controlled by the coordinate activation of naturally-occurring (n)Tregs – limiting early inflammation at the sites of infection – and pathogen-induced (i)Tregs, which regulated the expression of adaptive Th immunity in secondary lymphoid organs (Romani and Puccetti 2006). Because both the recovery of *C. albicans* from the gastrointestinal tract and the detection of underlying T_H1 reactivity such as DTH and lympho-proliferation can fluctuate in healthy subjects, it is likely that Tregs mediate tolerance to the fungus at the site of colonization. This may have allowed fungal persistence and the occurrence of memory immunity to a commensal. CMC, although encompassing a variety of clinical entities (Lilic 2002), has been associated with auto-immune polyendocrinopathy-candidiasis-ectodermal dystrophy a condition in which Treg induction is defective (Ryan et al. 2005).

The different Treg cell populations may have the capacity to influence the emergence or function of one another. This was best illustrated in murine aspergillosis where, early in infection, inflammation was controlled by the expansion, activation, and local recruitment of nTregs suppressing neutrophils; whereas late in infection, and similarly in allergy, tolerogenic iTregs inhibited T_H2 cells and prevented allergy to the fungus (Montagnoli et al. 2006). The level of inflammation and IFN-γ in the early stage set the subsequent adaptive stage by conditioning the indoleamine 2,3-dioxygenase (IDO)-dependent tolerogenic program of DCs and the subsequent activation and expansion of tolerogenic Tregs preventing allergy to the fungus. Therefore, regulatory mechanisms operating in the control of inflammation and allergy to the fungus are different but interdependent as the level of the inflammatory response early in infection may impact the susceptibility to allergy in continuous exposure to the fungus. The unifying mechanism linking nTregs to tolerogenic respiratory Tregs in response to the fungus is consistent with the revisited "hygiene hypothesis" of allergy in infections. This means an early reduction in microbial burden may predispose to allergy and simultaneously provide mechanistic explanations for the significance of the variable level of IFN-γ seen in allergic diseases and asthma and for the paradoxical worsening effect on allergy of T_H1 cells.

Collectively, these observations suggest that the capacity of Tregs to inhibit aspects of innate and adaptive immunity is pivotal in their regulatory function and further support the concept of 'protective tolerance' to fungi; implying that a host's immune defense may be adequate for protection without necessarily eliminating fungal pathogens (which would impair immune memory) or causing an unacceptable level of tissue damage (Romani and Puccetti 2006).

2 IDO as a Mammalian Regulator of Resistance and Tolerance to Fungi

The enzyme IDO has a complex role in immuno-regulation in infection, pregnancy, auto-immunity, transplantation, and neoplasia (Grohmann et al. 2003). IDO catalyzes the first and limiting step in the kynurenine pathway of tryptophan catabolism. Initially recognized in infection because of antimicrobial activity ('tryptophan starvation' of intracellular parasites), IDO is more importantly and widely involved in immune homeostasis of the mammalian host, and may even represent an evasion mechanism for microbes that establish commensalism or chronic infection (Zelante et al. 2009a, b). Within a conceptual framework from different areas of immune regulation, the many disparate functions of IDO in infection are now being reconciled, emphasizing a complex role for a metabolic pathway that appears to be conserved – though evolved in function – through the last 600 million years of evolution.

IDO and the downstream enzymes in the metabolic pathway of tryptophan degradation have a central role in chronic infection (Zelante et al. 2009b). Present in innate immune cells such as macrophages, neutrophils, and epithelial cells, IDO is a key player in the suppression of acute inflammatory responses. Furthermore, the activation of the amino acid starvation response was recently reported to inhibit T_H17 differentiation in vivo (Sundrud et al. 2009) and its expression by plasmacytoid dendritic cells (pDC) enables the onset of tolerance in adaptive immunity. Thus, in addition to direct effector activities largely involving tryptophan deprivation, IDO and the other enzymes of the metabolic pathway contribute immuno-active molecules to the generation of regulatory Tregs with anti-inflammatory and tolerogenic activities. These findings establish a mutual interaction between DCs and Tregs for the upkeep of immunological tolerance and predict that in certain infectious settings, any direct IDO antimicrobial activity resulting from tryptophan starvation could be somewhat blunted by the induction of tolerogenic responses that would allow pathogens to take advantage of the 'immune privilege' normally reserved to mammalian hosts (Mellor and Munn 2004).

In experimental fungal infections, IDO and kynurenines pivotally contribute to the delicate inflammatory/anti-inflammatory balance by providing the host with immune mechanisms adequate for protection without necessarily eliminating fungal pathogens or causing an unacceptable level of tissue damage (Romani and Puccetti 2006; Zelante et al. 2009b). More recently, while capable of inducing the (forkhead box P3) Foxp3-encoding gene transcriptionally, tryptophan catabolites were also found to suppress the gene encoding RORγt (retinoid-related orphan receptor gamma t) and the T_H17 lineage specification factor (De Luca et al. 2007). Thus, in their capacity to induce Tregs and inhibit T_H17, IDO and kynurenines pivotally contribute to cell lineage decision in experimental fungal infections and revealed an unexpected potential in the control of inflammation, allergy, and T_H17-driven inflammation in these infections. In this context, the T_H17 pathway, which downregulates tryptophan catabolism, may instead favor pathology and serve to accommodate the seemingly paradoxical association of chronic inflammation with fungal diseases (Romani et al. 2008a, b).

As already mentioned, the IDO mechanism has revealed an unexpected potential in the control of inflammation, not only in infection, but also in airway fungal allergy, a condition in which tolerogenic DCs could have a protective function. IDO expression is paradoxically upregulated in patients with allergy or auto-immune inflammation, a finding suggesting the occurrence of a homeostatic mechanism to halt ongoing inflammation. The induction of IDO could be an important mechanism underlying the anti-inflammatory action of corticosteroids (Grohmann et al. 2007).

The implications for IDO in immuno-regulation of fungal infections are manifold. As *C. albicans* is a commensal of the human gastrointestinal and genito-urinary tracts, and IFN-γ is an important mediator of protective immunity to the fungus, the IFN-γ/IDO axis may accommodate fungal persistence in a host environment rich in IFN-γ. Because IFN-γ is a potent IDO activator, this suggests the existence of an IFN-γ/IDO-dependent pathway leading to sequential T_H1/Treg cell activation in infection. In its ability to induce T_H1 immunity within a regulatory environment and to prevent T_H17 development, IDO expression may correlate with the occurrence of local tolerogenic responses. Alternatively, the high levels of IL-10 production may be a consequence of IDO activation by the fungus, impairing anti-fungal T_H1 immunity and favoring persistent infection. The fact that hyphae, more than yeasts, activate the expression of IDO, further suggests that differential sensing of fungal morphotypes through distinct recognition receptors may promote distinct immune responses and that fungal hyphae may also promote tolerance and thus contribute to commensalism and eventually to immuno-evasion. Because germinating *Aspergillus* conidia promote inflammatory responses by subverting tolerance (Montagnoli et al. 2006), the central role of tolerance at the fungus/host interface and the pivotal role of tryptophan catabolism in the tolerant state are both emphasized. Ultimately, the manipulation of Tregs by fungi gives further support to the notion that microbiota may function as a major regulator of immunological tolerance both locally and at distant sites.

3 T_H17 Cells

Accumulating data supports a role for T_H17 cells and T_H17 cytokines in inflammatory processes and in animal models of auto-immunity or inflammation (Littman and Rudensky 2010; Miossec et al. 2009; Korn et al. 2009; Zhou et al. 2009). T_H17 cells have an important function in the host defense against extracellular pathogens, but they also have become notorious for their role in the pathogenesis of many autoimmune and allergic disorders. T_H17 cells are a separate lineage of effector T_H cells contributing to immune pathogenesis previously attributed to the T_H1 lineage. They produce unique cytokines (IL-17, IL-17F, IL-21, and IL-22) and express transcription factors distinct from T_H1 and T_H2 cells. Naive mouse and human CD4+ T cells activated in the presence of TGF-β and IL-6 express the transcription factor RORγt and become T_H17 cells that are stabilized by DC derived IL-23 and amplified by IL-1 and IL-21. IL-6, IL-21, and IL-23 all induce phosphorylation of a signal transducer and activator of transcription 3 that has multiple binding sites on the IL-17A promoter. In contrast, TGF-β with IL-2 upregulates expression of the transcription factor Foxp3 and activates iTregs, which suppress immune responses. Several lines of evidence further support the notion of a reciprocal relationship between FoxP3+ Tregs and T_H17 cells. Since TGF-β can induce both RORγt and FoxP3, the level of

TGF-β and the presence of cytokines such as IL-6 and IL-21 dictate whether expression of RORγt/RORα or FoxP3 predominate, and depending on the level of free transcription factors, determine whether T cells will differentiate into the Treg or T_H17 phenotype. Interestingly though, T_H17 are found early during the initiation of an immune response and have been reported to be involved in a broad range of both T_H1- and T_H2-dominated immune responses (Dong 2008).

Emerging data on the mechanism by which T_H17 cells induce tissue inflammation suggests that T_H17 cells first infiltrate the site of tissue inflammation and then recruit other pro-inflammatory effector T cells (including T_H1 cells) and innate cells (including neutrophils) to the site of tissue inflammation. As IL-17 receptors are widely expressed on parenchyma/tissue cells and IL-17 induces production of IL-1, IL-6, TNF, matrix metalloproteinases, IL-8 and chemokines, these mediators coordinate infiltration of other cell types to the site of inflammation and mediate massive tissue inflammation at the site where IL-17 is abundantly produced.

T_H17 cells are induced in fungal infections through Toll-like receptors (TLR) and non-TLR dependent signaling (Heninger et al. 2006; Kleinschek et al. 2006; De Luca et al. 2007; Leibundgut-Landmann et al. 2007; Zelante et al. 2007; Armstrong-James et al. 2009; van de Veerdonk et al. 2009; Zhang et al. 2009; Rizzetto et al. 2010). T_H17 are present in the human T cell memory repertoire, similar to *C. albicans* (Acosta-Rodriguez et al. 2007; Fenoglio et al. 2009) and *A. fumigatus* (Bozza et al. 2009; Chai et al. 2010). Defective T_H17 cell differentiation has been linked to CMC in patients with primary immuno-deficiencies (Milner et al. 2008). Although recent evidence supports the importance of the Dectin-1/IL-17 axis in human mucocutaneous fungal infections (Ferwerda et al. 2009), both positive and negative effects on immune resistance have been attributed to T_H17 and IL-17 receptor (IL-17R) signaling in experimental fungal infections (Huang et al. 2004; De Luca et al. 2007; Conti et al. 2009). Thus, the role of IL-17 and T_H17 cells in immunity vs. pathology in fungal infections and diseases remains controversial (Zelante et al. 2009a, b). The high susceptibility to systemic (Huang et al. 2004) and oral (Conti et al. 2009) candidiasis in condition of defective IL-17RA signaling has been taken to indicate the essential role of IL-17RA signaling in host defense to *Candida*. This is thought to be through the ability of IL-17 to mobilize neutrophils and induce β-defensin-3. However, exogenous IL-17 administration failed to rescue the T_H17 deficiency and actually caused severe adverse reactions (Conti et al. 2009). IL-17A is dispensable for protection in gastric candidiasis and the neutralization of IL-17A greatly reduced the fungal burden and ameliorated the systemic and the gastrointestinal infections in both IL-17RA KO. IL-17A was elevated in IL-17F KO mice and contributed to susceptibility to the infection in these mice, a finding suggesting that IL-17F by inhibiting IL-17A may exert protective effects in candidiasis. However, the finding that IL-17A blockade increased resistance in IL-17F KO mice clearly indicated that neither cytokine is essential in infection (De Luca et al. 2010). It is likely that the protective vs. disease-promoting effect of the IL-17/T_H17 pathway may depend on the stage and site of infection. Being early, IL-17 is able to exert some forms of antifungal resistance via IL-22 (see below), defensins, and neutrophils while the failure to downregulate microbe-induced expression of IL-17

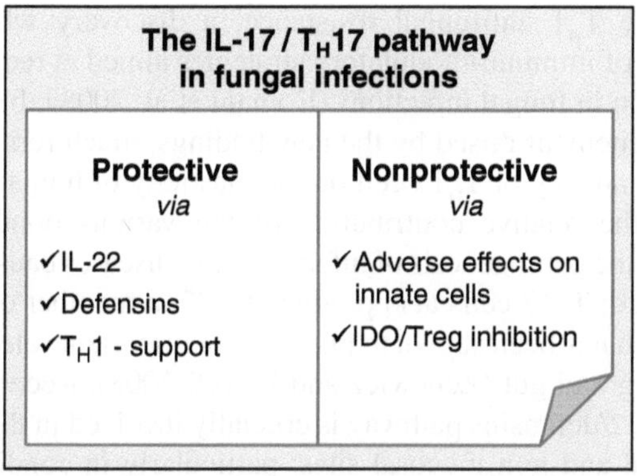

Fig. 3 Beneficial and harmful effects induced by the IL-17/T_H17 pathway in fungal infections. Under conditions of limited inflammation, naturally occurring IL-22+T_H17 cells may fulfill the role of a protective response that exploits primitive anti-fungal effector defense mechanisms of resistance. In contrast, persistent or exuberant inflammation resulting from dysregulated actions of multiple cytokines, including IL-17A, could ultimately promote the activation of pathogenic T_H17 leading to chronic inflammation and persistent infection through IDO inhibition. IDO, indoleamine 2,3-dioxygenase. See text for details

could eventually be one major link connecting infection with chronic inflammation (Fig. 3).

The mechanisms that linked inflammation to chronic infection have been credited to the offending potential of IL-17A that although promoting neutrophil recruitment, impeded the timely restriction of neutrophil inflammatory potential, preventing optimal protection from occurring (Romani et al. 2008a, b). IL-17A also activated the inflammatory program of neutrophils by counteracting the IFN-γ-dependent activation of IDO, which is known to limit the inflammatory status of neutrophils as well as by inducing the release of metalloproteinases and oxidants, and likely accounts for the high inflammatory pathology and tissue destruction associated with T_H17 cell activation. Evidence indicates that the detrimental side effect of the inflammatory action of an unopposed IL-23/IL-17 pathway (restrained by IDO) occurs through a mechanism leading to the sequential generation of regulatory and anti-inflammatory $V\gamma4^+$ γδ and $CD25^+$ αβ T cells (Romani et al. 2008a, b). This suggests that the intersection of γδ cells (present in all vertebrates) with tryptophan catabolism (conserved through the past 600 million years of evolution) might represent a milestone in the evolution of the immune system, combining the innate and acquired immune systems in the proper control of infection (Romani et al. 2008b).

These new findings provide a molecular connection between the failure to resolve inflammation and lack of antifungal immune resistance and points to strategies for immune therapy of fungal infections that will limit inflammation to stimulate an effective immune response. More generally, the T_H17 pathway could be involved in the immuno-pathogenesis of chronic fungal diseases where persistent fungal antigens may maintain immunological dysreactivity. As a matter of fact, IL-17 neutralization increased fungal clearance, ameliorated inflammatory pathology, and restored protective T_H1 antifungal resistance; a discovery which points to the therapeutic utility of immuno-modulatory strategies aimed at reducing T_H17-driven hyper-inflammation in fungal infections (Romani et al. 2008a, b).

Despite the excitement raised by the new findings, much remains to be learned, including the dependency of T_H17 cell on the plasticity of human CD4+ T cell differentiation and the relative contribution of the various populations of IL-17-producing cells to the pathogenesis of infections and diseases caused by the different fungi. In this regard, T_H17 cells also produce IL-22, a member of the IL-10 family of cytokines, which has been shown to play a more important role than IL-17 in host defense in the lung and gut (Zenewicz and Flavell 2008). Recent findings suggest that the IL-23/IL-22/defensins pathway is crucially involved in the control of fungal growth at mucosal and non mucosal sites, particularly in conditions of T_H1 deficiency (see below). Thus, further tweaking the T_H17 model, T_H17 may exert its protective role in fungal infections through IL-22.

4 IL-22 Defines a Novel Immune Pathway of Anti-fungal Resistance

Although IL-23 is known to maintain T_H17 cells in vivo (McGeachy et al. 2009), IL-23 can also operate independently of T_H17. The finding that IL-23 was protective in condition of IFN-γ deficiency (Zelante et al. 2007) suggests that IL-23 may regulate the production of cytokines other than IL-17A/IL-17F. It has recently been described that IL-23 is crucial for the production of IL-22 (Aujla et al. 2008; Zheng et al. 2008). IL-22, but not IL-17, provided protective immunity to systemic infection with attenuated *Salmonella enterica* serovar enteritidis in the absence of IL-12 (Schulz et al. 2008). IL-22 is primarily expressed by T_H17 CD4+ T cells and is highly upregulated during chronic inflammatory diseases and auto-immunity (Laurence et al. 2008; Vivier et al. 2009). IL-22 receptor expression is absent on immune cells, but is instead restricted to the tissues, providing signaling directions from the immune system to the tissues. However, the role of IL-22 in inflammatory responses has been filled with perplexing data suggesting both pro and anti-inflammatory functions. In several animal models of bacterial infections, IL-22 was found to be more important than IL-17 for host defense at mucosal sites by inducing the production of specific effector molecules (Aujla et al. 2008; Zheng et al. 2008).

Recent evidence indicates that IL-22 has a more crucial role than IL-17A or F in mucosal host defense to *Candida* (Fig. 4). The IL-23/IL-22 pathway is crucially involved in the control of fungal growth at mucosal and non-mucosal sites, particularly in conditions of T_H1 deficiency such as in IFN-γ- or IL-17R-deficient mice (De Luca et al. 2010). Thus, by expanding T_H17 effector cells (Zelante et al. 2007) and IL-22 production by innate cells (De Luca et al. 2010), IL-23 has a dual role in infection. IL-22 directly targeted gut epithelial cells to induce STAT3 phosphorylation, the Reg family proteins, and together with IL-17, S100A8 and S100A9, which are known to have potent anti-candidal activity and anti-inflammatory effects (Kolls et al. 2008).

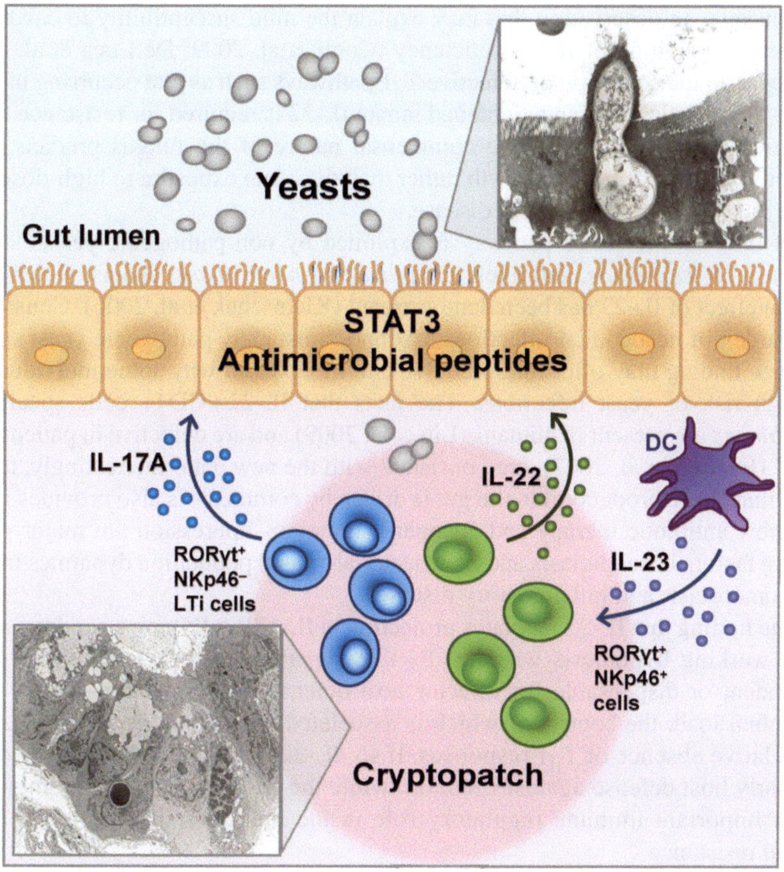

Fig. 4 IL-22 mediates first-line defence in mucosal candidiasis. Electron microscopy shows that *Candida albicans* penetrates gut epithelial cells with yeasts eventually residing within gut crypts (in the *insets*). Concomitantly, dendritic cell (DC)-derived IL-23 promotes the production IL-22 from cryptopatch CD3–NKp46+ expressing retinoic-acid related orphan receptor (ROR)γt, IL-23R, aryl hydrocarbon receptor, CCR5 and CCR7, indicating their similarity to gut RORγt+NKp46+ cells and cryptopatch lymphoid tissue inducer-like cells that are expanded by commensals. In contrast, NKp46-cells strongly express the IL-17A specific transcripts (from De Luca et al. 2010). IL-22 targets epithelial cells for STAT3 activation and, together with IL-17A, for antimicrobial peptide production

The new finding support a model in which the IL-23/IL-22 axis controls the initial fungal growth and tissue homeostasis likely exploiting primitive anti-fungal effector defense mechanisms (Xu et al. 2005), while adaptive T_H1/Treg cells prevent fungal dissemination and provide memory and tolerance (Romani and Puccetti 2006; Zelante et al. 2009a). The two pathways are reciprocally regulated and compensate each other in the relative absence of either one, consistent with the theme that adaptive immunity depends on innate immunity, and innate immunity requires adaptive regulation. With a functional T_H1 pathway, the IL-23/IL-22 axis is even dispensable, an observation that may explain the mild susceptibility to candidiasis as seen in condition of IL-22-deficiency (Conti et al. 2009; De Luca et al. 2010). However, in the condition of defective T_H1 pathways such as that occurring in IFN-γ or IL-17RA deficiency, an heightened innate IL-22 is required for resistance to low, infective doses. Ultimately, the commensal nature of the fungus predicts that a defective control of fungal growth rather than the acute exposure to high-dose challenge is associated with human disease.

Of interest, the IL-22 pathway is exploited by non-pathogenic yeasts such as *C. krusei*, *S. cerevisiae*, and likely *Cryptococcus neoformans*, against which a protective effect of IL-23 has been demonstrated (Kleinschek et al. 2006). Considering the burden of nosocomial infections by opportunistically pathogenic yeast species, the new finding may offer interpretative clues to explain why some individuals are at high risk of yeast infections. The facts that IL-22+CD4+ cells specific for *C. albicans* are present in humans (Liu et al. 2009) and are defective in patients with CMC (Eyerich et al. 2007), are consistent with the new data. Interestingly, the statistic that IL-22 production in the gut is driven by commensals also provides insight into how antibiotic therapy and iatrogenic immuno-suppression are major predisposing factors in candidiasis and how bacterial-fungal population dynamics impacts gut homeostasis and inflammatory diseases.

The finding that IL-22 mediates protection in IL-17RA KO mice suggests a plausible working hypothesis whereby IL-17R signaling in the gut is either organ-dependent or dispensable for effector host defense and protection against a low microbial load, the control of which is associated with IL-23 dependent IL-22 in the relative absence of T_H1 responses. If so, IL-22 may play a crucial role during the early host defense against *Candida*, while the IL-17/T_H17 pathway may serve as an important immune regulatory role in the induction of optimal T_H1 antifungal resistance.

These new findings help to further explain the susceptibility to bacterial and fungal infections in patients with autosomal dominant hyper-IgE syndrome (AD-HIES) (Ma et al. 2008; Milner et al. 2008). Due to dominant negative mutations of STAT3 (Holland et al. 2007), AD-HIES patients have a defective T_H17 that is likely amplified on epithelia where STAT3 mutations compromise the IL-22 effects. Thus, the defective T_H1 in AD-HIES patients (Netea et al. 2005) and the low IL-22 production in CMC (Eyerich et al. 2007) are clinical features consistent with the T_H1-supporting role and IL-22 production by T_H17 cells (Lin et al. 2009b; Romagnani et al. 2009). As a matter of fact, the epigenetic state of both *Tbet* and *Gata3* in in vitro polarized T_H17 shows marks of bivalency and T_H17 displays a large degree of plasticity with conversion to both T_H1 and T_H2 (Hirota et al. 2010). In line with this notion, T_H1 activation was indeed defective in IL-17R-deficient mice with candidiasis (Lin et al. 2009a; De Luca et al. 2010).

Also relevant to this matter is the observation that the contribution of STAT3-dependent mechanisms in susceptibility to fungal infections is dependent upon the cell types. STAT3 ablation in myeloid dendritic cells in vivo contributed to susceptibility in mucosal candidiasis and aspergillosis by subverting the tolerogenic response to fungi (Bonifazi et al. 2009, 2010). As a matter of fact, defective T_H17

responses in AD-HIES patients also occurred independently of STAT3 mutations (van de Veerdonk et al. 2010). This data would suggest that STAT3-dependent mechanisms, in addition to the lack of T_H17 per se, may contribute to susceptibility of fungal infections by subverting tolerance.

5 Conclusions

5.1 A Change of Strategy

The new discoveries in the field of fungal immunology have offered grounds for a better comprehension of cells and immune pathways that are amenable to manipulation in patients with or at risk of fungal infections. Our increased understanding of the functions that dictate development and role of T_H17 cells, as well as our knowledge of how T_H17/Tregs regulate each other and immune and non-immune cells provides guidelines for the design of immuno-modulatory therapies that limit inflammation in order to stimulate an effective immune response. The potential pathogenicity resulting from inappropriate T_H17 responses could be one reason for the complex degree of regulation of the T_H17 program. Improved understandings, control, and activity of T_H17 responses could provide vital information for the treatment of multiple inflammatory disorders. Tryptophan metabolites and T_H17 inhibitors are likely candidates as potent regulators capable of taming overzealous or heightened inflammatory host responses to the benefit of pathogen control and host survival (Romani et al. 2008a, b).

5.2 A Change of View

Infectious agents can induce auto-immune diseases but paradoxically suppress allergic and auto-immune disorders. A central question here will be to determine whether fungal exposure/colonization contributes to the burst of pathogenic autoimmunity or alternatively, whether dysregulation precedes or promotes diseases caused by fungi. Naturally occurring IL-22$^+$ T_H17 cells are highly enriched at mucosal sites, where continuous exposure to ubiquitous fungi occurs (through inhalation or via commensalism). This demands a sophisticated degree of adaptation to which integrated innate and adaptive immune responses contribute. The occurrence of IL-22 + T_H17 cells, employing ancient effector mechanisms of immunity, may represent a primitive mechanism of resistance against fungi under a condition of limited inflammation. However, in their ability to subvert the inflammatory program through the activation of the IL-23/T_H17 axis, fungi may eventually lead to immune dysregulation, including allergy and auto-immunity (Romani 2008a, b). The exploitation of the IFN-γ/IDO axis for functional specialization of anti-fungal effectors and regulatory mechanisms may have allowed commensal or ubiquitous fungi to co-evolve with the mammalian immune system to survive in conditions of high-threat inflammation and may represent a mechanism whereby dysregulated immunity is prevented in a manner similar to "protective" microbiota of the gastrointestinal tract. Together with the evidence for the critical role of IL-22 as a first line of defense in candidiasis and the reciprocal regulation with the T_H1 pathway, the above considerations suggest that protective immunity to fungi could be made up of a staged response involving an early, IL-22-dominated response followed by aT_H1/Treg response that will prevent fungal dissemination and supply memory. This implies that functionally distinct modules of immunity evolved to provide resistance and tolerance to commensal or airborne fungi in mammals.

Acknowledgments This study was supported by the Specific Targeted Research Project "ALLFUN" (FP7-HEALTH-2009) and the Italian project PRIN prot.2007KLCKP8_004.

References

Acosta-Rodriguez EV, Napolitani G et al. (2007). Interleukins 1beta and 6 but not transforming growth factor-beta are essential for the differentiation of interleukin 17-producing human T helper cells. Nat Immunol 8: 942–949.

Armstrong-James DP, Turnbull SA et al. (2009). Impaired interferon-gamma responses, increased interleukin-17 expression, and a tumor necrosis factor-alpha transcriptional program in invasive aspergillosis. J Infect Dis 200: 1341–1351.

Aujla SJ, Chan YR et al. (2008). IL-22 mediates mucosal host defense against Gram-negative bacterial pneumonia. Nat Med 14: 275–281.

Bach JF (2002). The effect of infections on susceptibility to auto-immune and allergic diseases. N Engl J Med 347: 911–920.

Bonifazi P, D'Angelo C et al. (2010). Intranasally delivered siRNA targeting PI3K/Akt/mTOR inflammatory pathways protects from aspergillosis. Mucosal Immunol 3: 193–205.

Bonifazi P, Zelante T et al (2009). Balancing inflammation and tolerance *in vivo* through dendritic cells by the commensal *Candida albicans*. Mucosal Immunol 2: 362–374.

Bozza S, Clavaud C et al. (2009). Immune Sensing of *Aspergillus fumigatus* Proteins, Glycolipids, and Polysaccharides and the Impact on Th Immunity and Vaccination. J Immunol 183: 2407–2414.

Cavassani KA, Campanelli AP et al. (2006). Systemic and local characterization of regulatory T cells in a chronic fungal infection in humans. J Immunol 177: 5811–5818.

Chai LY, van de Veerdonk F et al. (2010). Anti-*Aspergillus* human host defence relies on type 1 T helper (Th1), rather than type 17 T helper (Th17), cellular immunity. Immunology 130: 46–54.

Conti HR, Shen F et al (2009). Th17 cells and IL-17 receptor signaling are essential for mucosal host defense against oral candidiasis. J Exp Med 206: 299–311.

Corvino CL, Mamoni RL et al. (2007). Serum interleukin-18 and soluble tumour necrosis factor receptor 2 are associated with disease severity in patients with paracoccidioidomycosis. Clin Exp Immunol 147: 483–490.

De Luca A, Montagnoli C et al. (2007). Functional yet balanced reactivity to *Candida albicans* requires TRIF, MyD88, and IDO-dependent inhibition of Rorc. J Immunol 179: 5999–6008.

De Luca A, Zelante T et al. (2010). IL-22 defines a novel developmental pathway of antifungal resistance. Mucosal Immunol 3: 361–73.

Deepe GS Jr. and Gibbons RS (2008). TNF-alpha antagonism generates a population of antigen-specific CD4+CD25+ T cells that inhibit protective immunity in murine histoplasmosis. J Immunol 180: 1088–1097.

Dong C (2008). TH17 cells in development: an updated view of their molecular identity and genetic programming. Nat Rev Immunol 8: 337–348.

Eyerich K, Rombold S et al. (2007). Altered, but not diminished specific T cell response in chronic mucocutaneous candidiasis patients. Arch Dermatol Res 299: 475–481.

Fenoglio D, Poggi A et al. (2009). Vdelta1 T lymphocytes producing IFN-gamma and IL-17 are expanded in HIV-1-infected patients and respond to *Candida albicans*. Blood 113: 6611–6618.

Ferwerda B, Ferwerda G et al. (2009). Human dectin-1 deficiency and mucocutaneous fungal infections. N Engl J Med 361: 1760–1767.

Grohmann U, Fallarino F et al. (2003). Tolerance, DCs and tryptophan: much ado about IDO. Trends Immunol 24: 242–248.

Grohmann U, Volpi C et al. (2007). Reverse signaling through GITR ligand enables dexamethasone to activate IDO in allergy. Nat Med 13: 579–586.

Heninger E, Hogan LH et al. (2006). Characterization of the *Histoplasma capsulatum*-induced granuloma. J Immunol 177: 3303–3313.

Hirota K, Martin B et al. (2010). Development, regulation and functional capacities of Th17 cells. Semin Immunopathol 32: 3–16.

Holland SM, DeLeo FR et al. (2007). STAT3 mutations in the hyper-IgE syndrome. N Engl J Med 357: 1608–1619.

Hori S, Carvalho TL et al. (2002). CD25+CD4+ regulatory T cells suppress CD4+ T cell-mediated pulmonary hyperinflammation driven by *Pneumocystis carinii* in immuno-deficient mice. Eur J Immunol 32: 1282–1291.

Huang W, Na L et al. (2004). Requirement of interleukin-17A for systemic anti-*Candida albicans* host defense in mice. J Infect Dis 190: 624–631.

Kivity S, Agmon-Levin N et al. (2009). Infections and auto-immunity–friends or foes? Trends Immunol 30: 409–414.

Kleinschek MA, Muller U et al. (2006). IL-23 enhances the inflammatory cell response in *Cryptococcus neoformans* infection and induces a cytokine pattern distinct from IL-12. J Immunol 176: 1098–1106.

Kolls JK, McCray PB Jr et al. (2008). Cytokine-mediated regulation of antimicrobial proteins. Nat Rev Immunol 8: 829–835.

Korn T, Bettelli E et al. (2009). IL-17 and Th17 Cells. Annu Rev Immunol 27: 485–517.

Laurence A, O'Shea JJ et al. (2008). Interleukin-22: a sheep in wolf's clothing. Nat Med 14: 247–249.

Lazar-Molnar E, Gacser A et al. (2008). The PD-1/PD-L costimulatory pathway critically affects host resistance to the pathogenic fungus *Histoplasma capsulatum*. Proc Natl Acad Sci USA 105: 2658–2663.

Leibundgut-Landmann S, Gross O et al. (2007). Syk- and CARD9-dependent coupling of innate immunity to the induction of T helper cells that produce interleukin 17. Nat Immunol 8: 630–638.

Levin NA (2009). Beyond spaghetti and meatballs: skin diseases associated with the *Malassezia* yeasts. Dermatol Nurs 21: 7–13, 51; quiz 14.

Lilic D (2002). New perspectives on the immunology of chronic mucocutaneous candidiasis. Curr Opin Infect Dis 15: 143–147.

Lin L, Ibrahim AS et al. (2009). Th1-Th17 cells mediate protective adaptive immunity against *Staphylococcus aureus* and *Candida albicans* infection in mice. PLoS Pathog 5: e1000703.

Lin Y, Ritchea S et al. (2009). Interleukin-17 is required for T helper 1 cell immunity and host resistance to the intracellular pathogen *Francisella tularensis*. Immunity 31: 799–810.

Littman DR and Rudensky AY (2010) Th17 and regulatory T cells in mediating and restraining inflammation. Cell 140: 845–858.

Liu Y, Yang B et al. (2009). Memory IL-22-producing CD4+ T cells specific for *Candida albicans* are present in humans. Eur J Immunol 39: 1472–1479.

Loures FV, Pina A et al. (2010). Toll-like receptor 4 signaling leads to severe fungal infection associated with enhanced pro-inflammatory immunity and impaired expansion of regulatory T cells. Infect Immun 78: 1078–1088.

Ma HL, Liang S et al. (2008). IL-22 is required for Th17 cell-mediated pathology in a mouse model of psoriasis-like skin inflammation. J Clin Invest 118: 597–607.

McGeachy MJ, Chen Y et al. (2009). The interleukin 23 receptor is essential for the terminal differentiation of interleukin 17-producing effector T helper cells in vivo. Nat Immunol 10: 314–324.

McKinley L, Logar AJ et al. (2006). Regulatory T cells dampen pulmonary inflammation and lung injury in an animal model of pneumocystis pneumonia. J Immunol 177: 6215–626.

Mellor AL and Munn DH (2004). IDO expression by dendritic cells: tolerance and tryptophan catabolism. Nat Rev Immunol 4: 762–774.

Milner JD, Brenchley JM et al. (2008). Impaired T(H)17 cell differentiation in subjects with autosomal dominant hyper-IgE syndrome. Nature 452: 773–776.

Miossec P, Korn T et al. (2009). Interleukin-17 and type 17 helper T cells. N Engl J Med 361: 888–898.

Montagnoli C, Bacci A et al. (2002). B7/CD28-dependent CD4+CD25+ regulatory T cells are essential components of the memory-protective immunity to *Candida albicans*. J Immunol 169: 6298–6308.

Montagnoli C, Fallarino F et al. (2006). Immunity and tolerance to *Aspergillus* involve functionally distinct regulatory T cells and tryptophan catabolism. J Immunol 176: 1712–1723.

Moreira, A. P., K. A. Cavassani, et al. (2008). CCR5-dependent regulatory T cell migration mediates fungal survival and severe immuno-suppression. J Immunol 180: 3049–3056.

Netea MG, Kullberg BJ et al. (2005). Severely impaired IL-12/IL-18/IFNgamma axis in patients with hyper IgE syndrome. Eur J Clin Invest 35: 718–721.

O'Garra A and Vieira P (2004). Regulatory T cells and mechanisms of immune system control. Nat Med 10: 801–805.

Rizzetto L, Kuka M et al. (2010). Differential IL-17 production and mannan recognition contribute to fungal pathogenicity and commensalism. J Immunol 184: 4258–4268.

Romagnani S, Maggi E et al. (2009). Properties and origin of human Th17 cells. Mol Immunol 47:3–7.

Romani L (2001). Overview of the fungal pathogens. Immunology of Infectious Diseases. S. H. E. Kaufman, A. Sher and R. Ahmed. Washington D.C., ASM Press: 25–37.

Romani L (2004). Immunity to fungal infections. Nat Rev Immunol 4: 1–23.

Romani L (2008a). Cell mediated immunity to fungi: a reassessment. Med Mycol 46: 515–529.

Romani L (2008b). Parasites and auto-immunity: the case of fungi. Autoimmun Rev 8: 129–133.

Romani L, Fallarino F et al. (2008a). Defective tryptophan catabolism underlies inflammation in mouse chronic granulomatous disease. Nature 451: 211–215.

Romani L and Puccetti P (2006). Protective tolerance to fungi: the role of IL-10 and tryptophan catabolism. Trends Microbiol 14: 183–189.

Romani L and Puccetti P (2007). Controlling pathogenic inflammation to fungi. Expert Rev Anti Infect Ther 5: 1007–1017.

Romani L and Puccetti P (2008b). Immune regulation and tolerance to fungi in the lungs and skin. Chem Immunol Allergy 94: 124–137.

Romani L, Zelante T et al. (2008b). IL-17 and therapeutic kynurenines in pathogenic inflammation to fungi. J Immunol 180: 5157–5162.

Ryan KR, Lawson CA et al. (2005). CD4+CD25+ T-regulatory cells are decreased in patients with auto-immune polyendocrinopathy candidiasis ectodermal dystrophy. J Allergy Clin Immunol 116: 1158–1159.

Schneider DS and Ayres JS (2008). Two ways to survive infection: what resistance and tolerance can teach us about treating infectious diseases. Nat Rev Immunol 8: 889–895.

Schulz SM, Kohler G et al. (2008). Protective immunity to systemic infection with attenuated *Salmonella enterica* serovar enteritidis in the absence of IL-12 is associated with IL-23-dependent IL-22, but not IL-17. J Immunol 181: 7891–7901.

Shoham S and Levitz SM (2005). The immune response to fungal infections. Br J Haematol 129: 569–582.

Singh N and Perfect JR (2007). Immune reconstitution syndrome associated with opportunistic mycoses. Lancet Infect Dis 7: 395–401.

Sundrud MS, Koralov SB et al. (2009). Halofuginone inhibits TH17 cell differentiation by activating the amino acid starvation response. Science 324: 1334–1338.

van de Veerdonk FL, Marijnissen R et al. (2010). Milder clinical hyperimmunoglobulin E syndrome phenotype is associated with partial interleukin-17 deficiency. Clin Exp Immunol 159: 57–64.

van de Veerdonk FL, Marijnissen RJ et al. (2009). The macrophage mannose receptor induces IL-17 in response to *Candida albicans*. Cell Host Microbe 5: 329–340.

Vivier E, Spits H et al. (2009). Interleukin-22-producing innate immune cells: new players in mucosal immunity and tissue repair? Nat Rev Immunol 9: 229–234.

Xu T, Logsdon NJ et al. (2005). Structure of insect-cell-derived IL-22. Acta Crystallogr D Biol Crystallogr 61: 942–950.

Zelante T, De Luca A et al. (2007). IL-23 and the Th17 pathway promote inflammation and impair anti-fungal immune resistance. Eur J Immunol 37: 2695–2706.

Zelante T, De Luca A et al. (2009a). IL-17/Th17 in anti-fungal immunity: what's new? Eur J Immunol 39: 645–648.

Zelante T, Fallarino F et al. (2009b). Indoleamine 2,3-dioxygenase in infection: the paradox of an evasive strategy that benefits the host. Microbes Infect 11: 133–141.

Zenewicz LA and Flavell RA (2008). IL-22 and inflammation: leukin' through a glass onion. Eur J Immunol 38: 3265–3268.

Zhang Y, Wang F et al. (2009). Robust Th1 and Th17 immunity supports pulmonary clearance but cannot prevent systemic dissemination of highly virulent *Cryptococcus neoformans* H99. Am J Pathol 175: 2489–2500.

Zheng Y, Valdez PA et al. (2008). Interleukin-22 mediates early host defense against attaching and effacing bacterial pathogens. Nat Med 14: 282–289.

Zhou L, Chong MM et al. (2009). Plasticity of CD4+ T cell lineage differentiation. Immunity 30: 646–655.

T_H17 Cells in Organ Transplantation Rejection and Tolerance

Julia Tsang, Ling Lu, and Shuiping Jiang

Abstract The emergence of T_H17 cells as a distinct subset of effector T cell has led to a revised model of the adaptive immune system. In experimental and clinical transplantation, the T_H17-producing cytokine, interleukin (IL)-17, is evident in allograft rejection. It is conceivable that T_H17 cells could play a specific role in pathogenic process of allograft rejection. This chapter summarizes the current spectrum of T_H17 cells in transplant rejection. The tolerance and regulation of T_H17 response in the allogeneic context is also discussed.

1 Introduction

Allograft rejection is the major problem in solid organ transplantation, leading to the loss of graft function. It is an immunological process resulting from the recognition of alloantigens by recipient T cells. Among the T cells, $CD4^+$ T helper (Th) cells play a pivotal role in mediating the rejection process. The mainstay of controlling the rejection processes after transplantation is the administration of immuno-suppressive drugs. The current immuno-suppressive drugs target different events that inhibit T cell responses. Despite the advances made in past years, the long term graft outcome remains unsatisfactory. The knowledge of how T cells mediate allo-response in transplantation is an important basis for the development of new strategies for therapeutic intervention.

Th cells are heterogenous with regard to their protective functions, and different Th subsets induce different effector mechanisms. More than 20 years ago, two main subsets of Th cells with different functions and different patterns of cytokine secretion

S. Jiang (✉)
Department of Nephrology and Transplantation, Guy's Hospital, King's College London, London, UK
e-mail: sj4774@hotmail.com

S. Jiang (ed.), *T_H17 Cells in Health and Disease*,
DOI 10.1007/978-1-4419-9371-7_17,

were identified, namely T_H1 and T_H2 cells. The T_H1/T_H2 paradigm has helped us to understand the complex principles of Th cell biology and to comprehend different immune reactions. While the original T_H1/T_H2 model is enlightening and gives insight into the functions of helper T cells, it is becoming clear that it is far too simple to define its entire role or actions. The characterization of a novel Th subtype, namely T_H17, has cast further doubt on the basic T_H1/T_H2 paradigm. T_H17 cells are considered developmentally distinct from T_H1 and T_H2 cells and were initially described as a pathogenic population implicated in auto-immune disease. Since then, many studies have pointed to the role of T_H17 cells in a broad spectrum of diseases. Nonetheless, their relevance to graft rejection is only just beginning to be understood.

This chapter will focus on the recent evidence which suggests the relevance of T_H17 cells in graft rejection. Given the emerging role of T_H17 during the rejection process, therapeutic approaches targeting T_H17 cells and the possible actions of regulatory T cells (Tregs) on T_H17 responses will be discussed.

2 Th Subsets in Transplantation Rejection

The immune system evolved to protect multicellular organisms from a variety of pathogens. It has the ability to specifically recognize and eliminate foreign pathogens without harming the body. T cell antigen recognition is central to the immune functions that eliminate and neutralize foreign invaders. Once activated, $CD4^+$ T cells differentiate into different subsets and mediate other cell types to direct diverse effector mechanisms. The micro-environment in which naïve T cells encounter cognate antigen is critical in determining their differentiation, with factors such as local cytokine milieu and co-stimulation activity of antigen presenting cells (APCs) influencing the lineage selection. The functional properties of different subsets are characterized by the cytokines they secrete. T_H1 cells defined by their production of interferon gamma (IFNγ) were considered as crucial players in mediating graft rejection. T_H2 are defined by their secretion of IL-4, IL-5, IL-9, IL-10, and IL-13. T_H2 cells were thought to blunt the severity of allograft rejection by inhibiting T_H1 response, but in fact, they have been demonstrated to be another effector on promoting the rejection process. Notably, there is no redundancy in the Th cell differentiation and they regulate each other. IL-12 is crucial for T_H1 cell differentiation through signal transducer and activator of transcription (stat)-4 and the activation of a unique T_H1 transcriptional factor, T-bet, which upregulates IFNγ and downregulates IL-4 and IL-5. In contrast, IL-4 induces T_H2 cell differentiation through stat-6 and the activation of GATA-3, the T_H2 transcriptional factor. More recently, a third subset of Th cells were discovered which are characterized by their ability to produce IL-17, namely T_H17 cells. They express retinoic acid related orphan (ROR)γt as their lineage-specific transcriptional factor. T_H17 cells originally have been linked to auto-immune disease, but recent evidence also suggests their role in transplantation rejection.

Given that graft rejection appears to be a predominantly cell-mediated immune response, T_H1 cells are vastly held responsible for the process. By provision of T_H1 cytokines, they are able to activate cytotoxic response of $CD8^+$ T cells against alloantigens. Interestingly, IFNγ production by activated CD8 cells acts as a feedback loop to amplify T_H1 response. In addition, T_H1 cells are also capable of mediating the macrophage-dependent delayed-type hypersensitivity and the synthesis of complement-fixing immunoglobulin (Ig) G2a antibody by B cells. In addition to function as helper cells, T_H1 cells can also function as cytotoxic cells by the expression of Fas-ligand (FasL) on their surface. In clinical transplantation, T cells isolated from renal biopsies of patients with graft rejection were primed to high IFNγ production, suggesting the relevance of T_H1 responses in acute graft rejection (D'Elios et al. 1997). Although much clinical data has highlighted the association of T_H1 cytokine/T-bet to the rejection process (Atalar et al. 2009), surprisingly, it has been shown that knockout mice of IFNγ or IL-12 had accelerated the rejection process. One of the explanations was that T_H1 cytokine also play a role in the immuno-regulation process. In agreement with the hypothesis, IFNγ is shown to be important for the regulatory function of Tregs which are key players in the induction and maintenance of transplantation tolerance. It could also be explained by the fact that T_H1 cannot solely be responsible for mediating graft rejection (Wang et al. 2000).

In the absence of T_H1 response, allotransplant can skew the rejection towards T_H2 response. Mice rejected fully mismatched cardiac graft displayed the T_H1 cytokines IL-2 and IFNγ together with anti-donor $CD8^+$ CTL in the grafts. The T_H2 cytokines IL-4, IL-5, and IL-10 could not be detected. Nevertheless, when recipients were depleted of $CD8^+$ cells or had IL-12 blocked by various means, there was no prolongation of graft survival. Rejection still occurred, but showed a strong augmentation of T_H2 pathway (Chan et al. 1995; Piccotti et al. 1997). A similar observation was observed in clinical small bowel transplantation (Wu et al. 2006). When the T_H1 pathway of $CD4^+$ T cell differentiation was inhibited after profound T cell depletion, a predominant T_H2 response was revealed. More direct evidence for a role of allograft rejection of T_H2 responses comes from the studies using T_H2 polarized cell lines. Adoptive transfer of T_H2 cell lines in immuno-deficient animals induced allograft rejection. Moreover, the response from T_H2 cell lines was equally strong as T_H1 cell as shown by the fact that T_H2 alloreactive T cell rejected islet graft with similar efficiency to T_H1 cells (Barbara et al. 2000). Graft infiltration of eosinophils has been shown to be involved in the T_H2 mediated allograft rejection. Eosinophils were recruited and activated within the allograft through the combined action of IL-4, IL-5, and IL-13 produced by alloreactive T_H2 cells. IL-5 plays an essential role in the differentiation and proliferation of eosinophils in the bone marrow (Hogan and Foster 1996; Kopf et al. 1996). IL-4 and IL-13 upregulate the expression of VCAM-1 on endothelial cells, a critical adhesion molecule for eosinophils (Fukushi et al. 2000; Le Moine et al. 2002). Several studies in clinical transplant patients have suggested a role of T_H2 cytokines in chronic rejection. In renal transplant patients, development of chronic rejection was associated with a high IL-4 producing phenotype (Uboldi de Capei et al. 2004).

The T_H1/T_H2 paradigm has long been regarded as the basis for immunologists to perceive the adaptive immune response. However, contradictory data from autoimmunity showed that neutralization of T_H1 response, either by neutralizing antibodies or in knockout mice, did not result in the improvement of disease, but instead worsened disease progression. It has been suggested that the T_H1/T_H2 paradigm was not sufficient to fully explain the disease pathogenesis, implying that an important piece of puzzle was missing. Followed by the discovery that IL-23 promoted T_H17 differentiation, this new member of Th subset has been cast as a major player in autoimmunity. It has been established that IL-17 plays a role during the pathogenesis of a variety of auto-immune diseases, including experimental auto-immune encephalomyelitis (EAE), rheumatoid arthritis (RA), inflammatory bowel disease (IBD). It is interesting to note that in the setting of IFNγ (Miura et al. 2003) or IL-4 deficiency (Surquin et al. 2005), a massive neutrophil infiltration, a hallmark of T_H17 immune response, was found in allograft. This suggests that T_H17 cells play a role in allograft rejection.

3 The Evidence of T_H17 Cells in Allograft Rejection

Despite the recent emergence of T_H17 cells in adaptive immunity, IL-17 has long been associated with the allograft rejection process in both human and animal studies. In clinical studies, IL-17 is specifically present in kidneys undergoing rejection. More than 10 years ago, Van Kooten et al. showed the expression of IL-17 in kidney biopsies from patients suffering with graft rejection, whereas pre-transplant biopsies and normal kidneys were negative (Van Kooten et al. 1998). IL-17 expression was not only limited to human renal allograft during rejection, but was also found in urinary sediment. There was a 100% detectable rate of IL-17 mRNA expression in the mononuclear cells of urinary sediment of patients with rejection (Loong et al. 2002). The same group showed that IL-17 protein expression was found in human renal allograft biopsy tissues during subclinical (borderline) rejection, and none in biopsy tissues not showing any signs of rejection (Loong et al. 2002). Hence, IL-17 protein expression could serve as an early indicator for allograft rejection. Similar results were demonstrated in a rat renal allograft model. IL-17 cytokine was found scattered around in the borderline-changed rejected rat renal allografts without evidence of a serum creatinine increase, but was undetectable both in normal controls and in renal transplant tissues without signs of rejection (Hsieh et al. 2001). In this model, the kinetics of IL-17 mRNA expression was documented during the rejection process. IL-17 mRNA could be found as early as day 2 and peaked on postoperative day 5 (Loong et al. 2002). In addition to renal transplantation, IL-17 has been implicated in lung transplant rejection. IL-17 mRNA and protein levels were elevated in bronchoalveolar lavage (BAL) in lung transplanted patients with bronchiolitis obliterans syndrome (BOS) or acute rejection, compared to those without rejection (Vanaudenaerde et al. 2006). The elevated level of IL-17 was detectable before the airway invasion of neutrophils. Although the increase had disappeared 90 days after transplantation, the increased neutrophils persisted, implying the long lasting effect of IL-17. Further study from the same group also found that patients experiencing BOS had higher levels of T_H17 differentiating cytokines (IL-1β, IL-6, and IL-23), effector cytokine (IL-17), and IL-17-derived neutrophil-attracting chemokine (IL-8), highlighting the involvement of the IL-23/IL-17 pathway in lung transplantation (Vanaudenaerde et al. 2008). Lung allograft rejection is associated with immunity to Type V collagen (Haque et al. 2002). Interestingly, in rat and human lung transplants, type V collagen reactive lymphocytes were found to be IL-17 producing cells. Strong reactivity of these T_H17 cells was correlated to substantial increased incident and severity of BOS (Burlingham et al. 2007; Yoshida et al. 2006). More recently, IL-17 has also been implicated in acute rejection in liver transplantation. During the time course of serum levels of IL-23 and IL-17 during hepatic allograft rejection, a significant higher serum IL-17 level was found in the rejection group than in the non-rejection group (Fabrega et al. 2009).

Neutrophils, which are rapidly recruited by IL-17, have been described as effectors in both human and experimental allograft rejection. Human neutrophil infiltrates were frequently observed in biopsies of rejected allografts, and these infiltrates could predict the severity in human cardiac rejection (Healy et al. 2006). In mice, neutrophils were found to act as effector cells in allograft rejection, particularly in the settings where T_H1 and T_H2 pathways were blocked. In fact, neutrophil depletion significantly delayed skin graft rejection in IL-4 deficient mice (Surquin et al. 2005) and in cardiac allograft in IFNγ deficient mice (Miura et al. 2003). Taken together, this demonstrates the effector role of neutrophils during rejection. It is worth noting that the loss of T_H17 suppressive effects by IFNγ and IL-4 could result in increased IL-17-mediated neutrophil recruitment. Recently, Burrel et al. demonstrated that in T-bet deficient animals, IL-17 producing $CD8^+$ cells mediated allograft rejection, which is characterized by massive neutrophil infiltrates and was resistant to co-stimulatory blockade (Burrell et al. 2008). Altogether, this data suggests the presence of IL-17 producing cells and their effects during allograft rejection.

So far, only a few studies have provided more direct evidence on the causal relationship between IL-17 and rejection. In an in vitro study, Antonysamy et al. showed that IL-17R:Fc fusion protein could inhibit T-cell alloresponses in a mouse mixed lymphocyte reaction (Antonysamy et al. 1999a). They suggested that it acts by inhibiting the effect of IL-17 on promoting maturation of DC progenitors, thus its allo-stimulatory capacity (Antonysamy et al. 1999b). Administration of the same fusion protein into animals received a mismatched cardiac allograft could significantly extend graft survival (Antonysamy et al. 1999a; Tang et al. 2001). The effect of neutralization of IL-17 in inhibiting graft rejection could also be observed in T-bet deficient recipients (Yuan et al. 2008). In the absence of T_H1-mediated alloimmune responses, T_H17 cells mediate an aggressive pro-inflammatory response culminating in severe accelerated allograft rejection and vasculopathy. Anti-IL17 antibody inhibits the accelerated rejection. Interestingly, it appears that T_H17 cells mainly act at the site of the graft. Local expression of soluble IL-17 receptor-immunoglobulin in the graft by gene transfer could attenuate cytokine responses and leukocyte infiltration in rat cardiac allografts (Li et al. 2006). These findings are in line with the fact that IL-17 producing memory $CD4^+$ T cell express the graft/inflammatory site homing chemokine receptors, CCR4 and CCR6 (Acosta-Rodriguez et al. 2007). Most studies on IL-17 antagonism showed only a delayed rejection rather than complete protection, thus raising the possibility whether IL-17 alone is sufficient to mediate alloimmune responses. Indeed, it has been shown that in vitro generated T_H17 cells are sufficient to drive alloresponse leading to graft versus host disease (GVHD) in allogeneic bone marrow transplant setting (Iclozan et al. 2010).

The early detection of IL-17 during allograft rejection implicates that T_H17 plays a role in the early phase of rejection. In fact, it has been shown that IL-17 expression was enhanced significantly in inflamed transplants and in draining lymph nodes at the early stage of allocorneal rejection before the upregulation of T_H1-producing IFNγ (Chen et al. 2009a). Also, the inhibitory effect of IL-17 antagonism by IL17R:Fc on mononuclear cell infiltration in the intimal and medial compartments of rat cardiac allograft also occurred at an early time point (Tang et al. 2001). It is likely that T_H17 cells play a more prominent role in acute rejection. In agreement with this notion, TLR activation by CpG at early transplantation has shown to induce acute rejection which is IL-17 dependent (Chen et al. 2009b). However, its effect on chronic rejection is less clear. An earlier report showed that allografts harvested later showed intimal smooth muscle cell proliferation and luminal narrowing after IL-17 blockade, indicating that T_H17 cells have less effect on chronic graft rejection (Tang et al. 2001). In contrast, a recent report has shown that neutralization of IL-17 not only prevented vascular inflammation, but also suppressed the development of chronic vasculopathy (CAV) (Yuan et al. 2008). A correlation between short-term graft survival and the presence of T_H17 cells that produce IL-17 and IL-21 has been observed. There is a link between the expressions of activation-induced cytidine deaminase (the key enzyme of the germinal center reaction) and IL-21, suggesting that T_H17 could exert their deleterious effect by promoting lymphoid neogenesis, namely the organization of inflammatory effectors into ectopic germinal centers in which a local humoral immune response is elicited (Deteix et al. 2010). Moreover, IL-17 has been shown to be involved in the development of graft fibrosis in recipients transiently depleted of $CD4^+$ T cells (Faust et al. 2009). The apparent contrasting results could be explained by the difference in the transplantation models. The latter was using a single class II disparate model in T-bet$^{-/-}$ mice. Therefore, the initial vascular inflammation which precedes the pathological changes of impending CAV can be more efficiently controlled and consequently, the development of CAV minimized.

Taken together, T_H17 cells could be an alternative player in addition to T_H1 and T_H2 to mediate the allograft rejection process. Its role during early rejection is evident but the precise mechanism in transplantation immunity needs further investigation.

4 The Development of T_H17 Cells

Unlike T_H1/T_H2 differentiation, the development of T_H17 is less defined. Multiple signals derived from antigen presenting cells (APCs), the local environment, and $CD4^+$ T cells cooperate to induce full commitment into functional T_H17 cells.

Rather than T-bet and GATA-3 for T_H1 and T_H2 differentiation, factors inducing T_H17 differentiation activate the expression of RORγt (Ivanov et al. 2006).

Initially, it was reported that the IL-12 family member IL-23, a heterodimer consisting of the IL-12/23 p40 and the unique IL-23p19 sub-unit, was an important factor for the development of T_H17 cells (Aggarwal et al. 2003). However, subsequent studies showed that the effects were limited to memory T cells (Chen et al. 2007b). Since IL-23 receptor expression is low in naive T cells, IL-23 is unable to initiate T_H17 differentiation of naïve T cells. Therefore, it is unlikely that IL-23 would be the dominant factor for T_H17 cell differentiation. Further studies have demonstrated that the combination of IL-6 and transforming growth factor (TGF)-β was the key inducer for T_H17 cell differentiation of naive murine T cells (Bettelli et al. 2006; Veldhoen et al. 2006). IL-6 activates stat-3 via Janus family kinases. Stat-3 signaling is paramount for the activation of the T_H17 specific transcriptional factor, RORγt, leading to T_H17 cell differentiation (Yang et al. 2007). IL-6 alone induces only low level of RORγt expression because it also activates suppressor of cytokine signaling (SOCS-3) expression (Suto et al. 2008). TGF-β works together with IL-6 by inhibiting SOCS-3 and enhancing stat-3 (Qin et al. 2009). Pro-inflammatory cytokines IL-1β and IL-2 could also favor murine T_H17 cell differentiation. IL-1 receptor 1 expression in T cells, which was induced by IL-6, was necessary for the induction of EAE and for early T_H17 cell differentiation in vivo. IL-1 can enhance the TGF-β/IL-6-induced IL-17 production together with IL-23 (Sutton et al. 2006) and independently (Kryczek et al. 2007). IL-21 is cytokine-induced after IL-6 stimulation and is produced by differentiated T_H17 cells. Through stat-3 signaling, IL-21 acts in an autocrine fashion to upregulate its own expression, and inhibiting IFNγ production of T_H1 cells and inducing IL-23R expression (Chen et al. 2007a). Interestingly, IL-21 can replace IL-6 in T_H17 differentiation (Weaver et al. 2006). It acts as a positive feedback loop to further amplify the T_H17 cell development.

In contrast to murine T_H17 differentiation, factors to induce T_H17 development in human are more elusive, especially the role of TGF-β. This could be because in different reports, they have used various markers and methods to isolate T cells, and different conditions for cell activation and different culture mediums. Earlier works showed that IL-1β, IL-6, and/or IL-23 were sufficient to induce T_H17 cells, but unlike the mouse system, TGF-β has an inhibitory effect (Acosta-Rodriguez et al. 2007; Chen et al. 2007b; Evans et al. 2007). Nonetheless, more recent works have confirmed that similar cytokine pathways are involved in T_H17 development in both mice and humans. Using serum-free conditions which derived TGF-β, Manel et al. showed the essential role of exogenous TGF-β on human T_H17 development. TGF-β, along with IL-1 and IL-23 were necessary to induce IL-17 expression in naive human $CD4^+$ T cells from cord blood. However, TGF-β upregulated RORγt expression but simultaneously inhibited its ability to induce IL-17 expression. The addition of those inflammatory cytokines relieved this inhibition and increased RORγt-directed IL-17 expression (Manel et al. 2008). It is interesting to note that the effect of TGF-β on the induction of IL-17 in naïve human cells seems to be dose-dependent. TGF-β would only promote T_H17 differentiation at an optimal dose, while at a higher concentration it had a suppressive effect (Volpe et al. 2008; Yang et al. 2008a). Furthermore, the effect of TGF-β on T_H17 differentiation seems to differ from naïve and memory cell types. Memory T cells are more susceptible to the suppressive effect of TGF-β than naïve T cells. The optimal concentration of TGF-β that has shown to be necessary for IL-17 production in naïve T cells could inhibit IL-17 production by memory T cells (de Jong et al. 2010). Interestingly, $CD161^+$ T cells from umbilical cord blood have been identified as precursors of human IL-17 secreting cells. $CD161^+$ T cells can differentiate into T_H17 in response to IL-1β and IL-23, even in serum free conditions. The TGF-β-independent T_H17 differentiation could be due to the $CD161^+$ cells showing constitutively expression of RORγt, IL-23R, and CCR6, but not in the $CD161^-$ fraction (Cosmi et al. 2008). Although TGF-β had no direct effect on the expression of RORγt in $CD161^+$ T cells, it increased the relative proportions of $CD161^+$ T cells differentiating into IL-17 producing cells, while it inhibited IFNγ producing cells. This data implies the indirect role of TGF-β in T_H17 expansion as it is related to the suppression of T_H1 development (Santarlasci et al. 2009).

5 Treg Cell Plasticity and T_H17 Cell Differentiation

The dependence of TGF-β signaling on T_H17 development implies that T_H17 cells could be developmentally related to Tregs. Instead of acting as suppressors of T_H17 cell response, Tregs could favor the development of T_H17 cells. This would be a concern particularly in transplantation, as Tregs play an important role in the induction and maintenance of transplantation tolerance. Tregs could promote T_H17 development at least in two different fashions. On one hand, Tregs could represent a reservoir of T_H17 precursors. Tregs have plasticity in their early development. T_H17 and Tregs share a common dependence on TGF-β signaling. Studies on the intracellular signaling mechanism by which TGF-β regulates the development of both cell subsets showed that ERK and/or JNK pathways were involved in regulating Treg cell development, whereas the p38 pathway predominately modulated T_H17 cells (Lu et al. 2010). The developments of Tregs and T_H17 cells are also closely linked and depend on the balance between TGF-β and IL-6 signaling. TGF-β positively regulates both RORγt and FOXP3 transcription factors in naïve T cells. In the presence of TGF-β alone, naïve T cells failed to upregulate appreciable levels of IL-17, and progressively extinguished RORγt as they differentiated into Tregs. However, in the presence of IL-6 and TGF-β, naïve T cells differentiated into T_H17 cells by inhibiting FOXP3 expression (Bettelli et al. 2006; Zhou et al. 2008). Several reports showed that differentiated Tregs could also possess plasticity and become IL-17 producing cells. In the presence of IL-6, murine Tregs were able to express IL-17 upon activation and downregulate FOXP3 expression (Osorio et al. 2008; Radhakrishnan et al. 2008; Xu et al. 2007). It has been reported that the conversion of Tregs into IL-17 producing cells could be facilitated by activated dendritic cells (DCs). Dendritic cells activated after toll like receptor (TLR)-4 ligation (Xu et al. 2007), or after B7-DC crossing linking with a novel immune modulator B7-DCXAb (Radhakrishnan et al. 2008) were able to reprogram Tregs into T_H17 effectors. This reprogram was dependent on IL-6. The reprogrammed Tregs ceased to express IL-10 and TGF-β, and failed to suppress T cell responses (Radhakrishnan et al. 2008). Tregs would also convert into IL-17 producing cells upon co-culture with DCs activated via dectin-1, a C-type lectin receptor involved in fungal recognition. IL-23 produced by DCs was shown to drive this conversion (Osorio et al. 2008). Likewise, Beriou and colleagues showed that human Tregs could produce IL-17 in the presence of the pro-inflammatory cytokines IL-1β and IL-6 (Beriou et al. 2009). On the other hand, Treg would facilitate T_H17 cell differentiation by its production of TGF-β. Tregs can substitute TGF-β, and together with IL-6 or TLR activated DCs, they were able to facilitate the differentiation of IL-17-producing T cells (Veldhoen et al. 2006). Moreover, the production of TGF-β by Tregs could favor T_H17 cell differentiation indirectly by negatively regulating the T_H1 and T_H2 lineage transcription factors: T-bet and GATA3. Several in vitro studies have demonstrated that the addition of Tregs inhibited T_H1 and T_H2 cytokine production by T cells but promoted IL-17 synthesis upon stimulation by allogeneic DC (Benghiat et al. 2008; Santarlasci et al. 2009). Taken together, this raised concerns that under inflammation conditions, the suppressive effect of Tregs would be limited and that it could become an inducer of T_H17 response instead.

6 Other Factors Affecting T_H17 Cell Differentiation

Several other cytokines are also instrumental in T_H17 cell differentiation. IL-23 has been shown to be insufficient for T_H17 cell differentiation, but it is essential for the maintenance and/or expansion of T_H17 cells. It seems that IL-23 is required for fully differentiated T_H17 effector function. In the absence of IL-23, T cells reactivated in the presence of IL-6 and TGF-β were able to produce high amount of IL-17, but failed to upregulate the pro-inflammatory chemokines crucial for its pathogenesis. In addition, these IL-17 producing cells also produced IL-10, which has potent anti-inflammatory activities. The production of IL-10 could limit its pathogenic potential and the stability of the newly committed T_H17 cells. By contrast, naïve T cells stimulated in the presence of IL-6 and TGF-β together with IL-23 expressed both IL-17 and pro-inflammatory chemokines, but not IL-10 (McGeachy et al. 2007). The production of IL-23 by T_H17 cells acts in synergy with TGF-β to further amplify RORγt expression and the production of IL-17 in a stat-3 dependent manner (Yang et al. 2008b; Zhou et al. 2007), while IL-27, another IL-12 related cytokine, has shown to inhibit T_H17 development. Mice lacking IL-27 signaling had an increased T_H17 response (Batten et al. 2006; Stumhofer et al. 2006), but IL-27 is able to enhance the production of IL-10 by $CD4^+$ T cells (Stumhofer et al. 2007). Thus, IL-27 has been considered as an anti-inflammatory cytokine that could be a target for regulating T_H17 responses. However, the role of IL-27 in allograft rejection remains to be elucidated. T_H17 differentiation could be antagonized by IFNγ and IL-4, the key cytokines promoting T_H1 and T_H2 differentiation respectively. In the absence of IFNγ and IL-4, T_H17 differentiation pathway becomes prominent and can act as effector cells to mediate graft rejection (Miura et al. 2003; Surquin et al. 2005; Yi et al. 2009). Very recently, it had been identified that IFNγ can mediate T_H17 differentiation through differential regulation of osteopontin and IL-27 expression in DCs. IFNγ induces IL-27, while inhibiting OPN expression in DCs both in vitro and in vivo, and that engagement of IFNγR expressed by DCs leads to suppression of IL-17 production while inducing IL-10 from T cells (Murugaiyan et al. 2010). IL-2, a growth factor for T cells, was shown to constrain T_H17 generation through a stat-5 mediated pathway (Laurence et al. 2007). Stat-5 suppresses IL-17 expression by direct binding to the IL-17 promoter. Addition of IL-2 to T cell cultures in the presence of TGF-β and IL-6 was able not only to block the conversion of naïve T cells into T_H17 cells, but also to drive the generation of Tregs (Laurence et al. 2007).

7 T_H17 Cell Differentiation During Organ Transplantation

As discussed above, IL-17 production is detected at the early stage of post-transplantation and T_H17 development requires the presence of pro-inflammatory conditions. It is reasonable to argue that early events during the transplantation procedure result in production of inflammatory mediators that drive T_H17 cell differentiation. The earliest event that affects the local cytokine milieu and the context of how naïve T cells recognize alloantigens is the transplantation procedure itself. Each allograft must undergo ischemic reperfusion injury (IRI) during the transplantation process. As a consequence, it stimulates the release of numerous inflammatory mediators, including IL-1α, prostaglandin E2 (PGE2), ATP, and high mobility group protein (HMGB1). These mediators act directly and indirectly on the balance of T_H17 response. For example, PGE_2 is able to directly promote the differentiation, expansion, and pro-inflammatory function of human and mouse T_H17 cells (Boniface et al. 2009; Yao et al. 2009). Indirectly, these mediators act via their ability to induce production of pro-inflammatory cytokines, such as IL-1, IL-6, and TGF-β by resident macrophages (Rao and Pober 2008). As a result, the pro-inflammatory cytokine expression leads to the induction of T_H17 cells (Zhang et al. 2010). In fact, mice which received anti-IL-6 antibody treatment showed attenuated renal dysfunction caused by IRI, and had reduced neutrophil infiltration, one of the hallmark effector mechanisms of T_H17 response (Patel et al. 2005). In line with this observation, lung ischemia reperfusion injury has shown to induce type V collagen-specific T cells. As mentioned above, these cells mediated lung allograft rejection and produced IL-17 (Yoshida et al. 2006). The severity of tracheal obliteration and IL-17 mRNA level was significantly suppressed in tracheal allografts after IL-6 neutralization with anti-IL-6 antibody (Nakagiri et al. 2010). In addition to IRI, peri-operative infection could also shape the Th responses and modify the mechanism of rejection. Infections with micro-organisms of different origins can also trigger TLR activation. Some of the microbial products preferentially induced IL-12 expression and promoted T_H1 response, while others activated DCs to produce IL-23, promoting T_H17 response (Basso et al. 2009). Indeed, CpG mediated rejection was shown to be dependent on IL-17 (Chen et al. 2009b). During the allograft rejection process, T_H17 responses developed when T_H1 responses were impaired (Yuan et al. 2008). Thus, it is conceivable that there could be a risk of using treatments specifically targeting on T_H1 or T_H2 response, as it could promote T_H17 mediated rejection instead.

8 Regulation of T_H17 Cells by Immuno-Suppressive Agents

Given the clinical effectiveness of the current immuno-suppressive modalities in organ transplantation, it appears that T_H17 cells should also be susceptible. Although only limited information is available, some studies have suggested that current immuno-suppressive regimes which are effective for T_H1 response are also effective in controlling T_H17 cells. Cyclosporin A (CsA), a calcineurin inhibitor commonly used as part of the immuno-suppressive regimen, was able to inhibit IL-17 production (Liu et al. 2009; Zhang et al. 2008; Ziolkowska et al. 2000). Usage of $CD4^+$ T cells isolated from RA patients with CsA suppressed the IL-17 production and inhibited T_H17 cell differentiation (Cho et al. 2007; Zhang et al. 2008). Similar results were demonstrated in psoriasis patients upon treatment with CsA (Lowes et al. 2008). The efficacy of CsA-mediated suppression of T_H17 cells could correlate with clinical disease activity as shown in Vogt–Koyanagi–Harada (VDH) syndrome (Liu et al. 2009). Although CsA is an effective treatment to control both T_H17 and T_H1 responses, it has been shown to interfere with tolerance induction in experimental models.

Other immuno-suppressive agents have also been reported to control T_H17 responses. In addition to CsA, dexamethasone (DEX) was able to inhibit IL-17 production. However, its effect on T_H17 seems to be less effective than that of CsA. An in vitro analysis of PBMCs from VKH patients has shown that both CsA and DEX inhibited the production of IFNγ efficiently. IL-17 secretion was completely blocked by the presence of low doses CsA, while only partial blockade was observed with high doses of DEX (Liu et al. 2009). Rapamycin has also shown to be a much stronger inhibitor of IL-17 production than DEX, although its effect on T_H1 response is less pronounced (Yang et al. 2009). Rapamycin does not have the same detrimental effect on Treg cell development as CsA does. Rapamycin was able to promote the differentiation of Tregs and inhibit the generation of T_H17 cells, while CsA inhibited the differentiation of both Tregs and T_H17 cells (Kopf et al. 2007). Thus, Rapamycin seems to a desirable agent in controlling graft rejection. It is interesting to see whether other immuno-suppressive agents, such as mycophenolate mofetil (Abatacept), would have an effect on T_H17 cell differentiation.

Although the current immuno-suppressive regimens have improved significantly, there are many side effects associated with the long-term use of immuno-suppressive agents. The avoidance of long-term immuno-suppression by achieving immunological tolerance would be the ultimate solution to improve long-term allograft survival. Tolerogenic strategies should not only suppress T_H1/T_H2 responses, but also T_H17 responses. The development of T_H17 cells could render resistance to tolerance induction. One study has reported that $CD8^+$ T_H17 cells were capably of mediating co-stimulation blockade-resistant rejection in T-bet$^{-/-}$ allograft recipients. Depleting $CD8^+$ cells, neutralizing IL-17, or the T_H17-inducing cytokine IL-6, ablated the T_H17 response and reversed co-stimulation blockade-resistant graft rejection (Burrell et al. 2008). TLR9 signaling by exogenous CpG at the time of transplantation is sufficient to abrogate anti-CD154-mediated acceptance of fully mismatched cardiac allografts. The effect was previously attributed to the production of IL-6 which altered the inhibitory effect of Tregs. However, recent evidence suggested that T_H17 response is also responsible for the resistance of tolerance induction. Injection of anti-IL-17 antibody reversed the abrogative effect of CpG administration on anti-CD154 mediated cardiac allograft acceptance in IL-6 deficient mice (Chen et al. 2009b). Another factor which is thought to be necessary for anti-CD154-mediated tolerance is the development of Tregs (Lee et al. 2005). In fact, the suppressive activity of Tregs has been associated with the induction of transplantation tolerance in many experimental models. The idea of inducing tolerance through the therapeutic administration of Tregs has received much attention in recent years. Infusion of Tregs is currently considered as potential therapy to control allograft rejection or GVHD in clinical transplantation. Given the close interplay between T_H17 and Tregs as discussed above, the effect of Tregs on T_H17 development could be an important issue.

9 Problems Associated with Tregs-Based Therapy in Controlling T_H17 Responses

The dependence of T_H17 differentiation on TGF-β in the presence of pro-inflammatory cytokine implies that the pathogenic potential of the adoptive transferred Tregs. In local or systemic inflammation during rejection, administration of Tregs induces T_H17 activities (Korn et al. 2007b; Lohr et al. 2006). Indeed, adoptive transfer of Tregs pretreated with IL-6 did not confer protection in mice with a chronic lupus-like syndrome (Zheng et al. 2008), while administration of Tregs stimulated by pre-activated dendritic cells induced pathology in a murine model of diabetes (Radhakrishnan et al. 2008). In addition to promoting T_H17 differentiation, adoptive transferred Tregs become IL-17 producing cells due to plasticity in differentiation. Yang et al. showed that after adoptive transfer into lymphopenic hosts, a significant percentage of Tregs lose FOXP3 expression following immunization with cognate antigen in complete fraud adjuvant (Yang et al. 2008b). Similar results were shown by Hori and colleagues. They demonstrated that ex-FOXP3 positive cells produced strong pro-inflammatory cytokines, including IL-17 and IFNγ (Komatsu et al. 2009). These observations suggest the possibility of shifting the Treg-T_H17 axis towards T_H17 conversion. However, it is unclear whether any circumstances under which the effect of promoting T_H17 cells by Tregs would dominate over their protective function, making them pathogenic in the context of transplantation. Further investigation into the stability of Tregs is critical to the development of effective therapy for organ transplantation rejection.

The other issue of regulation of T_H17 cells by Tregs is that they appear to have limited effect on T_H17 cells in contrast to prominent suppression on T_H1 and T_H2 responses. In a mouse model of auto-immune gastritis (AIG), freshly isolated polyclonal Tregs suppressed T_H1 and T_H2 driven AIG. In contrast, the T_H17-mediated disease could only be moderately ameliorated by adoptive transferring of Tregs during the early phase of disease progression. The adoptive transfer of Tregs at later stages failed to protect mice from the disease at all (Stummvoll et al. 2008). In addition, Tregs isolated from the central nervous system of mice with EAE suppressed IFNγ production efficiently by CNS derived T effectors in co-culture, but were unable to suppress their production of IL-17 (O'Connor et al. 2007). Consistent data from human studies have also shown that Tregs failed to exert suppressive function on T_H17 cells at the level of proliferation (Annunziato et al. 2007) and cytokine production (Evans et al. 2007). Several lines of evidence have pointed to the conclusion that T_H17 cells are more resistant to regulation by Tregs. Like ex vivo and cytokine-induced T_H17 cells, the capacity of pCCL. RORC2-transduced T cells to produce IL-17 was not suppressed by Tregs (Crome et al. 2009). This implies that the T_H17 lineage transcriptional factor contributed to the resistance of T_H17 cells to Treg-mediated suppression. Nonetheless, it appears that Tregs have some capacity for controlling T_H17 development. Depletion of Tregs could heighten the production of IL-17 and IL-6 in male mice during EAE (Reddy et al. 2005).

10 Therapeutic Strategies for Tregs-Based Therapy in Controlling T_H17 Responses

In the model of AIG, it has been demonstrated that there is a difference in therapeutic efficiency based on the different timings which Tregs were administrated. T_H17 mediated AIG is prevented by adoptive transferred polyclonal Tregs early on, but not at later stages. Similar results have been observed in a study of T_H17 driven GVHD. When Tregs were co-transferred together with effector cells, there was a decrease in the total number of IL-17-producing cells. However, a delayed administration of Tregs failed to prevent GVHD (Lohr et al. 2006). This data indicates that the timing of Treg administration is crucial for suppressing T_H17 response. Tregs have the ability to block the initial activation of T_H17 response, but have limited effect on established T_H17 immunity, implying that early intervention by Tregs at the time of transplantation is an important strategy.

Several lines of studies have shown that antigen specific Tregs are more effective than polyclonal population in controlling auto-immune disease. Antigen specific Tregs not only blocked the initial response of auto-immunity, but were also effective in controlling ongoing disease when administrated during advanced disease progression. Given the effectiveness of disease control by antigen specific Tregs, it could be argued that antigen specificity should also able to control T_H17 response. In supporting this notion, Huter et al. has shown that induced Ag-specific Tregs (iTregs) were potent inhibitors of T_H17-mediated AIG induction and even exerted considerable therapeutic effects when administered after the inflammatory response in the target organ had already been initiated. In contrast, both polyclonal natural Tregs (nTregs) and polyclonal iTregs were ineffective in preventing the induction of AIG (Stummvoll et al. 2008). Thus, Ag-specific iTregs are capable of both preventing and treating disease induced by fully differentiated T_H17 effector cells. Apostolou et al. have shown that Foxp3 expression was stable in Tregs generated in vivo, regardless of whether they are iTregs or nTregs (Apostolou et al. 2008). Interestingly, it has been shown that the generation of stable murine iTregs in vivo is specific for one epitope of the male HY antigen and is able to regulate immune responses to other HY epitopes, illustrating the concept of linked suppression (Apostolou et al. 2008).

Several studies have highlighted that different subtypes of Tregs have differential effects in controlling T_H17 activity. It has been demonstrated that CD39, an ecto-nucleotidase which hydrolyzes ATP, is expressed on a subset of both mouse and human nTregs. This $CD4^+CD25^{high}CD39^+$ Treg subset was shown to be important for controlling graft rejection. Tregs from $CD39^{-/-}$ mice had reduced suppressive capacity in vitro and those Tregs failed to prevent allograft rejection in vivo (Deaglio et al. 2007). Furthermore, recent data has suggested that $CD39^+$ Tregs were also effective in controlling both T_H17 and T_H1 responses, in contrast to the $CD39^-$ counterpart which were only able to suppress T_H1 cells. $CD39^+$ Tregs have a lower potential of become IL-17 producing cells than the $CD39^-$ Tregs (Fletcher et al. 2009). Of note, ATP has been shown to promote T_H17 cells. By the removal of ATP, the ecto-nucleotidase activity of CD39 reduces IL-17 production and inhibits T_H17 cells in vivo (Atarashi et al. 2008). An analog of adenosine, the final metabolite of ATP breakdown, which binds to the adenosine 2A receptor on T cells, effectively inhibits IL-17 as well as IFNγ production and T cell proliferation. Therefore, the selection of Treg sub-types, namely $CD39^+$ Tregs, are important in suppressing both T_H1 and T_H17 cells.

Finally, the suppressive activity of Treg on T_H17 response could be improved with adjunctive therapy by preserving or enhancing Treg function, particularly in an inflammatory milieu. One consideration for Tregs to be effective in T_H17 mediated pathology was to stabilize Treg populations during therapeutic intervention. Given the reciprocal control of T_H17 and Treg differentiation (Bettelli et al. 2006), both subsets require TGF-β signaling, but the concomitant presence of inflammatory cytokines antagonize suppression and promote T_H17 responses. Strategies that could skew the response towards protection rather than T_H17 inflammation are desirable. One possibility would be the use of anti-IL-6 agents to block T_H17 pathway in order to enable effective Treg function. In support of this hypothesis, IL-6 blockade has been shown to restrain the development of both T_H1 and T_H17 response following immunization (Serada et al. 2008). IL-6 knockout mice were more resistant to EAE after immunization with myelin oligodendrocyte glycoprotein, and it resulted in the expansion of Tregs in the absence of T_H17 response (Korn et al. 2007a). Systemic administration of anti-IL-6 receptor antibodies was able to ameliorate auto-immune uveitis by suppressing both systemic and regional T_H17 responses (Yoshimura et al. 2009). Other pro-inflammatory cytokines that have been shown to overcome Treg mediated suppression include TNFα and IL-15, and they have suggested promoting T_H17 responses. Both TNFα and IL-15 could be possible targets for adjunctive therapy (Korn et al. 2007b; van Amelsfort et al. 2007). On the other hand, immunosuppressive agents that are able to diminish the inflammatory response without compromising Treg functions could be candidates to be used together with Treg therapy. For example, Rapamycin was able to promote Treg development and function while suppressing effector responses (Battaglia et al. 2005; Strauss et al. 2007; Valmori et al. 2006). It has been suggested that Rapamycin could function as additional therapy along with Tregs (Tsang et al. 2008).

11 Conclusion

It is clear that there are many distinct effector pathways that can lead to graft rejection. Multiple cytokines are involved and they interact with each other to mediate distinct pathways. The discovery of the newly identified T_H17 subset has opened a new avenue to elucidate the graft rejection process, hence providing new therapeutic interventions. Continuing investigations in unraveling the factors affecting T_H17 differentiation in transplantation will facilitate the development of novel immunosuppressive protocols and immune monitoring of allograft rejection process. With regard to immuno-suppression and tolerance strategies, the reciprocal relation between Tregs and T_H17 cells needs further investigation. Special attention should be paid to procedures such as ischemia-reperfusion injury and peri-operative infections which influence the rejection mechanisms. With our better understanding of Th mediated immune responses in graft rejection, it is possible to revise strategies to achieve clinical transplantation tolerance in the future.

References

Acosta-Rodriguez, E.V., Rivino, L., Geginat, J., Jarrossay, D., Gattorno, M., Lanzavecchia, A., Sallusto, F., and Napolitani, G. (2007). Surface phenotype and antigenic specificity of human interleukin 17-producing T helper memory cells. Nat Immunol *8*, 639–646.

Aggarwal, S., Ghilardi, N., Xie, M.H., de Sauvage, F.J., and Gurney, A.L. (2003). Interleukin-23 promotes a distinct CD4 T cell activation state characterized by the production of interleukin-17. J Biol Chem *278*, 1910–1914.

Annunziato, F., Cosmi, L., Santarlasci, V., Maggi, L., Liotta, F., Mazzinghi, B., Parente, E., Fili, L., Ferri, S., Frosali, F., *et al.* (2007). Phenotypic and functional features of human Th17 cells. J Exp Med *204*, 1849–1861.

Antonysamy, M.A., Fanslow, W.C., Fu, F., Li, W., Qian, S., Troutt, A.B., and Thomson, A.W. (1999a). Evidence for a role of IL-17 in alloimmunity: a novel IL-17 antagonist promotes heart graft survival. Transplant Proc *31*, 93.

Antonysamy, M.A., Fanslow, W.C., Fu, F., Li, W., Qian, S., Troutt, A.B., and Thomson, A.W. (1999b). Evidence for a role of IL-17 in organ allograft rejection: IL-17 promotes the functional differentiation of dendritic cell progenitors. J Immunol *162*, 577–584.

Apostolou, I., Verginis, P., Kretschmer, K., Polansky, J., Hühn, J., von Boehmer, H. (2008). Peripherally induced Treg: Mode, stability, and role in specific tolerance. J. Clin. Immunol *28*, 619–624.

Atalar, K., Afzali, B., Lord, G., and Lombardi, G. (2009). Relative roles of Th1 and Th17 effector cells in allograft rejection. Curr Opin Organ Transplant *14*, 23–29.

Atarashi, K., Nishimura, J., Shima, T., Umesaki, Y., Yamamoto, M., Onoue, M., Yagita, H., Ishii, N., Evans, R., Honda, K., *et al.* (2008). ATP drives lamina propria T(H)17 cell differentiation. Nature *455*, 808–812.

Barbara, J.A., Turvey, S.E., Kingsley, C.I., Spriewald, B.M., Hara, M., Witzke, O., Morris, P.J., and Wood, K.J. (2000). Islet allograft rejection can be mediated by CD4+, alloantigen experienced, direct pathway T cells of TH1 and TH2 cytokine phenotype. Transplantation *70*, 1641–1649.

Basso, A.S., Cheroutre, H., and Mucida, D. (2009). More stories on Th17 cells. Cell Res *19*, 399–411.

Battaglia, M., Stabilini, A., and Roncarolo, M.G. (2005). Rapamycin selectively expands CD4+CD25+FoxP3+ regulatory T cells. Blood *105*, 4743–4748.

Batten, M., Li, J., Yi, S., Kljavin, N.M., Danilenko, D.M., Lucas, S., Lee, J., de Sauvage, F.J., and Ghilardi, N. (2006). Interleukin 27 limits auto-immune encephalomyelitis by suppressing the development of interleukin 17-producing T cells. Nat Immunol *7*, 929–936.

Benghiat, F.S., Craciun, L., De Wilde, V., Dernies, T., Kubjak, C., Lhomme, F., Goldman, M., and Le Moine, A. (2008). IL-17 production elicited by allo-major histo-compatibility complex class II recognition depends on CD25posCD4pos T cells. Transplantation *85*, 943–949.

Beriou, G., Costantino, C.M., Ashley, C.W., Yang, L., Kuchroo, V.K., Baecher-Allan, C., and Hafler, D.A. (2009). IL-17-producing human peripheral regulatory T cells retain suppressive function. Blood *113*, 4240–4249.

Bettelli, E., Carrier, Y., Gao, W., Korn, T., Strom, T.B., Oukka, M., Weiner, H.L., and Kuchroo, V.K. (2006). Reciprocal developmental pathways for the generation of pathogenic effector TH17 and regulatory T cells. Nature *441*, 235–238.

Boniface, K., Bak-Jensen, K.S., Li, Y., Blumenschein, W.M., McGeachy, M.J., McClanahan, T.K., McKenzie, B.S., Kastelein, R.A., Cua, D.J., and de Waal Malefyt, R. (2009). Prostaglandin E2 regulates Th17 cell differentiation and function through cyclic AMP and EP2/EP4 receptor signaling. J Exp Med *206*, 535–548.

Burlingham, W.J., Love, R.B., Jankowska-Gan, E., Haynes, L.D., Xu, Q., Bobadilla, J.L., Meyer, K.C., Hayney, M.S., Braun, R.K., Greenspan, D.S. et al. (2007). IL-17-dependent cellular immunity to collagen type V predisposes to obliterative bronchiolitis in human lung transplants. J Clin Invest *117*, 3498–3506.

Burrell, B.E., Csencsits, K., Lu, G., Grabauskiene, S., and Bishop, D.K. (2008). CD8+ Th17 mediate costimulation blockade-resistant allograft rejection in T-bet-deficient mice. J Immunol *181*, 3906–3914.

Chan, S.Y., DeBruyne, L.A., Goodman, R.E., Eichwald, E.J., and Bishop, D.K. (1995). *In vivo* depletion of CD8+ T cells results in Th2 cytokine production and alternate mechanisms of allograft rejection. Transplantation *59*, 1155–1161.

Chen, H., Wang, W., Xie, H., Xu, X., Wu, J., Jiang, Z., Zhang, M., Zhou, L., and Zheng, S. (2009a). A pathogenic role of IL- 17 at the early stage of corneal allograft rejection. Transpl Immunol *21*, 155–161.

Chen, L., Ahmed, E., Wang, T., Wang, Y., Ochando, J., Chong, A.S., and Alegre, M.L. (2009b). TLR signals promote IL-6/IL-17-dependent transplant rejection. J Immunol *182*, 6217–6225.

Chen, Z., Laurence, A., and O'Shea, J.J. (2007a). Signal transduction pathways and transcriptional regulation in the control of Th17 differentiation. Semin Immunol *19*, 400–408.

Chen, Z., Tato, C.M., Muul, L., Laurence, A., and O'Shea, J.J. (2007b). Distinct regulation of interleukin-17 in human T helper lymphocytes. Arthritis Rheum *56*, 2936–2946.

Cho, M.L., Ju, J.H., Kim, K.W., Moon, Y.M., Lee, S.Y., Min, S.Y., Cho, Y.G., Kim, H.S., Park, K.S., Yoon, C.H., *et al.* (2007). Cyclosporine A inhibits IL-15-induced IL-17 production in CD4+ T cells via downregulation of PI3K/Akt and NF-kappaB. Immunol Lett *108*, 88–96.

Cosmi, L., De Palma, R., Santarlasci, V., Maggi, L., Capone, M., Frosali, F., Rodolico, G., Querci, V., Abbate, G., Angeli, R., *et al.* (2008). Human interleukin 17-producing cells originate from a CD161+CD4+ T cell precursor. J Exp Med *205*, 1903–1916.

Crome, S.Q., Wang, A.Y., Kang, C.Y., and Levings, M.K. (2009). The role of retinoic acid-related orphan receptor variant 2 and IL-17 in the development and function of human CD4+ T cells. Eur J Immunol *39*, 1480–1493.

D'Elios, M.M., Josien, R., Manghetti, M., Amedei, A., de Carli, M., Cuturi, M.C., Blancho, G., Buzelin, F., del Prete, G., and Soulillou, J.P. (1997). Predominant Th1 cell infiltration in acute rejection episodes of human kidney grafts. Kidney Int *51*, 1876–1884.

de Jong, E., Suddason, T., and Lord, G.M. (2010). Translational Mini-Review Series on Th17 Cells: Development of mouse and human T helper 17 cells. Clin Exp Immunol Clin Exp Immunol *159*, 148–58.

Deaglio, S., Dwyer, K.M., Gao, W., Friedman, D., Usheva, A., Erat, A., Chen, J.F., Enjyoji, K., Linden, J., Oukka, M., *et al.* (2007). Adenosine generation catalyzed by CD39 and CD73 expressed on regulatory T cells mediates immune suppression. J Exp Med *204*, 1257–1265.

Deteix, C., Attuil-Audenis, V., Duthey, A., Patey, N., McGregor, B., Dubois, V., Caligiuri, G., Graff-Dubois, S., Morelon, E., and Thaunat, O. (2010). Intragraft Th17 infiltrate promotes lymphoid neogenesis and hastens clinical chronic rejection. J Immunol *184*, 5344–5351.

Evans, H.G., Suddason, T., Jackson, I., Taams, L.S., and Lord, G.M. (2007). Optimal induction of T helper 17 cells in humans requires T cell receptor ligation in the context of Toll-like receptor-activated monocytes. Proc Natl Acad Sci USA *104*, 17034–17039.

Fabrega, E., Lopez-Hoyos, M., San Segundo, D., Casafont, F., and Pons-Romero, F. (2009). Changes in the serum levels of interleukin-17/interleukin-23 during acute rejection in liver transplantation. Liver Transpl *15*, 629–633.

Faust, S.M., Lu, G., Marini, B.L., Zou, W., Gordon, D., Iwakura, Y., Laouar, Y., and Bishop, D.K. (2009). Role of T cell TGFbeta signaling and IL-17 in allograft acceptance and fibrosis associated with chronic rejection. J Immunol *183*, 7297–7306.

Fletcher, J.M., Lonergan, R., Costelloe, L., Kinsella, K., Moran, B., O'Farrelly, C., Tubridy, N., and Mills, K.H. (2009). CD39+Foxp3+ regulatory T Cells suppress pathogenic Th17 cells and are impaired in multiple sclerosis. J Immunol *183*, 7602–7610.

Fukushi, J., Ono, M., Morikawa, W., Iwamoto, Y., and Kuwano, M. (2000). The activity of soluble VCAM-1 in angiogenesis stimulated by IL-4 and IL-13. J Immunol *165*, 2818–2823.

Haque, M.A., Mizobuchi, T., Yasufuku, K., Fujisawa, T., Brutkiewicz, R.R., Zheng, Y., Woods, K., Smith, G.N., Cummings, O.W., Heidler, K.M., *et al.* (2002). Evidence for immune responses to a self-antigen in lung transplantation: role of type V collagen-specific T cells in the pathogenesis of lung allograft rejection. J Immunol *169*, 1542–1549.

Healy, D.G., Watson, R.W., O'Keane, C., Egan, J.J., McCarthy, J.F., Hurley, J., Fitzpatrick, J., and Wood, A.E. (2006). Neutrophil transendothelial migration potential predicts rejection severity in human cardiac transplantation. Eur J Cardiothorac Surg *29*, 760–766.

Hogan, S.P., and Foster, P.S. (1996). Cellular and molecular mechanisms involved in the regulation of eosinophil trafficking in vivo. Med Res Rev *16*, 407–432.

Hsieh, H.G., Loong, C.C., Lui, W.Y., Chen, A., and Lin, C.Y. (2001). IL-17 expression as a possible predictive parameter for subclinical renal allograft rejection. Transpl Int *14*, 287–298.

Iclozan, C., Yu, Y., Liu, C., Liang, Y., Yi, T., Anasetti, C., and Yu, X.Z. (2010). Th17 cells are sufficient but not necessary to induce acute graft-versus-host disease. Biol Blood Marrow Transplant *16*, 170–8

Ivanov, II, McKenzie, B.S., Zhou, L., Tadokoro, C.E., Lepelley, A., Lafaille, J.J., Cua, D.J., and Littman, D.R. (2006). The orphan nuclear receptor RORgammat directs the differentiation program of pro-inflammatory IL-17+ T helper cells. Cell *126*, 1121–1133.

Komatsu, N., Mariotti-Ferrandiz, M.E., Wang, Y., Malissen, B., Waldmann, H., and Hori, S. (2009). Heterogeneity of natural Foxp3+ T cells: a committed regulatory T-cell lineage and an uncommitted minor population retaining plasticity. Proc Natl Acad Sci USA *106*, 1903–1908.

Kopf, H., de la Rosa, G.M., Howard, O.M., and Chen, X. (2007). Rapamycin inhibits differentiation of Th17 cells and promotes generation of FoxP3+ T regulatory cells. Int Immunopharmacol *7*, 1819–1824.

Kopf, M., Brombacher, F., Hodgkin, P.D., Ramsay, A.J., Milbourne, E.A., Dai, W.J., Ovington, K.S., Behm, C.A., Kohler, G., Young, I.G., *et al.* (1996). IL-5-deficient mice have a developmental defect in CD5+ B-1 cells and lack eosinophilia but have normal antibody and cytotoxic T cell responses. Immunity *4*, 15–24.

Korn, T., Bettelli, E., Gao, W., Awasthi, A., Jager, A., Strom, T.B., Oukka, M., and Kuchroo, V.K. (2007a). IL-21 initiates an alternative pathway to induce pro-inflammatory T(H)17 cells. Nature *448*, 484–487.

Korn, T., Reddy, J., Gao, W., Bettelli, E., Awasthi, A., Petersen, T.R., Backstrom, B.T., Sobel, R.A., Wucherpfennig, K.W., Strom, T.B., *et al.* (2007b). Myelin-specific regulatory T cells accumulate in the CNS but fail to control auto-immune inflammation. Nat Med *13*, 423–431.

Kryczek, I., Wei, S., Vatan, L., Escara-Wilke, J., Szeliga, W., Keller, E.T., and Zou, W. (2007). Cutting edge: opposite effects of IL-1 and IL-2 on the regulation of IL-17+ T cell pool IL-1 subverts IL-2-mediated suppression. J Immunol *179*, 1423–1426.

Laurence, A., Tato, C.M., Davidson, T.S., Kanno, Y., Chen, Z., Yao, Z., Blank, R.B., Meylan, F., Siegel, R., Hennighausen, L. et al. (2007). Interleukin-2 signaling via STAT5 constrains T helper 17 cell generation. Immunity *26*, 371–381.

Le Moine, A., Goldman, M., and Abramowicz, D. (2002). Multiple pathways to allograft rejection. Transplantation *73*, 1373–1381.

Lee, I., Wang, L., Wells, A.D., Dorf, M.E., Ozkaynak, E., and Hancock, W.W. (2005). Recruitment of Foxp3+ T regulatory cells mediating allograft tolerance depends on the CCR4 chemokine receptor. J Exp Med *201*, 1037–1044.

Li, J., Simeoni, E., Fleury, S., Dudler, J., Fiorini, E., Kappenberger, L., von Segesser, L.K., and Vassalli, G. (2006). Gene transfer of soluble interleukin-17 receptor prolongs cardiac allograft survival in a rat model. Eur J Cardiothorac Surg *29*, 779–783.

Liu, X., Yang, P., Lin, X., Ren, X., Zhou, H., Huang, X., Chi, W., Kijlstra, A., and Chen, L. (2009). Inhibitory effect of Cyclosporin A and corticosteroids on the production of IFN-gamma and IL-17 by T cells in Vogt-Koyanagi-Harada syndrome. Clin Immunol *131*, 333–342.

Lohr, J., Knoechel, B., Wang, J.J., Villarino, A.V., and Abbas, A.K. (2006). Role of IL-17 and regulatory T lymphocytes in a systemic auto-immune disease. J Exp Med *203*, 2785–2791.

Loong, C.C., Hsieh, H.G., Lui, W.Y., Chen, A., and Lin, C.Y. (2002). Evidence for the early involvement of interleukin 17 in human and experimental renal allograft rejection. J Pathol *197*, 322–332.

Lowes, M.A., Kikuchi, T., Fuentes-Duculan, J., Cardinale, I., Zaba, L.C., Haider, A.S., Bowman, E.P., and Krueger, J.G. (2008). Psoriasis vulgaris lesions contain discrete populations of Th1 and Th17 T cells. J Invest Dermatol *128*, 1207–1211.

Lu, L., Wang, J., Zhang, F., Chai, Y., Brand, D., Wang, X., Horwitz, D.A., Shi, W., and Zheng, S.G. (2010). Role of SMAD and non-SMAD signals in the development of Th17 and regulatory T cells. J Immunol *184*, 4295–4306.

Manel, N., Unutmaz, D., and Littman, D.R. (2008). The differentiation of human T(H)-17 cells requires transforming growth factor-beta and induction of the nuclear receptor RORgammat. Nat Immunol *9*, 641–649.

McGeachy, M.J., Bak-Jensen, K.S., Chen, Y., Tato, C.M., Blumenschein, W., McClanahan, T., and Cua, D.J. (2007). TGF-beta and IL-6 drive the production of IL-17 and IL-10 by T cells and restrain T(H)-17 cell-mediated pathology. Nat Immunol *8*, 1390–1397.

Miura, M., El-Sawy, T., and Fairchild, R.L. (2003). Neutrophils mediate parenchymal tissue necrosis and accelerate the rejection of complete major histocompatibility complex-disparate cardiac allografts in the absence of interferon-gamma. Am J Pathol *162*, 509–519.

Murugaiyan, G., Mittal, A., and Weiner, H.L. (2010). Identification of an IL-27/osteopontin axis in dendritic cells and its modulation by IFN-{gamma} limits IL-17-mediated auto-immune inflammation. Proc Natl Acad Sci USA.

Nakagiri, T., Inoue, M., Morii, E., Minami, M., Sawabata, N., Utsumi, T., Kadota, Y., Ideguchi, K., Tokunaga, T., and Okumura, M. (2010). Local IL-17 production and a decrease in peripheral blood regulatory T cells in an animal model of bronchiolitis obliterans. Transplantation *89*, 1312–1319.

O'Connor, R.A., Malpass, K.H., and Anderton, S.M. (2007). The inflamed central nervous system drives the activation and rapid proliferation of Foxp3+ regulatory T cells. J Immunol *179*, 958–966.

Osorio, F., LeibundGut-Landmann, S., Lochner, M., Lahl, K., Sparwasser, T., Eberl, G., and Reis e Sousa, C. (2008). DC activated via dectin-1 convert Treg into IL-17 producers. Eur J Immunol *38*, 3274–3281.

Patel, N.S., Chatterjee, P.K., Di Paola, R., Mazzon, E., Britti, D., De Sarro, A., Cuzzocrea, S., and Thiemermann, C. (2005). Endogenous interleukin-6 enhances the renal injury, dysfunction, and inflammation caused by ischemia/reperfusion. J Pharmacol Exp Ther *312*, 1170–1178.

Piccotti, J.R., Chan, S.Y., Li, K., Eichwald, E.J., and Bishop, D.K. (1997). Differential effects of IL-12 receptor blockade with IL-12 p40 homodimer on the induction of CD4+ and CD8+ IFN-gamma-producing cells. J Immunol *158*, 643–648.

Qin, H., Wang, L., Feng, T., Elson, C.O., Niyongere, S.A., Lee, S.J., Reynolds, S.L., Weaver, C.T., Roarty, K., Serra, R., *et al.* (2009). TGF-beta promotes Th17 cell development through inhibition of SOCS3. J Immunol *183*, 97–105.

Radhakrishnan, S., Cabrera, R., Schenk, E.L., Nava-Parada, P., Bell, M.P., Van Keulen, V.P., Marler, R.J., Felts, S.J., and Pease, L.R. (2008). Reprogrammed FoxP3+ T regulatory cells become IL-17+ antigen-specific auto-immune effectors *in vitro* and *in vivo*. J Immunol *181*, 3137–3147.

Rao, D.A., and Pober, J.S. (2008). Endothelial injury, alarmins, and allograft rejection. Crit Rev Immunol *28*, 229–248.

Reddy, J., Waldner, H., Zhang, X., Illes, Z., Wucherpfennig, K.W., Sobel, R.A., and Kuchroo, V.K. (2005). Cutting edge: CD4+CD25+ regulatory T cells contribute to gender differences in susceptibility to experimental auto-immune encephalomyelitis. J Immunol *175*, 5591–5595.

Santarlasci, V., Maggi, L., Capone, M., Frosali, F., Querci, V., De Palma, R., Liotta, F., Cosmi, L., Maggi, E., Romagnani, S. et al. (2009). TGF-beta indirectly favors the development of human Th17 cells by inhibiting Th1 cells. Eur J Immunol *39*, 207–215.

Serada, S., Fujimoto, M., Mihara, M., Koike, N., Ohsugi, Y., Nomura, S., Yoshida, H., Nishikawa, T., Terabe, F., Ohkawara, T., *et al.* (2008). IL-6 blockade inhibits the induction of myelin antigen-specific Th17 cells and Th1 cells in experimental auto-immune encephalomyelitis. Proc Natl Acad Sci U S A *105*, 9041–9046.

Strauss, L., Whiteside, T.L., Knights, A., Bergmann, C., Knuth, A., and Zippelius, A. (2007). Selective survival of naturally occurring human CD4+CD25+Foxp3+ regulatory T cells cultured with rapamycin. J Immunol *178*, 320–329.

Stumhofer, J.S., Laurence, A., Wilson, E.H., Huang, E., Tato, C.M., Johnson, L.M., Villarino, A.V., Huang, Q., Yoshimura, A., Sehy, D. et al. (2006). Interleukin 27 negatively regulates the development of interleukin 17-producing T helper cells during chronic inflammation of the central nervous system. Nat Immunol *7*, 937–945.

Stumhofer, J.S., Silver, J.S., Laurence, A., Porrett, P.M., Harris, T.H., Turka, L.A., Ernst, M., Saris, C.J., O'Shea, J.J., and Hunter, C.A. (2007). Interleukins 27 and 6 induce STAT3-mediated T cell production of interleukin 10. Nat Immunol *8*, 1363–1371.

Stummvoll, G.H., DiPaolo, R.J., Huter, E.N., Davidson, T.S., Glass, D., Ward, J.M., and Shevach, E.M. (2008). Th1, Th2, and Th17 effector T cell-induced auto-immune gastritis differs in pathological pattern and in susceptibility to suppression by regulatory T cells. J Immunol *181*, 1908–1916.

Surquin, M., Le Moine, A., Flamand, V., Rombaut, K., Demoor, F.X., Salmon, I., Goldman, M., and Abramowicz, D. (2005). IL-4 deficiency prevents eosinophilic rejection and uncovers a role for neutrophils in the rejection of MHC class II disparate skin grafts. Transplantation *80*, 1485–1492.

Suto, A., Kashiwakuma, D., Kagami, S., Hirose, K., Watanabe, N., Yokote, K., Saito, Y., Nakayama, T., Grusby, M.J., Iwamoto, I., *et al.* (2008). Development and characterization of IL-21-producing CD4+ T cells. J Exp Med *205*, 1369–1379.

Sutton, C., Brereton, C., Keogh, B., Mills, K.H., and Lavelle, E.C. (2006). A crucial role for interleukin (IL)-1 in the induction of IL-17-producing T cells that mediate auto-immune encephalomyelitis. J Exp Med *203*, 1685–1691.

Tang, J.L., Subbotin, V.M., Antonysamy, M.A., Troutt, A.B., Rao, A.S., and Thomson, A.W. (2001). Interleukin-17 antagonism inhibits acute but not chronic vascular rejection. Transplantation *72*, 348–350.

Tsang, J.Y., Tanriver, Y., Jiang, S., Xue, S.A., Ratnasothy, K., Chen, D., Stauss, H.J., Bucy, R.P., Lombardi, G., and Lechler, R. (2008). Conferring indirect allospecificity on CD4+CD25+ Tregs by TCR gene transfer favors transplantation tolerance in mice. J Clin Invest *118*, 3619–3628.

Uboldi de Capei, M., Dametto, E., Fasano, M.E., Messina, M., Pratico, L., Rendine, S., Segoloni, G., and Curtoni, E.S. (2004). Cytokines and chronic rejection: a study in kidney transplant long-term survivors. Transplantation *77*, 548–552.

Valmori, D., Tosello, V., Souleimanian, N.E., Godefroy, E., Scotto, L., Wang, Y., and Ayyoub, M. (2006). Rapamycin-mediated enrichment of T cells with regulatory activity in stimulated CD4+ T cell cultures is not due to the selective expansion of naturally occurring regulatory T cells but to the induction of regulatory functions in conventional CD4+ T cells. J Immunol *177*, 944–949.

van Amelsfort, J.M., van Roon, J.A., Noordegraaf, M., Jacobs, K.M., Bijlsma, J.W., Lafeber, F.P., and Taams, L.S. (2007). Pro-inflammatory mediator-induced reversal of CD4+,CD25+ regulatory T cell-mediated suppression in rheumatoid arthritis. Arthritis Rheum *56*, 732–742.

Van Kooten, C., Boonstra, J.G., Paape, M.E., Fossiez, F., Banchereau, J., Lebecque, S., Bruijn, J.A., De Fijter, J.W., Van Es, L.A., and Daha, M.R. (1998). Interleukin-17 activates human renal epithelial cells *in vitro* and is expressed during renal allograft rejection. J Am Soc Nephrol *9*, 1526–1534.

Vanaudenaerde, B.M., De Vleeschauwer, S.I., Vos, R., Meyts, I., Bullens, D.M., Reynders, V., Wuyts, W.A., Van Raemdonck, D.E., Dupont, L.J., and Verleden, G.M. (2008). The role of the IL23/IL17 axis in bronchiolitis obliterans syndrome after lung transplantation. Am J Transplant *8*, 1911–1920.

Vanaudenaerde, B.M., Dupont, L.J., Wuyts, W.A., Verbeken, E.K., Meyts, I., Bullens, D.M., Dilissen, E., Luyts, L., Van Raemdonck, D.E., and Verleden, G.M. (2006). The role of interleukin-17 during acute rejection after lung transplantation. Eur Respir J *27*, 779–787.

Veldhoen, M., Hocking, R.J., Atkins, C.J., Locksley, R.M., and Stockinger, B. (2006). TGFbeta in the context of an inflammatory cytokine milieu supports de novo differentiation of IL-17-producing T cells. Immunity *24*, 179–189.

Volpe, E., Servant, N., Zollinger, R., Bogiatzi, S.I., Hupe, P., Barillot, E., and Soumelis, V. (2008). A critical function for transforming growth factor-beta, interleukin 23 and pro-inflammatory cytokines in driving and modulating human T(H)-17 responses. Nat Immunol *9*, 650–657.

Wang, H., DeVries, M.E., Deng, S., Khandaker, M.H., Pickering, J.G., Chow, L.H., Garcia, B., Kelvin, D.J., Zhong, R. (2000). The axis of interleukin 12 and gamma interferon regulates acute vascular xenogeneic rejection. Nat Med *6*, 549–55.

Weaver, C.T., Harrington, L.E., Mangan, P.R., Gavrieli, M., and Murphy, K.M. (2006). Th17: an effector CD4 T cell lineage with regulatory T cell ties. Immunity *24*, 677–688.

Wu, T., Bond, G., Martin, D., Nalesnik, M.A., Demetris, A.J., and Abu-Elmagd, K. (2006). Histopathologic characteristics of human intestine allograft acute rejection in patients pre-treated with thymoglobulin or alemtuzumab. Am J Gastroenterol *101*, 1617–1624.

Xu, L., Kitani, A., Fuss, I., and Strober, W. (2007). Cutting edge: regulatory T cells induce CD4+CD25-Foxp3- T cells or are self-induced to become Th17 cells in the absence of exogenous TGF-beta. J Immunol *178*, 6725–6729.

Yang, K., Wen, J., Liu, X., Kijlstra, A., Chen, L., Chi, W., Zhou, H., Huang, X., and Yang, P. (2009). Inhibitory effect of rapamycin and dexamethasone on production of IL-17 and IFN-gamma in Vogt-Koyanagi-Harada patients. Br J Ophthalmol *93*, 249–253.

Yang, L., Anderson, D.E., Baecher-Allan, C., Hastings, W.D., Bettelli, E., Oukka, M., Kuchroo, V.K., and Hafler, D.A. (2008a). IL-21 and TGF-beta are required for differentiation of human T(H)17 cells. Nature *454*, 350–352.

Yang, X.O., Panopoulos, A.D., Nurieva, R., Chang, S.H., Wang, D., Watowich, S.S., and Dong, C. (2007). STAT3 regulates cytokine-mediated generation of inflammatory helper T cells. J Biol Chem *282*, 9358–9363.

Yang, X.O., Pappu, B.P., Nurieva, R., Akimzhanov, A., Kang, H.S., Chung, Y., Ma, L., Shah, B., Panopoulos, A.D., Schluns, K.S., *et al.* (2008b). T helper 17 lineage differentiation is programmed by orphan nuclear receptors ROR alpha and ROR gamma. Immunity *28*, 29–39.

Yao, C., Sakata, D., Esaki, Y., Li, Y., Matsuoka, T., Kuroiwa, K., Sugimoto, Y., and Narumiya, S. (2009). Prostaglandin E2-EP4 signaling promotes immune inflammation through Th1 cell differentiation and Th17 cell expansion. Nat Med *15*, 633–640.

Yi, T., Chen, Y., Wang, L., Du, G., Huang, D., Zhao, D., Johnston, H., Young, J., Todorov, I., Umetsu, D.T., *et al.* (2009). Reciprocal differentiation and tissue-specific pathogenesis of Th1, Th2, and Th17 cells in graft-versus-host disease. Blood *114*, 3101–3112.

Yoshida, S., Haque, A., Mizobuchi, T., Iwata, T., Chiyo, M., Webb, T.J., Baldridge, L.A., Heidler, K.M., Cummings, O.W., Fujisawa, T., *et al.* (2006). Anti-type V collagen lymphocytes that express IL-17 and IL-23 induce rejection pathology in fresh and well-healed lung transplants. Am J Transplant *6*, 724–735.

Yoshimura, T., Sonoda, K.H., Ohguro, N., Ohsugi, Y., Ishibashi, T., Cua, D.J., Kobayashi, T., Yoshida, H., and Yoshimura, A. (2009). Involvement of Th17 cells and the effect of anti-IL-6 therapy in auto-immune uveitis. Rheumatology (Oxford) *48*, 347–354.

Yuan, X., Paez-Cortez, J., Schmitt-Knosalla, I., D'Addio, F., Mfarrej, B., Donnarumma, M., Habicht, A., Clarkson, M.R., Iacomini, J., Glimcher, L.H. et al. (2008). A novel role of CD4 Th17 cells in mediating cardiac allograft rejection and vasculopathy. J Exp Med *205*, 3133–3144.

Zhang, C., Zhang, J., Yang, B., and Wu, C. (2008). Cyclosporin A inhibits the production of IL-17 by memory Th17 cells from healthy individuals and patients with rheumatoid arthritis. Cytokine *42*, 345–352.

Zhang, N., Kruger, B., Lal, G., Luan, Y., Yadav, A., Zang, W., Grimm, M., Waaga-Gasser, A.M., Murphy, B., Bromberg, J.S. et al. (2010). Inhibition of TLR4 signaling prolongs Treg-dependent murine islet allograft survival. Immunol Lett *127*, 119–125.

Zheng, S.G., Wang, J., and Horwitz, D.A. (2008). Cutting edge: Foxp3+CD4+CD25+ regulatory T cells induced by IL-2 and TGF-beta are resistant to Th17 conversion by IL-6. J Immunol *180*, 7112–7116.

Zhou, L., Ivanov, II, Spolski, R., Min, R., Shenderov, K., Egawa, T., Levy, D.E., Leonard, W.J., and Littman, D.R. (2007). IL-6 programs T(H)-17 cell differentiation by promoting sequential engagement of the IL-21 and IL-23 pathways. Nat Immunol *8*, 967–974.

Zhou, L., Lopes, J.E., Chong, M.M., Ivanov, II, Min, R., Victora, G.D., Shen, Y., Du, J., Rubtsov, Y.P., Rudensky, A.Y. et al. (2008). TGF-beta-induced Foxp3 inhibits T(H)17 cell differentiation by antagonizing RORgammat function. Nature *453*, 236–240.

Ziolkowska, M., Koc, A., Luszczykiewicz, G., Ksiezopolska-Pietrzak, K., Klimczak, E., Chwalinska-Sadowska, H., and Maslinski, W. (2000). High levels of IL-17 in rheumatoid arthritis patients: IL-15 triggers in vitro IL-17 production via cyclosporin A-sensitive mechanism. J Immunol *164*, 2832–2838.

T_H17 Cells, Proteins Associated with T_H17 Polarization, and Their Role in Graft vs. Host Disease

Jonathan S. Serody, Timothy P. Moran, and Bruce R. Blazar

Abstract Graft vs. host disease (GVHD) is mediated by donor T cells that recognize minor and major MHC disparities in the host and initiate an inflammatory cascade the results in tissue damage in specific GVHD target organs. Over the past 20 years, multiple different T cell subsets under control of specific transcription factors have been described. Of these, the T_H1 pathway in which T cells generate IFNγ under control of the transcription factor T-bet has been most closely associated with GVHD. Quite recently, several new T cell subsets have been discovered. T_H17 cells are a recently defined subset in which the cells are differentiated in the presence of IL-6, TGF-β1, and IL-1β and expanded by the cytokine IL-23. RORγt and RORα are the critical transcription factors necessary for the generation of T_H17 cells. Here, we have reviewed the current data on the roles of T_H17 cells and the proteins critical for their generation in the pathogenesis of acute and chronic GVHD. Both data from murine models and clinical studies have been evaluated. While a definitive role for T_H17 cells remains to be established, current data suggests that T_H1/T_H17 cells together may be critical mediators of tissue pathology during GVHD.

1 Introduction

Allogeneic stem cell transplantation (allo-SCT) is the preferred treatment for patients with relapsed leukemia, myelodysplastic syndrome, bone marrow failure syndromes, and congenital immuno-deficiency diseases, and is increasingly used for the treatment of patients with low-grade lymphoid malignancies, specific types of myeloproliferative diseases, and inherited mucopolysaccharidoses (AS and

B.R. Blazar (✉)
Department of Pediatrics, Division of Blood and Marrow Transplantation,
University of Minnesota Cancer Center, Minneapolis, MN 55455, USA
e-mail: blaza001@umn.edu

S. Jiang (ed.), *T_H17 Cells in Health and Disease*,
DOI 10.1007/978-1-4419-9371-7_18, © Springer Science+Business Media, LLC 2011

Cowan 1997; Attal et al. 1994; Bensinger et al. 1997; Burnett and Eden 1997; Butturini and Gale 1994; Forman 1998; Gribben 1997; Khouri and Champlin 2004; Levitt and Lin 1996; Smith 2003; Storb et al. 1986; Tabbara and Ingram 2003; Vlachos and Lipton 1996). It is estimated that 25,000 patients will undergo an allo-SCT in 2009. However, approximately 100,000 individuals could benefit from allo-SCT if they had a suitable donor (Horowitz CIBMTR, personal communication). The primary reason that HLA identical donors are preferred for allo-SCT is the occurrence of graft vs. host disease (GVHD), which is mediated by donor T cells that recognize major or minor histocompatibility complex (MHC) disparities present in the recipient (Blazar et al. 1997; Chao and Schlegel 1995; Ferrara et al. 1996; Klingebiel and Schlegel 1998; Krenger and Ferrara 1996; Lazarus et al. 1997; Mielcarek et al. 2003). Because the incidence and severity of GVHD is directly related to matching MHC proteins between donor and recipient, most transplant centers do not offer transplants to individuals with a two antigen or greater mismatch with the donor. Thus, approaches that limit the occurrence of GVHD are critical for the greater adoption of allo-SCT in the treatment of specific diseases.

2 GVHD Biology

The central tenets for GVHD biology were described by Billingham over 35 years ago (Billingham 1966). These indicated that; (1) the host must be incapable of rejecting the graft, (2) the graft must contain immuno-competent cells, and (3) there must be incompatibilities in transplantation antigens between donor and host. Subsequently, the immuno-competent cells were shown to be T cells that were contained in the donor marrow or much later in the peripheral blood stem cell fraction. These naïve T cells mediate GVHD in phases as described by Ferrara and colleagues (Ferrara et al. 1996; Krenger et al. 1997). The conditioning phase is mediated by the damage of host tissue in the pre-transplant conditioning regimen that leads to the upregulation of inflammatory mediators and cell adhesion molecules and the enhanced expression by recipient antigen presenting cells (APCs) of MHC proteins and co-stimulatory molecules. In the activation phase, donor T cells interact with host APCs, leading to activation and differentiation toward a presumed T helper 1- T-cytotoxic 1 (T_H1/Tc1) pathway and subsequent migration of these cells to target tissue affected during acute GVHD. Finally, the effector phase involves target organ damage mediated by cytolytic and other effector mechanisms such as the production of TNF-α, perforin and granzyme, reactive oxygen species (Wysocki et al. 2005), and Fas/FasL (CD95/CD95L) interactions.

This model would suggest several aspects of GVHD biology that have not been supported by subsequent evaluations. If acute GVHD is mediated by T_H1/Tc1 T cells, the infusion of T cells incapable of generating IFN-γ should prevent, or at least diminish the severity of acute GVHD. However, this has not been found. Murphy et al. evaluated whether donor T cells incapable of generating IFN-γ could mediate GVHD in a murine model (Murphy et al. 1998). IFNKO ($H2^b$) donor cells were given to lethally irradiated B10.BR ($H2^k$) recipient mice. The median survival for recipient mice given IFNKO donor cells was 21 days compared to 38 days for the use of wild type (WT) T cells. This finding was duplicated using BALB/c IFNKO donor cells given to lethally irradiated B6 recipients demonstrating that this finding was not strain specific. Finally, the authors demonstrated a similar effect using anti-IFN-γ mAb. This data suggested that not only was IFN-γ not critical for the generation of acute GVHD, but that its absence exacerbated GVHD lethality. The mechanism behind this finding was not clear, although the authors did show that T cells from IFNKO mice generated more IL-2 after stimulation.

3 Signaling Proteins and GVHD

A second approach to evaluating the function of different T cell subsets in GVHD was performed using T cells lacking signaling proteins critical to the generation of IFN-γ, and as a result T_H1 T cells. The transcription factor *STAT4* is critical for signaling through the T_H1 polarizing cytokine, IL-12 (Jacobson et al. 1995), and as a result *STAT4* knockout mice are deficient in the generation of T_H1 T cell responses (Kaplan et al. 1996). Nikolic et al. evaluated the induction of acute GVHD using donors unable to generate *STAT4* (T_H1 impaired) or *STAT6* (T_H2 impaired) (Nikolic et al. 2000). They found that *STAT4* $^{-/-}$ donor T cells mediated acute GVHD with delayed kinetics, compared to WT or *STAT6* $^{-/-}$ T cells. Interestingly, recipients of *STAT4* knockout cells demonstrated extensive cutaneous changes with marked dermal infiltration and lysis along the epidermal/dermal junction, similar to that recently described after the infusion of T_H17 cells generated ex vivo (Carlson et al. 2009). This severe skin pathology was not found using *STAT6* knockout donors. The authors hypothesized that acute GVHD induced in the absence of *STAT4* was mediated by T_H2 T cells.

There is still controversy regarding the ability of T_H2 T cells to mediate acute GVHD. Krenger et al. evaluated the ability of T_H2 polarized T cells to mediate acute GVHD (Krenger et al. 1995). They found that culturing $CD4^+$ T cells in the presence of IL-4 generated T_H2 polarized donor cells. Subsequently, they injected these donor T_H2 cells to class II MHC-disparate bm12 recipients. Recipients of T_H2 T cells generated significantly less IFN-γ compared to those given T cells cultured without IL-4. Production of the pro-inflammatory cytokine, TNF-α, was also decreased after the receipt of T_H2 T cells and was not significantly induced by LPS. Injection of LPS post-transplantation led to death in all hosts receiving naïve $CD4^+$ T cells, but only in a fraction of those receiving T_H2 T cells. Interestingly, the addition of T_H2 T cells to naïve T cells led to a marked decrease in the generation of IFN-γ and GVHD pathology, suggesting that T_H2 cells modulate GVHD responses. This work has questioned whether T_H2 cells can mediate acute GVHD and has led to a significant interest in identifying new populations of T cells capable of mediating GVHD.

4 T_H17 T Cells

Approximately, 20 years ago, seminal work from Mosmann and Coffman demonstrated that CD4⁺ T cell clones generated specific cytokines (Mosmann et al. 1986) (Fig. 1). T cells that generated IFN-γ, IL-2, IL-3, and GM-CSF were termed T_H1 cells, while those that generated IL-4 and IL-5 were termed T_H2 cells. T_H2 cells were shown to generate IL-9 and IL-13 in addition to IL-4 and IL-5 (McKenzie and Zurawski 1995). Follow up work (Zheng and Flavell 1997) (Szabo et al. 2000) demonstrated that there were transcription factors that were specific for the individual T cell lineages; GATA-3 was shown to be critical for the generation of T_H2 $CD4^+$ T cells, while T-bet and eomesodermin (Pearce et al. 2003) were needed for the production of T_H1 $CD4^+$ T cells and $CD8^+$ effector function. Until approximately 13 years ago, these were the only two groups of $CD4^+$ T cells that were recognized (Fig. 1).

In 1995, Sakaguchi and his group identified a group of thymically derived T cells that expressed CD4 and CD25 and whose removal by neonatal thymectomy led to multi-organ system auto-immunity (Sakaguchi et al. 1995). Subsequent investigators including those from Sakaguchi's group have found that these T cells, termed regulatory T cells (T_{reg}), express a specific transcription factor, Foxp3, that is critical

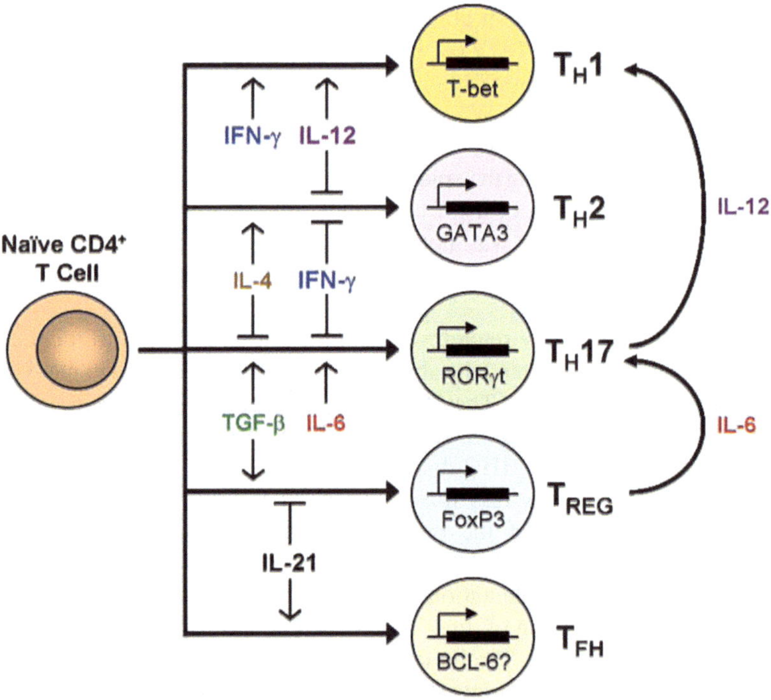

Fig. 1 Polarization of CD4± T cells may proceed along multiple pathways: The critical cytokine and transcription factors important for the polarization of CD4+ T cells is shown. The roles of cytokines in enhancing or blocking polarization along particular pathways is shown

for the function of T_{reg} cells (Fontenot et al. 2003, 2005; Fontenot and Rudensky 2004; Hori et al. 2003). T_{reg} cells express CD4, CD25, and a number of other surface molecules including, but not limited, to GITR, CD103, and LAG-3. They block the proliferation and effector function of $CD4^+$ and $CD8^+$ T cells via a contact-dependent process in vitro that may be critically dependent on the expression of CTLA-4 (Suri-Payer et al. 1998; Thornton and Shevach 1998; Wing et al. 2008). In vivo, the generation of IL-10 and TGF-β1 by these cells may be essential (Bach and Francois Bach 2003; Wood and Sakaguchi 2003). In addition to the thymically derived population of T_{reg} cells, there is a population of inducible T_{reg} cells. This population of peripheral $CD4^+$ T cells is generated by the production of TGF-β1 (Zhou and Levitsky 2007) (Fig. 1).

A fourth population of T cells was recently identified based on the finding of differences in the biological response of mice lacking the p40 and p35 chains of IL-12 (Becher et al. 2003; Cooper et al. 2002; Oppmann et al. 2000). Ensuing investigators found that the p40 chain can also pair with another protein, p19, to generate a cytokine termed IL-23. Initial work suggested that the generation of IL-23 was important in the production of IL-17 by $CD4^+$ T cells (T_H17 cells). Subsequent work has shown that TGF-β1, IL-6 and/or IL-21, and IL-1β are the critical cytokines for the generation of T_H17 T cells (Bettelli et al. 2006). Following this finding, Ivanov et al. demonstrated that the gene *Rorc* (retinoid-related orphan receptor gamma) was a critical transcription factor in the generation of T_H17 T cells (Ivanov et al. 2006). Yang et al. demonstrated that RORγt can pair with RORα in T_H17 cells (Yang et al. 2008).

One initial question for GVHD biologists was whether a cell type distinct from T_H1/Tc1 T cells could mediate GVHD. If another cell type could initiate GVHD, did it mediate this alone or in concert with T_H1/Tc1 T cells? Is GVHD induced in the liver, lungs, GI tract, or skin, and is it mediated by the same population of T cells, or is there organ-specific populations of T cells responsible for organ-specific GVHD? These issues will be discussed in this chapter, although much of this work is still speculative. We will present separate sections on the function of T_H17 cells and proteins associated with the T_H17 program using animal models from correlative clinical associations, and which will be discussed in the latter part of this chapter.

5 Function of T_H17 Cells in Acute GVHD

5.1 Cytokines Important in the Generation of T_H17 Cells and Their Impact on GVHD

T_H17 cell generation can be divided into different phases of the response based on timing and function. Polarization of T_H17 cells is critical early in the response and this is mediated by specific cytokines such IL-6, IL-21, TGF-β1, and IL-1β,

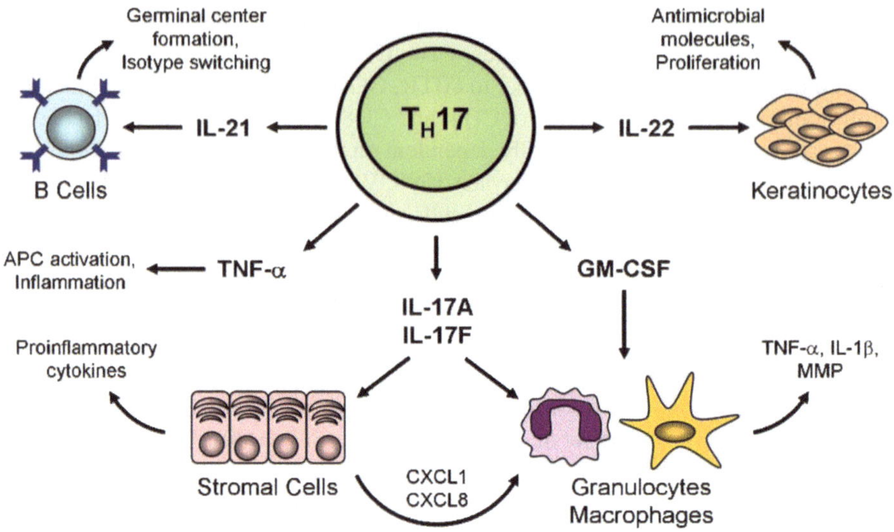

Fig. 2 Effector pathways mediated by T_H17 cells: The mechanisms by which T_H17 cells mediate immune activity at the tissue level is shown. The critical roles for IL-17A, IL-17F, IL-21, IL-22, TNF-α and GM-CSF generated by T_H17 cells is shown

depending on the species (mouse or human) and the absence of cytokines that polarize toward T_H1, T_H2, or T_{reg} cells such as IFN-γ and IL-12, IL-4, and IL-2 (Fig. 1). Expansion of T_H17 cells depends on the presence of IL-23 and generation of the transcription factors RORγt and/or RORα. The effector function of T_H17 cells depends on the generation of cytokines that mediate inflammatory responses such as IL-17A, IL-17F, IL-21, IL-22, and TNF-α (Fig. 2). The recruitment of T_H17 cells is dependent on the expression of CCL20 locally and its interaction with CCR6 expressed on T_H17 cells.

Each of these areas has been investigated in GVHD models and we will discuss these in the following sections. However, several caveats need to be established. First, this is a very complex area in which several of the proteins have a multitude of effects (IL-6, TGF-β1) on the pathogenesis of GVHD and thus studies probing these proteins and pathways may influence GVHD without necessarily impacting on T_H17 cells. Second, several of these proteins and pathways have their own effects (IFN-γ, IL-12, IL-4), as well as specific effects on the generation of T_H17 cells. Separating these effects in some cases is difficult, and in others impossible. This is made more complex by the activity of proteins like IFN-γ, which blocks the generation of T_H17 cells, but also is generated by T_H17 cells in a lymphopenic environment. Third, the effector activity of T_H17 cells is complex with a number of cytokines generated by these cells, and blocking the function of one of these, such as IL-17A, does not reflect on the global activity of T_H17 cells. Some cytokines, such as TNF-α, are made by multiple different phenotypic T cells, including T_H17 cells. Finally, several of these cytokines impact on local tissues to generate proteins or peptides that may independently be important in GVHD biology.

5.2 IL-6 and TGF-β1

Recent work has indicated that IL-6 and TGF-β1 are critical in the generation of murine T_H17 cells while TGF-β1, IL-21, and IL-1β are critical for the generation of human T_H17 cells, skewing T cells away from the T_{reg} lineage (Fig. 1). One area of investigation into the pathophysiology of GVHD and T_H17 cells is the impact of IL-6 and TGF-β1. IL-6 is produced by T cells, B cells, monocytes, DCs, keratinocytes, fibroblasts, and endothelial cells. Patients with elevated plasma levels of IL-6 (Barak et al. 1995; Imamura et al. 1994), as well as those with a recipient or donor IL-6 genotype that results in increased IL-6 production, (Cavet et al. 2001; Socie et al. 2001) have an increased incidence and severity of GVHD. Membrane-bound IL-6R is expressed on hematopoietic and hepatic cells. Anti-IL-6R antibody has been used to treat auto-immune diseases such as rheumatoid arthritis and colitis (Atreya et al. 2000). Recently, Drobyski and colleagues have reported that IL-6 and IL-6R levels are significantly increased in the sera and in GVHD target organs (Chen et al. 2009). Blockade of the IL-6/IL-6R signaling complex using mAbs resulted in reduced severity of GVHD. IL-6 was shown to inhibit the reconstitution of T_{regs} during GVHD via both thymic-dependent and thymic-independent pathways, although the magnitude of this reduction was not sufficient to preclude a beneficial effect of IL-6/IL-6R pathway blockade on GVHD-induced mortality. IL-6 also is known to inhibit the generation of iTregs induced by TGF-β by blocking Foxp3 expression (Dominitzki et al. 2007). Consistent with this observation, IL-6/IL-6R pathway blockades increase the frequency of iTregs in the colon, as has been documented for IL-21/IL-21R pathway blockades (see below). T_H17 cells were reduced in frequency in the spleen, liver, and lung. Thus, IL-6/IL-6R signaling represents an important decision point between iTregs and T_H17 cells.

TGF-β1 has potent immuno-regulatory properties implicated in the induction of alloantigen-specific tolerance. G-CSF stimulated donor T cells have been shown to diminish the incidence of acute GVHD in rodents and humans. G-CSF augments TGF-β production by donor $CD4^+$ T cells. Conversely, TGF-β neutralization accelerates acute GVHD lethality that is associated with impaired donor T cell expansion and proliferation and cytolytic capacity (Banovic et al. 2005). Notably, neutralization of TGF-β1 post-BMT attenuated histologic features of chronic GVHD. In rodents, IL-10 and TGF-β collaborate to induce a regulatory T cell subset that is highly immune suppressive and can inhibit GVHD lethality (Zeller et al. 1999). In humans, IL-10 has been shown to be essential in forming T regulatory type I cells (Groux et al. 1996; Roncarolo et al. 2001) that can produce TGF-β and has been linked to transplantation tolerance in mixed chimeric SCID recipients, implicating IL-10 and/or TGF-β as immune suppressive cytokines that may inhibit acute GVHD generation. In vivo, TGF-β1 and IL-10 have additive properties in protecting murine recipients of G-CSF treated donor T cells against lethal acute GVHD (Banovic et al. 2005).

NK cells also have been demonstrated to be both beneficial and deleterious in murine BMT, depending on their activation status. Activated NK cells produce TGFβ1 (Murphy et al. 1992a, b). Murphy and co-workers have shown that that the adoptive transfer of activated NK cells of donor type improved allogeneic marrow engraftment without inducing GVHD in mice and can directly inhibit GVHD while retaining graft vs. tumor effects in tumor-bearing mice (Asai et al. 1998). Treatment with anti-TGF-β1 mAb significantly shortened the survival compared with mice receiving NK cells without mAb. Moreover, mice receiving mAB alone without allogeneic donor splenocytes had a significant increase in GVHD vs. those that host production of TGF-β was also important in regulating GVHD. Together, this data demonstrates that TGF-β was at least partially responsible for the protection resulting from the activated NK cells.

5.3 IL-23 and Its Effects on GVHD

IL-23 is generated by APCs including monocytes, macrophages, and dendritic cells after stimulation of pattern recognition receptors such as Toll-like receptors. IL-23 binds to the IL-23 receptor, which is a heterodimer composed of the IL-12 receptor β1 and a novel chain, the IL-23 receptor. IL-23 mediates its effects by signaling predominantly via STAT3. The activity of IL-23 is not critical for the generation of T_H17 cells, but is absolutely required for the expansion of T_H17 cells via interaction with the IL-23 receptor on T cells, whose expression is governed by the expansion of RORγt.

Several groups have suggested a role for IL-23 in the pathogenesis of acute GVHD, although the mechanisms responsible vary. Das et al. evaluated the activity of IL-23 in a murine transplant model in which recipient mice received bone marrow and spleen cells unable to generate IL-23 due to the loss of the p19 chain (Das et al. 2009a). There was a marked improvement in survival of mice given grafts unable to generate IL-23 with substantial reductions in the generation of IL-17 and TNF-α and less impressive decreases in IL-1β and IL-6. Despite the significant overall improvement of recipient mice given bone marrow and spleen cells unable to generate IL-23, the only pathological difference was found in the colon. This effect was independent of the infusion of wild type or IL-23 null T cells, suggesting that donor APCs were the critical source of IL-23. Fascinatingly, despite the known role that IL-23 plays in the expansion of T_H17 cells, there were very few T_H17 cells found in the colon of BALB/c mice given wild type or IL-23 $^{-/-}$ bone marrow. However, in the absence of IL-23 there was a significant reduction of $CD4^+$ IFN-γ-producing T cells in the colon. A follow-up presentation by this group in 2009 demonstrated that the absence of IL-23 in donor bone marrow cells did not impact on the anti-tumor activity after allogeneic stem cell transplantation (Das et al. 2009b). Additionally, this group demonstrated that an anti-p19 mAb was capable of mediating the same colonic protection as the infusion of bone marrow unable to generate IL-23, suggesting a new translatable approach to the prevention of acute GVHD. These findings were supported by a follow-up manuscript from Thompson et al. that demonstrated diminished GVHD after the infusion of spleen cells and bone marrow unable to generate p19, although that group correlated the diminished GVHD with decreased generation of protein and mRNA for IL-17A systemically (Thompson et al. 2010).

5.4 T_H17 Transcription Factors and Signaling Proteins

Two transcription factors have been shown to be critically important for the generation of T_H17 cells; RORγt (*Rorc*) and RORα via the generation of the signaling protein, *STAT3*. The function of RORγt has been briefly evaluated by two separate groups. Iclozan et al. (2010) evaluated the function of ex vivo expanded T_H17 cells in the pathogenesis of acute GVHD and as part of that study infused T cells from RORγt $^{-/-}$ mice obtained from the Jackson Laboratory into lethally irradiated Balb/c recipient mice. $1–2 \times 10^6$ $CD4^+$ WT or RORγt $^{-/-}$ donor T cells were given with TCD bone marrow cells. No difference in median survival or weight loss was found comparing recipients of WT or RORγt $^{-/-}$ donor T cells. In contrast, there has been one presentation that showed that donor T cells which could not generate RORγt had diminished GVHD using a lower dose of $CD4^+$ T cells (Kappel et al. 2008). Our group has found that the infusion of donor T cells from RORγt $^{-/-}$ mice into lethally irradiated B6D2 recipients, profoundly diminished the incidence and severity of acute GVHD. This suggests that the role of RORγt may either be model-dependent or dependent on the cell inoculum used to induce GVHD.

STAT3 has been evaluated in the biology of GVHD (Lu et al. 2008). Proliferating alloreactive T cells demonstrated early increased expression of phosphorylated STAT3. Inhibition of STAT3 using cucurbitacin I, which inhibits STAT3 in addition to p53, was used to treat T cells for 1 h ex vivo prior to their administration to induce GVHD. Cucurbitacin-I treated T cells led to diminished GVHD with over 50% of recipient mice surviving compared to none of the control animals. This effect was associated with diminished $CD8^+$ donor T cells and an increase in the percentage of Foxp3-expressing regulatory T cells. Unfortunately, the authors did not correlate the effects of cucurbitacin-I on the in vivo transcription of STAT3.

6 T_H17 Effector Cell Function

6.1 IL-17A

IL-17 is a pro-inflammatory cytokine that induces the production of IL-6, g-CSF, GM-CSF, PGE2, and chemokines critical for the recruitment of neutrophils such as CXCL1, CXCL5, and CXCL8 (Broxmeyer 1996; Kolls and Linden 2004). IL-17, which is generated predominantly by αβ and γδ T cells mediates its activity by binding to the IL-17 receptor found on all hematopoietic cells and in endothelial, epithelial, and fibroblast cells. The IL-17 receptor differs from other cytokine receptors in that it does not mediate activity via the STAT/JAK pathway but via ACT1 and TRAF6, which eventually leads to activation of NF-κB.

Two groups have recently attempted to directly evaluate the function of IL-17A in the pathology of acute GVHD using mouse models. The initial description of a role for IL-17A in the pathophysiology of GVHD was described by Yi et al. using mice unable to generate IL-17A (Yi et al. 2008). For their work, they infused wild type T cell-depleted bone marrow supplemented with splenocytes from either wild type or B6 mice unable to generate IL-17 to MHC. Completely mismatched BALB/c recipients were given 800 cGy of irradiation for conditioning. They demonstrated greater weight loss and more severe diarrhea in recipients of IL-17 $^{-/-}$ donor splenocytes, compared to those given wild type splenocytes. Recipient mice given IL-17 null splenocytes had much greater inflammation in the colon, liver, and lungs compared to those given wild type splenocytes. The mechanisms mediating these findings appeared to be the enhanced generation of pro-inflammatory cytokines, especially IFN-γ, in the absence of IL-17. Increased amounts of IFN-γ were found in recipient mice given IL-17 null splenocytes and blocking the function of IFN-γ. Using antibody treatment led to diminished GVHD in recipients of IL-17 null splenocytes. Thus, this study suggested that IL-17 production may be important in limiting the generation of IFN-γ, and this group's in vitro work proposed that this was mediated by the influence of IL-17 on the ability of APCs to polarize a T_H1 response. What is important to note however, is that T_H17 cells generate a significant number of cytokines in addition to IL-17A, including IL-17F, IL-21, IL-22, and TNF-α. T_H17 cells may be potent inducers of GVHD even if IL-17A has an anti-inflammatory effect.

The role of IL-17A in the pathogenesis may not be straightforward, as Kappel et al. performed similar work using the same strain of donor mice unable to generate IL-17 and found completely disparate results (Kappel et al. 2009). This group used 850 cGy of irradiation split into two doses with the administration of 5×10^5 $CD4^+$ T cells, the typical dose used to generate lethal GVHD using B6 donors and BALB/c recipients. They found that recipient mice given this number of $CD4^+$ IL-17 null T cells had improved survival with a modest to moderate delay in lethality, dependent on the model employed. Interestingly, the infusion of whole T cells did not lead to enhanced GVHD, as was revealed in the earlier group. To perhaps provide more confusion, Kappel et al. found that the infusion of CD4+ T cells unable to generate IL-17 led to diminished production of IFN-γ-producing immune cells on day 7 post transplantation. As a result, it is currently not clear what role that IL-17A plays in the pathophysiology of acute GVHD.

6.2 *IL-17 and Chronic GVHD*

Chen et al. (2010) have evaluated the function of IL-17 in an activated syngeneic transplant model using RAG donors that recapitulates some of the clinical findings of chronic GVHD (Chen et al. 2010). At the time of syngeneic transplantation, recipient mice were treated with anti-IL-17A monoclonal antibody. This treatment had no effect on GVHD-induced pathology, cellular expansion post transplantation, or on the number of T_H17 cells generated post transplantation. Similar results were found using IL-17 $^{-/-}$ donor cells, although in transplants using knockout donor T cells, there was a marked decrease in the number of IL-17-producing T cells, which did not impact survival or the incidence/severity of chronic GVHD.

6.3 IL-21

Interleukin-21 (IL-21) is produced by $CD4^+$ T cells (especially T_H17-producing cells), T-follicular helper cells (Tfh), natural killer T (NKT) cells (Parrish-Novak et al. 2000), and signals through the IL-21R complex composed of the γ_CR and the IL-21R. IL-21R is expressed on hematopoietic and epithelial cells and promotes the activation, differentiation, maturation, or expansion of NK cells, B cells, $CD8^+$, $CD4^+$ T cells, DCs, and macrophages, (Davis et al. 2007; Leonard and Spolski 2005; Monteleone et al. 2008) and facilitates auto-immunity in some (Korn et al. 2007; Liu et al. 2008; Spolski et al. 2008), but not all (Coquet et al. 2008) by supporting immunoglobulin production and T_H17 cell-mediated pathogenesis. Because IL-21 augments T_H17 cell differentiation, indirect evidence for the role of IL-21 in GVHD pathogenesis may be derived from GVHD studies, as discussed above. IL-21 has been described variably as an inhibitor (Wurster et al. 2002) or enhancer (Monteleone et al. 2005) of T_H1 differentiation. Because IL-21 supports T_H17 cell survival at the expense of T_{regs} (Bettelli et al. 2006) and by inhibiting naive T-cell conversion into $CD4^+25^+FoxP3^+$ regulatory T cells (termed inducible Tregs, iTregs) (Fantini et al. 2007), and limiting the suppression of Teffs by T_{regs}, and augmenting T_H17 responses, (Peluso et al. 2007) IL-21 may increase GVHD lethality.

Recent studies have shown that IL-21 blockade by genetic (using IL-21 or IL-21R$^{-/-}$ donor cells) or pharmacological methods (anti-IL-21 mAb or a decoy receptor for IL-21) led to a reduced GVHD-induced lethality. In one study, a reduced frequency of IFN-γ secreting $CD4^+$ and $CD8^+$ lamina propria (LP) cells was found. Because no reduction in the mean GVHD scores were noted in the other organs, yet the frequency of $CD4^+$ and $CD8^+$ T cells producing IFN-γ were significantly reduced in the liver and the spleen, this data pointed to the possibility of a more local (gastrointestinal) influence of IL-21 on GVHD lethality. Although IFN-γ-producing T_H1 cells are known to cause gastrointestinal injury during GVHD (Burman et al. 2007; Nikolic et al. 2000), seemingly conflicting data exists for the relationship between IL-21 and IFN-γ production (Wurster et al. 2002),(Strengell et al. 2002). Our data indicates that during GVHD, IL-21 blockade reduced the generation of IFN-γ-secreting cells (Strengell et al. 2002), and increased the frequency of IL-4-producing cells in the colon. Consistent with the limited role of T_H17 cells in gastrointestinal GVHD (Carlson et al. 2009; Kappel et al. 2009), a difference in the frequency of T_H17 cells relative to the presence of IL-21 was not observed, and the reduced frequency of IFN-γ-secreting LP T-cells alone would not seem to explain the reduced GVHD-lethality witnessed since prior studies have indicated that IFN-$\gamma^{-/-}$ T-cells accelerated while IL-4$^{-/-}$ T-cells reduced GVHD lethality. We favor the possibility that the balance between cytokine-secreting cells in a given site is more important than the frequency of a single cytokine, although the reduced frequency of IFN-γ secreting cells may be reflective of a more immune suppressive environment which is not conducive to Teff generation. This can be seen in instances where ex vivo expanded T_{regs} suppress IFN-γ production by Teffs during GVHD (Trenado et al. 2006).

By using $CD4^+FoxP3^-$ donor T-cells, we were able to demonstrate that IL-21 production by donor T-cells impedes iT_{reg} generation. In the absence of IL-21, a significant decrease in T_{reg} frequency was seen in LP and hepatic T cells. The overall higher T_{reg} frequencies in the LP vs. the liver may account for the greater reduction in pathological scores of the gastrointestinal tract seen in the context of IL-21 blockade. A requirement for FoxP3 expression and function in donor T-cells for ameliorating GVHD in recipients of $IL\text{-}21^{-/-}$ vs. wt T-cells was seen in studies in which FoxP3-deficient *scurfy* T-cells isolated from chimeras that also contained congenic $FoxP3^+$ wt T cells were unable to reduce GVHD lethality following adoptive transfer into allogeneic recipients treated with anti-IL-21 Ab. The lack of a beneficial effect of IL-21 blockade in mice receiving scurfy cells supports the hypothesis that $FoxP3^+$ T_{regs} are necessary for optimal GVHD inhibition and suggests that other T_{reg} independent mechanisms, such the direct effects of IL-21 on epithelial tissue injury or cytokine skewing, are unlikely explanations for the beneficial effects of IL-21 blockade on GVHD. IL-21 blockade, while allowing significant GVHD attenuation, did not preclude effective GVL responses. Thus, IL-21 neutralization is an attractive strategy for acute GVHD prevention with GVL retention that should be explored for clinical application. The effect of IL-21 neutralization on chronic GVHD has not been reported to date.

IL-21 may have beneficial effects post-BMT in recipients with GVHD. For example, IL-21 is critical for germinal center (GC) formation and GC B cell antibody and pathogen responses. Long term initiating cells (LTis), which produce IL-21 and respond to IL-21, seed the developing lymph node and Peyer's patches during fetal development and initiate the formation of both organs (Blum and Pabst 2006; Ivanov et al. 2006). All types of LTis require expression of the RORγt for their differentiation and function and deficient mice lacking RORγt do not develop lymph nodes, Peyer's patches, and other gut-associated lymphoid tissues. We have shown that conditioning of BMT recipients by TBI injures the stromal cell compartment in secondary lymphoid tissue (Kelly et al. 2008) associated with a marked reduction in LTi cells. Therefore, long-term IL-21 neutralization, as might be needed to prevent or treat GVHD, could impair antibody responses if LTis and IL-21 prove to be beneficial in organogenesis of secondary lymphoid organs post-BMT. It is also possible that active GVHD may be injurious to secondary lymphoid organogenesis. Future studies will be needed to determine the effects of IL-21 neutralization on chronic GVHD, lymphoid organogenesis, and pathogen responses.

6.4 *IL-17A, IL-17F and TNF-α*

Our group has taken another approach to evaluate whether T_H17 cells can mediate lethal acute GVHD (Carlson et al. 2008). Initially, we demonstrated that the administration of anti-IFN-γ mAb greatly increased the number of T_H17 cells present in secondary lymphoid tissue, the liver, and lung. However, anti-IL-23 therapy did not block the generation of T_H17 cells after anti-IFN-γ mAb and we were not able to evaluate the function of the T_H17 axis given the large number of cytokines generated by these cells including IL-17A, IL-17F, IL-21, IL-22, TNF-α, and others. To circumvent this problem, we generated a method for the expansion of a near homogeneous population of T_H17 cells that could be infused to lethally irradiated recipients. By sorting naïve CD4⁺ T cells and culturing them in the presence of the cytokines TGF-β1, IL-6, IL-1β, TNF-α, and mAb specific for IFN-γ and IL-2, we could generate approximately 50–65% T_H17 cells in culture. Expanding these cells in the presence of IL-23 allowed us to get a population that was 95% T_H17 cells, with less than 2% contaminating IFN-γ-producing T cells. Infusion of these ex vivo expanded B6 T_H17 cells and led to lethal GVHD when given to haplo-identical B6D2 recipient mice. All recipients developed severe skin, GI tract, and lung pathology. The skin pathology was the most severe seen by our group, with very high, statistically significant, pathology scores and a large number of mice developing substantial fur loss. Interestingly, we found that the skin manifestations depended on the generation of IL-17 as anti-IL-17 mAb treatment abrogated fur loss in B6D2 recipients but had no effect on overall survival, which was improved by blocking TNF-α. Finally, we found that T_H17 cells initially generated significant quantities of IL-17F, but that this did not continue. T_H17 cells isolated 2 weeks post transplant from recipient mice did not generate IL-17F, suggesting that this cytokine was not critical to the overall inflammatory response induced by T_H17 cells. A second group used a similar approach to confirm these findings (Iclozan et al. 2009). They demonstrated that T_H17 cells generated ex vivo mediated substantial GVHD with earlier mortality than the administration of T_H1 T cells, while CD4⁺ T cells from B6 donors were given to lethally irradiated Balb/c mice. They found substantial tissue damage in the liver, small intestine, colon, and lung using T_H17 cells. The lung inflammation was significantly greater than that found using ex vivo generated T_H1 cells. T_H17 cells expanded more rapidly in the lymphopenic post transplant environment compared to T_H1 cells. However, this group found data contrary to that found with our group, a MHC complete mismatch model, T cells unable to generate RORγt could still mediate acute GVHD.

A second interesting finding came from our work infusing ex vivo expanded T_H17 cells. We generated eGFP-expressing T_H17 cells and found that these cells gradually lost expression of IL-17A, but increased the expression of IFN-γ. This led to the generation of T cells that produced both IL-17A and IFN-γ initially post transplantation. However, infusion of T_H17 cells unable to generate IFN-γ did not abrogate the pathology found using T_H17 cells, signifying that the production of IFN-γ was not critical for T_H17-mediated tissue inflammation. Subsequently, we found that IL-12 could reprogram T_H17 cells to the T_H1 pathway. This data demonstrated that the T_H17 pathway is not fixed and that classic T_H17 cells may provide a component of the IFN-γ generated post transplantation.

Since the publication of these findings, our group has evaluated for the presence of T_H17 cells from clinical tissues in patients with acute and chronic GVHD. Although quite preliminary, we have found substantial expression in lesional skin samples of the T_H17 cytokines, IL-17A, IL-22, and the transcription factor RORγt. Clearly, more work is needed to determine if T_H17 cells are critical for clinical GVHD induced in the skin and GI tract.

6.5 T_H17 Cell Migration

T_H17 cells express the chemokine receptor, CCR6, and migrate in response to the chemokine ligand, CCL20. While a specific role for CCR6 or CCL20 in the migration of T_H17 cells in GVHD models has not been performed, the absence of CCR6 on donor T cells was found to diminish GVHD in a MHC minor mismatch model with significantly diminished GVHD in the GI tract and skin (Varona et al. 2005). Interestingly, this was associated with diminished production of IFN-γ, which may have been mediated by impaired trafficking of T_H17 cells.

7 T_H1 and T_H2 Cell Impact on T_H17 Cells and GVHD

Stem cell transplantation differs from other forms of organ transplantation with the requirement for conditioning therapy to allow hematopoietic donor stem cells to engraft (Fig. 3). Typically, this conditioning therapy involves the use of total body irradiation and/or chemotherapy (Anagnastopoulos et al. 2004; Basara et al. 2002; Blaise et al. 2004; Mielcarek et al. 2003; Mohty et al. 2003). Conditioning therapy has profound effects on the generation of pro-inflammatory cytokines and the upregulation of co-stimulatory proteins on host antigen-presenting cells that may be

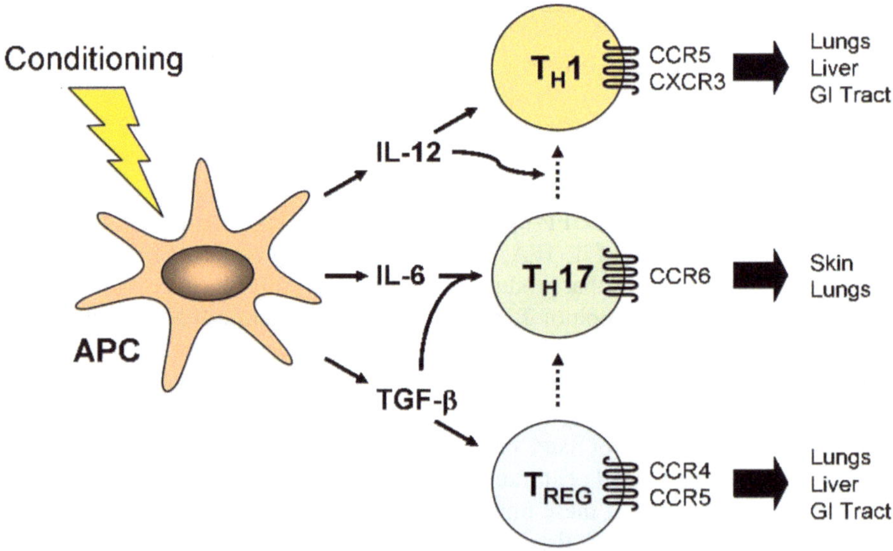

Fig. 3 Effects of conditioning therapy given prior to bone marrow or stem cell transplantation on the generation of pro-inflammatory cytokines by host antigen-presenting cells (APCs): three major cytokines are generated by host APCs which can initiate the polarization of T cells along a Th1 axis mediated by IL-12, a T_H17 axis mediated by IL-6 and a T_{reg} or iT_{reg} axis mediated by TGF-β1

critical in the initiation of acute GVHD (Ferrara et al. 1996; Hill and Ferrara 2000; Krenger et al. 1997). Our group evaluated the generation of cytokines in the first 24 h after conditioning by host APCs (Serody et al. 2008) and found substantial production of the p40 and p35 chains of IL-12 and IL-6, with much less expression of IL-23 p19, and no expression of IL-13. Thus, early conditioning effects favor the generation of T_H1 and T_H17 cells and limits T_H2 cell expansion. Early IFN-γ production blocks the generation of T_H17 cells, and it is quite conceivable that the more intense the conditioning therapy, the greater the expansion of T_H1 T cells at the expense of T_H17 cells.

As indicated, investigators have clearly shown that T cells unable to generate or respond to the T_H1 cytokine can mediate GVHD, leading investigators to evaluate the activity of a number of different T cell subsets in inducing GVHD in preclinical models. Recently, Tawara et al. investigated the function of T_H2 cells in the generation of acute GVHD in murine models (Tawara et al. 2008). For this work, they isolated donor T cells from mice genetically unable to generate IL-4, IL-5, IL9, and IL-13 by homologous recombination in which the Il4 gene cluster containing Il5 and Il13 was disrupted using standard homologous recombination and crossed with Il-9 knockout mice. BALB/c wild type or quadruple knockout mice were used as a source of donor T cells that were given with wild type T cell depleted BALB/c bone marrow cells. The median survival rate was statistically poorer for recipient mice given the Quad-deficient T cells, which correlated with higher clinical GVHD scores, and worsened GI tract GVHD pathology in the first week post-transplant. Interestingly, when either IL-4 or STAT6 was targeted alone, donor T cells from those animals caused either an improved outcome (IL-4) or a similar outcome with wild type T cells (STAT6). The enhanced GVHD found after the infusion of Quad4 knockout T cells correlated with enhanced production in vitro by those T cells of IFN-γ, and IL-17. Thus, this work suggested that the absence of T_H2 cytokine production by T cells exacerbated GVHD by the increased generation of T_H1 and/or T_H17 cytokines.

Simultaneous work done by our group investigated the generation of T_H17 cells in mice that had received depletion antibodies targeting IFN-γ. We had confirmed the data originally published by Murphy et al. indicating that blocking IFN-γ led to earlier and more lethal GVHD. When we evaluated recipient mice treated twice weekly post transplant with anti-IFN-γ, we found an increased percentage of T_H17 cells in the liver, lung, and lymphoid tissue compared to mice receiving controlled antibodies, with an even greater number of T_H17 cells found at these sites (Carlson and Serody 2007).

8 Idiopathic Pneumonia Syndrome

Clinical acute GVHD is characterized by involvement of the GI tract, liver, and skin. In mice, acute GVHD can involve the lung, and while classic acute GVHD is not typically seen here, an aggressive inflammatory reaction termed idiopathic pneumonia syndrome is not uncommon (Zhu et al. 2008). This syndrome is characterized by the presence of interstitial and alveolar pneumonitis in the absence of an infectious etiology (Afessa and Peters 2008). IPS appears to be linked to conditioning-related changes in cytokine expression in the lung and the presence of alloreactive T cells and perhaps infectious sequelae such as lipopolysaccharide (Panoskaltsis-Mortari et al. 1997). In animal models, IPS is often an advanced occurrence in GVHD, peaking at 6–10 weeks post transplant.

Cytokines, especially IFN-γ, are critical to the occurrence of IPS. Burman et al. recently evaluated the function of IFN-γ in GVHD and IPS (Burman et al. 2007). They similarly found that the infusion of g-CSF mobilized donor cells unable to generate IFN-γ, and led to enhanced GVHD lethality. The infusion of donor cells lacking the IFN-γ receptors led to diminished GVHD with an improvement in survival and a marked decrease in the clinical score of GVHD. In the absence of IFN-γ, recipient mice developed significant inflammation in the lungs consistent with IPS. Similarly, when this group transplanted Balb/c splenocytes into B6-recipient mice lacking the IFN-γ receptor again, they found that these animals developed accelerated lethality due to IPS. Thus, this data demonstrated that donor IFN-γ is critical for the prevention of IPS and that it mediates this effect by interacting with host cells. The mechanism behind this finding was not clear, although this group demonstrated that IFN-γ was not functioning via the generation of nitric oxide or indolamine 2–3 dioxygenase.

A follow-up manuscript from Mauermann et al. (2008) may have provided an explanation to the paradoxical findings of Burman. Using a different haplo-identical transplant model, they demonstrated that in the absence of IFN-γ, IPS required the infusion of $CD4^+$ T cells with only minor contribution when $CD8^+$ T cells from IFN-$\gamma^{-/-}$ mice were given. Bronchoalveolar lavage after the infusion of wild type or IFN-$\gamma^{-/-}$ $CD4^+$ T cells demonstrated an increase in neutrophils in mice receiving IFN-$\gamma^{-/-}$ $CD4^+$ T cells. To identify the relevant cell population mediating IPS, they focused on T_H17 cells and previous data regarding the function of these cells in the recruitment of neutrophils. IL-17A was found from the $CD4^+$ T cell population after the infusion of IFN-$\gamma^{-/-}$ donor cells. Moreover, anti-IL-17A treatment led to an improvement in survival and a reduction in IPS. Additionally, etanercept, a soluble TNF-α receptor, treatment-ameliorated pulmonary complications induced by the administration of IFN-$\gamma^{-/-}$ donor cells, perhaps suggesting a role for TNF-α produced by T_H17 cells in this process.

Recently, Yi et al. have revisited the function of T_H2 cells in the pathophysiology of IPS and GVHD (Yi et al. 2009). After the administration of IFN-$\gamma^{-/-}$ donor $CD4^+$ T cells to lethally irradiated BALB/c recipients, they confirmed enhanced severity and lethality of GVHD, which was correlated with enhanced skin and lung disease. Next, they infused IFN-γ/IL-$17^{-/-}$ donor T cells to lethally irradiated BALB/c recipients and found much less GI tract and skin disease, but a similar degree of lung inflammation. They went on to show that blocking IL-4 in vivo impacted the severity of IPS, but found no difference in the number of T_H2 cells when using recipients lacking IL-4R, which also did not get IPS. These facts suggest that IFN-γ may be important in blocking the generation of T_H17 cells and preventing IPS, T_H17 cells may mediate IPS partly via the production of TNF-α and finally, that T_H2 T cells play a role in IPS. Yi et al. showed that mice given T cells which polarize toward a T_H2 response post-transplantation have significantly less pathology due to GVHD in the skin, liver, and GI tract.

9 Clinical Support for T_H17/IL-23 Function in GVHD

While murine studies are helpful in our understanding of clinical transplantation, they serve as a model to investigate clinically relevant pathways and approaches to prevent or treat GVHD. Our group and others have been interested in determining if there is an organ specific role for T_H17 cells during clinical GVHD. We have done a preliminary analysis of skin biopsy samples from patients with acute and chronic GVHD, searching for the expression of cytokine mRNA. We found that mRNA for *Il17*, *Il21*, and *Il22* were increased from lesional skin, especially in the setting of cGVHD. Dander et al. analyzed the frequency of T_H17 cells in the peripheral blood of patients with GVHD compared to healthy donors (Dander et al. 2009). The T_H17 population was increased in patients with acute GVHD, compared to healthy donors or allograft recipients without GVHD. This correlated with increased levels of IL-17 in the plasma, but only in patients with active GVHD. Interestingly, we found that patients with active GVHD had a decreased percentage of Foxp3-expressing regulatory T cells, and that in a small subset of these patients the decrease correlated with an increase of T_H17 cells. We then evaluated for the presence of T cells in the skin from six patients and found that all of the $CD3^+$ T cells were generating IL-17 and IFN-γ, similar to the cytokine expression from T cells isolated from the liver. Although it is hard to draw firm conclusions from anecdotal longitudinal assessments, our group used IL-17 ELISPOT assays to demonstrate that increases in this cell population correlated with the occurrence of acute GVHD.

A second approach to evaluating the role of the T_H17 axis has involved analysis for SNPs associated with GVHD in IL-23 or the IL-23 receptor (IL-23R). Previous work had shown a very strong association between a SNP present in the IL-23R (G to A change at 1,142 leading to a substitution of glycine for arginine, R381G, and Crohn's disease). (Baldassano et al. 2007; Duerr et al. 2006). This SNP is located in the IL-23 receptor chain of the IL-23 receptor. Elmaagacli et al. evaluated for the G>A change in a cohort of 221 transplant recipients from Germany and their HLA identical donors (Elmaagacli et al. 2008). The gene variant occurred in 13.1% of recipients and 11.3% of donors. Severe grades III-IV acute GVHD was found less frequently in recipients transplanted with donor cells from the IL-23R SNP (22% vs. 4%). Recipients with the G>A change also had less frequent severe GVHD (3.5% vs. 22%). In this evaluation, all but one donor and recipient were heterozygous for the SNP. A second unit undergoing unrelated transplantation was also evaluated, in which all donors and recipients who had the SNP were heterozygous. Again, recipients transplanted with donors having the IL-23R SNP had a significant decrease in grades III-IV acute GVHD compared to those receiving donor cells with the WT SNP (3.8% vs. 25.5%). However, in this analysis there was no beneficial effect if only the recipients had the IL-23R G>A change. No difference was found for the incidence of chronic GVHD and the IL-23R SNP. Overall survival was not impacted by the presence of the IL-23R polymorphism.

Two other groups have evaluated the interaction between polymorphisms in the IL-23R and outcomes post transplantation. Gruhn et al. evaluated another group from Germany, this time focusing on children and young adults (Gruhn et al. 2009). The IL-23R SNP was found in 7.8% of patients and 11% of donors with none of the patients being homozygous for this. Patients receiving donor grafts with the G>A SNP had a significantly decreased risk of acute GVHD (5% vs. 33.3% for grade II or greater), however, in this analysis, the frequency of severe GVHD was not affected by the occurrence of the polymorphism in the patient. When this group confined their analyses to patients under the age of 16, there was still a significantly decreased incidence of acute GVHD when the donor had the G>A polymorphism in the IL-23R. There was no association with the G>A polymorphism and the incidence of chronic GVHD or overall survival rate found.

One recent manuscript has questioned whether the G>A polymorphism in the IL-23R is implicated in a diminished risk of acute GVHD. Nguyen et al. evaluated a larger patient cluster done in association with the National Marrow Donor Program (Nguyen et al. 2010). 390 patients that underwent T-replete transplants for the treatment of acute or chronic leukemia or myelodysplastic syndrome primarily utilizing myeloablative conditioning were included. Two SNPs in the IL-23 R, including the G>A SNP, and a second polymorphism associated with ankylosing spondylitis were evaluated. 13% of the donors and 16% of the recipients had the variant SNPs. No association between either of the SNPs and the incidence or severity of acute GVHD was found. Thus, it is not clear at this time if these differences are due to differences in the transplantation approaches, methods used to prevent GVHD, or perhaps ethnic differences in the activity of the IL-23 axis between these groups.

10 Conclusion

GVHD is a complex process in which the pathophysiology is dependent on conditioning-induced changes to host APCs interacting with donor cells. We know that T cells are critical for the pathogenesis of GVHD, although the phenotype of T cell that mediates this complex process is not well understood. Because of previous data which indicates that blocking IFN-γ exacerbates acute GVHD, investigators over the past decade have searched for other subsets of T cells that could mediate acute GVHD. To that end, recent work has shown that T_H17 cells are capable of mediating inflammation characteristics of acute GVHD with a particular tropism for inducing inflammation of the skin, GI tract, liver, and lungs. Multiple studies have linked proteins associated with T_H17 cells in the pathogenesis of acute GVHD, although in the majority of these studies, a direct relationship between GVHD incidence or severity and T_H17 cell activity was not established. However, our group and others have shown that conditioning therapy leads to a marked increase in the generation of IL-6 in the host, which provides a favorable environment for the generation of T_H17 cells. As T_H17 cells generate characteristic cytokines like IL-17A, IL-17F, IL-21, IL-22, and TNF-α, research continues to determine the roles of these cytokines in the pathogenesis of acute GVHD. We have discussed recently evaluated work and although some of the data is paradoxical, we can conclude that blocking the function of cytokines produced by T_H17 cells ameliorates the pathology associated with acute GVHD. Some of the cytokines, like IL-17A, may have organ-specific activity and could be future targets for the treatment of severe cutaneous GVHD. These studies suggest several potential clinical approaches to limit acute GVHD, including blocking the functions of IL-6, IL-23, or IL-21 cytokines, which are critical for T_H17 polarization or effector function. Additionally, data from our group and others, has implicated the cytokine TNF-α in the pathogenesis of acute GVHD induced by T_H17 cells and clinical studies have demonstrated benefit in blocking the function of TNFα in patients with acute GVHD, especially involving the GI tract or skin (Couriel et al. 2004).

In the future, we anticipate a significant number of clinical studies which will attempt to correlate the presence of T_H17 cells or proteins involved in T_H17 polarization with the occurrence and/or severity of acute or chronic GVHD. A number of inhibitors of cytokines generated by T_H17 cells, such as IL-17A and IL-22 are being targeted by pharmaceutical companies for the treatment of diseases like psoriasis and could be evaluated for the treatment of patients with acute GVHD involving the skin. Additionally, inhibitors of IL-12 and IL-23 are currently available for the treatment of auto-immune disease (Griffiths et al. 2010), and could be tested for the prevention of acute GVHD. It is quite possible that work investigating the role of T_H17 cells in the pathogenesis of acute and chronic GVHD may provide a new understanding to the tissue-specific tropism of GVHD and result in new avenues for diagnosis and/or treatment.

References

Afessa, B., and Peters, S.G. (2008). Non-infectious pneumonitis after blood and marrow transplant. Curr Opin Oncol *20*, 227–233.

Anagnastopoulos, A., Aleman, A., and Giralt, S. (2004). Non-myeloablative reduced-intensity transplantation in multiple myeloma. Semin Oncol *31*, 33–36.

AS, O.M., and Cowan, M.J. (1997). Bone marrow transplantation for inherited diseases [see comments]. Current Opinion in Oncology *9*, 126–130.

Asai, O., Longo, D.L., Tian, Z.G., Hornung, R.L., Taub, D.D., Ruscetti, F.W., and Murphy, W.J. (1998). Suppression of graft versus host disease and amplification of graft versus tumor effects by activated natural killer cells after allogeneic bone marrow transplantation. J Clin Invest *101*, 1835–1842.

Atreya, R., Mudter, J., Finotto, S., Mullberg, J., Jostock, T., Wirtz, S., Schutz, M., Bartsch, B., Holtmann, M., Becker, C. et al. (2000). Blockade of interleukin 6 trans signaling suppresses T-cell resistance against apoptosis in chronic intestinal inflammation: evidence in Crohn's disease and experimental colitis in vivo. Nat Med *6*, 583–588.

Attal, M., Canal, P., Schlaifer, D., Chatelut, E., Dezeuze, A., Huguet, F., Payen, C., Pris, J., and Laurent, G. (1994). Escalating dose of mitoxantrone with high-dose cyclophosphamide, carmustine, and etoposide in patients with refractory lymphoma undergoing autologous bone marrow transplantation. Journal of Clinical Oncology *12*, 141–148.
Bach, J.F., and Francois Bach, J. (2003). Regulatory T cells under scrutiny. Nat Rev Immunol *3*, 189–198.
Baldassano, R.N., Bradfield, J.P., Monos, D.S., Kim, C.E., Glessner, J.T., Casalunovo, T., Frackelton, E.C., Otieno, F.G., Kanterakis, S., Shaner, J.L. et al. (2007). Association of variants of the interleukin-23 receptor gene with susceptibility to pediatric Crohn's disease. Clin Gastroenterol Hepatol *5*, 972–976.
Banovic, T., MacDonald, K.P., Morris, E.S., Rowe, V., Kuns, R., Don, A., Kelly, J., Ledbetter, S., Clouston, A.D., and Hill, G.R. (2005). TGF-beta in allogeneic stem cell transplantation: friend or foe? Blood *106*, 2206–2214.
Barak, V., Levi-Schaffer, F., Nisman, B., and Nagler, A. (1995). Cytokine dysregulation in chronic graft versus host disease. Leuk Lymphoma *17*, 169–173.
Basara, N., Roemer, E., Kraut, L., Guenzelmann, S., Schmetzer, B., Kiehl, M.G., and Fauser, A.A. (2002). Reduced intensity preparative regimens for allogeneic hematopoietic stem cell transplantation: a single center experience. Bone Marrow Transplant *30*, 651–659.
Becher, B., Durell, B.G., and Noelle, R.J. (2003). IL-23 produced by CNS-resident cells controls T cell encephalitogenicity during the effector phase of experimental auto-immune encephalomyelitis. J Clin Invest *112*, 1186–1191.
Bensinger, W.I., Buckner, D., and Gahrton, G. (1997). Allogeneic stem cell transplantation for multiple myeloma. Hematology - Oncology Clinics of North America *11*, 147–157.
Bettelli, E., Carrier, Y., Gao, W., Korn, T., Strom, T.B., Oukka, M., Weiner, H.L., and Kuchroo, V.K. (2006). Reciprocal developmental pathways for the generation of pathogenic effector TH17 and regulatory T cells. Nature *441*, 235–238.
Billingham, R.E. (1966). The biology of graft versus host reactions. In The Harvey Lectures (New York Academic Press), pp. 21–78.
Blaise, D., Bay, J.O., Faucher, C., Michallet, M., Boiron, J.M., Choufi, B., Cahn, J.Y., Gratecos, N., Sotto, J.J., Francois, S, et al. (2004). Reduced-intensity preparative regimen and allogeneic stem cell transplantation for advanced solid tumors. Blood *103*, 435–441.
Blazar, B.R., Korngold, R., and Vallera, D.A. (1997). Recent advances in graft versus host disease (GVHD) prevention. Immunological Reviews *157*, 79–109.
Blum, K.S., and Pabst, R. (2006). Keystones in lymph node development. J Anat *209*, 585–595.
Broxmeyer, H.E. (1996). Is interleukin 17, an inducible cytokine that stimulates production of other cytokines, merely a redundant player in a sea of other biomolecules? J Exp Med *183*, 2411–2415.
Burman, A.C., Banovic, T., Kuns, R.D., Clouston, A.D., Stanley, A.C., Morris, E.S., Rowe, V., Bofinger, H., Skoczylas, R., Raffelt, N. et al. (2007). IFNgamma differentially controls the development of idiopathic pneumonia syndrome and GVHD of the gastrointestinal tract. Blood *110*, 1064–1072.
Burnett, A.K., and Eden, O.B. (1997). The treatment of acute leukemia. Lancet *349*, 270–275.
Butturini, A., and Gale, R.P. (1994). Allogeneic bone marrow transplantation for leukemia. Current Opinion in Hematology *1*, 402–405.
Carlson, M.J., and Serody, J.S. (2007). TH17 cells mediate extensive skin pathology in acute GVHD. In American Society of Hematology (Atlanta Georgia).
Carlson, M.J., West, M.L., Coghill, J.M., Panoskaltsis-Mortari, A., Blazar, B.R., and Serody, J.S. (2008). In vitro differentiated TH17 cells mediate lethal acute graft versus host disease with severe cutaneous and pulmonary pathology.
Carlson, M.J., West, M.L., Coghill, J.M., Panoskaltsis-Mortari, A., Blazar, B.R., and Serody, J.S. (2009). In vitro-differentiated TH17 cells mediate lethal acute graft versus host disease with severe cutaneous and pulmonary pathologic manifestations. Blood *113*, 1365–1374.
Cavet, J., Dickinson, A.M., Norden, J., Taylor, P.R., Jackson, G.H., and Middleton, P.G. (2001). Interferon-gamma and interleukin-6 gene polymorphisms associate with graft versus host disease in HLA-matched sibling bone marrow transplantation. Blood *98*, 1594–1600.

Chao, N.J., and Schlegel, P.G. (1995). Prevention and treatment of graft versus host disease. Annals of the New York Academy of Sciences *770*, 130–140.

Chen, X., Das, R., Komorowski, R., Beres, A., Hessner, M.J., Mihara, M., and Drobyski, W.R. (2009). Blockade of interleukin-6 signaling augments regulatory T-cell reconstitution and attenuates the severity of graft versus host disease. Blood *114*, 891–900.

Chen, X., Das, R., Komorowski, R., van Snick, J., Uyttenhove, C., and Drobyski, W.R. (2010). Interleukin 17 is not required for auto-immune-mediated pathologic damage during chronic graft versus host disease. Biol Blood Marrow Transplant *16*, 123–128.

Cooper, A.M., Kipnis, A., Turner, J., Magram, J., Ferrante, J., and Orme, I.M. (2002). Mice lacking bioactive IL-12 can generate protective, antigen-specific cellular responses to mycobacterial infection only if the IL-12 p40 subunit is present. J Immunol *168*, 1322–1327.

Coquet, J.M., Chakravarti, S., Smyth, M.J., and Godfrey, D.I. (2008). Cutting edge: IL-21 is not essential for Th17 differentiation or experimental auto-immune encephalomyelitis. J Immunol *180*, 7097–7101.

Couriel, D., Saliba, R., Hicks, K., Ippoliti, C., de Lima, M., Hosing, C., Khouri, I., Andersson, B., Gajewski, J., Donato, M. et al. (2004). Tumor necrosis factor-alpha blockade for the treatment of acute GVHD. Blood *104*, 649–654.

Dander, E., Balduzzi, A., Zappa, G., Lucchini, G., Perseghin, P., Andre, V., Todisco, E., Rahal, D., Migliavacca, M., Longoni, D, et al. (2009). Interleukin-17-producing T-helper cells as new potential player mediating graft versus host disease in patients undergoing allogeneic stem-cell transplantation. Transplantation *88*, 1261–1272.

Das, R., Chen, X., Komorowski, R., Hessner, M.J., and Drobyski, W.R. (2009a). Interleukin-23 secretion by donor antigen-presenting cells is critical for organ-specific pathology in graft versus host disease. Blood *113*, 2352–2362.

Das, R., Komorowski, R., Hessner, M., Subramaniam, H., Cua, D., Huettner, C., and Drobyski, W.R. (2009b). Blockade of IL-23 signaling results in targeted protection fo the colon and allows for separation of graft versus host and graft versus leukemia effects. In American Society of Hematology (New Orleans), p. 231.

Davis, I.D., Skak, K., Smyth, M.J., Kristjansen, P.E., Miller, D.M., and Sivakumar, P.V. (2007). Interleukin-21 signaling: functions in cancer and autoimmunity. Clin Cancer Res *13*, 6926–6932.

Dominitzki, S., Fantini, M.C., Neufert, C., Nikolaev, A., Galle, P.R., Scheller, J., Monteleone, G., Rose-John, S., Neurath, M.F., and Becker, C. (2007). Cutting edge: trans-signaling via the soluble IL-6R abrogates the induction of FoxP3 in naive CD4+CD25 T cells. J Immunol *179*, 2041–2045.

Duerr, R.H., Taylor, K.D., Brant, S.R., Rioux, J.D., Silverberg, M.S., Daly, M.J., Steinhart, A.H., Abraham, C., Regueiro, M., Griffiths, A, et al. (2006). A genome-wide association study identifies IL23R as an inflammatory bowel disease gene. Science *314*, 1461–1463.

Elmaagacli, A.H., Koldehoff, M., Landt, O., and Beelen, D.W. (2008). Relation of an interleukin-23 receptor gene polymorphism to graft versus host disease after hematopoietic-cell transplantation. Bone Marrow Transplant *41*, 821–826.

Fantini, M.C., Rizzo, A., Fina, D., Caruso, R., Becker, C., Neurath, M.F., Macdonald, T.T., Pallone, F., and Monteleone, G. (2007). IL-21 regulates experimental colitis by modulating the balance between Treg and Th17 cells. Eur J Immunol *37*, 3155–3163.

Ferrara, J.L., Cooke, K.R., Pan, L., and Krenger, W. (1996). The immuno-pathophysiology of acute graft versus host-disease. Stem Cells *14*, 473–489.

Fontenot, J.D., Gavin, M.A., and Rudensky, A.Y. (2003). Foxp3 programs the development and function of CD4+CD25+ regulatory T cells. Nat Immunol *4*, 330–336.

Fontenot, J.D., Rasmussen, J.P., Williams, L.M., Dooley, J.L., Farr, A.G., and Rudensky, A.Y. (2005). Regulatory T cell lineage specification by the forkhead transcription factor foxp3. Immunity *22*, 329–341.

Fontenot, J.D., and Rudensky, A.Y. (2004). Molecular aspects of regulatory T cell development. Semin Immunol *16*, 73–80.

Forman, S.J. (1998). Stem cell transplantation in acute leukemia. Current Opinion in Oncology *10*, 10–16.

Gribben, J.G. (1997). Bone marrow transplantation for low-grade B-cell malignancies [see comments]. Current Opinion in Oncology *9*, 117–121.

Griffiths, C.E., Strober, B.E., van de Kerkhof, P., Ho, V., Fidelus-Gort, R., Yeilding, N., Guzzo, C., Xia, Y., Zhou, B., Li, S. et al. (2010). Comparison of ustekinumab and etanercept for moderate-to-severe psoriasis. N Engl J Med *362*, 118–128.

Groux, H., Bigler, M., de Vries, J.E., and Roncarolo, M.G. (1996). Interleukin-10 induces a long-term antigen-specific anergic state in human CD4+ T cells [see comments]. Journal of Experimental Medicine *184*, 19–29.

Gruhn, B., Intek, J., Pfaffendorf, N., Zell, R., Corbacioglu, S., Zintl, F., Beck, J.F., Debatin, K.M., and Steinbach, D. (2009). Polymorphism of interleukin-23 receptor gene but not of NOD2/CARD15 is associated with graft versus host disease after hematopoietic stem cell transplantation in children. Biol Blood Marrow Transplant *15*, 1571–1577.

Hill, G.R., and Ferrara, J.L. (2000). The primacy of the gastrointestinal tract as a target organ of acute graft versus host disease: rationale for the use of cytokine shields in allogeneic bone marrow transplantation. Blood *95*, 2754–2759.

Hori, S., Nomura, T., and Sakaguchi, S. (2003). Control of regulatory T cell development by the transcription factor Foxp3. Science *299*, 1057–1061. Epub 2003 Jan 1059.

Iclozan, C., Yu, Y., Liu, C., Liang, Y., Yi, T., Anasetti, C., and Yu, X.Z. (2009). Th17 cells are sufficient but not necessary to induce acute graft versus host disease. Biol Blood Marrow Transplant.

Iclozan, C., Yu, Y., Liu, C., Liang, Y., Yi, T., Anasetti, C., and Yu, X.Z. (2010). T helper17 cells are sufficient but not necessary to induce acute graft versus host disease. Biol Blood Marrow Transplant *16*, 170–178.

Imamura, M., Hashino, S., Kobayashi, H., Kubayashi, S., Hirano, S., Minagawa, T., Tanaka, J., Fujii, Y., Kobayashi, M., Kasai, M. et al. (1994). Serum cytokine levels in bone marrow transplantation: synergistic interaction of interleukin-6, interferon-gamma, and tumor necrosis factor-alpha in graft versus host disease. Bone Marrow Transplant *13*, 745–751.

Ivanov, II, McKenzie, B.S., Zhou, L., Tadokoro, C.E., Lepelley, A., Lafaille, J.J., Cua, D.J., and Littman, D.R. (2006). The orphan nuclear receptor RORgammat directs the differentiation program of pro-inflammatory IL-17+ T helper cells. Cell *126*, 1121–1133.

Jacobson, N.G., Szabo, S.J., Weber-Nordt, R.M., Zhong, Z., Schreiber, R.D., Darnell, J.E., Jr., and Murphy, K.M. (1995). Interleukin 12 signaling in T helper type 1 (Th1) cells involves tyrosine phosphorylation of signal transducer and activator of transcription (Stat)3 and Stat4. J Exp Med *181*, 1755–1762.

Kaplan, M.H., Sun, Y.L., Hoey, T., and Grusby, M.J. (1996). Impaired IL-12 responses and enhanced development of Th2 cells in Stat4-deficient mice. Nature *382*, 174–177.

Kappel, L.W., Goldberg, G.L., Ivanov, I.I., Na, I.-K., King, C., Suh, D., Smith, O.M., Ligh, C., Littman, D.R., and Brink, M.R.M.v.d. (2008). IL-17 is required for CD4-mediated GVHD. In Tandem BMT Meetings (San Diego CA).

Kappel, L.W., Goldberg, G.L., King, C.G., Suh, D.Y., Smith, O.M., Ligh, C., Holland, A.M., Grubin, J., Mark, N.M., Liu, C, et al. (2009). IL-17 contributes to CD4-mediated graft versus host disease. Blood *113*, 945–952.

Kelly, R.M., Highfill, S.L., Panoskaltsis-Mortari, A., Taylor, P.A., Boyd, R.L., Hollander, G.A., and Blazar, B.R. (2008). Keratinocyte growth factor and androgen blockade work in concert to protect against conditioning regimen-induced thymic epithelial damage and enhance T-cell reconstitution after murine bone marrow transplantation. Blood *111*, 5734–5744.

Khouri, I.F., and Champlin, R.E. (2004). Non-myeloablative stem cell transplantation for lymphoma. Semin Oncol *31*, 22–26.

Klingebiel, T., and Schlegel, P.G. (1998). GVHD: overview on pathophysiology, incidence, clinical and biological features. Bone Marrow Transplantation *21*, S45–49.

Kolls, J.K., and Linden, A. (2004). Interleukin-17 family members and inflammation. Immunity *21*, 467–476.

Korn, T., Bettelli, E., Gao, W., Awasthi, A., Jager, A., Strom, T.B., Oukka, M., and Kuchroo, V.K. (2007). IL-21 initiates an alternative pathway to induce pro-inflammatory T(H)17 cells. Nature *448*, 484–487.

Krenger, W., and Ferrara, J.L. (1996). Graft versus host disease and the Thl/Th2 paradigm. Immunologic Research *15*, 50–73.

Krenger, W., Hill, G.R., and Ferrara, J.L. (1997). Cytokine cascades in acute graft versus host disease. Transplantation *64*, 553–558.

Krenger, W., Snyder, K.M., Byon, J.C., Falzarano, G., and Ferrara, J.L. (1995). Polarized type 2 alloreactive CD4+ and CD8+ donor T cells fail to induce experimental acute graft versus host disease. J Immunol *155*, 585–593.

Lazarus, H.M., Vogelsang, G.B., and Rowe, J.M. (1997). Prevention and treatment of acute graft versus host disease: the old and the new. A report from the Eastern Cooperative Oncology Group (ECOG). Bone Marrow Transplantation *19*, 577–600.

Leonard, W.J., and Spolski, R. (2005). Interleukin-21: a modulator of lymphoid proliferation, apoptosis and differentiation. Nat Rev Immunol *5*, 688–698.

Levitt, L., and Lin, R. (1996). Biology and treatment of adult acute lymphoblastic leukemia. Western Journal of Medicine *164*, 143–155.

Liu, R., Bai, Y., Vollmer, T.L., Bai, X.F., Jee, Y., Tang, Y.Y., Campagnolo, D.I., Collins, M., Young, D.A., La Cava, A., and Shi, F.D. (2008). IL-21 receptor expression determines the temporal phases of experimental auto-immune encephalomyelitis. Exp Neurol *211*, 14–24.

Lu, S.X., Alpdogan, O., Lin, J., Balderas, R., Campos-Gonzalez, R., Wang, X., Gao, G.J., Suh, D., King, C., Chow, M, et al. (2008). STAT-3 and ERK 1/2 phosphorylation are critical for T-cell alloactivation and graft versus host disease. Blood *112*, 5254–5258.

Mauermann, N., Burian, J., von Garnier, C., Dirnhofer, S., Germano, D., Schuett, C., Tamm, M., Bingisser, R., Eriksson, U., and Hunziker, L. (2008). Interferon-gamma regulates idiopathic pneumonia syndrome, a Th17+CD4+ T-cell-mediated graft versus host disease. Am J Respir Crit Care Med *178*, 379–388.

McKenzie, A.N., and Zurawski, G. (1995). Interleukin-13: characterization and biologic properties. Cancer Treat Res *80*, 367–378.

Mielcarek, M., Martin, P.J., Leisenring, W., Flowers, M.E., Maloney, D.G., Sandmaier, B.M., Maris, M.B., and Storb, R. (2003). Graft versus host disease after non-myeloablative versus conventional hematopoietic stem cell transplantation. Blood *102*, 756–762. Epub 2003 Mar 2027.

Mohty, M., Bay, J.O., Faucher, C., Choufi, B., Bilger, K., Tournilhac, O., Vey, N., Stoppa, A.M., Coso, D., Chabannon, C. et al. (2003). Graft versus host disease following allogeneic transplantation from HLA-identical sibling with antithymocyte globulin-based reduced-intensity preparative regimen. Blood *102*, 470–476.

Monteleone, G., Monteleone, I., Fina, D., Vavassori, P., Del Vecchio Blanco, G., Caruso, R., Tersigni, R., Alessandroni, L., Biancone, L., Naccari, G.C. et al. (2005). Interleukin-21 enhances T-helper cell type I signaling and interferon-gamma production in Crohn's disease. Gastroenterology *128*, 687–694.

Monteleone, G., Pallone, F., and MacDonald, T.T. (2008). Interleukin-21: a critical regulator of the balance between effector and regulatory T-cell responses. Trends Immunol *29*, 290–294.

Mosmann, T.R., Cherwinski, H., Bond, M.W., Giedlin, M.A., and Coffman, R.L. (1986). Two types of murine helper T cell clone. I. Definition according to profiles of lymphokine activities and secreted proteins. J Immunol *136*, 2348–2357.

Murphy, W.J., Bennett, M., Kumar, V., and Longo, D.L. (1992a). Donor-type activated natural killer cells promote marrow engraftment and B cell development during allogeneic bone marrow transplantation. J Immunol *148*, 2953–2960.

Murphy, W.J., Keller, J.R., Harrison, C.L., Young, H.A., and Longo, D.L. (1992b). Interleukin-2-activated natural killer cells can support hematopoiesis in vitro and promote marrow engraftment in vivo. Blood *80*, 670–677.

Murphy, W.J., Welniak, L.A., Taub, D.D., Wiltrout, R.H., Taylor, P.A., Vallera, D.A., Kopf, M., Young, H., Longo, D.L., and Blazar, B.R. (1998). Differential effects of the absence of interferon-gamma and IL-4 in acute graft versus host disease after allogeneic bone marrow transplantation in mice. J Clin Invest *102*, 1742–1748.

Nguyen, Y.P., Al-Lehibi, A., Gorbe, E., Li, E., Haagenson, M., Want, T., Spellman, S., Lee, S.J., and Davidson, N.O. (2010). Insufficient evidence for association fo NOD2/CARD15 or other

inflammatory bowel disease-associated markers on GVHD incidence or other adverse outcomes in T-replete, unrelated donor transplantation. Blood *115*, 3625–1321.
Nikolic, B., Lee, S., Bronson, R.T., Grusby, M.J., and Sykes, M. (2000). Th1 and Th2 mediate acute graft versus host disease, each with distinct end-organ targets. The Journal of clinical investigation *105*, 1289–1298.
Oppmann, B., Lesley, R., Blom, B., Timans, J.C., Xu, Y., Hunte, B., Vega, F., Yu, N., Wang, J., Singh, K. et al. (2000). Novel p19 protein engages IL-12p40 to form a cytokine, IL-23, with biological activities similar as well as distinct from IL-12. Immunity *13*, 715–725.
Panoskaltsis-Mortari, A., Taylor, P.A., Yaeger, T.M., Wangensteen, O.D., Bitterman, P.B., Ingbar, D.H., Vallera, D.A., and Blazar, B.R. (1997). The critical early pro-inflammatory events associated with idiopathic pneumonia syndrome in irradiated murine allogeneic recipients are due to donor T cell infusion and potentiated by cyclophosphamide. J Clin Invest *100*, 1015–1027.
Parrish-Novak, J., Dillon, S.R., Nelson, A., Hammond, A., Sprecher, C., Gross, J.A., Johnston, J., Madden, K., Xu, W., West, J. et al. (2000). Interleukin 21 and its receptor are involved in NK cell expansion and regulation of lymphocyte function. Nature *408*, 57–63.
Pearce, E.L., Mullen, A.C., Martins, G.A., Krawczyk, C.M., Hutchins, A.S., Zediak, V.P., Banica, M., DiCioccio, C.B., Gross, D.A., Mao, C.A. et al. (2003). Control of effector CD8+ T cell function by the transcription factor Eomesodermin. Science *302*, 1041–1043.
Peluso, I., Fantini, M.C., Fina, D., Caruso, R., Boirivant, M., MacDonald, T.T., Pallone, F., and Monteleone, G. (2007). IL-21 counteracts the regulatory T cell-mediated suppression of human CD4+ T lymphocytes. J Immunol *178*, 732–739.
Roncarolo, M.G., Bacchetta, R., Bordignon, C., Narula, S., and Levings, M.K. (2001). Type 1 T regulatory cells. Immunol Rev *182*, 68–79.
Sakaguchi, S., Sakaguchi, N., Asano, M., Itoh, M., and Toda, M. (1995). Immunologic self-tolerance maintained by activated T cells expressing IL-2 receptor alpha-chains (CD25). Breakdown of a single mechanism of self-tolerance causes various auto-immune diseases. J Immunol *155*, 1151–1164.
Serody, J., Wysocki, C.A., and Moran, T.P. (2008). Recruitment of donor T cells to secondary lymphoid tissue is enhanced by conditioning therapy and donor T cells/marrow. In American Society of Hematology (San Francisco), p. 3506.
Smith, S.M. (2003). Non-myeloablative transplantation for lymphoma. Curr Treat Options Oncol *4*, 261–268.
Socie, G., Loiseau, P., Tamouza, R., Janin, A., Busson, M., Gluckman, E., and Charron, D. (2001). Both genetic and clinical factors predict the development of graft versus host disease after allogeneic hematopoietic stem cell transplantation. Transplantation *72*, 699–706.
Spolski, R., Kashyap, M., Robinson, C., Yu, Z., and Leonard, W.J. (2008). IL-21 signaling is critical for the development of type I diabetes in the NOD mouse. Proc Natl Acad Sci USA *105*, 14028–14033.
Storb, R., Deeg, H.J., Farewell, V., Doney, K., Appelbaum, F., Beatty, P., Bensinger, W., Buckner, C.D., Clift, R., Hansen, J. et al. (1986). Marrow transplantation for severe aplastic anemia: methotrexate alone compared with a combination of methotrexate and cyclosporine for prevention of acute graft versus host disease. Blood *68*, 119–125.
Strengell, M., Sareneva, T., Foster, D., Julkunen, I., and Matikainen, S. (2002). IL-21 up-regulates the expression of genes associated with innate immunity and Th1 response. J Immunol *169*, 3600–3605.
Suri-Payer, E., Amar, A.Z., Thornton, A.M., and Shevach, E.M. (1998). CD4+CD25+ T cells inhibit both the induction and effector function of autoreactive T cells and represent a unique lineage of immunoregulatory cells. J Immunol *160*, 1212–1218.
Szabo, S.J., Kim, S.T., Costa, G.L., Zhang, X., Fathman, C.G., and Glimcher, L.H. (2000). A novel transcription factor, T-bet, directs Th1 lineage commitment. Cell *100*, 655–669.
Tabbara, I.A., and Ingram, R.M. (2003). Non-myeloablative therapy and allogeneic hematopoietic stem cell transplantation. Exp Hematol *31*, 559–566.
Tawara, I., Maeda, Y., Sun, Y., Lowler, K.P., Liu, C., Toubai, T., McKenzie, A.N., and Reddy, P. (2008). Combined Th2 cytokine deficiency in donor T cells aggravates experimental acute graft versus host disease. Exp Hematol *36*, 988–996.

Thompson, J.S., Chu, Y., Glass, J.F., and Brown, S.A. (2010). Absence of IL-23p19 in donor allogeneic cells reduces mortality from acute GVHD. Bone Marrow Transplant *45*, 712–22.

Thornton, A.M., and Shevach, E.M. (1998). CD4+CD25+ immunoregulatory T cells suppress polyclonal T cell activation in vitro by inhibiting interleukin 2 production. J Exp Med *188*, 287–296.

Trenado, A., Sudres, M., Tang, Q., Maury, S., Charlotte, F., Gregoire, S., Bonyhadi, M., Klatzmann, D., Salomon, B.L., and Cohen, J.L. (2006). Ex vivo-expanded CD4+CD25+ immunoregulatory T cells prevent graft versus host disease by inhibiting activation/differentiation of pathogenic T cells. J Immunol *176*, 1266–1273.

Varona, R., Cadenas, V., Gomez, L., Martinez, A.C., and Marquez, G. (2005). CCR6 regulates CD4+ T-cell-mediated acute graft versus host disease responses. Blood *106*, 18–26.

Vlachos, A., and Lipton, J.M. (1996). Bone marrow failure in children. Current Opinion in Pediatrics *8*, 33–41.

Wing, K., Onishi, Y., Prieto-Martin, P., Yamaguchi, T., Miyara, M., Fehervari, Z., Nomura, T., and Sakaguchi, S. (2008). CTLA-4 control over Foxp3+ regulatory T cell function. Science *322*, 271–275.

Wood, K.J., and Sakaguchi, S. (2003). Regulatory T cells in transplantation tolerance. Nat Rev Immunol *3*, 199–210.

Wurster, A.L., Rodgers, V.L., Satoskar, A.R., Whitters, M.J., Young, D.A., Collins, M., and Grusby, M.J. (2002). Interleukin 21 is a T helper (Th) cell 2 cytokine that specifically inhibits the differentiation of naive Th cells into interferon gamma-producing Th1 cells. The Journal of experimental medicine *196*, 969–977.

Wysocki, C.A., Panoskaltsis-Mortari, A., Blazar, B.R., and Serody, J.S. (2005). Leukocyte migration and graft versus host disease. Blood *105*, 4191–4199.

Yang, X.O., Pappu, B.P., Nurieva, R., Akimzhanov, A., Kang, H.S., Chung, Y., Ma, L., Shah, B., Panopoulos, A.D., Schluns, K.S. et al. (2008). T helper 17 lineage differentiation is programmed by orphan nuclear receptors ROR alpha and ROR gamma. Immunity *28*, 29–39.

Yi, T., Chen, Y., Wang, L., Du, G., Huang, D., Zhao, D., Johnston, H., Young, J., Todorov, I., Umetsu, D.T. et al. (2009). Reciprocal differentiation and tissue-specific pathogenesis of Th1, Th2, and Th17 cells in graft versus host disease. Blood *114*, 3101–3112.

Yi, T., Zhao, D., Lin, C.L., Zhang, C., Chen, Y., Todorov, I., LeBon, T., Kandeel, F., Forman, S., and Zeng, D. (2008). Absence of donor Th17 leads to augmented Th1 differentiation and exacerbated acute graft versus host disease. Blood *112*, 2101–2110.

Zeller, J.C., Panoskaltsis-Mortari, A., Murphy, W.J., Ruscetti, F.W., Narula, S., Roncarolo, M.G., and Blazar, B.R. (1999). Induction of CD4+ T cell alloantigen-specific hyporesponsiveness by IL-10 and TGF-beta. J Immunol *163*, 3684–3691.

Zheng, W., and Flavell, R.A. (1997). The transcription factor GATA-3 is necessary and sufficient for Th2 cytokine gene expression in CD4 T cells. Cell *89*, 587–596.

Zhou, G., and Levitsky, H.I. (2007). Natural regulatory T cells and de novo-induced regulatory T cells contribute independently to tumor-specific tolerance. J Immunol *178*, 2155–2162.

Zhu, K.E., Hu, J.Y., Zhang, T., Chen, J., Zhong, J., and Lu, Y.H. (2008). Incidence, risks, and outcome of idiopathic pneumonia syndrome early after allogeneic hematopoietic stem cell transplantation. Eur J Haematol *81*, 461–466.

Interplay of Pathogenic T_H1/T_H17 Cells and Regulatory T Cells in Auto-immune Disease: A Tale of Yin and Yang

Xuebin Liu, Stewart Leung, Lei Fang, Xi Chen, Taylor Guo, and Jingwu Zhang

Abstract The propagation and regulation of an immune response is driven by effector and regulatory T cells in a network and fashion resembling the "yin" and "yang" concept in the traditional Chinese philosophy. The shift of the balance between "yin" and "yang" determines the direction of the response towards inflammation or its resolution. In auto-immune disease, the break of tolerance to self-antigens leads to differentiation and propagation of auto-reactive effector T cells and the restoration of the balance is a logical approach to effective treatments. In this review, we describe the characteristics and development of pathogenic T_H1 and T_H17 cells and T_{reg} cells which regulate them in auto-immune disease. The emphasis is given to the crucial roles of cytokines in influencing lineage differentiation and function as well as interactions of these T cell subsets. We discuss current immunotherapeutic strategies involving cytokine or cytokine receptor antibodies for the treatment of auto-immune diseases and the potential of traditional Chinese medicine in restoring the balance of "Yin" and "Yang" in a disease setting.

1 Introduction

$CD4^+$ T helper (T_H) cells are essential effectors of the immune response and inflammation. For more than two decades, the T_H1/T_H2 paradigm, originally introduced by Mosmann and Coffman, has been used to explain most of the phenomena related to adaptive immunity (Mosmann and Coffman 1989). This paradigm has shifted following the recent discovery of T_H17 and regulatory T (T_{reg}) cells, two novel T subsets that are thought to be critically involved in mediating and regulating auto-immune responses (Sakaguchi 2004; Harrington et al. 2005; Park et al. 2005; Afzali et al. 2007). These opposite but interactive roles between pathogenic T_H1/T_H17 and T_{reg}

J. Zhang (✉)
Department of Neuroimmunology, GlaxoSmithKline Research and Development Center, 898 Halei Road, Building 3, Zhangjiang Pudong, Shanghai 201203, China
e-mail: jingwu.z.zang@gsk.com

S. Jiang (ed.), *T_H17 Cells in Health and Disease*,
DOI 10.1007/978-1-4419-9371-7_19, © Springer Science+Business Media, LLC 2011

cells are perhaps best understood by the principle of "yin" and "yang" that represents nature's autonomy, whereby the former functions as mediator, the latter acts as a regulator of inflammation. An in-depth understanding of the precise pathological mechanisms of these pathogenic T cell subsets is urgently needed to identify novel targets and develop new specific treatment approaches for auto-immune diseases, where interplay or interaction of T_H1/T_H17 and T_{reg} cells are controlled by their cytokine environment. Pathogenic T_H1/T_H17 cells function as the "yang", secreting pro-inflammatory cytokines including IFN-γ and IL-17. They have been shown to promote inflammatory responses and contribute to the pathogenesis of multiple human auto-immune disorders such as multiple sclerosis (MS), and rheumatoid arthritis (RA), experimental auto-immune encephalomyelitis (EAE), and collagen-induced arthritis (CIA) (Hwang et al. 2004; Komiyama et al. 2006; Weaver et al. 2007). In contrast, T_{reg} cells, acting as the "yin", are crucial for the maintenance of localized immuno-suppression and normally keep pathogenic T_H1/T_H17 cells in check (Corthay 2009). T_{reg} cells were found to be deficient in patients with MS and RA, potentially contributing to a shift of the immune system toward a pro-inflammatory state (Costantino et al. 2008).

2 Roles of T_H1/T_H17 and T_{reg} Cells in Auto-immune Diseases

2.1 Differentiation and Function of T_H1 and T_H17 Cells in Auto-immune Diseases

The pathogenesis of auto-immune diseases, including MS and RA, arises from the break of tolerance towards self-antigens and the development of auto-aggressive effector T cells infiltrating into the target tissues. MS is a chronic inflammatory disease characterized by progressive demyelination and axon injury within the central nervous system (CNS), leading to numerous neurological and mental symptoms (Issazadeh et al. 1995; Bielekova et al. 2000; Sospedra and Martin 2005; McFarland and Martin 2007). A key event in the pathophysiology of MS is the increased migration of T lymphocytes across the blood–brain barrier into the brain, spinal cord, or both. Previously, T_H1 cells have been regarded as the main pathogenic T cells driving auto-immune tissue damage, which is attributed to the activation of macrophages by T_H1-producing cytokine, IFN-γ (Sospedra and Martin 2005). CD4+ T helper cells invading the CNS at the peak of EAE express IFN-γ (Renno et al. 1995). In addition, treatment of mice with IL-12 promotes development of IFN-γ-producing T_H1 cells and aggravates the clinical course of CIA, another organ-specific auto-immune disease model (Germann et al. 1996). Paradoxically, IFN-$\gamma^{-/-}$ mice were found highly susceptible to many organ-specific auto-immune diseases including EAE, and blocking IFN-γ or its signaling pathway exacerbated auto-immune disease severity in EAE and CIA (Billiau et al. 1988; Murphy et al. 2003). Furthermore, the cytokine IL-23, but not IL-12 was shown to be crucial for mounting an auto-reactive

T cell response in the CNS in EAE (Cua et al. 2003; Langrish et al. 2005). In addition to the paradoxical properties of IFN-γ (Chen et al. 2006), the findings led to the discovery of a new lineage of T helper subset, IL-17-producing T_H17 cells. Given that IL-23 and IL-12 specifically promote T_H17 and T_H1 differentiation and that T_H17 differentiation is strongly inhibited by IFN-γ, it is indicated that T_H17 cells are in fact responsible for the induction of auto-immune response in EAE.

There is evidence that T_H17 cells are activated during the disease process and responsible for recruiting other inflammatory cell types to mediate CNS pathology in EAE and potentially in MS. Elevated levels of IL-17 have been reported in the serum, peripheral blood mononuclear cells (PBMC), and cerebrospinal fluid (CSF) of MS patients (Matusevicius et al. 1999; Lock et al. 2002; Tzartos et al. 2008). Vaknin-Dembinsky et al. reported that dendritic cells (DC) derived from MS patients secrete higher levels of IL-23 but similar amounts of IL-12 as compared to that of healthy controls (Vaknin-Dembinsky et al. 2008). Recent studies suggest that targeting T_H17 cells is efficacious in T cell-mediated auto-immune conditions such as EAE (Serada et al. 2008). These observations suggest that T_H17 cells are at least as important as, if not more important than, T_H1 cells in driving the inflammatory process in tissue-specific auto-immunity. In the past 4 years, the importance of T_H17 cells in the pathogenesis of auto-immune disease has been demonstrated in several well-characterized auto-immune animal models. However, the new findings on T_H17 cells do not necessarily dismiss the important role of T_H1 cells in auto-immune disease. Rather, T_H1 and T_H17 cells work in unison through different pathways to induce auto-immune pathology. There is evidence that adoptive transfer of MOG-specific T_H1 cell lines or clones, in the absence of IL-17-expressing T cells, is able to induce EAE in naïve mouse models (Pettinelli and McFarlin 1981). Moreover, only partial protection is conferred in EAE in IL-17$^{-/-}$ mice or by neutralizing IL-17 (Komiyama et al. 2006). It is now accepted that T_H1 and T_H17 cells are both involved in the pathogenesis of EAE and MS (Kroenke et al. 2008; Stromnes et al. 2008). In our own study, analysis of MOG-specific T cells from spinal cord of mice with EAE further revealed the presence of substantial numbers of T cells co-expressing IL-17 and IFN-γ (Liu et al. 2010). The findings are consistent with the recent reports that have identified double positive cells such as T_H1-like cells (Acosta-Rodriguez et al. 2007b). However, it remains to be investigated as to whether these IL-17$^+$ double positive T_H1-like cells are encephalitogenic in EAE.

2.2 T_{reg} Cells in Auto-immune Diseases

T_{reg} cells co-express CD4 and CD25 and its signature transcription factor Forkhead box protein 3 (Foxp3) (Curotto de Lafaille and Lafaille 2009). They are capable of actively suppressing effector cells and dampening a wide spectrum of immune responses, including those associated with auto-immune diseases (Sakaguchi 2004). Thymic-derived naturally occurring T_{reg} (nT_{reg}) cells are critical for maintaining self-tolerance and preventing auto-immunity and tissue injury by inhibiting the proliferation and effector functions of auto-reactive pathogenic T_H1/T_H17 cells (Sakaguchi et al. 1995; Stummvoll et al. 2008). In addition to nT_{reg} cells, another type of Foxp3$^+$CD4$^+$ regulatory T cells, termed inducible regulatory T (iT_{reg}) cells, originate from the peripheral immune system. They differentiate from naive CD4+ T cells in the presence of TGF-β through the induction of Foxp3 expression (Chen et al. 2003). iT_{reg} cells can also act as a negative regulator for the control of activation and effector functions of self-antigen reactive T cells and maintain peripheral tolerance (Huter et al. 2008). In our hands, iTreg cells are potent regulators for autoimmune inflammation upon adoptive transfer (unpublished data).

It is widely reported that auto-reactive T cells isolated from patients with autoimmune disease have a lower threshold of activation (Anderton 2006). Recent studies have suggested that CD4$^+$CD25^{hi} T_{reg} cells are functionally impaired in patients with MS (Fletcher et al. 2009). Interestingly, it has been shown that T_{reg} cells isolated from the circulation of patients with MS, although normal in frequency, have impaired inhibitory functions against effector T cell proliferation in an in vitro co-culture system (Viglietta et al. 2004; Kumar et al. 2006). The functional impairment of T_{reg} cells is also reported in EAE where myelin-specific functional T_{reg} cells accumulate in the CNS but fail to control auto-immune inflammation (Korn et al. 2007b). In RA patients, reduced numbers of CD4$^+$CD25^{hi} T_{reg} cells were observed in peripheral blood (Sarkar and Fox 2007). It has been reported that T_{reg} cells derived from RA patients can neither regulate pro-inflammatory cytokine secretion by pathogenic T cells and monocytes or suppress auto-reactive T cells (Ehrenstein et al. 2004; Valencia et al. 2006). It should be noted that the frequency of CD4$^+$CD25^{hi} T_{reg} cells is higher in rheumatoid synovium than in blood of patients with RA (Cao et al. 2003, 2004; van Amelsfort et al. 2004; Mottonen et al. 2005; Korn et al. 2007a, b; Chen et al. 2009). Our recent study demonstrated that T_{reg} cells accumulate in the rheumatoid synovium and co-exist with pathogenic T_H17 and T_H1 cells. However, these Treg cells were functionally deficient, which was attributable to altered post-translational modification of Foxp3 (unpublished). Taken together, these findings suggest that functionally impaired T_{reg} cells and increased numbers of auto-pathogenic T_H17 cells contribute to the break in tolerance and shift of the immune system towards a pro-inflammatory state.

3 Cytokine Environment and Signaling Pathways for Differentiation and Maintenance of T_H1/T_H17 and T_{reg} Cells

3.1 T Subsets Differentiation Pathway

Efficient host defense against pathogens is achieved through complex signaling between the innate immune and adaptive immune systems. In response to the type of pathogen presented by conventional antigen presenting cells (APC) and in the presence of a cytokine milieu produced by activated APC, naive CD4$^+$ T cells can

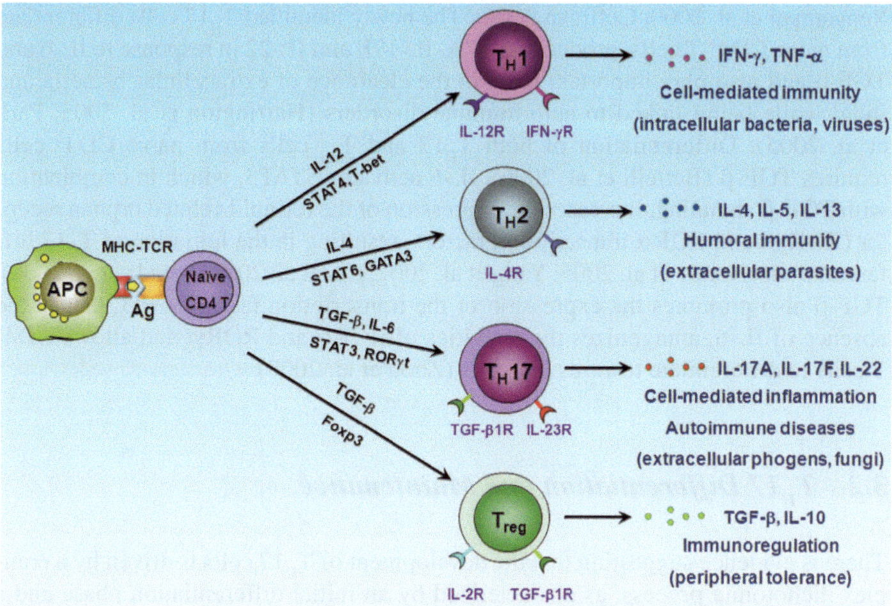

Fig. 1 Model for T helper cell differentiation from naive CD4+ T cells. In the presence of IL-12, differentiation of naive CD4+ T cells into T_H1 cells requires activation of the master regulator transcription factor T-bet through STAT1 and STAT4. T_H1 cells produce IFN-γ and are involved in cell-mediated immunity against intracellular bacteria and viruses. IL-4 promotes the activation of STAT6 and GATA3, which are responsible for T_H2 cell differentiation. T_H2 cells are important in humoral immunity against parasites, an action that is mediated through their production of IL-4, IL-5, and IL-13. The combination of TGF-β and pro-inflammatory cytokines such as IL-6 and IL-23, drives the differentiation of naive CD4+ T cells into IL-17-producing T helper cells (T_H17) through the regulation of STAT3 and RORγt. T_H17 cells play a critical role in host protection against extracellular pathogens and in inflammatory auto-immune diseases. In addition, TGF-β can induce differentiation of naive CD4+ T cells into Foxp3+ T_{reg} cells, which produce TGF-β and IL-10 and act as modulators of immune responses

differentiate into specific T cell lineages, T helper type 1 cells (T_H1), T_H2, T_H17, and T_{reg} cells, as categorized by the expression of specific transcription factors and distinct effector function with their signature cytokine profiles (Fig. 1). In the presence of IFN-γ and IL-12, naive CD4 T cells differentiate towards T_H1 cells in a process that is dependent on the activities of STAT1, STAT4 and the lineage-specific transcription factor T-bet. T_H1 cells produce large quantities of IFN-γ and play a critical role in protective immunity against intracellular pathogens through the activation of macrophages (Mosmann and Coffman 1989; Szabo et al. 2000; Coffman 2006). IL-4 promotes T_H2 differentiation through activation of STAT6 and lineage-specific transcription factor GATA3. T_H2 cells produce IL-4, IL-5 IL-13, and IL-25, which are important for the orchestration of humoral immune responses clearing extracellular pathogens and parasites through the induction of immuno-globulin class switching to IgG1 and IgE. (Mosmann et al. 1986; Zheng and Flavell 1997;

Rengarajan et al. 2000; Coffman 2006). The newly identified T_H17 cells differentiate from naive $CD4^+$ T cells, produce IL-17A, IL-17F and IL-22 in response to IL-6 and TGF-β and also play important roles in the clearance of extracellular bacteria and fungi while being linked to auto-immune disorders (Harrington et al. 2005; Park et al. 2005). Differentiation of both T_H17 and T_{reg} cells from naive $CD4^+$ cells requires TGF-β (Bettelli et al. 2006). IL-6 activates STAT3, which in combination with TGF-β signaling, increases the expression of the retinoid-related orphan receptor (ROR)γt and RORα transcription factors, resulting in the initiation of T_H17 differentiation (Ivanov et al. 2006; Yang et al. 2007; Liu et al. 2008; Serada et al. 2008). TGF-β also promotes the expression of the transcription factor Foxp3, that in the absence of IL-6, antagonizes the activities of RORα and RORγt and allows $CD4^+$ T cells to differentiate towards T_{reg} cells (Zhou et al. 2008).

3.2 T_H17 Differentiation and Maintenance

There is evidence suggesting that the development of T_H17 cells is driven by a complex dichotomic process, as characterized by an initial differentiation phase and a later phase of survival and expansion of committed T_H17 cells. Several cytokines such as TGF-β, IL-6, IL-1β, IL-21, and IL-23, have been demonstrated to regulate/induce T_H17 cell differentiation, critically contributing to solid outcome of autoimmune disease (de Jong et al. 2002; Aggarwal et al. 2003; Korn et al. 2007a, b, 2009; Vang et al. 2008). IL-23, which shares the p40 unit with IL-12, was the first cytokine described to upregulate IL-17 expression in $CD4^+$ T cells. Cua et al. dissected the participation and contribution of IL-12 and IL-23 in EAE induction by using animals with gene disruptions for each of the sub-units forming IL-12 and IL-23, namely p19, p35 and p40. They found that p19 or p40-deficient mice were resistant to EAE induction and unable to generate IL-17-producing $CD4^+$ T cells (T_H17 cells), while mice deficient only in IL-12 ($p35^{-/-}$) were highly susceptible to EAE induction (Cua et al. 2003). Similar results were shown in a CIA model (Murphy et al. 2003). Resistance seems to correlate with the absence of IL-17-producing $CD4^+$ T cells, despite normal induction of antigen-specific IFN-γ-producing T_H1 cells. In contrast, IL-12-deficient $p35^{-/-}$ mice developed more IL-17-producing $CD4^+$ T cells. It was further demonstrated that T_H17 cell differentiation from truly naive $CD4^+$ T cells was driven by a combination of the pro-inflammatory cytokine IL-6 and TGF-β in an in vitro system (McGeachy et al. 2007). In vivo analysis of IL-6 KO mice also demonstrated an important role of IL-6 in the differentiation of T_H17 cells since these animals are resistant to the development of EAE and do not develop a competent T_H17 cell response (Bettelli et al. 2006). Moreover, the T_H1 and T_H2 cytokines, IFN-γ and IL-4, were found to negatively regulate the generation of T_H17 cells in vitro and in vivo. Murine T_H17 lymphocytes differentiate from naive $CD4^+$ cells in a specific cytokine environment, which includes TGF-β and IL-6 or IL-21 (Mangan et al. 2006; Korn et al. 2007a, b), whereas human T_H17 cell development requires TGF-β, IL-1β and IL-2 in combination with IL-6, IL-21, or IL-23 (Acosta-Rodriguez et al. 2007a, b; Wilson et al. 2007; Manel et al. 2008; Yang et al. 2008a). Independent studies demonstrated that TGF-β and IL-6 are key drivers of the initiation of T_H17 differentiation, and IL-1β and TNF-α participate in T_H17 differentiation and can amplify this process (Skarica et al. 2009; Sutton et al. 2009). IL-21, another cytokine produced by T_H17 cells, has been shown to provide an additional autocrine amplification signal (Korn et al. 2007a, b). In fact, IL-6 or IL-21 can induce T_H17 cells to produce more IL-21 (Zhou et al. 2007) and disruption of the IL-21 pathway results in reduced T_H17 differentiation (Korn et al. 2007a, b; Nurieva et al. 2007). Although IL-23 plays an important role in T_H17-mediated inflammatory auto-immune diseases, it does not participate in T_H17 initiation, due to the lack of IL-23R expression on naive $CD4^+$ T cells (Stritesky et al. 2008; Chen et al. 2009). As IL-23R expression is turn on in differentiating T_H17 cells, it can be further upregulated by IL-23 itself and other cytokines, such as IL-21. IL-23 further amplifies the induction of T_H17 cells through its receptor by stabilizing the STAT3 signal, given that IL-6, IL-21, and IL-23 all utilize the Jak-Stat pathway and activate STAT3 (Cho et al. 2006; Yang et al. 2007; Bayer et al. 2008). Thus, IL-23 is proposed to function in the late stage of T_H17 cell differentiation.

It is crucial to understand in vivo differentiation, maintenance of T_H17 cells, and the underlying mechanisms in the context of an auto-immune process. The basic mechanisms of IL-23 in vivo maintenance of differentiated T_H17 cells remain poorly understood. Cua et al. recently indicated that IL-23 may play an important role in the terminal differentiation of T_H17 cells through its effect on re-expression of IL-7R on T_H17 cells (Chen et al. 2009). We recently discovered that the second phase of T_H17 cell development, namely survival and expansion of committed T_H17 cells, is controlled by IL-7 signaling through STAT5. The study was prompted by genome-wide single nucleotide polymorphism scanning in a large population of MS patients and controls, suggesting that polymorphisms in IL-7R are important susceptibility factors for MS (Gregory et al. 2007; Hafler et al. 2007; Lundmark et al. 2007). In addition, there are several reports showing that T-cell reactivity against myelin basic protein is augmented in those MS patients with elevated IL-7 and IL-7R levels (Bielekova et al. 1999; Traggiai et al. 2001; Vudattu et al. 2009).

IL-7 is primarily produced by non-hematopoietic stromal and epithelial cells, including fibroblastic reticular cells in the T-cell zone of lymphoid organs, bone marrow stromal cells, thymic epithelial cells, and liver and intestinal epithelial cells (Lee and Surh 2005; Mazzucchelli and Durum 2007). To a lesser degree, DC and macrophages also produce IL-7 and it is also expressed in some organs, including the brain. The IL-7R is expressed on most mature T cells at high levels. It is also expressed on dendritic cells, monocytes, and subsets of developing B cells and T cells, but not on mature T_{reg} and B cells. IL-7 binds to IL-7R and activates the JAK/STAT5 and PI3K-AKT signaling pathways (Palmer et al. 2008). As a T-cell growth factor, IL-7 plays a role in the regulation of peripheral homeostasis of the CD4 T cell pool (Sportes et al. 2008). Our study demonstrated that IL-7 is essential for survival and expansion of pathogenic T_H17 cells in EAE. IL-7 directly expanded effector T_H17 cells in EAE and human T_H17 cells from subjects with MS, whereas it was not required for T_H17 differentiation. IL-7R antagonism rendered differentiated T_H17 cells susceptible to apoptosis through altered expression levels of the pro-survival protein Bcl-2 and pro-apoptotic protein Bax, leading to decreased EAE severity (Liu et al. 2010). In particular, the role of IL-7 and IL-7R antagonism was examined in comparison to IL-6, IL-23, and other related cytokines, namely thymic stromal lymphopoietin (TSLP) and IL-15 (Liu et al. 2010). Taken together, it becomes clear that there are two distinct steps in the development of T_H17 cells, induction/terminal differentiation and survival/expansion, in which several cytokines are sequentially involved (IL-6, IL-21, IL-23, and IL-7) and play different roles. IL-6 is a major player in the initial T_H17 differentiation through the STAT3 pathway. IL-21 provides an additional autocrine amplification signal. IL-23 acts through stabilization of STAT3 to promote T_H17 terminal differentiation especially in the secondary lymphoid organs. In this regard, our data show that IL-7 is a key cytokine for the survival and in vivo expansion of committed T_H17 cells by signaling through a JAK/STAT-5 pathway (Fig. 2). However, there is a potential connection between IL-7 and IL-23 in T_H17 development, as IL-23 receptor is required for the re-expression of IL-7R in activated T_H17 cells (McGeachy et al. 2009). Our data clearly supports that IL-7 is superior to IL-23 for T_H17 cell expansion, which could not be blocked by IL-23 antibody (unpublished data). The results provide compelling evidence for

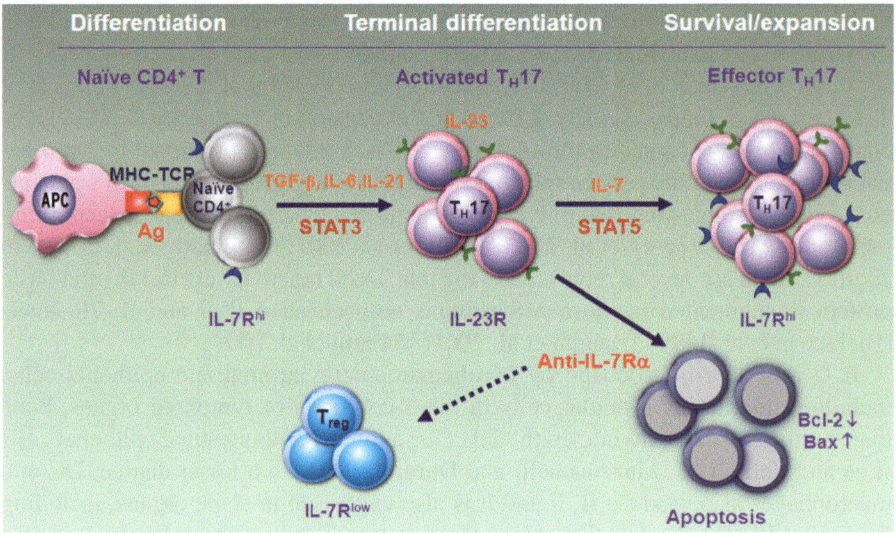

Fig. 2 Critical role of IL-7/IL-7R signaling in survival and expansion of differentiated T_H17 cells. T_H17 cell development is a dichotomic process that is regulated through a complex cytokine network. The differentiation of T_H17 cells is mainly mediated by STAT-3 signaling through cytokines such as IL-6, IL-21, and IL-23. There is dynamic expression of IL-7R on T_H17 cells in the course of T cell activation/differentiation. In the latter phase, where IL-7R is re-expressed, IL-7 is critically required to sustain survival and expansion of differentiated T_H17 cells through STAT-5 signaling. IL-7R antagonism renders differentiated T_H17 cells susceptible to apoptosis through altered expression of pro-apoptotic protein Bax and anti-apoptotic molecule, Bcl-2

a novel role of IL-7/IL-7R in pathogenic T_H17 cell development and function in EAE and support the rationale for IL-7R antagonism as a potential treatment for MS (see further discussion below) and perhaps other auto-immune conditions.

3.3 T_{reg} Development and Function

Inflammatory cytokine milieu not only promotes the T_H1 and T_H17 response but also affects T_{reg} induction and function. T_{reg} cells express the transcription factor Foxp3, which is not only a lineage specification factor but also a functional marker of T_{reg} cells (Fontenot et al. 2003; Hori et al. 2003; Khattri et al. 2003). Genetic defects in the *Foxp3* gene, which affect the development and function of T_{reg} cells, have been identified as being responsible for X-linked recessive inflammatory disease in *Scurfy* mice, and IPEX (immuno-dysregulation polyendocrinopathy enteropathy X-linked) syndrome in humans (Chatila et al. 2000; Bennett et al. 2001; Brunkow et al. 2001; Wildin et al. 2001). Furthermore, ectopic expression of Foxp3 is sufficient to confer conventional T cells with regulatory properties (Hori et al. 2003; Fontenot et al. 2005b; Wan and Flavell 2005). Multiple signals, such as TCR and co-stimulatory molecule CD28, are required for the thymic development of T_{reg} cells (Salomon et al. 2000; Bensinger et al. 2001; Tai et al. 2005). Common γ-chain (γ_c) cytokines IL-2, IL-7, and IL-15 are required for Foxp3 expression and T_{reg} cell development (Bayer et al. 2005, 2008; Soper et al. 2007; Lio and Hsieh 2008; Yang et al. 2008a, b). Gene knockout of γ_c in mice leads to complete absence of $Foxp3^+$ T_{reg} cells in thymus and peripheral immune system (Fontenot et al. 2005a, b). In this process, transcription factor STAT5 is critically involved in the activation of γ_c cytokine receptors and is shown to bind to Foxp3 promoter (Burchill et al. 2007; Yao et al. 2007). iT_{reg} cells that differentiate from naive $CD4^+$ T cells in the peripheral immune system upon activation in the presence of TGF-β seem less stable than nT_{reg} cells (Yang et al. 2008a, b; Zhou et al. 2008). TCR repertoire of nT_{reg} cells that are selected by high-avidity interactions in the thymus is mostly towards self-antigen, whereas that of iT_{reg} cells that differentiate from conventional $CD4^+$ is similar to naïve conventional $CD4^+$ T cells in the TCR range (Curotto de Lafaille and Lafaille 2009). Cytokines such as IL-4, IFNγ, IL-6 that induce differentiation of T helper cells mostly dampen iT_{reg} development (Zhou et al. 2009). Inhibition of TGF-β-induced T_{reg} cells by these cytokines or co-stimulation could be suppressed by retinoic acid (RA) (Benson et al. 2007; Mucida et al. 2007; Hill et al. 2008). $CD103^+$ dendritic cells isolated from the small intestine and the mesenteric lymph nodes, which produce both TGF-β and RA, efficiently promote iT_{reg} differentiation (Coombes et al. 2007; Sun et al. 2007).

$Foxp3^+$ T_{reg} cells control the immune response through multiple mechanisms. T_{reg} cells are potent suppressors of helper T cells. Many studies have demonstrated T_{reg}-mediated suppression by inhibiting the production of IL-2 mRNA in the responder T cells (Thornton and Shevach 1998). Although in vitro T_{reg} cells exert their suppressive function through cell–cell contact, it should not be ruled out that T_{reg} cells could secrete soluble factor. IL-35, a newly identified inhibitory cytokine, contributes to T_{reg} function by directly acting on responder T cells (Collison et al. 2007). Another candidate factor is galectin-1 which induces cell cycle arrest of responder cells (Garin et al. 2007). One other mechanism is cytolysis of target cells; T_{reg} cells can express granzyme A or B and kill target cells (Grossman et al. 2004; Gondek et al. 2005). Furthermore, T_{reg} cells could downregulate or inhibit upregulation of CD80/86 of APCs through CTLA-4 (Wing et al. 2008). Catalytic inactivation of extracellular ATP by CD39 expressed by T_{reg} cells represents a mechanism that prevents deleterious effect of ATP on APC function (Borsellino et al. 2007; Deaglio et al. 2007). Moreover, IL-10 and TGF-β have been shown to be important mediators of T_{reg}-mediated suppression of colitis, type 1 diabetes, and EAE (Asseman et al. 1999; Burkhart et al. 1999; Suri-Payer and Cantor 2001; Green et al. 2003; Zhang et al. 2004).

Recent studies have suggested that the reduced numbers and impaired regulatory function of T_{reg} cells are responsible for human auto-immune disorders, including type 1 diabetes, RA and MS (Ehrenstein et al. 2004; Viglietta et al. 2004; Lindley et al. 2005). Although the accumulation of T_{reg} cells in the rheumatoid synovium and CNS lesions of animals with EAE may represent the immune system's attempt to regulate local inflammation, it remains to be discovered how local auto-immune inflammation perturbs T_{reg}-mediated immune regulation. There are two possibilities: pro-inflammatory cytokines may hamper T_{reg} cell function and/or the inflammatory cytokine milieu may induce defects in T_{reg} cells, possibly through post-translational modifications of Foxp3. Recently, we demonstrated that the suppressive function of T_{reg} cells derived from rheumatoid synovium of RA patients is impaired. The inhibitory property of RA-derived synovial fluid on normal T_{reg} cells originating from healthy donors could only be neutralized with a TNF-α antibody (data unpublished). In addition, TNF-α inhibits the regulatory function of nT_{reg} cells and TGF-β1-induced $CD4^+CD25^+$ T_{reg} cells in vitro (Ehrenstein et al. 2004; Valencia et al. 2006). TNF-α-mediated inhibition of T_{reg} suppressive function is associated with altered post-translational modification of Foxp3 (data unpublished). The mechanism of TNF-α-mediated inhibition of T_{reg} cells involves signaling through TNFRII, which is constitutively expressed on resting T_{reg} cells and is up-regulated by TNF-α. Treatment with Infliximab® (a therapeutic TNF-α antibody) restored the function of T_{reg} cells and increased their expression of *Foxp3* mRNA and protein (Valencia et al. 2006). In addition to RA, it has also been reported that CNS-infiltrating $CD4^+Foxp3^-$ T effector cells, which produce IL-6, IL-21, and TNF-α, are resistant to T_{reg}-mediated suppression (Korn et al. 2007a, b; Peluso et al. 2007). Furthermore, a recent study by Tao et al. demonstrated that acetylation of the forkhead domain of Foxp3 is critical for optimal T_{reg} function and Foxp3 binding to the *IL-2* promoter (Tao et al. 2007). Treatment with histone deacetylase inhibitor (HDACi) in vitro promoted T_{reg} gene expression and enhanced T_{reg} suppressive function: treatment in vivo increased T_{reg}-mediated suppression of homeostatic proliferation and prolonged allograft survival in a T_{reg} dependent manner (Tao et al. 2007). It is conceivable that pro-inflammatory cytokines, such as TNF-α, could alter the post-translational modification state of Foxp3 protein, leading to impaired T_{reg} function and modulate inflammation in auto-immune disease.

3.4 Interplay Between T_H17 and T_{reg}

Distinct populations of T_H17 and T_{reg} cells coexist and are reciprocally regulated during differentiation in healthy tissue (Bettelli et al. 2006). Imbalances in the ratio of these lymphocytes have been implicated in a wide range of auto-immune disorders including MS and RA. Bettelli et al. have demonstrated a reciprocal developmental relationship between $Foxp3^+$ T_{reg} and T_H17 cells. TGF-β triggers the expression of Foxp3 in naive T cells, whereas TGF-β drives T_H17 differentiation from naive T cells in combination with IL-6. It is well established that inflammatory cytokines such as IL-6 and IL-21 inhibit TGF-β-induced T_{reg} generation (Bettelli et al. 2006; Fantini et al. 2007; Hill et al. 2008). Blocking of IL-6 signaling in vivo can generate $Foxp3^+$ T_{reg} cells and reduce EAE severity (Korn et al. 2008). Furthermore, Infliximab therapy leads to the generation of $CD4^+CD25^{hi}Foxp3^+$ T_{reg} cells that are $CD62L^{low}$ and suppress effector T cells through TGF-β and IL-10 (Nadkarni et al. 2007). Some recent findings showed that T_{reg} cells could be converted to IL-17-producing cells. Pro-inflammatory cytokines, such as IL-6, IL-1, and IL-21 were able to decrease the Foxp3 expression levels of committed iT_{reg} or nT_{reg} cells and induce IL-17 production by these cells in a TGF-β-dependent manner (Yang et al. 2008a, b). In addition, the $IL\text{-}17^+/Foxp3^+$ T cells were functionally impaired. Differentiation of both Treg and T_H17 cells requires TGF-β through different transcription factors, RORγt for T_H17 cells and Foxp3 for Treg cells. In combination with TCR signal, TGF-β is able to induce both RORγt and Foxp3 in CD4 T cells (Zhou et al. 2008). Foxp3 directly binds to RORγt and inhibits RORγt transcription activity through the LxxLL motif and the recruitment of TIP60-HDAC7 (Yang et al. 2008a, b). Over-expression of Foxp3 in naïve CD4 T cells results in the inhibition of T_H17 cell differentiation. IL-6, IL-21 and IL-23 hamper Foxp3-mediated inhibition of RORγt through Foxp3, thereby promoting T_H17 cell differentiation and trans-differentiation (Zhou et al. 2008). These findings indicate that the pro-inflammatory milieu could convert T_{reg} cells to T_H17 cells and shift the balance between immune regulation and inflammation towards inflammation. Thus, targeting these cytokines could have great therapeutic potential as these treatments would lead to restoration of the "yin" and "yang" of the immune system (Fig. 3).

In short, pro-inflammatory cytokines e.g. IL-6, IL-1b, and IL-21 not only support T_H17 differentiation, but also constrain T_{reg} development through affecting the balance between RORγt and Foxp3. In contrast, IL-2 and retinoic acid, which support T_{reg} differentiation, could perturb T_H17 development (Laurence et al. 2007; Mucida et al. 2007). Furthermore, cytokines secreted by T_H17 cells such as TNF-α and IL-21 could dampen T_{reg} function, while IL-10 and IL-35 are able to suppress the pro-inflammatory response. Taken together, these findings show that the persistence of inflammation at the site of pathology is significantly associated with a reduced level and impaired function of $CD4^+CD25^+Foxp3^+$ T_{reg} cells induced by pro-inflammatory cytokines. Therapeutic alteration of these cytokines holds great promise as a treatment for auto-immune disease.

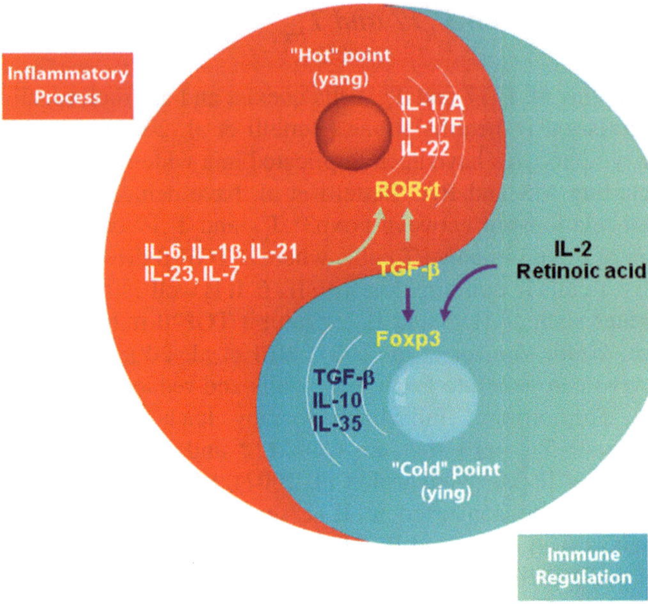

Fig. 3 "Yin" and "yang" interplay between T_H17 and T_{reg} cells in inflammatory auto-immunity. A model based on the traditional Chinese idea of "yin" and "yang" can be used to describe the dynamic regulation of T_H17 and T_{reg} cells. Cytokines play an important role in determining the balance between "yin", represented by T_{reg} cells, and "yang" represented by T_H17 cells. In the center of the circle is the cytokine TGF-β which drives the expression of the signature transcription factors RORγt for T_H17 and Foxp3 for T_{reg} cells. The reciprocal regulation of the two transcription factors is crucial to the dynamic operation of the system. Cytokines including IL-6, IL-23, and IL-7 drive the system towards "yang", which is further amplified by the effector cytokines, including IL-17. In contrast, IL-2 and hormones including retinoic acid drive the system towards "yin", which is further amplified by effector cytokines, including IL-10

4 Therapeutic Strategies for Auto-immune Diseases

The interplay between pathogenic and regulatory T cells provides excellent rationale for the development of effective therapeutic approaches to auto-immune diseases.

4.1 Cytokine Antibody Therapies

Many companies are developing anti-cytokine antibodies against pathways relevant to T_H17 and T_{reg} cells for the treatment of common human auto-immune diseases. These approaches are summarized in Table 1.

T_H17 and T_{reg} cells are important for the promotion and regulation of the pathology of many auto-immune diseases. Since IL-6 plays a critical role in the development

Table 1 Current clinical trials of antibodies targeting T_H17 and T_{reg} cells for the treatment of autoimmune disease

Cytokine	Example of therapeutic agent (company: drug name)	Stage of clinical trial	Therapeutic outcome	Reference
IL-6	Anti-IL-6 receptor (Roche: Actemra/Tocilizumab)		RA: efficacious	Oldfield et al. (2009)
IL-12/IL-23	Anti-P40 (Centocor: Stelara/Ustekinumab)	MS: discontinued Crohn's: phase IIb	MS: failed PS: efficacious	Papp et al. (2008)
	Anti-P40 (Abbot: ABT-874)	Crohn's: phase II PS: phase II MS: phase II		Ding et al. (2008), Menter (2009)
	Antagonist (SYNTA: Apilimod/STA5326)	RA: phase II Crohn's: phase II		Billich (2007), Sands et al. (2009)
IL-23	Anti-P19 (Shering Plough)	Phase I		
	Anti-IL-23 (Eli Lilly)	Phase I		
IL-1β	IL-1β Antagonist (Amgen: Kineret/Anakinra)		RA: approved	Waugh and Perry (2005)
	Anti-IL-1 (Amgen: AMG108)	RA: phase II		Waugh and Perry (2005)
IL-17	Anti-IL-17 (Eli Lilly:LY2439821)	RA: phase II PS: phase II		Steinman (2010); van den Berg and Miossec (2009)
	Anti-IL-17 (Norvatis: AIN457)	RA: phase II Crohn's: phase II		(Steinman (2010); van den Berg and Miossec 2009)
	Anti-IL-17 (Amgen)	RA: phase Ib/IIa		Steinman (2010)
TNF-α	Soluble receptor (Amgen: Enbrel/etanercept)		RA: approved MS: failed	Dhillon et al. (2007), Stubgen (2008)
	Anti-TNF-α (Centocor: Remicade/Infliximab)	Crohn's: phase III	RA: approved MS: failed	Maini et al. (2004), Stubgen (2008)
	Anti-TNF-α (Abbot: Humira/Adalimumab)	Crohn's: phase III completed	RA: approved MS: failed	Burmester et al. (2007), Stubgen (2008)
IL-2	Anti-IL-2 receptor (Biogen-Idec: Zenapax/ Dacliqumab)	MS: phase II completed		Martin (2008)
BAFF/April	Soluble receptor (Zymogenetics/Merck: Atacicept)		MS: failed	Gatto (2008)

MS multiple sclerosis; *RA* rheumatoid arthritis; *PS* psoriasis; *Crohn's*: Crohn's disease

of the T_H17 response and the inhibition of T_{reg} functions, many trials have provided strong evidence that inhibition of the biological activity of IL-6 and/or its receptors represents a promising approach for the treatment of IL-6-associated inflammatory diseases. In fact, antibodies against IL-6 receptor (IL-6R) and IL-6 are being developed by a number of companies. Elevated IL-6 and IL-17 levels in the joints of RA patients have been documented (Chabaud et al. 1999). An anti-IL-6 receptor monoclonal antibody, known as Tocilizumab, is efficacious and has been approved for the treatment of RA (Mircic and Kavanaugh 2009). However, targeting of the IL-6R may lead to potential toxicity issues as it is expressed on a number of normal cells, including hepatocytes. Indeed, elevated cholesterol levels have been reported in patients treated with anti-IL-6R. Targeting IL-6 is also a possibility for the treatment of MS, but there is currently no known clinical trial. One challenge to consider is that neutralization of IL-6 signaling may not be sufficient to block the expansion of committed pathogenic T_H17 cells as seen in EAE model (Serada et al. 2008).

The common sub-unit of IL-12 and IL-23, p40, has been targeted by at least two companies. Their p40-specific antibodies are particularly efficacious in the treatment of psoriasis and there are ongoing clinical trials for Crohn's disease (Elliott et al. 2009). The results suggested an important contribution of T_H17 cells to the pathology of psoriasis (Steinman 2010). However, the outcome of a clinical trial of a p40 mAb (Ustekinumab) for MS was unexpected. The treated MS subjects did not show a reduction in brain lesions, suggesting that blocking the differentiation of both T_H1 and T_H17 cells may not be sufficient for effective MS therapy. Perhaps, in an established disease course, a majority of pathogenic T_H subsets have already matured, rendering these blocking agents specific for T_H differentiation ineffective. Agents specific for IL-23 targeting the p19 sub-unit are currently in phase I trial. These agents may be suitable for the treatment of psoriasis and Crohn's disease, but it is uncertain whether they will be successful in MS. Therapeutic agents against IL-6 and IL-23 pathways that target T_H17 differentiation may not work in MS, and it has been suggested by EAE experiments where anti-IL-6 or anti-IL-23 treatments were only efficacious when they were applied at the time of immunization (the prevention protocol), but not at the peak of disease (the therapeutic protocol) (Chen et al. 2006; Serada et al. 2008). These observations further confirm that there are two distinct steps in the development of T_H17 cells; differentiation and survival/expansion, in which IL-6 and IL-23 are only involved in the first differentiation phase. However, once pathogenic T_H17 cells are committed, targeting the signal pathway that relates only to early stage of the development of T_H17 cells may not be effective in controlling an on-going process of the disease.

Some companies are developing direct blockers for IL-17 and have moved into phase II trials in patients with RA and psoriasis (Garber 2009). Since T_H17 and T_H1 cells secrete cytokines that are important for defense against pathogens, directly targeting IL-17 is believed to be safer and should reduce the risk of developing severe infections. However, the therapeutic potential of anti-IL-17 may also be limited, because T_H17 and T_H1 cells will continue to thrive in the subject. The anti-inflammatory benefits of anti-IL-17 therapies will have to be compared to those of anti-TNF-α and anti-IL-1 in the future.

Overall, anti-cytokine agents targeting T_H17 cell development are generally promising for many auto-immune diseases with the exception of MS. MS cannot be treated with anti-TNF-α antibodies or soluble receptors targeting general inflammation, since these agents have been associated with monophasic CNS demyelination or worsening of known MS (Fromont et al. 2009; Moreland 2009). The recent halting of a trial of the BAFF/April soluble receptor was due to an observed increase in MS disease activity, suggests that targeting B cells in MS patients is equally unsuccessful (Gatto 2008). The outcomes of the unsuccessful MS trials will prompt researchers to consider alternative therapeutic approaches. In this regard, targeting mature instead of differentiating pathogenic T_H subsets could be a viable alternative. An antibody against the IL-2R sub-unit CD25 is currently in phase II trial for MS (Bielekova et al. 2009). Interestingly, IL-2R and IL-7R were identified as genes associated with MS susceptibility by genome-wide association studies. These two unappealing cytokine receptors have roles in the survival and expansion of pathogenic T_H1 and T_H17 cells (Liu et al. 2010; Amadi-Obi et al. 2007). T_{reg} cells responsible for resolving inflammation express very low levels of IL-7R and will be unaffected by antibodies against IL-7 and IL-7R, providing an extra benefit to this approach.

4.2 *Use of Traditional Chinese Medicine in the Treatment of Auto-immune Diseases*

As alluded to earlier, the overall treatment strategy for auto-immune diseases is to restore the dynamic balance of the "yin" and the "yang" inherent to the immune system. This notion fits with the time-honored philosophy underpinning traditional Chinese medicine (TCM) which aims to restore the body's homeostasis through the combination of several ingredients with "yin" or "yang" effects. It should come as no surprise that many single chemical entities isolated from clinically-efficacious herbal products possess anti-inflammatory or immuno-regulatory properties. These active components and their derivatives can be used as chemical probes for mechanistic and structural-activity relationship studies towards the identification of potentially novel biological mechanisms and pharmacological targets. In recent years, we have been active in this area and investigated a number of these compounds. We found that instead of behaving like general immuno-suppressants, there are compounds from the TCM collection that target distinct immune/inflammation pathways. One such compound is SM933, a derivative of Artemisinin, which is approved for the treatment of malaria. We found that SM933 could control EAE inflammation through both T_H2 deviation and selective inhibition of the proliferation of activated, but not resting, T cells (Wang et al. 2007). We further demonstrated that the immuno-regulatory properties were associated with alteration in the NF-κB and IFN-γ/Rig-G/JAB-1 pathways, the latter of which represents a novel signaling node that controls cell cycle progression (Wang et al. 2007). Another example is berbamine, an alkaloid derivative from *Berberis vulgaris L*, that also reduces EAE severity (Ren et al. 2008). Detailed analyses indicate that in addition to regulating the co-stimulatory function of APC, berbamine selectively inhibits T_H1, but not T_H17 cells, by promoting STAT4 degradation through upregulation of its ubiquitin E3 ligase, SLIM (Ren et al. 2008). Interestingly, another isoquinoline alkaloid from *Berberis* plants, berberine, was found to suppress EAE through the inhibition of T_H17, and to a lesser extent T_H1 cell differentiation (Cui et al. 2009) (data unpublished). We propose to take advantage of the insights we have gained from molecular studies of T_H cell differentiation, as well as the immuno-modulatory properties of natural compounds, in the quest for novel pathway-based immuno-modulatory therapeutics.

5 Conclusion

In summary, it is important to maintain an appropriate balance between T_{reg} cells (yin) and effector T_H17/T_H1 cells (yang) that can ensure effective immunity while avoiding pathological auto-immunity. The pro-inflammatory cytokine milieu produced during an immune response plays an important role in driving the imbalance between these two T cell subsets and the outcome of inflammatory auto-immune diseases. Cytokines such as IL-6 and IL-23 promote the "yang" and others such as TGF-β and IL-10 promote the "yin" functions of the immune system. The dynamic nature of an immune response was recently highlighted by the demonstration of interconversion between T_H17 and T_{reg} cells under the influence of a particular cytokine micro-environment. A better understanding of the molecular mechanisms regulating T_{reg} and T_H17 cells in vitro and in vivo will create opportunities for the development of therapeutic approaches that could be used to regulate the T_{reg} and T_H17 cell balance in human inflammatory diseases. The Western medicine approach usually targets just a single cytokine to inhibit pathogenic T cells using antibodies, while the TCM approach places an emphasis on the restoration of the balance of "yin" and "yang" through multiple therapeutic ingredients to increase the number of T_{reg} cells and limit the pathogenic T_H1/T_H17 cell populations. A combination of the two approaches may result in better therapies for auto-immune diseases in the future.

References

Acosta-Rodriguez, E. V., G. Napolitani, et al. (2007a). "Interleukins 1beta and 6, but not transforming growth factor-beta are essential for the differentiation of interleukin 17-producing human T helper cells." *Nat Immunol* **8**(9): 942–9.

Acosta-Rodriguez, E. V., L. Rivino, et al. (2007b). "Surface phenotype and antigenic specificity of human interleukin 17-producing T helper memory cells." *Nat Immunol* **8**(6): 639–46.

Afzali, B., G. Lombardi, et al. (2007). "The role of T helper 17 (Th17) and regulatory T cells (Treg) in human organ transplantation and auto-immune disease." *Clin Exp Immunol* **148**(1): 32–46.

Aggarwal, S., N. Ghilardi, et al. (2003). "Interleukin-23 promotes a distinct CD4 T cell activation state characterized by the production of interleukin-17." *J Biol Chem* **278**(3): 1910–4.

Amadi-Obi, A., C. R. Yu, et al. (2007). "TH17 cells contribute to uveitis and scleritis and are expanded by IL-2 and inhibited by IL-27/STAT1." *Nat Med* **13**(6): 711–8.

Anderton, S. M. (2006). "Avoiding auto-immune disease--T cells know their limits." *Trends Immunol* **27**(5): 208–14.

Asseman, C., S. Mauze, et al. (1999). "An essential role for interleukin 10 in the function of regulatory T cells that inhibit intestinal inflammation." *J Exp Med* **190**(7): 995–1004.

Bayer, A. L., J. Y. Lee, et al. (2008). "A function for IL-7R for CD4+CD25+Foxp3+ T regulatory cells." *J Immunol* **181**(1): 225–34.

Bayer, A. L., A. Yu, et al. (2005). "Essential role for interleukin-2 for CD4(+)CD25(+) T regulatory cell development during the neonatal period." *J Exp Med* **201**(5): 769–77.

Bennett, C. L., J. Christie, et al. (2001). "The immune dysregulation, polyendocrinopathy, enteropathy, X-linked syndrome (IPEX) is caused by mutations of FOXP3." *Nat Genet* **27**(1): 20–1.

Bensinger, S. J., A. Bandeira, et al. (2001). "Major histocompatibility complex class II-positive cortical epithelium mediates the selection of CD4(+)25(+) immuno-regulatory T cells." *J Exp Med* **194**(4): 427–38.

Benson, M. J., K. Pino-Lagos, et al. (2007). "All-trans retinoic acid mediates enhanced T reg cell growth, differentiation, and gut homing in the face of high levels of co-stimulation." *J Exp Med* **204**(8): 1765–74.

Bettelli, E., Y. Carrier, et al. (2006). "Reciprocal developmental pathways for the generation of pathogenic effector TH17 and regulatory T cells." *Nature* **441**(7090): 235–8.

Bielekova, B., B. Goodwin, et al. (2000). "Encephalitogenic potential of the myelin basic protein peptide (amino acids 83–99) in multiple sclerosis: results of a phase II clinical trial with an altered peptide ligand." *Nat Med* **6**(10): 1167–75.

Bielekova, B., T. Howard, et al. (2009). "Effect of anti-CD25 antibody daclizumab in the inhibition of inflammation and stabilization of disease progression in multiple sclerosis." *Arch Neurol* **66**(4): 483–9.

Bielekova, B., P. A. Muraro, et al. (1999). "Preferential expansion of auto-reactive T lymphocytes from the memory T-cell pool by IL-7." *J Neuroimmunol* **100**(1–2): 115–23.

Billiau, A., H. Heremans, et al. (1988). "Enhancement of experimental allergic encephalomyelitis in mice by antibodies against IFN-gamma." *J Immunol* **140**(5): 1506–10.

Billich, A. (2007). "Drug evaluation: apilimod, an oral IL-12/IL-23 inhibitor for the treatment of auto-immune diseases and common variable immuno-deficiency." *IDrugs* **10**(1): 53–9.

Borsellino, G., M. Kleinewietfeld, et al. (2007). "Expression of ectonucleotidase CD39 by Foxp3+ Treg cells: hydrolysis of extracellular ATP and immune suppression." *Blood* **110**(4): 1225–32.

Brunkow, M. E., E. W. Jeffery, et al. (2001). "Disruption of a new forkhead/winged-helix protein, scurfin, results in the fatal lymphoproliferative disorder of the scurfy mouse." *Nat Genet* **27**(1): 68–73.

Burchill, M. A., J. Yang, et al. (2007). "IL-2 receptor beta-dependent STAT5 activation is required for the development of Foxp3+ regulatory T cells." *J Immunol* **178**(1): 280–90.

Burkhart, C., G. Y. Liu, et al. (1999). "Peptide-induced T cell regulation of experimental autoimmune encephalomyelitis: a role for IL-10." *Int Immunol* **11**(10): 1625–34.

Burmester, G. R., X. Mariette, et al. (2007). "Adalimumab alone and in combination with disease-modifying anti-rheumatic drugs for the treatment of rheumatoid arthritis in clinical practice: the Research in Active Rheumatoid Arthritis (ReAct) trial." *Ann Rheum Dis* **66**(6): 732–9.

Cao, D., V. Malmstrom, et al. (2003). "Isolation and functional characterization of regulatory CD25brightCD4+ T cells from the target organ of patients with rheumatoid arthritis." *Eur J Immunol* **33**(1): 215–23.

Cao, D., R. van Vollenhoven, et al. (2004). "CD25brightCD4+ regulatory T cells are enriched in inflamed joints of patients with chronic rheumatic disease." *Arthritis Res Ther* **6**(4): R335–46.

Chabaud, M., J. M. Durand, et al. (1999). "Human interleukin-17: A T cell-derived pro-inflammatory cytokine produced by the rheumatoid synovium." *Arthritis Rheum* **42**(5): 963–70.

Chatila, T. A., F. Blaeser, et al. (2000). "JM2, encoding a fork head-related protein, is mutated in X-linked auto-immunity-allergic dysregulation syndrome." *J Clin Invest* **106**(12): R75–81.

Chen, W., W. Jin, et al. (2003). "Conversion of peripheral CD4+CD25- naive T cells to CD4+CD25+ regulatory T cells by TGF-beta induction of transcription factor Foxp3." *J Exp Med* **198**(12): 1875–86.
Chen, X., L. Fang, et al. (2009). "Thymic regulation of auto-immune disease by accelerated differentiation of Foxp3+ regulatory T cells through IL-7 signaling pathway." *J Immunol* **183**(10): 6135–44.
Chen, Y., C. L. Langrish, et al. (2006). "Anti-IL-23 therapy inhibits multiple inflammatory pathways and ameliorates auto-immune encephalomyelitis." *J Clin Invest* **116**(5): 1317–26.
Cho, M. L., J. W. Kang, et al. (2006). "STAT3 and NF-kappaB signal pathway is required for IL-23-mediated IL-17 production in spontaneous arthritis animal model IL-1 receptor antagonist-deficient mice." *J Immunol* **176**(9): 5652–61.
Coffman, R. L. (2006). "Origins of the T(H)1-T(H)2 model: a personal perspective." *Nat Immunol* **7**(6): 539–41.
Collison, L. W., C. J. Workman, et al. (2007). "The inhibitory cytokine IL-35 contributes to regulatory T-cell function." *Nature* **450**(7169): 566–9.
Coombes, J. L., K. R. Siddiqui, et al. (2007). "A functionally specialized population of mucosal CD103+ DCs induces Foxp3+ regulatory T cells via a TGF-beta and retinoic acid-dependent mechanism." *J Exp Med* **204**(8): 1757–64.
Corthay, A. (2009). "How do regulatory T cells work?" *Scand J Immunol* **70**(4): 326–36.
Costantino, C. M., C. Baecher-Allan, et al. (2008). "Multiple sclerosis and regulatory T cells." *J Clin Immunol* **28**(6): 697–706.
Cua, D. J., J. Sherlock, et al. (2003). "Interleukin-23 rather than interleukin-12 is the critical cytokine for auto-immune inflammation of the brain." *Nature* **421**(6924): 744–8.
Cui, G., X. Qin, et al. (2009). "Berberine differentially modulates the activities of ERK, p38 MAPK, and JNK to suppress Th17 and Th1 T cell differentiation in type 1 diabetic mice." *J Biol Chem* **284**(41): 28420–9.
Curotto de Lafaille, M. A. and J. J. Lafaille (2009). "Natural and adaptive foxp3+ regulatory T cells: more of the same or a division of labor?" *Immunity* **30**(5): 626–35.
de Jong, B. A., T. W. Huizinga, et al. (2002). "Production of IL-1beta and IL-1Ra as risk factors for susceptibility and progression of relapse-onset multiple sclerosis." *J Neuroimmunol* **126**(1–2): 172–9.
Deaglio, S., K. M. Dwyer, et al. (2007). "Adenosine generation catalyzed by CD39 and CD73 expressed on regulatory T cells mediates immune suppression." *J Exp Med* **204**(6): 1257–65.
Dhillon, S., K. A. Lyseng-Williamson, et al. (2007). "Etanercept: a review of its use in the management of rheumatoid arthritis." *Drugs* **67**(8): 1211–41.
Ding, C., J. Xu, et al. (2008). "ABT-874, a fully human monoclonal anti-IL-12/IL-23 antibody for the potential treatment of auto-immune diseases." *Curr Opin Investig Drugs* **9**(5): 515–22.
Ehrenstein, M. R., J. G. Evans, et al. (2004). "Compromised function of regulatory T cells in rheumatoid arthritis and reversal by anti-TNFalpha therapy." *J Exp Med* **200**(3): 277–85.
Elliott, M., J. Benson, et al. (2009). "Ustekinumab: lessons learned from targeting interleukin-12/23p40 in immune-mediated diseases." *Ann N Y Acad Sci* **1182**: 97–110.
Fantini, M. C., A. Rizzo, et al. (2007). "IL-21 regulates experimental colitis by modulating the balance between Treg and Th17 cells." *Eur J Immunol* **37**(11): 3155–63.
Fletcher, J. M., R. Lonergan, et al. (2009). "CD39+Foxp3+ regulatory T Cells suppress pathogenic Th17 cells and are impaired in multiple sclerosis." *J Immunol* **183**(11): 7602–10.
Fontenot, J. D., M. A. Gavin, et al. (2003). "Foxp3 programs the development and function of CD4+CD25+ regulatory T cells." *Nat Immunol* **4**(4): 330–6.
Fontenot, J. D., J. P. Rasmussen, et al. (2005a). "A function for interleukin 2 in Foxp3-expressing regulatory T cells." *Nat Immunol* **6**(11): 1142–51.
Fontenot, J. D., J. P. Rasmussen, et al. (2005b). "Regulatory T cell lineage specification by the forkhead transcription factor foxp3." *Immunity* **22**(3): 329–41.
Fromont, A., J. De Seze, et al. (2009). "Inflammatory demyelinating events following treatment with anti-tumor necrosis factor." *Cytokine* **45**(2): 55–7.
Garber, K. (2009). "Immunology's quiet upheaval." *Nat Biotechnol* **27**(8): 687–9.

Garin, M. I., C. C. Chu, et al. (2007). "Galectin-1: a key effector of regulation mediated by CD4+CD25+ T cells." *Blood* **109**(5): 2058–65.

Gatto, B. (2008). "Atacicept, a homodimeric fusion protein for the potential treatment of diseases triggered by plasma cells." *Curr Opin Investig Drugs* **9**(11): 1216–27.

Germann, T., H. Hess, et al. (1996). "Characterization of the adjuvant effect of IL-12 and efficacy of IL-12 inhibitors in type II collagen-induced arthritis." *Ann N Y Acad Sci* **795**: 227–40.

Gondek, D. C., L. F. Lu, et al. (2005). "Cutting edge: contact-mediated suppression by CD4+CD25+ regulatory cells involves a granzyme B-dependent, perforin-independent mechanism." *J Immunol* **174**(4): 1783–6.

Green, E. A., L. Gorelik, et al. (2003). "CD4+CD25+ T regulatory cells control anti-islet CD8+ T cells through TGF-beta-TGF-beta receptor interactions in type 1 diabetes." *Proc Natl Acad Sci USA* **100**(19): 10878–83.

Gregory, S. G., S. Schmidt, et al. (2007). "Interleukin 7 receptor alpha chain (IL7R) shows allelic and functional association with multiple sclerosis." *Nat Genet* **39**(9): 1083–91.

Grossman, W. J., J. W. Verbsky, et al. (2004). "Human T regulatory cells can use the perforin pathway to cause autologous target cell death." *Immunity* **21**(4): 589–601.

Hafler, D. A., A. Compston, et al. (2007). "Risk alleles for multiple sclerosis identified by a genome-wide study." *N Engl J Med* **357**(9): 851–62.

Harrington, L. E., R. D. Hatton, et al. (2005). "Interleukin 17-producing CD4+ effector T cells develop via a lineage distinct from the T helper type 1 and 2 lineages." *Nat Immunol* **6**(11): 1123–32.

Hill, J. A., J. A. Hall, et al. (2008). "Retinoic acid enhances Foxp3 induction indirectly by relieving inhibition from CD4+CD44hi Cells." *Immunity* **29**(5): 758–70.

Hori, S., T. Nomura, et al. (2003). "Control of regulatory T cell development by the transcription factor Foxp3." *Science* **299**(5609): 1057–61.

Huter, E. N., G. A. Punkosdy, et al. (2008). "TGF-beta-induced Foxp3+ regulatory T cells rescue scurfy mice." *Eur J Immunol* **38**(7): 1814–21.

Hwang, S. Y., J. Y. Kim, et al. (2004). "IL-17 induces production of IL-6 and IL-8 in rheumatoid arthritis synovial fibroblasts via NF-kappaB- and PI3-kinase/Akt-dependent pathways." *Arthritis Res Ther* **6**(2): R120–8.

Issazadeh, S., M. Mustafa, et al. (1995). "Interferon gamma, interleukin 4 and transforming growth factor beta in experimental auto-immune encephalomyelitis in Lewis rats: dynamics of cellular mRNA expression in the central nervous system and lymphoid cells." *J Neurosci Res* **40**(5): 579–90.

Ivanov, II, B. S. McKenzie, et al. (2006). "The orphan nuclear receptor RORgammat directs the differentiation program of pro-inflammatory IL-17+ T helper cells." *Cell* **126**(6): 1121–33.

Khattri, R., T. Cox, et al. (2003). "An essential role for Scurfin in CD4+CD25+ T regulatory cells." *Nat Immunol* **4**(4): 337–42.

Komiyama, Y., S. Nakae, et al. (2006). "IL-17 plays an important role in the development of experimental auto-immune encephalomyelitis." *J Immunol* **177**(1): 566–73.

Korn, T., E. Bettelli, et al. (2007a). "IL-21 initiates an alternative pathway to induce pro-inflammatory T(H)17 cells." *Nature* **448**(7152): 484–7.

Korn, T., E. Bettelli, et al. (2009). "IL-17 and Th17 Cells." *Annu Rev Immunol* **27**: 485–517.

Korn, T., M. Mitsdoerffer, et al. (2008). "IL-6 controls Th17 immunity in vivo by inhibiting the conversion of conventional T cells into Foxp3+ regulatory T cells." *Proc Natl Acad Sci USA* **105**(47): 18460–5.

Korn, T., J. Reddy, et al. (2007b). "Myelin-specific regulatory T cells accumulate in the CNS but fail to control auto-immune inflammation." *Nat Med* **13**(4): 423–31.

Kroenke, M. A., T. J. Carlson, et al. (2008). "IL-12- and IL-23-modulated T cells induce distinct types of EAE based on histology, CNS chemokine profile, and response to cytokine inhibition." *J Exp Med* **205**(7): 1535–41.

Kumar, M., N. Putzki, et al. (2006). "CD4+CD25+FoxP3+ T lymphocytes fail to suppress myelin basic protein-induced proliferation in patients with multiple sclerosis." *J Neuroimmunol* **180**(1–2): 178–84.

Langrish, C. L., Y. Chen, et al. (2005). "IL-23 drives a pathogenic T cell population that induces auto-immune inflammation." *J Exp Med* **201**(2): 233–40.

Laurence, A., C. M. Tato, et al. (2007). "Interleukin-2 signaling via STAT5 constrains T helper 17 cell generation." *Immunity* **26**(3): 371–81.

Lee, S. K. and C. D. Surh (2005). "Role of interleukin-7 in bone and T-cell homeostasis." *Immunol Rev* **208**: 169–80.

Lindley, S., C. M. Dayan, et al. (2005). "Defective suppressor function in CD4(+)CD25(+) T-cells from patients with type 1 diabetes." *Diabetes* **54**(1): 92–9.

Lio, C. W. and C. S. Hsieh (2008). "A two-step process for thymic regulatory T cell development." *Immunity* **28**(1): 100–11.

Liu, X., Y. S. Lee, et al. (2008). "Loss of STAT3 in CD4+ T cells prevents development of experimental auto-immune diseases." *J Immunol* **180**(9): 6070–6.

Liu, X., S. Leung, et al. (2010). "Crucial role of interleukin-7 in T helper type 17 survival and expansion in auto-immune disease." *Nat Med* **16**(2): 191–7.

Lock, C., G. Hermans, et al. (2002). "Gene-microarray analysis of multiple sclerosis lesions yields new targets validated in auto-immune encephalomyelitis." *Nat Med* **8**(5): 500–8.

Lundmark, F., K. Duvefelt, et al. (2007). "Variation in interleukin 7 receptor alpha chain (IL7R) influences risk of multiple sclerosis." *Nat Genet* **39**(9): 1108–13.

Maini, R. N., F. C. Breedveld, et al. (2004). "Sustained improvement over two years in physical function, structural damage, and signs and symptoms among patients with rheumatoid arthritis treated with infliximab and methotrexate." *Arthritis Rheum* **50**(4): 1051–65.

Manel, N., D. Unutmaz, et al. (2008). "The differentiation of human T(H)-17 cells requires transforming growth factor-beta and induction of the nuclear receptor RORgammat." *Nat Immunol* **9**(6): 641–9.

Mangan, P. R., L. E. Harrington, et al. (2006). "Transforming growth factor-beta induces development of the T(H)17 lineage." *Nature* **441**(7090): 231–4.

Martin, R. (2008). "Humanized anti-CD25 antibody treatment with daclizumab in multiple sclerosis." *Neurodegener Dis* **5**(1): 23–6.

Matusevicius, D., P. Kivisakk, et al. (1999). "Interleukin-17 mRNA expression in blood and CSF mononuclear cells is augmented in multiple sclerosis." *Mult Scler* **5**(2): 101–4.

Mazzucchelli, R. and S. K. Durum (2007). "Interleukin-7 receptor expression: intelligent design." *Nat Rev Immunol* **7**(2): 144–54.

McFarland, H. F. and R. Martin (2007). "Multiple sclerosis: a complicated picture of autoimmunity." *Nat Immunol* **8**(9): 913–9.

McGeachy, M. J., K. S. Bak-Jensen, et al. (2007). "TGF-beta and IL-6 drive the production of IL-17 and IL-10 by T cells and restrain T(H)-17 cell-mediated pathology." *Nat Immunol* **8**(12): 1390–7.

McGeachy, M. J., Y. Chen, et al. (2009). "The interleukin 23 receptor is essential for the terminal differentiation of interleukin 17-producing effector T helper cells in vivo." *Nat Immunol* **10**(3): 314–24.

Menter, A. (2009). "The status of biologic therapies in the treatment of moderate to severe psoriasis." *Cutis* **84**(4 Suppl): 14–24.

Mircic, M. and A. Kavanaugh (2009). "The clinical efficacy of tocilizumab in rheumatoid arthritis." *Drugs Today (Barc)* **45**(3): 189–97.

Moreland, L. W. (2009). "Cytokines as targets for anti-inflammatory agents." *Ann N Y Acad Sci* **1182**: 88–96.

Mosmann, T. R., H. Cherwinski, et al. (1986). "Two types of murine helper T cell clone. I. Definition according to profiles of lymphokine activities and secreted proteins." *J Immunol* **136**(7): 2348–57.

Mosmann, T. R. and R. L. Coffman (1989). "TH1 and TH2 cells: different patterns of lymphokine secretion lead to different functional properties." *Annu Rev Immunol* **7**: 145–73.

Mottonen, M., J. Heikkinen, et al. (2005). "CD4+ CD25+ T cells with the phenotypic and functional characteristics of regulatory T cells are enriched in the synovial fluid of patients with rheumatoid arthritis." *Clin Exp Immunol* **140**(2): 360–7.

Mucida, D., Y. Park, et al. (2007). "Reciprocal TH17 and regulatory T cell differentiation mediated by retinoic acid." *Science* **317**(5835): 256–60.

Murphy, C. A., C. L. Langrish, et al. (2003). "Divergent pro- and antiinflammatory roles for IL-23 and IL-12 in joint auto-immune inflammation." *J Exp Med* **198**(12): 1951–7.
Nadkarni, S., C. Mauri, et al. (2007). "Anti-TNF-alpha therapy induces a distinct regulatory T cell population in patients with rheumatoid arthritis via TGF-beta." *J Exp Med* **204**(1): 33–9.
Nurieva, R., X. O. Yang, et al. (2007). "Essential autocrine regulation by IL-21 in the generation of inflammatory T cells." *Nature* **448**(7152): 480–3.
Oldfield, V., S. Dhillon, et al. (2009). "Tocilizumab: a review of its use in the management of rheumatoid arthritis." *Drugs* **69**(5): 609–32.
Palmer, M. J., V. S. Mahajan, et al. (2008). "Interleukin-7 receptor signaling network: an integrated systems perspective." *Cell Mol Immunol* **5**(2): 79–89.
Papp, K. A., R. G. Langley, et al. (2008). "Efficacy and safety of ustekinumab, a human interleukin-12/23 monoclonal antibody, in patients with psoriasis: 52-week results from a randomised, double-blind, placebo-controlled trial (PHOENIX 2)." *Lancet* **371**(9625): 1675–84.
Park, H., Z. Li, et al. (2005). "A distinct lineage of CD4 T cells regulates tissue inflammation by producing interleukin 17." *Nat Immunol* **6**(11): 1133–41.
Peluso, I., M. C. Fantini, et al. (2007). "IL-21 counteracts the regulatory T cell-mediated suppression of human CD4+ T lymphocytes." *J Immunol* **178**(2): 732–9.
Pettinelli, C. B. and D. E. McFarlin (1981). "Adoptive transfer of experimental allergic encephalomyelitis in SJL/J mice after in vitro activation of lymph node cells by myelin basic protein: requirement for Lyt 1+ 2- T lymphocytes." *J Immunol* **127**(4): 1420–3.
Ren, Y., L. Lu, et al. (2008). "Novel immuno-modulatory properties of berbamine through selective down-regulation of STAT4 and action of IFN-gamma in experimental auto-immune encephalomyelitis." *J Immunol* **181**(2): 1491–8.
Rengarajan, J., S. J. Szabo, et al. (2000). "Transcriptional regulation of Th1/Th2 polarization." *Immunol Today* **21**(10): 479–83.
Renno, T., M. Krakowski, et al. (1995). "TNF-alpha expression by resident microglia and infiltrating leukocytes in the central nervous system of mice with experimental allergic encephalomyelitis. Regulation by Th1 cytokines." *J Immunol* **154**(2): 944–53.
Sakaguchi, S. (2004). "Naturally arising CD4+ regulatory t cells for immunologic self-tolerance and negative control of immune responses." *Annu Rev Immunol* **22**: 531–62.
Sakaguchi, S., N. Sakaguchi, et al. (1995). "Immunologic self-tolerance maintained by activated T cells expressing IL-2 receptor alpha-chains (CD25). Breakdown of a single mechanism of self-tolerance causes various auto-immune diseases." *J Immunol* **155**(3): 1151–64.
Salomon, B., D. J. Lenschow, et al. (2000). "B7/CD28 co-stimulation is essential for the homeostasis of the CD4+CD25+ immunoregulatory T cells that control auto-immune diabetes." *Immunity* **12**(4): 431–40.
Sands, B. E., E. W. Jacobson, et al. (2009). "Randomized, double-blind, placebo-controlled trial of the oral interleukin-12/23 inhibitor apilimod mesylate for treatment of active Crohn's disease." *Inflamm Bowel Dis*.
Sarkar, S. and D. A. Fox (2007). "Regulatory T cell defects in rheumatoid arthritis." *Arthritis Rheum* **56**(3): 710–3.
Serada, S., M. Fujimoto, et al. (2008). "IL-6 blockade inhibits the induction of myelin antigen-specific Th17 cells and Th1 cells in experimental auto-immune encephalomyelitis." *Proc Natl Acad Sci USA* **105**(26): 9041–6.
Skarica, M., T. Wang, et al. (2009). "Signal transduction inhibition of APCs diminishes th17 and Th1 responses in experimental auto-immune encephalomyelitis." *J Immunol* **182**(7): 4192–9.
Soper, D. M., D. J. Kasprowicz, et al. (2007). "IL-2Rbeta links IL-2R signaling with Foxp3 expression." *Eur J Immunol* **37**(7): 1817–26.
Sospedra, M. and R. Martin (2005). "Immunology of multiple sclerosis." *Annu Rev Immunol* **23**: 683–747.
Sportes, C., F. T. Hakim, et al. (2008). "Administration of rhIL-7 in humans increases in vivo TCR repertoire diversity by preferential expansion of naive T cell subsets." *J Exp Med* **205**(7): 1701–14.

Steinman, L. (2010). "Mixed results with modulation of TH-17 cells in human auto-immune diseases." *Nat Immunol* **11**(1): 41–4.
Stritesky, G. L., N. Yeh, et al. (2008). "IL-23 promotes maintenance but not commitment to the Th17 lineage." *J Immunol* **181**(9): 5948–55.
Stromnes, I. M., L. M. Cerretti, et al. (2008). "Differential regulation of central nervous system auto-immunity by T(H)1 and T(H)17 cells." *Nat Med* **14**(3): 337–42.
Stubgen, J. P. (2008). "Tumor necrosis factor-alpha antagonists and neuropathy." *Muscle Nerve* **37**(3): 281–92.
Stummvoll, G. H., R. J. DiPaolo, et al. (2008). "Th1, Th2, and Th17 effector T cell-induced autoimmune gastritis differs in pathological pattern and in susceptibility to suppression by regulatory T cells." *J Immunol* **181**(3): 1908–16.
Sun, C. M., J. A. Hall, et al. (2007). "Small intestine lamina propria dendritic cells promote de novo generation of Foxp3 T reg cells via retinoic acid." *J Exp Med* **204**(8): 1775–85.
Suri-Payer, E. and H. Cantor (2001). "Differential cytokine requirements for regulation of autoimmune gastritis and colitis by CD4(+)CD25(+) T cells." *J Auto-immun* **16**(2): 115–23.
Sutton, C. E., S. J. Lalor, et al. (2009). "Interleukin-1 and IL-23 induce innate IL-17 production from gammadelta T cells, amplifying Th17 responses and autoimmunity." *Immunity* **31**(2): 331–41.
Szabo, S. J., S. T. Kim, et al. (2000). "A novel transcription factor, T-bet, directs Th1 lineage commitment." *Cell* **100**(6): 655–69.
Tai, X., M. Cowan, et al. (2005). "CD28 co-stimulation of developing thymocytes induces Foxp3 expression and regulatory T cell differentiation independently of interleukin 2." *Nat Immunol* **6**(2): 152–62.
Tao, R., E. F. de Zoeten, et al. (2007). "Deacetylase inhibition promotes the generation and function of regulatory T cells." *Nat Med* **13**(11): 1299–307.
Thornton, A. M. and E. M. Shevach (1998). "CD4+CD25+ immuno-regulatory T cells suppress polyclonal T cell activation in vitro by inhibiting interleukin 2 production." *J Exp Med* **188**(2): 287–96.
Traggiai, E., T. Biagioli, et al. (2001). "IL-7-enhanced T-cell response to myelin proteins in multiple sclerosis." *J Neuroimmunol* **121**(1–2): 111–9.
Tzartos, J. S., M. A. Friese, et al. (2008). "Interleukin-17 production in central nervous system-infiltrating T cells and glial cells is associated with active disease in multiple sclerosis." *Am J Pathol* **172**(1): 146–55.
Vaknin-Dembinsky, A., G. Murugaiyan, et al. (2008). "Increased IL-23 secretion and altered chemokine production by dendritic cells upon CD46 activation in patients with multiple sclerosis." *J Neuroimmunol* **195**(1–2): 140–5.
Valencia, X., G. Stephens, et al. (2006). "TNF downmodulates the function of human CD4+CD25hi T-regulatory cells." *Blood* **108**(1): 253–61.
van Amelsfort, J. M., K. M. Jacobs, et al. (2004). "CD4(+)CD25(+) regulatory T cells in rheumatoid arthritis: differences in the presence, phenotype, and function between peripheral blood and synovial fluid." *Arthritis Rheum* **50**(9): 2775–85.
van den Berg, W. B. and P. Miossec (2009). "IL-17 as a future therapeutic target for rheumatoid arthritis." *Nat Rev Rheumatol* **5**(10): 549–53.
Vang, K. B., J. Yang, et al. (2008). "IL-2, -7, and -15, but not thymic stromal lymphopoeitin, redundantly govern CD4+Foxp3+ regulatory T cell development." *J Immunol* **181**(5): 3285–90.
Viglietta, V., C. Baecher-Allan, et al. (2004). "Loss of functional suppression by CD4+CD25+ regulatory T cells in patients with multiple sclerosis." *J Exp Med* **199**(7): 971–9.
Vudattu, N. K., S. Kuhlmann-Berenzon, et al. (2009). "Increased numbers of IL-7 receptor molecules on CD4+CD25-CD107a+T-cells in patients with auto-immune diseases affecting the central nervous system." *PLoS One* **4**(8): e6534.
Wan, Y. Y. and R. A. Flavell (2005). "Identifying Foxp3-expressing suppressor T cells with a bicistronic reporter." *Proc Natl Acad Sci USA* **102**(14): 5126–31.
Wang, Z., J. Qiu, et al. (2007). "Anti-inflammatory properties and regulatory mechanism of a novel derivative of artemisinin in experimental auto-immune encephalomyelitis." *J Immunol* **179**(9): 5958–65.

Waugh, J. and C. M. Perry (2005). "Anakinra: a review of its use in the management of rheumatoid arthritis." *BioDrugs* **19**(3): 189–202.
Weaver, C. T., R. D. Hatton, et al. (2007). "IL-17 family cytokines and the expanding diversity of effector T cell lineages." *Annu Rev Immunol* **25**: 821–52.
Wildin, R. S., F. Ramsdell, et al. (2001). "X-linked neonatal diabetes mellitus, enteropathy and endocrinopathy syndrome is the human equivalent of mouse scurfy." *Nat Genet* **27**(1): 18–20.
Wilson, N. J., K. Boniface, et al. (2007). "Development, cytokine profile and function of human interleukin 17-producing helper T cells." *Nat Immunol* **8**(9): 950–7.
Wing, K., Y. Onishi, et al. (2008). "CTLA-4 control over Foxp3+ regulatory T cell function." *Science* **322**(5899): 271–5.
Yang, L., D. E. Anderson, et al. (2008a). "IL-21 and TGF-beta are required for differentiation of human T(H)17 cells." *Nature* **454**(7202): 350–2.
Yang, X. O., R. Nurieva, et al. (2008b). "Molecular antagonism and plasticity of regulatory and inflammatory T cell programs." *Immunity* **29**(1): 44–56.
Yang, X. O., A. D. Panopoulos, et al. (2007). "STAT3 regulates cytokine-mediated generation of inflammatory helper T cells." *J Biol Chem* **282**(13): 9358–63.
Yao, Z., Y. Kanno, et al. (2007). "Non-redundant roles for Stat5a/b in directly regulating Foxp3." *Blood* **109**(10): 4368–75.
Zhang, X., D. N. Koldzic, et al. (2004). "IL-10 is involved in the suppression of experimental autoimmune encephalomyelitis by CD25+CD4+ regulatory T cells." *Int Immunol* **16**(2): 249–56.
Zheng, W. and R. A. Flavell (1997). "The transcription factor GATA-3 is necessary and sufficient for Th2 cytokine gene expression in CD4 T cells." *Cell* **89**(4): 587–96.
Zhou, L., M. M. Chong, et al. (2009). "Plasticity of CD4+ T cell lineage differentiation." *Immunity* **30**(5): 646–55.
Zhou, L., Ivanov, II, et al. (2007). "IL-6 programs T(H)-17 cell differentiation by promoting sequential engagement of the IL-21 and IL-23 pathways." *Nat Immunol* **8**(9): 967–74.
Zhou, L., J. E. Lopes, et al. (2008). "TGF-beta-induced Foxp3 inhibits T(H)17 cell differentiation by antagonizing RORgammat function." *Nature* **453**(7192): 236–40.

The Role of Interleukin-17 in Systemic Lupus Erythematosus

José C. Crispín, Vasileios C. Kyttaris, and George C. Tsokos

Abstract Altered cytokine production has long been recognized to be involved in the immune pathogenesis of systemic lupus erythematosus (SLE). Interleukin 17 (IL-17) has recently been recognized as a contribution to the advancement of the auto-immune response and the expression of tissue injury. Increased production of IL-17 has been documented in humans with SLE and therefore validated its contribution to increased auto-antibody production and kidney inflammation. However, suppression of IL-17 production in lupus-prone mice results in disease improvement.

1 Introduction

Systemic lupus erythematosus (SLE) is an auto-immune disease that affects numerous organs and systems in a diverse pattern. Although clinically heterogeneous, patients with SLE share a common failure in tolerance mechanisms that leads to the development of an immune response against ubiquitous self-antigens (Crispin et al. 2010b). Tissue injury is thought to occur as a consequence of the chronic inflammatory response through various pathways that include the deposition of immune complexes as well as cellular infiltration and release of cytokines in target organs.

The clinical diagnosis of SLE relies on the identification of markers of systemic tolerance failure (e.g. anti-nuclear and anti-double stranded DNA (dsDNA) antibodies) associated to the presence of target organ involvement (e.g. glomerulonephritis, skin rash). Most of what we know about cellular and molecular pathogenesis of SLE in humans is based on in vitro work performed with peripheral blood circulating lymphocytes of clinically diverse individuals with SLE. On the other hand, studies

G.C. Tsokos (✉)
Division of Rheumatology, Beth Israel Deaconess Medical Center, Harvard Medical School, 330 Brookline Avenue, CLS-937, Boston, MA 02115, USA
e-mail: gtsokos@bidmc.harvard.edu

S. Jiang (ed.), *T_H17 Cells in Health and Disease*,
DOI 10.1007/978-1-4419-9371-7_20,

performed in animal models have allowed us to dissect pathogenic pathways able to induce auto-immunity and organ damage that are similar, to a certain degree, to what is observed in the human disease (Liu and Mohan 2006). Congruent information from studies using human cells and studies in lupus-prone mice may enable definitive conclusions and support for advancement to clinical therapeutic trials.

IL-17 is known for its powerful capacity to induce inflammation (Korn et al. 2007). However, it is also involved in the regulation of B cell function by T cells (Doreau et al. 2009). IL-17 is produced at increased levels by T cells from patients with SLE and animals with lupus-like diseases (Crispin et al. 2008; Zhang et al. 2009). Its ability to promote local inflammation and drive B cell responses indicates that it may be involved in several aspects of SLE pathogenesis, both in the underlying systemic auto-immune response and in the instigation of the inflammatory response in target organs. In this chapter we will review the evidence that links IL-17 to SLE and discuss the mechanisms whereby it promotes lupus.

2 IL-17 in Human SLE

Production of IL-17 is increased in patients with SLE (Table 1). Serum levels of IL-17 are abnormally high in patients with SLE when compared to healthy individuals (Cheng et al. 2009; Doreau et al. 2009; Wong et al. 2000, 2008; Zhao et al. 2010). Moreover, a higher percentage of T cells from peripheral blood express IL-17 (Crispin et al. 2008; Wong et al. 2008; Yang et al. 2009).

Evidence of IL-17 production in target organs (kidney and skin) of patients with SLE has been documented by immuno-fluorescence (Crispin et al. 2008) and immuno-histochemistry (Yang et al. 2009). Also, IL-17A mRNA has been found in the kidney (Wang et al. 2009) and urine sediment (Kwan et al. 2009) of patients with lupus nephritis. This data indicates that IL-17 is a cytokine abundantly produced in patients with SLE.

IL-17 is produced by neutrophils, NK cells, and several T cell types including $CD4^+$, $CD8^+$, double negative (DN; $CD4^-CD8^-$), and TCR-$\gamma\delta$ (Korn et al. 2007).

Table 1 Evidence for augmented IL-17 production in human SLE

Finding	References
Increased IL-17 in SLE sera	Cheng et al. (2009), Doreau et al. (2009) Wong et al. (2000, 2008), Zhao et al. (2010)
Increased frequency of IL-17-producing T cells in SLE peripheral blood	Crispin et al. (2008), Wong et al. (2008), Yang et al. (2009)
Presence of IL-17-producing T cells in SLE target organs	Crispin et al. (2008), Wang et al. (2010), Yang et al. (2009)
IL-17 mRNA in urinary sediment of patients with SLE	Kwan et al. (2009)

Among $CD4^+$ T cells, IL-17 is primarily a product of two effector cell types, T_H17 and follicular helper T cells (T_{FH}) (Johnston et al. 2009; Korn et al. 2009; Nurieva et al. 2009). T_H17 cells are generated when naïve T cells are primed in the presence of TGF-β and inflammatory cytokines (e.g. IL-1β, IL-6, IL-21, IL-23) (Bettelli et al. 2006; Yang et al. 2008). Besides IL-17, these cells also produce IL-17F, IL-21, and IL-22 (Korn et al. 2009). T_{FH} are cells specialize in providing help to B cells, mainly through IL-21 and CD40L (Johnston et al. 2009; Nurieva et al. 2009). Although they produce IL-17, their role in this context has not been thoroughly investigated.

Evidence suggests that the T_H17 subset is abnormally expanded in patients with SLE. Accordingly, a higher fraction of $CD4^+$ T cells produce IL-17 in these patients (Crispin et al. 2008; Wong et al. 2008; Yang et al. 2009). This might be caused by increased differentiation of naïve T cells into T_H17 cells, or by higher survival and/or expansion of the T_H17 pool in patients with SLE. Although this question has not been directly addressed, indirect evidence suggests that several factors may facilitate T_H17 differentiation in patients with SLE. Production of certain cytokines involved in the generation of T_H17 cells, namely IL-6, IL-21, and IL-23, are increased in patients with SLE (Huang et al. 2007; Linker-Israeli et al. 1991; Wong et al. 2008, 2010). On the other hand, production of IL-2, a cytokine able to stop differentiation of T_H17 cells (Stockinger 2007) is decreased (Alcocer-Varela and Alarcon-Segovia 1982). Plasmacytoid dendritic cells (pDC), specialized antigen-presenting cells, have been associated to SLE pathogenesis because they produce large quantities of type I interferons upon stimulation via Toll-like receptors (TLR) (Banchereau and Pascual 2006). These cells may induce the generation of T_H17 cells in an IL-1β- and IL-23-dependent manner when stimulated through TLR able to detect nucleic acids (Kattah et al. 2008; Lombardi et al. 2009; Yu et al. 2010). This mechanism may prove to be especially significant in patients with SLE since they have circulating immune complexes that contain RNA and DNA.

Numbers of TCR-$\alpha\beta^+$ DN ($CD4^-CD8^-$) T cells are expanded in patients with SLE (Anand et al. 2002; Crispin et al. 2008; Shivakumar et al. 1989). These cells produce pro-inflammatory cytokines (including IL-17 and IL-1β) (Crispin et al. 2008; Crispin and Tsokos 2009) and are able to provide help to B cells (Dean et al. 2002; Shivakumar et al. 1989). They also represent a large portion of kidney-infiltrating T cells in patients with lupus nephritis (Crispin et al. 2008). At least a fraction of TCR-$\alpha\beta^+$ DN T cells derived from $CD8^+$ T cells lose the expression of CD8 following activation (Crispin and Tsokos 2009). The circumstances that drive this conversion are not known. However, data indicates that in mice, it may depend on IL-23 (Kyttaris et al. 2010).

In summary, diverse cytokine abnormalities common in patients with SLE may skew the differentiation of T cells into IL-17-producing $CD4^+$ and DN T cells. This phenomenon could promote the auto-immune process by increasing the activation of immune cells and stimulating B cell proliferation and antibody production (Doreau et al. 2009). Release of IL-17 in target organs by infiltrating T cells probably contributes to local tissue injury by instigating the inflammatory response.

3 IL-17 in Lupus Animal Models

In order to assess the pathogenic significance of the T_H17 cell subpopulation, in vivo studies using animal models of lupus have to be undertaken. One of the most useful mouse models of lupus is the MRL/*lpr* which is characterized by an exuberant immunologic response with lymphadenopathy, nephritis, and dermatitis. MRL/*lpr* mice do not express the Fas molecule, leading to deficient apoptosis and abnormal accumulation of lymphocytes. The majority of these accumulating lymphocytes are $B220^+$ T cells that do not bear the CD4 and CD8 markers (DN T cells).

Early studies suggested that the auto-immune response in MRL/*lpr* mice greatly depended on the production of interferon gamma by T cells. Blocking IFN-γ with an IFN-γ receptor/Fc fusion molecule could abrogate nephritis in these animals (Lawson et al. 2000; Prud'homme et al. 1995). The exact driving force behind the production of IFN-γ was unclear though. One of the first stimuli that were implicated in the generation of the IFN-γ producing T cells in lupus prone mice was IL-12. IL-12 is a heterodimeric cytokine consisting of p40 and p35 sub-units and is produced by antigen-presenting cells. Its role in lupus nephritis was shown by production of IL-12-deficient lupus prone mice. Indeed, these MRL/*lpr* IL-12p40 deficient mice had significantly milder nephritis than MRL/*lpr* mice (Kikawada et al. 2003). This was accompanied by a decrease in the manufacture of intra-renal IFN-γ in the IL-12p40 deficient mice. Somewhat different results were seen when animals lacking the p35 sub-unit of the IL-12 were used as an inducible lupus model (Calvani et al. 2003). These mice were injected with the hydrocarbon oil pristane, a molecule that results in the development of lupus in roughly one third of the cases when injected into mice. IL-12p35-deficient mice injected with pristane did not develop nephritis but still had "nephritogenic anti-bodies" (anti-dsDNA, anti-chromatin). This data pointed to the fact that deficiency of p40 and p35 may not result in exactly the same phenotype. The role of IL-12 in the development of lupus in murine models was further complicated by the fact that macrophages derived from some auto-immunity-prone mouse strains (MRL/++ and New Zealand Black/White F1) produced far less IL-12 when activated than normal mice (Alleva et al. 1998).

These early reports suggested that although IL-12 may be playing a role in the development of lupus nephritis, especially in the abnormal accumulation of IFN-γ producing T cells in the kidneys, there may be another IL-12-related factor that would explain the discrepancies. Although IFN-γ is important in the pathogenesis of lupus nephritis, the fact that the majority of T cells ($CD3^+CD4^-CD8$) do not produce IFN-γ, suggested alternative mediators of pathology in lupus.

As IL-23 shares the p40 sub-unit with IL-12, it became a prime candidate to explain some of the observed discrepancies between the different studies in lupus-prone mice, and in particular, the protection against lupus that IL-12/23 p40 deficiency confers. Along the same lines, since IL-23 is a pivotal cytokine for the maintenance of the effector T_H17 cells, the cytokine duet IL-23/IL-17 could be an important mediator of lupus pathology in lieu or parallel to IL-12/IFN-γ.

T cells from Fas deficient lupus-prone MRL/*lpr* and B6/*lpr* mice were found to express IL-17A (Zhang et al. 2009), with the expression levels increasing as the mice aged and became sicker. Most of the IL-17A$^+$ cells, contrary to conventional CD4$^+$ T$_H$17, were found to be CD3$^+$CD4$^-$CD8, and not only populated the secondary lymphoid organs, but also invaded the interstitium of the kidneys. Two lines of evidence suggested that these cells pathogenic potential was dependent on IL-23. First, the IL-23 receptor was highly expressed in the lymphocytes from B6/*lpr* animals and paralleled the expression of IL-17A. Secondly, incubation of lymph node cells from lupus-prone mice with IL-23 and injection into Rag1$^{-/-}$ lymphocyte-deficient mice led to the development of nephritis in the recipient mice, with deposition of immunoglobulin and complement to the glomeruli. Incubation of lymph node cells from lupus prone mice with IL-23 led to a significant increase of double negative T cells, primarily because of increase in their proliferation. Other studies in MRL/*lpr* also reached similar conclusions that IL-17$^+$ T cells infiltrated the kidneys of mice that have developed nephritis (Wang et al. 2008).

The pathogenic role of IL-23 in murine lupus was further supported by a series of experiments in IL-23 receptor deficient B6/*lpr* mice (Kyttaris et al. 2010). As compared to IL-23R$^{+/+}$B6/*lpr* mice, these did not develop secondary lymphoid organ hyperplasia and had similar numbers of DN T cells as normal B6 mice would. The production of IL-17A, IFN-γ, immunolgobulin, and auto-antibodies was significantly reduced in these animals. As a consequence of the blunted auto-immune response, the IL-23R$^{-/-}$B6/*lpr* mice did not develop glomerulonephritis. This data suggests a trophic effect of IL-23 on CD3$^+$CD4$^-$CD8$^-$IL-17A$^+$ cells that significantly contributed to the lupus pathogenesis.

IL-23 seems to play a pivotal role in the development of lupus, not only in animals with expanded DN T cell population, but also in other models with lupus-like conditions. One of these models was the Ro52-deficient mouse (Espinosa et al. 2009). Ro52/Trim21 is a nuclear protein that is thought to ubiquitinate interferon-related factors, and its deficiency led to the development of lupus-like disease. Ro52 deficient mice exhibited severe dermatitis (induced by ear tags) and systemic inflammation, including nephritis. The inflammatory response was characterized by the production of IL-23, IL-17, and auto-antibodies (ANA, anti-dsDNA). Importantly though, mice that were deficient for both Ro52 and IL-23p19 did not develop autoimmunity as did Ro52$^{-/-}$IL-23p19$^{+/+}$ mice. In another lupus model, the NZM 2328 mice, deficient of both tumor necrosis factor receptors 1 and 2 led to accelerated nephritis with activation of the IL-17 pathway (Jacob et al. 2009). Therefore, the IL-23/IL-17 pathway has been proven to be involved in the development of lupus-like auto-immunity, and in particular nephritis in mice.

Clearly, more studies are needed to ascertain the exact stimulus that leads to the production of IL-23 and the mechanism by which IL-23 leads to the development of lupus nephritis in these animal models. Interferon effect or Toll like receptor engagement by nucleic acid containing complexes, both significant contributors to SLE pathogenesis, may be the underlying causes for the over-production of IL-23. One interesting study showed dendritic cells that phagocytose apoptotic blebs could induce the maturation of T cells into a T$_H$17 phenotype (Fransen et al. 2009), suggesting that increased apoptosis and inefficient handling of the apoptotic material in SLE may result in aberrant T cell activation.

The influx of T_H17 cells in the kidneys of lupus prone mice may be directed by the aberrant production of chemokines that bind to receptors such as the CXCR3. Indeed CXCR3–/–MRL/*lpr* mice displayed much milder nephritis, with a significant decrease in infiltrating T_H1 (IFN-γ) and T_H17 cells in their kidneys (Steinmetz et al. 2009).

The downstream effects of IL-23-dependent T_H17 cells seem to be mediated by both cellular and humoral factors. As mentioned above, IL-17A$^+$ cells infiltrate the kidneys of lupus prone mice, possibly coordinating the abnormal immune response in situ via production of pro-inflammatory cytokines. At the same time, incubation of lymph node cells from lupus-prone mice with IL-23 but not control PBS, pre-injection into lymphocyte deficient mice, led to the deposition of immunoglobulin and supplement to the glomeruli of the recipient mice (Zhang et al. 2009). This latter observation suggested that IL-23 may directly or indirectly stimulate the production of nephritogenic antibodies. In another murine model of lupus (BXD2) that develops an exuberant humoral response, IL-17 has been implicated in the creation of germinal centers and the production of auto-antibodies. T cells that are IL-17$^+$ and B cells that express the IL-17 receptor on their surface co-localize in the germinal centers in the spleens of these animals. IL-17 deficient BXD2 mice had a significant decrease in auto-antibody production while overexpression of IL-17 in young healthy mice lead to the acceleration of the disease and the creation of germinal centers (Hsu et al. 2008). It is also possible that IL-23 induces the production by IL-17$^+$ T cells of IL-21 and/or CD40L, two molecules that are produced by extrafollicular T helper cells and have been shown to be important in auto-antibody production in lupus-prone mice (Odegard et al. 2008).

4 Conclusion

A number of cytokines have previously been reported to get abnormally expressed in lupus-prone mice and in patients with SLE. Specifically, interferons alpha and gamma have been atypically expressed and contribute to the manifestation of autoimmunity and lupus nephritis. Their significance has led to exploitation as therapeutic targets. Similarly, IL-6 has been found to be expressed in increased amounts in humans and mice with SLE and antibodies directed against its receptor are already in clinical trials (Crispin et al. 2010a). On the other hand, decreased expression of IL-2 has been coined to abnormal T cell function and the expression of auto-immunity (Setoguchi et al. 2005).

IL-17 and its upstream modulators have been recently recognized to be involved in the expression of auto-immune pathology in patients and relevant animal models of rheumatoid arthritis (Kirkham et al. 2006), multiple sclerosis (Matusevicius et al. 1999), inflammatory bowel disease (Duerr et al. 2006) and psoriasis (Krueger et al. 2007). In this chapter we provided a brief critical summary of the information generated during the last 2 years about the role of IL-17 in the expression of SLE pathology. In brief, both humans and mice with lupus display an increased number of IL-17 producing cells which includes $CD4^+$ and the expanded $CD3^+CD4^-CD8^-$ cells. The most important information is the fact that IL-17 producing cells are present in the inflamed kidneys of humans with lupus nephritis (Crispin et al. 2008) and mice with nephritis (Zhang et al. 2009). Although it is possible that non-IL-17-producing T cells may enter the kidney where they acquire IL-17 producing capability, because increased numbers of IL-17-producing cells are found in the periphery, we assume that these cells enter the tissues. This is important because the levels of IL-17-producing cells can be used as heralds of impeding tissue injury to urge the initiation of treatment.

The information summarized here demonstrates that genetic reduction of IL-17 production leads to suppression of auto-antibody levels and nephritis (Kyttaris et al. 2010). Infusion of cells producing increased amounts of IL-17 into $Rag1^{-/-}$ mice causes increased production of auto-antibodies and lupus nephritis (Zhang et al. 2009). This strongly urges the therapeutic use of the inhibition of IL-17 production or IL-17 action in the treatment of patients with SLE and lupus nephritis.

References

Alcocer-Varela, J., and Alarcon-Segovia, D. (1982). Decreased production of and response to interleukin-2 by cultured lymphocytes from patients with systemic lupus erythematosus. J Clin Invest *69*, 1388–1392.

Alleva, D.G., Kaser, S.B., and Beller, D.I. (1998). Intrinsic defects in macrophage IL-12 production associated with immune dysfunction in the MRL/++ and New Zealand Black/White F1 lupus-prone mice and the Leishmania major-susceptible BALB/c strain. J Immunol *161*, 6878–6884.

Anand, A., Dean, G.S., Quereshi, K., Isenberg, D.A., and Lydyard, P.M. (2002). Characterization of CD3+ CD4- CD8- (double negative) T cells in patients with systemic lupus erythematosus: activation markers. Lupus *11*, 493–500.

Banchereau, J., and Pascual, V. (2006). Type I interferon in systemic lupus erythematosus and other auto-immune diseases. Immunity. *25*, 383–392.

Bettelli, E., Carrier, Y., Gao, W., Korn, T., Strom, T.B., Oukka, M., Weiner, H.L., and Kuchroo, V.K. (2006). Reciprocal developmental pathways for the generation of pathogenic effector TH17 and regulatory T cells. Nature *441*, 235–238.

Calvani, N., Satoh, M., Croker, B.P., Reeves, W.H., and Richards, H.B. (2003). Nephritogenic autoantibodies but absence of nephritis in Il-12p35-deficient mice with pristane-induced lupus. Kidney Int. *64*, 897–905.

Cheng, F., Guo, Z., Xu, H., Yan, D., and Li, Q. (2009b). Decreased plasma IL22 levels, but not increased IL17 and IL23 levels, correlate with disease activity in patients with systemic lupus erythematosus. Ann Rheum Dis *68*, 604–606.

Crispin, J.C., Kyttaris, V.C., Terhorst, C., and Tsokos, G.C. (2010a). T cells as therapeutic targets in systemic lupus erythematosus. Nat Rev Rheumatol *6*, 317–25

Crispin, J.C., Liossis, S.N., Kis-Toth, K., Lieberman, L., Kyttaris, V.C., Juang, Y.T., and Tsokos, G.C. (2010b). Pathogenesis of human systemic lupus erythematosus: recent advances. Trends Mol Med *16*, 47–57.

Crispin, J.C., Oukka, M., Bayliss, G., Cohen, R.A., Van Beek, C.A., Stillman, I.E., Kyttaris, V.C., Juang, Y.T., and Tsokos, G.C. (2008). Expanded double negative T cells in patients with systemic lupus erythematosus produce IL-17 and infiltrate the kidneys. J Immunol *181*, 8761–8766.

Crispin, J.C., and Tsokos, G.C. (2009). Human TCR-$\alpha\beta^+$ CD4$^-$ CD8$^-$ T cells can derive from CD8$^+$ T cells and display an inflammatory effector phenotype. J Immunol *183*, 4675–4681.

Dean, G.S., Anand, A., Blofeld, A., Isenberg, D.A., and Lydyard, P.M. (2002). Characterization of CD3+ CD4- CD8- (double negative) T cells in patients with systemic lupus erythematosus: production of IL-4. Lupus *11*, 501–507.

Doreau, A., Belot, A., Bastid, J., Riche, B., Trescol-Biemont, M.C., Ranchin, B., Fabien, N., Cochat, P., Pouteil-Noble, C., Trolliet, P., et al. (2009). Interleukin 17 acts in synergy with B cell-activating factor to influence B cell biology and the pathophysiology of systemic lupus erythematosus. Nat Immunol *10*, 778–785.

Duerr, R.H., Taylor, K.D., Brant, S.R., Rioux, J.D., Silverberg, M.S., Daly, M.J., Steinhart, A.H., Abraham, C., Regueiro, M., Griffiths, A., et al. (2006). A genome-wide association study identifies IL23R as an inflammatory bowel disease gene. Science *314*, 1461–1463.

Espinosa, A., Dardalhon, V., Brauner, S., Ambrosi, A., Higgs, R., Quintana, F.J., Sjostrand, M., Eloranta, M.L., Ni, G.J., Winqvist, O., et al. (2009). Loss of the lupus auto-antigen Ro52/Trim21 induces tissue inflammation and systemic auto-immunity by dysregulating the IL-23-Th17 pathway. J Exp Med *206*, 1661–1671.

Fransen, J.H., Hilbrands, L.B., Ruben, J., Stoffels, M., Adema, G.J., van, d., V, and Berden, J.H. (2009). Mouse dendritic cells matured by ingestion of apoptotic blebs induce T cells to produce interleukin-17. Arthritis Rheum *60*, 2304–2313.

Hsu, H.C., Yang, P., Wang, J., Wu, Q., Myers, R., Chen, J., Yi, J., Guentert, T., Tousson, A., Stanus, A.L., et al. (2008). Interleukin 17-producing T helper cells and interleukin 17 orchestrate autoreactive germinal center development in auto-immune BXD2 mice. Nat. Immunol *9*, 166–175.

Huang, X., Hua, J., Shen, N., and Chen, S. (2007). Dysregulated expression of interleukin-23 and interleukin-12 sub-units in systemic lupus erythematosus patients. Mod. Rheumatol. *17*, 220–223.

Jacob, N., Yang, H., Pricop, L., Liu, Y., Gao, X., Zheng, S.G., Wang, J., Gao, H.X., Putterman, C., Koss, M.N., et al. (2009). Accelerated pathological and clinical nephritis in systemic lupus erythematosus-prone New Zealand Mixed 2328 mice doubly deficient in TNF receptor 1 and TNF receptor 2 via a Th17-associated pathway. J Immunol *182*, 2532–2541.

Johnston, R.J., Poholek, A.C., DiToro, D., Yusuf, I., Eto, D., Barnett, B., Dent, A.L., Craft, J., and Crotty, S. (2009). Bcl6 and Blimp-1 are reciprocal and antagonistic regulators of T follicular helper cell differentiation. Science *325*, 1006–1010.

Kattah, M.G., Wong, M.T., Yocum, M.D., and Utz, P.J. (2008). Cytokines secreted in response to Toll-like receptor ligand stimulation modulate differentiation of human Th17 cells. Arthritis Rheum. *58*, 1619–1629.

Kikawada, E., Lenda, D.M., and Kelley, V.R. (2003). IL-12 deficiency in MRL-Fas(lpr) mice delays nephritis and intrarenal IFN-gamma expression, and diminishes systemic pathology. J Immunol *170*, 3915–3925.

Kirkham, B.W., Lassere, M.N., Edmonds, J.P., Juhasz, K.M., Bird, P.A., Lee, C.S., Shnier, R., and Portek, I.J. (2006). Synovial membrane cytokine expression is predictive of joint damage progression in rheumatoid arthritis: a two-year prospective study (the DAMAGE study cohort). Arthritis Rheum. *54*, 1122–1131.

Korn, T., Bettelli, E., Oukka, M., and Kuchroo, V.K. (2009). IL-17 and Th17 Cells. Annu. Rev Immunol *27*, 485–517.

Korn, T., Oukka, M., Kuchroo, V., and Bettelli, E. (2007). Th17 cells: effector T cells with inflammatory properties. Semin. Immunol. *19*, 362–371.

Krueger, G.G., Langley, R.G., Leonardi, C., Yeilding, N., Guzzo, C., Wang, Y., Dooley, L.T., and Lebwohl, M. (2007). A human interleukin-12/23 monoclonal antibody for the treatment of psoriasis. N. Engl. J Med *356*, 580–592.

Kwan, B.C., Tam, L.S., Lai, K.B., Lai, F.M., Li, E.K., Wang, G., Chow, K.M., Li, P.K., and Szeto, C.C. (2009). The gene expression of type 17 T-helper cell-related cytokines in the urinary sediment of patients with systemic lupus erythematosus. Rheumatology. (Oxford) *48*, 1491–1497.

Kyttaris, V.C., Zhang, Z., Kuchroo, V.K., Oukka, M., and Tsokos, G.C. (2010). IL-23 receptor deficiency prevents the development of lupus nephritis in B6/lpr mice. J Immunol *184*, 4605–9.

Lawson, B.R., Prud'homme, G.J., Chang, Y., Gardner, H.A., Kuan, J., Kono, D.H., and Theofilopoulos, A.N. (2000). Treatment of murine lupus with cDNA encoding IFN-gammaR/Fc. J Clin. Invest *106*, 207–215.

Linker-Israeli, M., Deans, R.J., Wallace, D.J., Prehn, J., Ozeri-Chen, T., and Klinenberg, J.R. (1991). Elevated levels of endogenous IL-6 in systemic lupus erythematosus. A putative role in pathogenesis. J Immunol *147*, 117–123.

Liu, K., and Mohan, C. (2006). What do mouse models teach us about human SLE? Clin. Immunol. *119*, 123–130.

Lombardi, V., Van, O.L., Horiot, S., and Moingeon, P. (2009). Human dendritic cells stimulated via TLR7 and/or TLR8 induce the sequential production of Il-10, IFN-gamma, and IL-17A by naive CD4+ T cells. J Immunol *182*, 3372–3379.

Matusevicius, D., Kivisakk, P., He, B., Kostulas, N., Ozenci, V., Fredrikson, S., and Link, H. (1999). Interleukin-17 mRNA expression in blood and CSF mononuclear cells is augmented in multiple sclerosis. Mult. Scler. *5*, 101–104.

Nurieva, R.I., Chung, Y., Martinez, G.J., Yang, X.O., Tanaka, S., Matskevitch, T.D., Wang, Y.H., and Dong, C. (2009). Bcl6 mediates the development of T follicular helper cells. Science *325*, 1001–1005.

Odegard, J.M., Marks, B.R., DiPlacido, L.D., Poholek, A.C., Kono, D.H., Dong, C., Flavell, R.A., and Craft, J. (2008). ICOS-dependent extra-follicular helper T cells elicit IgG production via IL-21 in systemic auto-immunity. J Exp Med *205*, 2873–2886.

Prud'homme, G.J., Kono, D.H., and Theofilopoulos, A.N. (1995). Quantitative polymerase chain reaction analysis reveals marked overexpression of interleukin-1 beta, interleukin-1 and interferon-gamma mRNA in the lymph nodes of lupus-prone mice. Mol Immunol *32*, 495–503.

Setoguchi, R., Hori, S., Takahashi, T., and Sakaguchi, S. (2005). Homeostatic maintenance of natural Foxp3(+) CD25(+) CD4(+) regulatory T cells by interleukin (IL)-2 and induction of auto-immune disease by IL-2 neutralization. J Exp Med *201*, 723–735.

Shivakumar, S., Tsokos, G.C., and Datta, S.K. (1989). T cell receptor alpha/beta expressing double-negative (CD4-/CD8-) and CD4+ T helper cells in humans augment the production of pathogenic anti-DNA auto-antibodies associated with lupus nephritis. J Immunol *143*, 103–112.

Steinmetz, O.M., Turner, J.E., Paust, H.J., Lindner, M., Peters, A., Heiss, K., Velden, J., Hopfer, H., Fehr, S., Krieger, T., et al. (2009). CXCR3 mediates renal Th1 and Th17 immune response in murine lupus nephritis. J Immunol *183*, 4693–4704.

Stockinger, B. (2007). Good for Goose, but not for Gander: IL-2 interferes with Th17 differentiation. Immunity *26*, 278–279.

Wang, Y., Ito, S., Chino, Y., Goto, D., Matsumoto, I., Murata, H., Tsutsumi, A., Hayashi, T., Uchida, K., Usui, J., et al. (2009). Laser microdissection-based analysis of cytokine balance in the kidneys of patients with lupus nephritis. Clin. Exp. Immunol. *159*, 1–10.

Wang, Y., Ito, S., Chino, Y., Iwanami, K., Yasukochi, T., Goto, D., Matsumoto, I., Hayashi, T., Uchida, K., and Sumida, T. (2008). Use of laser microdissection in the analysis of renal-infiltrating T cells in MRL/lpr mice. Mod. Rheumatol. *18*, 385–393.

Wong, C.K., Ho, C.Y., Li, E.K., and Lam, C.W. (2000). Elevation of pro-inflammatory cytokine (IL-18, IL-17, IL-12) and Th2 cytokine (IL-4) concentrations in patients with systemic lupus erythematosus. Lupus *9*, 589–593.

Wong, C.K., Lit, L.C., Tam, L.S., Li, E.K., Wong, P.T., and Lam, C.W. (2008). Hyperproduction of IL-23 and IL-17 in patients with systemic lupus erythematosus: implications for Th17-mediated inflammation in auto-immunity. Clin Immunol *127*, 385–393.

Wong, C.K., Wong, P.T., Tam, L.S., Li, E.K., Chen, D.P., and Lam, C.W. (2010). Elevated Production of B Cell Chemokine CXCL13 is Correlated with Systemic Lupus Erythematosus Disease Activity. J Clin Immunol *30*, 45–52.

Yang, J., Chu, Y., Yang, X., Gao, D., Zhu, L., Yang, X., Wan, L., and Li, M. (2009). Th17 and natural Treg cell population dynamics in systemic lupus erythematosus. Arthritis Rheum *60*, 1472–1483.

Yang, L., Anderson, D.E., Baecher-Allan, C., Hastings, W.D., Bettelli, E., Oukka, M., Kuchroo, V.K., and Hafler, D.A. (2008). IL-21 and TGF-beta are required for differentiation of human T(H)17 cells. Nature *454*, 350–352.
Yu, C.F., Peng, W.M., Oldenburg, J., Hoch, J., Bieber, T., Limmer, A., Hartmann, G., Barchet, W., Eis-Hubinger, A.M., and Novak, N. (2010). Human plasmacytoid dendritic cells support Th17 cell effector function in response to TLR7 ligation. J Immunol *184*, 1159–1167.
Zhang, Z., Kyttaris, V.C., and Tsokos, G.C. (2009). The role of IL-23/IL-17 axis in lupus nephritis. J Immunol *183*, 3160–3169.
Zhao, X.F., Pan, H.F., Yuan, H., Zhang, W.H., Li, X.P., Wang, G.H., Wu, G.C., Su, H., Pan, F.M., Li, W.X., et al. (2010). Increased serum interleukin 17 in patients with systemic lupus erythematosus. Mol. Biol Rep. *37*, 81–85.

IL-17 in B Cell Biology and Systemic Lupus Erythematosus

Agnès Doreau-Bastid, Jeremy Bastid, Alexandre Belot, Jean-François Eliaou, and Nathalie Bonnefoy-Berard *

Abstract Recent data emphasizes the role of interleukin 17 (IL-17) in several chronic inflammatory and auto-immune diseases, systemic lupus erythematosus, in particular (SLE). It has been reported that in addition to its well-known pro-inflammatory effect, IL-17 directly regulates B cell migration within germinal centers. Furthermore, we demonstrated that IL-17 acts in synergy with B cell activating factor (BAFF) to directly regulate key aspects of B cell biology such as survival, proliferation, and differentiation into immunoglobulin (Ig)-secreting cells. We proposed and discussed herein the hypothesis that IL-17 might play a pivotal role in the pathophysiology of auto-immune diseases such as SLE because of its combined actions on both T cell driven inflammation and B cell biology. In these settings, IL-17 could be considered as the link between the T and the B cell components contributing to SLE pathogenesis. The IL-17 pathway may thus represent a new attractive target with high therapeutic potential in the treatment of SLE.

---

* Jean-François Eliaou and Nathalie Bonnefoy-Berard equally contribute to the manuscript and therefore share last authorship.

N. Bonnefoy-Berard (✉)
Université de Lyon, Lyon, France
and
INSERM, U851, 21 Avenue Tony Garnier, 69007 Lyon, France
and
Université Lyon1, IFR128, Lyon, France
e-mail: nathalie.bonnefoy-berard@inserm.fr

S. Jiang (ed.), *T_H17 Cells in Health and Disease*,
DOI 10.1007/978-1-4419-9371-7_21, © Springer Science+Business Media, LLC 2011

1 Introduction

Interleukin 17A (hereafter named IL-17) is a potent pro-inflammatory cytokine mainly produced by the T_H17 $CD4^+$ T cell sub-population, but also by several other cell types like natural killer cells, natural killer T cells, double negative CD3 T cells, γδ-T cells, and neutrophils (Korn et al. 2007). Although IL-17-producing cells play an important role in the initiation of defense against infection by bacteria and fungi, they are also viewed as a key inducers and organizers of tissue inflammation (Miossec et al. 2009). Increased production of IL-17 is associated with the pathogenesis of several chronic inflammatory and auto-immune diseases, including psoriasis, rheumatoid arthritis, multiple sclerosis, Crohn's disease (Yen et al. 2006; Boniface et al. 2008; Zhang et al. 2010), and more recently SLE (Wong et al. 2000, 2008; Pan et al. 2008; Pene et al. 2008; Crispin et al. 2008; Kwan et al. 2009; Wang et al. 2009; Doreau et al. 2009; Yang et al. 2009). The pathogenic effect of IL-17 in such auto-immune diseases is mainly due to its capacity to induce the release of pro-inflammatory chemokines, cytokines, and growth factors. As exemplified in rheumatoid arthritis, IL-17 also favors the release of matrix metalloproteinases, leading to the recruitment and activation of neutrophils in inflamed joints, but to the destruction of the extracellular matrix and bone resorption (Miossec et al. 2009). We and others (Hsu et al. 2008; Doreau et al. 2009) recently identified a novel function of IL-17 as a direct regulator of B cell biology, suggesting that deregulated production of IL-17 could lead not only to T cell-mediated tissue injury, but also to B cell dysfunctions and subsequent production of auto-reactive antibodies. This review will focus on addressing the new aspect of IL-17 and B cell biology and will discuss the importance of these findings regarding SLE pathophysiology.

2 IL-17 and B Cell Biology

2.1 IL-17: A Novel Regulator of B Cell Survival

B cell ontogenesis is a highly regulated process that relies on a progressive and continued maturation of B cells from bone marrow (BM) to peripheral blood and lymph nodes. This maturation process is achieved with the concomitant "silencing" of auto-reactive clones. Three mechanisms have been proposed for the removal of auto-reactive B cells: deletion by apoptosis, induction of anergy, and receptor editing. In healthy donors, more than 75% of the antibodies produced by early immature B cells are reactive with several self-antigens (Wardemann and Nussenzweig 2007). The majority of these auto-reactive B cell clones are removed from the nascent repertoire at the first central tolerance checkpoint before the surface IgM-positive immature B cell stage (Wardemann et al. 2003). Another removal step occurs early after the BM exits for new emigrant B cells, called transitional B cells, which recognize peripheral self-antigens that are not expressed in the BM. Failure to remove auto-reactive clones at either checkpoint results in persistence of self-reactive B cells and thus increased susceptibility of auto-immunity.

While assessing the effects of IL-17 on B cells, we demonstrated that IL-17 directly increases B cell viability through the induction of the NF-κB pathway, as previously demonstrated for B cell-activating factor (BAFF) (Mackay and Browning 2002). More importantly, both cytokines display a strong synergy when combined together and are able to maintain naive and memory B cells alive for more than a week. In addition to its effect on spontaneous B cell apoptosis, we showed that IL-17 also alters the clonal deletion process that contributes to elimination of self-reactive B cells. Using the B104 human-lymphoma cell line that undergoes apoptosis following BCR engagement as cellular model of clonal deletion (Ishigami et al. 1992), we demonstrated that IL-17 efficiently inhibits anti-IgM-mediated cell death, an effect enhanced when IL-17 is combined with BAFF. In pathological and inflammatory settings, IL-17 itself, or in combination with BAFF, may induce a defect in B cell tolerance. Strikingly, among the whole B cell population, the transitional B cells (which are the target of peripheral tolerance checkpoint and normally undergo peripheral tolerance process) where reported to be elevated in SLE (Sims et al. 2005). Interestingly, these transitional B cells express IL-17 and BAFF receptors (Fig. 1) and the combination of both cytokines efficiently protects this sub-population from cell death (Doreau et al. unpublished data). Whether there is a causative link with IL-17 and BAFF levels in patients, the anti-apoptotic effect of both cytokines and the increased proportion of transitional B cells in SLE patients are now under investigation in our laboratory.

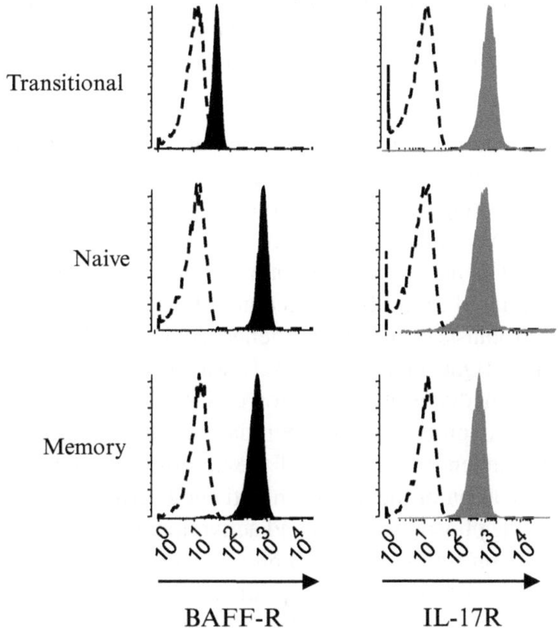

Fig. 1 Expression of IL-17-R and BAFF-R on B cell sub-populations. Representative histograms showed expression of IL-17R and BAFF-R on transitional, naïve and memory human B cells

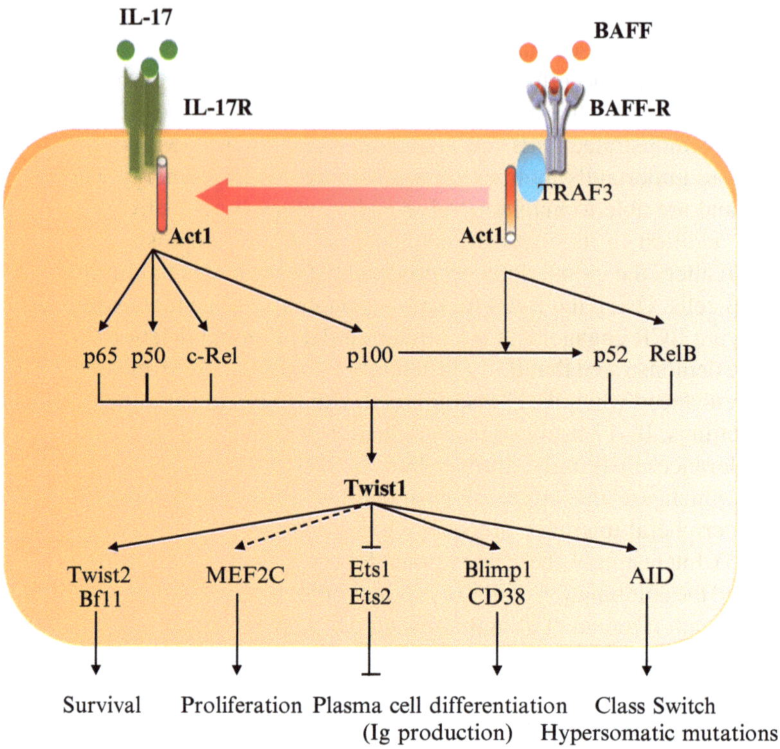

Fig. 2 Schematic representation of IL-17R and BAFF-R signaling pathways in B cells. Activation of both classical and non-classical NF-κB pathway following IL-17R and BAFF-R stimulation in B cells leads to the activation of Twist-1 transcription factor that controls transcription of genes involved in B cell survival, proliferation, and differentiation

One of the interesting observations in this work was the strong synergy revealed between IL-17 and BAFF cytokines. The synergistic effect of IL-17 and BAFF in promoting survival relies on two molecular mechanisms. These mechanisms exploit the Act1 adaptator protein and the NF-κB transcription factor pathway (Fig. 2). Act1 is an essential adaptor molecule in IL-17-mediated signaling that is directly recruited to the IL-17 receptor (IL-17R) through SEFIR (for similar expression to fibroblasts growth factor genes, IL-17R and Toll-IL-1R) domain homotypic interactions (Qian et al. 2007). Act1 was recently identified as a U-box E3 ubiquitin ligase that mediates ubiquitination of TRAF6 on Lys63 (Liu et al. 2009). Rather than leading to proteoasomal degradation, this modification of TRAF6 leads to Lys63-dependent recruitment of the ubiquitin-binding proteins TAB2 and TAB3, and the activation of the ubiquitin-activated serine kinase TAK1. Activation of TAK1 leads to the phosphorylation and activation of members of the IκB kinase complex, which then phosphorylates IκB proteins bound to NF-κB. This phosphorylation of the IκB proteins favors their ubiquitination and proteasomal degradation, which enables the translocation of active NF-κB sub-units into the nucleus. Interestingly, Act1 also binds BAFF receptor through a multiprotein complex in a SEFIR-independent manner and inhibits BAFF signaling on NF-κB (Qian et al. 2004). However two very striking molecular mechanisms contradict this apparent IL-17 and BAFF opposite effects. We showed that when both receptors are stimulated, Act1 binds preferentially IL-17 receptor to the detriment of BAFF receptor, enhancing both pathways and leading to a strong NF-κB activation. Moreover, IL-17 binding to its receptor is sufficient to induce the dissociation and the recruitment of Act1 molecules already fixed to BAFF receptor, highlighting an unexpected and mysterious mode of action. Thus, after the binding of IL-17 and BAFF to their receptors, Act 1 is preferentially diverted to the IL-17 receptor, enforcing its pathway while decreasing its inhibitory effect on BAFF signaling pathway, leading to a synergistic activation of the NF-κB transcription factor. The second mechanism underlying synergy between the two cytokines is their impact on the different sub-units of the NF-κB transcription factor. Although both cytokines lead to NF-κB sub-units translocation into the nucleus, they display distinct targets. IL-17 activates the so-called canonical NF-κB pathway by promoting nuclear translocation of the NF-kB sub-units p65, p50 and cRel, whereas BAFF activates the non-canonical pathway by triggering nuclear translocation of the NF-κB sub-units p52 and RelB. Furthermore, we demonstrated that the IL-17 "feeds" the BAFF pathway by generating p100, which will get processed into p52 if BAFF binds to its receptor (Doreau et al. 2009).

As a consequence of the NF-κB activation, IL-17 and BAFF stimulation induces the NF-κB target Twist-1, an anti-apoptotic bHLH transcription factor (Maestro et al. 1999). By using knockdown assays, it was possible to demonstrate that IL-17 and BAFF-mediated survival of B cells completely relies on Twist-1 induction. However, Twist-1 induction is an early and transient step, as the protein is not detectable anymore beyond 8 h post-cytokine stimulation of B cells. Twist-1-mediated survival indeed relies on the induction of two of its anti-apoptotic target genes, Twist-2 and Bfl-1, which are known regulators of the p53 pathway and the Bcl-2 family members. Thus, Twist-1 and its two target genes, Twist-2 and Bfl-1, appear as key mediators of IL-17 and/or BAFF-mediated survival in B cells (Doreau et al. 2009).

2.2 *IL-17: Impact on B Cell Proliferation and Differentiation*

It was shown that extensive naive B cell proliferation, isotypic switch, and differentiation to immunoglobulin (Ig)-secreting cells requires BCR cross-linking by antigens followed by cognate interaction with helper T cells and an additional signal delivered by the triggering of any of the Toll-like receptors (TLR) expressed by B cells (Ruprecht and Lanzavecchia 2006). While each cytokine alone has no effect with regards to B cell proliferation or differentiation, the combination of IL-17 and BAFF can substitute for either TLR or CD40 signal. Therefore, cognate T cell help through CD40 signaling or TLR triggering is not required for B cells activated in the presence of IL-17 and BAFF to acquire all the plasma cell markers and produce IgM, IgG, and IgA (Doreau et al. 2009). The effect of IL-17 and BAFF in combination differs from the effect of IL-21, but can also promote survival and proliferation of human B cells that respond to antigen and receive T cell help signal (Ettinger et al. 2007). As the IL-17 and BAFF combination impedes apoptosis, but also promotes B cell proliferation and differentiation into Ig-secreting cells, one can speculate that such combination of cytokines provide a favorable environment for the maintenance and activation of auto-reactive B cells and the production of auto-reactive antibodies. Reinforcing a direct role for IL-17 in humoral response and the emergence of pathogenic auto-antibodies, Mountz and colleagues (Hsu et al. 2008) proposed that T_H17 T cells and IL-17 participate in the formation of auto-reactive germinal centers (GCs) by arresting B cell migration at these sites. The capacity of IL-17 to lower B cell motility and prolonged retention of B cells in the GCs combined with its protective effect on B cell apoptosis (demonstrated towards human B cells only) suggest that IL-17 may play a major role in humoral auto-immunity.

Many clues indicate that the proliferation and differentiation processes induced in the presence of IL-17 and BAFF are controlled by Twist-1. Twist-1 knockdown in B cells completely abrogates the IL-17 and BAFF-mediated differentiation into effector cells. Furthermore, B cell proliferation requires MEF2C (Wilker et al. 2008), which is a known target of Twist-1 (Cripps et al. 1998). Taken together, this data shows that in contrast to physiological B cell activation, Twist-1 induction is a signature of pathological activation of B cells in chronic inflammatory settings characterized by high levels of IL-17 and BAFF.

3 IL-17 in SLE Pathophysiology

3.1 IL-17-Mediated B Cell Dysfunctions: Impact on SLE Pathophysiology

Although auto-antibody producing B cells appear to play a central role in the pathogenesis of SLE, dysregulation of the innate immune system, activation of monocytes, dendritic cells, or neutrophils, as well as overproduction of inflammatory cytokines and impaired T cell tolerance have also been proposed as key events in the development of this auto-immune disease. Besides the demonstration using normal B cells, using IL-17 by itself or in combination with BAFF directly affects key processes that may impact on B cell tolerance. Several related studies reported an overproduction of IL-17 and other T_H17 related cytokines (IL-21, IL-23) in SLE (Wong et al. 2000, 2008; Kwan et al. 2009; Wang et al. 2009; Doreau et al. 2009; Yang et al. 2009). These observations suggest that IL-17 may represent a major component of the inflammatory environment in SLE, and that deregulated production of IL-17 may represent a novel mechanism for the alteration of B cells responses in SLE patients. Supporting such a hypothesis, we demonstrated that sera from SLE patients with high IL-17 and/or BAFF concentrations significantly increased normal B cell survival compared to sera from healthy individuals. Complete abrogation of the pro-survival effect of SLE sera displaying high concentrations of IL-17 and BAFF can be achieved by blockade of both IL-17 and BAFF with antagonistic antibodies. Furthermore, in the presence of BCR and CD40 stimuli, SLE patients' sera containing elevated concentrations of IL-17 and BAFF efficiently sustained B cell proliferation and differentiation of naive and memory B cells. Again, IL-17 and BAFF-specific antagonists can abrogate these effects.

Several murine models also suggest that SLE may involve the IL-17/T_H17 pathway (Garrett-Sinha et al. 2008; Nalbandian et al. 2009). Among those, the BXD2 mice that display SLE-like auto-immune diseases, including arthritis and glomerulonephritis, support a role for IL-17 in the development of germinal center (GC) B cell-derived auto-antibodies by directly down-modulating the chemotactic response of murine B cells to CXCL12, and leading to their prolonged retention in the GC (Hsu et al. 2008). These results provide evidences that besides acting as a mediator of inflammation, IL-17 also acts as a direct regulator of B cell functions. Thus, deregulated expression of IL-17 may severely impaired B cell homeostasis. An interesting finding is the recent demonstration that BAFF can promote proliferation of T_H17 cells (Lai Kwan Lam et al. 2008), suggesting that BAFF may amplify the production of IL-17.

Further strengthening the potential role of IL-17 in SLE pathogenesis in humans, we and others reported that serum concentration of IL-17 positively correlates with the SLEDAI score (Wong et al. 2008; Doreau et al. 2009) and anti-dsDNA titers (Doreau et al. 2009). The frequency of IL-17-producing T cells is significantly increased in peripheral blood of SLE patients (Wong et al. 2008; Crispin et al. 2008; Yang et al. 2009) at the site of disease activity (Crispin et al. 2008; Yang et al. 2009). Finally, the number of IL-17 producing cells was reported to further augment during SLE flare, especially in patients with vasculitis (Yang et al. 2009).

3.2 IL17: A Link Between T and B Cell Dysfunctions in SLE

The predominance of the B or T cell component in the contribution to SLE pathogenesis has been a matter of intense debate until now (Groom and Mackay 2008). During the past decade, advocates of the T cell or B cell involvement struggled against each other by providing clues from animal models and clinical observations as well as either T cell or B cell specific treatment efficacy.

For the defenders of the T cell dominance, implicating defects in T cell activation and tolerance, the presence of T cell tissue infiltrates, ectopic germinal center-like lymphoid formations, and the difficulty to precisely address the pathogenic role of auto-antibodies, constituted strong arguments. The efficacy of steroids and immuno-suppressants (that are widely used to treat SLE patients) was also interpreted as a hallmark of the T cell dominance in SLE pathophysiology, as these treatments suppress inflammation, but more specifically reduce T cell activation. According to this view, the minor contribution of the B cell component, only considered as a consequence of the T cell activation, consisted in the T cell help-mediated production of high-affinity, isotype-switched auto-antibodies leading to immune complex deposits and enhanced inflammation.

On the other hand, results obtained from animal models and clinical trials using B cell-depleting agents such as rituximab brought strong clues of a predominant involvement of the B cell component in SLE pathogenesis. Hypothesis of exclusive B cell related mechanisms has therefore emerged to explain certain forms of the disease. T cell-independent activation and maturation of auto-reactive B cells via triggering of TLRs and increased production of BAFF is the pivot of this pathophysiological model (Leadbetter et al. 2003; Groom et al. 2007).

The demonstration that IL-17 has a major impact on B cell biology might reconcile the two apparent opposite views aiming to explain the pathophysiology of SLE. Within this context, it is important to note that IL-17, alone or in synergy with BAFF, induces survival, proliferation, and differentiation of potentially auto-reactive transitional B cells in humans and leads to the production of high-affinity auto-reactive antibodies. Because of its pro-inflammatory properties, together with its major impact on various target-cells such as macrophages, dendritic cells, endothelial cells, fibroblasts, osteoblasts, chondrocytes and now B cells, IL-17 appears as forefront cytokine in the scenario driving auto-immune diseases (especially SLE) (Fig. 3). Clear demonstration of such a pivotal role in SLE pathogenesis will come from the use of inhibitors of IL-17 in clinical trials to precisely assess the impact on B cell biology.

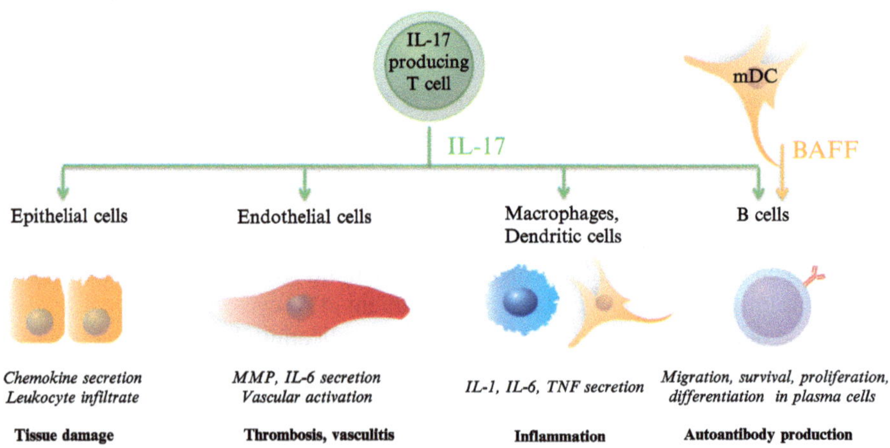

Fig. 3 Effects of IL-17 produced by T cells on main cellular components involved in SLE pathophysiology. For each cell type, biological effects in response to IL-17 stimulation and consequences in SLE pathophysiology are indicated

References

Boniface K, Blom B, Liu Y J, and de Waal Malefyt R (2008) From interleukin-23 to T-helper 17 cells: human T-helper cell differentiation revisited. Immunol Rev 226: 132–146

Cripps R M, Black B L, Zhao B et al. (1998) The myogenic regulatory gene Mef2 is a direct target for transcriptional activation by Twist during Drosophila myogenesis. Genes Dev 12: 422–434

Crispin J C, Oukka M, Bayliss G et al. (2008) Expanded double negative T cells in patients with systemic lupus erythematosus produce IL-17 and infiltrate the kidneys. J Immunol 181: 8761–8766

Doreau, A., Belot, A., Bastid, et al. (2009) Interleukin 17 acts in synergy with B cell-activating factor to influence B cell biology and the pathophysiology of systemic lupus erythematosus. Nat Immunol 10: 778–785

Ettinger R, Sims G P, Robbins et al. (2007) IL-21 and BAFF/BLyS synergize in stimulating plasma cell differentiation from a unique population of human splenic memory B cells. J Immunol 178: 2872–2882

Garrett-Sinha L A, John S, Gaffen S L (2008) IL-17 and the Th17 lineage in systemic lupus erythematosus. Curr Opin Rheumatol 20: 519–525

Groom J, Mackay F (2008) B cells flying solo. Immunol Cell Biol 86: 40–46

Groom J R, Fletcher C A, Walters S N et al. (2007) BAFF and MyD88 signals promote a lupus-like disease independent of T cells. J Exp Med 204: 1959–1971

Hsu H C, Yang P, Wang J et al. (2008) Interleukin 17-producing T helper cells and interleukin 17 orchestrate auto-reactive germinal center development in auto-immune BXD2 mice. Nat Immunol 9: 166–175

Ishigami T, Kim K M, Horiguchi Y et al. (1992) Anti-IgM antibody-induced cell death in a human B lymphoma cell line, B104, represents a novel programmed cell death. J Immunol 148: 360–368

Korn T, Oukka M, Kuchroo V, Bettelli E (2007) Th17 cells: effector T cells with inflammatory properties. Sem Immunol 19: 362–371

Kwan B C, Tam L S, Lai K B et al. (2009) The gene expression of type 17 T-helper cell-related cytokines in the urinary sediment of patients with systemic lupus erythematosus. Rheumatology 48: 1491–1497

Lai Kwan Lam Q, King Hung Ko O, Zheng B J et al. (2008) Local BAFF gene silencing suppresses Th17-cell generation and ameliorates auto-immune arthritis. Proc Natl Acad Sci USA 105: 14993–14998

Leadbetter E A, Rifkin I R, Marshak-Rothstein A (2003) Toll-like receptors and activation of autoreactive B cells. Curr Dir Autoimmun 6: 105–122

Liu C, Qian W, Qian Y et al. (2009) Act1, a U-box E3 ubiquitin ligase for IL-17 signaling. Sci Signal 2: ra63.

Mackay F, Browning J L (2002) BAFF: a fundamental survival factor for B cells. Nat Rev Immunol 2: 465–475

Maestro R, Dei Tos A P, Hamamori Y, et al. (1999) Twist is a potential oncogene that inhibits apoptosis. Genes Dev 13: 2207–2217

Miossec P, Korn T, Kuchroo V K (2009) Interleukin-17 and type 17 helper T cells. N Engl J Med 361: 888–898

Nalbandian A, Crispin J C, Tsokos G C (2009) Interleukin-17 and systemic lupus erythematosus: current concepts. Clin Exp Immunol 157: 209–215

Pan H F, Ye D Q, Li X P (2008) Type 17 T-helper cells might be a promising therapeutic target for systemic lupus erythematosus. Nat Clin Pract 4: 352–353

Pene J, Chevalier S, Preisser L et al. (2008) Chronically inflamed human tissues are infiltrated by highly differentiated Th17 lymphocytes. J Immunol 180: 7423–7430

Qian Y, Liu C, Hartupee J et al. (2007) The adaptor Act1 is required for interleukin 17-dependent signaling associated with auto-immune and inflammatory disease. Nat Immunol 8: 247–256

Qian Y, Qin J, Cui G et al. (2004) Act1, a negative regulator in CD40- and BAFF-mediated B cell survival. Immunity 21: 575–587

Ruprecht C R, Lanzavecchia A (2006) Toll-like receptor stimulation as a third signal required for activation of human naive B cells. Eur J Immunol 36: 810–816

Sims G P, Ettinger R, Shirota Y et al. (2005). Identification and characterization of circulating human transitional B cells. Blood 105: 4390–4398

Wang Y, Ito S, Chino Y et al. (2009) Laser microdissection-based analysis of cytokine balance in the kidneys of patients with lupus nephritis. Clin Exp Immunol 159: 1–10

Wardemann H, Nussenzweig M C (2007) B-cell self-tolerance in humans. Adv Immunol 95: 83–110

Wardemann H, Yurasov S, Schaefer A et al. (2003) Predominant autoantibody production by early human B cell precursors. Science 301: 1374–1377

Wilker P R, Kohyama M, Sandau M Met al. (2008) Transcription factor Mef2c is required for B cell proliferation and survival after antigen receptor stimulation. Nat Immunol 9: 603–612

Wong C K, Ho C Y, Li E K et al. (2000) Elevation of pro-inflammatory cytokine (IL-18, IL-17, IL-12) and Th2 cytokine (IL-4) concentrations in patients with systemic lupus erythematosus. Lupus 9: 589–593

Wong C K, Lit L C, Tam L S, et al. (2008) Hyperproduction of IL-23 and IL-17 in patients with systemic lupus erythematosus: implications for Th17-mediated inflammation in auto-immunity. Clin Immunol 127: 385–393

Yang J, Chu Y, Yang X et al. (2009) Th17 and natural Treg cell population dynamics in systemic lupus erythematosus. Arthritis Rheum 60: 1472–1483

Yen D, Cheung J, Scheerens H et al. (2006) IL-23 is essential for T cell-mediated colitis and promotes inflammation via IL-17 and IL-6. J Clin Invest 116: 1310–1316

Zhang L, Yang X Q, Cheng J, et al. (2010) Increased Th17 cells are accompanied by FoxP3(+) Treg cell accumulation and correlated with psoriasis disease severity. Clin Immunol 135: 108–17

IL-17 and T_H17 Cells in Human Rheumatoid Arthritis

Pierre Miossec

Abstract The field of IL-17 biology has vastly increased recently. In 2005–2006, the cell source of IL-17 named T_H17 cells was identified in mice and then in humans. The IL-17 protein had been discovered 10 years earlier as a T cell derived cytokine. These new findings on T_H17 cells have led to an update of the classification of T cells into effector T cell subsets.

The discovery of IL-17 was immediately linked to rheumatoid arthritis (RA). The first indication that T cells producing IL-17 could be a different subset was also first shown using T cell clones from RA synovial fluid and synovium. These findings have led to the concept of IL-17 targeting for treatment of various inflammatory conditions, including RA.

We will review these findings focusing on human RA as the prototypic chronic inflammatory disease where inflammation leads to matrix destruction. However, some of these observations also apply to other diseases and such concepts are leading to new therapeutic applications.

1 Identification of IL-17 with a First Link to Rheumatoid Arthritis

IL-17 was discovered under the name of CTLA-8 as a gene product without clear function. IL-17 was described in 1995/1996 as a pro-inflammatory cytokine produced by T cells. The key experiment was demonstration that the addition of IL-17 to fibroblast-like cells increased IL-6 and other pro-inflammatory cytokine production, immediately showing the link to inflammation (Fossiez et al. 1996; Yao et al. 1995a, b). The link to RA was also established because synoviocytes obtained

P. Miossec (✉)
Department of Immunology and Rheumatology, Hôspital Edouard Herriot,
69437 Lyon, Cedex 03, France
e-mail: miossec@univ-lyon1.fr

S. Jiang (ed.), *T_H17 Cells in Health and Disease*,
DOI 10.1007/978-1-4419-9371-7_22,

from the synovium membrane of RA patients were used in these first experiments. These results suggested a potential contribution of IL-17 to RA pathogenesis.

At the same time, IL-17 was shown to induce IL-8, the key chemokine to attract neutrophils (Fossiez et al. 1996). The effect of IL-17 on neutrophil induction and maturation indicated its role in the acute mechanisms of host defense. In addition, neutrophils are also present at the site of chronic inflammation. In the case of RA, high numbers of neutrophils are found in synovial fluid, but to a lesser degree in the synovium. Increased survival of neutrophils contributes to the chronic nature of RA (Turrel-Davin et al. 2010).

2 IL-17 and the IL-17 Family Members in Rheumatoid Arthritis

IL-17 now referred to as IL-17A, is the founding member of the IL-17 family, which includes IL-17A to F (Aggarwal and Gurney 2002). IL-17F shows a 5% sequence homology with IL-17A. IL-17A and F are produced as dimers, including a significant proportion of A/F heterodimers. IL-17F has similar effects as IL-17A, but usually to a lesser degree (Zrioual et al. 2008). However, when combined with TNF on RA synoviocytes, a synergistic effect is observed, almost as potent as the effect seen with IL-17A.

Staining of the RA synovium with antibodies against IL-17A and IL-17F indicated that the two cytokines were expressed locally, even with the indication of a more frequent expression of IL-17F (Zrioual et al. 2009). Using results from microarrays, over expression of both IL-17A and IL-17F was observed in RA whole blood when compared to control levels. Compared to IL-17A, IL-17F alone induced the expression of only 27 genes, compared to 165 genes for IL-17A in RA synoviocytes. No gene was increased by IL-17A and decreased by IL-17A, or the opposite. Of the 27 genes induced by IL-17F, no gene was identified to be specific of IL-17-F. However, these effects could still vary in vivo. In the mouse, IL-17F is not needed to initiate experimental encephalitis, because it acts as a pro-inflammatory cytokine in acute experimental colitis. IL-17F KO mice are protected, whereas those with IL-17A deficiency show increased colon damage (Yang et al. 2008).

3 IL-17 Receptors and Rheumatoid Arthritis

A receptor for IL-17 (IL-17R) was defined in 1995–1996 simultaneously as IL-17 was discovered. Its rather low affinity for IL-17 suggested the presence of additional chains (Yao et al. 1995a, b, 1997). Systematic sequence analysis identified other proteins with a partial homology with the IL-17R (Toy et al. 2006). Today, the first member of the IL-17R family is the original IL-R receptor renamed IL-17RA. The second important member is IL-17RC. The physical association of these two receptors has been shown, although it is still unclear whether these are two chains of a single receptor, or two different receptors in close association. It was previously proposed that IL-17A could be the receptor for IL-17A and IL-17RC the receptor for IL-17F. It appears now that IL-17RA and RC interact together for an optimal response, in particular when IL-17A is combined with TNF on RA synoviocytes. The inhibition of the two IL-17 receptors is needed to constrain the response of RA synoviocytes to the combination of IL-17A with TNF (Zrioual et al. 2008).

In order to explain the synergistic interactions between IL-17A and TNF, pre-incubation studies with RA synoviocytes indicated that IL-17A was acting first in order to increase the response to TNF (Zrioual et al. 2008). This is linked to the specific induction of the type II TNF receptor, leading to an increased sensitivity to TNF. There was no effect on the type I TNF receptor or on IL-17RA or IL-17RC, showing the specificity of the effect on the type II TNF receptor.

Using results from microarrays, overexpression of both IL-17RA and IL-17RC was observed in RA-whole blood and linked to RA disease activity (Zrioual et al. 2008). Their expression is rather ubiquitous and the two receptors are highly expressed by synoviocytes and synovium cells. Blood cells express variable levels of IL-17R, high in B cells and monocytes, but low in T cells. Using siRNA, down-regulation of either IL-17RA or IL-17RC in RA synoviocytes led to a near complete abrogation of IL-6 activation, in line with results in the mouse (Toy et al. 2006).

4 Contribution of IL-17 to Rheumatoid Arthritis

RA is characterized by chronic inflammation of the synovium membrane of joints. Local cell interactions induce pro-inflammatory cytokine production, which in turn activates the release of proteases leading to bone and cartilage destruction (Firestein 2003). Concordant results using mouse and human models of RA showed that IL-17 is involved in the pro-inflammatory patterns associated with joint inflammation (Koenders et al. 2006a, b, Hillyer et al. 2009; Melis et al. 2010) . In the mouse, a single injection of IL-17 into a normal knee is sufficient to induce cartilage damage (Chabaud et al. 2001). The continuous administration of IL-17 by gene over-expression into a normal knee induces massive damage with extensive inflammatory cell migration, bone erosions, and cartilage degradation (Lubberts et al. 2002). The results indicate that the local release of IL-17 could reproduce the key features of human RA. Conversely, inhibition with specific factors including blocking anti-IL-17A antibodies and IL-17RA-soluble receptors provided protection from inflammation and destruction (Lubberts et al. 2000). Addition of IL-4 through gene therapy acting more broadly on other cytokines could also inhibit IL-17 production and action and protect from bone destruction by reducing the key interactions between RANK and RANK ligands (Lubberts et al. 2000).

The link between IL-17 and human RA was reinforced when it was shown that pieces of RA synovium could produce bioactive IL-17. This activity was measured with a bioassay where supernatants of RA synovium pieces induced a massive production of IL-6 by RA synoviocytes (Chabaud et al. 1999). Such production was reduced by two thirds with a monoclonal antibody blocking human IL-17. This robust production of IL-6 using supernatants containing IL-17 was in contrast with the limited effect of recombinant IL-17 alone. The same findings were obtained with human bone explants obtained during RA wrist surgery. Here again, high levels of functional IL-17 were detected, indicating the role of IL-17-producing cells from bone marrow in juxta-articular bone destruction (Chabaud et al. 1999; Sato et al. 2006). This effect was associated with RANK-ligand expression by osteoblasts and synoviocytes interacting with RANK expressing cells, and not only osteoclasts but also mature dendritic cells (Page and Miossec 2005; Sato et al. 2006). In addition to synovium inflammation, IL-17 induces inflammation associated bone resorption (Sato et al. 2006) and contributes to similar conditions such as joint prosthesis loosening and periodontal disease (Oda et al. 2003; Andersson et al. 2007). Expression of IL-17 in RA synovium biopsies was associated with high disease activity and severity, with a link between higher IL-17 expression and increased joint damage (Kirkham et al. 2006).

Similar findings were obtained in juvenile idiopathic arthritis with high levels of IL-17 in synovial fluid (Nistala et al. 2008). This induces production of other cytokines and enzymes by synoviocytes (Agarwal et al. 2008). In the same condition, an enrichment of IL-17 producing cells was observed with an opposite relationship with the number of regulatory T cells (Nistala et al. 2008).

The role of IL-17 in disease chronicity was confirmed by showing that in the mouse collagen arthritis model, the IL-17 effect was dependent on the presence of TNF at the early phase, whereas at a later stage, the disease was mostly IL-17 driven, and not TNF dependent anymore (Koenders et al. 2006a, b). In RA synoviocytes, IL-17 may contribute to chronicity by reducing cell sensitivity to apoptotic signals. This was extended in an arthritis model in mice lacking the IL-17RA receptor, which had reduced disease activity and synovium infiltrate, but enhanced apoptotic activity.

5 T_H17 Pathway in Rheumatoid Arthritis

5.1 T_H17 Cells and Rheumatoid Arthritis

The most recent discovery was the identification of the cell source of IL-17 (Harrington et al. 2006; Miossec 2007). This led a major change in the classification of T cells into T_H1 and T_H2 cells with the addition of the T_H17 cells. However, early studies had already shown in 1999 that T cells from RA synovium and synovium fluid could produce functional IL-17 alone or in association with IFNγ, but not with IL-4 (Aarvak et al. 1999). These results were extended by the demonstration of the frequent co-expression of IFNγ and IL-17 in blood cells. However, in situ immuno-staining of the RA synovium showed two isolated populations of T cells producing either IFNγ or IL-17 (Page et al. 2004). Double positive cells were rarely seen.

It remains to be shown whether these cells with such a secreting pattern are still able to further differentiate or have reached a final stage. Of interest in this context, was the demonstration that cytokine-secreting T cells have a particular morphology with a plasma cell appearance, a pattern that can be induced in vitro and is associated with the loss of TCR and CD, but not CD4 (Page and Miossec 2004). As for B cells, the plasma cell morphology of the IFNγ and IL-17-producing cells strongly suggests that this is a fixed pattern related to a final stage of differentiation.

In addition, T_H17 cells are involved in cell interactions through the expression of RANKL. Such RANK-RANKL interaction is the final bridge whereby osteoblasts activate osteoclasts and lead to bone destruction. Similar interactions are found between fibroblasts and dendritic cells (Page and Miossec 2005).

Although the production of IL-17A has been linked with the T_H17 subset, other T cell subsets have been shown to produce IL-17, such as γδ T cells and NK T cells. Cells from a different lineage could also contribute. In the context of RA synovium, mast cells have been shown to express IL-17 (Hueber et al. 2010). The same may also apply to neutrophils. Their respective contribution to the total amount of IL-17 present locally in the RA joint remains to be clarified.

5.2 Other T_H17-Derived Cytokines and Rheumatoid Arthritis

T_H17 cells have been described for the production of IL-17A, but other cytokines are produced by T_H17 cells, including IL-17F, IL-22, IL-21, TNF, and IL-6.

IL-22 is a member of the IL-10 family that synergizes with IL-17A or IL-17F to regulate genes associated with innate skin immunity (Liang et al. 2006). In particular, IL-22 was shown to mediate dermal inflammation (Zheng et al. 2007). As such, IL-22 has been linked to the pathogenesis of psoriasis. Skin biopsies from patients with psoriasis show a high expression of IL-17, as well as IL-23 and IL-22 (Zaba et al. 2007). Increased frequency of IL-22 positive T cells was also seen in RA blood (Colin et al. 2010). In an animal model, IL-23 was found to promote T_H17 differentiation and IL-22 production (Mus et al. 2010).

IL-21 is a key cytokine for the amplification of the T_H17 pathway. IL-21 induces IL-17 production through a Stat3 pathway. Since IL-17 is produced at high levels by T_H17 cells, this involves autocrine amplification. IL-21 expression correlates with the presence of T_H17 cells in synovial fluid and blood (Niu et al. 2010).

5.3 Governing Effects of Effector and Regulatory T Cell Subsets

The inhibitory effect of the key T_H2 cytokine IL-4 on IL-17 function and production was established quickly when IL-17 was discovered. This was shown in both mouse and human models. For instance, treatment of mouse arthritis with prolonged expression of IL-4 was able to control disease activity and destruction with an inhibitory effect on IL-17 production (Lubberts et al. 2000). The inhibitory effect of IFNγ on the T_H17 pathway was established early in the mouse. The reverse was recently shown in human cells. In fact, IL-17 inhibits the T_H1 pathway through a specific means on the IL-12 receptor-specific β2 chain, with no effect on the common β1 chain and the IL-23 specific chain (Toh et al. 2009). IL-17 can hinder the T_H1 pathway by curbing the response to IL-12, leading to reduced IFNγ production. These results have been extended to the RA model. Reduced expression and production of IFNγ was observed in RA PBMC compared to controls. In addition, RA cells showed a very sensitive effect to the inhibitory of IL-17. IL-17 contributes to the systemic T_H1 defect observed in RA and linked to disease activity. This may explain the increased sensitivity of RA patients to infections with intra-cellular pathogens, namely tuberculosis.

Whereas the combination of IL-6 and TGFβ favors the T_H17 pathway, TGFβ alone favors the T cell regulatory pathway (Bettelli et al. 2006, 2008). Since IL-6 is highly produced during RA inflammation, this acts through a positive feedback to favor persistence of inflammation. IL-6 itself inhibits the induction of Foxp3 by TGFβ. This balance has been well studied in various mouse models of chronic inflammation and extended to the human situation (Bettelli et al. 2006; Ogura et al. 2008).

6 IL-17 Targeting for Treatment of Rheumatoid Arthritis

Cytokine targeting in RA has shown major efficacy with a TNF-soluble receptor and monoclonal antibodies against TNF and the IL-6 receptor. Some of the tools based on similar concepts are now available to target the IL-17 pathway.

6.1 IL-17 Specific Inhibitors

The two major options are targeting of the ligand or the receptor (Dong 2008). Tools ready for human applications include monoclonal antibodies against IL-17A and IL-17RA. Additional targets from recent research on the IL-17 and IL-17R families could include IL-17A, IL-17F or both, and for the receptor, IL-17RA, IL-17RC, or both.

The first positive results from a clinical trial were when anti-IL-17 antibodies were obtained in patients with psoriasis. A single IV injection was able to induce prolonged control of skin lesions. Two phase II clinical trials have been performed in RA. Both indicated an effect on signs and symptoms with no obvious adverse events (Genovese et al. 2010). Additional trials in RA and other inflammatory conditions are starting.

6.2 Use of Non-IL-17 Anti-cytokine Inhibitors

Some inhibitors can act on pathways upstream of IL-17 and/or on other cytokines in addition to IL-17. Since IL-23 is implicated in T_H17 differentiation and IL-17 production, the inhibition of IL-23 is a way to control the T_H17 pathway by acting upstream. Treatment with a monoclonal antibody against p40, which is common to IL-12 and IL-23, has shown efficacy in psoriasis and Crohn's disease, but not clearly in RA (Mannon et al. 2004; Krueger et al. 2007). Specific inhibitors of IL-23 remain to be tested and compared with the p40 inhibitors. Similarly, specific inhibitors of IL-21 and IL-22 will have to be compared with the IL-17 inhibitors.

Some cytokines have broad anti-IL-17 properties. IL-4 inhibits the production and functions of IL-17 as well as TNF, IL-1, and IL-6 (Lubberts et al. 2000; Sarkar et al. 2006). IL-25/IL-17E is a member of the IL-17 family (Kleinschek et al. 2007), with anti-inflammatory properties on T_H17 cells (Owyang et al. 2006). All these possibilities remain to be tested.

6.3 Acting on Regulatory Pathways

New approaches could use the effects of T_H17 cells on the function of regulatory T cells (Bettelli et al. 2007). IL-17, IL-6, and IL-1 not only promote inflammation, but also inhibit regulatory T cell functions. In RA patients, TNFα inhibitors may control the disease by inhibiting TNFα and by correcting part of the defective regulatory T cell functions (Nadkarni et al. 2007). Because of the key role of IL-6 in inducing T_H17 cells, another option is targeting the IL-6 receptor with a monoclonal antibody (Yokota et al. 2005). Prolonged control of inflammation through reduction of IL-6 may allow restoration of the regulatory properties of T cells.

6.4 Limitations of IL-17 Targeting

Targeting cytokines may interfere with immune defense. Inhibition of TNF was associated with an increased risk of tuberculosis reactivation. For IL-17, the link with neutrophil biology was apparent from the first results. IL-17 appears to be critical for neutrophil activation and migration (Chung et al. 2003). IL-17 is a strong inducer of IL-8, a key chemokine for neutrophils. Thus, inhibition of IL-17 increased the frequency and severity of extra-cellular bacterial infections and fungi. In the mouse, inhibition of IL-17 has been associated with increased mortality from bacterial lung infections (Dubin and Kolls 2007).

7 Conclusion

The story of IL-17 and of the T_H17 subset has been associated from the very beginning with RA. Of importance, is the strong interaction between IL-17 and TNF leading to enhanced destruction and long-lasting chronicity. Some of these effects could apply to other inflammatory arthritis such as spondylarthropathies and psoriasis arthritis. Clinical trials with IL-17 inhibitors are now including these conditions (Miossec et al. 2009).

References

Aarvak T, Chabaud M, Miossec P, et al. (1999) IL-17 is produced by some pro-inflammatory Th1/Th0 cells but not by Th2 cells. J Immunol 162:1246–1251.

Agarwal S, Misra R and Aggarwal A (2008) Interleukin 17 levels are increased in juvenile idiopathic arthritis synovial fluid and induce synovial fibroblasts to produce pro-inflammatory cytokines and matrix metalloproteinases. J Rheumatol 35:515–519.

Aggarwal S and Gurney AL (2002) IL-17: prototype member of an emerging cytokine family. J Leukoc Biol 71:1–8.

Andersson MK, Lundberg P, Ohlin A, et al. (2007) Effects on osteoclast and osteoblast activities in cultured mouse calvarial bones by synovial fluids from patients with a loose joint prosthesis and from osteoarthritis patients. Arthritis Res Ther 9:R18.

Bettelli E, Carrier Y, Gao W, et al. (2006) Reciprocal developmental pathways for the generation of pathogenic effector TH17 and regulatory T cells. Nature 441:235–238.

Bettelli E, Korn T, Oukka M, et al. (2008) Induction and effector functions of T(H)17 cells. Nature 453:1051–1057.

Bettelli E, Oukka M and Kuchroo VK (2007) T(H)-17 cells in the circle of immunity and autoimmunity. Nat Immunol 8:345–350.

Chabaud M, Durand JM, Buchs N, et al. (1999) Human interleukin-17: A T cell-derived proinflammatory cytokine produced by the rheumatoid synovium. Arthritis Rheum 42:963–970.

Chabaud M, Lubberts E, Joosten L, et al. (2001) IL-17 derived from juxta-articular bone and synovium contributes to joint degradation in rheumatoid arthritis. Arthritis Res 3:168–177.

Chung DR, Kasper DL, Panzo RJ, et al. (2003) CD4+ T cells mediate abscess formation in intra-abdominal sepsis by an IL-17-dependent mechanism. J Immunol 170:1958–1963.

Colin EM, Asmawidjaja PS, van Hamburg JP, et al. (2010) 1,25-dihydroxyvitamin D3 modulates Th17 polarization and interleukin-22 expression by memory T cells from patients with early rheumatoid arthritis. Arthritis Rheum 62:132–142.

Dong C (2008) IL-23/IL-17 biology and therapeutic considerations. J Immunotoxicol 5:43–46.

Dubin PJ and Kolls JK (2007) IL-23 mediates inflammatory responses to mucoid Pseudomonas aeruginosa lung infection in mice. Am J Physiol Lung Cell Mol Physiol 292:L519-528.

Firestein GS (2003) Evolving concepts of rheumatoid arthritis. Nature 423:356–361.

Fossiez F, Djossou O, Chomarat P, et al. (1996) T cell interleukin-17 induces stromal cells to produce pro-inflammatory and hematopoietic cytokines. J Exp Med 183:2593–2603.

Genovese MC, Van den Bosch F, Roberson SA, et al. (2010) LY2439821, a humanized anti-interleukin-17 monoclonal antibody, in the treatment of patients with rheumatoid arthritis: A phase I randomized, double-blind, placebo-controlled, proof-of-concept study. Arthritis Rheum 62:929–939.

Harrington LE, Mangan PR and Weaver CT (2006) Expanding the effector CD4 T-cell repertoire: the Th17 lineage. Curr Opin Immunol 18:349–356.

Hillyer P, Larche MJ, Bowman EP, et al. (2009) Investigating the role of the interleukin-23/-17A axis in rheumatoid arthritis. Rheumatology (Oxford) 48:1581–1589.

Hueber AJ, Asquith DL, Miller AM, et al. (2010) Mast cells express IL-17A in rheumatoid arthritis synovium. J Immunol 184:3336–3340.

Kirkham BW, Lassere MN, Edmonds JP, et al. (2006) Synovial membrane cytokine expression is predictive of joint damage progression in rheumatoid arthritis: a two-year prospective study (the DAMAGE study cohort). Arthritis Rheum 54:1122–1131.

Kleinschek MA, Owyang AM, Joyce-Shaikh B, et al. (2007) IL-25 regulates Th17 function in auto-immune inflammation. J Exp Med 204:161–170.

Koenders MI, Joosten LA, van den Berg WB (2006a) Potential new targets in arthritis therapy: interleukin (IL)-17 and its relation to tumor necrosis factor and IL-1 in experimental arthritis. Ann Rheum Dis 65 Suppl 3:iii29-iii33.

Koenders MI, Lubberts E, van de Loo FA, et al. (2006b) Interleukin-17 acts independently of TNF-alpha under arthritic conditions. J Immunol 176:6262–6269.

Krueger GG, Langley RG, Leonardi C, et al. (2007) A human interleukin-12/23 monoclonal antibody for the treatment of psoriasis. N Engl J Med 356:580–592.

Liang SC, Tan XY, Luxenberg DP, et al. (2006) Interleukin (IL)-22 and IL-17 are co-expressed by Th17 cells and cooperatively enhance expression of antimicrobial peptides. J Exp Med 203:2271–2279.

Lubberts E, Joosten LA, Chabaud M, et al. (2000) IL-4 gene therapy for collagen arthritis suppresses synovial IL-17 and osteoprotegerin ligand and prevents bone erosion. J Clin Invest 105:1697–1710.

Lubberts E, Joosten LA, van de Loo FA, et al. (2002) Overexpression of IL-17 in the knee joint of collagen type II immunized mice promotes collagen arthritis and aggravates joint destruction. Inflamm Res 51:102–104.

Mannon PJ, Fuss IJ, Mayer L, et al. (2004) Anti-interleukin-12 antibody for active Crohn's disease. N Engl J Med 351:2069–2079.

Melis L, Vandooren B, Kruithof E, et al. (2010) Systemic levels of IL-23 are strongly associated with disease activity in rheumatoid arthritis but not spondyloarthritis. Ann Rheum Dis 69:618–623.

Miossec P (2007) Interleukin-17 in fashion, at last: ten years after its description, its cellular source has been identified. Arthritis Rheum 56:2111–2115.

Miossec P, Korn T and Kuchroo VK (2009) Interleukin-17 and type 17 helper T cells. N Engl J Med 361:888–898.

Mus AM, Cornelissen F, Asmawidjaja PS, et al. (2010) Interleukin-23 promotes Th17 differentiation by inhibiting T-bet and FoxP3 and is required for elevation of interleukin-22, but not interleukin-21, in auto-immune experimental arthritis. Arthritis Rheum 62:1043–1050.

Nadkarni S, Mauri C and Ehrenstein MR (2007) Anti-TNF-alpha therapy induces a distinct regulatory T cell population in patients with rheumatoid arthritis via TGF-beta. J Exp Med 204:33–39.

Nistala K, Moncrieffe H, Newton KR, et al. (2008) Interleukin-17-producing T cells are enriched in the joints of children with arthritis, but have a reciprocal relationship to regulatory T cell numbers. Arthritis Rheum 58:875–887.

Niu X, He D, Zhang X, et al. (2010) IL-21 regulates Th17 cells in rheumatoid arthritis. Hum Immunol 71:334–341.

Oda T, Yoshie H and Yamazaki K (2003) Porphyromonas gingivalis antigen preferentially stimulates T cells to express IL-17 but not receptor activator of NF-kappaB ligand in vitro. Oral Microbiol Immunol 18:30–36.

Ogura H, Murakami M, Okuyama Y, et al. (2008) Interleukin-17 promotes autoimmunity by triggering a positive-feedback loop via interleukin-6 induction. Immunity 29:628–636.

Owyang AM, Zaph C, Wilson EH, et al. (2006) Interleukin 25 regulates type 2 cytokine-dependent immunity and limits chronic inflammation in the gastrointestinal tract. J Exp Med 203:843–849.

Page G and Miossec P (2004) Paired synovium and lymph nodes from rheumatoid arthritis patients differ in dendritic cell and chemokine expression. J Pathol 204:28–38.
Page G and Miossec P (2005) RANK and RANKL expression as markers of dendritic cell-T cell interactions in paired samples of rheumatoid synovium and lymph nodes. Arthritis Rheum 52:2307–2312.
Page G, Sattler A, Kersten S, et al. (2004) Plasma cell-like morphology of Th1-cytokine-producing cells associated with the loss of CD3 expression. Am J Pathol 164:409–417.
Sarkar S, Tesmer LA, Hindnavis V, et al. (2006) Interleukin-17 as a molecular target in immune-mediated arthritis: Immuno-regulatory properties of genetically modified murine dendritic cells that secrete interleukin-4. Arthritis Rheum 56:89–100.
Sato K, Suematsu A, Okamoto K, et al. (2006) Th17 functions as an osteoclastogenic helper T cell subset that links T cell activation and bone destruction. J Exp Med 203:2673–2682.
Toh ML, Kawashima M, Zrioual S, et al. (2009) IL-17 inhibits human Th1 differentiation through IL-12Rbeta2 downregulation. Cytokine 2009 48:226–30.
Toy D, Kugler D, Wolfson M, et al. (2006) Cutting Edge: Interleukin 17 Signals through a Heteromeric Receptor Complex. J Immunol 177:36–39.
Turrel-Davin F, Tournadre A, Pachot A, et al. (2010) FoxO3a involved in neutrophil and T cell survival is overexpressed in rheumatoid blood and synovial tissue. Ann Rheum Dis 69:755–760.
Yang XO, Chang SH, Park H, et al. (2008) Regulation of inflammatory responses by IL-17F. J Exp Med 205:1063–1075.
Yao Z, Fanslow WC, Seldin MF, et al. (1995a) Herpesvirus Saimiri encodes a new cytokine, IL-17, which binds to a novel cytokine receptor. Immunity 3:811–821.
Yao Z, Painter SL, Fanslow WC, et al. (1995b) Human IL-17: a novel cytokine derived from T cells. J Immunol 155:5483–5486.
Yao Z, Spriggs MK, Derry JM, et al. (1997) Molecular characterization of the human interleukin (IL)-17 receptor. Cytokine 9:794–800.
Yokota S, Miyamae T, Imagawa T, et al. (2005) Therapeutic efficacy of humanized recombinant anti-interleukin-6 receptor antibody in children with systemic-onset juvenile idiopathic arthritis. Arthritis Rheum 52:818–825.
Zaba LC, Cardinale I, Gilleaudeau P, et al. (2007) Amelioration of epidermal hyperplasia by TNF inhibition is associated with reduced Th17 responses. J Exp Med *204*, 3183–3194.
Zheng Y, Danilenko DM, Valdez P, et al. (2007) Interleukin-22, a T(H)17 cytokine, mediates IL-23-induced dermal inflammation and acanthosis. Nature 445:648–651.
Zrioual S, Ecochard R, Tournadre A, et al. (2009) Genome-wide comparison between IL-17A- and IL-17F-induced effects in human rheumatoid arthritis synoviocytes. J Immunol 182:3112–3120.
Zrioual S, Toh ML, Tournadre A, et al. (2008) IL-17RA and IL-17RC Receptors Are Essential for IL-17A-Induced ELR+ CXC Chemokine Expression in Synoviocytes and Are Overexpressed in Rheumatoid Blood. J Immunol 180:655–663.

The Role of the IL-23/T_H17 Immune Pathway in the Pathogenesis of Arthritis

Erik Lubberts

Abstract Studies with IL-23-deficient mice revealed the critical role of IL-23 and in particular IL-17-producing cells, including T_H17 cells, in the initiation of experimental auto-immune arthritis. Interestingly, the recognition of T_H17 cells and T_H17 cytokines at the very early stage of rheumatoid arthritis (RA) fits with the concept that the IL-23/T_H17 immune pathway is important in the development of this auto-immune disease. However, the exact biological role of this immune pathway in the development of RA and other arthropathies needs further examination. The first data from clinical trials using anti-IL-17A antibody treatment in psoriatic arthritis and RA is promising. Whether the regulation of T_H17 cell activity or specific combinations of T_H17 cytokines will have additional value compared to neutralizing IL-17A activity or TNF needs to be elucidated. However, first evidence from in vitro studies showed that the inhibition of the T_H17 pathway in addition to anti-TNF suppressed the pro-inflammatory feedback loop between T_H17 cells and synovial fibroblasts of patients with early RA. This approach may help to reach the ultimate goal of permanent remission or even prevent the development of this crippling disease. In addition, further understanding of the plasticity of T cell subsets in the pathogenesis of chronic destructive arthritis especially at different stages of the disease will be essential to understanding the T cell biology in RA. Moreover, understanding T cell plasticity due to different therapies may further improve treatment of patients with chronic destructive arthritis.

E. Lubberts (✉)
Department of Rheumatology, Erasmus MC, University Medical Center,
's-Gravendijkwal 230, 3015 CE, Rotterdam, The Netherlands
and
Department of Immunology, Erasmus MC, University Medical Center,
's-Gravendijkwal 230, 3015 CE, Rotterdam, The Netherlands
e-mail: E.Lubberts@erasmusmc.nl

S. Jiang (ed.), *T_H17 Cells in Health and Disease*,
DOI 10.1007/978-1-4419-9371-7_23,

1 Introduction

Rheumatoid arthritis (RA) is an auto-immune disease characterized by chronic inflammation of the joint synovium by activated inflammatory leukocytes (such as T helper lymphocytes, monocytes), synovial hyperplasia, neo-angiogenesis, and progressive destruction of cartilage and bone. This disease affects 1–2% of the population worldwide, most commonly middle-aged women. The etiology of RA is unknown but pro-inflammatory cytokines play a central role. Regulation of the cytokine imbalance might represent a solid way to control this disease. The pro-inflammatory cytokines TNFalpha and IL-1 play a crucial role in the pathogenesis of arthritis, driving enhanced production of cytokines, chemokines, and degradative enzymes (Arend and Dayer 1995). Clinical studies revealed efficacy after blocking TNFalpha. However, a subset of patients does not respond to anti-TNF and this treatment does not cure the disease. Therefore, cytokines or factors other than IL-1 and TNFalpha may also participate in the pro-inflammatory cytokine cascade.

T cells represent a large proportion of the inflammatory cells invading the synovial tissue. T cell activation and migration into the synovium occurs as an early consequence of disease, and these cells adopt a pro-inflammatory phenotype. Considerable evidence now supports a role for T cells in the initiation and perpetuation of the chronic inflammation prevalent in RA. Interestingly, the vast majority of these are (memory) T cells producing IL-17 that is up-regulated in early disease and is thought to contribute to the inflammation associated with RA (Chabaud et al. 1999; Raza et al. 2005).

Although the levels of IL-17A in sera of RA patients is hard to detect, elevated levels of this T cell cytokine have been demonstrated in synovial fluid of these patients (Chabaud et al. 1999; Kotake et al. 1999; Ziolkowska et al. 2000). Similar to the human situation, in the auto-immune collagen-induced arthritis (CIA) murine model, it is hard to detect IL-17A levels in the serum of the arthritic mice. In contrast, elevated levels of IL-17A have been found in inflamed synovium (Lubberts et al. 2001). IL-17A blocking experiments have shown the importance of this T cell cytokine as pro-inflammatory in the pathogenesis of CIA (Lubberts et al. 2001, 2004; Bush et al. 2002).

IL-17A induces the production of pro-inflammatory mediators, such as IL-1 and TNFalpha from several joint cells including synovial fibroblast, macrophages, and chondrocytes (Yao et al. 1995a, b; Fossiez et al. 1996; Jovanovic et al. 1998) (Fig. 1). In addition, IL-17A induces receptor activator of NF-kB ligand (RANKL) expression that is crucial for osteoclastogenesis and bone resorption (Fig. 1). IL-17A can synergize with these cytokines (IL-1, TNF, RANKL), but has direct activity as well (Miossec 2003; Lubberts et al. 2005a, b). Furthermore, IL-17A is able to promote cartilage destruction and bone erosion in experimental arthritis (Lubberts et al. 2001, 2002). When IL-17A is combined with other cytokines already thought to be important in arthritic disease, even more marked tissue destruction occurs (Chabaud and Miossec 2001, Lubberts et al. Unpublished observations). Moreover, IL-17 receptor signaling has been identified as a critical pathway in turning an acute

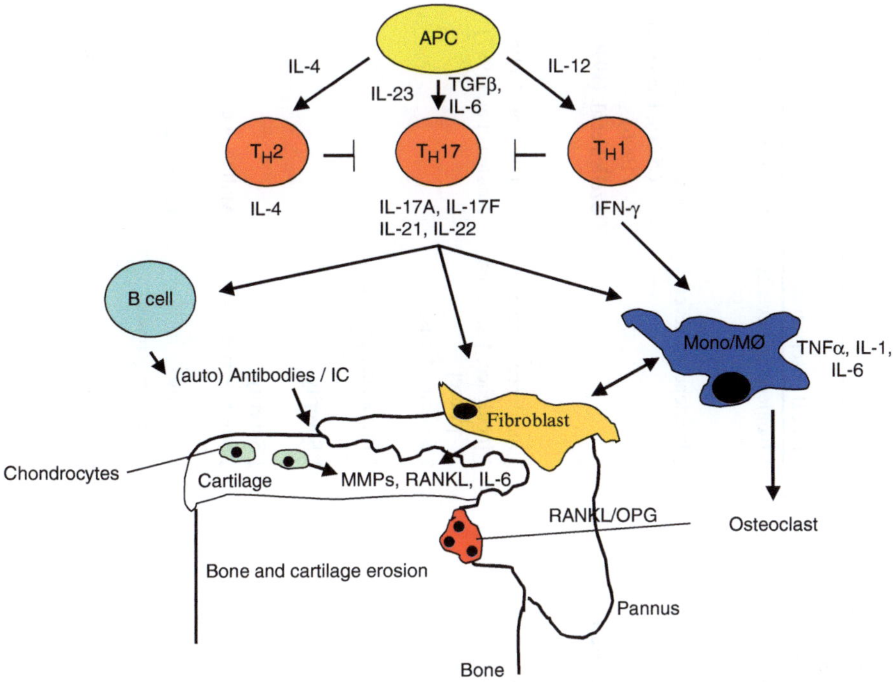

Fig. 1 Schematic overview of the role of IL-17/T_H17 in relation with other key cytokines and the cellular pathways of synovitis and concomitant joint destruction. (Adapted from Lubberts 2008)

synovitis into a chronic destructive arthritis (Lubberts et al. 2005a, b). These observations strongly implicate IL-17A as an important mediator of arthritis. The identification of six IL-17 family members (IL-17A-F) may extend the role of this novel cytokine family in the pathogenesis of chronic destructive joint inflammation (Lubberts 2003).

2 The Role of IL-23 in the Pathogenesis of Arthritis

2.1 The Discovery of IL-23 and the Importance of This Cytokine in Regulating CD4+ IL-17A+ T Cells in Experimental Arthritis

In early studies, the role of IL-12 in experimental disease models was often studied by using IL-12p40-knock-out mice, or neutralizing antibodies raised against p40. At present, it is known that such approaches not only target IL-12 but also IL-23. Table 1 summarizes relevant studies in which the role of IL-12 or IL-23 in different

Table 1 Overview of the in vivo role of IL-12 and IL-23 cytokines in different models of arthritis

Model	Treatment/target	Phase of disease	Effect (–, diminished) (+, exacerbation)	Clinical observation	References
CIA, DBA/1	recIL-12 (50–1,000 ng)	With immunization (CII/IFA) and with booster (CII/PBS)	+++	↑ Incidence, ↑ severity	Germann et al. (1995)
CIA, DBA/1	recIL-12 (1 μg)	With immunization (CII/CFA) and with booster (CII/PBS)	−−	↓ Incidence, ↓ severity	Oppmann et al. (2000), Hess et al. (1996)
CIA, C57BL/6 or B10.Q	recIL-12 (40–1,000 ng)	With immunization (CII/IFA) and with booster (CII/PBS)	No effect	–	Szeliga et al. (1996)
CIA, DBA/1	IL-12p40$^{-/-}$	Knock-out	−−	↓ Incidence, ↓ severity	McIntyre et al. (1996)
CIA, DBA/1	recIL-12 (100 ng)	Before onset (days 28–32)	++	↑ Incidence, ↑ severity, ↑ onset	Joosten et al. (1997)
CIA, DBA/1	LPS/anti-IL-12 Ab (polyclonal, 200 μg)	Before onset (days 28, 30, 32)	−−	↑ Severity, ↑ onset	
CIA, Bom	Anti-IL-12 Ab (as above)	Before onset (days 28, 30, 32)	−−	↓ Incidence, ↓ onset	
CIA, Bom	Anti-IL-12 Ab (as above)	Around onset (days 35, 37, 39)	++	↑ Severity	
CIA, Bom	recIL-12 (100 ng)	Around onset (d35-41)	−−	↓ Severity	
CIA, DBA/1	Anti-IL-12p40 Ab (10F6, 500 μg)	From immunization to onset (twice a week)	−−	↓ Severity, = incidence	Malfait et al. (1998)
CIA, DBA/1	AdIL-12p70	Before onset (day 25)	+	↑ Severity	Parks et al. (1998)
	Anti-IL-12 Ab (antiserum)	Before onset (days 26, 28, 30, 32)	–	↓ Severity, ↓ onset	
CIA, DBA/1	Anti-IL-12p40 Ab in WT	After immunization (days 0, 7, 14)	−−−	No onset	Matthys et al. (1998)
	Anti-IL-12p40 Ab in IFN-γ$^{-/-}$	After immunization (days 0, 7, 14)	−−−	No onset	

	Anti-IL-12p40 Ab in IFN-$\gamma^{-/-}$ (C17.8, 200 µg)	After onset (days 24, 31, 28)	---	↓ Severity	
CIA, DBA/1	recIL-12 (5 ng)	Around onset (days 24–33)	++	↑ Severity	Kasama et al. (1999)
	recIL-12 (500 ng)	Around onset (days 24–33)	0/–	↓ Severity, ↓ onset	
CIA, DBA/1	CII-specific $CD4^+$ T cells transduced with IL-12p40	Before booster (d20)	---	↓ Severity	Nakajima et al. (2001)
CIA, C57BL/6	IL-12p35$^{-/-}$	Knock-out	No effect	Similar to WT	Murphy et al. (2003)
	IL-12p40$^{-/-}$		---	No onset	
	IL-23p19$^{-/-}$		---	No onset	
CIA (rats)	Anti-IL-23 Ab (Polyclonal, Santa Cruz)	After onset (days 14, 16, 18, 20, 22)	–	↓ Severity	Yago et al. (2007)
IL-1Ra$^{-/-}$	AdIL-23p40p19	At 8 weeks of age	+	Arthritis acceleration, ↑ pannus formation	Ju et al. (2008)
PG (acute)	IL-23p19$^{-/-}$	Knock-out	–	↓ Joint swelling	Cornelissen and Lubberts (Unpublished)
AIA (primary)	IL-23p19$^{-/-}$	Knock-out	--	↓ Joint swelling, inflammation	Cornelissen et al. (2009a, b)
AIA (flare)	IL-23p19$^{-/-}$	Knock-out	--	↓ Joint swelling	Cornelissen and Lubberts (Unpublished)
AIA (flare)	Anti-IL-23p19 Ab	Before flare-induction	--	↓ Joint swelling	Cornelissen and Lubberts (Unpublished)

Ab Antibody, *Ad* Adenovirus (transduced with specified gene), *AIA* Antigen-Induced Arthritis (methylated BSA), *CII* Collagen type 2, *CFA* Complete Freund's adjuvant, *CIA* Collagen-induced arthritis, *IFA* Incomplete Freund's adjuvant, *LPS* Lipopolysaccharide, *PBS* Phosphate-buffered saline, *PG* Peptidoglycan, *Rec* recombinant. (Adapted from Cornelissen et al. 2009a, b)

experimental arthritis models was investigated. In the auto-immune collagen-induced arthritis (CIA) model, the effect of targeting p40 resulted in a clear reduction in disease score, regardless at which stage of disease stage the treatment was started.

This indicates that IL-12 and/or IL-23 have a prominent role at multiple stages of disease. From these studies it is not clear whether IL-12 or IL-23 act independently or if they act synergistically at different stages of disease. In CIA, targeting IL-12p35 specifically did not ameliorate the disease (Murphy et al. 2003). On the other hand, IL-12 administrated around the time of CIA onset exacerbated collagen arthritis (Joosten et al. 1997). Overall, the exact role of IL-12 in CIA is still not fully understood, mostly since there are no studies wherein IL-12p35 is therapeutically targeted. In CIA, it is clear that the IL-23/T_H17 axis is critical for the development of auto-immune arthritis. Mice that were deficient for IL-23 were completely protected against the development of CIA. Interestingly, these IL-23 deficient mice had no IL-17+ CD4 T cells, but did have normal antigen-specific IFN-gamma producing T_H1 cells (Murphy et al. 2003). This data shows that IL-17-producing CD4+ T cells, but not IFN-gamma-producing CD4+ T cells are critical for the initiation of CIA. We found that in the T cell-dependent antigen-induced monoarthritis (AIA) model that although lack of IL-23 did not prevent the early onset of AIA, these mice were fully protected against progression of joint inflammation into a destructive synovitis (Cornelissen et al. 2009a, b). Lower proportions of IL-17-positive TCR gammadelta and IL-17-positive CD4+ T cells were noted in the spleen, as well as LN and inflamed synovium. Interestingly, although monoarthritis induction in IL-23 deficient mice did not result in a lower proportion of IFN-gamma-positive CD4+ T cells in the spleen and lymph node, a lower proportion of these cells was detected in the target organ, the joint (Cornelissen et al. 2009a, b). IL-23 could be required to promote T_H1 and T_H17 effector responses, especially at the target organ/tissue of inflammation, and further studies are needed to understand this phenomenon. Additionally, in an adjuvant-free acute arthritis model induced with peptidoglycan, IL-23p19$^{-/-}$ mice developed significantly milder arthritis (Cornelissen and Lubberts Unpublished observations). This information suggests that IL-23 plays an essential role in the development of arthritis caused by different etiological mechanisms.

2.2 *IL-23 Regulates IL-17A and IL-17F Expression in TCRgammadelta T Cells in Experimental Arthritis*

Next to CD4$^+$ TCRαβ T cells, CD4$^-$ TCRγδ (γδ) T cells are able to produce pro-inflammatory cytokines, like IL-17 and IFN-γ. It has been shown that IL-17 production by γδ T cells can be induced by IL-23 (Stark et al. 2005; Shibata et al. 2007). The production of IL-17 by γδ T cells might be an important mechanism during the early inflammatory response. Evidence for the role of γδ T cells in the first line of defense was provided by the finding that in the lung and skin, γδ T cells are the largest RORγt$^+$ population, producing high levels of IL-17 (Lochner et al. 2008).

Although γδ T cells are able to produce IL-17, their role in auto-immune diseases is not well understood. Increased numbers of γδ T cells are found in the RA synovium (Holoshitz et al. 1989; Kjeldsen-Kragh et al. 1990). The role of γδ T cells in experimental arthritis has been investigated in a number of studies and their opposed role in arthritis etiology and pathogenesis differs between these reports. The role of γδ T cells in experimental arthritis was investigated by using TCRδ$^{-/-}$ mice, which completely lack γδ T cells. These mice showed no difference in CIA incidence and severity, suggesting a minor role for γδ T cells in the development of CIA (Corthay et al. 1999). However, treatment with anti-TCRγδ mAb (UC7-13D5) before injection of type II collagen delayed both the onset and the severity of CIA. In contrast, treatment of arthritic mice led to a more severe form of CIA. These opposing effects may be caused by the antibody depleting or blocking non-activated γδ T cells in non-diseased mice, but inducing activated γδ T cells in diseased mice (Peterman et al. 1993). In addition, in rat adjuvant arthritis, depletion of γδ T cells resulted in an aggravation of joint destruction (Pelegri et al. 1996). However, it was shown that treatment with the UC7-13D5 and GL3 mAbs led to the internalization of the γδ TCR, making γδ T cells 'invisible' for assessment by flow cytometry, which is a frequently used method to evaluate the efficacy of antibody-based depletion (Koenecke et al. 2009). This might imply that although significant biological effects were observed (Peterman et al. 1993), the question remains what the true fate was of the targeted γδ T cells is. It has been shown that γδ T cells are able to produce high levels of IL-17 during CIA and that the numbers of IL-17-producing γδ T cells were equal to that of CD4$^+$ IL-17-producing cells (Roark et al. 2007). The IL-17 producing γδ T cells in CIA preferentially expressed the Vγ4/Vδ4 γδ TCR and upon treatment with a depleting anti-Vγ4 antibody, it was shown that the Vγ4$^+$ γδ T cells contributed to CIA (Roark et al. 2007). This indicates that a specific subtype of γδ T cells which produces IL-17 are involved in CIA development (Roark et al. 2007, 2008). IL-23, but not IL-6, was required for IL-17A production from NKT cells and γδ T cells, as these cells constitutively express IL-23R and RORγt, unlike naive CD4$^+$ T cells (Shibata et al. 2007; Lochner et al. 2008; Rachitskaya et al. 2008). In line with these observations, we noticed that during arthritic conditions in antigen-induced arthritis (AIA), relatively high proportions of IL-17 producing γδ T cells are present (Cornelissen et al. 2009a, b). The expression of IL-17 by these γδ T was associated with a high RORγt transcription level, which was significantly higher compared to the level in CD4$^+$ T cells (Cornelissen et al. 2009a, b). To demonstrate the essential role of IL-23 in the induction of these IL-17-producing γδ T cells, we induced AIA in IL-23p19$^{-/-}$ mice. In these mice, IL-17-producing γδ T cells were absent while IFN-γ producing γδ T cells were still present (Cornelissen et al. 2009a, b).

Recently, it has been shown that thymic ligand-naïve γδ T cells are already able to produce IL-17, whereas ligand-experienced γδ T cells produce IFN-γ (Jensen et al. 2008). Directly after immunization, these IL-17-producing γδ T cells were found in the draining lymph nodes, suggesting that they might play a role in the onset of an inflammatory response (Jensen et al. 2008). This adds new insight into a potential role of γδ T cells during inflammation and in auto-immune diseases.

3 Role of T_H17 Cells in the Pathogenesis of Arthritis

3.1 Co-expression of T_H1 and T_H17 Cells in Rheumatoid Arthritis

Rheumatoid arthritis (RA) is characterized by synovial inflammation and destruction of joint cartilage and bone and mediated by persistent synthesis of pro-inflammatory cytokines and matrix metalloproteinases (MMP) (Arend and Dayer 1995). T cell activation and migration occurs as an early consequence of RA and these cells adopt a pro-inflammatory phenotype. Classically, auto-immune diseases such as RA were thought to be a T_H1, and not a T_H2 associated disorder (Firestein 2003). At present, it is unclear whether RA is a T_H1 and/or T_H17-mediated disease (Lubberts 2008). It has been shown that IL-17A is produced by some pro-inflammatory T_H1/T_H0 cells isolated from synovial membranes and from synovial fluid of RA patients. In joints of patients with established RA, T_H1 cells were predominantly observed (Aarvak et al. 1999). With enhanced IL-17A levels in the joint, IL-17A producing CCR6+ memory T cells have been identified in synovial fluid of RA patients (Yamada et al. 2008). We detected a relatively high percentage of IL-17A-producing CCR+ memory T cells in PBMC from treatment naïve RA patients (Colin et al. 2010, Van Hamburg et al. 2011). Of note, the T_H1 IFN-γ and the T_H17 IL-17 cytokines are often co-expressed in human memory $CD4^+CD45^+RO^+$ T cells from treatment naïve RA patients (Colin et al. 2010).

3.2 T_H17 Cells Found in Different Arthropathies

Although relatively high numbers of T_H17 cells have been shown in PBMC from treatment naïve RA patients (Colin et al. 2010; Van Hamburg et al. 2011), the proportion of these cells in PBMC and SFMC in established RA varies between no T_H17 to low T_H17 cells (Table 2). Yamada et al. (2008) showed that IL-17 positive cells were detected in CD45RO+ CD4 T cells in RA and most IL-17 positive cells produced neither IFN-gamma nor IL-4, but TNF (Yamada et al. 2008). The frequency of T_H17 cells was neither increased in RA nor correlated with the DAS28 disease activity score. Unexpectedly, the frequency of T_H17 cells was significantly decreased in the joints compared with PBMC of the same patients with RA, whereas T_H1 cells were more abundant in the joints than in PBMC in these established RA patients (Yamada et al. 2008). Increased levels of IL-17A, IL-6, TGF-beta, and IFN-gamma concentrations in sera and synovial fluid of reactive arthritis (ReA) and undifferentiated spondyloarthropathy (uSpA) compared to RA have been shown, suggesting that T_H1 and T_H17 cells could be the major agents in inflammation in ReA/uSpA (Singh et al. 2007) (Table 2).

Table 2 The IL-23/IL-17 axis in different arthropathies

Arthropathy	Systemic	At site of inflammation	References
Early RA	T_H17: IL-17A, IL-17F, IL-22, TNF	ND	Colin et al. (2010)
	T_H1: IFN-gamma		
Established RA	T_H17	T_H17↓	Yamada et al. (2008)
	T_H1↓	T_H1↑	
	No T_H17	IL-17A↑	Jandus et al. (2008), Chabaud et al. (1999), Kotake et al. (1999), Ziolkowska et al. (2000)
ReA/undiff. spondylarthropathy	IL-17A	IL-17A	Singh et al. (2007)
	IL-6	IL-6	
	TGF-beta	TGF-beta	
	IFN-gamma	IFN-gamma	
Seronegative spondylaarthritides: – Psoriatic arthritis – Ankylosing spondylitis	T_H17	ND	Jandus et al. (2008)
Systemic sclerosis	IL-17A mRNA, IL-17A levels	ND	Kurasawa et al. (2000), Radstake et al. (2009)
	CD4, CD45RA, CD45RO: IL-23R+ & IL-17A+		
	IL-6		
	IL-23		
	IL-1alpha		
SLE	No IL-17A	ND	Kurasawa et al. (2000)
	No IL-17A		Meyers et al. (2006)
	IL-17A		Wong et al. (2009)
	IL-17A		Sullivan et al. (2000)
Lyme arthritis	ND	IL-17+ cells	Codolo et al. (2008)
Wegener's granulomatosis	T_H17	ND	Abdulahad et al. (2008)
	T_H2		
Sjogren's syndrome	IL-17A/IL-23	IL-17+	Nguyen et al. (2008)
		IL-23+	
		IL-18	Sakai et al. (2008)
		T_H17	
Juvenile idiopathic arthritis	IL-17+ T cells	IL-17+ T cells↑	Nistala et al. (2008)

Adapted from Lubberts 2010
ND: not determined

Jandus et al. (2008) found increased numbers of circulating T_H17 cells in the peripheral blood of patients with seronegative spondylarthritides (psoriatic arthritis and ankylosing spondylitis), but not in patients with RA (Table 2). In addition, T_H17 cells from the spondylarthritis patients showed advanced differentiation and were polyfunctional in terms of T cell receptor-driven cytokine production (Jandus et al. 2008).

Subclinical gut inflammation is common in spondylarthritis. A strong and significant up-regulation of IL-23p19 transcript was found in the terminal ileum in patients with AS and patients with Crohn's disease (CD). IL-23 was abundantly produced by infiltrating monocyte-like cells in inflamed mucosa from AS and CD patients (Ciccia et al. 2009). Notably, Paneth cells were identified as a major source of IL-23 in patients with AS, patients with CD, and normal controls (Ciccia et al. 2009). It has been shown that Paneth cells can also be a source of IL-17A in a TNF-induced experimental shock model in mice (Takahashi et al. 2008). However, unlike CD, in AS patients, IL-23 was not associated with up-regulation of IL-17 and the IL-17-inducing cytokines IL-6 and IL-1beta, indicating that overexpression of IL-23, but not IL-17, is a pivotal feature of sub-clinical gut inflammation in AS (Ciccia et al. 2009).

In patients with systemic sclerosis (SSc), increased IL-17A messenger RNA was expressed in unstimulated PBMC and lymphocytes from the skin and lungs of SSc patients, but not in similar samples from patients with systemic lupus erythematosus (SLE), polymyositis/dermatomyositis, or from healthy donors (Table 2). IL-17 levels were also increased in the serum of SSc patients, but not in that of SLE patients or healthy donors. IL-17 overexpression was significantly related to the early stage of SSc, but not to other clinical features of SSc (Kurasawa et al. 2000). Furthermore, Radstake et al. (2009) showed that CD4, CD45RA and CD45RO cells from SSc patients highly expressed the IL-23 receptor, which was associated with high IL-17 expression as well (Radstake et al. 2009). In contrast, IFN-gamma and TGF-beta were selectively upregulated in subsets of SSc patients. In addition, circulating levels of IL-17 inducing cytokines IL-6, IL-23, and IL-1alpha were increased in all subsets of SSc patients. The combination of IL-17, IFN-gamma, and TGFbeta levels in CD45RO and CD45RA cells from SSc patients can be useful to distinguish between limited cutaneous SSc (lcSSc), early diffuse cutaneous SSc (edcSSc), or late diffuse cutaneous SSc (ldSSc) (Radstake et al. 2009).

Data implicating IL-17 in lupus has become available (Sullivan et al. 2000, Wong et al. 2009) with some evidence arguing against a role for IL-17 in human lupus (Kurasawa et al. 2000, Meyers et al. 2006) (Table 2). In the BXD2 mouse model for lupus-like auto-immunity, a dramatic upregulation of serum IL-17 and numbers of T_H17 cells has been demonstrated (Hsu et al. 2008). BXD2 mice form spontaneous germinal centers which are IL-17-dependent. Treatment with IL-17 promotes the secretion of both IgM and IgG auto-antibodies, which is reduced by crossing BXD2 mice to IL-17R KO mice (Hsu et al. 2008). A potential role for IL-17-producing T cells has been shown in Ets-1 knockout mice, another mouse model of lupus (Moisan et al. 2007; Wang et al. 2005), and in the spontaneous mouse model of lupus, the New Zealand Black (NZB) x SWR F1 cross (SNF1 mice) (Jang et al. 2009). These studies reveal that there is increasing evidence in both humans and mouse models that IL-17-producing cells play a role in SLE progression, although T_H17 does not always seem necessary.

Human Lyme arthritis is caused by Borrelia burgdorferi and characterized by an inflammatory infiltrate that consists mainly of neutrophils and T cells. Anti-NapA (neutrophil-activating protein A of B burgdorferi) antibodies were found in 48% of the patients with Lyme arthritis, but were undetectable in the healthy controls (Codolo et al. 2008). T cells from the synovial fluid of patients with Lyme arthritis produced IL-17 in response to NapA. Moreover, NapA was able to induce the expression of IL-23 in neutrophils and monocytes, as well as the expression of IL-6, IL-1beta, and TGF-beta in monocytes via Toll-like receptor (TLR) 2 (Codolo et al. 2008). Therefore, NapA of B burgdorferi is able to elicit a synovial fluid T_H17 cell response that might play a crucial role in the pathogenesis of Lyme arthritis (Codolo et al. 2008) (Table 2). In addition, IL-23 is required for the development of arthritis in mice vaccinated and challenged with Borrelia species (Kotloski et al. 2008).

In patients with Wegener's granulomatosis (WG) in remission, the percentage of T_H17 cells and T_H2 cells within the activated CD69+ and CD4+ T cell population were significantly increased, while no difference was found in T_H1 cells compared with the percentage in healthy controls (Table 2). Increased percentages of T_H17 cells in response to tetanus toxoid (TT) and staphylococcal enterotoxin B (SEB) were found both in antineutrophil cytoplasmic antibody (ANCA) positive and in ANCA-negative patients, while an increased frequency of PR3-specific T_H17 cells was restricted to ANCA-positive patients. Therefore, a skewed T_H17 response found in ANCA-positive WG patients following stimulation with the autoantigen PR3 suggests that IL-17 is involved in disease pathogenesis and could constitute a new therapeutic target for WG (Abdulahad et al. 2008).

Also in Sjogren's syndrome (SS), an upregulation has been shown of the T_H17/IL-23 system at the time of disease (Nguyen et al. 2008) (Table 2). Salivary gland biopsy specimens from SS patients revealed strong positive staining for both IL-17 and IL-23 within lymphocytic foci and diffuse staining on epithelial tissues. In sera and saliva from SS patients, IL-17 and IL-6 were present at varying levels (Nguyen et al. 2008).

In addition, IL-18 and T_H17 cells detected in the salivary glands in SS patients are associated with the pathogenesis of SS in the salivary glands (Sakai et al. 2008).

IL-17+ T cells were also detected in the joints of children with juvenile idiopathic arthritis (JIA) and these cells were enriched in the joint, compared to the blood of JIA patients (Nistala et al. 2008) (Table 2). Of note, IL-17+ T cell numbers were higher in patients with extended oligoarthritis, the more severe subtype of JIA, as compared with patients with persistent oligoarthritis, the milder subtype (Nistala et al. 2008). Within the joint, there was an inverse relationship between IL-17+ T cells and FoxP3+ Treg cells (Nistala et al. 2008).

These studies suggest that IL-17+ T cells contribute to the pathogenesis of different arthropathies and that the stage of the disease and the site of expression might be important for the role of these IL-17+ T cells in the pathogenesis of the disease. Furthermore, the balance between T_H17/T_H1 and T_H17/Treg may be critical in disease outcome.

3.3 CCR6+ T_H17, But Not CCR6– T_H1 Cells from Patients with Early RA are Potent Activators of RA Synovial Fibroblasts

T_H17 cells have been detected in different arthropathies but their functional role in the human disease has not been established. Both T_H1 and T_H17 cells have been recognized in RA but it remains unclear whether T_H1 and/or T_H17 cells drive disease chronicity (Lubberts 2008). In experimental arthritis models, T_H17 cells have been recognized as important contributors to the inflammatory processes (Lubberts et al. 2005a, b; Lubberts 2008) and preferential recruitment of CCR6-expressing T_H17 cells to inflamed joints via CCL20 in rheumatoid arthritis and its animal model has been shown (Hirota et al. 2007). However, in EAE and EAU experimental auto-immune models, evidence also suggests the potential of T_H1 cells as pathogenic mediators (Kroenke et al. 2008; Luger et al. 2008). T_H17-dependent experimental arthritis models were described as being induced by a specific antigen with complete Freund's adjuvant (CFA), as well as spontaneous arthritis models (Hirota et al. 2007; Murphy et al. 2003; Nakae et al. 2003; Van Hamburg et al. 2009) and T_H17-independent/T_H1 dependent arthritis models exist (Doodes et al. 2008). For example, the development of proteoglycan-induced arthritis has been shown to be IL-17-independent and the severity of arthritis is dependent on the production of IFN-gamma (Doodes et al. 2008). Both IFN-gamma and IL-17 have the potential to induce PGIA, although the strength of the IFN-gamma response regulated the contribution of each of these T helper effector cytokines to disease (Doodes et al. 2010).

Whether this is also the case in human arthritis needs further clarification, including what the functional role of these T cell subsets is with focus on different stages of the disease.

T cells and the cytokines IL-17A and TNF-alpha have been shown to activate RA synovial fibroblast (RASF), resulting in the expression of pro-inflammatory cytokines such as IL-6 and IL-8, mediators of joint inflammation (Chabaud et al. 1999; Miossec 2003; Bombara et al. 1993; Brennan and McInnes 2008; McInnes et al. 2000; Parsonage et al. 2008; Tran et al. 2007). In addition, an IL-17A-triggered positive-feedback loop of IL-6 signaling in fibroblasts has been described in experimental arthritis and discussed, but not shown for human T_H17-mediated auto-immune disorders (McInnes et al. 2000; Ogura et al. 2008). However, characterization of the human T cell population that activates RASF and is responsible for IL-6 and IL-8 cytokine induction has not been established. We identified IL-17A+ TNF+ T_H17 cells in PBMC from treatment naïve early RA patients. These cells are identified as T_H17 cells since they express CCR6, IL-17F, IL-22, IL-26, RORc, CCL20, low T-bet, low FoxP3, and low IFN-gamma. Their functional potential was examined using co-cultures with early RA synovial fibroblasts. Primary CCR6+ T_H17, but not CCR6- T_H1 memory T cells were potent inducers of IL-6, IL-8, and MMP-1 and MMP-3 by the synovial fibroblast in these co-cultures. Of interest, IL-17A expression was increased in these co-cultures with synovial fibroblasts indicating the induction of a pro-inflammatory feedback loop. This T_H17-RASF pro-inflammatory feedback loop might be an important mechanism driving persistent arthritis (Van Hamburg et al. 2011). Blocking experiments revealed that both TNF and IL-17A needed to be blocked for optimal suppressive effect, especially at the level of MMP-1 and MMP-3, indicating additional value of blocking T_H17 activity to anti-TNF neutralization in early RA (Van Hamburg et al. 2011). This is the first data indicating potential pathogenicity of T_H17 cells in human arthritis. Interestingly, data from first clinical trials with anti-IL-17A antibody treatment in psoriatic arthritis patients and RA patients looks promising. Since additional neutralization of the T_H17 cytokine IL-17A to TNF-alpha was needed to suppress the mechanism that drives the pro-inflammatory feedback loop in co-cultures of cells from treatment naïve early RA patients, blocking T_H17 activity in addition to anti-TNF in early RA and potentially other T_H17-mediated disorders may be essential to reach the ultimate goals. These goals are to prevent the development of persistent and destructive arthritis at the very early stages of the disease, and to reach permanent remission.

4 IL-23 and Pathogenic T_H17 Cells

The role of IL-1, IL-6, IL-23, TGF-β, and TNF-α in the differentiation of mouse and human T_H17 cells has been discovered (Korn et al. 2009). However, the role of IL-23 in the effector function of T_H17 cells is not clearly understood. The essential role for IL-23 in the pathogenic function of T_H17 cells was shown in EAE (McGeachy et al. 2007). In this experiment, T_H17 cells, which were polarized under IL-23 polarizing conditions, were potent inducers of EAE, while the pathogenic function in TGF-β/IL-6 polarized T_H17 cells was completely abrogated. A possible explanation for this discrepancy could be the inhibited induction of IL-10 in the IL-23 polarized cells. TGF-β/IL-6 polarized T_H17 cells modulated the induction of EAE by IL-23-polarized T_H17 cells (McGeachy et al. 2007). This data indicates that IL-23 is essential for the pathogenic function of T_H17 cells. We have shown that IL-23 and TGF-β/IL-6 differentially regulate T helper-specific transcription factors during T_H17 development in auto-immune experimental arthritis, which may explain the pathogenic potential of IL-23 (Mus et al. 2010). In CD4+ T cells from non-immunized and type II collagen immunized DBA/1 mice, IL-23 increased the expression of IL-17A, IL-17F, and RORγt. In contrast to TGF-β/IL-6, IL-23 inhibiting the T_H1 and regulatory T cell-specific transcription factors T-bet and FoxP3. Of note, the IL-23-induced increase in IL-17A and IL-17F levels was prevented in T-bet deficient mice, indicating a functional link between IL-23 and T-bet under the T_H17 culture condition (Mus et al. 2010).

5 Role of the T_H17 Cytokines in Arthritis

5.1 *IL-17F*

Apart from the well studied effects of IL-17A in arthritis (Miossec 2003; Lubberts et al. 2002; Lubberts 2008; Iwakura et al. 2008), little is known about the specific role of IL-17F in arthritis. It has been shown that IL-17F is an important regulator of inflammatory responses and seems to function differently than IL-17A in immune responses and diseases (Yang et al. 2008a, b). Although IL-17F has many biologically overlapping effects with IL-17A, IL-17F is less potent in activating synovial fibroblast (Zrioual et al. 2009). IL-17F has been shown to be potentially destructive to cartilage in vitro (Hymowitz et al. 2001). IL-17F played only marginal roles in collagen-induced arthritis and arthritis in IL-1RA-deficient mice (Ishigame et al. 2009). In contrast, both IL-17F and IL-17A were involved in host defense against mucoepithelial infection by Staphylococcus aureus and Citrobacter rodentium (Ishigame et al. 2009). IL-17A was produced mainly in T cells, whereas IL-17F was produced in T cells, innate immune cells, and epithelial cells (Ishigame et al. 2009). Further studies are needed to clarify the role of IL-17F in arthritis and the interaction of IL-17F with other T_H17 cytokines.

5.2 *IL-21*

IL-21 was recently reported to play an important role in the generation of T_H17 cells. IL-21 was shown to be potently induced by IL-6 (Zhou et al. 2007). IL-21 can be produced by T_H17 cells (Korn et al. 2007; Nurieva et al. 2007) and can also be involved in T_H17 polarization, since in the absence of IL-6 and IL-21 in combination with TGF-beta could function as an alternative signal for the induction of T_H17 cells (Korn et al. 2007).

In arthritis, IL-21R-deficient (IL-21R^{-}/$^{-}$)K/BxN mice were completely refractory in the development of spontaneous arthritis (Jang et al. 2009). These mice contained fewer CD4+ T cells with a reduced proportion of homeostatically proliferating cells, fewer follicular Th cells, and, surprisingly, more T_H17 cells than the control mice (Jang et al. 2009). Moreover, these mice also failed to develop IgG1+ memory B cells and auto-antigen-specific IgG1 antibodies secreting cells (Jang et al. 2009). This data suggests that IL-21 forms a positive feedback autocrine loop involving homeostatically activated CD4+ cells and is essential in the development of auto-immune arthritis by mechanisms dependent on follicular Th cells development, auto-reactive B cell maturation, and RANKL induction, but independent of T_H17 cell function (Jang et al. 2009). Blocking IL-21 in CIA by in vivo administration of soluble IL-21R-Fc fusion protein delayed the onset and progression of arthritis (Jang et al. 2009). In addition, in MLR/lpr mice, a spontaneous model for SLE, blocking IL-21 attenuated disease (Herber et al. 2007). Genetic association of IL-21 polymorphisms with SLE has been suggested (Sawalha et al. 2008). In addition, a polymorphism within IL-21R confers a risk for SLE (Webb et al. 2009). Moreover, highly suggestive evidence has been provided for IL-2/IL-21 loci as a risk factor for RA (Daha et al. 2009). These statistics suggest IL-21 to be an interesting therapeutic target in arthritis.

5.3 IL-22

IL-22 is a cytokine belonging to the IL-10 super family (Renauld 2003) and has been shown to act as an effector cytokine of the T_H17 lineage (Liang et al. 2006). IL-22 is primarily produced by activated T cells and natural killer cells (Wolk et al. 2006). High levels of IL-22 were expressed both in the lining and the sublining layers of RA synovial tissues (Ikeuchi et al. 2005). The majority of IL-22-positive cells were synovial fibroblasts and macrophages. IL-22R1 expression was also expressed in both the lining and the sublining layers of RA synovial tissues (Ikeuchi et al. 2005). IL-22 significantly increased proliferation of RA synovial fibroblast and the production of MCP-1 in vitro (Ikeuchi et al. 2005).

In line with other inflammatory models such as EAE (Kreymborg et al. 2007), we found that IL-23 is essential for IL-22 production in polarized CD4+ T cells from non-immunized and type II collagen-immunized DBA/1 mice (Mus et al. 2010). IL-22 deficient mice were less susceptible to CIA than wild type mice, as evidenced by their decreased incidence of arthritis and decreased pannus formation (Geboes et al. 2009). Remarkably, less severe CIA in IL-22 deficient mice was associated with increased production of CII-specific and total IgG antibodies, whereas cellular CII responses were unchanged (Geboes et al. 2009). In vitro, IL-22 was found to promote osteoclastogenesis (Geboes et al. 2009). Although this proposes a pro-inflammatory role of endogenous IL-22 in arthritis, promoting osteoclastogenesis and regulating antibody production, the uncoupling between low incidence and higher antibody production in IL-22 deficient mice during CIA is unclear and needs further investigation.

In addition to IL-22 produced by T_H17 cells, new evidence exist that IL-22 can be produced by effector T cells without IL-17A by a subset of human skin-homing memory T cells (Duhen et al. 2009; Trifari et al. 2009). These so called "T_H22" cells can also be found in RA (Van Hamburg et al. 2011). However, the function of these cells in the pathogenesis of RA is unknown.

6 Plasticity Between T_H17, T_H1 and Treg Cells in Arthritis

T_H17 cells appear to be associated with Treg cells, which are indicated by the practice of a common inducer; TGF-β, an overlapping chemokine receptor profile, and the expression of the T_H17 associated transcription factor RORγt (Van Hamburg et al. 2009;

Ivanov et al. 2006). In mice, it has been shown that Treg cells can be converted to IL-17 producing T cells (Radhakrishnan et al. 2008; Xu et al. 2007; Yang et al. 2008a, b). Additionally, human Treg cells, defined as CD4+CD25highFoxP3+-CD127–CD27+, were reported to differentiate to IL-17-producing T cells. This was accompanied by an up-regulation of RORγt and CCR6 expression (Koenen et al. 2008).

Next to plasticity between Treg and T_H17 cells, new data has emerged that T_H17 cells are closely related to T_H1 cells. In the mouse, differentiated T_H17 cells responded rapidly in vitro to IL-12, by upregulating the expression of IFN-γ and downregulating IL-17 expression. The development of IFN-γ-producing effector T cells from IL-17-producing progenitor cells is inhibited in the presence of TGF-β, which is important for the maintenance of IL-17 expression by T_H17 polarized cells (Lee et al. 2009; Lexberg et al. 2008; Martin-Orozco et al. 2009).

Recently, a role for epigenetic regulation of effector T cell plasticity was found. For example, the gene encoding for T-bet, the master regulator of T_H1 differentiation, was found to be in an active state, according to histone methylation marks, in both T_H17 and Treg cells. This implies that T_H17 and Treg cells still have the potential to upregulate the expression of T-bet and to differentiate towards T_H1 cells (Wei et al. 2009).

Associations between T cell subsets such as T_H17, T_H1, and Treg cells have been found in inflammation and progression of RA. In joint inflammation of RA, high levels of IL-17 can be found, but hardly any IFN-γ expression can be detected in the joint synovium (Chabaud et al. 1999; Raza et al. 2005). This was in line with studies in mice, which indicate the importance for IL-17 in the induction of arthritis (Lubberts et al. 2005a, b). In contrast, IFN-γ cytokine levels were detected in the joint synovium in later stages of RA, and IFN-γ expression was detected in joint infiltrating CD4 T cells and in CD4 T cell clones, obtained from inflamed synovium (Aarvak et al. 1999; Dolhain et al. 1996; Morita et al. 1998). Increased levels of IFN-γ were detected in lymph nodes and synovium at later time points when adjuvant arthritis was induced in rats. In contrast, IL-17 was induced shortly after initiation and declined in time (Bush et al. 2001). In a 2 years predictive study, it became evident that IFN-γ was playing a role in reducing joint progression (Kirkham et al. 2006). Inhibition of IFN-γ in advanced stages exacerbated the disease (Kirkham et al. 2006).

Next to T_H1 cells, an inverse correlation between IL-17-producing and FoxP3+ Treg cells has been identified. In children with juvenile idiopathic arthritis (JIA), the balance between IL-17+ cells and Treg cells may be critical for the disease outcome (Nistala et al. 2008). In addition, Treg cells are present in higher numbers in inflamed joints in patients with a mild RA phenotype, as compared to patients with a more severe disease (de Kleer et al. 2004; Ruprecht et al. 2005).

These observations indicated that plasticity between Treg, T_H1 and T_H17 subsets exist and that differentiation of these subsets was not completely restricted to separate lineages. All together, these recent advances provide a completely new concept of T cell plasticity that may be relevant to the arthritis process.

7 Modulation of the IL-23/IL-17 Axis in the Pathogenesis of Arthritis

Since T_H17 cells and T_H17 cytokines can have a pro-inflammatory effect in arthritis, it will be interesting to discover how to regulate the activity of these cells/cytokines. It has been shown that overexpression of the T_H2-specific transcription factor, GATA3, in T cells can modulate T_H17 cell differentiation and protects against severe joint inflammation and bone erosion in experimental arthritis (Van Hamburg et al. 2009). Moreover, GATA3 overexpression resulted in reduced gene expression of the T_H17-associated transcription factor RORgammat (Van Hamburg et al. 2009).

Treatment of mice with CIA with anti-IL-6R mAb on day 0 markedly suppressed the induction of T_H17 cells and arthritis development, but treatment with this antibody on day 14 failed to suppress both T_H17 differentiation and arthritis. Treatment of mice with TNFR-Fc from day 0 to day 14 suppressed neither T_H17 differentiation nor arthritis, but treatment from day 21 to day 35 successfully ameliorated arthritis without inhibiting T_H17 induction (Fujimoto et al. 2008). This study indicates that the protective effect of IL-6 blockade, but not TNF blockade in CIA correlated with the inhibition of T_H17 differentiation and suggests that IL-6 blockade in RA in human likely involves a therapeutic mechanism distinct from that of TNF blockade (Fujimoto et al. 2008). Notley et al. (2008) showed that TNF blockade using TNFR-Fc fusion protein or anti-TNFmonoclonal antibody reduced collagen arthritis severity, but unexpectedly expanded populations of T_H1 and T_H17 cells, which were shown by adoptive transfer to be pathogenic (Notley et al. 2008). T_H1 and T_H17 cell populations were also expanded in CII-immunized TNFR p55KO, but not p75KO. The expansion of T_H1/T_H17 cells was abrogated by the blockade of p40 (Notley et al. 2008). Although TNF blockade increased numbers of T_H1 an T_H17 cells in LN, it inhibited their accumulation in the joint, thereby providing an explanation for the paradox that anti-TNF therapy ameliorates arthritis despite increasing numbers of pathogenic T cells (Notley et al. 2008).

Lai Kwan Lam et al. showed that local BAFF gene silencing suppresses T_H7 cell generation and ameliorates auto-immune arthritis (Lai Kwan Lam et al. 2008). TNF superfamily member B cell-activating factor (BAFF) plays an important role in humoral immunity and in auto-immune diseases, including RA. Local BAFF gene targeting inhibited pro-inflammatory cytokine expression, suppressed generation of plasma cells and T_H17 cells, and markedly ameliorated joint pathology (Lai Kwan Lam et al. 2008). This study revealed a previously unrecognized role for BAFF in promoting the expansion of T_H17 cells and demonstrated IL-17 as a crucial effector cytokine for BAFF-mediated pro-inflammatory effects during collagen-induced arthritis development (Lai Kwan Lam et al. 2008).

Apart from regulation of T_H17 cells in experimental mouse models for arthritis, Colin et al. showed modulatory effects of 1,25-dihydroxyvitamin D3 (1,25-$(OH)_2D_3$) on the memory T_H17 activity in treatment naïve early RA patients (Colin et al. 2010). 1,25-$(OH)_2D_3$ reduced the levels of IL-17A and IFN-gamma and increased IL-4 in stimulated PBMC from treatment naïve early RA patients (Colin et al. 2010).

Interestingly, 1,25-$(OH)_2D_3$, in contrast to dexamethasone, directly modulated human T_H17 polarization accompanied with suppression of IL-17A, IL-17F, TNF-alpha, and IL-22 production by FACS sorted memory T cells from these early RA patients (Colin et al. 2010).

Acknowledgement I would like to acknowledge all members of the Lubberts lab for their excellent input and discussions.

References

Aarvak T, Chabaud M, Kallberg E (1999) Change in the Th1/Th2 phenotype of memory T-cell clones from rheumatoid arthritis synovium. Scand J Immunol 50:1–9.

Abdulahad WH, Stegeman CA, Limburg PC et al (2008) Skewed distribution of Th17 lymphocytes in patients with Wegener's granulomatosis in remission. Arthritis Rheum 58:2196–2205.

Arend WP, Dayer J-M (1995) Inhibition of the production and effects of interleukin-1 and tumor necrosis factor alpha in rheumatoid arthritis. Arthritis Rheum 38:151–60.

Bombara MP, Webb DL, Conrad P et al (1993) Cell contact between T cells and synovial fibroblasts causes induction of adhesion molecules and cytokines. J Leuk Biol 54:399–406.

Brennan FM, McInnes IB (2008) Evidence that cytokines play a role in rheumatoid arthritis. J Clin Invest 118:3537–3545.

Bush KA, Farmer KM, Walker JS et al (2002) Reduction of joint inflammation and bone erosion in rat adjuvant arthritis by treatment with interleukin-17 receptor IgG1 Fc fusion protein. Arthritis Rheum 46:802–805.

Bush KA, Walker JS, Lee CS et al (2001) Cytokine expression and synovial pathology in the initiation and spontaneous resolution phases of adjuvant arthritis: interleukin-17 expression is upregulated in early disease. Clin Exp Immunol 123:487–495.

Chabaud M, Durand JM, Buchs N et al (1999) Human interleukin-17: a T cell-derived pro-inflammatory cytokine produced by the rheumatoid synovium. Arthritis Rheum 42:963–70.

Chabaud M, Miossec P (2001) The combination of tumor necrosis factor alpha blockade with interleukin-1 and interleukin-17 blockade is more effective for controlling synovial inflammation and bone resorption in an ex vivo model. Arthritis Rheum 44:1293–03.

Ciccia F, Bombardieri M, Principato A et al (2009) Overexpression of interleukin-23, but not interleukin-17, as an immunologic signature of subclinical intestinal inflammation in ankylosing spondylitis. Arthritis Rheum 60:955–965.

Codolo G, Amedei A, Steere AC et al (2008) Borrelia burgdorferi NapA-driven Th17 cell inflammation in lyme arthritis. Arthritis Rheum 58:3609–3617.

Colin EM, Asmawidjaja PS, van Hamburg JP et al (2010) 1,25-dihydroxyvitamin D3 modulates Th17 polarization and interleukin-22 expression by memory T cells from patients with early rheumatoid arthritis. Arthritis Rheum 62:132–42.

Cornelissen F, Mus AM, Asmawidjaja PS et al (2009) Interleukin-23 is critical for full-blown expression of a non-auto-immune destructive arthritis and regulates interleukin-17A and RORgammat in gammadelta T cells. Arthritis Res Ther 11:R194. Epub 2009 Dec 17.

Cornelissen F, Van Hamburg JP, Lubberts E (2009) The IL-12/IL-23 axis and its role in the Th17 cell development, pathology and plasticity in arthritis. Curr Opin Invest Drugs 10:452–462.

Corthay A, Johansson A, Vestberg M et al (1999) Collagen-induced arthritis development requires alpha beta T cells but not gamma delta T cells: studies with T cell-deficient (TCR mutant) mice. Int Immunol 11:1065–1073.

Daha NA, Kurreeman FA, Marques RB et al (2009) Confirmation of STAT4, IL2/IL21, and CTLA4 polymorphisms in rheumatoid arthritis. Arthritis Rheum 60:1255–1260.

de Kleer IM, Wedderburn LR, Taams LS et al (2004) CD4+CD25bright regulatory T cells actively regulate inflammation in the joints of patients with the remitting form of juvenile idiopathic arthritis. J Immunol 172:6435–6443.
Dolhain RJ, van der Heiden AN, ter Haar NT et al (1996) Shift toward T lymphocytes with a T helper 1 cytokine-secretion profile in the joints of patients with rheumatoid arthritis. Arthritis Rheum 39:1961–1969.
Doodes PD, Cao Y, Hamel KM et al (2008) Development of proteoglycan-induced arthritis is independent of IL-17. J Immunol 181:329–337
Doodes PD, Cao Y, Hamel KM et al (2010) IFN-gamma regulates the requirement for IL-17 in proteoglycan-induced arthritis. J Immunol 184:1552–1559.
Duhen T, Geiger R, Jarrossay D et al (2009) Production of interleukin 22 but not interleukin 17 by a subset of human skin-homing memory T cells. Nature Immunol 10:857–863.
Firestein GS (2003) Evolving concepts of rheumatoid arthritis. Nature 423:356–361.
Fossiez F, Djossou O, Chomarat P et al (1996) T cell interleukin-17 induces stromal cells to produce pro-inflammatory and hematopoietic cytokines. J Exp Med 183:2593–03.
Fujimoto M, Serada S, Mihara M et al (2008) Interleukin-6 blockade suppresses auto-immune arthritis in mice by the inhibition of inflammatory Th17 responses. Arthritis Rheum 58: 3710–3719.
Geboes L, Dumoutier L, Kelchtermans H et al (2009) Pro-inflammatory role of the Th17 cytokine interleukin-22 in collagen-induced arthritis in C57BL/6 mice. Arthritis Rheum 60:390–395.
Germann T, Szeliga J, Hess H et al (1995) Administration of interleukin 12 in combination with type II collagen induces severe arthritis in DBA/1 mice. PNAS 92:4823–4827.
Herber D, Brown TP, Liang S et al (2007) IL-21 has a pathogenic role in a lupus-prone mouse model and its blockade with IL-21R.Fc reduces disease progression. J Immunol 178:3822–3830.
Hess H, Gately MK, Rude E et al (1996) High doses of interleukin-12 inhibit the development of joint disease in DBA/1 mice immunized with type II collagen in complete Freund's adjuvant. Eur J Immunol 26:187–191.
Hirota K, Yoshitomi H, Hashimoto M et al (2007) Preferential recruitment of CCR6-expressing Th17 cells to inflamed joints via CCL20 in rheumatoid arthritis and its animal model. J Exp Med 204:2803–2812.
Holoshitz J, Koning F, Coligan JE et al (1989) Isolation of CD4- CD8- mycobacteria-reactive T lymphocyte clones from rheumatoid arthritis synovial fluid. Nature 339:226–229.
Hsu HC, Yang P, Wang J et al (2008) Interleukin 17-producing T helper cells and interleukin 17 orchestrate autoreactive germinal center development in auto-immune BXD2 mice. Nature Immunol 9:166–175.
Hymowitz SG, Filvaroff EH, Yin JP et al (2001) IL-17s adopt a cystine knot fold: structure and activity of a novel cytokine, IL-17F, and implications for receptor binding. EMBO J 20: 5332–5341.
Ikeuchi H, Kuroiwa T, Hiramatsu N et al (2005) Expression of interleukin-22 in rheumatoid arthritis: potential role as a pro-inflammatory cytokine. Arthritis Rheum 52:1037–1046.
Ishigame H, Kakuta S, Nagai T et al (2009) Differential roles of interleukin-17A and -17F in host defense against mucoepithelial bacterial infection and allergic responses. Immunity 30:108–119.
Ivanov, II, McKenzie BS, Zhou L et al (2006) The orphan nuclear receptor RORgammat directs the differentiation program of pro-inflammatory IL-17+ T helper cells. Cell 126:1121–1133.
Iwakura Y, Nakae S, Saijo S et al (2008) The roles of IL-17A in inflammatory immune responses and host defense against pathogens. Immunol Rev 226:57–79.
Jandus C, Bioley G, Rivals JP et al (2008) Increased numbers of circulating polyfunctional Th17 memory cells in patients with seronegative spondylarthritides. Arthritis Rheum 58:2307–2317.
Jang E, Cho SH, Park H et al (2009) A positive feedback loop of IL-21 signaling provoked by homeostatic CD4+CD25- T cell expansion is essential for the development of arthritis in auto-immune K/BxN mice. J Immunol 182:4649–4656.
Jensen KD, Su X, Shin S et al (2008) Thymic selection determines gammadelta T cell effector fate: antigen-naive cells make interleukin-17 and antigen-experienced cells make interferon gamma. Immunity 29:90–100.

Joosten LA, Lubberts E, Helsen MM et al (1997) Dual role of IL-12 in early and late stages of murine collagen type II arthritis. J Immunol 159:4094–4102.
Jovanovic DV, DiBattista JA, Martel-Pelletier J et al (1998) IL-17 stimulates the production and expression of pro-inflammatory cytokines, IL-1beta and TNFalpha, by human macrophages. J Immunol 160:3513–21.
Ju JH, Cho ML, Moon YM et al (2008) IL-23 induces receptor activator of NF-kappaB ligand expression on CD4+ T cells and promotes osteoclastogenesis in an auto-immune arthritis model. J Immunol 181:1507–1518.
Kasama T, Yamazaki J, Hanaoka R et al (1999) Biphasic regulation of the development of murine type II collagen-induced arthritis by interleukin-12: possible involvement of endogenous interleukin-10 and tumor necrosis factor alpha. Arthritis Rheum 42:100–109.
Kirkham BW, Lassere MN, Edmonds JP et al (2006) Synovial membrane cytokine expression is predictive of joint damage progression in rheumatoid arthritis: A two-year prospective study (the DAMAGE study cohort). Arthritis Rheum 54:1122–1131.
Kjeldsen-Kragh J, Quayle A, Kalvenes C et al (1990) T gamma delta cells in juvenile rheumatoid arthritis and rheumatoid arthritis. In the juvenile rheumatoid arthritis synovium the T gamma delta cells express activation antigens and are predominantly V delta 1+, and a significant proportion of these patients have elevated percentages of T gamma delta cells. Scand J Immunol 32:651–659.
Koenecke C, Chennupati V, Schmitz S et al (2009) In vivo application of mAb directed against the gammadelta TCR does not deplete but generates "invisible" gammadelta T cells. Eur J Immunol 39:372–379.
Koenen HJ, Smeets RL, Vink PM et al (2008) Human CD25highFoxp3pos regulatory T cells differentiate into IL-17-producing cells. Blood 112:2340–2352.
Korn T, Bettelli E, Gao W et al (2007) IL-21 initiates an alternative pathway to induce proinflammatory T(H)17 cells. Nature 448:484–487.
Korn T, Bettelli E, Oukka M et al (2009) IL-17 and Th17 Cells. Annu Rev Immunol 27:485–517.
Kotake S, Udagawa N, Takahashi N et al (1999) IL-17 in synovial fluids from patients with rheumatoid arthritis is a potent stimulator of osteoclastogenesis. J Clin Invest 103:1345–1352.
Kotloski NJ, Nardelli DT, Peterson SH et al (2008) Interleukin-23 is required for development of arthritis in mice vaccinated and challenged with Borrelia species. Clin Vaccine Immunol 15:1199–1207.
Kreymborg K, Etzensperger R, Dumoutier L et al (2007) IL-22 is expressed by Th17 cells in an IL-23-dependent fashion, but not required for the development of auto-immune encephalomyelitis. J Immunol 179:8098–8104.
Kroenke MA, Carlson TJ, Andjelkovic AV et al (2008) IL-12- and IL-23-modulated T cells induce distinct types of EAE based on histology, CNS chemokine profile, and response to cytokine inhibition. J Exp Med 205:1535–1541.
Kurasawa K, Hirose K, Sano H (2000) Increased interleukin-17 production in patients with systemic sclerosis. Arthritis Rheum 43:2455–2463
Lai Kwan Lam Q, King Hung Ko O, Zheng BJ et al (2008) Local BAFF gene silencing suppresses Th17-cell generation and ameliorates auto-immune arthritis. PNAS 105:14993–14998.
Lee YK, Turner H, Maynard CL et al (2009) Late developmental plasticity in the T helper 17 lineage. Immunity 30:92–107.
Lexberg MH, Taubner A, Forster A et al (2008) Th memory for interleukin-17 expression is stable in vivo. Eur J Immunol 38:2654–2664.
Liang SC, Tan XY, Luxenberg DP et al (2006) Interleukin (IL)-22 and IL-17 are co-expressed by Th17 cells and cooperatively enhance expression of antimicrobial peptides. J Exp Med 203:2271–2279.
Lochner M, Peduto L, Cherrier M et al (2008) In vivo equilibrium of pro-inflammatory IL-17+ and regulatory IL-10+ Foxp3+ RORgamma t+ T cells. J Exp Med 205:1381–1393.
Lubberts E (2003) The role of IL-17 and family members in the pathogenesis of arthritis. Curr Opin Investig Drugs 4:572–7.
Lubberts E (2008) IL-17/Th17 targeting: on the road to prevent chronic destructive arthritis? Cytokine 41:84–91.
Lubberts E (2010) Th17 cytokines and arthritis. Semin Immunopathol 32:43–53

Lubberts E, Joosten LA, Oppers B et al (2001) IL-1-independent role of IL-17 in synovial inflammation and joint destruction during collagen-induced arthritis. J Immunol 167:1004–1013.

Lubberts E, Joosten LAB, Van de Loo FA et al (2002) Overexpression of IL-17 in the knee joint of collagen type II immunized mice promotes collagen arthritis and aggravates joint destruction. Inflam Res 51:102–4.

Lubberts E, Koenders MI, Oppers-Walgreen B et al (2004) Treatment with a neutralizing anti-murine interleukin-17 antibody after the onset of collagen-induced arthritis reduces joint inflammation, cartilage destruction, and bone erosion. Arthritis Rheum 50:650–659.

Lubberts E, Koenders MI, van den Berg WB (2005) The role of T-cell interleukin-17 in conducting destructive arthritis: lessons from animal models. Arthritis Res Ther 7:29–37.

Lubberts E, Schwarzenberger P, Huang W et al (2005) Requirement of IL-17 receptor signaling in radiation-resistant cells in the joint for full progression of destructive synovitis. J Immunol 175:3360–8.

Luger D, Silver PB, Tang J et al (2008) Either a Th17 or a Th1 effector response can drive autoimmunity: conditions of disease induction affect dominant effector category. J Exp Med 205:799–810.

Malfait AM, Butler DM, Presky DH et al (1998) Blockade of IL-12 during the induction of collagen-induced arthritis (CIA) markedly attenuates the severity of the arthritis. Clin Exp Immunol 111:377–383.

Martin-Orozco N, Chung Y, Chang SH et al (2009) Th17 cells promote pancreatic inflammation but only induce diabetes efficiently in lymphopenic hosts after conversion into Th1 cells. Eur J Immunol 39:216–224.

Matthys P, Vermeire K, Mitera T et al (1998) Anti-IL-12 antibody prevents the development and progression of collagen-induced arthritis in IFN-gamma receptor-deficient mice. Eur J Immunol 28:2143–2151.

McGeachy MJ, Bak-Jensen KS, Chen Y et al (2007) TGF-beta and IL-6 drive the production of IL-17 and IL-10 by T cells and restrain T(H)-17 cell-mediated pathology. Nat Immunol 8:1390–1397.

McInnes IB, Leung BP, Liew FY (2000) Cell-cell interactions in synovitis. Interactions between T lymphocytes and synovial cells. Arthritis Res 2:374–378.

McIntyre KW, Shuster DJ, Gillooly KM et al (1996) Reduced incidence and severity of collagen-induced arthritis in interleukin-12-deficient mice. Eur J Immunol 26:2933–2938.

Meyers JA, Mangini AJ, Nagai T et al (2006) Blockade of TLR9 agonist-induced type I interferons promotes inflammatory cytokine IFN-gamma and IL-17 secretion by activated human PBMC. Cytokine 35:235–246.

Miossec P (2003) Interleukin-17 in rheumatoid arthritis: if T cells were to contribute to inflammation and destruction through synergy. Arthritis Rheum 48:594–601.

Moisan J, Grenningloh R, Bettelli E et al (2007) Ets-1 is a negative regulator of Th17 differentiation. J Exp Med 204:2825–2835.

Morita Y, Yamamura M, Kawashima M et al (1998) Flow cytometric single-cell analysis of cytokine production by CD4+ T cells in synovial tissue and peripheral blood from patients with rheumatoid arthritis. Arthritis Rheum 41:1669–1676.

Murphy CA, Langrish CL, Chen Y et al (2003) Divergent pro and anti-inflammatory roles for IL-23 and IL-12 in joint auto-immune inflammation. J Exp Med 198:1951–1957.

Mus AMC, Cornelissen F, Asmawidjaja PS et al (2010) Interleukin-23 promotes Th17 differentiation by inhibiting T-bet and FoxP3 and is required for interleukin-22, but not interleukin-21, in auto-immune experimental arthritis. Arthritis Rheum 62:1043–1050.

Nakae S, Saijo S, Horai R et al (2003) IL-17 production from activated T cells is required for the spontaneous development of destructive arthritis in mice deficient in IL-1 receptor antagonist. PNAS 100:5986–5990.

Nakajima A, Seroogy CM, Sandora MR et al (2001) Antigen-specific T cell-mediated gene therapy in collagen-induced arthritis. J Clin Invest 107:1293–1301.

Nguyen CQ, Hu MH, Li Y et al (2008) Salivary gland tissue expression of interleukin-23 and interleukin-17 in Sjogren's syndrome: findings in humans and mice. Arthritis Rheum 58:734–743.

Nistala K, Moncrieffe H, Newton KR et al (2008) Interleukin-17-producing T cells are enriched in the joints of children with arthritis, but have a reciprocal relationship to regulatory T cell numbers. Arthritis Rheum 58:875–887.
Notley CA, Inglis JJ, Alzabin S et al (2008) Blockade of tumor necrosis factor in collagen-induced arthritis reveals a novel immune-regulatory pathway for Th1 and Th17 cells. J Exp Med 205:2491–2497.
Nurieva R, Yang XO, Martinez G et al (2007) Essential autocrine regulation by IL-21 in the generation of inflammatory T cells. Nature 448:480–483.
Ogura H, Murakami M, Okuyama Y et al (2008) Interleukin-17 promotes auto-immunity by triggering a positive-feedback loop via interleukin-6 induction. Immunity 29:628–636.
Oppmann B, Lesley R, Blom B et al (2000) Novel p19 protein engages IL-12p40 to form a cytokine, IL-23, with biological activities similar as well as distinct from IL-12. Immunity 13:715–725.
Parks E, Strieter RM, Lukacs NW et al (1998) Transient gene transfer of IL-12 regulates chemokine expression and disease severity in experimental arthritis. J Immunol 160:4615–4619.
Parsonage G, Filer A, Bik M et al (2008) Prolonged, granulocyte-macrophage colony-stimulating factor-dependent, neutrophil survival following rheumatoid synovial fibroblast activation by IL-17 and TNFalpha. Arthritis Res Ther 10:R47.
Pelegri C, Kuhnlein P, Buchner E et al (1996) Depletion of gamma/delta T cells does not prevent or ameliorate, but rather aggravates, rat adjuvant arthritis. Arthritis Rheum 39:204–215.
Peterman GM, Spencer C, Sperling AI et al (1993) Role of gamma delta T cells in murine collagen-induced arthritis. J Immunol 151:6546–6558.
Rachitskaya AV, Hansen AM, Horai R et al (2008) Cutting edge: NKT cells constitutively express IL-23 receptor and RORgammat and rapidly produce IL-17 upon receptor ligation in an IL-6-independent fashion. J Immunol 180:5167–5171.
Radhakrishnan S, Cabrera R, Schenk EL et al (2008) Reprogrammed FoxP3+ T regulatory cells become IL-17+ antigen-specific auto-immune effectors in vitro and in vivo. J Immunol 181:3137–3147.
Radstake TR, van Bon L, Broen J et al (2009) The pronounced Th17 profile in systemic sclerosis (SSc) together with intracellular expression of TGFbeta and IFNgamma distinguishes SSc phenotypes. PloS One 4:e5903.
Raza K, Falciani F, Curnow SJ et al (2005) Early rheumatoid arthritis is characterized by a distinct and transient synovial fluid cytokine profile of T cell and stromal cell origin. Arthritis Res Ther 7:R784-795.
Renauld JC (2003) Class II cytokine receptors and their ligands: key antiviral and inflammatory modulators. Nature Rev 3:667–676.
Roark CL, French JD, Taylor MA et al (2007) Exacerbation of Collagen-Induced Arthritis by Oligoclonal, IL-17-Producing {gamma}{delta} T Cells. J Immunol 179:5576–5583.
Roark CL, Simonian PL, Fontenot AP et al (2008) gammadelta T cells: an important source of IL-17. Curr Opin Immunol 20:353–357.
Ruprecht CR, Gattorno M, Ferlito F et al (2005) Co-expression of CD25 and CD27 identifies FoxP3+ regulatory T cells in inflamed synovia. J Exp Med 201:1793–1803.
Sakai A, Sugawara Y, Kuroishi T et al (2008) Identification of IL-18 and Th17 cells in salivary glands of patients with Sjogren's syndrome, and amplification of IL-17-mediated secretion of inflammatory cytokines from salivary gland cells by IL-18. J Immunol 181:2898–2906.
Sawalha AH, Kaufman KM, Kelly JA et al (2008) Genetic association of interleukin-21 polymorphisms with systemic lupus erythematosus. Ann Rheum Dis 67:458–461.
Shibata K, Yamada H, Hara H et al (2007) Resident V{delta}1+ {gamma}{delta} T Cells Control Early Infiltration of Neutrophils after Escherichia coli Infection via IL-17 Production. J Immunol 178:4466–4472.
Singh R, Aggarwal A, Misra R (2007) Th1/Th17 cytokine profiles in patients with reactive arthritis/undifferentiated spondyloarthropathy. J Rheum 34:2285–2290.
Stark MA, Huo Y, Burcin TL et al (2005) Phagocytosis of apoptotic neutrophils regulates granulopoiesis via IL-23 and IL-17. Immunity 22:285–294.
Sullivan KE, Piliero LM, Dharia T et al (2000) 3' polymorphisms of ETS1 are associated with different clinical phenotypes in SLE. Human mutation 16:49–53.

Szeliga J, Hess H, Rude E et al (1996) IL-12 promotes cellular but not humoral type II collagen-specific Th 1-type responses in C57BL/6 and B10.Q mice and fails to induce arthritis. Int Immunol 8:1221–1227.
Takahashi N, Vanlaere I, de Rycke R et al (2008) IL-17 produced by Paneth cells drives TNF-induced shock. J Exp Med 205:1755–1761.
Tran CN, Lundy SK, White PT et al (2007) Molecular interactions between T cells and fibroblast-like synoviocytes: role of membrane tumor necrosis factor-alpha on cytokine-activated T cells. Am J Pathol 171:1588–1598.
Trifari S, Kaplan CD, Tran EH et al (2009) Identification of a human helper T cell population that has abundant production of interleukin 22 and is distinct from T(H)-17, T(H)1 and T(H)2 cells. Nature Immunol 10:864–871.
Van Hamburg JP, Mus AM, de Bruijn MJ et al (2009) GATA-3 protects against severe joint inflammation and bone erosion and reduces differentiation of Th17 cells during experimental arthritis. Arthritis Rheum 60:750–759.
Van Hamburg JP, Asmawidjaja PS, Davelaar N et al (2011) Th17 cells, but not Th1 cells, from patients with early rheumatoid arthritis are potent inducers of matrix metalloproteinases and proinflammatory cytokines upon synovial fibroblast interaction, including autocrine interleukin-17A production. Arthritis Rheum 63:73–83.
Wang D, John SA, Clements JL et al (2005) Ets-1 deficiency leads to altered B cell differentiation, hyper-responsiveness to TLR9 and auto-immune disease. Internat Immunol 17:1179–1191.
Webb R, Merrill JT, Kelly JA et al (2009) A polymorphism within IL21R confers risk for systemic lupus erythematosus. Arthritis Rheum 60:2402–2407.
Wei G, Wei L, Zhu J et al (2009) Global mapping of H3K4me3 and H3K27me3 reveals specificity and plasticity in lineage fate determination of differentiating CD4+ T cells. Immunity 30:155–167.
Wolk K, Witte E, Wallace E et al (2006) IL-22 regulates the expression of genes responsible for antimicrobial defense, cellular differentiation, and mobility in keratinocytes: a potential role in psoriasis. Eur J Immunol 36:1309–1323.
Wong CK, Ho CY, Li EK et al (2009) Elevation of pro-inflammatory cytokine (IL-18, IL-17, IL-12) and Th2 cytokine (IL-4) concentrations in patients with systemic lupus erythematosus. Lupus 9:589–593.
Xu L, Kitani A, Fuss I et al (2007) Cutting edge: regulatory T cells induce CD4+CD25-Foxp3- T cells or are self-induced to become Th17 cells in the absence of exogenous TGF-beta. J Immunol 178:6725–6729.
Yago T, Nanke Y, Kawamoto M et al (2007) IL-23 induces human osteoclastogenesis via IL-17 in vitro, and anti-IL-23 antibody attenuates collagen-induced arthritis in rats. Arthritis Res Ther 9:R96.
Yamada H, Nakashima Y, Okazaki K et al (2008) Th1 but not Th17 cells predominate in the joints of patients with rheumatoid arthritis. Ann Rheum Dis 67:1299–1304.
Yang XO, Chang SH, Park H et al (2008) Regulation of inflammatory responses by IL-17F. J Exp Med 205:1063–1075.
Yang XO, Pappu BP, Nurieva R et al (2008) T helper 17 lineage differentiation is programmed by orphan nuclear receptors ROR alpha and ROR gamma. Immunity 28:29–39.
Yao Z, Fanslow WC, Seldin MF et al (1995) Herpesvirus Saimiri encodes a new cytokine, IL-17, which binds to a novel cytokine receptor. Immunity 3:811–21.
Yao Z, Painter SL, Fanslow WC et al (1995) Human IL-17: a novel cytokine derived from T cells. J Immunol 155:5483–86.
Zhou L, Ivanov, II, Spolski R et al (2007) IL-6 programs T(H)-17 cell differentiation by promoting sequential engagement of the IL-21 and IL-23 pathways. Nature Immunol 8:967–974.
Ziolkowska M, Koc A, Luszczykiewicz G et al (2000) High levels of IL-17 in rheumatoid arthritis patients: IL-15 triggers in vitro IL-17 production via cyclosporin A-sensitive mechanism. J Immunol 164:2832–2838.
Zrioual S, Ecochard R, Tournadre A et al (2009) Genome-wide comparison between IL-17A- and IL-17F-induced effects in human rheumatoid arthritis synoviocytes. J Immunol 182:3112–3120.

The Roles of IL-22 and Its Related Family Members in the Pathogenesis of Psoriasis

Patricia Valdez and Wenjun Ouyang

Abstract IL-22 is a T_H17 cytokine. It belongs to IL-10 family of cytokines that also includes IL-10, IL-19, IL-20, IL-24, and IL-26. IL-26 is also produced by T_H17 cells, while IL-24 is a T_H2 cytokine. All of these cytokines can also be produced by other leukocytes. Cytokine networks play essential roles in the pathogenesis of psoriasis. In psoriatic skin, the expression of IL-19, IL-20, IL-22, IL-24 and IL-26 is elevated. Infiltrating immune cells are the primary cellular sources. However, the receptors for these cytokines are expressed mainly on epithelial cells, including keratinocytes, but not on leukocytes. In psoriatic skin, these cytokines induce epidermal keratinocytes to display many pathogenic features, including hyperplasia, abnormal differentiation, and overexpression of psoriasin and other psoriatic markers. These cytokines mediate the crosstalk between infiltrating immune cells and epidermal keratinocytes.

1 Introduction

IL-10 was first defined in 1989 (Moore et al. 2001) as a pleiotropic cytokine that was identified by its ability to inhibit the functions of T_H1 cells. IL-10 is produced by a spectrum of leukocytes and exerts broad immune regulatory functions. It has taken almost 10 years for people to realize that IL-10 is actually only one member of a large group of cytokines that share structural homology (Pestka et al. 2004). The completion of the human genome sequencing has prompted many researchers to use bio-informatic approaches to identify novel genes based on their sequence homologies. Many new cytokine genes have been identified in the late 1990s and

W. Ouyang (✉)
Immunology Department, Genentech, Inc., South San Francisco, CA 94080, USA
e-mail: ouyang.wenjun@gene.com

S. Jiang (ed.), *T_H17 Cells in Health and Disease*,
DOI 10.1007/978-1-4419-9371-7_24, © Springer Science+Business Media, LLC 2011

Fig. 1 The percentage of identify of IL-20 sub-family cytokines

	IL-10	IL-19	IL-20	IL-22	IL-24	IL-26
IL-10	100.0	18.0	24.7	24.1	19.8	24.7
IL-19		100.0	40.3	15.8	26.2	20.4
IL-20			100.0	21.9	31.4	20.4
IL-22				100.0	22.8	21.0
IL-24					100.0	20.1
IL-26						100.0

early 2000s. IL-19, IL-20, IL-22, IL-24, IL-26, IL-28, and IL-29 are among these cytokines discovered during this period. These cytokines are classified as IL-10 family cytokines based on their structural similarity to IL-10. IL-28 and IL-29 share many biological functions with type I IFNs, and have also been referred to as IFNλ (Sheppard et al. 2003; Dumoutier et al. 2004). IL-19, IL-20, IL-22, IL-24, and IL-26 target epithelial cells and play similar biological roles. Because the functions of these 5 cytokines are largely different from IL-10, IL-28, and IL-29, we have sub-grouped these cytokines into the IL-20 sub-family (Sa et al. 2007). Here we focus our discussion on the discovery of these five cytokines and their potential pathogenic functions in psoriasis (Fig. 1).

2 The Discovery of IL-19, IL-20, IL-22, IL-24, and IL-26

IL-19 was identified through a genomic search for IL-10 homologues (Gallagher et al. 2000). Human IL-19 shares similar genomic exon and intron structure, and 18% amino acid identity with human IL-10. Human IL-19 has two different splicing forms. The shorter form of IL-19 has the consensus signal sequence and is easily secreted from cells when overexpressed. The longer splicing form includes an additional 38 amino acids at the 5′ end, which severely limits its secretion from cells (Gallagher et al. 2000). Structural analysis of recombinant human IL-19 revealed that it forms a unique seven helical bundled monomer, which is different from the intercalated dimer structure of IL-10 (Chang et al. 2003). IL-20 was first cloned by Blumberg and colleagues by mining EST databases (Blumberg et al. 2001). They revealed that IL-20 has 176 amino acids and shares sequence homology with other IL-10 family cytokines. Both IL-19 and IL-20 share six conserved cysteine residues, whereas IL-10 only has four conserved cysteine residues. This difference has a significant impact on the structure of IL-20. Similar to IL-19, IL-20 also does not form an intercalating dimer (Zdanov 2004). Both IL-19 and IL-20 are located within a conserved cytokine cluster on chromosome 1 in human and mice, and are closely linked to IL-10 (Jones and Flavell 2005). IL-24 is another gene that is located within this 195 kb cytokine cluster. In fact, it is the first gene among this group of cytokines to be identified. IL-24 was originally cloned as melanoma differentiation-associated (MDA) cDNA 7 from cultured human melanoma cells after treatment with IFNβ and mezerein by subtractive hybridization method (Jiang et al. 1995). This gene was also cloned from T_H2 cells, and named as FISP, because it is preferentially expressed by T_H2 cells and induced by IL-4 (Schaefer et al. 2001). Both names, MDA7 and FISP, were later replaced by the official name, IL-24, due to its conserved cytokine structure (Caudell et al. 2002).

The other two members of the IL-20 cytokine sub-family, IL-22 and IL-26, do not reside within the IL-10 cytokine cluster (Dumoutier et al. 2000a, b). Both genes are located on human chromosome 12. IL-22 is only 90 kb away from the IFNγ gene.

IL-26 is located in between IL-22 and IFNγ, and is 27 kb downstream of the IL-22 gene (Dumoutier et al. 2000a, b). IL-22 was first named as IL-TIF to represent IL-10 related T cell-derived inducible factor (Dumoutier et al. 2000a, b). It was isolated using a cDNA subtractive method from an IL-9 treated lymphoma cell line. IL-22 shares 22% identity with IL-10. The active form of IL-22 is also a monomer. IL-26 was originally identified as AK155 (Knappe et al. 2000). AK155 is induced in herpesvirus saimiri transformed T lymphocytes. It was cloned from these cells by subtractive hybridization. AK155/IL-26 shares 24% homology with human IL-10. However, there is no mouse IL-26 that has been identified. Unlike IL-19, IL-20, and IL-22, IL-26 can form a dimer upon being released from cells. Another unique feature of IL-26 is its ability to interact with heparin, which results in the association of IL-26 with cell surfaces (Hor et al. 2004). Heparin blocks the function of IL-26. The biological significance of this phenomenon is currently unclear.

3 Identification of the Receptors for This Group of Cytokines

The first receptor for this group of cytokines was identified by Gurney and colleagues from Genentech, Inc (Xie et al. 2000). They showed that IL-22 can bind and signal through a heterodimeric receptor complex, composed of IL-22R and IL-10Rβ chains. Human IL-22R has 574 amino acids, and shares sequence homology to IL-10Ra. These findings were subsequently confirmed by Pestka's group (Kotenko et al. 2001a, b). At almost the same time, IL-20 and its receptors were cloned (Blumberg et al. 2001). IL-20 binds to two orphan receptors of the class II cytokine receptor family. They were named IL-20Rα (IL-20R1) and IL-20Rβ (IL-20R2). It was soon realized that this receptor complex also serves as a functional receptor for IL-19 and IL-24 (Dumoutier et al. 2001a, b; ParrishNovak et al. 2002; Wang et al. 2002). IL-20Rβ chain can also pair with IL-22R, which serves as an additional functional receptor for both IL-20 and IL-24 (Dumoutier et al. 2001a, b). Interestingly, IL-19 cannot signal through this particular receptor pair. To make the story more complicated, the receptor for IL-26 was identified as the IL-20Rα and IL-10Rβ complex (Hor et al. 2004; Sheikh et al. 2004). As a result, IL-20Rβ is required for

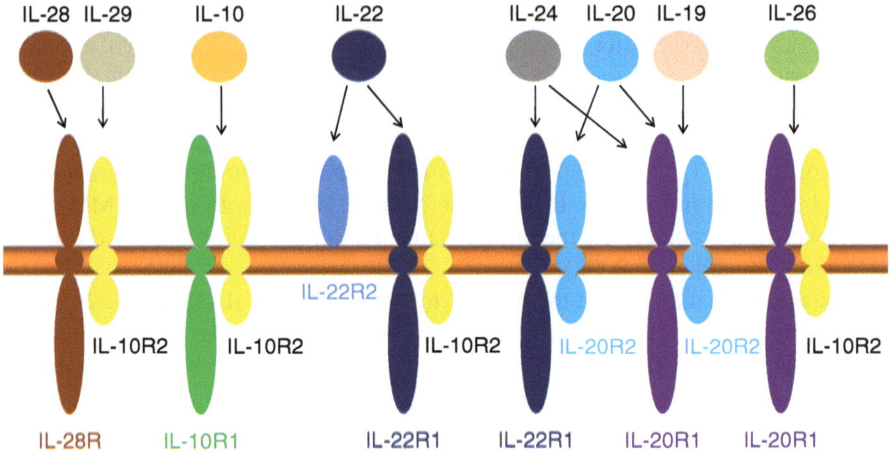

Fig. 2 Receptors for IL-10 family cytokines

the functions of IL-19, IL-20, and IL-24 (Fig. 2). IL-20Rα is used by IL-19, IL-20, IL-24, and IL-26. IL-22R can deliver signals for IL-20, IL-22, and IL-24. In addition, IL-10Rβ is necessary for IL-10, IL-22, IL-26, as well as for the interferon λ family cytokines, IL-28 and IL-29 (Fig. 2). IL-10Rα is still the unique receptor for IL-10. Consequently, the complicated receptor and ligand interactions will require more careful studies to dissect the functions of each individual cytokine. In addition to all the membrane bound receptors, a soluble receptor, IL-22BP or IL-22R2, has been cloned (Dumoutier et al. 2001a, b; Kotenko et al. 2001a, b; Xu et al. 2001). IL-22BP shares structural similarity to the extracellular domains of IL-22R and IL-20Rα, but it does not contain a trans-membrane domain or intracellular tail. In vitro, IL-22BP blocks the function of IL-22.

Like IL-10, all IL-20 sub-family cytokines activate Stat molecules upon binding to their receptors (Xie et al. 2000; Blumberg et al. 2001; Dumoutier et al. 2001a, b). Stat3 is the dominant Stat molecule that is activated by these cytokines. IL-22 has been reported to be able to activate Stat1 and Stat5 as well (Xie et al. 2000; Lejeune et al. 2002). In addition to Stat pathways, IL-22 can also activate MAPK pathways (Lejeune et al. 2002). In a rat hepatoma cell line, IL-22 induced phosphorylation of ERK1/2, JNK and p38 kinase. Many studies have shown that IL-24 can induce apoptosis of various tumor cells (Sarkar et al. 2007). Some studies suggested that this function of IL-24 is independent of its ability to activate Stat3. The exact signaling molecules that mediate this phenomenon, however, are still unclear (Kreis et al. 2007, 2008).

4 T_H17 and Other Cellular Sources of IL-22

CD4+ helper T cells are diverse in function and have been categorized into subsets according to these functions, the most studied thus far being T_H1, T_H2, and T_H17. T_H1 cells have been describe as helpers for cell-mediated immunity, largely producing

IFNγ, and T_H2 cells have been described as helpers for humeral immunity and helminth infection, largely producing IL-4. More recently the T_H17 subset has been described as mediators of both infection and inflammation, producing IL-17A, IL-17F, IL-22, and IL-21 (Ouyang et al. 2008; Korn et al. 2009).

The first evidence for a distinct T cell subset expressing IL-17 came from the observation that splenocytes stimulated with microbial lipo-peptide produced large amounts of IL-17 compared to T_H1 or T_H2-type cytokines (InfanteDuarte et al. 2000). In an effort to further understand the regulation of IL-17, it was found that a newly recognized cytokine, IL-23, enhanced IL-17 production from memory cells (Aggarwal et al. 2003). IL-23 exists as a heterodimer of p19 and p40, which is shared with IL-12. For many years it was believed that T_H1 cells were the sole mediators of Mog peptideinduced EAE, but soon it was recognized that IL-23 played a larger role than IL-12 (Cua et al. 2003). T cells stimulated with IL-23 could induce much more severe disease compared to T cells stimulated with IL-12, suggesting IL-17 producing T cells could mediate inflammatory disease (Langrish et al. 2005). Efforts to determine the conditions required to generate T_H17 cells from naïve T cell precursors in vitro were met with some controversy. It has generally been accepted that T_H17 cells can be generated in vitro from naive cells via TGFβ and IL-6, plus IL-23 or IL-21 (Bettelli et al. 2006; Mangan et al. 2006; Veldhoen et al. 2006; Korn et al. 2007). Given the role of T_H17 cells in the inflammatory response, the products mediating this inflammatory response became very interesting. IL-22 was soon described as a product of T_H17 cells (Chung et al. 2006; Liang et al. 2006; Zheng et al. 2007). It soon becomes apparent that IL-22 and IL-17 are differentially regulated as IL-17 requires both TGFb and IL-6 whereas IL-22 only requires IL-6 and is inhibited by TGFb (Zheng et al. 2007). Based on this finding, it was predicted that IL-22, but not IL-17 producing T cells could exist. The first in vivo evidence for independent expression of IL-17 and IL-22 came from analysis of atopic dermatitis skin biopsies (Nograles et al. 2009). The presence of IL-22 producing T cells that did not produce IL-17 positively correlated with disease severity. IL-22-expressing T cells were also found in skin from psoriasis, atopic eczema, and allergic contact dermatitis (Eyerich et al. 2009; Kagami et al. 2009).

Recently, T cells exclusively producing IL-22 were isolated from human PBMCs which express CD4, CCR6, and the skin homing receptors CCR10 and CCR4 (Duhen et al. 2009; Trifari et al. 2009). IL-6 and TNF appear sufficient to differentiate these cells from naïve human T cells (Duhen et al. 2009). Given their expression of skin homing receptors, it makes sense that IL-22-producing T cells could play an important role is skin surveillance. Further evidence for a role in skin came from a study that showed both memory and naïve CD4 T cells that are activated by human Langerhan's cells induced high IL-22 production (Fujita et al. 2009). It is possible that LCs might induce the same recently described IL-22-producing T cells.

Although much work initially focused on T cell production of IL-22, T cells are not the only source of IL-22. The first evidence of an important role for innate production of IL-22 came from a study in which IL-22 was found to mediate protection from infection with the attaching/effacing gut bacterium, Citrobacter rodentium (Zheng et al. 2008). In this study, it was found that IL-22 production was still induced in Rag2/mice during infection, and blocking IL-22 led to early mortality. NK cells, γδ T cells, monocytes and CD8 T cells were initially described as nonCD4 T cell sources of IL-22 (Wolk et al. 2002; Zheng et al. 2007). Subsequent studies revealed that a cell type known as the lymphoidtissue inducer (LTi) cell also produced IL-22 (Cupedo et al. 2009; Takatori et al. 2009). LTi cells are required to form secondary lymphoid organs, including Peyer's patches. They express lymphotoxin and in humans they can express NKcell markers including CD56 (Cupedo et al. 2009). Human LTi cells are found in the mesenteric lymph node and tonsils and respond to IL-23 stimulation by expressing both IL-22 and IL-17 (Colonna 2009; Cupedo et al. 2009). Mouse LTi and LTilike cells can be found in the spleen and intestinal lamina propria where they express lymphotoxin, CD4, CD127, and CCR6. Like human LTi, mouse LTi respond to IL-23 stimulation by producing both IL-22 and IL-17.

Another cell type that shares some similarities with LTi cells is the newly identified NK cell subset marked by NKp44 expression in humans and NKp46 expression in mice. Unlike traditional NK cells, NKp46+ and NKp44+ cells do not produce perforin or IFNγ (Colonna 2009). Mouse NKp46+ cells are lineage negative and express CD127 and cKit, similar to LTi cells. These IL-22 producing cells were found in the Peyer's patches and intestinal lamina propria, possibly in or around crypto patches (SatohTakayama et al. 2008; Luci et al. 2009; Sanos et al. 2009). In addition, Nkp46+ cells that are NK1.1 low or intermediate also express high levels of Rorγt and IL-22, most likely induced by the presence of microbial flora in the gut. A large fraction of IL-22 producing cells isolated from the small intestine lamina propria of Citrobacter rodentium infected mice were NKp46+, providing more evidence for the induction of IL-22 production by bacteria (Cella et al. 2009). Human NKp44+ cells express CCR and are both CD56+ and CD56. Unlike mouse Nkp46+ cells which secrete only IL-22, these cells secrete both IL-22 and IL-17 in response to IL-23. Similar to Nkp46 cells, they are found in peyer's patches, tonsils, and intestinal lamina propria during bacterial infection (Cella et al. 2009).

Dendritic cells were also found to produce IL-22 in response to IL-23 (Zheng et al. 2008). Using IL-23 RGFP reporter mice, it was shown that γδ T cells, macrophage, dendritic cells, a subset of macrophage, and LTilike cells all express the IL-23 receptor. In addition, lymph node myeloid cells expressing IL-23R responded to IL-23 by producing IL-22 (Awasthi et al. 2009). IL-22 was found to be produced by dendritic cells in the gut, suggesting dendritic cells could play an important role in protection from infection (Zheng et al. 2008; Awasthi et al. 2009; Pickert et al. 2009).

Finally, γδ cells were shown to express IL-23R and produce IL-17, IL-21, and IL-22 in response to IL-23 plus IL-1b (Martin et al. 2009; Sutton et al. 2009). These cells promoted CD4 T cell production of IL-17 and increase susceptibility to EAE, suggesting a role for gamma delta T cells in amplification of the T_H17 cell response (Sutton et al. 2009). Although γδ T cells produce both IL-17 and IL-22, it was shown that IL-22 expression was dependent on AHR while IL-17 was not, providing further evidence for differential regulation of IL-22 and IL-17 (Martin et al. 2009).

5 Cellular Sources and Regulation of the IL-20 Sub-family of Cytokines

Given its similarity to IL-10, the initial search for the cellular sources of IL-19 was focused on IL-10 producing cells, such as monocytes. Indeed, LPS treated monocytes produce both IL-10 and IL-19 in vitro (Gallagher et al. 2000). In vivo, IL-19 mRNA is also induced in the liver upon LPS challenge (Wegenka et al. 2007).Cytokines such as GMCSF, IL-4 and IL-13 can also augment IL-19 expression from monocytes (Gallagher et al. 2000). Recently, IL-19 mRNA has also been demonstrated to be induced in various epithelial cells and fibroblasts. For example, extracellular HIVI Tat protein induces both IL-19 and IL-20 expression from MCF7 and AV3 cell lines (Bettaccini et al. 2005). Triggering adenosine receptors by 5(Nethylcarboxamido) adenosine (NECA) induces IL-19 production from human bronchial epithelial (HBE) cells (Zhong et al. 2006). In addition, IL-4, IL-13, IL-17 and IL-22 stimulate the expression of IL-19 from HBE cells (Huang et al. 2008) (Aujla et al. 2008). IL-1β can also augment IL-19 message level from keratinocytes (Yano et al. 2008). The majority of data from these studies, however, relied on the measurement of IL-19 transcript levels. Further confirmation of these results at the protein level will be required.

The regulation of IL-20 resembles that of IL-19. IL-20 is primarily produced by myeloid cells (Wolk et al. 2002). This is not unexpected given its close link with IL-19 on chromosome 1. LPS stimulation induces IL-20 production from monocytes (Wolk et al. 2002), primary glial cells (Hosoi et al. 2004), and macrophage cell lines (Hosoi et al. 2004). Monocyte-derived dendritic cells (DCs) maintain their ability to express IL-20, as well as IL-19. However, monocyte-derived macrophages terminate IL-20 expression, despite retaining their low IL-19 expression (Wolk et al. 2008). Neither cell population produces IL-24 or IL-22. In addition to leukocytes, keratinocytes are a potential source of IL-20. Keratinocytes stimulated with ultraviolet B light (Hunt et al. 2006), IL-1β (Kunz et al. 2006; Otkjaer et al. 2007) and IL-22 augment the transcription of IL-20 (Sa et al. 2007; Tohyama et al. 2009; Wolk et al. 2009a, b). NFκB and p38 pathways are required for IL-20 expression in keratinocytes upon IL-1β treatment (Otkjaer et al. 2007).

Although IL-24 resides within the same locus of IL-19 and IL-20, the production and regulation of IL-24 is somewhat different from that of IL-19 and IL-20. IL-24 can be produced by myeloid cells upon LPS stimulation (Wolk et al. 2002). However, T cells, and in particular T_H2 cells are also a potential source of IL-24 (Schaefer et al. 2001). IL-4 plays an important role in the induction of IL-24 production from T_H2 cells. Similar regulation in T_H2 cells has not been observed for IL-19 and IL-20. Cells of non-hematopoietic origin can also produce IL-24. Similar to IL-19 and IL-20, IL-24 transcript levels are also augmented in keratinocytes upon IL-1β treatment (Yano et al. 2008). Many tumor cells, especially melanoma cells, are reported to be the potential sources of IL-24 (Fisher 2005).

As discussed above, human IL-26 is closely linked to IL-22. IL-26 is mainly produced by T cells, and has also been recently identified as an effector cytokine for human T_H17 cells (Pene et al. 2008).

In summary, we are still at the very early stages of understanding the production and regulation of this family of cytokines. The regulatory pathways, especially during disease processes, for IL-19, IL-20, and IL-24 need to be more fully elucidated in the future.

6 Biology of IL-20 Sub-family Cytokines

One of the key differences between IL-20 sub-family cytokines and IL-10 is in their cellular targets. IL-10 is an immune regulatory cytokine that directly acts on various immune cells (Moore et al. 2001). IL-10Rα is primarily expressed by leukocytes, whereas IL-10Rβ chain is ubiquitously expressed. After the discovery of the receptors for IL-20 sub-family cytokines, the expression of these receptors has been investigated. To date, overwhelming data suggest that IL-20Rα, IL-20Rβ, and IL-22R are mainly expressed on cells of epithelial origin or on fibroblasts, but not on leukocytes (Wolk et al. 2002; Ouyang et al. 2008). These data point to the biological functions of the IL-20 sub-family cytokines on epithelial cells.

Keratinocytes, synovial cells and liver hepatocytes are all reported to respond to IL-19 (Kunz et al. 2006; Sa et al. 2007; Wegenka et al. 2007; Sakurai et al. 2008). As we will discuss later, the functions of IL-19 on keratinocytes are similar to those of other family cytokines. The potential unique functions of IL-19 on other epithelial cells remain unclear. Some studies claim that IL-19 can induce various immune responses from leukocytes. For example, IL-19 has been reported to induce IL-6 and TNFα from monocytes (Liao et al. 2002), and IL-10 from DCs (Jordan et al. 2005), but it also promotes T_H2 cytokine production from both murine and human T cells (Liao et al. 2004; Oral et al. 2006). However, these studies failed to provide convincing evidence that the receptors for IL-19 are present on these cells. It is possible that there are additional receptors, other than IL-20Rα and IL-20Rβ, used by IL-19. Future studies will be required to confirm these findings.

IL-20 and IL-24 have diverse biological functions. We will focus our discussion on their roles in keratinocytes and inflammatory skin diseases in the next section. In addition, studies also support the potential functions of IL-20 in hematopoiesis and angiogenesis (Liu et al. 2003; Tritsaris et al. 2007). In vitro, IL-20 specifically increases multi-potential progenitors from CD34+ bone marrow cell culture (Liu et al. 2003). IL-20 also promotes corneal angiogenesis, endothelial cell sprouting and tube formation (Tritsaris et al. 2007). Although there are studies demonstrating IL-20 production from immune cells, the expression of IL-20 receptors on these cells has not been confirmed (Hummelshoj et al. 2006; Oral et al. 2006). There is large body of literature on the functions of IL-24 in repression of various tumor cells (Fisher 2005). The exact mechanisms of IL-24 tumor repression are still controversial (Kreis et al. 2007, 2008). However, IL-24 shares many similar biological functions with other family members on keratinocytes. Due to a lack of the murine homology, the biological function of human IL-26 is still largely unknown.

The importance of IL-22 in infectious diseases and auto-immune diseases has been recently revealed (Ouyang et al. 2008). Many in vitro studies on epithelial cells treated with IL-22 indicated that IL-22 is a pro-inflammatory cytokine (Wolk et al. 2004; Boniface et al. 2005; Sa et al. 2007) (Aggarwal et al. 2001; Nagalakshmi et al. 2004; Whittington et al. 2004; Aujla et al. 2008). IL-22 induces not only pro-inflammatory cytokines and chemokines, but also various antimicrobial peptides, hinting at its role in controlling infections (Wolk et al. 2004; Boniface et al. 2005). In addition, IL-22 also alters the proliferation and differentiation properties of various cells, especially keratinocytes (Boniface et al. 2005; Sa et al. 2007). Furthermore, IL-22 promotes tissue remodeling and healing through the induction of proteinases and protease inhibitors (Wolk et al. 2006; Sa et al. 2007). Recently, Koll's group and our group revealed the crucial roles of IL-22 in host defense against extracellular bacterial infections, and protection of the integrity of mucosal epithelial layers (Aujla et al. 2008; Zheng et al. 2008). During Klebsiella pneumonia infection in the lung and Citrobacter rodentium infection in the colon, IL-22 is required to help the host establish a sufficient first line of epithelial defense, without which the bacteria can quickly spread systemically and result in mortality. The functions of IL-22 in these models include the induction of antimicrobial peptides, such as Reg family proteins and lipocalin 2, and restoration of epithelial integrity. These studies are among the first to reveal the essential functions of IL-20 sub-family cytokines in infectious diseases, suggesting similar roles for other family members. In addition, several groups have demonstrated the role of IL-22 in preventing tissue damage in auto-immune diseases (Pan et al. 2004; Radaeva et al. 2004; Zenewicz et al. 2007; Sugimoto et al. 2008). IL-22 protects the liver from damage in ConA induced hepatitis (Radaeva et al. 2004; Zenewicz et al. 2007). IL-22 also helps to reduce disease severity in several models of IBD (Sugimoto et al. 2008). In a TCRα knockout model of colitis and DSS induced colitis, IL-22 can enhance colon mucus secretion and boost the function of goblet cells. In conclusion, IL-22 and other IL-20 sub-family cytokines are important effector cytokines that mediate essential physiological functions in host defense of infection, in auto-immune diseases, and in other immune processes.

7 IL-20 Sub-family Cytokines in the Pathogenesis of Psoriasis

Psoriasis is an auto-immune inflammatory skin disease (Nickoloff and Nestle 2004; Lowes et al. 2007). The first evidence providing the link for this group of cytokines with psoriasis came from the study of IL-20 transgenic mice (Blumberg et al. 2001). Mice overexpressing IL-20 developed neonatal lethality and were smaller in size. Strikingly, the skin of these transgenic mice was abnormal with a tight, shiny, and wrinkled appearance. Histological analyses further revealed that these mice had a thickened epidermis with hyperkerotosis and a compact stratum corneum, all of which are features seen in psoriatic skin (Blumberg et al. 2001).

Based on this breakthrough, overwhelming data has accumulated in the past few years to further support the pathogenic roles of this group of cytokines in psoriasis (Ouyang et al. 2008).

The potential genetic associations of IL-19, IL-20, IL-24 and IL-20 receptors with psoriasis have been investigated. In one study, IL-20 1053 G allele was reported to be associated with psoriasis (Kingo et al. 2004). The same group further showed that some haplotypes of IL-19, IL-20 and IL-24 genes were associated with an increased risk of psoriasis (Koks et al. 2004, 2005). In addition, while there is no association of any individual SNP within IL-20Rα and IL-20Rβ loci with psoriasis, some haplotypes of these two genes render either an association or protective effect for psoriasis (Kingo et al. 2008). These findings, however, should be confirmed in independent psoriatic populations.

IL-20 sub-family cytokines are upregulated in psoriatic skin (Romer et al. 2003). The transcripts of IL-19, and IL-20 and IL-24 have been detected by in situ hybridization in psoriatic skin (Romer et al. 2003). The upregualtion of these cytokines was specific for lesional, but not uninvolved, psoriatic skin. The expression of IL-24 was detected primarily in infiltrating mononuclear cells in the dermis. However, the message for both IL-19 and IL-20 was located in the epidermis, especially confined to the basal and suprabasal keratinocytes. As we discussed above, keratinocytes upregulate IL-19 or IL-20 upon stimulation with IL-1β, IL-22, or UVB light (Hunt et al. 2006; Kunz et al. 2006; Otkjaer et al. 2007; Sa et al. 2007; Tohyama et al. 2009; Wolk et al. 2009a, b). A study from Krueger's group, however, suggested that CD68+/CD11c+ myeloid derived cells are prominent sources of IL-20 in psoriatic skin (Wang et al. 2006). This finding is also consistent with earlier in vitro results showing that monocytes produce IL-19, IL-20, and IL-24 when stimulated with LPS (Wolk et al. 2002). Therefore, whether keratinocytes, myeloid cells, or both are the primary cellular sources of these cytokines during the pathogenesis of psoriasis needs to be further investigated. Nevertheless, these augmented IL-19, IL-20, and IL-24 levels in psoriatic skin are significantly reduced when disease symptoms are ameliorated upon treatment with cyclosporine A, Calcipotriol, IL-4, or alefacept (Ghoreschi et al. 2003; Romer et al. 2003; Otkjaer et al. 2005; Wang et al. 2006). A recent study utilizing a human xenograft model showed that IL-20 blockade helped resolve psoriasis lesions and prevent psoriasis induction when psoriatic skin was transferred to immune-deficient mice (Stenderup et al. 2009). IL-20 receptors are expressed on keratinocytes and are detected in psoriatic skin (Blumberg et al. 2001; Romer et al. 2003). An early study suggested that IL-20 receptors are augmented in psoriatic skin (Blumberg et al. 2001). Two subsequent studies, however, have shown that compared to expression levels in normal or non-lesional skin, expression levels of IL-20 receptors are actually decreased in lesional psoriatic skin, suggesting a potential negative feedback loop for the regulation of these receptors upon ligand engagement (Otkjaer et al. 2005; Wang et al. 2006).

IL-22 is also upregulated in lesional psoriatic skin (Wolk et al. 2004; Boniface et al. 2007; Zaba et al. 2007). IL-22 is an effector cytokine of T_H17 cells (Liang et al. 2006; Zheng et al. 2007). T cells isolated from psoriatic skin produce high level of

IL-22, as well as IL-17 (Boniface et al. 2007; Lowes et al. 2008; Pene et al. 2008). Recent studies have demonstrated that IL-23 plays an essential role in regulating IL-22 expression in various leukocytes, including in T_H17 cells (Ouyang et al. 2008). IL-23 is composed of p19 and p40 sub-units (Kastelein et al. 2007), and the expression of both sub-units is elevated in lesional psoriatic skin (Lee et al. 2004). Importantly, an antip40 antibody has shown impressive clinical efficacy in the treatment of psoriasis (Krueger et al. 2007). In addition, antiTNFa therapies, such as etanercept, which are associated with reduced T_H17 responses including the reduction of IL-22, can significantly improve the symptoms of psoriatic patients, (Zaba et al. 2007) (Caproni et al. 2009).

Many in vitro studies have helped to reveal the mechanisms downstream of the IL-20 sub-family cytokines in the pathogenesis of psoriasis. Early in vitro studies focused on the functions of IL-22 on primary human keratinocytes. An early study demonstrated that IL-22 induced β defensin 2 and β defensin 3 from keratinocytes (Wolk et al. 2004). A follow up microarray study of IL-22-treated keratinocytes examined 1,126 genes (Wolk et al. 2006). Approximately 1% of these genes responded to the treatment of IL-22 in keratinocytes, including genes involved in antimicrobial defense, cell differentiation and cell mobility (Wolk et al. 2006). Morel and colleagues adopted a more powerful in vitro system to study the biology of this group of cytokines on human skin. They used the reconstituted human epidermis (RHE) culture system to explore the biology of IL-22 (Boniface et al. 2005). In this 3D culture system with stratified epidermal keratinocytes, IL-22 promoted the hyper-proliferation of keratinocytes, hypogranulosis, and the expression of many psoriatic markers such as psoriasin (S100A7), all of which are features observed in psoriatic skin (Boniface et al. 2005). Inspired by this research, we decided to use this system to systematically study the biology of IL-20 sub-family cytokines, not only because the RHE system closely resembled human epidermis, but also because IL-20 and IL-22 receptors are unstable and expressed at a low level on monolayer cultures of primary keratinocytes (Valdez et al. 2007). Although IL-20, IL-22, and IL-24 can promote the proliferation of monolayer keratinocytes, their responses are not as robust and induce much less gene expression in monolayer culture than is induced in the RHE culture system. Therefore, the RHE system is perhaps the most ideal system to study the biology of IL-20 sub-family cytokines in epidermal keratinocytes. We noticed that IL-19, IL-20, IL-22, and IL-24 all had very similar biology, although to a differing extents in RHE, despite the different receptor usages by these cytokines. All of these cytokines induced acanthosis, Stat3 activation, and expression of many psoriatic markers in RHE. We also performed genome wide microarray analysis with chips containing 54,675 probes. The gene expression profiles further confirmed our conclusion that this group of cytokines induced very similar biology in epidermal keratinocytes. However, the potencies of these cytokines were clearly different, with IL-22 having the greatest effect, followed by IL-24, IL-20, and IL-19. IL-26 failed to induce any noticeable biology in our study. The reason for this observation remains unclear. Compared to previous studies, our data revealed many additional biological effects of these cytokines on epidermal keratinocytes. Most importantly, the IL-20 family cytokines induced a large group of chemokines, such as IL-8/CXCL8, CXCL7, and CXCL1, in RHE. These chemokines may play essential roles in recruiting various immune cells during the pathogenesis of psoriasis. In addition, these cytokines also promoted the production of HBEGF and VEGF, both of which may also participate in psoriasis. Finally, the genes regulated by IL-20 sub-family cytokines in the RHE system were highly correlated with the genes regulated in psoriatic skin. In conclusion, all of the in vitro data supported the pathogenic functions of IL-20 sub-family cytokines in the onset of psoriasis.

Studies using in vivo pre-clinical models have further corroborated this notion. As previously discussed, mice overexpressing IL-20 developed a skin phenotype resembling that observed in psoriatic skin, by stimulating abnormal proliferation and differentiation of the epidermis (Blumberg et al. 2001). Similar results have also been observed in IL-22 and IL-24 transgenic mice (He and Liang 2010; Wolk et al. 2009a, b). This data indicates that in vivo, IL-20 and IL-22 exert similar biological functions as they do in vitro. Since IL-23 is essential for IL-22 production in vitro, we hypothesized that the pathogenic functions of IL-23 in psoriasis might be partially mediated by IL-22. Injection of IL-23 into mouse ears leads to an inflammatory skin phenotype, including dermal inflammation and epidermal acanthosis (Kopp et al. 2003). We found that IL-23 also stimulated IL-22 production, as well as IL-17 production in the ear skin (Zheng et al. 2007). However, we did not detect the upregulation of IL-19, IL-20, or IL-24 in this system. Ablation of IL-22, both in IL-22 deficient mice and by using an IL-22 neutralizing antibody, dramatically reduced the ear thickness and epidermal acanthosis induced by IL-23. In addition, the neutrophilic infiltration and dermal inflammation were decreased. Our study, for the first time, established that IL-22 linked IL-23 and epidermal keratinocyte hypo-proliferation. Psoriasis is characterized by immune cell infiltration in both dermis and epidermis, as well as epidermal keratinocyte hyper-proliferation. The mechanisms of hyper-proliferation induced by immune cell activation are not well understood. Our data suggested that IL-22, as well as the other IL-20 sub-family cytokines, may provide the missing links between the immune system and epidermal keratinocytes in the pathogenesis of psoriasis (Zheng et al. 2007). Interestingly, in a modified model, IL-23 was injected in the back of the mouse instead of the ear. In this model, IL-23 also induced the expression of IL-19 and IL-24, but not IL-20, in the skin (Chan et al. 2006). Indeed, in this model, IL-23 failed to induce an inflammatory skin phenotype in IL-20Rβ deficient mice, which lack IL-19, IL-20, and IL-24 signaling. The mechanisms underlying the differences observed in both studies are unclear. IL-22 is also upregulated in the second model, but its role has not been examined. A recent study further demonstrated that IL-22 is also required for the development of inflammatory skin disease in the CD45Rb high T cell transfer model of psoriasis (Ma et al. 2008). This is also an IL-23 dependent model, since antibody blocking p40 inhibits the development of disease. In summary, these in vivo studies have established the essential roles of IL-20 sub-family cytokines in psoriasis.

8 Conclusions

Great progress has been made in the past few years in understanding the biological functions of the IL-20 sub-family of cytokines. These cytokines are produced by various immune cells and regulate the biological functions of epithelial cells. The crosstalk mediated by these cytokines unquestionably plays an essential role in various immune responses, especially during infectious and auto-immune diseases. There are, however, still many unanswered questions. For example, how are these cytokines regulated in vivo during the inflammatory processes? What are the unique or redundant roles of each individual cytokine in diseases, given their overlapping receptor usage? Answers to these questions will help to develop future therapies for both auto-immune and infectious diseases.

References

Aggarwal, S., N. Ghilardi, et al. (2003). "Interleukin-23 promotes a distinct CD4 T cell activation state characterized by the production of interleukin-17." *J Biol Chem* **278**(3): 1910–4.

Aggarwal, S., M. H. Xie, et al. (2001). "Acinar cells of the pancreas are a target of interleukin-22." *J Interferon Cytokine Res* **21**(12): 1047–53.

Aujla, S. J., Y. R. Chan, et al. (2008). "IL-22 mediates mucosal host defense against Gram-negative bacterial pneumonia." *Nat Med* **14**(3): 275–81.

Awasthi, A., L. Riol-Blanco, et al. (2009). "Cutting edge: IL-23 receptor gfp reporter mice reveal distinct populations of IL-17-producing cells." *J Immunol* **182**(10): 5904–8.

Bettaccini, A. A., A. Baj, et al. (2005). "Proliferative activity of extracellular HIV-1 Tat protein in human epithelial cells: expression profile of pathogenetically relevant genes." *BMC Microbiol* **5**(1): 20.

Bettelli, E., Y. Carrier, et al. (2006). "Reciprocal developmental pathways for the generation of pathogenic effector TH17 and regulatory T cells." *Nature* **441**(7090): 235–8.

Blumberg, H., D. Conklin, et al. (2001). "Interleukin 20: discovery, receptor identification, and role in epidermal function." *Cell* **104**(1): 9–19.

Boniface, K., F. X. Bernard, et al. (2005). "IL-22 inhibits epidermal differentiation and induces proinflammatory gene expression and migration of human keratinocytes." *J Immunol* **174**(6): 3695–702.

Boniface, K., E. Guignouard, et al. (2007). "A role for T cell-derived interleukin 22 in psoriatic skin inflammation." *Clin Exp Immunol* **150**(3): 407–15.

Caproni, M., E. Antiga, et al. (2009). "Serum levels of IL-17 and IL-22 are reduced by etanercept, but not by acitretin, in patients with psoriasis: a randomized-controlled trial." *J Clin Immunol* **29**(2): 210–4.

Caudell, E. G., J. B. Mumm, et al. (2002). "The protein product of the tumor suppressor gene, melanoma differentiation-associated gene 7, exhibits immunostimulatory activity and is designated IL-24." *J Immunol* **168**(12): 6041–6.

Cella, M., A. Fuchs, et al. (2009). "A human natural killer cell subset provides an innate source of IL-22 for mucosal immunity." *Nature* **457**(7230): 722–5.

Chan, J. R., W. Blumenschein, et al. (2006). "IL-23 stimulates epidermal hyperplasia via TNF and IL- 20R2-dependent mechanisms with implications for psoriasis pathogenesis." *J Exp Med* **203**(12): 2577–87.

Chang, C., E. Magracheva, et al. (2003). "Crystal structure of interleukin-19 defines a new sub-family of helical cytokines." *J Biol Chem* **278**(5): 3308–13.

Chung, Y., X. Yang, et al. (2006). "Expression and regulation of IL-22 in the IL-17-producing CD4+ T lymphocytes." *Cell Res* **16**(11): 902–7.

Colonna, M. (2009). "Interleukin-22-producing natural killer cells and lymphoid tissue inducer-like cells in mucosal immunity." *Immunity* **31**(1): 15–23.

Cua, D. J., J. Sherlock, et al. (2003). "Interleukin-23 rather than interleukin-12 is the critical cytokine for autoimmune inflammation of the brain." *Nature* **421**(6924): 744–8.

Cupedo, T., N. K. Crellin, et al. (2009). "Human fetal lymphoid tissue-inducer cells are interleukin 17-producing precursors to RORC+ CD127+ natural killer-like cells." *Nat Immunol* **10**(1): 66–74.

Duhen, T., R. Geiger, et al. (2009). "Production of interleukin 22 but not interleukin 17 by a subset of human skin-homing memory T cells." *Nat Immunol* **10**(8): 857–63.

Dumoutier, L., C. Leemans, et al. (2001a). "Cutting edge: STAT activation by IL-19, IL-20 and mda-7 through IL-20 receptor complexes of two types." *J Immunol* **167**(7): 3545–9.

Dumoutier, L., D. Lejeune, et al. (2001b). "Cloning and characterization of IL-22 binding protein, a natural antagonist of IL-10-related T cell-derived inducible factor/IL-22." *J Immunol* **166**(12): 7090–5.

Dumoutier, L., J. Louahed, et al. (2000a). "Cloning and characterization of IL-10-related T cell-derived inducible factor (IL-TIF), a novel cytokine structurally related to IL-10 and inducible by IL-9." *J Immunol* **164**(4): 1814–9.

Dumoutier, L., A. Tounsi, et al. (2004). "Role of the interleukin (IL)-28 receptor tyrosine residues for antiviral and antiproliferative activity of IL-29/interferon-lambda 1: similarities with type I interferon signaling." *J Biol Chem* **279**(31): 32269–74.

Dumoutier, L., E. Van Roost, et al. (2000b). "IL-TIF/IL-22: genomic organization and mapping of the human and mouse genes." *Genes Immun* **1**(8): 488–94.

Eyerich, S., K. Eyerich, et al. (2009). "Th22 cells represent a distinct human T cell subset involved in epidermal immunity and remodeling." *J Clin Invest* **119**(12): 3573–85.

Fisher, P. B. (2005). "Is mda-7/IL-24 a "magic bullet" for cancer?" *Cancer Res* **65**(22): 10128–38.

Fujita, H., K. E. Nograles, et al. (2009). "Human Langerhans cells induce distinct IL-22-producing CD4+ T cells lacking IL-17 production." *Proc Natl Acad Sci USA* **106**(51): 21795–800.

Gallagher, G., H. Dickensheets, et al. (2000). "Cloning, expression and initial characterization of interleukin-19 (IL-19), a novel homologue of human interleukin-10 (IL-10)." *Genes Immun* **1**(7): 442–50.

Ghoreschi, K., P. Thomas, et al. (2003). "Interleukin-4 therapy of psoriasis induces Th2 responses and improves human autoimmune disease." *Nat Med* **9**(1): 40–6.

He, M. and P. Liang (2010) "IL-24 Transgenic Mice: In Vivo Evidence of Overlapping Functions for IL-20, IL-22, and IL-24 in the Epidermis." *J Immunol* **184**(4):1793–8.

Hor, S., H. Pirzer, et al. (2004). "The T-cell lymphokine interleukin-26 targets epithelial cells through the interleukin-20 receptor 1 and interleukin-10 receptor 2 chains." *J Biol Chem* **279**(32): 33343–51.

Hosoi, T., S. Wada, et al. (2004). "Bacterial endotoxin induces IL-20 expression in the glial cells." *Brain Res Mol Brain Res* **130**(1-2): 23–9.

Huang, F., S. Wachi, et al. (2008). "Potentiation of IL-19 expression in airway epithelia by IL-17A and IL- 4/IL-13: important implications in asthma." *J Allergy Clin Immunol* **121**(6): 1415–21, 1421 e1-3.

Hummelshoj, L., L. P. Ryder, et al. (2006). "The role of the interleukin-10 subfamily members in immunoglobulin production by human B cells." *Scand J Immunol* **64**(1): 40–7.

Hunt, D. W., W. A. Boivin, et al. (2006). "Ultraviolet B light stimulates interleukin-20 expression by human epithelial keratinocytes." *Photochem Photobiol* **82**(5): 1292–300.

Infante-Duarte, C., H. F. Horton, et al. (2000). "Microbial lipopeptides induce the production of IL-17 in Th cells." *J Immunol* **165**(11): 6107–15.

Jiang, H., J. J. Lin, et al. (1995). "Subtraction hybridization identifies a novel melanoma differentiation associated gene, mda-7, modulated during human melanoma differentiation, growth and progression." *Oncogene* **11**(12): 2477–86.

Jones, E. A. and R. A. Flavell (2005). "Distal enhancer elements transcribe intergenic RNA in the IL-10 family gene cluster." *J Immunol* **175**(11): 7437–46.
Jordan, W. J., J. Eskdale, et al. (2005). "Human IL-19 regulates immunity through auto-induction of IL-19 and production of IL-10." *Eur J Immunol* **35**(5): 1576–82.
Kagami, S., H. L. Rizzo, et al. (2009). "Circulating Th17, Th22, and Th1 Cells Are Increased in Psoriasis." *J Invest Dermatol* 130(5): 1373–83.
Kastelein, R. A., C. A. Hunter, et al. (2007). "Discovery and biology of IL-23 and IL-27: related but functionally distinct regulators of inflammation." *Annu Rev Immunol* **25**: 221–42.
Kingo, K., S. Koks, et al. (2004). "Polymorphisms in the interleukin-20 gene: relationships to plaque-type psoriasis." *Genes Immun* **5**(2): 117–21.
Kingo, K., R. Mossner, et al. (2008). "Association analysis of IL20RA and IL20RB genes in psoriasis." *Genes Immun* **9**(5): 445–51.
Knappe, A., S. Hor, et al. (2000). "Induction of a novel cellular homolog of interleukin-10, AK155, by transformation of T lymphocytes with herpesvirus saimiri." *J Virol* **74**(8): 3881–7.
Koks, S., K. Kingo, et al. (2004). "Combined haplotype analysis of the interleukin-19 and -20 genes: relationship to plaque-type psoriasis." *Genes Immun* **5**(8): 662–7.
Koks, S., K. Kingo, et al. (2005). "Possible relations between the polymorphisms of the cytokines IL-19, IL-20 and IL-24 and plaque-type psoriasis." *Genes Immun* **6**(5): 407–15.
Kopp, T., P. Lenz, et al. (2003). "IL-23 production by cosecretion of endogenous p19 and transgenic p40 in keratin 14/p40 transgenic mice: evidence for enhanced cutaneous immunity." *J Immunol* **170**(11): 5438–44.
Korn, T., E. Bettelli, et al. (2007). "IL-21 initiates an alternative pathway to induce proinflammatory T(H)17 cells." *Nature* **448**(7152): 484–7.
Korn, T., E. Bettelli, et al. (2009). "IL-17 and Th17 Cells." *Annu Rev Immunol* **27**: 485–517.
Kotenko, S. V., L. S. Izotova, et al. (2001a). "Identification of the functional interleukin-22 (IL-22) receptor complex: the IL-10R2 chain (IL-10Rbeta) is a common chain of both the IL-10 and IL-22 (IL-10-related T cell-derived inducible factor, IL-TIF) receptor complexes." *J Biol Chem* **276**(4): 2725–32.
Kotenko, S. V., L. S. Izotova, et al. (2001b). "Identification, cloning, and characterization of a novel soluble receptor that binds IL-22 and neutralizes its activity." *J Immunol* **166**(12): 7096–103.
Kreis, S., D. Philippidou, et al. (2008). "IL-24: a classic cytokine and/or a potential cure for cancer?" *J Cell Mol Med.* **12**(6A): 2505–10.
Kreis, S., D. Philippidou, et al. (2007). "Recombinant interleukin-24 lacks apoptosis-inducing properties in melanoma cells." *PLoS ONE* **2**(12): e1300.
Krueger, G. G., R. G. Langley, et al. (2007). "A human interleukin-12/23 monoclonal antibody for the treatment of psoriasis." *N Engl J Med* **356**(6): 580–92.
Kunz, S., K. Wolk, et al. (2006). "Interleukin (IL)-19, IL-20 and IL-24 are produced by and act on keratinocytes and are distinct from classical ILs." *Exp Dermatol* **15**(12): 991–1004.
Langrish, C. L., Y. Chen, et al. (2005). "IL-23 drives a pathogenic T cell population that induces autoimmune inflammation." *J Exp Med* **201**(2): 233-40.
Lee, E., W. L. Trepicchio, et al. (2004). "Increased expression of interleukin 23 p19 and p40 in lesional skin of patients with psoriasis vulgaris." *J Exp Med* **199**(1): 125–30.
Lejeune, D., L. Dumoutier, et al. (2002). "Interleukin-22 (IL-22) activates the JAK/STAT, ERK, JNK, and p38 MAP kinase pathways in a rat hepatoma cell line. Pathways that are shared with and distinct from IL-10." *J Biol Chem* **277**(37): 33676–82.
Liang, S. C., X. Y. Tan, et al. (2006). "Interleukin (IL)-22 and IL-17 are coexpressed by Th17 cells and cooperatively enhance expression of antimicrobial peptides." *J Exp Med* **203**(10): 2271–9.
Liao, S. C., Y. C. Cheng, et al. (2004). "IL-19 induced Th2 cytokines and was up-regulated in asthma patients." *J Immunol* **173**(11): 6712–8.
Liao, Y. C., W. G. Liang, et al. (2002). "IL-19 induces production of IL-6 and TNF-alpha and results in cell apoptosis through TNF-alpha." *J Immunol* **169**(8): 4288–97.
Liu, L., C. Ding, et al. (2003). "Selective enhancement of multipotential hematopoietic progenitors in vitro and in vivo by IL-20." *Blood* **102**(9): 3206–9.

Lowes, M. A., A. M. Bowcock, et al. (2007). "Pathogenesis and therapy of psoriasis." *Nature* **445**(7130): 866–73.
Lowes, M. A., T. Kikuchi, et al. (2008). "Psoriasis vulgaris lesions contain discrete populations of Th1 and Th17 T cells." *J Invest Dermatol* **128**(5): 1207–11.
Luci, C., A. Reynders, et al. (2009). "Influence of the transcription factor RORgammat on the development of NKp46+ cell populations in gut and skin." *Nat Immunol* **10**(1): 75–82.
Ma, H. L., S. Liang, et al. (2008). "IL-22 is required for Th17 cell-mediated pathology in a mouse model of psoriasis-like skin inflammation." *J Clin Invest* **118**(2): 597–607.
Mangan, P. R., L. E. Harrington, et al. (2006). "Transforming growth factor-beta induces development of the T(H)17 lineage." *Nature* **441**(7090): 231–4.
Martin, B., K. Hirota, et al. (2009). "Interleukin-17-producing gammadelta T cells selectively expand in response to pathogen products and environmental signals." *Immunity* **31**(2): 321–30.
Moore, K. W., R. de Waal Malefyt, et al. (2001). "Interleukin-10 and the interleukin-10 receptor." *Annu Rev Immunol* **19**: 683–765.
Nagalakshmi, M. L., A. Rascle, et al. (2004). "Interleukin-22 activates STAT3 and induces IL-10 by colon epithelial cells." *Int Immunopharmacol* **4**(5): 679–91.
Nickoloff, B. J. and F. O. Nestle (2004). "Recent insights into the immunopathogenesis of psoriasis provide new therapeutic opportunities." *J Clin Invest* **113**(12): 1664–75.
Nograles, K. E., L. C. Zaba, et al. (2009). "IL-22-producing T22 T cells account for upregulated IL-22 in atopic dermatitis despite reduced IL-17-producing TH17 T cells." *J Allergy Clin Immunol* **123**(6): 1244–52 e2.
Oral, H. B., S. V. Kotenko, et al. (2006). "Regulation of T cells and cytokines by the interleukin-10 (IL-10)-family cytokines IL-19, IL-20, IL-22, IL-24 and IL-26." *Eur J Immunol* **36**(2): 380–8.
Otkjaer, K., K. Kragballe, et al. (2005). "The dynamics of gene expression of interleukin-19 and interleukin-20 and their receptors in psoriasis." *Br J Dermatol* **153**(5): 911–8.
Otkjaer, K., K. Kragballe, et al. (2007). "IL-20 gene expression is induced by IL-1beta through mitogenactivated protein kinase and NF-kappaB-dependent mechanisms." *J Invest Dermatol* **127**(6): 1326–36.
Ouyang, W., J. K. Kolls, et al. (2008). "The biological functions of T helper 17 cell effector cytokines in inflammation." *Immunity* **28**(4): 454–67.
Pan, H., F. Hong, et al. (2004). "Hydrodynamic gene delivery of interleukin-22 protects the mouse liver from concanavalin A-, carbon tetrachloride-, and Fas ligand-induced injury via activation of STAT3." *Cell Mol Immunol* **1**(1): 43–9.
Parrish-Novak, J., W. Xu, et al. (2002). "Interleukins 19, 20, and 24 signal through two distinct receptor complexes. Differences in receptor-ligand interactions mediate unique biological functions." *J Biol Chem* **277**(49): 47517–23.
Pene, J., S. Chevalier, et al. (2008). "Chronically inflamed human tissues are infiltrated by highly differentiated Th17 lymphocytes." *J Immunol* **180**(11): 7423-30.
Pestka, S., C. D. Krause, et al. (2004). "Interleukin-10 and related cytokines and receptors." *Annu Rev Immunol* **22**: 929–79.
Pickert, G., C. Neufert, et al. (2009). "STAT3 links IL-22 signaling in intestinal epithelial cells to mucosal wound healing." *J Exp Med* **206**(7): 1465–72.
Radaeva, S., R. Sun, et al. (2004). "Interleukin 22 (IL-22) plays a protective role in T cell-mediated murine hepatitis: IL-22 is a survival factor for hepatocytes via STAT3 activation." *Hepatology* **39**(5): 1332–42.
Romer, J., E. Hasselager, et al. (2003). "Epidermal overexpression of interleukin-19 and -20 mRNA in psoriatic skin disappears after short-term treatment with cyclosporine a or calcipotriol." *J Invest Dermatol* **121**(6): 1306–11.
Sa, S. M., P. A. Valdez, et al. (2007). "The effects of IL-20 subfamily cytokines on reconstituted human epidermis suggest potential roles in cutaneous innate defense and pathogenic adaptive immunity in psoriasis." *J Immunol* **178**(4): 2229–40.
Sakurai, N., T. Kuroiwa, et al. (2008). "Expression of IL-19 and its receptors in RA: potential role for synovial hyperplasia formation." *Rheumatology (Oxford)* **47**(6): 815–20.

Sanos, S. L., V. L. Bui, et al. (2009). "RORgammat and commensal microflora are required for the differentiation of mucosal interleukin 22-producing NKp46+ cells." *Nat Immunol* **10**(1): 83–91.

Sarkar, D., I. V. Lebedeva, et al. (2007). "Melanoma differentiation associated gene-7 (mda-7)/IL-24: a 'magic bullet' for cancer therapy?" *Expert Opin Biol Ther* **7**(5): 577–86.

Satoh-Takayama, N., C. A. Vosshenrich, et al. (2008). "Microbial flora drives interleukin 22 production in intestinal NKp46+ cells that provide innate mucosal immune defense." *Immunity* **29**(6): 958–70.

Schaefer, G., C. Venkataraman, et al. (2001). "Cutting edge: FISP (IL-4-induced secreted protein), a novel cytokine-like molecule secreted by Th2 cells." *J Immunol* **166**(10): 5859–63.

Sheikh, F., V. V. Baurin, et al. (2004). "Cutting edge: IL-26 signals through a novel receptor complex composed of IL-20 receptor 1 and IL-10 receptor 2." *J Immunol* **172**(4): 2006–10.

Sheppard, P., W. Kindsvogel, et al. (2003). "IL-28, IL-29 and their class II cytokine receptor IL-28R." *Nat Immunol* **4**(1): 63–8.

Stenderup, K., C. Rosada, et al. (2009). "Interleukin-20 plays a critical role in maintenance and development of psoriasis in the human xenograft transplantation model." *Br J Dermatol* **160**(2): 284–96.

Sugimoto, K., A. Ogawa, et al. (2008). "IL-22 ameliorates intestinal inflammation in a mouse model of ulcerative colitis." *J Clin Invest* **118**(2): 534–44.

Sutton, C. E., S. J. Lalor, et al. (2009). "Interleukin-1 and IL-23 induce innate IL-17 production from gammadelta T cells, amplifying Th17 responses and autoimmunity." *Immunity* **31**(2): 331–41.

Takatori, H., Y. Kanno, et al. (2009). "Lymphoid tissue inducer-like cells are an innate source of IL-17 and IL-22." *J Exp Med* **206**(1): 35–41.

Tohyama, M., Y. Hanakawa, et al. (2009). "IL-17 and IL-22 mediate IL-20 subfamily cytokine production in cultured keratinocytes via increased IL-22 receptor expression." *Eur J Immunol* **39**(10): 2779–88.

Trifari, S., C. D. Kaplan, et al. (2009). "Identification of a human helper T cell population that has abundant production of interleukin 22 and is distinct from T(H)-17, T(H)1 and T(H)2 cells." *Nat Immunol* **10**(8): 864–71.

Tritsaris, K., M. Myren, et al. (2007). "IL-20 is an arteriogenic cytokine that remodels collateral networks and improves functions of ischemic hind limbs." *Proc Natl Acad Sci U S A* **104**(39): 15364–9.

Veldhoen, M., R. J. Hocking, et al. (2006). "TGFbeta in the context of an inflammatory cytokine milieu supports de novo differentiation of IL-17-producing T cells." *Immunity* **24**(2): 179–89.

Wang, F., E. Lee, et al. (2006). "Prominent production of IL-20 by CD68+/CD11c+ myeloid-derived cells in psoriasis: Gene regulation and cellular effects." *J Invest Dermatol* **126**(7): 1590–9.

Wang, M., Z. Tan, et al. (2002). "Interleukin 24 (MDA-7/MOB-5) signals through two heterodimeric receptors, IL-22R1/IL-20R2 and IL-20R1/IL-20R2." *J Biol Chem* **277**(9): 7341–7.

Wegenka, U. M., N. Dikopoulos, et al. (2007). "The murine liver is a potential target organ for IL-19, IL-20 and IL-24: Type I Interferons and LPS regulate the expression of IL-20R2." *J Hepatol* **46**(2): 257–65.

Whittington, H. A., L. Armstrong, et al. (2004). "Interleukin-22: a potential immunomodulatory molecule in the lung." *Am J Respir Cell Mol Biol* **31**(2): 220–6.

Wolk, K., H. S. Haugen, et al. (2009a). "IL-22 and IL-20 are key mediators of the epidermal alterations in psoriasis while IL-17 and IFN-gamma are not." *J Mol Med* **87**(5): 523–36.

Wolk, K., S. Kunz, et al. (2002). "Cutting edge: immune cells as sources and targets of the IL-10 family members?" *J Immunol* **168**(11): 5397–402.

Wolk, K., S. Kunz, et al. (2004). "IL-22 increases the innate immunity of tissues." *Immunity* **21**(2): 241–54.

Wolk, K., E. Witte, et al. (2006). "IL-22 regulates the expression of genes responsible for antimicrobial defense, cellular differentiation, and mobility in keratinocytes: a potential role in psoriasis." *Eur J Immunol* **36**(5): 1309–23.

Wolk, K., E. Witte, et al. (2009b). "The Th17 cytokine IL-22 induces IL-20 production in keratinocytes: a novel immunological cascade with potential relevance in psoriasis." *Eur J Immunol* **39**(12): 3570–81.
Wolk, K., K. Witte, et al. (2008). "Maturing dendritic cells are an important source of IL-29 and IL-20 that may cooperatively increase the innate immunity of keratinocytes." *J Leukoc Biol* **83**(5): 1181–93.
Xie, M. H., S. Aggarwal, et al. (2000). "Interleukin (IL)-22, a novel human cytokine that signals through the interferon receptor-related proteins CRF2-4 and IL-22R." *J Biol Chem* **275**(40): 31335–9.
Xu, W., S. R. Presnell, et al. (2001). "A soluble class II cytokine receptor, IL-22RA2, is a naturally occurring IL-22 antagonist." *Proc Natl Acad Sci USA* **98**(17): 9511–6.
Yano, S., T. Banno, et al. (2008). "Transcriptional responses of human epidermal keratinocytes to cytokine interleukin-1." *J Cell Physiol* **214**(1): 1–13.
Zaba, L. C., I. Cardinale, et al. (2007). "Amelioration of epidermal hyperplasia by TNF inhibition is associated with reduced Th17 responses." *J Exp Med* **204**(13): 3183–94.
Zdanov, A. (2004). "Structural features of the interleukin-10 family of cytokines." *Curr Pharm Des* **10**(31): 3873–84.
Zenewicz, L. A., G. D. Yancopoulos, et al. (2007). "Interleukin-22 but not interleukin-17 provides protection to hepatocytes during acute liver inflammation." *Immunity* **27**(4): 647–59.
Zheng, Y., D. M. Danilenko, et al. (2007). "Interleukin-22, a T(H)17 cytokine, mediates IL-23-induced dermal inflammation and acanthosis." *Nature* **445**(7128): 648–51.
Zheng, Y., P. A. Valdez, et al. (2008). "Interleukin-22 mediates early host defense against attaching and effacing bacterial pathogens." *Nat Med* **14**(3): 282–9.
Zhong, H., Y. Wu, et al. (2006). "A2B adenosine receptors induce IL-19 from bronchial epithelial cells, resulting in TNF-alpha increase." *Am J Respir Cell Mol Biol* **35**(5): 587–92.

IL-23/T_H17 Pathway in Psoriasis and Inflammatory Skin Diseases

Kristine E. Nograles, Batya Davidovici, and James G. Krueger

Abstract Psoriasis is a highly prevalent and heritable skin disease that affects the population worldwide. While the importance of activated T cells in disease pathogenesis has long been appreciated, the central role of the IL-23/T_H17 pathway has only recently been realized. In this chapter, we discuss the contributions of T_H17 cells in psoriasis, as well as in other inflammatory skin diseases.

1 Introduction

Psoriasis is a chronic relapsing inflammatory disease of the skin that results in great morbidity for those severely affected. Although common, its clinical manifestations are diverse, and its genetics complex. Thus, psoriasis pathogenesis is not yet fully understood, although great strides have been made recently.

Severe psoriasis, which occurs in about 30% of patients, is a disabling disease affecting the physical and emotional wellbeing of patients. Recent studies indicate that severe, long-standing psoriasis is associated with co-morbidities, as well as a shorter lifespan due to cardiovascular mortality. In addition, patients with psoriasis, like those with other major medical disorders, have reduced levels of employment and income. The combined cost of long-term therapy and social burden of this disease have a major impact not only on the patient, but also on health care systems.

J.G. Krueger (✉)
Laboratory for Investigative Dermatology, The Rockefeller University,
1230 York Avenue, New York, NY 10065, USA
e-mail: jgk@rockefeller.edu

S. Jiang (ed.), *T_H17 Cells in Health and Disease*,
DOI 10.1007/978-1-4419-9371-7_25,

2 Epidemiology

A literature search of epidemiologic studies from around the world reveals that the estimated prevalence of psoriasis is 0.6–4.8%. However, caution is needed when reviewing these figures. Disparities such as variations in sampling techniques, definitions of prevalence (e.g., point vs. lifetime prevalence), case definitions (e.g., self report vs. physician diagnosis), ages, and populations can be observed.

A recent population-based survey in the United States estimates the lifetime prevalence of self-reported psoriasis among adults to be 2.2% (Stern et al. 2004). Similarly, the General Practice Research Database (GPRD), broadly representative of the general UK population, pegs the prevalence of psoriasis at 1.5% (Gelfand et al. 2005) However, ethnic factors appear to influence the prevalence of psoriasis, which ranges from no cases in the Samoan population to 12% in Arctic Kasach'ye (Farber and Nall 1998). The influence of ethnicity becomes particularly evident within the United States where the prevalence among blacks (0.45–0.7%) (Kenney 1971) is far lower than the remainder of the U.S. population (1.4–4.6%) (Christophers 2001).

Psoriasis is a highly heritable disease, although the exact pattern of inheritance is still unknown (Henseler and Christophers 1985). As many as half of the siblings of persons with psoriasis develop the disease when both parents are affected, compared to 16% when only one parent has psoriasis, and 8% when neither parent is affected (Watson et al. 1972). The concordance rate for monozygotic twins is around 70%, compared with approximately 20% for dizygotic twins (Farber et al. 1974; Brandrup et al. 1982). Importantly, recent SNP-based genome-wide association studies (GWAS) have implicated genes encoding IL-23 (IL23A, IL12B) and IL-23 receptor (IL23R) in psoriasis susceptibility (Nair et al. 2008, 2009).

3 Clinical Manifestations and Histopathologic Features

Psoriasis is heterogeneous and may have many clinical variations. In fact, patients may have mixed types of lesions, or have alterations in lesion morphology during the course of their disease. For example, guttate psoriasis, an acute eruption of small (<1 cm) papules of psoriasis, transforms into large plaque psoriasis in approximately 70% of patients (Williams et al. 1976). Similarly, large plaque-type psoriasis may become pustular and vice versa. The various clinical presentations for psoriasis have been extensively reviewed by The International Psoriasis Council, and interested readers are directed there (Griffiths et al. 2007).

The great majority (~90%) of patients, however, have chronic plaque psoriasis characterized by red, scaly, discoid lesions varying in size from 0.5 cm in diameter to large confluent areas on the trunk and limbs (Griffiths et al. 2004) (Fig. 1). A sharp line of demarcation delineating a plaque from clinically normal, uninvolved skin can be observed. The lesions may occur singly at predisposed sites (e.g. extensor aspects of knees and elbows) or disseminated (generalized) over the body. Other

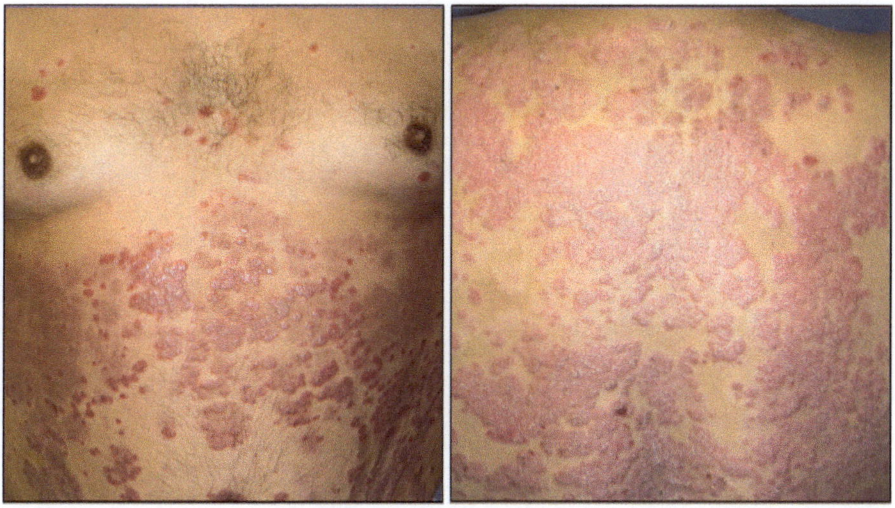

Fig. 1 A clinical picture (trunk, front and back) of a patient with moderate-to-severe psoriasis vulgaris demonstrates the wide distribution of red, raised, scaly discoid plaques that are sharply demarcated from normal, non-lesional skin. On the patient's back, the plaques coalesce and cover most of the skin surface

areas that may be affected include the intertriginous areas (i.e. submammary, groins, axillae, genitalia and natal cleft), scalp, palms and soles, and nails.

Several defining histologic changes can be observed as lesions develop. These include (1) a thickened epidermis (acanthosis) from premature maturation of keratinocytes and rapid keratinocyte proliferation, (2) reduced or absent granular layer (hypogranulosis) and retention of nuclei by corneocytes (parakeratosis) as a result of aberrant differentiation of keratinocytes, (3) increased numbers of tortuous capillaries as well as marked dilation of blood vessels in the papillary dermis causing visible erythema, and (4) a dense inflammatory infiltrate composed of $CD4^+$ T helper cells and antigen presenting dendritic cells (DCs) in the dermis, and $CD8^+$ T cells and neutrophils in the epidermis (Fig. 2).

Most of the T cells found in psoriasis lesions are memory cells (CD45RO+) that express cutaneous lymphocyte antigen (CLA) (Morganroth et al. 1991; Fuhlbrigge et al. 1997). These cells also express high levels of CCR6 chemokine receptor (Homey et al. 2000; Teraki et al. 2004) that allow for chemotaxis towards elevated CCL20 gradients as observed in psoriasis lesions. CD4+ and CD8+ T cells also display markers for persistent activation such as IL-2R and HLA-DR (Ferenczi et al. 2000).

Myeloid dermal dendritic cells (DCs), marked by CD11c and HLA-DR expression, are also markedly increased in psoriasis lesions (Fig. 2) (Abrams et al. 2000; Gottlieb et al. 2005; Lowes et al. 2005). Mature DCs, marked by CD83, DC-LAMP and BDCA-1, are found in dermal aggregates in close proximity to CD3 T cells (Lowes et al. 2005; Zaba et al. 2009b), while immature DCs (BDCA-1 negative, DC-LAMP low), are scattered across the upper reticular dermis and dermal papillae (Fuentes-Duculan et al. 2009). Among the immature DC population is a subset of

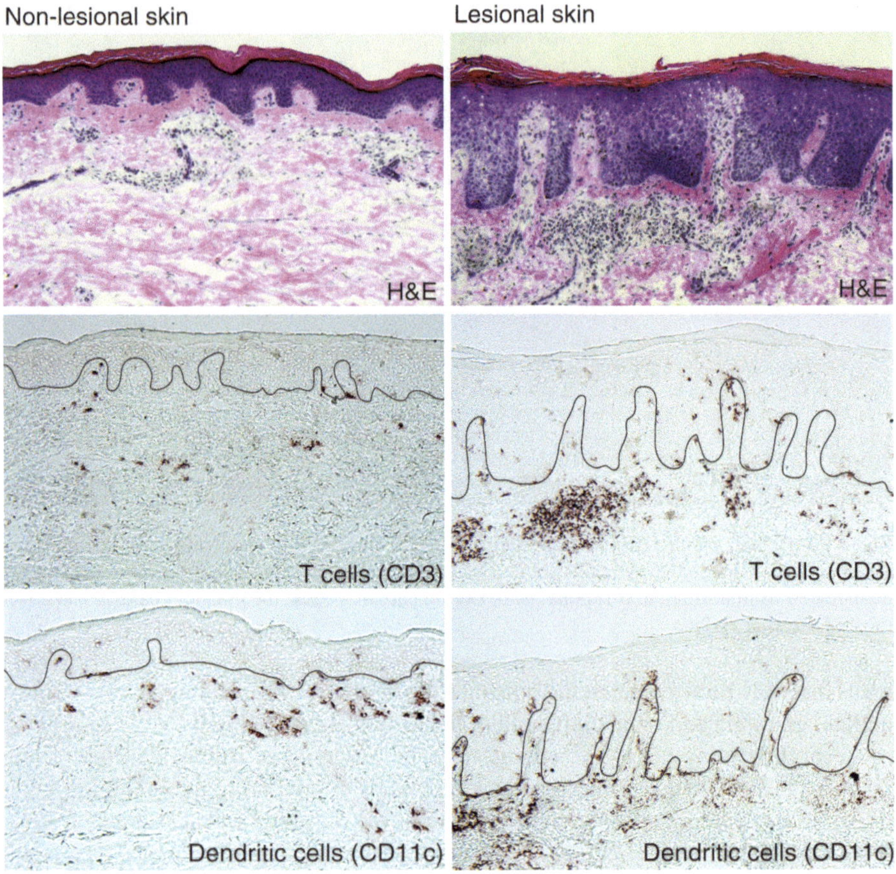

Fig. 2 Histologic comparison of lesional and non-lesional skin. H&E sections demonstrate striking differences between lesional and non-lesional skin, including a markedly thickened epidermis (acanthosis, A) and an increased inflammatory infiltrate in the dermis (I). Among these dermal inflammatory cells are T cells, identified by the pan-T-cell marker, CD3+, and dendritic cells, identified by CD11c+

inflammatory DCs known as tumor necrosis factor (TNF) and inducible nitric oxide synthase (iNOS) producing DCs (TIP-DCs) (Lowes et al. 2005; Fuentes-Duculan et al. 2009). Although immature DCs comprise a majority of DCs that infiltrate the psoriasis lesion, both populations can induce T cell proliferation and polarize them towards the type 1 and type 17 phenotypes (Fuentes-Duculan et al. 2009). Myeloid DCs in psoriasis express IL-23 (Lee et al. 2004; Zaba et al. 2007), a cytokine that promotes the expansion and stimulation of T_H17 cells (Di Cesare et al. 2009).

Another type of dendritic cells that are activated and increased in psoriasis is the plasmacytoid DC subset (pDCs, marked by BDCA-2) (Nestle et al. 2005). These cells produce high levels of IFNα, and may play a vital role in the initiation of psoriasis lesions (Lande et al. 2007).

4 Immunologic Basis of Psoriasis

As the epidermal alterations are the most striking clinical and histologic features of psoriasis, it comes to no surprise that this disease was previously considered to be of aberrant keratinocyte proliferation and differentiation. In fact, various forms of treatment, i.e. anthralin, vitamin D analogues, photochemotherapy and methotrexate were all thought to be effective because of their anti-proliferative effects. However, more recent evidence has shown that the epidermal changes in psoriasis occur in response to activated T cells. Moreover, we have come to appreciate that the above treatment modalities also have anti-inflammatory effects.

Perhaps the strongest argument that the development of psoriasis lesions is T cell mediated comes from clinical improvement with targeted therapy, such as calcineurin antagonists (cyclosporin and tacrolimus) that have major inhibitory effect on T cell activation, and DAB_{389}IL-2, a fusion protein that causes activated T lymphocytes to undergo apoptosis (Ellis et al. 1986; Gottlieb et al. 1995). While it can be argued that cyclosporine may also directly affect keratinocyte proliferation, DAB_{389}IL-2 does not, thus providing a direct pathogenic link between activated T lymphocytes and epidermal changes in psoriasis (Gottlieb et al. 1995). Although more selective T-cell antagonists, e.g. DAB_{389}IL-2, alefacept and efalizumab, can indeed reverse psoriasis, recent studies using IL-12/IL-23 inhibitors more strongly implicate the IL-23/T_H17 axis in this disease.

Many of the infiltrating T cells in psoriasis lesions are type 1 CD4 helper T cells (T_H1) and type 1 CD8 cytotoxic T cells (Tc1) that secrete the inflammatory cytokines IFNγ and TNFα upon activation (Szabo et al. 1998; Austin et al. 1999). Indeed, many genes downstream of IFNγ are upregulated in psoriasis lesions compared to non-lesional skin (Lew et al. 2004a, b). IFNγ induces the expression of numerous inflammatory molecules in keratinocytes, including CXCL9/MIG, CXCL10/IP-10 and CXCL11/I-TAC that are key chemotactic factors for recruiting more type 1 cells that contribute to further inflammation in psoriasis (Nograles et al. 2008). Much of this biology is described in past reports (Lew et al. 2004a, b), and will not be further discussed here. Recently, IFNγ has been shown to stimulate DCs to produce IL-1 and IL-23 that promote T_H17 differentiation and activation.(Kryczek et al. 2008) (Fig. 2).

5 T_H17 Cells in Psoriasis

Psoriasis research has recently focused on interleukin-17 producing helper T cells, T_H17. These cells are developmentally and functionally distinct from the classic T_H1 and T_H2 lineages (Bettelli et al. 2007), and are thought to have evolved primarily to provide immunity against pathogens in epithelial surfaces. If T_H17 cell differentiation is impaired, as in hyper IgE syndrome, recurrent C. albicans and Staph infections are observed (Milner et al. 2008). Similarly, autoantibodies against T_H17 cytokines may cause chronic mucocutaneous candidiasis (Puel et al. 2010).

The growth, survival, and effector functions of T_H17 cells are dependent on IL-23 (Zhou et al. 2007), a heterodimeric cytokine composed of p19 (encoded by the IL23A gene), and p40 (shared with IL-12, and encoded by IL12B) sub-units that binds to a receptor complex encoded by IL23R and IL12RB1. Recent genome-wide association studies have identified IL23A, IL12B and IL23R to be psoriasis-associated gene signals, implicating the IL-23/T_H17 pathway to psoriasis pathogenesis (Liu et al. 2008; Nair et al. 2008, 2009). Indeed, IL-23 and T_H17 cells were found to be markedly abundant in psoriasis lesions (Lowes et al. 2008; Lee et al. 2004), perhaps as a direct effect of genetic variations in regulatory regions of the above-mentioned genes. Accordingly, T_H17-associated cytokines, IL-17A (Li et al. 2004), IL-17F (Guttman-Yassky et al. 2008), IL-22 (Boniface et al. 2007), and IL-21 (Caruso et al. 2009), are markedly elevated in psoriasis lesions compared to non-lesional or normal skin.

5.1 IL-17 Specific Effects

Members of the IL-17 cytokine family include IL-17A, IL-17B, IL-17C, IL-17D, IL-17E and IL-17F. By far, the best characterized of these are IL-17A and IL-17F. While both IL-17A and IL-17F form covalent homodimers, recent findings have shown that they can also form IL-17A-IL-17F heterodimeric complexes (Wright et al. 2007; Chang and Dong 2007). Which dimeric complex is dominant in psoriasis lesions is currently unknown, although transcripts for both IL-17A and IL-17F are elevated in psoriatic lesions (Guttman-Yassky et al. 2008). Because IL-17A has been studied more extensively than IL-17F, and both share many biological properties, the contributions of IL-17A in psoriasis will be the subject of further discussion.

The IL-17 receptor family comprises five receptor sub-units, IL-17RA to IL-17RE. It is not entirely clear how these sub-units interact to form functional receptor complexes, but IL-17RA appears to be the common signaling sub-unit used by at least four of the IL-17 cytokine family members (Gaffen 2009). To elicit IL-17A and IL-17F responses, IL-17RA pairs with IL-17RC (Gaffen 2009).

IL-17RA mRNA is expressed ubiquitously, with particularly high levels in haematopoietic cells (Ishigame et al. 2009). In contrast, IL-17RC mRNA is selectively expressed in non-hematopoietic tissues such as the colon, small intestine, and lung (Kuestner et al. 2007; Ishigame et al. 2009). This may provide explanation for the preferential sensitivity of epithelial surfaces to IL-17A. Indeed, CD11c+ DCs and CD163+ macrophages within the psoriatic lesion express IL-17RA but exhibit minimal response to IL-17 stimulation (unpublished data). In contrast, keratinocytes express both IL-17RA and IL-17RC, and are highly sensitive to IL-17 (Nograles et al. 2008).

IL-17 stimulation induces keratinocytes to upregulate ELR+ CXC chemokines, including IL-8, CXCL1, CXCL3, CXCL5 and CXCL6, to promote neutrophil influx into the psoriatic lesion (Fig. 3) (Nograles et al. 2008). This upregulation can be further enhanced by the presence of IFNγ and TNFα (Fig. 4) (Teunissen et al. 1998; Albanesi et al. 1999). In addition to these pro-inflammatory chemokines, IL-17 induces keratinocytes to express CCL20 (Nograles et al. 2008; Harper et al. 2009).

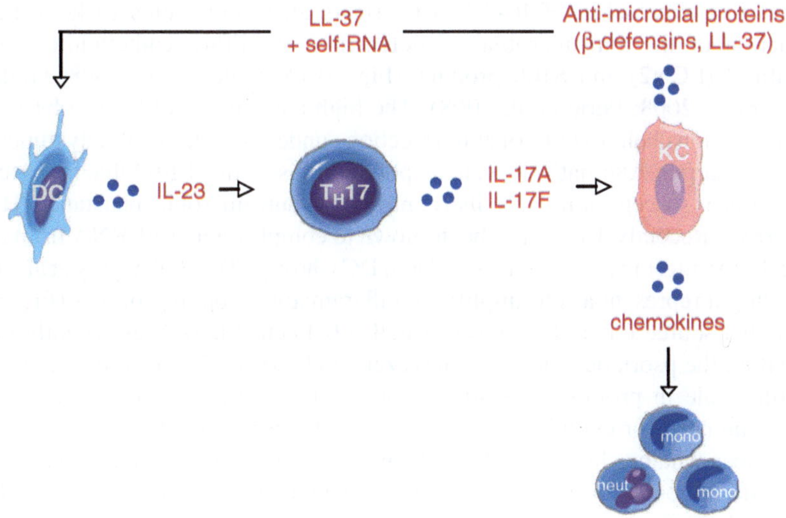

Fig. 3 Model for T_H17 activity in psoriasis. IL-23 producing dendritic cells stimulate T_H17 cells to secrete IL-17A and IL-17F. These cytokines trigger keratinocyte production of a number of chemokines that drive mononuclear cells and neutrophils into the psoriatic lesion. In addition, keratinocytes also secrete anti-microbial proteins in response to IL-17 stimulation. These anti-microbial proteins, in coordination with self-RNA from apoptotic cells, may further activate dendritic cells creating a self-amplifying inflammatory loop

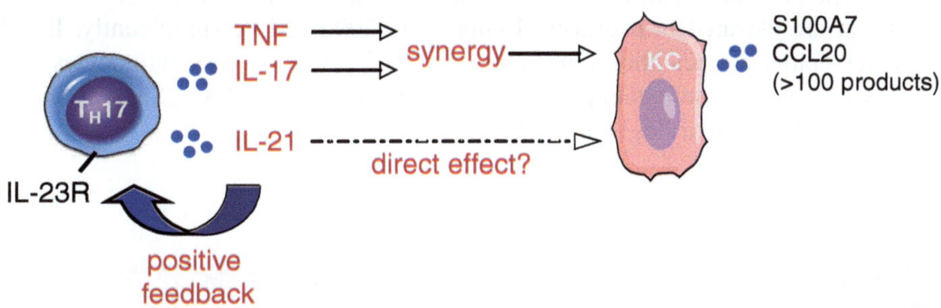

Fig. 4 Synergistic and autocrine activity in T_H17 cells. IL-17 and TNFa work synergistically to induce pro-inflammatory keratinocyte products. Another T_H17-associated cytokine, IL-21, induces the upregulation of IL-23R, enabling T_H17 cells to respond to IL-23 stimulation. It is still unclear whether IL-21 may have a direct effect of keratinocytes

This drives CCR6-bearing cells, including memory T cells, DCs and T_H17 cells themselves, into the psoriatic lesion. The influx of CCR6+ cells into the skin may be critical, as CCR6-null mice do not develop lesions even with IL-23 stimulation (Hedrick et al. 2009). In addition, IL-17 can induce skin fibroblasts to produce IL-6, a cytokine that commits naïve T cells to the T_H17 lineage, potentially activating a positive feedback loop that perpetuates T_H17 inflammation (Fossiez et al. 1996).

The over-expression of IL-17 in psoriasis also contributes to keratinocyte production of the anti-microbials β-defensin 2 (DEFB4), cathelicidin (LL37), lipocalin-2 (LCN2) and S100 proteins (Fig. 3) (Nograles et al. 2008; Guttman-Yassky et al. 2008; Peric et al. 2008). The high anti-microbial levels observed in psoriasis lesions help to control skin infections under conditions of a dysfunctional epidermal barrier. Alternatively, genetic polymorphisms in DEFB4 that is associated with psoriasis susceptibility may also contribute to anti-microbial resistance (Hollox et al. 2008). Recently, LL-37 has been shown to complex with self-RNA from apoptotic cells to trigger the activation of myeloid DCs through TLR8 (Ganguly et al. 2009), which might represent a self-amplifying inflammatory loop in psoriasis (Fig. 3).

Another source for IL-17 is $CD8^+$ T cells (Tc17) that have been recently identified within the psoriatic epidermis (Kryczek et al. 2008). These cells may have an important role in promoting psoriatic epidermal response, as their contributions obviate the need for cytokines to diffuse from the dermis (Elder et al. 2010). It is still unclear whether human Tc17 cells are influenced by the same conditions as T_H17 cells, although murine models suggest that they might also be driven by IL-23 (Ciric et al. 2009).

5.2 IL-22 Specific effects

Another cytokine commonly associated with T_H17 cells is IL-22. This cytokine works synergistically with IL-17 to enhance the expression of anti-microbial peptides that are elevated in psoriasis (Liang et al. 2006). More significantly, IL-22 mediates epidermal acanthosis and abnormal differentiation of keratinocytes that are key pathologic findings in psoriasis (Fig. 5) (Zheng et al. 2007; Sa et al. 2007; Liang et al. 2006; Nograles et al. 2008).

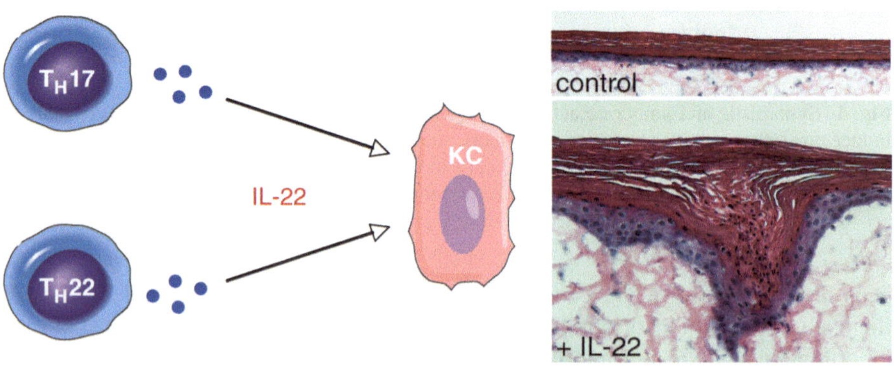

Fig. 5 Model for IL-22 activity in psoriasis. T_H17 cells, as well as T_H22 cells, contribute to the elevated levels of IL-22 in lesional skin. This cytokine promotes epidermal acanthosis and parakeratosis that are hallmarks of psoriasis histology

IL-22 production is commonly attributed to T_H17 cells based on early studies utilizing murine models.(Liang et al. 2006; Zheng et al. 2007) Accordingly, we found that ~40% of IL-22-producing T helper cells in psoriasis are T_H17 cells (Nograles et al. 2009). However, the majority of IL-22 contributors in psoriasis lesions does not co-express IL-17 or the T_H1 cytokine, IFNγ, and represent a unique subset of helper T cells (Nograles et al. 2009). These IL-22 producing T helper cells, T_H22, co-express CCR6 and skin homing receptors CCR4 and CCR10 (Trifari et al. 2009; Duhen et al. 2009), thus may presumably respond to elevated CCL20 levels in psoriatic skin. Potentially, different DC subsets in psoriasis lesions might regulate T_H17 vs. T_H22 activation. $CD11c^+$ dermal DCs have been shown to stimulate T_H17 cells, while epidermal langerhans cells (LCs) have been shown to preferentially stimulate T_H22 responses (Fujita et al. 2009). Cytokine products from DCs that can stimulate IL-22 production include IL-23 and IL-12 (Volpe et al. 2009).

5.3 *IL-21 Specific Effects*

T_H17 cells also secrete IL-21 (Nurieva et al. 2007; Wei et al. 2007), an autocrine factor that induces naïve $CD4^+$ T cells to upregulate IL-23R, IL-17 and IL-21 (Fig. 4) (Wei et al. 2007; Zhou et al. 2007; Manel et al. 2008). While IL-23 plays a key role in inflammatory disease, IL-23 receptor (IL-23R) is not present on naïve T cells. By upregulating IL-23R, IL-21 enables these cells to respond to IL-23, promoting T_H17 commitment and expansion *in vivo* (Spolski and Leonard 2008). When IL-21 is deficient, the generation of T_H17 cells is impaired, making this cytokine a potential target for the treatment of T_H17-mediated diseases including psoriasis (Nurieva et al. 2007). Interestingly, a region of chromosome 4q27 that harbors the IL21 gene has recently been identified as a locus for psoriasis susceptibility (Liu et al. 2008).

IL-21 and IL-21-producing $CD4^+$ T cells are increased in psoriasis lesions compared to non-lesional skin (Caruso et al. 2009). When injected intradermally into mice, IL-21 causes epidermal hyperplasia that is not blocked by IL-22-specific antibodies (Caruso et al. 2009). This would imply a direct effect of IL-21 on keratinocytes, instead of an indirect one via the T_H17 pathway. This may be possible as the IL-21 receptor (IL-21R) is expressed not only on leukocytes, but also on keratinocytes (Spolski and Leonard 2008). However, the common cytokine gamma chain, γ_c, a required signaling component of IL-21R (Habib et al. 2002), is only variably expressed in keratinocytes (Corrigall et al. 2001). Thus, whether IL-21 may indeed directly trigger keratinocyte responses warrants further investigation (Fig. 4).

5.4 *Psoriasis Co-morbidities and T_H17 Cells*

It is now increasingly being recognized that psoriasis is not merely a skin disease, but is probably associated with other co-morbidities such as psoriatic arthritis,

Crohn's disease (an inflammatory bowel disease), the metabolic syndrome (insulin resistance, hypertension, central obesity, and combined hyperlipidemia) (Lopez-Candales 2001), and cardio-vascular diseases (CVD). Some of these co-morbidities may develop in direct relationship to the duration and severity of psoriasis and therefore are frequently found in patients with severe and or long-lasting disease. Furthermore, recently, studies based on large cohorts of psoriasis patients showed that severe psoriasis was association with increased mortality and CVD was found to be the most common cause of death (Gelfand et al. 2007).

Psoriasis and atherosclerosis have similar histological characteristics involving T cells, macrophages and monocytes. In particular, the extravasation of leukocytes through the endothelium is characteristic of both psoriatic and atherosclerotic plaques. Unstable psoriatic and atherosclerotic plaques both have also an increased percentage of activated T cells expressing a Th-1 pattern of cytokines including local and systemic expression of adhesion molecules, and endothelins. As was extensively discussed, T_H17 cells play an important role in the pathogenesis of psoriasis and broadly activate inflammation in a variety of organ systems (Arican et al. 2005; Sabat et al. 2007). Interestingly, IL-17 is elevated in the sera of patients with unstable CVD (Hashmi and Zeng 2006) and is preferentially expressed in animal models of aged coronary arteries that are susceptible to ischemia (Csiszar et al. 2003). Thus, T_H17 cells may also play an important role in the pathogenesis of atherosclerosis.

The association between psoriasis and Crohn's disease might be explained by genetic mechanisms, as recent studies have revealed many overlapping genetic loci common in both diseases (Duerr et al. 2006; Rioux et al. 2007). Variations in the IL23R gene were found in both diseases (Duerr et al. 2006; Cargill et al. 2007), once more demonstrating the importance of the IL-23/T_H17 pathway in both diseases.

5.5 Downregulation of the T_H17 Pathway Correlates with Successful Treatment

Cyclosporine, long considered the "gold standard" for the treatment of moderate-to-severe-psoriasis, is a calcineurin antagonist that broadly inhibits T cell activation. It is, therefore, not surprising that cyclosporine treatment is associated with down-regulation of both T_H1 and T_H17 pathways (Haider et al. 2008).

The vital role of the T_H17 pathway in psoriasis pathogenesis comes from the genomic analysis of psoriasis lesions during etanercept treatment (Zaba et al. 2007). While blockade of TNF is considered to be its primary action, treatment success with etanercept is highly dependent on the inactivation of myeloid dendritic cells and T_H17 immune response (Zaba et al. 2009a). Thus, non-response to etanercept treatment is associated with a failure to down-regulate the T_H17 pathway (Zaba et al. 2009a). Indeed, etanercept is associated with a rapid reduction of IL-23 and T_H17 cell products (IL-17, IL-22, CCL20 and DEFB4). By contrast, T_H1 cellular products are reduced late in disease resolution (Zaba et al. 2007).

The anti-IL-12p40 antibody was developed to prevent the biological activity of both IL-12 and IL-23. Early studies utilizing this antibody for psoriasis treatment, however, show significant suppression of p40 and IL-23p19, but not IL-12p35 (Toichi et al. 2006). This implies that the p40 antibody targets the IL-23/T_H17 pathway more efficiently than the IL-12/T_H1 pathway. If the IL-23/T_H17 pathway is truly upstream in the course of disease development, then selective disruption of IL-23 signaling should lead to reduced T_H17 T cells and long-lasting improvement in psoriasis. Indeed, the treatment of patients with anti-p40 monoclonal antibody (Ustekinumab) has led to significant, and sustained, clinical improvement in psoriasis (Leonardi et al. 2008; Papp et al. 2008).

Given that psoriasis is effectively treated with tumor necrosis factor (TNF) antagonists, some link between TNF and the IL23/T_H17 pathway should be mentioned. TNF is a likely amplifier of the T_H17 pathway by stimulating CD11c+ dermal DCs to produce IL-23. Accordingly, anti-TNF-α agents have been shown to modulate IL-23p19 and IL-12p40 mRNA levels, as well as decrease IL-17 production and inflammatory infiltrates in the psoriatic skin (Zaba et al. 2007).

6 Involvement of the IL-23/T_H17 Pathway in Other Inflammatory Skin Diseases

6.1 Atopic Dermatitis

Atopic dermatitis (AD) is another common chronic inflammatory skin disorder that results from an interaction between genetic susceptibility, skin barrier defects, and immunological factors (Leung and Bieber 2003). A potential involvement of T_H17 cells during onset of disease has been previously suggested, based on the presence of IL-17+ cells in early lesions, but not in chronic ones (Toda et al. 2003; Koga et al. 2008). However, there is decreased expression of IL-17 and IL-17-induced innate defense molecules in chronic AD lesions compared with psoriasis (Guttman-Yassky et al. 2008). The decreased activity of the T_H17 pathway might explain the relative deficiency of skin-derived anti-microbial proteins in AD (de Jongh et al. 2005). In contrast, there is upregulated expression of IL-22 independent of T_H17 cells in AD, which may provide explanation for the reactive epidermal hyperplasia observed in the chronic phase of the disease. (Nograles et al. 2009)

6.2 Allergic Contact Dermatitis

Allergic contact dermatitis (ACD) is a delayed-type hypersensitivity reaction to sensitizing haptens. This reaction is T-cell mediated, with T_H1 and Tc1 cells that are thought to be the primary effector cells. Recent reports, however, show that IL-23 and T_H17 cells are increased after allergen-challenge (Larsen et al. 2009; Zhao et al. 2009), suggesting a role for this pathway in eliciting an allergic skin reaction.

7 Conclusions

Although many factors that contribute to the initiation of psoriatic lesions remain obscure, psoriasis can serve as a model for studying interactions of immigrating immune cells with resident epithelial and mesenchymal cells. This disease vividly emphasizes the importance of T cells and DCs in its pathogenesis. In addition, it illustrates how advances in our understanding of molecular immunopathogenesis can be translated into innovative therapies. Accordingly, the recently FDA-approved anti-p40 monoclonal antibody (Ustekinumab) which has led to significant and sustained clinical improvement in psoriasis and implies the pivotal role of Th-17/IL-23 pathway in the pathogenesis of psoriasis. Ongoing clinical trials will further establish whether specific blockade of IL-23 can collapse the activation of T_H17 and T_H22 cells in psoriasis, or whether targeted blockade of either IL-17 or IL-22 will be sufficient for disease resolution.

Acknowledgement This publication was made possible by grant number 5UL1RR024143 from the National Center for Research Resources (NCRR), a component of the National Institutes of Health (NIH), and NIH Roadmap for Medical Research. JGK has received research grants from Amgen, Centocor, Wyeth and Eli Lilly for the study of therapeutic agents related to the IL-23/T_H17 pathway.

References

Abrams, J. R., S. L. Kelley, E. Hayes, T. Kikuchi, M. J. Brown, S. Kang, M. G. Lebwohl, C. A. Guzzo, B. V. Jegasothy, P. S. Linsley, and J. G. Krueger. 2000. Blockade of T lymphocyte costimulation with cytotoxic T lymphocyte-associated antigen 4-immunoglobulin (CTLA4Ig) reverses the cellular pathology of psoriatic plaques, including the activation of keratinocytes, dendritic cells, and endothelial cells. *J Exp Med* 192 (5):681–94.

Albanesi, C., A. Cavani, and G. Girolomoni. 1999. IL-17 is produced by nickel-specific T lymphocytes and regulates ICAM-1 expression and chemokine production in human keratinocytes: synergistic or antagonist effects with IFN-gamma and TNF-alpha. *J Immunol* 162 (1):494–502.

Arican, O., M. Aral, S. Sasmaz, and P. Ciragil. 2005. Serum levels of TNF-alpha, IFN-gamma, IL-6, IL-8, IL-12, IL-17, and IL-18 in patients with active psoriasis and correlation with disease severity. *Mediators Inflamm* 2005 (5):273–9.

Austin, L. M., M. Ozawa, T. Kikuchi, I. B. Walters, and J. G. Krueger. 1999. The majority of epidermal T cells in Psoriasis vulgaris lesions can produce type 1 cytokines, interferon-gamma, interleukin-2, and tumor necrosis factor-alpha, defining TC1 (cytotoxic T lymphocyte) and TH1 effector populations: a type 1 differentiation bias is also measured in circulating blood T cells in psoriatic patients. *J Invest Dermatol* 113 (5):752–9.

Bettelli, E., T. Korn, and V. K. Kuchroo. 2007. Th17: the third member of the effector T cell trilogy. *Curr Opin Immunol* 19 (6):652–7.

Boniface, K., E. Guignouard, N. Pedretti, M. Garcia, A. Delwail, F. X. Bernard, F. Nau, G. Guillet, G. Dagregorio, H. Yssel, J. C. Lecron, and F. Morel. 2007. A role for T cell-derived interleukin 22 in psoriatic skin inflammation. *Clin Exp Immunol* 150 (3):407–15.

Brandrup, F., N. Holm, N. Grunnet, K. Henningsen, and H. E. Hansen. 1982. Psoriasis in monozygotic twins: variations in expression in individuals with identical genetic constitution. *Acta Derm Venereol* 62 (3):229–36.

Cargill, M., S. J. Schrodi, M. Chang, V. E. Garcia, R. Brandon, K. P. Callis, N. Matsunami, K. G. Ardlie, D. Civello, J. J. Catanese, D. U. Leong, J. M. Panko, L. B. McAllister, C. B. Hansen, J. Papenfuss, S. M. Prescott, T. J. White, M. F. Leppert, G. G. Krueger, and A. B. Begovich. 2007. A large-scale genetic association study confirms IL12B and leads to the identification of IL23R as psoriasis-risk genes. *Am J Hum Genet* 80 (2):273–90.

Caruso, R., E. Botti, M. Sarra, M. Esposito, C. Stolfi, L. Diluvio, M. L. Giustizieri, V. Pacciani, A. Mazzotta, E. Campione, T. T. Macdonald, S. Chimenti, F. Pallone, A. Costanzo, and G. Monteleone. 2009. Involvement of interleukin-21 in the epidermal hyperplasia of psoriasis. *Nat Med* 15 (9):1013–5.

Chang, S. H., and C. Dong. 2007. A novel heterodimeric cytokine consisting of IL-17 and IL-17F regulates inflammatory responses. *Cell Res* 17 (5):435–40.

Christophers, E. 2001. Psoriasis--epidemiology and clinical spectrum. *Clin Exp Dermatol* 26 (4):314–20.

Ciric, B., M. El-behi, R. Cabrera, G. X. Zhang, and A. Rostami. 2009. IL-23 drives pathogenic IL-17-producing CD8+ T cells. *J Immunol* 182 (9):5296–305.

Corrigall, V. M., M. Arastu, S. Khan, C. Shah, M. Fife, T. Smeets, P. P. Tak, and G. S. Panayi. 2001. Functional IL-2 receptor beta (CD122) and gamma (CD132) chains are expressed by fibroblast-like synoviocytes: activation by IL-2 stimulates monocyte chemoattractant protein-1 production. *J Immunol* 166 (6):4141–7.

Csiszar, A., Z. Ungvari, A. Koller, J. G. Edwards, and G. Kaley. 2003. Aging-induced proinflammatory shift in cytokine expression profile in coronary arteries. *FASEB J* 17 (9):1183–5.

de Jongh, G. J., P. L. Zeeuwen, M. Kucharekova, R. Pfundt, P. G. van der Valk, W. Blokx, A. Dogan, P. S. Hiemstra, P. C. van de Kerkhof, and J. Schalkwijk. 2005. High expression levels of keratinocyte antimicrobial proteins in psoriasis compared with atopic dermatitis. *J Invest Dermatol* 125 (6):1163–73.

Di Cesare, A., P. Di Meglio, and F. O. Nestle. 2009. The IL-23/Th17 axis in the immunopathogenesis of psoriasis. *J Invest Dermatol* 129 (6):1339–50.

Duerr, R. H., K. D. Taylor, S. R. Brant, J. D. Rioux, M. S. Silverberg, M. J. Daly, A. H. Steinhart, C. Abraham, M. Regueiro, A. Griffiths, T. Dassopoulos, A. Bitton, H. Yang, S. Targan, L. W. Datta, E. O. Kistner, L. P. Schumm, A. T. Lee, P. K. Gregersen, M. M. Barmada, J. I. Rotter, D. L. Nicolae, and J. H. Cho. 2006. A genome-wide association study identifies IL23R as an inflammatory bowel disease gene. *Science* 314 (5804):1461–3.

Duhen, T., R. Geiger, D. Jarrossay, A. Lanzavecchia, and F. Sallusto. 2009. Production of interleukin 22 but not interleukin 17 by a subset of human skin-homing memory T cells. *Nat Immunol* 10 (8):857–63.

Elder, J. T., A. T. Bruce, J. E. Gudjonsson, A. Johnston, P. E. Stuart, T. Tejasvi, J. J. Voorhees, G. R. Abecasis, and R. P. Nair. 2010. Molecular Dissection of Psoriasis: Integrating Genetics and Biology. *J Invest Dermatol* 130 (5):1213–26.

Ellis, C. N., D. C. Gorsulowsky, T. A. Hamilton, J. K. Billings, M. D. Brown, J. T. Headington, K. D. Cooper, O. Baadsgaard, E. A. Duell, T. M. Annesley, and et al. 1986. Cyclosporine improves psoriasis in a double-blind study. *JAMA* 256 (22):3110–6.

Farber, E. M., M. L. Nall, and W. Watson. 1974. Natural history of psoriasis in 61 twin pairs. *Arch Dermatol* 109 (2):207–11.

Farber, E.M., and M.L. Nall. 1998. Epidemiology: natural history and genetics. In *Psoriasis*, edited by H. H. Roenigk and H. I. Maibach. New York: Marcel Dekker.

Ferenczi, K., L. Burack, M. Pope, J. G. Krueger, and L. M. Austin. 2000. CD69, HLA-DR and the IL-2R identify persistently activated T cells in psoriasis vulgaris lesional skin: blood and skin comparisons by flow cytometry. *J Autoimmun* 14 (1):63–78.

Fossiez, F., O. Djossou, P. Chomarat, L. Flores-Romo, S. Ait-Yahia, C. Maat, J. J. Pin, P. Garrone, E. Garcia, S. Saeland, D. Blanchard, C. Gaillard, B. Das Mahapatra, E. Rouvier, P. Golstein, J. Banchereau, and S. Lebecque. 1996. T cell interleukin-17 induces stromal cells to produce proinflammatory and hematopoietic cytokines. *J Exp Med* 183 (6):2593–603.

Fuhlbrigge, R. C., J. D. Kieffer, D. Armerding, and T. S. Kupper. 1997. Cutaneous lymphocyte antigen is a specialized form of PSGL-1 expressed on skin-homing T cells. *Nature* 389 (6654):978–81.

Fujita, H., K.E. Nograles, T. Kikuchi, J. Gonzalez, J.A. Carucci, and J.G. Krueger. 2009. Human Langerhans cells induce distinct IL-22-producing CD4+ T cells lacking IL-17 production. *Proc Natl Acad Sci USA* 106 (51):21795–800.

Gaffen, S. L. 2009. Structure and signalling in the IL-17 receptor family. *Nat Rev Immunol* 9 (8):556–67.

Ganguly, D., G. Chamilos, R. Lande, J. Gregorio, S. Meller, V. Facchinetti, B. Homey, F. J. Barrat, T. Zal, and M. Gilliet. 2009. Self-RNA-antimicrobial peptide complexes activate human dendritic cells through TLR7 and TLR8. *J Exp Med* 206 (9):1983–94.

Gelfand, J. M., A. B. Troxel, J. D. Lewis, S. K. Kurd, D. B. Shin, X. Wang, D. J. Margolis, and B. L. Strom. 2007. The risk of mortality in patients with psoriasis: results from a population-based study. *Arch Dermatol* 143 (12):1493–9.

Gelfand, J. M., R. Weinstein, S. B. Porter, A. L. Neimann, J. A. Berlin, and D. J. Margolis. 2005. Prevalence and treatment of psoriasis in the United Kingdom: a population-based study. *Arch Dermatol* 141 (12):1537–41.

Gottlieb, A. B., F. Chamian, S. Masud, I. Cardinale, M. V. Abello, M. A. Lowes, F. Chen, M. Magliocco, and J. G. Krueger. 2005. TNF inhibition rapidly down-regulates multiple proinflammatory pathways in psoriasis plaques. *J Immunol* 175 (4):2721–9.

Gottlieb, S. L., P. Gilleaudeau, R. Johnson, L. Estes, T. G. Woodworth, A. B. Gottlieb, and J. G. Krueger. 1995. Response of psoriasis to a lymphocyte-selective toxin (DAB389IL-2) suggests a primary immune, but not keratinocyte, pathogenic basis. *Nat Med* 1 (5):442–7.

Griffiths, C. E., E. Christophers, J. N. Barker, R. J. Chalmers, S. Chimenti, G. G. Krueger, C. Leonardi, A. Menter, J. P. Ortonne, and L. Fry. 2007. A classification of psoriasis vulgaris according to phenotype. *Br J Dermatol* 156 (2):258–62.

Griffiths, C.E.M., R.D.R. Camp, and J.N.W.N. Barker. 2004. Psoriasis. In *Rook's Textbook of Dermatology*, edited by T. Burns, S. Breathnach, N. Cox and C. Griffiths. Oxford: Blackwell Publishing.

Guttman-Yassky, E., M. A. Lowes, J. Fuentes-Duculan, L. C. Zaba, I. Cardinale, K. E. Nograles, A. Khatcherian, I. Novitskaya, J. A. Carucci, R. Bergman, and J. G. Krueger. 2008. Low expression of the IL-23/Th17 pathway in atopic dermatitis compared to psoriasis. *J Immunol* 181 (10):7420–7.

Habib, T., S. Senadheera, K. Weinberg, and K. Kaushansky. 2002. The common gamma chain (gamma c) is a required signaling component of the IL-21 receptor and supports IL-21-induced cell proliferation via JAK3. *Biochemistry* 41 (27):8725–31.

Haider, A. S., M. A. Lowes, M. Suarez-Farinas, L. C. Zaba, I. Cardinale, A. Khatcherian, I. Novitskaya, K. M. Wittkowski, and J. G. Krueger. 2008. Identification of cellular pathways of "type 1," Th17 T cells, and TNF- and inducible nitric oxide synthase-producing dendritic cells in autoimmune inflammation through pharmacogenomic study of cyclosporine A in psoriasis. *J Immunol* 180 (3):1913–20.

Harper, E. G., C. Guo, H. Rizzo, J. V. Lillis, S. E. Kurtz, I. Skorcheva, D. Purdy, E. Fitch, M. Iordanov, and A. Blauvelt. 2009. Th17 cytokines stimulate CCL20 expression in keratinocytes in vitro and in vivo: implications for psoriasis pathogenesis. *J Invest Dermatol* 129 (9):2175–83.

Hashmi, S., and Q. T. Zeng. 2006. Role of interleukin-17 and interleukin-17-induced cytokines interleukin-6 and interleukin-8 in unstable coronary artery disease. *Coron Artery Dis* 17 (8):699–706.

Hedrick, M. N., A. S. Lonsdorf, A. K. Shirakawa, C. C. Richard Lee, F. Liao, S. P. Singh, H. H. Zhang, A. Grinberg, P. E. Love, S. T. Hwang, and J. M. Farber. 2009. CCR6 is required for IL-23-induced psoriasis-like inflammation in mice. *J Clin Invest* 119 (8):2317–29.

Henseler, T., and E. Christophers. 1985. Psoriasis of early and late onset: characterization of two types of psoriasis vulgaris. *J Am Acad Dermatol* 13 (3):450–6.

Hollox, E. J., U. Huffmeier, P. L. Zeeuwen, R. Palla, J. Lascorz, D. Rodijk-Olthuis, P. C. van de Kerkhof, H. Traupe, G. de Jongh, M. den Heijer, A. Reis, J. A. Armour, and J. Schalkwijk. 2008. Psoriasis is associated with increased beta-defensin genomic copy number. *Nat Genet* 40 (1):23–5.

Homey, B., M. C. Dieu-Nosjean, A. Wiesenborn, C. Massacrier, J. J. Pin, E. Oldham, D. Catron, M. E. Buchanan, A. Muller, R. deWaal Malefyt, G. Deng, R. Orozco, T. Ruzicka, P. Lehmann, S. Lebecque, C. Caux, and A. Zlotnik. 2000. Up-regulation of macrophage inflammatory protein-3 alpha/CCL20 and CC chemokine receptor 6 in psoriasis. *J Immunol* 164 (12):6621–32.

Ishigame, H., S. Kakuta, T. Nagai, M. Kadoki, A. Nambu, Y. Komiyama, N. Fujikado, Y. Tanahashi, A. Akitsu, H. Kotaki, K. Sudo, S. Nakae, C. Sasakawa, and Y. Iwakura. 2009. Differential roles of interleukin-17A and -17F in host defense against mucoepithelial bacterial infection and allergic responses. *Immunity* 30 (1):108–19.

Kenney, J.A. 1971. Psoriasis in the American black. In *Psoriasis: Proceedings of the International Symposium*. Stanford, California: Stanford University Press.

Koga, C., K. Kabashima, N. Shiraishi, M. Kobayashi, and Y. Tokura. 2008. Possible pathogenic role of Th17 cells for atopic dermatitis. *J Invest Dermatol* 128 (11):2625–30.

Kryczek, I., A. T. Bruce, J. E. Gudjonsson, A. Johnston, A. Aphale, L. Vatan, W. Szeliga, Y. Wang, Y. Liu, T. H. Welling, J. T. Elder, and W. Zou. 2008. Induction of IL-17+ T cell trafficking and development by IFN-gamma: mechanism and pathological relevance in psoriasis. *J Immunol* 181 (7):4733–41.

Kuestner, R. E., D. W. Taft, A. Haran, C. S. Brandt, T. Brender, K. Lum, B. Harder, S. Okada, C. D. Ostrander, J. L. Kreindler, S. J. Aujla, B. Reardon, M. Moore, P. Shea, R. Schreckhise, T. R. Bukowski, S. Presnell, P. Guerra-Lewis, J. Parrish-Novak, J. L. Ellsworth, S. Jaspers, K. E. Lewis, M. Appleby, J. K. Kolls, M. Rixon, J. W. West, Z. Gao, and S. D. Levin. 2007. Identification of the IL-17 receptor related molecule IL-17RC as the receptor for IL-17F. *J Immunol* 179 (8):5462–73.

Lande, R., J. Gregorio, V. Facchinetti, B. Chatterjee, Y. H. Wang, B. Homey, W. Cao, B. Su, F. O. Nestle, T. Zal, I. Mellman, J. M. Schroder, Y. J. Liu, and M. Gilliet. 2007. Plasmacytoid dendritic cells sense self-DNA coupled with antimicrobial peptide. *Nature* 449 (7162):564–9.

Larsen, J. M., C. M. Bonefeld, S. S. Poulsen, C. Geisler, and L. Skov. 2009. IL-23 and T(H)17-mediated inflammation in human allergic contact dermatitis. *J Allergy Clin Immunol* 123 (2):486–92.

Lee, E., W. L. Trepicchio, J. L. Oestreicher, D. Pittman, F. Wang, F. Chamian, M. Dhodapkar, and J. G. Krueger. 2004. Increased expression of interleukin 23 p19 and p40 in lesional skin of patients with psoriasis vulgaris. *J Exp Med* 199 (1):125–30.

Leonardi, C. L., A. B. Kimball, K. A. Papp, N. Yeilding, C. Guzzo, Y. Wang, S. Li, L. T. Dooley, and K. B. Gordon. 2008. Efficacy and safety of ustekinumab, a human interleukin-12/23 monoclonal antibody, in patients with psoriasis: 76-week results from a randomised, double-blind, placebo-controlled trial (PHOENIX 1). *Lancet* 371 (9625):1665–74.

Leung, D. Y., and T. Bieber. 2003. Atopic dermatitis. *Lancet* 361 (9352):151–60.

Lew, W., A. M. Bowcock, and J. G. Krueger. 2004. Psoriasis vulgaris: cutaneous lymphoid tissue supports T-cell activation and "Type 1" inflammatory gene expression. *Trends Immunol* 25 (6):295–305.

Lew, W., E. Lee, and J. G. Krueger. 2004. Psoriasis genomics: analysis of proinflammatory (type 1) gene expression in large plaque (Western) and small plaque (Asian) psoriasis vulgaris. *Br J Dermatol* 150 (4):668–76.

Li, J., D. Li, and Z. Tan. 2004. The expression of interleukin-17, interferon-gamma, and macrophage inflammatory protein-3 alpha mRNA in patients with psoriasis vulgaris. *J Huazhong Univ Sci Technolog Med Sci* 24 (3):294–6.

Liang, S. C., X. Y. Tan, D. P. Luxenberg, R. Karim, K. Dunussi-Joannopoulos, M. Collins, and L. A. Fouser. 2006. Interleukin (IL)-22 and IL-17 are coexpressed by Th17 cells and cooperatively enhance expression of antimicrobial peptides. *J Exp Med* 203 (10):2271–9.

Liu, Y., C. Helms, W. Liao, L. C. Zaba, S. Duan, J. Gardner, C. Wise, A. Miner, M. J. Malloy, C. R. Pullinger, J. P. Kane, S. Saccone, J. Worthington, I. Bruce, P. Y. Kwok, A. Menter, J. Krueger, A. Barton, N. L. Saccone, and A. M. Bowcock. 2008. A genome-wide association study of psoriasis and psoriatic arthritis identifies new disease loci. *PLoS Genet* 4 (3):e1000041.

Lopez-Candales, A. 2001. Metabolic syndrome X: a comprehensive review of the pathophysiology and recommended therapy. *J Med* 32 (5–6):283–300.

Lowes, M. A., F. Chamian, M. V. Abello, J. Fuentes-Duculan, S. L. Lin, R. Nussbaum, I. Novitskaya, H. Carbonaro, I. Cardinale, T. Kikuchi, P. Gilleaudeau, M. Sullivan-Whalen, K. M. Wittkowski, K. Papp, M. Garovoy, W. Dummer, R. M. Steinman, and J. G. Krueger. 2005. Increase in TNF-alpha and inducible nitric oxide synthase-expressing dendritic cells in psoriasis and reduction with efalizumab (anti-CD11a). *Proc Natl Acad Sci U S A* 102 (52):19057–62.

Lowes, M. A., T. Kikuchi, J. Fuentes-Duculan, I. Cardinale, L. C. Zaba, A. S. Haider, E. P. Bowman, and J. G. Krueger. 2008. Psoriasis vulgaris lesions contain discrete populations of Th1 and Th17 T cells. *J Invest Dermatol* 128 (5):1207–11.

Manel, N., D. Unutmaz, and D. R. Littman. 2008. The differentiation of human T(H)-17 cells requires transforming growth factor-beta and induction of the nuclear receptor RORgammat. *Nat Immunol* 9 (6):641–9.

Milner, J. D., J. M. Brenchley, A. Laurence, A. F. Freeman, B. J. Hill, K. M. Elias, Y. Kanno, C. Spalding, H. Z. Elloumi, M. L. Paulson, J. Davis, A. Hsu, A. I. Asher, J. O'Shea, S. M. Holland, W. E. Paul, and D. C. Douek. 2008. Impaired T(H)17 cell differentiation in subjects with autosomal dominant hyper-IgE syndrome. *Nature* 452 (7188):773–6.

Morganroth, G. S., L. S. Chan, G. D. Weinstein, J. J. Voorhees, and K. D. Cooper. 1991. Proliferating cells in psoriatic dermis are comprised primarily of T cells, endothelial cells, and factor XIIIa+ perivascular dendritic cells. *J Invest Dermatol* 96 (3):333–40.

Nair, R. P., K. C. Duffin, C. Helms, J. Ding, P. E. Stuart, D. Goldgar, J. E. Gudjonsson, Y. Li, T. Tejasvi, B. J. Feng, A. Ruether, S. Schreiber, M. Weichenthal, D. Gladman, P. Rahman, S. J. Schrodi, S. Prahalad, S. L. Guthery, J. Fischer, W. Liao, P. Y. Kwok, A. Menter, G. M. Lathrop, C. A. Wise, A. B. Begovich, J. J. Voorhees, J. T. Elder, G. G. Krueger, A. M. Bowcock, and G. R. Abecasis. 2009. Genome-wide scan reveals association of psoriasis with IL-23 and NF-kappaB pathways. *Nat Genet* 41 (2):199–204.

Nair, R. P., A. Ruether, P. E. Stuart, S. Jenisch, T. Tejasvi, R. Hiremagalore, S. Schreiber, D. Kabelitz, H. W. Lim, J. J. Voorhees, E. Christophers, J. T. Elder, and M. Weichenthal. 2008. Polymorphisms of the IL12B and IL23R genes are associated with psoriasis. *J Invest Dermatol* 128 (7):1653–61.

Nestle, F. O., C. Conrad, A. Tun-Kyi, B. Homey, M. Gombert, O. Boyman, G. Burg, Y. J. Liu, and M. Gilliet. 2005. Plasmacytoid predendritic cells initiate psoriasis through interferon-alpha production. *J Exp Med* 202 (1):135–43.

Nograles, K. E., L. C. Zaba, A. Shemer, J. Fuentes-Duculan, I. Cardinale, T. Kikuchi, M. Ramon, R. Bergman, J. G. Krueger, and E. Guttman-Yassky. 2009. IL-22-producing "T22" T cells account for upregulated IL-22 in atopic dermatitis despite reduced IL-17-producing TH17 T cells. *J Allergy Clin Immunol* 123 (6):1244–52 e2.

Nograles, K.E., L. C. Zaba, E. Guttman-Yassky, J. Fuentes-Duculan, M. Suarez-Farinas, I. Cardinale, A. Khatcherian, J. Gonzalez, K.C. Pierson, T.R. White, C. Pensabene, I. Coats, I. Novitskaya, M. A. Lowes, and J.G. Krueger. 2008. Th17 cytokines interleukin (IL)-17 and IL-22 modulate distinct inflammatory and keratinocyte-response pathways. *Br J Dermatol* 159 (5):1092–102.

Nurieva, R., X. O. Yang, G. Martinez, Y. Zhang, A. D. Panopoulos, L. Ma, K. Schluns, Q. Tian, S. S. Watowich, A. M. Jetten, and C. Dong. 2007. Essential autocrine regulation by IL-21 in the generation of inflammatory T cells. *Nature* 448 (7152):480–3.

Papp, K. A., R. G. Langley, M. Lebwohl, G. G. Krueger, P. Szapary, N. Yeilding, C. Guzzo, M. C. Hsu, Y. Wang, S. Li, L. T. Dooley, and K. Reich. 2008. Efficacy and safety of ustekinumab, a human interleukin-12/23 monoclonal antibody, in patients with psoriasis: 52-week results from a randomised, double-blind, placebo-controlled trial (PHOENIX 2). *Lancet* 371 (9625):1675–84.

Peric, M., S. Koglin, S. M. Kim, S. Morizane, R. Besch, J. C. Prinz, T. Ruzicka, R. L. Gallo, and J. Schauber. 2008. IL-17A enhances vitamin D3-induced expression of cathelicidin antimicrobial peptide in human keratinocytes. *J Immunol* 181 (12):8504–12.

Puel, A., R. Doffinger, A. Natividad, M. Chrabieh, G. Barcenas-Morales, C. Picard, A. Cobat, M. Ouachee-Chardin, A. Toulon, J. Bustamante, S. Al-Muhsen, M. Al-Owain, P. D. Arkwright, C. Costigan, V. McConnell, A. J. Cant, M. Abinum, M. Polak, P. F. Bougneres, D. Kumararatne, L. LMarodi, A. Nahum, C. Roifman, S. Blanche, A. Fischer, C. Bodemer, L. Abel, D. Lilic, and J. L. Casanova. 2010. Autoantibodies against IL-17A, IL-17F and IL-22 in patients with chronic mucocutaneous candidiasis and autoimmune polyendocrine syndrome type I. *J Exp Med* 207 (2):291–7.

Rioux, J. D., R. J. Xavier, K. D. Taylor, M. S. Silverberg, P. Goyette, A. Huett, T. Green, P. Kuballa, M. M. Barmada, L. W. Datta, Y. Y. Shugart, A. M. Griffiths, S. R. Targan, A. F. Ippoliti, E. J. Bernard, L. Mei, D. L. Nicolae, M. Regueiro, L. P. Schumm, A. H. Steinhart, J. I. Rotter, R. H. Duerr, J. H. Cho, M. J. Daly, and S. R. Brant. 2007. Genome-wide association study identifies new susceptibility loci for Crohn disease and implicates autophagy in disease pathogenesis. *Nat Genet* 39 (5):596–604.

Sa, S. M., P. A. Valdez, J. Wu, K. Jung, F. Zhong, L. Hall, I. Kasman, J. Winer, Z. Modrusan, D. M. Danilenko, and W. Ouyang. 2007. The effects of IL-20 subfamily cytokines on reconstituted human epidermis suggest potential roles in cutaneous innate defense and pathogenic adaptive immunity in psoriasis. *J Immunol* 178 (4):2229–40.

Sabat, R., S. Philipp, C. Hoflich, S. Kreutzer, E. Wallace, K. Asadullah, H. D. Volk, W. Sterry, and K. Wolk. 2007. Immunopathogenesis of psoriasis. *Exp Dermatol* 16 (10):779–98.

Spolski, R., and W. J. Leonard. 2008. Interleukin-21: basic biology and implications for cancer and autoimmunity. *Annu Rev Immunol* 26:57–79.

Stern, R. S., T. Nijsten, S. R. Feldman, D. J. Margolis, and T. Rolstad. 2004. Psoriasis is common, carries a substantial burden even when not extensive, and is associated with widespread treatment dissatisfaction. *J Investig Dermatol Symp Proc* 9 (2):136–9.

Szabo, S. K., C. Hammerberg, Y. Yoshida, Z. Bata-Csorgo, and K. D. Cooper. 1998. Identification and quantitation of interferon-gamma producing T cells in psoriatic lesions: localization to both CD4+ and CD8+ subsets. *J Invest Dermatol* 111 (6):1072–8.

Teraki, Y., A. Miyake, R. Takebayashi, and T. Shiohara. 2004. Homing receptor and chemokine receptor on intraepidermal T cells in psoriasis vulgaris. *Clin Exp Dermatol* 29 (6):658–63.

Teunissen, M. B., C. W. Koomen, R. de Waal Malefyt, E. A. Wierenga, and J. D. Bos. 1998. Interleukin-17 and interferon-gamma synergize in the enhancement of proinflammatory cytokine production by human keratinocytes. *J Invest Dermatol* 111 (4):645–9.

Toda, M., D. Y. Leung, S. Molet, M. Boguniewicz, R. Taha, P. Christodoulopoulos, T. Fukuda, J. A. Elias, and Q. A. Hamid. 2003. Polarized in vivo expression of IL-11 and IL-17 between acute and chronic skin lesions. *J Allergy Clin Immunol* 111 (4):875–81.

Toichi, E., G. Torres, T. S. McCormick, T. Chang, M. A. Mascelli, C. L. Kauffman, N. Aria, A. B. Gottlieb, D. E. Everitt, B. Frederick, C. E. Pendley, and K. D. Cooper. 2006. An anti-IL-12p40 antibody down-regulates type 1 cytokines, chemokines, and IL-12/IL-23 in psoriasis. *J Immunol* 177 (7):4917–26.

Trifari, S., C. D. Kaplan, E. H. Tran, N. K. Crellin, and H. Spits. 2009. Identification of a human helper T cell population that has abundant production of interleukin 22 and is distinct from T(H)-17, T(H)1 and T(H)2 cells. *Nat Immunol* 10 (8):864–71.

Volpe, E., M. Touzot, N. Servant, M. A. Marloie-Provost, P. Hupe, E. Barillot, and V. Soumelis. 2009. Multiparametric analysis of cytokine-driven human Th17 differentiation reveals a differential regulation of IL-17 and IL-22 production. *Blood* 114 (17):3610–4.

Watson, W., H. M. Cann, E. M. Farber, and M. L. Nall. 1972. The genetics of psoriasis. *Arch Dermatol* 105 (2):197–207.

Wei, L., A. Laurence, K. M. Elias, and J. J. O'Shea. 2007. IL-21 is produced by Th17 cells and drives IL-17 production in a STAT3-dependent manner. *J Biol Chem* 282 (48):34605–10.

Williams, R. C., A. W. McKenzie, J. H. Roger, and V. C. Joysey. 1976. HL-A antigens in patients with guttate psoriasis. *Br J Dermatol* 95 (2):163–7.

Wright, J. F., Y. Guo, A. Quazi, D. P. Luxenberg, F. Bennett, J. F. Ross, Y. Qiu, M. J. Whitters, K. N. Tomkinson, K. Dunussi-Joannopoulos, B. M. Carreno, M. Collins, and N. M. Wolfman. 2007. Identification of an interleukin 17F/17A heterodimer in activated human CD4+ T cells. *J Biol Chem* 282 (18):13447–55.

Zaba, L. C., I. Cardinale, P. Gilleaudeau, M. Sullivan-Whalen, M. Suarez Farinas, J. Fuentes-Duculan, I. Novitskaya, A. Khatcherian, M. J. Bluth, M. A. Lowes, and J. G. Krueger. 2007. Amelioration of epidermal hyperplasia by TNF inhibition is associated with reduced Th17 responses. *J Exp Med* 204 (13):3183–94.

Zaba, L. C., M. Suarez-Farinas, J. Fuentes-Duculan, K. E. Nograles, E. Guttman-Yassky, I. Cardinale, M. A. Lowes, and J. G. Krueger. 2009a. Effective treatment of psoriasis with etanercept is linked to suppression of IL-17 signaling, not immediate response TNF genes. *J Allergy Clin Immunol* 124 (5):1022–10 e1-395.

Zaba, L.C., J. Fuentes-Duculan, N.J. Eungdamrong, M.V. Abello, I. Novitskaya, K.C. Pierson, J. Gonzalez, J.G. Krueger, and M.A. Lowes. 2009b. Psoriasis is characterized by accumulation of immunostimulatory and Thl/Th17 Cell-Polarizing Myeloid Dendritic Cells. *J Invest Dermatol* 129:79–88.

Zhao, Y., A. Balato, R. Fishelevich, A. Chapoval, D. L. Mann, and A. A. Gaspari. 2009. Th17/Tc17 infiltration and associated cytokine gene expression in elicitation phase of allergic contact dermatitis. *Br J Dermatol* 161 (6):1301–6.

Zheng, Y., D. M. Danilenko, P. Valdez, I. Kasman, J. Eastham-Anderson, J. Wu, and W. Ouyang. 2007. Interleukin-22, a T(H)17 cytokine, mediates IL-23-induced dermal inflammation and acanthosis. *Nature* 445 (7128):648–51.

Zhou, L., Ivanov, II, R. Spolski, R. Min, K. Shenderov, T. Egawa, D. E. Levy, W. J. Leonard, and D. R. Littman. 2007. IL-6 programs T(H)-17 cell differentiation by promoting sequential engagement of the IL-21 and IL-23 pathways. *Nat Immunol* 8 (9):967–74.

T_H17 Cells in the Tumor Micro-environment

Cailin Moira Wilke and Weiping Zou

Abstract Although it has been established that T_H17 cells play pivotal roles in the development and pathogenesis of many auto-immune diseases, their function in cancer is not yet as well-defined. Current research has documented both pro and anti-tumor roles for T_H17 cells or more precisely, T_H17-associated cytokines. Here we will examine those functions, as well as population kinetics, phenotype, and sources of T_H17 cells within the tumor micro-environment.

1 T_H17 Distribution in Tumor-Bearing Hosts

Many groups have undertaken studies of T_H17 in patients with various cancers, including ovarian (Charles et al. 2009; Kryczek et al. 2007b, 2009a; Miyahara et al. 2008), breast (Horlock et al. 2009), colon (Kryczek et al. 2009a), gastric (Zhang et al. 2008), hepatocellular (Kryczek et al. 2009a; Zhang et al. 2009), prostate cancer (Derhovanessian et al. 2009; Sfanos et al. 2008), small cell lung carcinoma (Koyama et al. 2008), non-small cell lung carcinoma (Chen et al. 2009), renal-cell carcinoma (Inozume et al. 2009; Kryczek et al. 2009a), lymphoma (Yang et al. 2009), melanoma (Kryczek et al. 2009a; von Euw et al. 2009), and myeloma (Dhodapkar et al. 2008). It should be noted that several of these investigations, while instructive, have focused on T_H17 populations in the peripheral blood rather than within the tumors themselves. Our laboratory has examined T_H17 distribution extensively within ovarian cancer patients (Kryczek et al. 2007b, 2009a) and we

W. Zou (✉)
Department of Surgery, School of Medicine, University of Michigan, C560B MSRB II, 1150 W. Medical Center Dr., Ann Arbor, MI 48109-0669, USA
e-mail: wzou@med.umich.edu

S. Jiang (ed.), *T_H17 Cells in Health and Disease*,
DOI 10.1007/978-1-4419-9371-7_26,

have found that the prevalence of T_H17 in the tumor-draining lymph nodes (TDLN) and peripheral blood is comparable to that in the blood of healthy donors. While T_H17 cells make up a proportionally small population of blood lymphocytes, experimental evidence from our group and others demonstrates that they may be found in higher numbers within the tumor micro-environment, due either to induction or recruitment. Su et al. (2010) found significantly higher numbers of T_H17 cells expanded from tumor-infiltrating lymphocyte (TIL) populations than in normal tissue from patients with breast cancer, melanoma, and colon cancer. Mouse studies have also confirmed the presence of T_H17 cells in tumors, although in smaller proportions than other T cell subsets (Kryczek et al. 2007b).

2 T_H17 Phenotype and Recruitment into the Tumor Micro-environment

Our group has observed that T_H17 cells in the tumor lack high levels of the markers HLA-DR, CD25, and granzyme B, and therefore do not seem to be conventional effector cells. They also do not express programmed cell death 1 (PD-1) and forkhead box P3 (FoxP3), implying that they likely do not participate in immune suppression through these pathways (Kryczek et al. 2009a). It appears that T_H17 cells are a subset of T cells distinct from regulatory T cells and other effector T cells in the tumor. Several laboratories have contributed knowledge regarding the cytokine profile of T_H17 cells within the tumor micro-environment. Our group has found that these T_H17 cells secrete high levels of granulocyte-macrophage colony stimulating factor (GM-CSF), tumor necrosis factor alpha (TNFα), interleukin-2 (IL-2), and interferon gamma (IFNγ) (Kryczek et al. 2009a). In addition to the data obtained from ovarian cancer patients, other investigators have observed similar cytokine profiles of T_H17 cells in patients with cancers of the skin, pancreas, intestine, and liver, as well as in mouse tumor models (Sharma et al. 2009). Interestingly, Su et al. documented a slightly different cytokine profile in T_H17 expanded from TILs in various cancers; high amounts of IL-8 and TNFα, but no IL-2. Because this phenotype is more closely aligned with that of T_H17 cells in healthy tissue (Liu and Rohowsky-Kochan 2008), it suggests that there may be a difference in phenotype between T_H17 cells freshly isolated from the tumor and those expanded from tumor-associated precursors. Conversely, but perhaps less likely, it may imply differential T_H17 profiles in different malignancies. Muranski et al. (2008) and Hinrichs et al. (2009) found that T_H17-polarized mouse CD4+ and CD8+ T cells express IFNγ, mirroring the IFNγ our group has observed in T_H17 from ovarian cancer-associated ascites. It seems that tumor-associated T_H17 expresses a cytokine profile similar to that of T_H17 cells in virus-response settings (Almeida et al. 2007; Arens et al. 2008; Precopio et al. 2007). The data obtained from studies of cancer patients suggest that tumor-associated T_H17 cells may promote anti-tumor immunity through the action of their cytokine products.

Our group has also found that human tumor-associated T_H17 cells express high levels of several tissue homing molecules, including CXCR4 and CCR6, c-type lectin-like receptor CD161, and the CD49 integrin isoforms c, d, and e. In contrast, these T_H17 do not have elevated expression of CCR2, CCR5, or CCR7 (Kryczek et al. 2009a). It is possible that these surface molecules aid in cell recruitment and retention in inflammatory environments, and since many reports have documented the similarities between a tumor and wound-healing or inflammatory environments (Schafer and Werner 2008), they may recruit cells to the tumor micro-environment as well. Interestingly, multiple groups have reported elevated levels of CXCL12 (Kryczek et al. 2005; Zou et al. 2001) and CCL20 (Bell et al. 1999), the ligands for CXCR4 and CCR6 in the tumor micro-environment. CXCL12 is both strongly chemotatic for lymphocytes (Bleul et al. 1996) and plays a large role in induction of angiogenesis (Zheng et al. 2007). CCL20 is also a chemotactic agent for lymphocytes and dendritic cells, and weakly attracts neutrophils (Hieshima et al. 1997). Although these chemokines and receptors may aid T_H17 recruitment to the tumor micro-environment, it is important to note that both CCR6 and CD161 expression have been observed in high levels on T_H17 from healthy donors and on other tissue lymphocytes and dendritic cells from inflammatory environments (Annunziato et al. 2007; Cosmi et al. 2008; Kleinschek et al. 2009). These molecules may not, then, be used as specific markers for T_H17 identification or selection. A recent study documented roles for toll-like receptor (TLR) and Nod2 signaling-induced monocyte chemotactic protein-1 (MCP-1, CCL2) and chemokine ligand 5 (CCL5, RANTES) for in vitro chemotaxis of tumor-infiltrating T_H17 clones (Su et al. 2010). CCL2 recruits monocytes, memory T cells, and dendritic cells to sites of tissue injury and infection (Carr et al. 1994; Xu et al. 1996), while CCL5 can induce chemotaxis of T cells and other leukocytes. The authors suggest that both tumor cells and tumor-associated fibroblasts serve as the sources for these chemokines (Su et al. 2010); however, these experiments must first be validated with primary cells. If either of these molecules is involved in T_H17 recruitment to the tumor, CCL5 may be the more likely candidate – recall that the receptor for CCL2, CCR2, is not found on tumor-infiltrating T_H17 cells.

3 Induction of Tumor-Associated T_H17 Cells

It is necessary to understand where tumor-associated T_H17 cells arise. One important avenue involves induction in the tumor micro-environment itself. We and others have explored this phenomenon, and documented a pivotal role for micro-environmental antigen-presenting cells (APCs). The best T_H17 inducers are tumor-associated macrophages (TAMs) and myeloid dendritic cells (mDC); both populations induced IL-17 production from memory (CD45RO+ CD25+) but not naïve T cells. Plasmacytoid DCs (pDC) were not efficient at this induction. Interestingly, TAMs, which outnumber mDC in several human cancers (Kryczek et al. 2006; Zou et al. 2001) are superior to both normal macrophages and mDCs in inducing IL-17 production (Kryczek et al. 2008b, 2009a). TAMs may serve as the primary means of human T_H17 induction in the tumor micro-environment. It seems logical, then, that TAMs secrete higher amounts of interleukin-1 beta (IL-1β) than macrophages from healthy donors. Several reports have documented that IL-1β, but not interleukin-1 alpha (IL-1α), interleukin-6 (IL-6), transforming growth factor beta (TGFβ), or interleukin-23 (IL-23) is crucial for human tumor-associated myeloid APC-mediated T_H17 induction. In line with this observation, our group has observed negligible levels of IL-1α and IL-23 in human tumor ascites. Because IL-23, IL-1β, and IL-1α are crucial in expansion of T_H17 memory cells in psoriatic patients (and in vitro) (Kryczek et al. 2008a, b), it is likely that the expansion mechanisms involved in cancer patients may be different than those implicated in patients with auto-immune disease. The debate regarding the importance of IL-6 and TGFβ in human T_H17 development is ongoing. Many studies suggest that TGFβ is crucial; however, large amounts of TGFβ polarize T cells towards a T regulatory cell (Treg) phenotype and suppress T_H17 development (Acosta-Rodriguez et al. 2007; Manel et al. 2008; Wilson et al. 2007; Yang et al. 2008). Interestingly, TGFβ and IL-6 are often found in high levels in the tumor micro-environment (Zou 2005). If TGFβ and IL-6 are necessary in T_H17 promotion, it stands to reason that there should be high T_H17 numbers within the tumor. However, as we have noted above, intra-tumoral T_H17 numbers are limited in both mouse and human patients. Two pieces of evidence support a developmental mechanism that does not rely on IL-6 and TGFβ: first, numbers of T_H17 cells (and levels of IL-17) do not correlate with ovarian tumor levels of IL-6 and TGFβ, and second, only blockade of IL-1β, not IL-6 or TGFβ, reduced T_H17 induction by myeloid APCs isolated from ovarian cancer patients (Kryczek et al. 2009a). It seems then, that IL-1β is the crucial mediator of T_H17 induction in the ovarian tumor micro-environment. Several mouse studies from the past few years also document a key role for this cytokine in T_H17 development (Chung et al. 2009; Gulen et al. 2010; Kryczek et al. 2007a; Sutton et al. 2006).

Interestingly, in addition to induction stimulated by myeloid APCs, it is probable that T_H17 increase their own frequency at the tumor site via production of CCL20. As we mentioned above, CCL20 signaling through CCR6 is profoundly chemotactic for lymphocytes and DCs (Martin-Orozco et al. 2009b). Augmenting DC trafficking towards the tumor increases the possibility of IL-1β secretion at the tumor site and thus enhances the likelihood of T_H17 induction. Additionally, because human T_H17 cells also express high levels of CCR6 (Kryczek et al. 2008a, 2009a) and migrate well towards CCL20 (Kryczek et al. 2008a), they may promote their own induction at and retention within the tumor micro-environment.

4 Interaction with Other T cell Subsets

T_H17 within the tumor micro-environment will interact with numerous other cell types. One of the most important intra-tumoral relationships for T_H17 is with Treg cells. Studies from our laboratory have documented an inverse correlation between

T_H17 and Treg numbers within both human and mouse tumors (Kryczek et al. 2007b, 2009a), prompting consideration of whether there may be a dynamic relationship between the two T cell subsets. Fascinatingly, three fairly recent reports have demonstrated that mature Treg cells in the periphery may be converted to T_H17 cells depending upon both the signals they receive and their surrounding cytokine milieu (Bettelli et al. 2006; Deknuydt et al. 2009; Mucida et al. 2007). IL-6 and IL-1β, both prototypical inflammatory cytokines, seem to be the major determinants of this conversion. In the study by Bettelli et al. (2006) IL-6 completely inhibited the generation of Foxp3+ Treg cells induced by TGFβ; instead, together with TGFβ, IL-6 promoted the differentiation of naive T cells into T_H17 cells. Deknuydt et al. (2009) demonstrated that IL-1β (in the presence or absence of IL-6) could induce conversion of both naïve ($CD45RA^+CCR7^+$) and memory ($CD45RA^-$) Treg cells into T_H17 cells. The existence of FoxP3+ IL-17+ Treg cells in human peripheral blood and lymphoid tissue (Beriou et al. 2009; Voo et al. 2009) supports the possibility of conversion, although at present it is unknown whether these cells arise from Treg cells or T_H17. These cells do retain their suppressive function along with FoxP3 and CD25 expression, but also produce IL-17 and express RORγt, the T_H17 lineage-specific transcription factor. Notably, despite the high levels of IL-6 observed in some cancers (Kryczek et al. 2000; Zou 2005), the number of T_H17 cells within the tumor remains limited; it may be that an as-yet unidentified mechanism is subverting micro-environmental IL-6. A report of note from Mucida et al. demonstrated that IL-6 was unable to induce IL-17 production from FoxP3+ cells in the presence of retinoic acid, a Vitamin A metabolite that enhances TGFβ signaling while simultaneously impeding IL-6 signaling (Mucida et al. 2007). Currently, it remains unknown whether retinoic acid in the tumor micro-environment has any effect on intra-tumoral Tregs.

To continue our investigation into the limited numbers of T_H17 in the tumor micro-environment, we examined other mechanisms of possible T_H17 suppression. We found that tumor-associated Tregs express high levels of CD39, an ectonucleotidase that converts ATP into the nucleoside adenosine (Kryczek et al. 2009a). Adenosine can then inhibit T_H17 cell function by binding to one of their adenosine type 1 purinergic G protein-coupled cell surface receptors [likely A2a (Ohta et al. 2006; Ohta and Sitkovsky 2001)] and, through the adenosinergic pathway, suppress their IL-17 production. It has been suggested by other groups that this pathway may be used by mouse Treg cells to suppress both T cell activation (Borsellino et al. 2007; Deaglio et al. 2007) and T_H17-mediated protective immunity (Kao et al. 2009). Additionally, Chaudhry et al. (2009) reported that mouse Tregs can suppress T_H17-mediated responses in vivo in a STAT3-dependent fashion; selective ablation of STAT3 in Tregs lead to CD4+ T cell activation and an increase in T_H17 responses. If Tregs are so efficient, then, at suppression of T_H17 development and function, and if T_H17 cells provide protective immunity to tissues, then recruitment (Curiel et al. 2004; Enarsson et al. 2006), retention, and direction of Tregs by the tumor itself may serve as an efficient means of evading the immune system.

Another important relationship within the tumor environment is that of T_H17 cells with T_H1 cells. Evidence reviewed by Muranski et al. suggests that these populations are developmentally, phenotypically, and functionally linked within the tumor micro-environment (Muranski and Restifo 2009a). IFNγ, the signature T_H1 cytokine, is expressed by populations of human T_H17 cells in tumors (Kryczek et al. 2009a), in settings of auto-immune disease (Kryczek et al. 2008a), and in T_H17-polarized mouse cells in vitro (Muranski et al. 2008). It appears that these double-positive cells may develop either from the T_H1 or T_H17 lineage (Lee et al. 2009). Certain experimental evidence supports the idea of a T_H17-like precursor in this paradigm: first, when antigen-specific IL-17^+CD8^+ cells are adoptively transferred into antigen-bearing hosts, the cells switch to become IFNγ+ CD8+ cells (Yen et al. 2009). Second, T_H17 cells can re-differentiate into T_H1 cells in lymphopenic mice (Bending et al. 2009; Martin-Orozco et al. 2009a; Nurieva et al. 2009). Because cancer patients mimic lymphopenic hosts (especially after cytotoxic chemotherapy) and we have found substantial numbers of IL-17+ IFNγ+ cells within ovarian cancer patients(Kryczek et al. 2009a), it is possible that T_H17 may initially express minimal IFNγ but then be converted slowly to a more T_H1-like phenotype. Conversely, we have observed that T_H1 cells in acute inflammatory responses may be converted to T_H17 cells (more often found in settings of chronic inflammation) via IFNγ-dependent production of IL-1 and IL-23 in local DCs (Kryczek et al. 2008b). This data supports the notion that T_H1 cells are the precursor and may pass through a double-positive stage during their conversion towards a T_H17 phenotype. Either (or both) concepts are possible, and as is often the case with immune cells, their developmental pathway may depend upon the conditions within the local environment. It is important to note that IFNγ in mice can prevent T_H17 development from naïve T cells(Harrington et al. 2005; Kimura et al. 2007; Kryczek et al. 2008a), and recall that not every phenomenon observed in human or mouse is applicable to the other. In summary, it is possible that the interaction between T_H1 and T_H17 cells is beneficial to the cancer-bearing host and serves an anti-tumor role. We explore the anti-tumor effects of T_H17 in the following section.

5 The Role of T_H17 in Anti-tumor Immunity

The debate as to whether T_H17 cells fulfill an anti- or pro-tumor function is ongoing (Bronte 2008; Munn 2009; Muranski and Restifo 2009b; Murugaiyan and Saha 2009). Of course, it is possible that both may be true, and simply depend on the environment of the T_H17 population in question. We will first review the evidence that T_H17 cells play a positive role in anti-tumor immunity. Supporting this claim are the following pieces of information: first, we have established that tumor-infiltrating T_H17 cells display a multi-effector cytokine profile – much like T_H17 cells in patients with viral infections(Almeida et al. 2007; Precopio et al. 2007) – suggesting that they are functional effector T cells. Secondly, T_H17 and Treg numbers are inversely correlated within the tumor micro-environment (Curiel et al. 2004; Kryczek et al. 2009a). Finally, T_H17 are positively associated with IFNγ+ effector T cells, cytotoxic CD8+ T cells, and NK cells in the same tumor environment (Kryczek et al. 2009a, b; Martin-Orozco et al. 2009b). Interestingly, the majority of mouse data corroborates an anti-tumor role for T_H17 cells. Experiments in mice have demonstrated that transgenic T cells polarized to a T_H17 phenotype via IL-6 and TGFβ are capable of eliminating tumors (Hinrichs et al. 2009; Muranski et al. 2008). Transgenic experiments with addition or deletion of IL-17 in mice have also been instructive: IL-17A knockout mice have faster tumor growth and more lung metastases than wild-type (WT) mice, and forced expression of IL-17 in tumor cells suppresses tumor progression(Benchetrit et al. 2002; Hirahara et al. 2001; Kryczek et al. 2009b; Martin-Orozco et al. 2009b). Improved anti-tumor immunity has also resulted from immunotherapies directed towards increasing T_H17 activity – these include blocking indoleamine 2,3-dioxygenase (IDO) (Sharma et al. 2009), use of IL-7 as an adjuvant (Pellegrini et al. 2009), and induction of hsp70 (Kottke et al. 2007).

Several groups have investigated the relationship between intra-tumoral T_H17 cells and tumor progression and/or patient survival. In prostate cancer patients, Sfanos et al. (2008) identified an inverse correlation between T_H17 differentiation stage and tumor progression. T_H17 induction is achieved in melanoma patients upon vaccination with antibody specific to cytotoxic T lymphocyte antigen 4 (CTLA4), and the level of IL-17 in tumor-associated ascites in these patients correlates positively with survival (von Euw et al. 2009). In our own ovarian cancer studies, we have determined that the level of IL-17 within patient ascites is significantly associated with improved survival (78 months in patients with greater than median IL-17 levels, 27 months in patients with lower than median IL-17 levels(Kryczek et al. 2009a)). Additionally, we found that tumor ascites IL-17 levels were significant predictors of patient death; patients with higher IL-17 levels had a significantly reduced death hazard. These studies support a protective role for T_H17 in tumor immunity.

As we documented above, T_H17 cells do not express granzyme B or perforin, and they do not have direct effects on proliferation or apoptosis of primary ovarian cancer cells (Kryczek et al. 2009a; Yen et al. 2009). It seems, then, that T_H17 cells exert their anti-tumor effects in an indirect fashion. They may recruit other effector cells to the tumor itself, as shown in studies by Galon, Sato, and Zhang (Galon et al. 2006; Sato et al. 2005; Zhang et al. 2003). We have found a positive correlation between intra-tumoral IL-17 levels and numbers of IFNγ-secreting effector T cells. Here, IL-17 and IFNγ synergized to induce production of CXCL9 and CXCL10, T_H1-type chemokines, which in turn contributed to effector T cell migration toward the tumor itself. CXCL9 is a T cell chemo-attractant, while CXCL10 is chemotactic for monocytes and macrophages, T cells, NK cells, and dendritic cells. CXCL10 also promotes T cell adhesion to endothelial cells and inhibits angiogenesis (Angiolillo et al. 1995; Dufour et al. 2002). Studies in our lab found that levels of CXCL9 and 10 within ascites from ovarian cancer patients correlated directly with infiltrating micro-environmental NK and CD8+ T cells (Kryczek et al. 2009a). T_H17 action within the tumor may also be related to DC recruitment to the tumor and tumor-draining lymph nodes. The laboratories of Shimon Sakaguchi and later, Chen Dong, reported that T_H17 cells can produce large amounts of CCL20 (Hirota et al. 2007; Yamazaki et al. 2008). Intra-tumoral T_H17

production of CCL20 could, via binding of DC CCR6, lead to trafficking of DC to the tumor micro-environment. In the presence of these DCs, CD8+ T cells are effectively primed, activated, and can then enact powerful anti-tumor immunity.

6 Pro-tumor Effects of T_H17-Associated Signature Cytokines

Other investigations have revealed pro-tumorigenic activities of T_H17-associated signature cytokines. The most notable mouse studies investigated the effects of IL-17 on stromal cells, vascular endothelial cells, and tumor cells themselves. In immuno-deficient nude and SCID mice, investigators have shown that exogenous IL-17 can induce angiogenesis and therefore contribute to tumor growth through vascularization (Numasaki et al. 2003, 2005; Tartour et al. 1999). Interestingly, studies in immuno-competent mice have quite different results, as summarized in the previous section (Benchetrit et al. 2002; Hirahara et al. 2001). Nam et al. demonstrated that IL-17 could suppress apoptosis of mouse 4T1 breast and CT26 colon cancer cell lines (as well as the human breast MDA MB231 cell line) but not others (human breast Hs 578T and melanoma MDA MB435, mouse mammary EMT6) in vitro. Interestingly, this laboratory also found that knockdown of the IL-17 receptor in 4T1 mouse mammary cancer cells enhanced apoptosis and decreased tumor growth in vivo (Nam et al. 2008). This suggests that the constituents of the tumor micro-environment may have profound consequences on how T_H17 cells and their signature cytokines affect the tumor. It is also necessary to note the likelihood that exogenous and endogenous IL-17 (including T_H17-derived IL-17) may have different effects; within the tumor micro-environment itself, the producers, targets, and local concentrations of IL-17 are crucial. The bioavailability of this cytokine no doubt greatly influences its own downstream effects. A 2008 study by Zhang et al. (2008) documented increased numbers of T_H17 cells in tumor-draining lymph nodes of gastric cancer patients with more advanced disease. Along the same lines, a report by Derhovanessian et al. (2009) demonstrated an inverse correlation between pretreatment frequency of circulating CD4+ IL-17+ cells and time to disease progression; prostate cancer patients with higher pretreatment T_H17 numbers experienced quicker declines in health than those with lower T_H17 numbers. Finally, Zhang et al. (2009) recently published a study investigating the prognostic potential of T_H17 cells patients with hepatocellular carcinoma (HCC). The investigators determined that intra-tumoral IL-17-producing cell density, which correlated proportionally with tumor micro vessel density, could serve as an independent prognostic factor for patient survival and was negatively associated with patient outcome. In non-small cell lung cancer (NSCLC) patients, high levels of IL-17 in situ within the tumor correlated with significantly poorer patient survival and higher blood vessel density, indicating that T_H17 cells may promote cancer progression through induction of angiogenesis (Chen et al. 2009). This study parallels an earlier report documenting a pro-angiogenic role for IL-17 (Numasaki et al. 2003). Many groups have shown that IL-17 can induce expression or enhance the effects of several angiogenic factors from fibroblasts and endothelial cells, including keratinocyte-derived chemokine (KC), prostaglandin E2 (PGE2), and vascular endothelial cell growth factor (VEGF) (Honorati et al. 2006; Numasaki et al. 2004; Takahashi et al. 2005). A recent study of *Bacteroides fragilis* – a common human commensal bacterium – showed that it induced T_H17-mediated colitis and subsequent colonic tumors in WT mice via activation of signal transducer and activator of transcription-3 (STAT3). Interestingly, antibody-mediated blockade of IL-17 and IL-23R inhibited colitis, colonic hyperplasia and tumor formation (Wu et al. 2009). Another study from the same year by Wang et al. (2009) showed that two transferred tumors had reduced growth in IL-$17^{-/-}$ mice but drastically accelerated growth in IFN$\gamma^{-/-}$ mice – this correlates well with studies from our laboratory demonstrating the importance of a T_H1-type anti-tumor response (Kryczek et al. 2009a). The Wang study showed that IL-17 induced IL-6 production by tumor and stromal cells, subsequently activating the oncogenic transcription factor STAT3. STAT3 can upregulate pro-survival and pro-angiogenic genes such as MYC, Cyclin D1, Survivin, and VEGF (Yu et al. 2007). A very recent study published by He et al. (2010) also investigated the effects of IL-17 signaling upon the mouse tumor environment. This article reported significantly inhibited growth of three different tumor cell lines in IL-17 receptor knockout (IL-17R$^{-/-}$) mice; interestingly, in accordance with the Wang study discussed above, IFNγR$^{-/-}$ mice had accelerated tumor growth. Mice lacking both the IL-17R and the IFNγR had significantly reduced tumor growth when compared to wild-type mice. He et al. also found that systemic administration of IL-17A accelerated tumor growth, while treatment with antiIL-17A significantly reduced tumor growth. Interestingly, the lack of IL-17R was associated with an increase in intra-tumoral CD8+ T cells and a decrease in the presence of myeloid-derived suppressor cells (MDSC) within the tumor micro-environment. Finally, the investigators found that co-injection of MDSC isolated from tumor-bearing IL-17R–/– mice had no effect on tumor growth, while co-injection of MDSC from tumor-bearing WT mice significantly increased it. Experiments with IL-17-treated MDSC yielded similar results.

It is important to mention the role of IL-23 when discussing the pro-tumorigenic capacity of T_H17 cells. IL-23 is an APC-produced IL-12 family member which promotes the survival and expansion of T_H17 cells. A study by Langrowski et al. (2006) reported that IL-23$^{-/}$mice are resistant to chemically-induced tumors, and that this resistance is associated with reduced skin expression of matrix metalloprotease 9 (MMP9), reduced expression of the angiogenic marker CD31, lower vessel density, and increased levels of intra-tumoral CD8+ T cell infiltration. Langrowski et al. (2007) subsequently discussed additional pro-tumor effects of IL-23: it stimulated intra-tumoral neutrophil and macrophage infiltration and antagonized IL-12 and IFNγ, both necessary cytokines for cytotoxic immune responses. Given the close relationship between IL-23 and T_H17 cells, it is enticing to postulate that T_H17 cells (or their development) play some role in the pro-tumor effect of IL-23 in this model. Despite this study, the anti-tumor properties of IL-23 have been rather well-documented: over-expression of IL-23 in tumors and vaccination with IL-23-transduced DCs had led to robust CD8+ T cell infiltration of tumors and subsequent inhibition of tumor growth (Hu et al. 2006; Oniki et al. 2006; Overwijk et al. 2006; Yuan et al. 2006). In mice, treatment with IL-23 leads to suppression of a pre-existing fibrosarcoma and better survival (Kaiga et al. 2007). Thus, it appears that IL-23 may – like IL-17 – have both pro- and anti-tumor capabilities, dependent upon the cell populations and immune responses occurring within the surrounding environment.

7 Conclusion

Within this chapter, we have discussed the T_H17 phenotypes of cancer patients and mouse models, the recruitment of T_H17 to and the development and expansion of T_H17 within the tumor micro-environment, and the associations of T_H17 cells with other tumor-associated immune cell populations. We have reviewed the current reports on pro- and anti-tumor effects of T_H17 cells and their signature cytokines, and discussed investigations into the prognostic significance of T_H17 populations within patient tumors and peripheral organs. It is important to understand that T_H17, like many other immune cells, may not have a consistent role within multiple tumor environments, but rather may serve protective or detrimental functions to the host depending on the type, the carcinogen, location, and developmental stage of the tumor in question. Current literature supports a pro-tumor role for T_H17 cells – or more precisely, their signature cytokines – in chronic inflammation-associated cancers (including cancers induced by chemical carcinogens or chronic viral infections), and an anti-tumor function for T_H17 cells in most human carcinomas which are not associated with chronic infection or inflammation. Future studies will help elucidate these differences.

References

Acosta-Rodriguez, E.V., Napolitani, G., Lanzavecchia, A., and Sallusto, F. (2007). Interleukins 1beta and 6 but not transforming growth factor-beta are essential for the differentiation of interleukin 17-producing human T helper cells. Nat Immunol *8*, 942–949.

Almeida, J.R., Price, D.A., Papagno, L., Arkoub, Z.A., Sauce, D., Bornstein, E., Asher, T.E., Samri, A., Schnuriger, A., Theodorou, I., *et al.* (2007). Superior control of HIV-1 replication by CD8+ T cells is reflected by their avidity, polyfunctionality, and clonal turnover. J Exp Med *204*, 2473–2485.

Angiolillo, A.L., Sgadari, C., Taub, D.D., Liao, F., Farber, J.M., Maheshwari, S., Kleinman, H.K., Reaman, G.H., and Tosato, G. (1995). Human interferon-inducible protein 10 is a potent inhibitor of angiogenesis in vivo. J Exp Med *182*, 155–162.

Annunziato, F., Cosmi, L., Santarlasci, V., Maggi, L., Liotta, F., Mazzinghi, B., Parente, E., Fili, L., Ferri, S., Frosali, F., *et al.* (2007). Phenotypic and functional features of human Th17 cells. J Exp Med *204*, 1849–1861.

Arens, R., Wang, P., Sidney, J., Loewendorf, A., Sette, A., Schoenberger, S.P., Peters, B., and Benedict, C.A. (2008). Cutting edge: murine cytomegalovirus induces a polyfunctional CD4 T cell response. J Immunol *180*, 6472–6476.

Bell, D., Chomarat, P., Broyles, D., Netto, G., Harb, G.M., Lebecque, S., Valladeau, J., Davoust, J., Palucka, K.A., and Banchereau, J. (1999). In breast carcinoma tissue, immature dendritic cells reside within the tumor, whereas mature dendritic cells are located in peritumoral areas. J Exp Med *190*, 1417–1426.

Benchetrit, F., Ciree, A., Vives, V., Warnier, G., Gey, A., Sautes-Fridman, C., Fossiez, F., Haicheur, N., Fridman, W.H., and Tartour, E. (2002). Interleukin-17 inhibits tumor cell growth by means of a T-cell-dependent mechanism. Blood *99*, 2114–2121.

Bending, D., De La Pena, H., Veldhoen, M., Phillips, J.M., Uyttenhove, C., Stockinger, B., and Cooke, A. (2009). Highly purified Th17 cells from BDC2.5NOD mice convert into Th1-like cells in NOD/SCID recipient mice. J Clin Invest *119*, 565–572.

Beriou, G., Costantino, C.M., Ashley, C.W., Yang, L., Kuchroo, V.K., Baecher-Allan, C., and Hafler, D.A. (2009). IL-17-producing human peripheral regulatory T cells retain suppressive function. Blood *113*, 4240–4249.

Bettelli, E., Carrier, Y., Gao, W., Korn, T., Strom, T.B., Oukka, M., Weiner, H.L., and Kuchroo, V.K. (2006). Reciprocal developmental pathways for the generation of pathogenic effector TH17 and regulatory T cells. Nature *441*, 235–238.

Bleul, C.C., Fuhlbrigge, R.C., Casasnovas, J.M., Aiuti, A., and Springer, T.A. (1996). A highly efficacious lymphocyte chemo-attractant, stromal cell-derived factor 1 (SDF-1). J Exp Med *184*, 1101–1109.

Borsellino, G., Kleinewietfeld, M., Di Mitri, D., Sternjak, A., Diamantini, A., Giometto, R., Hopner, S., Centonze, D., Bernardi, G., Dell'Acqua, M.L., *et al.* (2007). Expression of ectonucleotidase CD39 by Foxp3+ Treg cells: hydrolysis of extracellular ATP and immune suppression. Blood *110*, 1225–1232.

Bronte, V. (2008). Th17 and cancer: friends or foes? Blood *112*, 214.

Carr, M.W., Roth, S.J., Luther, E., Rose, S.S., and Springer, T.A. (1994). Monocyte chemo-attractant protein 1 acts as a T-lymphocyte chemo-attractant. Proc Natl Acad Sci USA *91*, 3652–3656.

Charles, K.A., Kulbe, H., Soper, R., Escorcio-Correia, M., Lawrence, T., Schultheis, A., Chakravarty, P., Thompson, R.G., Kollias, G., Smyth, J.F., *et al.* (2009). The tumor-promoting actions of TNF-alpha involve TNFR1 and IL-17 in ovarian cancer in mice and humans. J Clin Invest *119*, 3011–3023.

Chaudhry, A., Rudra, D., Treuting, P., Samstein, R.M., Liang, Y., Kas, A., and Rudensky, A.Y. (2009). CD4+ regulatory T cells control TH17 responses in a Stat3-dependent manner. Science *326*, 986–991.

Chen, X., Wan, J., Liu, J., Xie, W., Diao, X., Xu, J., Zhu, B., and Chen, Z. (2009). Increased IL17-producing cells correlate with poor survival and lymphangiogenesis in NSCLC patients. Lung Cancer *69*, 348–354.

Chung, Y., Chang, S.H., Martinez, G.J., Yang, X.O., Nurieva, R., Kang, H.S., Ma, L., Watowich, S.S., Jetten, A.M., Tian, Q., and Dong, C. (2009). Critical regulation of early Th17 cell differentiation by interleukin-1 signaling. Immunity *30*, 576–587.

Cosmi, L., De Palma, R., Santarlasci, V., Maggi, L., Capone, M., Frosali, F., Rodolico, G., Querci, V., Abbate, G., Angeli, R., *et al.* (2008). Human interleukin 17-producing cells originate from a CD161+CD4+ T cell precursor. J Exp Med *205*, 1903–1916.

Curiel, T.J., Coukos, G., Zou, L., Alvarez, X., Cheng, P., Mottram, P., Evdemon-Hogan, M., Conejo-Garcia, J.R., Zhang, L., Burow, M., *et al.* (2004). Specific recruitment of regulatory T cells in ovarian carcinoma fosters immune privilege and predicts reduced survival. Nat Med *10*, 942–949.

Deaglio, S., Dwyer, K.M., Gao, W., Friedman, D., Usheva, A., Erat, A., Chen, J.F., Enjyoji, K., Linden, J., Oukka, M., *et al.* (2007). Adenosine generation catalyzed by CD39 and CD73 expressed on regulatory T cells mediates immune suppression. J Exp Med *204*, 1257–1265.

Deknuydt, F., Bioley, G., Valmori, D., and Ayyoub, M. (2009). IL-1beta and IL-2 convert human Treg into T (H) 17 cells. Clin Immunol *131*, 298–307.

Derhovanessian, E., Adams, V., Hahnel, K., Groeger, A., Pandha, H., Ward, S., and Pawelec, G. (2009). Pretreatment frequency of circulating IL-17+ CD4+ T-cells, but not Tregs, correlates with clinical response to whole-cell vaccination in prostate cancer patients. Int J Cancer *125*, 1372–1379.

Dhodapkar, K.M., Barbuto, S., Matthews, P., Kukreja, A., Mazumder, A., Vesole, D., Jagannath, S., and Dhodapkar, M.V. (2008). Dendritic cells mediate the induction of polyfunctional human IL17-producing cells (Th17-1 cells) enriched in the bone marrow of patients with myeloma. Blood *112*, 2878–2885.

Dufour, J.H., Dziejman, M., Liu, M.T., Leung, J.H., Lane, T.E., and Luster, A.D. (2002). IFNgamma-inducible protein 10 (IP-10; CXCL10)-deficient mice reveal a role for IP-10 in effector T cell generation and trafficking. J Immunol *168*, 3195–3204.

Enarsson, K., Lundgren, A., Kindlund, B., Hermansson, M., Roncador, G., Banham, A.H., Lundin, B.S., and Quiding-Jarbrink, M. (2006). Function and recruitment of mucosal regulatory T cells in human chronic Helicobacter pylori infection and gastric adenocarcinoma. Clin Immunol *121*, 358–368.

Galon, J., Costes, A., Sanchez-Cabo, F., Kirilovsky, A., Mlecnik, B., Lagorce-Pages, C., Tosolini, M., Camus, M., Berger, A., Wind, P., *et al.* (2006). Type, density, and location of immune cells within human colorectal tumors predict clinical outcome. Science *313*, 1960–1964.

Gulen, M.F., Kang, Z., Bulek, K., Youzhong, W., Kim, T.W., Chen, Y., Altuntas, C.Z., Sass Bak-Jensen, K., McGeachy, M.J., Do, J.S., *et al.* (2010). The Receptor SIGIRR Suppresses Th17 Cell Proliferation via Inhibition of the Interleukin-1 Receptor Pathway and mTOR Kinase Activation. Immunity *32*, 54–66.

Harrington, L.E., Hatton, R.D., Mangan, P.R., Turner, H., Murphy, T.L., Murphy, K.M., and Weaver, C.T. (2005). Interleukin 17-producing CD4+ effector T cells develop via a lineage distinct from the T helper type 1 and 2 lineages. Nat Immunol *6*, 1123–1132.

He, D., Li, H., Yusuf, N., Elmets, C.A., Li, J., Mountz, J.D., and Xu, H. (2010). IL-17 promotes tumor development through the induction of tumor promoting micro-environments at tumor sites and myeloid-derived suppressor cells. J Immunol *184*, 2281–2288.

Hieshima, K., Imai, T., Opdenakker, G., Van Damme, J., Kusuda, J., Tei, H., Sakaki, Y., Takatsuki, K., Miura, R., Yoshie, O., and Nomiyama, H. (1997). Molecular cloning of a novel human CC chemokine liver and activation-regulated chemokine (LARC) expressed in liver. Chemotactic activity for lymphocytes and gene localization on chromosome 2. J Biol Chem *272*, 5846–5853.

Hinrichs, C.S., Kaiser, A., Paulos, C.M., Cassard, L., Sanchez-Perez, L., Heemskerk, B., Wrzesinski, C., Borman, Z.A., Muranski, P., and Restifo, N.P. (2009). Type 17 CD8+ T cells display enhanced antitumor immunity. Blood *114*, 596–599.

Hirahara, N., Nio, Y., Sasaki, S., Minari, Y., Takamura, M., Iguchi, C., Dong, M., Yamasawa, K., and Tamura, K. (2001). Inoculation of human interleukin-17 gene-transfected Meth-A fibrosarcoma cells induces T cell-dependent tumor-specific immunity in mice. Oncology *61*, 79–89.

Hirota, K., Yoshitomi, H., Hashimoto, M., Maeda, S., Teradaira, S., Sugimoto, N., Yamaguchi, T., Nomura, T., Ito, H., Nakamura, T., *et al.* (2007). Preferential recruitment of CCR6expressing Th17 cells to inflamed joints via CCL20 in rheumatoid arthritis and its animal model. J Exp Med *204*, 2803–2812.

Honorati, M.C., Neri, S., Cattini, L., and Facchini, A. (2006). Interleukin-17, a regulator of angiogenic factor release by synovial fibroblasts. Osteoarthritis Cartilage *14*, 345–352.

Horlock, C., Stott, B., Dyson, P.J., Morishita, M., Coombes, R.C., Savage, P., and Stebbing, J. (2009). The effects of trastuzumab on the CD4+CD25+FoxP3+ and CD4+IL17A+T-cell axis in patients with breast cancer. Br J Cancer *100*, 1061–1067.

Hu, J., Yuan, X., Belladonna, M.L., Ong, J.M., Wachsmann-Hogiu, S., Farkas, D.L., Black, K.L., and Yu, J.S. (2006). Induction of potent antitumor immunity by intra-tumoral injection of interleukin 23-transduced dendritic cells. Cancer Res *66*, 8887–8896.

Inozume, T., Hanada, K., Wang, Q.J., and Yang, J.C. (2009). IL-17 secreted by tumor reactive T cells induces IL-8 release by human renal cancer cells. J Immunother *32*, 109–117.

Kaiga, T., Sato, M., Kaneda, H., Iwakura, Y., Takayama, T., and Tahara, H. (2007). Systemic administration of IL-23 induces potent antitumor immunity primarily mediated through Th1-type response in association with the endogenously expressed IL-12. J Immunol *178*, 7571–7580.

Kao, J.Y., Zhang, M., Miller, M.J., Mills, J.C., Wang, B., Liu, M., Eaton, K.A., Zou, W., Berndt, B.E., Cole, T.S., *et al.* (2009). Helicobacter pylori Immune Escape Is Mediated by Dendritic Cell-Induced Treg Skewing and Th17 Suppression in Mice. Gastroenterology *138*, 1046–1054.

Kimura, A., Naka, T., and Kishimoto, T. (2007). IL-6-dependent and -independent pathways in the development of interleukin 17-producing T helper cells. Proc Natl Acad Sci USA *104*, 12099–12104.

Kleinschek, M.A., Boniface, K., Sadekova, S., Grein, J., Murphy, E.E., Turner, S.P., Raskin, L., Desai, B., Faubion, W.A., de Waal Malefyt, R., *et al.* (2009). Circulating and gut-resident human Th17 cells express CD161 and promote intestinal inflammation. J Exp Med *206*, 525–534.

Kottke, T., Sanchez-Perez, L., Diaz, R.M., Thompson, J., Chong, H., Harrington, K., Calderwood, S.K., Pulido, J., Georgopoulos, N., Selby, P., *et al.* (2007). Induction of hsp70-mediated Th17 autoimmunity can be exploited as immunotherapy for metastatic prostate cancer. Cancer Res *67*, 11970–11979.

Koyama, K., Kagamu, H., Miura, S., Hiura, T., Miyabayashi, T., Itoh, R., Kuriyama, H., Tanaka, H., Tanaka, J., Yoshizawa, H., *et al.* (2008). Reciprocal CD4+ T-cell balance of effector CD62Llow CD4+ and CD62LhighCD25+ CD4+ regulatory T cells in small cell lung cancer reflects disease stage. Clin Cancer Res *14*, 6770–6779.

Kryczek, I., Banerjee, M., Cheng, P., Vatan, L., Szeliga, W., Wei, S., Huang, E., Finlayson, E., Simeone, D., Welling, T.H., *et al.* (2009a). Phenotype, distribution, generation, and functional and clinical relevance of Th17 cells in the human tumor environments. Blood *114*, 1141–1149.

Kryczek, I., Bruce, A.T., Gudjonsson, J.E., Johnston, A., Aphale, A., Vatan, L., Szeliga, W., Wang, Y., Liu, Y., Welling, T.H., *et al.* (2008a). Induction of IL-17+ T cell trafficking and development by IFN-gamma: mechanism and pathological relevance in psoriasis. J Immunol *181*, 4733–4741.

Kryczek, I., Grybos, M., Karabon, L., Klimczak, A., and Lange, A. (2000). IL-6 production in ovarian carcinoma is associated with histiotype and biological characteristics of the tumour and influences local immunity. Br J Cancer *82*, 621–628.

Kryczek, I., Lange, A., Mottram, P., Alvarez, X., Cheng, P., Hogan, M., Moons, L., Wei, S., Zou, L., Machelon, V., *et al.* (2005). CXCL12 and vascular endothelial growth factor synergistically induce neoangiogenesis in human ovarian cancers. Cancer Res *65*, 465–472

Kryczek, I., Wei, S., Gong, W., Shu, X., Szeliga, W., Vatan, L., Chen, L., Wang, G., and Zou, W. (2008b). Cutting edge: IFN-gamma enables APC to promote memory Th17 and abate Th1 cell development. J Immunol *181*, 5842–5846.

Kryczek, I., Wei, S., Szeliga, W., Vatan, L., and Zou, W. (2009b). Endogenous IL-17 contributes to reduced tumor growth and metastasis. Blood *114*, 357–359.

Kryczek, I., Wei, S., Vatan, L., Escara-Wilke, J., Szeliga, W., Keller, E.T., and Zou, W. (2007a). Cutting edge: opposite effects of IL-1 and IL-2 on the regulation of IL-17+ T cell pool IL-1 subverts IL-2-mediated suppression. J Immunol *179*, 1423–1426.

Kryczek, I., Wei, S., Zou, L., Altuwaijri, S., Szeliga, W., Kolls, J., Chang, A., and Zou, W. (2007b). Cutting edge: Th17 and regulatory T cell dynamics and the regulation by IL2 in the tumor micro-environment. J Immunol *178*, 6730–6733.

Kryczek, I., Zou, L., Rodriguez, P., Zhu, G., Wei, S., Mottram, P., Brumlik, M., Cheng, P., Curiel, T., Myers, L., *et al.* (2006). B7-H4 expression identifies a novel suppressive macrophage population in human ovarian carcinoma. J Exp Med *203*, 871–881.

Langowski, J.L., Kastelein, R.A., and Oft, M. (2007). Swords into plowshares: IL-23 repurposes tumor immune surveillance. Trends Immunol *28*, 207–212.

Langowski, J.L., Zhang, X., Wu, L., Mattson, J.D., Chen, T., Smith, K., Basham, B., McClanahan, T., Kastelein, R.A., and Oft, M. (2006). IL-23 promotes tumour incidence and growth. Nature *442*, 461–465.

Lee, Y.K., Turner, H., Maynard, C.L., Oliver, J.R., Chen, D., Elson, C.O., and Weaver, C.T. (2009). Late developmental plasticity in the T helper 17 lineage. Immunity *30*, 92–107.

Liu, H., and Rohowsky-Kochan, C. (2008). Regulation of IL-17 in human CCR6+ effector memory T cells. J Immunol *180*, 7948–7957.

Manel, N., Unutmaz, D., and Littman, D.R. (2008). The differentiation of human T (H)-17 cells requires transforming growth factor-beta and induction of the nuclear receptor RORgammat. Nat Immunol *9*, 641–649.

Martin-Orozco, N., Chung, Y., Chang, S.H., Wang, Y.H., and Dong, C. (2009a). Th17 cells promote pancreatic inflammation but only induce diabetes efficiently in lymphopenic hosts after conversion into Th1 cells. Eur J Immunol *39*, 216–224.

Martin-Orozco, N., Muranski, P., Chung, Y., Yang, X.O., Yamazaki, T., Lu, S., Hwu, P., Restifo, N.P., Overwijk, W.W., and Dong, C. (2009b). T helper 17 cells promote cytotoxic T cell activation in tumor immunity. Immunity *31*, 787–798.

Miyahara, Y., Odunsi, K., Chen, W., Peng, G., Matsuzaki, J., and Wang, R.F. (2008). Generation and regulation of human CD4+ IL-17-producing T cells in ovarian cancer. Proc Natl Acad Sci USA *105*, 15505–15510.

Mucida, D., Park, Y., Kim, G., Turovskaya, O., Scott, I., Kronenberg, M., and Cheroutre, H. (2007). Reciprocal TH17 and regulatory T cell differentiation mediated by retinoic acid. Science *317*, 256–260.

Munn, D.H. (2009). Th17 cells in ovarian cancer. Blood *114*, 1134–1135.

Muranski, P., Boni, A., Antony, P.A., Cassard, L., Irvine, K.R., Kaiser, A., Paulos, C.M., Palmer, D.C., Touloukian, C.E., Ptak, K., *et al.* (2008). Tumor-specific Th17-polarized cells eradicate large established melanoma. Blood *112*, 362–373.

Muranski, P., and Restifo, N.P. (2009a). Adoptive immunotherapy of cancer using CD4 (+) T cells. Curr Opin Immunol *21*, 200–208.

Muranski, P., and Restifo, N.P. (2009b). Does IL-17 promote tumor growth? Blood *114*, 231–232

Murugaiyan, G., and Saha, B. (2009). Protumor vs antitumor functions of IL-17. J Immunol *183*, 4169–4175.

Nam, J.S., Terabe, M., Kang, M.J., Chae, H., Voong, N., Yang, Y.A., Laurence, A., Michalowska, A., Mamura, M., Lonning, S., *et al.* (2008). Transforming growth factor beta subverts the immune system into directly promoting tumor growth through interleukin-17. Cancer Res *68*, 3915–3923.

Numasaki, M., Fukushi, J., Ono, M., Narula, S.K., Zavodny, P.J., Kudo, T., Robbins, P.D., Tahara, H., and Lotze, M.T. (2003). Interleukin-17 promotes angiogenesis and tumor growth. Blood *101*, 2620–2627.

Numasaki, M., Lotze, M.T., and Sasaki, H. (2004). Interleukin-17 augments tumor necrosis factor-alpha-induced elaboration of proangiogenic factors from fibroblasts. Immunol Lett *93*, 39–43.

Numasaki, M., Watanabe, M., Suzuki, T., Takahashi, H., Nakamura, A., McAllister, F., Hishinuma, T., Goto, J., Lotze, M.T., Kolls, J.K., and Sasaki, H. (2005). IL-17 enhances the net angiogenic activity and in vivo growth of human non-small cell lung cancer in SCID mice through promoting CXCR-2-dependent angiogenesis. J Immunol *175*, 6177–6189.

Nurieva, R., Yang, X.O., Chung, Y., and Dong, C. (2009). Cutting edge: in vitro generated Th17 cells maintain their cytokine expression program in normal but not lymphopenic hosts. J Immunol *182*, 2565–2568.

Ohta, A., Gorelik, E., Prasad, S.J., Ronchese, F., Lukashev, D., Wong, M.K., Huang, X., Caldwell, S., Liu, K., Smith, P., *et al.* (2006). A2A adenosine receptor protects tumors from antitumor T cells. Proc Natl Acad Sci USA *103*, 13132–13137.

Ohta, A., and Sitkovsky, M. (2001). Role of G-protein-coupled adenosine receptors in downregulation of inflammation and protection from tissue damage. Nature *414*, 916–920.

Oniki, S., Nagai, H., Horikawa, T., Furukawa, J., Belladonna, M.L., Yoshimoto, T., Hara, I., and Nishigori, C. (2006). Interleukin-23 and interleukin-27 exert quite different antitumor and vaccine effects on poorly immunogenic melanoma. Cancer Res *66*, 6395–6404.

Overwijk, W.W., de Visser, K.E., Tirion, F.H., de Jong, L.A., Pols, T.W., van der Velden, Y.U., van den Boorn, J.G., Keller, A.M., Buurman, W.A., Theoret, M.R., *et al.* (2006). Immunological and antitumor effects of IL-23 as a cancer vaccine adjuvant. J Immunol *176*, 5213–5222.

Pellegrini, M., Calzascia, T., Elford, A.R., Shahinian, A., Lin, A.E., Dissanayake, D., Dhanji, S., Nguyen, L.T., Gronski, M.A., Morre, M., *et al.* (2009). Adjuvant IL-7 antagonizes multiple cellular and molecular inhibitory networks to enhance immunotherapies. Nat Med *15*, 528–536.

Precopio, M.L., Betts, M.R., Parrino, J., Price, D.A., Gostick, E., Ambrozak, D.R., Asher, T.E., Douek, D.C., Harari, A., Pantaleo, G., *et al.* (2007). Immunization with vaccinia virus induces polyfunctional and phenotypically distinctive CD8 (+) T cell responses. J Exp Med *204*, 1405–1416.

Sato, E., Olson, S.H., Ahn, J., Bundy, B., Nishikawa, H., Qian, F., Jungbluth, A.A., Frosina, D., Gnjatic, S., Ambrosone, C., *et al.* (2005). Intraepithelial CD8+ tumor-infiltrating lymphocytes and a high CD8+/regulatory T cell ratio are associated with favorable prognosis in ovarian cancer. Proc Natl Acad Sci USA *102*, 18538–18543.

Schafer, M., and Werner, S. (2008). Cancer as an overhealing wound: an old hypothesis revisited. Nat Rev Mol Cell Biol *9*, 628–638.

Sfanos, K.S., Bruno, T.C., Maris, C.H., Xu, L., Thoburn, C.J., DeMarzo, A.M., Meeker, A.K., Isaacs, W.B., and Drake, C.G. (2008). Phenotypic analysis of prostate-infiltrating lymphocytes reveals TH17 and Treg skewing. Clin Cancer Res *14*, 3254–3261.

Sharma, M.D., Hou, D.Y., Liu, Y., Koni, P.A., Metz, R., Chandler, P., Mellor, A.L., He, Y., and Munn, D.H. (2009). Indoleamine 2, 3-dioxygenase controls conversion of Foxp3+ Tregs to TH17-like cells in tumor-draining lymph nodes. Blood *113*, 6102–6111.

Su, X., Ye, J., Hsueh, E.C., Zhang, Y., Hoft, D.F., and Peng, G. (2010). Tumor micro-environments direct the recruitment and expansion of human Th17 cells. J Immunol *184*, 1630–1641.

Sutton, C., Brereton, C., Keogh, B., Mills, K.H., and Lavelle, E.C. (2006). A crucial role for interleukin (IL)-1 in the induction of IL-17-producing T cells that mediate auto-immune encephalomyelitis. J Exp Med *203*, 1685–1691.

Takahashi, H., Numasaki, M., Lotze, M.T., and Sasaki, H. (2005). Interleukin-17 enhances bFGF-, HGF- and VEGF-induced growth of vascular endothelial cells. Immunol Lett *98*, 189–193.

Tartour, E., Fossiez, F., Joyeux, I., Galinha, A., Gey, A., Claret, E., Sastre-Garau, X., Couturier, J., Mosseri, V., Vives, V., *et al.* (1999). Interleukin 17, a T-cell-derived cytokine, promotes tumorigenicity of human cervical tumors in nude mice. Cancer Res *59*, 3698–3704.

von Euw, E., Chodon, T., Attar, N., Jalil, J., Koya, R.C., Comin-Anduix, B., and Ribas, A. (2009). CTLA4 blockade increases Th17 cells in patients with metastatic melanoma. J Transl Med *7*, 35.

Voo, K.S., Wang, Y.H., Santori, F.R., Boggiano, C., Arima, K., Bover, L., Hanabuchi, S., Khalili, J., Marinova, E., Zheng, B., *et al.* (2009). Identification of IL-17-producing FOXP3+ regulatory T cells in humans. Proc Natl Acad Sci USA *106*, 4793–4798.

Wang, L., Yi, T., Kortylewski, M., Pardoll, D.M., Zeng, D., and Yu, H. (2009). IL-17 can promote tumor growth through an IL-6-Stat3 signaling pathway. J Exp Med *206*, 1457–1464.

Wilson, N.J., Boniface, K., Chan, J.R., McKenzie, B.S., Blumenschein, W.M., Mattson, J.D., Basham, B., Smith, K., Chen, T., Morel, F., *et al.* (2007). Development, cytokine profile and function of human interleukin 17-producing helper T cells. Nat Immunol *8*, 950–957.

Wu, S., Rhee, K.J., Albesiano, E., Rabizadeh, S., Wu, X., Yen, H.R., Huso, D.L., Brancati, F.L., Wick, E., McAllister, F., *et al.* (2009). A human colonic commensal promotes colon tumorigenesis via activation of T helper type 17 T cell responses. Nat Med *15*, 10161022.

Xu, L.L., Warren, M.K., Rose, W.L., Gong, W., and Wang, J.M. (1996). Human recombinant monocyte chemotactic protein and other C-C chemokines bind and induce directional migration of dendritic cells in vitro. J Leukoc Biol *60*, 365–371.

Yamazaki, T., Yang, X.O., Chung, Y., Fukunaga, A., Nurieva, R., Pappu, B., Martin-Orozco, N., Kang, H.S., Ma, L., Panopoulos, A.D., *et al.* (2008). CCR6 regulates the migration of inflammatory and regulatory T cells. J Immunol *181*, 8391–8401.

Yang, L., Anderson, D.E., Baecher-Allan, C., Hastings, W.D., Bettelli, E., Oukka, M., Kuchroo, V.K., and Hafler, D.A. (2008). IL-21 and TGF-beta are required for differentiation of human T(H)17 cells. Nature *454*, 350–352.

Yang, Z.Z., Novak, A.J., Ziesmer, S.C., Witzig, T.E., and Ansell, S.M. (2009). Malignant B cells skew the balance of regulatory T cells and TH17 cells in B-cell non-Hodgkin's lymphoma. Cancer Res *69*, 5522–5530.

Yen, H.R., Harris, T.J., Wada, S., Grosso, J.F., Getnet, D., Goldberg, M.V., Liang, K.L., Bruno, T.C., Pyle, K.J., Chan, S.L., *et al.* (2009). Tc17 CD8 T cells: functional plasticity and subset diversity. J Immunol *183*, 7161–7168.

Yu, H., Kortylewski, M., and Pardoll, D. (2007). Crosstalk between cancer and immune cells: role of STAT3 in the tumour micro-environment. Nat Rev Immunol *7*, 41–51.

Yuan, X., Hu, J., Belladonna, M.L., Black, K.L., and Yu, J.S. (2006). Interleukin-23-expressing bone marrow-derived neural stem-like cells exhibit antitumor activity against intracranial glioma. Cancer Res *66*, 2630–2638.

Zhang, B., Rong, G., Wei, H., Zhang, M., Bi, J., Ma, L., Xue, X., Wei, G., Liu, X., and Fang, G. (2008). The prevalence of Th17 cells in patients with gastric cancer. Biochem Biophys Res Commun *374*, 533–537.

Zhang, J.P., Yan, J., Xu, J., Pang, X.H., Chen, M.S., Li, L., Wu, C., Li, S.P., and Zheng, L. (2009). Increased intra-tumoral IL-17-producing cells correlate with poor survival in hepatocellular carcinoma patients. J Hepatol *50*, 980–989.

Zhang, L., Conejo-Garcia, J.R., Katsaros, D., Gimotty, P.A., Massobrio, M., Regnani, G., Makrigiannakis, A., Gray, H., Schlienger, K., Liebman, M.N., *et al.* (2003). Intra-tumoral T cells, recurrence, and survival in epithelial ovarian cancer. N Engl J Med *348*, 203–213.

Zheng, H., Fu, G., Dai, T., and Huang, H. (2007). Migration of endothelial progenitor cells mediated by stromal cell-derived factor-1alpha/CXCR4 via PI3K/Akt/eNOS signal transduction pathway. J Cardiovasc Pharmacol *50*, 274–280.

Zou, W. (2005). Immunosuppressive networks in the tumour environment and their therapeutic relevance. Nat Rev Cancer *5*, 263–274.

Zou, W., Machelon, V., Coulomb-L'Hermin, A., Borvak, J., Nome, F., Isaeva, T., Wei, S., Krzysiek, R., Durand-Gasselin, I., Gordon, A., *et al.* (2001). Stromal-derived factor-1 in human tumors recruits and alters the function of plasmacytoid precursor dendritic cells. Nat Med *7*, 1339–1346.

IL-17-Producing NKT Cells in Airway Inflammation

Dale T. Umetsu and Rosemarie H. DeKruyff

Abstract Allergic asthma, the most common form of asthma, is thought to be mediated by allergen-specific T_H2 cells, orchestrating a complex allergic inflammatory cascade. However, asthma is heterogeneous, and several other distinct forms of asthma have been recognized, and are associated with additional cellular and molecular pathways involving innate immunity, neutrophils, airway epithelial cells, and subsets of natural killer T (NKT) cells. A role for NKT cells in asthma has been suggested by extensive studies in several different animal models of asthma. This indicates that distinct subsets of NKT cells, some producing IL-17 and others inducing alternatively activated alveolar macrophages, function in concert with T_H2 cells or independently of adaptive immunity in causing asthma. These studies greatly expand our understanding the cellular and molecular mechanisms that drive the development of asthma.

1 Introduction

Bronchial asthma is a complex and heterogeneous disease caused by environmental factors in genetically susceptible individuals. Although many distinct environmental factors have been recognized by clinicians for hundreds of years to cause asthma (infection, exercise, air pollution, aspirin, and allergens), the treatment of and research regarding asthma over the past 25 years has focused primarily on the most common form of asthma; allergic asthma. In this form of asthma, allergen exposure, often beginning in childhood, drives allergen sensitization, the development of allergen-specific T_H2 cells, and the expansion of eosinophils, mast cells, and basophils,

D.T. Umetsu (✉)
Division of Immunology and Allergy, Children's Hospital, Harvard Medical School, Boston, MA 02115, USA
e-mail: Dale.Umetsu@childrens.harvard.edu

S. Jiang (ed.), *T_H17 Cells in Health and Disease*,
DOI 10.1007/978-1-4419-9371-7_27, © Springer Science+Business Media, LLC 2011

all producing IL-4, IL-5, IL-9, and IL-13. These cell types induce an adaptive allergic inflammatory response that results in the symptoms of wheezing, breathlessness, and chest tightness as well as airway hyper-reactivity (AHR), a cardinal feature of asthma. On the other hand, some forms of asthma develop independently of T_H2 driven allergic inflammation and in the absence of adaptive immunity, and therefore involve additional non-traditional pathways, including the involvement of IL-17 producing natural killer T (NKT) cells.

2 T_H2-Driven Allergic Inflammation

Since 1986, allergen-specific T_H2 cells, a component of the elegantly simple T_H1/T_H2 paradigm proposed by Bob Coffman and Tim Mosmann, has dominated the field of asthma. T_H2 cells produce IL-4, which enhances IgE production and differentiation of T_H2 cells, IL-5, which increases eosinophil growth and differentiation, IL-9, which enhances mast cell differentiation, and IL-13, which directly affects airway smooth muscle and mucus gland and induces AHR (Wills-Karp 1999; Busse and Lemanske 2001). T_H2 cells are thought be present in the lungs of most patients with asthma, particularly those with allergic asthma (Robinson et al. 1992), and the development of allergen sensitization in young children is a major risk factor for the later development of asthma (Wills-Karp and Ewart 1997; Martinez et al. 1995). In animal models of allergic asthma, depletion of $CD4^+$ T cells prevents the development of AHR (Gavett et al. 1994), suggesting that T_H2 cells are required for the development of asthma.

According to the initial T_H1/T_H2 paradigm, allergen-specific T_H1 cells inhibit the development of allergen-specific T_H2 cells and asthma. However, it became clear that the presence of allergen-specific T_H1 cells in the lungs provided a pro-inflammatory effect (Hansen et al. 1999), and that allergen-specific adaptive T_{Reg} cells producing IL-10 or TGF-β were better at preventing the development of asthma (Hansen et al. 2000; Oh et al. 2002). Furthermore, it became apparent that AHR and asthma could be inhibited by a spectrum of allergen-specific T_{Reg} cell types, including T_H2-like T_{Reg} cells producing IL-10 and expressing GATA-3 and Foxp3 (Akbari et al. 2001, 2002), as well as T_H1-like T_{Reg} cells producing IFN-γ and IL-10, and expressing T-bet and Foxp3 (Stock et al. 2004). Other allergen-specific T_{Reg} cells include those expressing membrane TGF-β (Ostroukhova et al. 2004), T_R1 cells (Akdis et al. 2004), *Mycobacterium* induced T_{Reg} cells (Zuany-Amorim et al. 2002), as well as natural T_{Reg} cells, expressing CD25 and Foxp3 (Lewkowich et al. 2005) and T_{Reg} cells induced with vitamin D and corticosteroids (Xystrakis et al. 2006; Barrat et al. 2002), were also found to be effective in preventing the development of allergen-induced AHR.

The various types of T_{Reg} cells of course, develop under different conditions. For example, exposure of mice to allergen into the respiratory tract induced allergen-specific tolerance (McMenamin and Holt 1993) associated with the development of allergen-specific adaptive T_{Reg} cells (Akbari et al. 2002; Ostroukhova et al. 2004).

Antigen in the respiratory tract activated dendritic cells (DCs) which then induced T_{Reg} cell development in an IL-10 and ICOS-dependent pathway. Plasmacytoid DCs (pDCs) have been shown to play an important role in the induction of T_{Reg} cells in the lung, since deletion of these cells resulted in worsening of airway inflammation (Lambrecht and Hammad 2009). In humans, subcutaneous administration of allergens similarly induced the development of antigen specific T_{Reg} cells producing IL-10, and was associated with improvement of allergic symptoms (Akdis et al. 1998, 2004). In mice, immunotherapy with the adjuvant *Listeria monocytogenes* with antigen was associated with the activation of $CD8\alpha^+$ DCs and the development of T_{Reg} cells expressing T-bet, Foxp3 and producing IFN-γ and IL-10 (Stock et al. 2004). Allergen-specific T_{Reg} can also be induced by exposure to TGF-β, which induces expression of Foxp3. The dose of TGF-β is critical since high doses can cause airway remodeling and fibrosis, whereas lower doses might induce T_{Reg} cells. For example, nursing pups of lactating female mice exposed to inhaled allergens develop antigen-specific T_{Reg} cells that suppressed AHR, but only if the lactating mothers expressed low levels of TGF-β in "breast" milk (Verhasselt et al. 2008). Furthermore, regulatory T_R1 cells can be induced in a pathway that involves corticosteroids and Vitamin D (Xystrakis et al. 2006; Barrat et al. 2002). These T_R1 cells, which do not express Foxp3, appear to be deficient in patients with severe, corticosteroid resistant asthma, and can be induced in T cells from such patients by Vitamin D exposure (Xystrakis et al. 2006). All of these T_{Reg} cell types are thought to be deficient in patients with asthma and allergy (Akdis et al. 2004).

3 Innate Immunity and Allergic Inflammation

While T_H2 adaptive immunity (and the lack of adaptive T_{Reg} cells) clearly plays an important role and can explain many features of asthma, this form of adaptive immunity cannot explain other aspects of asthma. Asthma is a heterogeneous disorder, and T_H2-mediated allergic asthma is only one of several forms. For example, many patients have a non-allergic form of asthma, associated with exercise, viral infection, exposure to air pollution, and airway neutrophils, rather than eosinophils. These patients have normal levels of IgE, and do not have allergen-specific T_H2 cells or increased allergen-specific IgE. Many non-T_H2 cytokines and factors, such as IFN-γ (Cho et al. 2005; Nakao et al. 2001), IL-17 (Bullens et al. 2006; Oboki et al. 2008; Pichavant et al. 2008), and non-T_H2 cells, such as neutrophils (Haldar and Pavord 2007; Kay 1989; Simpson et al. 2007) are present in the lungs of many asthmatics, particularly in patients with severe disease, or with corticosteroid-resistant asthma (Chanez et al. 2007; Green et al. 2002). The presence of allergen-specific T_H2 cells, for example in patients with allergic rhinitis, does not necessarily correlate with the development of asthma, as most patients with allergic rhinitis do not have asthma. This indicates that the development of allergen-specific T_H2 cells and allergen-specific IgE alone are not sufficient for the development of asthma. Finally, treatments that target or eliminate T_H2 cells and T_H2 cytokines (e.g., anti-IL-4, anti-IL-5 mAb, and IL-13 antagonists) have not been as effective as hoped in many clinical trials for asthma (Wenzel et al. 2007; Leckie et al. 2000), suggesting that other non-T_H2 factors critically regulate asthma.

4 Innate Pathophysiological Mechanisms in Asthma

Many features of asthma not explained by the T_H2 paradigm might involve the innate immune system. For example, a number of cell types including airway epithelial cells, basophils (Perrigoue et al. 2009), eosinophils (Shi et al. 2000), alternatively activated macrophages (Pesce et al. 2006), DCs, natural killer T (NKT) cells (Matangkasombut et al. 2009a), and most likely a newly described "natural helper cell" (Moro et al. 2010) play very important roles in mediating inflammation in both allergic and non-allergic asthma. The involvement of innate immunity in asthma is not surprising since the innate immune system can sense many environmental factors that affect the development of asthma, such as exposure to infections, including the hepatitis A virus (HAV) or *Helicobacter pylori* (Matricardi et al. 2000, 2002), endotoxins (Braun-Fahrlander et al. 2002), pets (Ownby et al. 2002), farm animals (Riedler et al. 2001), and bacteria in barns (Conrad et al. 2009). Exposure to such factors in the first years of life is associated with protection against the later development of allergic asthma, whereas exposure to air pollution and endotoxins (Trompette et al. 2009; Hammad et al. 2009) may enhance the predisposition towards the development of asthma. All of these environmental factors may be sensed by components of the innate immune system, such as TLRs expressed in the lung and on airway epithelial cells (Hammad et al. 2009). Airway epithelial cells produce large amounts of cytokines and chemokines that can modulate asthma. In particular, three cytokines produced by epithelial cells, TSLP, IL-25, and IL-33, have potent effects in the airways, even in the absence of adaptive immunity, and have recently received a great deal of interest.

5 Thymic Stromal Lymphopoietin

TSLP is a novel cytokine that has been suggested to mediate allergic inflammation at the epithelial-dendritic cell interface (Liu 2009). TSLP is secreted by epithelial cells in the skin, gut, and lung, and activates DCs, which preferentially induce T_H2 cell differentiation in T cells (Liu et al. 2007; Watanabe et al. 2005). *i*NKT cells express the TSLP receptor, and respond to TSLP by preferentially increasing IL-13, but not IFN-γ or IL-4 production (Nagata et al. 2007). Moreover, TSLP transgenic mice developed pulmonary eosinophilia, increased IgE production and allergen-induced AHR, but only when *i*NKT cells were present (Nagata et al. 2007), suggesting that TSLP is an important mediator in asthma.

IL-25, is a member of IL-17 family of cytokines (also known as IL-17E), is present in the lungs of some patients with asthma (Letuve et al. 2006) and greatly enhances

T_H2 response (Owyang et al. 2006). In a model of allergic asthma, IL-25 was expressed by lung epithelial cells and induced allergic airway inflammation as well as AHR (Fort et al. 2001; Angkasekwinai et al. 2007; Tamachi et al. 2006), while blocking IL-25 eliminated T_H2 response and AHR (Ballantyne et al. 2007). IL-17RB serves as a receptor for IL-25 and is expressed on non-B/non-T (NBNT) c-kit$^+$FcεRI$^-$ cells in mesenteric lymph nodes (Fallon et al. 2006). The c-kit$^+$FcεRI$^-$ cells are strikingly similar to a more recently described population of "natural helper cells", which respond to IL-2 with IL-25 or IL-33 by producing very large quantities of IL-5 and IL-13 (much more than CD4 T cells). The novel lymphocytes have the appearance of small lymphocytes and express c-Kit and Scal-1, but are Lin$^-$, and express IL-7RB and the IL-33 receptor (ST/2). These cells were found in human and mouse mesentery, along blood vessels, surrounded by adipose tissues (Moro et al. 2010). These cells produce very large amounts of T_H2 cytokines in response to IL-25 or IL-33, and regulate self-renewal of B1 cells. Infection with a helminth (Nippostrongylus brasiliensis) resulted in activation of these cells. Since these cells are present in naive mice and in RAG$^{-/-}$ mice, this population is called natural helper cells.

The receptor for IL-25, IL-17RB is also expressed by a subset of *i*NKT cells (Terashima et al. 2008; Stock et al. 2009). Importantly, the IL-17RB-expressing *i*NKT cells greatly enhanced AHR induced by IL-25 and sub-optimal doses of protein allergen, because this response occurred only in wildtype but not in Jα18$^{-/-}$ mice (Terashima et al. 2008). Furthermore, transferring IL-17RB$^+$, but not IL-17RB$^-$ *i*NKT cells into Jα18$^{-/-}$ mice restored allergen-induced AHR (Stock et al. 2009). The novel IL-17RB$^+$ *i*NKT cell subset expressed CD4, preferentially produced IL-13 and IL-4, and expressed CCR4 (Terashima et al. 2008), which are known to mediate the localization of *i*NKT cells into the airways (Meyer et al. 2007). This study suggests that CD4$^+$ IL-17RB$^+$ *i*NKT cells play a critical role in the pathogenesis of asthma induced by IL-25.

Another cytokine that affects *i*NKT cells is IL-33, a member of IL-1 family and a ligand for ST2, a receptor on T_H2 cells, some *i*NKT cells (Schmitz et al. 2005; Baekkevold et al. 2003), and natural helper cells (Moro et al. 2010). Administration of IL-33 into the lungs of mice induced AHR and goblet cell hyperplasia. The precise mechanism of this AHR response in not totally clear, particularly since this response could occur in the absence of adaptive immune cells (e.g., in RAG-2$^{-/-}$ mice), but may involve the direct activation of mast cells and basophils (Kondo et al. 2008; Bourgeois et al. 2009; Smithgall et al. 2008), as well as natural helper cells (Moro et al. 2010), and has not been directly examined.

6 NKT Cells in Asthma

Another cell type that is affected by both IL-25 and IL-33 which has been shown to have important effects in asthma, is NKT cells. NKT cells comprise a small subset of T cells that share characteristics of NK cells and conventional T cells, and whose potent functions in amplifying immune responses have only recently been appreciated

(Bendelac et al. 2007). NKT cells express a relatively unique transcription factor, PLZF, specific for NKT cells (Savage et al. 2008) and other innate or activated T cells (Savage et al. 2008; Kreslavsky et al. 2009), and an invariant T cell receptor (TCR), Vα14Jα18 in mice and Vα24 in humans. NKT cells are restricted by the MHC class I-like molecule, CD1d, and when activated rapidly produce large amounts of IL-4 and IFN-γ. The presence of a highly conserved TCR that functions as a pattern recognition receptor and the rapid production of cytokines by NKT cells suggest that NKT cells play an important role in innate immunity, consistent with many studies demonstrating an important regulatory role for NKT cells in autoimmunity, infectious diseases, cancer and asthma (Kronenberg 2005).

A role for NKT cells in asthma has been suggested in a number of mouse experimental models of asthma showing that the development of AHR requires the presence of different subsets of NKT cells. For example, allergen-induced AHR failed to occur in NKT cell deficient mice, but adoptive transfer of wild type NKT cells restored AHR in these NKT cell deficient mice (Riedler et al. 2001; Conrad et al. 2009). Normal T_H2 responses developed in the NKT cell deficient mice, indicating that NKT cells were not required for the development of T_H2 responses, but that T_H2 cells, which respond to protein allergens, were not sufficient in the absence of NKT cells for the development of AHR (Riedler et al. 2001; Conrad et al. 2009). Thus, in allergen-induced AHR, both NKT and allergen-specific T_H2 cells/adaptive immunity are required in experimental models of allergic asthma.

In several other distinct experimental models of asthma, the presence of NKT cells has also been shown to be required. For example, in a model of asthma induced by exposure to air pollution, a known trigger of asthma symptoms in patients (Aris et al. 1993; Gent et al. 2003), the presence of NKT cells was also required (Pichavant et al. 2008). In this model, mice were repeatedly exposed to ozone, a major component of air pollution, and developed severe AHR and airway inflammation (Pichavant et al. 2008). Unlike allergen-induced AHR, which is associated with airway eosinophilia, ozone-induced AHR was associated with airway neutrophilia and expansion of airway macrophages (Pichavant et al. 2008). Importantly, NKT cell deficient mice ($CD1d^{-/-}$ and $J\alpha18^{-/-}$ mice) failed to develop ozone-induced AHR, indicating that NKT cells were also required for the development of this form of AHR. Furthermore, IL-17 production was required for this form of experimental asthma, since $IL\text{-}17A^{-/-}$ mice failed to develop ozone-induced AHR. Importantly, a unique $NK1.1^{-}IL\text{-}17^{+}$ NKT cell subset, distinct from the NKT cells found in the lungs of mice developing allergen-induced AHR, was present in the lungs of ozone-exposed mice. These NKT cells have been shown to require RORγT for development, constitutively express IL-23 receptor (Rachitskaya et al. 2008), produce IL-22, and have been identified in situations associated with neutrophilic inflammation (Michel et al. 2007; Lee et al. 2008; Coquet et al. 2008; Goto et al. 2009), presumably because IL-17 is a potent neutrophil chemotactic factor. While T_H17 cells were also observed in the lungs of ozone-exposed mice, T_H17 cells were not required for the development of ozone-induced AHR since MHC class II deficient mice, which lack conventional $CD4^+$ T cells but have NKT cells, developed ozone-induced AHR.

Although IL-17 can also be also produced by $\gamma\delta$ T cells in the lungs, IL-17 producing $\gamma\delta$ T cells were not observed in the lungs of ozone-exposed mice. Finally, since ozone-induced AHR occurred in the absence of conventional $CD4^+$ T cells and adaptive immunity, it is clear that this form of experimental asthma was driven primarily by innate immunity and IL-17 producing NKT cells.

The observation that ozone-induced AHR required the production of IL-17 (Pichavant et al. 2008) was one of the first to demonstrate an important role of IL-17 in the development of AHR. IL-17 has been found in the lungs of patients with severe asthma, particularly asthma associated with airway neutrophils. Initially, IL-17 was thought to inhibit allergen-induced AHR (Schnyder-Candrian et al. 2006). However, in experimental asthma models, while IL-17A$^{-/-}$ mice developed severe allergen-induced AHR, they failed to develop ozone-induced AHR (Pichavant et al. 2008), indicating that IL-17A was required for ozone, but not for allergen-induced AHR. More recent studies suggest that IL-17A production can develop along with T_H2 cell mediated airways disease, and can potentiate allergic asthma (Wilson et al. 2009). The precise role of T_H17 cells versus NKT cells producing IL-17A is not yet clear. Nevertheless, the idea that IL-17A producing cells might be synergistic with T_H2 cells is not too surprising since patients with asthma rarely have only allergic asthma, but instead have symptoms induced by multiple environmental insults, including allergy, air pollution, and infection. Thus, human asthma is not only heterogeneous in terms of different forms of asthma, but is also extremely complex, due to combinations of forms that may develop and be synergistic in a given individual.

7 Virus-Induced AHR and Chronic Lung Disease

In a third mouse model of AHR, a requirement for NKT cells has also been demonstrated, in a form induced with a respiratory virus. Respiratory viral infections trigger asthma symptoms in virtually all patients with asthma, including patients with allergic and with non-allergic asthma (Gern and Busse 2000; Sigurs et al. 2005; Hamelin et al. 2006). In the mouse model, Kim et al. (2008a) demonstrated that infection with Sendai virus, which is related to human viruses that are clinically important (RSV and para-influenza virus), induced acute airway inflammation followed by a delayed but chronic phase of mucous cell metaplasia and increased AHR, resembling human chronic lung disease (asthma and COPD). Surprisingly, the AHR and airway inflammatory response required the presence of another distinct subset of NKT cells; $CD4^-$. This virus-induced AHR failed to occur in NKT cell deficient mice (CD1d$^{-/-}$ and Jα18$^{-/-}$ mice), indicating its dependence on NKT cells. In this model, NKT cells induced alternative activation of alveolar macrophages, which produced IL-13, and in turn drove increased mucus production and AHR. Since the virus-induced AHR response occurred in MHC class II$^{-/-}$ mice, innate immunity, rather than adaptive immunity was required for the development of this form of AHR.

8 The Many Pathways to Asthma

These studies of NKT cells in different experimental models of asthma suggest that several distinct subsets of NKT cells play a role in different forms of asthma and that some forms of asthma can develop in the absence of T_H2 cells and adaptive immunity. The precise mechanisms however, by which NKT cells might cause AHR in any of these models are not yet clear. One possible mechanism by which NKT cells function in asthma may be that some NKT cell subsets synergize with T_H2 cells (e.g., in allergen-induced AHR), possibly by licensing T_H2 cells to induce AHR (Akbari et al. 2003). T_H2 cells, but not NKT cells, respond to protein allergens and can induce a degree of eosinophilic inflammation (Das et al. 2006). However, allergic airway inflammation alone is not sufficient for asthma, and the frequency of allergen-specific T_H2 cells activated by a given allergen may be too low to induce AHR (a fundamental feature of asthma) without amplification by NKT cells. These results might explain why NKT cell deficient mice cannot develop allergen-induced AHR even though these mice clearly generate allergen-specific T_H2 responses, and suggest that T_H2 cells and NKT cells might have synergistic functions. This possibility is supported by the observation that lipids from some pollens (e.g., cypress tree pollen) can directly activate NKT cells (Agea et al. 2005), and might enhance the development of pollen sensitization by enhancing the development of pollen-specific T_H2 cells.

Another possibility is that T_H2 cells and $CD4^+$ IL-4/IL-13 producing NKT cells may have overlapping functions. For example, if artificially large numbers of T_H2 cells are placed in the lungs, as when large numbers of allergen-specific T_H2 cells are transferred into SCID or $RAG1^{-/-}$ mice, AHR can be induced by the T_H2 cells in the absence of NKT cells (Hansen et al. 1999). This idea is not so surprising, since AHR can be induced by the administration of artificially large quantities of recombinant cytokines such as rIL-13 (Wills-Karp et al. 1998), rIL-25, or rIL-33 (Kondo et al. 2008) into the lungs of mice, including $RAG1^{-/-}$ mice, which have no T cells or NKT cells. In the case of rIL-33 however, the effect may be mediated through IL-13 produced by IL-33-activated mast cells, basophils or natural helper cells. Therefore, there may be many pathways that lead to the development of AHR, some that require adaptive immunity, and some that entirely bypass T cells and require innate immunity, including subsets of NKT cells or natural helper cells.

In non-allergic forms of asthma that develop independent of T_H2 cells, NKT cells may play particularly important roles. Asthma associated with air pollution, bacterial infection, or severe corticosteroid resistant asthma, and subsets of NKT cells (e.g., NKT cells producing IL-17), may mediate the development of AHR. The activation of NKT cells in these instances may be induced by alveolar macrophages, other innate cells in the lungs (e.g., airway epithelial cells or dendritic cells), or by glycolipids from bacteria, such as *Sphingomonas paucimobilis*, *Staphlococcus aureus*, or *E. coli* (Kinjo et al. 2006; Meyer et al. 2006; Kim et al. 2008b). The lack of involvement of T_H2 cells and eosinophils in these forms of asthma may explain why these forms are resistant to corticosteroids, the most common therapy for asthma.

These pathways may be similar to virus-induced experimental asthma, where $CD4^-$ NKT cells in conjunction with alternatively activated alveolar macrophages (Kim et al. 2008a) may mediate the development of AHR independent of adaptive immunity, eosinophils and T_H2 cells. In humans, corticosteroid therapy is often ineffective for asthma symptoms induced by acute viral infection, presumably because of the development of $CD1d^+$ $IL\text{-}13R^+$ alternatively activated macrophages, interacting with activated NKT cells (Subrata et al. 2009).

9 NKT Cells in Human Asthma

While extensive studies of the regulation of asthma by NKT cells have been performed in mice, the translation of these studies to human asthma has be marked by controversy. This is primarily due to disagreement regarding the precise number of NKT cells that might be present in the lungs of patients with asthma (Akbari et al. 2006; Vijayanand et al. 2007). To date, at least 14 studies have been performed examining bronchoalveolar (BAL) fluid and/or endobronchial biopsies from patients with asthma for the presence of NKT cells (Matangkasombut et al. 2009a). Most of these (ten reports) showed that NKT cells are increased in number in the lungs of patients with asthma, while the other four studies concluded that NKT cells were not increased (Vijayanand et al. 2007; Mutalithas et al. 2007; Thomas et al. 2006; Bratke et al. 2007). While it is possible that technical difficulties in identifying NKT cells, use of sputum rather than BAL fluid samples, and lack of appropriate control populations in some studies are responsible, it is likely that the heterogeneous nature of asthma, particularly in terms of clinical severity and control has contributed to the apparent discrepancies. This has become evident in a recent study of patients with a very broad range of asthma clinical severity and symptom control. In this study, patients with severe, poorly controlled asthma had a consistent and very significant increase in the number of BAL fluid NKT cells compared to non-asthmatic controls, whereas only about 50% of patients with well-controlled (moderate to severe) asthma had detectable increases in the number of BAL fluid NKT cells compared to non-asthmatic controls (Matangkasombut et al. 2009b). Therefore, the number of pulmonary NKT cells appears to be quite variable, and many, though clearly not all patients with asthma have an increase in the number of pulmonary NKT cells. In contrast, normal individuals consistently have undetectable levels of pulmonary NKT cells.

The absence of detectable levels of NKT cells in the lungs of some asthma patients (particularly patients with mild or well-controlled asthma) suggests that in humans, understanding the biology of NKT cells will require more than simply counting the number of NKT cells in the lungs. Rather, performance of functional studies of NKT cells is more likely to be informative, though such studies in humans are much more difficult to perform than in mice. Several functional studies have been completed, such as one in which patients with moderate to severe allergic asthma were challenged with allergen. After such challenge, a significant increase in the number of pulmonary NKT cells developed, as observed in endobronchial biopsy specimens at 24 h, compared to baseline, and was associated with a significant increase in AHR (Reynolds et al. 2009). These results are consistent with allergen challenge studies in mice showing that the development of AHR required the presence of NKT cells (Akbari et al. 2003). A similar challenge in non-human primates (cynomolgus monkeys) with α-GalCer, selectively activating NKT cells, resulted in the development of significant AHR (Matangkasombut et al. 2008). Since monkeys are closely related to humans in terms of their genome, respiratory physiology and immunological responses, these results predict that human NKT cells indeed play an important functional role in regulating the development of human asthma. However, additional functional studies of NKT cells in the lungs of patients with asthma, along with assessments of NKT cell phenotypes and their cytokine profiles, are required to more fully understand the role of NKT cells in human asthma.

10 Summary

Studies in mice have clearly shown that T_H2 cells and NKT cells regulate the development of allergen-induced AHR. It has become clear that asthma is heterogeneous, and many pathways can lead to the development of asthma, involving both adaptive and innate immunity. These pathways to asthma involve not only allergen-specific T_H2 cells, which are clearly inhibited by allergen specific and natural T_{Reg} cells, but also by innate pathways involving NKT cells, neutrophils, IL-17, macrophages, epithelial cells, and natural helper cells. The mechanisms by which NKT cells interact with airway epithelial cells, macrophages, natural helper cells and T cells, the role of T_{Reg} cells in down modulating the innate pathways to asthma, and how these pathways interact in patients with asthma, will require much further studies, both in human and murine models.

References

Agea, E., A. Russano, O. Bistoni, R. Mannucci, I. Nicoletti, L. Corazzi, A. Postle, G. De Libero, S. Porcelli, and F. Spinozzi. 2005. Human CD1-restricted T cell recognition of lipids from pollens. *J Exp Med* 202:295–308.

Akbari, O., R.H. DeKruyff, and D.T. Umetsu. 2001. Pulmonary dendritic cells secreting IL-10 mediate T cell tolerance induced by respiratory exposure to antigen. *Nat Immunol.* 2:725–731.

Akbari, O., J.L. Faul, E.G. Hoyte, G.J. Berry, J. Wahlstrom, M. Kronenberg, R.H. DeKruyff, and D.T. Umetsu. 2006. CD4+ invariant T-cell-receptor+ natural killer T cells in bronchial asthma. *N Engl J Med* 354:1117–1129.

Akbari, O., G.J. Freeman, E.H. Meyer, E.A. Greenfield, T.T. Chang, A.H. Sharpe, G. Berry, R.H. DeKruyff, and D.T. Umetsu. 2002. Antigen-specific regulatory T cells develop via the ICOS-ICOS-Ligand pathway and inhibit allergen-induced airway hyperreactivity. *Nat Medicine* 8:1024–1032.

Akbari, O., P. Stock, E. Meyer, M. Kronenberg, S. Sidobre, T. Nakayama, M. Taniguchi, M.J. Grusby, R.H. DeKruyff, and D.T. Umetsu. 2003. Essential role of NKT cells producing IL-4 and IL-13 in the development of allergen-induced airway hyperreactivity. *Nat Medicine* 9:582–588.

Akdis, C.A., T. Blesken, M. Akdis, B. Wuthrich, and K. Blaser. 1998. Role of interleukin 10 in specific immunotherapy. *J Clin Invest.* 102:98–106.

Akdis, M., J. Verhagen, A. Taylor, F. Karamloo, C. Karagiannidis, R. Crameri, S. Thunberg, G. Deniz, R. Valenta, H. Fiebig, C. Kegel, R. Disch, C.B. Schmidt-Weber, K. Blaser, and C.A. Akdis. 2004. Immune responses in healthy and allergic individuals are characterized by a fine balance between allergen-specific T regulatory 1 and T helper 2 cells. *J Exp Med* 199:1567–1575.

Angkasekwinai, P., H. Park, Y.H. Wang, S.H. Chang, D.B. Corry, Y.J. Liu, Z. Zhu, and C. Dong. 2007. Interleukin 25 promotes the initiation of proallergic type 2 responses. *J Exp Med* 204:1509–1517.

Aris, R.M., D. Christian, P.Q. Hearne, K. Kerr, W.E. Finkbeiner, and J.R. Balmes. 1993. Ozone-induced airway inflammation in human subjects as determined by airway lavage and biopsy. *Am Rev Respir Dis* 148:1363–1372.

Baekkevold, E.S., M. Roussigne, T. Yamanaka, F.E. Johansen, F.L. Jahnsen, F. Amalric, P. Brandtzaeg, M. Erard, G. Haraldsen, and J.P. Girard. 2003. Molecular characterization of NF-HEV, a nuclear factor preferentially expressed in human high endothelial venules. *Am J Pathol* 163:69–79.

Ballantyne, S.J., J.L. Barlow, H.E. Jolin, P. Nath, A.S. Williams, K.F. Chung, G. Sturton, S.H. Wong, and A.N. McKenzie. 2007. Blocking IL-25 prevents airway hyperresponsiveness in allergic asthma. *J Allergy Clin Immunol* 120:1324–1331.

Barrat, F.J., D.J. Cua, A. Boonstra, D.F. Richards, C. Crain, H.F. Savelkoul, R. de Waal-Malefyt, R.L. Coffman, C.M. Hawrylowicz, and A. O'Garra. 2002. In vitro generation of interleukin 10-producing regulatory CD4(+) T cells is induced by immunosuppressive drugs and inhibited by T helper type 1 (Th1)- and Th2-inducing cytokines. *J Exp Med* 195:603–616.

Bendelac, A., P.B. Savage, and L. Teyton. 2007. The biology of NKT cells. *Annu Rev Immunol* 25:297–336.

Bourgeois, E., L.P. Van, M. Samson, S. Diem, A. Barra, S. Roga, J.M. Gombert, E. Schneider, M. Dy, P. Gourdy, J.P. Girard, and A. Herbelin. 2009. The pro-Th2 cytokine IL-33 directly interacts with invariant NKT and NK cells to induce IFN-gamma production. *Eur J Immunol* 39:1046–1055.

Bratke, K., P. Julius, and J.C. Virchow. 2007. Invariant natural killer T cells in obstructive pulmonary diseases. *N Engl J Med* 357:194; author reply 194–195.

Braun-Fahrlander, C., J. Riedler, U. Herz, W. Eder, M. Waser, L. Grize, S. Maisch, D. Carr, F. Gerlach, A. Bufe, R. Lauener, R. Schierl, H. Renz, D. Nowak, and E. von Mutius. 2002. Environmental exposure to endotoxin and its relation to asthma in school-age children. *N Engl J Med* 347:869–877.

Bullens, D.M., E. Truyen, L. Coteur, E. Dilissen, P.W. Hellings, L.J. Dupont, and J.L. Ceuppens. 2006. IL-17 mRNA in sputum of asthmatic patients: linking T cell driven inflammation and granulocytic influx? *Respir Res* 7:135.

Busse, W.W., and R.F. Lemanske, Jr. 2001. Asthma. *N Engl J Med* 344:350–362.

Chanez, P., S.E. Wenzel, G.P. Anderson, J.M. Anto, E.H. Bel, L.P. Boulet, C.E. Brightling, W.W. Busse, M. Castro, B. Dahlen, S.E. Dahlen, L.M. Fabbri, S.T. Holgate, M. Humbert, M. Gaga, G.F. Joos, B. Levy, K.F. Rabe, P.J. Sterk, S.J. Wilson, and I. Vachier. 2007. Severe asthma in adults: what are the important questions? *J Allergy Clin Immunol* 119:1337–1348.

Cho, S.H., L.A. Stanciu, S.T. Holgate, and S.L. Johnston. 2005. Increased interleukin-4, interleukin-5, and interferon-gamma in airway CD4+ and CD8+ T cells in atopic asthma. *Am J Respir Crit Care Med* 171:224–230.

Conrad, M.L., R. Ferstl, R. Teich, S. Brand, N. Blumer, A.O. Yildirim, C.C. Patrascan, A. Hanuszkiewicz, S. Akira, H. Wagner, O. Holst, E. von Mutius, P.I. Pfefferle, C.J. Kirschning, H. Garn, and H. Renz. 2009. Maternal TLR signaling is required for prenatal asthma protection by the nonpathogenic microbe Acinetobacter lwoffii F78. *J Exp Med* 206:2869–2877.

Coquet, J.M., S. Chakravarti, K. Kyparissoudis, F.W. McNab, L.A. Pitt, B.S. McKenzie, S.P. Berzins, M.J. Smyth, and D.I. Godfrey. 2008. Diverse cytokine production by NKT cell subsets and identification of an IL-17-producing CD4-NK1.1- NKT cell population. *Proc Natl Acad Sci USA* 105:11287–11292.

Das, J., P. Eynott, R. Jupp, A. Bothwell, L. Van Kaer, Y. Shi, and G. Das. 2006. Natural killer T cells and CD8+ T cells are dispensable for T cell-dependent allergic airway inflammation. *Nat Med* 12:1345–1346; author reply 1347.

Fallon, P.G., S.J. Ballantyne, N.E. Mangan, J.L. Barlow, A. Dasvarma, D.R. Hewett, A. McIlgorm, H.E. Jolin, and A.N. McKenzie. 2006. Identification of an interleukin (IL)-25-dependent cell population that provides IL-4, IL-5, and IL-13 at the onset of helminth expulsion. *J Exp Med* 203:1105–1116.

Fort, M.M., J. Cheung, D. Yen, J. Li, S.M. Zurawski, S. Lo, S. Menon, T. Clifford, B. Hunte, R. Lesley, T. Muchamuel, S.D. Hurst, G. Zurawski, M.W. Leach, D.M. Gorman, and D.M. Rennick. 2001. IL-25 induces IL-4, IL-5, and IL-13 and Th2-associated pathologies in vivo. *Immunity* 15:985–995.

Gavett, S.H., X. Chen, F. Finkelman, and M. Wills-Karp. 1994. Depletion of murine CD4+ T lymphocytes prevents antigen-induced airway hyperreactivity and pulmonary eosinophilia. *Am J Respir Cell Mol Biol* 10:587–593.

Gent, J.F., E.W. Triche, T.R. Holford, K. Belanger, M.B. Bracken, W.S. Beckett, and B.P. Leaderer. 2003. Association of low-level ozone and fine particles with respiratory symptoms in children with asthma. *JAMA* 290:1859–1867.

Gern, J.E., and W.W. Busse. 2000. The role of viral infections in the natural history of asthma. *J Allergy Clin Immunol* 106:201–212.

Goto, M., M. Murakawa, K. Kadoshima-Yamaoka, Y. Tanaka, K. Nagahira, Y. Fukuda, and T. Nishimura. 2009. Murine NKT cells produce Th17 cytokine interleukin-22. *Cell Immunol* 254:81–84.

Green, R.H., C.E. Brightling, G. Woltmann, D. Parker, A.J. Wardlaw, and I.D. Pavord. 2002. Analysis of induced sputum in adults with asthma: identification of subgroup with isolated sputum neutrophilia and poor response to inhaled corticosteroids. *Thorax* 57:875–879.

Haldar, P., and I.D. Pavord. 2007. Noneosinophilic asthma: a distinct clinical and pathologic phenotype. *The Journal of allergy and clinical immunology* 119:1043–1052; quiz 1053–1044.

Hamelin, M.E., G.A. Prince, A.M. Gomez, R. Kinkead, and G. Boivin. 2006. Human metapneumovirus infection induces long-term pulmonary inflammation associated with airway obstruction and hyperresponsiveness in mice. *J Infect Dis* 193:1634–1642.

Hammad, H., M. Chieppa, F. Perros, M.A. Willart, R.N. Germain, and B.N. Lambrecht. 2009. House dust mite allergen induces asthma via Toll-like receptor 4 triggering of airway structural cells. *Nat Med* 15:410–416.

Hansen, G., G. Berry, R.H. DeKruyff, and D.T. Umetsu. 1999. Allergen-specific Th1 cells fail to counterbalance Th2 cell-induced airway hyperreactivity but cause severe airway inflammation. *J Clin Invest* 103:175–183.

Hansen, G., J.J. McIntire, V.P. Yeung, G. Berry, G.J. Thorbecke, L. Chen, R.H. DeKruyff, and D.T. Umetsu. 2000. CD4+ Th Cells Engineered To Produce Latent TGF-b1 Reverse Allergen-Induced Airway Hyperreactivity and Inflammation. *J Clin Invest* 11:89–96.

Kay, A.B. 1989. Inflammatory cells in bronchial asthma. *J Asthma* 26:335–344.

Kim, E.Y., J.T. Battaile, A.C. Patel, Y. You, E. Agapov, M.H. Grayson, L.A. Benoit, D.E. Byers, Y. Alevy, J. Tucker, S. Swanson, R. Tidwell, J.W. Tyner, J.D. Morton, M. Castro, D. Polineni, G.A. Patterson, R.A. Schwendener, J.D. Allard, G. Peltz, and M.J. Holtzman. 2008. Persistent activation of an innate immune response translates respiratory viral infection into chronic lung disease. *Nat Med* 14:633–640.

Kim, S., S. Lalani, V.V. Parekh, T.L. Vincent, L. Wu, and L. Van Kaer. 2008. Impact of bacteria on the phenotype, functions, and therapeutic activities of invariant NKT cells in mice. *J Clin Invest* 118:2301–2315.

Kinjo, Y., E. Tupin, D. Wu, M. Fujio, R. Garcia-Navarro, M.R. Benhnia, D.M. Zajonc, G. Ben-Menachem, G.D. Ainge, G.F. Painter, A. Khurana, K. Hoebe, S.M. Behar, B. Beutler, I.A.

Wilson, M. Tsuji, T.J. Sellati, C.H. Wong, and M. Kronenberg. 2006. Natural killer T cells recognize diacylglycerol antigens from pathogenic bacteria. *Nat Immunol* 7:978–986.

Kondo, Y., T. Yoshimoto, K. Yasuda, S. Futatsugi-Yumikura, M. Morimoto, N. Hayashi, T. Hoshino, J. Fujimoto, and K. Nakanishi. 2008. Administration of IL-33 induces airway hyperresponsiveness and goblet cell hyperplasia in the lungs in the absence of adaptive immune system. *Int Immunol* 20:791–800.

Kreslavsky, T., A.K. Savage, R. Hobbs, F. Gounari, R. Bronson, P. Pereira, P.P. Pandolfi, A. Bendelac, and H. von Boehmer. 2009. TCR-inducible PLZF transcription factor required for innate phenotype of a subset of gammadelta T cells with restricted TCR diversity. *Proc Natl Acad Sci USA* 106:12453–12458.

Kronenberg, M. 2005. Toward an understanding of NKT cell biology: progress and paradoxes. *Annu Rev Immunol* 23:877–900.

Lambrecht, B.N., and H. Hammad. 2009. Biology of lung dendritic cells at the origin of asthma. *Immunity* 31:412–424.

Leckie, M.J., A. ten Brinke, J. Khan, Z. Diamant, B.J. O'Connor, C.M. Walls, A.K. Mathur, H.C. Cowley, K.F. Chung, R. Djukanovic, T.T. Hansel, S.T. Holgate, P.J. Sterk, and P.J. Barnes. 2000. Effects of an interleukin-5 blocking monoclonal antibody on eosinophils, airway hyperresponsiveness, and the late asthmatic response. *Lancet* 356:2144–2148.

Lee, K.A., M.H. Kang, Y.S. Lee, Y.J. Kim, D.H. Kim, H.J. Ko, and C.Y. Kang. 2008. A distinct subset of natural killer T cells produces IL-17, contributing to airway infiltration of neutrophils but not to airway hyperreactivity. *Cell Immunol* 251:50–55.

Letuve, S., S. Lajoie-Kadoch, S. Audusseau, M.E. Rothenberg, P.O. Fiset, M.S. Ludwig, and Q. Hamid. 2006. IL-17E upregulates the expression of proinflammatory cytokines in lung fibroblasts. *J Allergy Clin Immunol* 117:590–596.

Lewkowich, I.P., N.S. Herman, K.W. Schleifer, M.P. Dance, B.L. Chen, K.M. Dienger, A.A. Sproles, J.S. Shah, J. Kohl, Y. Belkaid, and M. Wills-Karp. 2005. CD4+CD25+ T cells protect against experimentally induced asthma and alter pulmonary dendritic cell phenotype and function. *J Exp Med* 202:1549–1561.

Liu, Y.J. 2009. TSLP in epithelial cell and dendritic cell cross talk. *Adv Immunol* 101:1–25.

Liu, Y.J., V. Soumelis, N. Watanabe, T. Ito, Y.H. Wang, W. Malefyt Rde, M. Omori, B. Zhou, and S.F. Ziegler. 2007. TSLP: an epithelial cell cytokine that regulates T cell differentiation by conditioning dendritic cell maturation. *Annu Rev Immunol* 25:193–219.

Martinez, F.D., A.L. Wright, L.M. Taussig, C.J. Holberg, M. Halonen, W.J. Morgan, and G.H. Associates. 1995. Asthma and wheezing in the first six years of life. *New Engl J Med* 332:133–138.

Matangkasombut, P., G. Marigowda, A. Irvine, I. Idris, M. Pichavant, H. Kim, Y. Yasumi, S. Wilson, R. DeKruyff, J.L. Faul, E. Israel, O. Akbari, and D.T. Umetsu. 2009. Natural Killer T Cells in the Lungs of Patients with Asthma. *J Allergy Clin Immunol* 123:1181–1185.

Matangkasombut, P., M. Pichavant, R.H. Dekruyff, and D.T. Umetsu. 2009. Natural killer T cells and the regulation of asthma. *Mucosal Immunol* 2:383–392.

Matangkasombut, P., M. Pichavant, T. Yasumi, C. Hendricks, P.B. Savage, R.H. Dekruyff, and D.T. Umetsu. 2008. Direct activation of natural killer T cells induces airway hyperreactivity in nonhuman primates. *J Allergy Clin Immunol* 121:1287–1289.

Matricardi, P., F. Rosmini, S. Riondino, M. Fortini, L. Ferrigno, M. Rapicetta, and S. Bonini. 2000. Exposure to foodborne and orofecal microbes versus airborne viruses in relation to atopy and allergic asthma: epidemiological study. *BMJ* 320:412–417.

Matricardi, P.M., F. Rosmini, V. Panetta, L. Ferrigno, and S. Bonini. 2002. Hay fever and asthma in relation to markers of infection in the United States. *J Allergy Clin Immunol* 110:381–387.

McMenamin, C., and P.G. Holt. 1993. The natural immune response to inhaled soluble protein antigens involves major histocompatibility complex (MHC) class I-restricted CD8+ T cell-mediated but MHC class II-restricted CD4+ T cell-dependent immune deviation resulting in selective suppression of immunoglobulin E production. *J Exp Med* 178:889–899.

Meyer, E.H., S. Goya, O. Akbari, G.J. Berry, P.B. Savage, M. Kronenberg, T. Nakayama, R.H. DeKruyff, and D.T. Umetsu. 2006. Glycolipid activation of invariant T cell receptor+ NK T cells is sufficient to induce airway hyperreactivity independent of conventional CD4+ T cells. *Proc Natl Acad Sci USA* 103:2782–2787.

Meyer, E.H., M.A. Wurbel, T.L. Staton, M. Pichavant, M.J. Kan, P.B. Savage, R.H. DeKruyff, E.C. Butcher, J.J. Campbell, and D.T. Umetsu. 2007. iNKT cells require CCR4 to localize to the airways and to induce airway hyperreactivity. *J Immunol* 179:4661–4671.

Michel, M.L., A.C. Keller, C. Paget, M. Fujio, F. Trottein, P.B. Savage, C.H. Wong, E. Schneider, M. Dy, and M.C. Leite-de-Moraes. 2007. Identification of an IL-17-producing NK1.1(neg) iNKT cell population involved in airway neutrophilia. *J Exp Med* 204:995–1001.

Moro, K., T. Yamada, M. Tanabe, T. Takeuchi, T. Ikawa, H. Kawamoto, J. Furusawa, M. Ohtani, H. Fujii, and S. Koyasu. 2010. Innate production of T(H)2 cytokines by adipose tissue-associated c-Kit(+)Sca-1(+) lymphoid cells. *Nature* 463:540–544.

Mutalithas, K., J. Croudace, C. Guillen, S. Siddiqui, D. Thickett, and A. Wardlaw. 2007. Bronchoalveolar lavage invariant natural killer T cells are not increased in asthma. *J Allergy Clin Immunol* 119:1274–1276.

Nagata, Y., H. Kamijuku, M. Taniguchi, S. Ziegler, and K. Seino. 2007. Differential role of thymic stromal lymphopoietin in the induction of airway hyperreactivity and Th2 immune response in antigen-induced asthma with respect to natural killer T cell function. *Int Arch Allergy Immunol* 144:305–314.

Nakao, F., K. Ihara, K. Kusuhara, Y. Sasaki, N. Kinukawa, A. Takabayashi, S. Nishima, and T. Hara. 2001. Association of IFN-gamma and IFN regulatory factor 1 polymorphisms with childhood atopic asthma. *J Allergy Clin immunol* 107:499–504.

Oboki, K., T. Ohno, H. Saito, and S. Nakae. 2008. Th17 and allergy. *Allergol Int* 57:121–134.

Oh, J.W., C.M. Seroogy, E.H. Meyer, O. Akbari, G. Berry, C.G. Fathman, R.H. DeKruyff, and D.T. Umetsu. 2002. CD4 T helper cells engineered to produce IL-10 reverse allergen-induced airway hyperreactivity and inflammation. *J Allergy Clin Immunol* 110:460–468.

Ostroukhova, M., C. Seguin-Devaux, T.B. Oriss, B. Dixon-McCarthy, L. Yang, B.T. Ameredes, T.E. Corcoran, and A. Ray. 2004. Tolerance induced by inhaled antigen involves CD4(+) T cells expressing membrane-bound TGF-beta and FOXP3. *J Clin Invest* 114:28–38.

Ownby, D., C. Johnson, and E. Peterson. 2002. Exposure to dogs and cats in the first year of life and risk of allergic sensitization at 6 to 7 years of age. *JAMA* 288:963–972.

Owyang, A.M., C. Zaph, E.H. Wilson, K.J. Guild, T. McClanahan, H.R. Miller, D.J. Cua, M. Goldschmidt, C.A. Hunter, R.A. Kastelein, and D. Artis. 2006. Interleukin 25 regulates type 2 cytokine-dependent immunity and limits chronic inflammation in the gastrointestinal tract. *J Exp Med* 203:843–849.

Perrigoue, J.G., S.A. Saenz, M.C. Siracusa, E.J. Allenspach, B.C. Taylor, P.R. Giacomin, M.G. Nair, Y. Du, C. Zaph, N. van Rooijen, M.R. Comeau, E.J. Pearce, T.M. Laufer, and D. Artis. 2009. MHC class II-dependent basophil-CD4+ T cell interactions promote T(H)2 cytokine-dependent immunity. *Nat Immunol* 10:697–705.

Pesce, J., M. Kaviratne, T.R. Ramalingam, R.W. Thompson, J.F. Urban, Jr., A.W. Cheever, D.A. Young, M. Collins, M.J. Grusby, and T.A. Wynn. 2006. The IL-21 receptor augments Th2 effector function and alternative macrophage activation. *J Clin Invest* 116:2044–2055.

Pichavant, M., S. Goya, E.H. Meyer, R.A. Johnston, H.Y. Kim, P. Matangkasombut, M. Zhu, Y. Iwakura, P.B. Savage, R.H. DeKruyff, S.A. Shore, and D.T. Umetsu. 2008. Ozone exposure in a mouse model induces airway hyperreactivity that requires the presence of natural killer T cells and IL-17. *J Exp Med* 205:385–393.

Rachitskaya, A.V., A.M. Hansen, R. Horai, Z. Li, R. Villasmil, D. Luger, R.B. Nussenblatt, and R.R. Caspi. 2008. Cutting edge: NKT cells constitutively express IL-23 receptor and RORgammat and rapidly produce IL-17 upon receptor ligation in an IL-6-independent fashion. *J Immunol* 180:5167–5171.

Reynolds, C., J. Barkans, P. Clark, H. Kariyawasam, D. Altmann, B. Kay, and R. Boyton. 2009. Natural killer T cells in bronchial biopsies from human allergen challenge model of allergic asthma. *J Allergy Clin Immunol* 124:860-862; author reply 862.

Riedler, J., C. Braun-Fahrlander, W. Eder, M. Schreuer, M. Waser, S. Maisch, D. Carr, R. Schierl, D. Nowak, and E. von Mutius. 2001. Exposure to farming in early life and development of asthma and allergy: a cross-sectional survey. *Lancet* 358:1129–1133.

Robinson, D.S., Q. Hamid, S. Ying, A. Tsicopoulos, J. Barkans, A.M. Bentley, C. Corrigan, S.R. Durham, and A.B. Kay. 1992. Predominant TH2-like bronchoalveolar T-lymphocyte population in atopic asthma. *N Engl J Med* 326:298–304.

Savage, A.K., M.G. Constantinides, J. Han, D. Picard, E. Martin, B. Li, O. Lantz, and A. Bendelac. 2008. The transcription factor PLZF directs the effector program of the NKT cell lineage. *Immunity* 29:391–403.

Schmitz, J., A. Owyang, E. Oldham, Y. Song, E. Murphy, T.K. McClanahan, G. Zurawski, M. Moshrefi, J. Qin, X. Li, D.M. Gorman, J.F. Bazan, and R.A. Kastelein. 2005. IL-33, an interleukin-1-like cytokine that signals via the IL-1 receptor-related protein ST2 and induces T helper type 2-associated cytokines. *Immunity* 23:479–490.

Schnyder-Candrian, S., D. Togbe, I. Couillin, I. Mercier, F. Brombacher, V. Quesniaux, F. Fossiez, B. Ryffel, and B. Schnyder. 2006. Interleukin-17 is a negative regulator of established allergic asthma. *J Exp Med* 203:2715–2725.

Shi, H.Z., A. Humbles, C. Gerard, Z. Jin, and P.F. Weller. 2000. Lymph node trafficking and antigen presentation by endobronchial eosinophils. *J Clin Invest* 105:945–953.

Sigurs, N., P.M. Gustafsson, R. Bjarnason, F. Lundberg, S. Schmidt, F. Sigurbergsson, and B. Kjellman. 2005. Severe respiratory syncytial virus bronchiolitis in infancy and asthma and allergy at age 13. *Am J Respir Crit Care Med* 171:137–141.

Simpson, J.L., T.V. Grissell, J. Douwes, R.J. Scott, M.J. Boyle, and P.G. Gibson. 2007. Innate immune activation in neutrophilic asthma and bronchiectasis. *Thorax* 62:211–218.

Smithgall, M.D., M.R. Comeau, B.R. Yoon, D. Kaufman, R. Armitage, and D.E. Smith. 2008. IL-33 amplifies both Th1- and Th2-type responses through its activity on human basophils, allergen-reactive Th2 cells, iNKT and NK cells. *Int Immunol* 20:1019–1030.

Stock, P., O. Akbari, G. Berry, G. Freeman, R. DeKruyff, and D.T. Umetsu. 2004. Induction of TH1-like regulatory cells that express Foxp3 and protect against airway hyperreactivity. Nat *Immunol.* 5:1149–1156.

Stock, P., V. Lombardi, V. Kohlrautz, and O. Akbari. 2009. Induction of airway hyperreactivity by IL-25 is dependent on a subset of invariant NKT cells expressing IL-17RB. *J Immunol* 182:5116–5122.

Subrata, L.S., J. Bizzintino, E. Mamessier, A. Bosco, K.L. McKenna, M.E. Wikstrom, J. Goldblatt, P.D. Sly, B.J. Hales, W.R. Thomas, I.A. Laing, P.N. LeSouef, and P.G. Holt. 2009. Interactions between innate antiviral and atopic immunoinflammatory pathways precipitate and sustain asthma exacerbations in children. *J Immunol* 183:2793–2800.

Tamachi, T., Y. Maezawa, K. Ikeda, S. Kagami, M. Hatano, Y. Seto, A. Suto, K. Suzuki, N. Watanabe, Y. Saito, T. Tokuhisa, I. Iwamoto, and H. Nakajima. 2006. IL-25 enhances allergic airway inflammation by amplifying a TH2 cell-dependent pathway in mice. *J Allergy Clin Immunol* 118:606–614.

Terashima, A., H. Watarai, S. Inoue, E. Sekine, R. Nakagawa, K. Hase, C. Iwamura, H. Nakajima, T. Nakayama, and M. Taniguchi. 2008. A novel subset of mouse NKT cells bearing the IL-17 receptor B responds to IL-25 and contributes to airway hyperreactivity. *J Exp Med* 205:2727–2733.

Thomas, S.Y., C.M. Lilly, and A.D. Luster. 2006. Invariant natural killer T cells in bronchial asthma (correspondence). *N Engl J Med* 354:2613–2616.

Trompette, A., S. Divanovic, A. Visintin, C. Blanchard, R.S. Hegde, R. Madan, P.S. Thorne, M. Wills-Karp, T.L. Gioannini, J.P. Weiss, and C.L. Karp. 2009. Allergenicity resulting from functional mimicry of a Toll-like receptor complex protein. *Nature* 457:585–588.

Verhasselt, V., V. Milcent, J. Cazareth, A. Kanda, S. Fleury, D. Dombrowicz, N. Glaichenhaus, and V. Julia. 2008. Breast milk-mediated transfer of an antigen induces tolerance and protection from allergic asthma. *Nat Med* 14:170–175.

Vijayanand, P., G. Seumois, C. Pickard, R.M. Powell, G. Angco, D. Sammut, S.D. Gadola, P.S. Friedmann, and R. Djukanovic. 2007. Invariant natural killer T cells in asthma and chronic obstructive pulmonary disease. *N Engl J Med* 356:1410–1422.

Watanabe, N., S. Hanabuchi, M.A. Marloie-Provost, S. Antonenko, Y.J. Liu, and V. Soumelis. 2005. Human TSLP promotes CD40 ligand-induced IL-12 production by myeloid dendritic cells but maintains their Th2 priming potential. *Blood* 105:4749–4751.

Wenzel, S., D. Wilbraham, R. Fuller, E.B. Getz, and M. Longphre. 2007. Effect of an interleukin-4 variant on late phase asthmatic response to allergen challenge in asthmatic patients: results of two phase 2a studies. *Lancet* 370:1422–1431.

Wills-Karp, M. 1999. Immunologic basis of antigen-induced airway hyperresponsiveness. *Annu Rev Immunol* 17:255–281.

Wills-Karp, M., and S.L. Ewart. 1997. The genetics of allergen-induced airway hyperresponsiveness in mice. *Am J Respir Crit Care Med* 156:S89–96.

Wills-Karp, M., J. Luyimbazi, X. Xu, B. Schofield, T.Y. Neben, C.L. Karp, and D.D. Donaldson. 1998. Interleukin-13: central mediator of allergic asthma. *Science* 282:2258–2261.

Wilson, R.H., G.S. Whitehead, H. Nakano, M.E. Free, J.K. Kolls, and D.N. Cook. 2009. Allergic sensitization through the airway primes Th17-dependent neutrophilia and airway hyperresponsiveness. *Am J Respir Crit Care Med* 180:720–730.

Xystrakis, E., S. Kusumakar, S. Boswell, E. Peek, Z. Urry, D.F. Richards, T. Adikibi, C. Pridgeon, M. Dallman, T.K. Loke, D.S. Robinson, F.J. Barrat, A. O'Garra, P. Lavender, T.H. Lee, C. Corrigan, and C.M. Hawrylowicz. 2006. Reversing the defective induction of IL-10-secreting regulatory T cells in glucocorticoid-resistant asthma patients. *J Clin Invest* 116:146–155.

Zuany-Amorim C, Sawicka E, Manlius C, Le Moine A, Brunet LR, Kemeny DM, Bowen G, Rook G, and Walker C. 2002. Suppression of airway eosinophilia by killed Mycobacterium vaccae-induced allergen-specific regulatory T-cells. *Nat Med* 8:8625–8629.

IL-17-Producing γδ T Cells in Auto-immune Disease

Rebecca L. O'Brien, Philip L. Simonian, Yafei Huang, Christina L. Roark, Niyun Jin, M. Kemal Aydintug, Jennifer L. Chain, Andrew P. Fontenot, and Willi K. Born

Abstract In both mice and humans, γδ T cells often represent a significant source of IL-17. Here we review the evidence that IL-17-producing γδ T cells contribute to auto-immune disease and discuss the role that they play in disease processes, including their influence on the development or activity of T_H17 αβ T cells. Although IL-17-producing γδ T cells clearly can exacerbate auto-immune disease, in some systems they have instead been shown to play a protective role. The ability to produce IL-22 as well may be critical for this protective role.

1 Introduction

IL-17 has been shown to be of critical importance to the host because it acts to deploy neutrophils needed for eliminating extracellular bacteria and fungi (Curtis and Way 2009). Despite this, IL-17 acts as a major player in a variety of auto-immune diseases, in which neutrophil recruitment and activation can also cause host tissue damage and destruction without the benefit of host protection, emphasizing the desirability of achieving a balance in immune mediators in preventing auto-immune attack. In addition, IL-17 has a number of other effects that contribute to particular auto-immune processes: (1) inducing osteoclasts to resorb bone, which promotes joint damage in rheumatoid arthritis (Kotake et al. 1999), (2) promoting the breakdown of the blood/brain barrier, which accelerates the development of encephalomyelitis in a model of multiple sclerosis (MS) (Huppert et al. 2010), and (3) stimulating IL-20 production with other cytokines, which promotes terminal

R.L. O'Brien (✉)
Integrated Department of Immunology, National Jewish Health,
1400 Jackson St., Denver, CO 80206, USA
and
University of Colorado Denver, Aurora, CO 80045, USA
e-mail: obrienr@njhealth.org

S. Jiang (ed.), *T_H17 Cells in Health and Disease*,

DOI 10.1007/978-1-4419-9371-7_28,

keratinocyte differentiation and exacerbates psoriasis (Tohyama et al. 2009; Wolk et al. 2009). Furthermore, because IL-17-producing CD4+ αβ TCR+ T_H17 cells develop from the same precursors as do CD4+ regulatory T cells (Tregs), the development of T_H17 cells also results in a reduction in T_H1 CD4+ T cell regulators (Afzali et al. 2010), which additionally aggravates inflammatory damage.

2 IL-17-Producing γδ T Cells Are Elicited in Many Diseases

γδ T cells that produce IL-17 have now been shown to expand in a wide variety of disease models in mice. They are strongly induced by bacterial or fungal infections with several different species, and in these models they are often a major source of IL-17 in the hosts, particularly at early time points following the initial infection (Roark et al. 2008; O'Brien et al. 2009). In some cases, IL-17-producing γδ T cells have been shown to be critical for an effective immune response against particular infectious agents, including *Mycobacterium bovis BCG* (Umemura et al. 2007), *Escherichia coli* (Shibata et al. 2007), and *Listeria monocytogenes* (Hamada et al. 2008; Meeks et al. 2009). IL-17-producing γδ T cells have also been identified in infectious diseases in humans, including patients with active *Mycobacterium tuberculosis* infections, in which they are two to three times more abundant than in normal controls (Peng et al. 2008). Responses by IL-17-producing γδ T cells have also been reported in models of auto-immunity (see below), and in hypersensitivity pneumonitis (Simonian et al. 2009), lung injury (Braun et al. 2008), and chronic granulomatous disease (Romani et al. 2008; O'Brien et al. 2009).

3 IL-17-Producing γδ T Cells Play a Role in Auto-immune Disease Models

In many auto-immune diseases and disease models, IL-17 has been shown to play a pathogenic role, and given their apparent penchant to produce IL-17, it is not surprising that γδ T cells have been shown to be involved in at least some of these. Collagen-induced arthritis (CIA) in mice is a model of rheumatoid arthritis; in which joint disease is induced in mice carrying particular $H\text{-}2^q$ MHC alleles (including DBA/1), following intradermal immunization with type II bovine collagen emulsified in Complete Freund's Adjuvant (CFA). Within about 40 days, the mice develop rheumatoid arthritis-like symptoms, consisting of swollen joints, anti-collagen antibodies, and inflammatory infiltrates in the synovia of the joints, and potential bone and/or cartilage destruction. We discovered that among γδ T cells in the draining lymph nodes of the arthritic joints of mice with CIA, as well as in the affected joints themselves, those that had acquired an activated phenotype frequently expressed a Vγ4Vδ4+ TCR (Roark et al. 2007). In fact, this relatively rare subset increased in number by an average of 60 times in mice with CIA, and about

80% of these expressed intracellular IL-17 (IL-17A) following non-specific stimulation (PMA/ionomycin). In comparison, among other cells from the same mice, less than 2% of lymph node γδ T cells expressing Vγ1, the most abundant γδ T cell type in lymphoid tissue, expressed IL-17 when identically stimulated. Moreover, nearly all of the responding Vγ4Vδ4+ cells had highly similar TCR junctional sequences. In particular, a δ chain junctional "motif" dominated, including the use of Dδ2 in one particular reading frame, preceded by a conserved arginine codon, and a total length limitation of 5–6 amino acids encoded by N, P, or D elements. A γ chain junctional motif was also apparent in most cases, consisting of a leucine codon, generated in several ways, and the inclusion of a tyrosine codon at beginning of the Jγ1 gene. This suggested that oligoclonal Vγ4Vδ4+ cells bearing closely related junctions were driven to expand in mice with CIA as a result of stimulation with a particular antigen. Because they were biased to produce IL-17, we examined the role of this TCR-limited Vγ4Vδ4+ subset in the disease process by depleting all cells expressing Vγ4 via injection of an anti-Vγ4 monoclonal antibody 4 days before the second injection of collagen/CFA. This normally induces a spike in the number of cells bearing the Vγ4Vδ4 TCR, but instead resulted in a reduction in the average disease severity score in the anti-Vγ4-treated group, compared to mice either sham-treated with non-specific polyclonal hamster IgG, or treated with an anti-Vγ1 monoclonal antibody. The Vγ4-depleted mice also showed a reduction in the overall incidence of disease, and a small drop in serum anti-collagen IgG, which mediates pathological effects in CIA. Because the numbers of CD4+ αβ T cells producing IL-17 (T_H17 cells) are normally similar to the numbers of Vγ4Vδ4+ IL-17 producing cells in mice with CIA, the depletion of Vγ4+ γδ T cells may have produced these effects in large part because of an overall reduction in IL-17. Finally, despite the apparent specificity of the γδ T cell response in CIA, these Vγ4Vδ4+ cells do not respond to type II collagen (unpublished results), and an obvious, although somewhat weaker response by the same cells can be elicited by intradermal immunization with CFA emulsified in PBS only (Roark et al. 2007). Therefore, the response by the Vγ4Vδ4+ γδ T cell subset in CIA is not critical for disease induction, but rather appears to be elicited by the inflammation induced during CIA, and acts to promote additional inflammatory damage.

A substantial response by IL-17-producing γδ T cells in CIA was recently confirmed by Ito et al. (2009). IL-17-producing γδ T cells were noted at several stages of disease in the joints, where the number of IL-17-producing γδ T cells was found to exceed that of T_H17 cells at most of the time points analyzed. They also found that IL-17-producing γδ cells can be strongly induced from naïve γδ T cells isolated from the joints by in vitro culture in the presence of IL-1β and IL-23, apparently without a need for TCR stimulation. IL-17-producing γδ cells expressed the chemokine receptor CCR6, which binds to the chemokine CCL20, known to be expressed by both epithelial cells and fibroblast-like synoviocytes. Whether γδ T cells detected in this study comprise the same Vγ4Vδ4 junctionally-limited subset that predominated in our own CIA study was not clear, and although PCR analysis suggested that they were not, this approach can be misleading because certain primers may amplify more efficiently than others, and multiple Vγs can be transcribed in the same cell. The authors did, however, confirm our findings that IL-17-producing γδ T cells elicited during CIA exacerbated disease by showing that when these cells are isolated either from the swollen joints of mice with CIA or draining lymph nodes and adoptively transferred by injection directly into the joints of mice with induced CIA, the treated mice showed significantly elevated arthritis scores, as compared to PBS-treated controls. Ito et al. also examined mice with SKG arthritis, in which joint inflammation develops as consequence of a point mutation in the ZAP-70 gene (Sakaguchi et al. 2003), but in these mice, very few IL-17-producing γδ T cells were detected. This result may not be meaningful because the developing γδ T cells in this strain could be defective due to the weak ZAP-70 signal, which appears to be important in γδ T cell development (Hayes et al. 2005). Furthermore, IL-17-producing γδ T cells were also recently implicated as playing an important role in a non-auto-immune arthritis model, antigen-induced arthritis (Cornelissen et al. 2009), in which an immune response to bovine serum albumin is elicited by prior sensitization, then followed by challenge via antigen injection into the joints. Thus, IL-17-producing γδ T cells may be important, not only in arthritis mediated by an auto-immune response, but in other types of arthritis as well. Ito et al. went on to examine 11 patients with rheumatoid arthritis for the presence of IL-17-producing γδ T cells. They did not observe IL-17-producing γδ T cells in either synovial fluid or tissue, although a small percentage of γδ T cells producing IFN-γ was detected, and αβ TCR+ T_H17 cells were predominant. Because of the small number of patients examined and the fact that all had been receiving anti-rheumatic drugs, a role for IL-17-producing γδ T cells in rheumatoid arthritis in humans was not ruled out by this study.

IL-17-producing γδ T cells have also been implicated in a mouse model of uveitis, in which an auto-immune attack on the uvea of the eye is induced by subcutaneous immunization with a synthetic peptide representing the N-terminus of interphotoreceptor retinoid-binding protein (IRBP) emulsified in CFA (Cui et al. 2009). When T cells from the spleen and draining lymph nodes of mice treated were isolated and cultured with APCs in IL-2 plus the IRBP peptide, over 90% expressed IFNγ, and virtually all of these expressed an αβ TCR. In contrast, when the T cells were instead cultured with APCs plus the IRBP peptide in the presence of IL-23, most of the proliferating T cells were found to express IL-17, and of these, about 17% were γδ T cells. Perhaps because subcutaneous CFA was used in the immunization protocol, the majority of the responding γδ T cells from these cultures were Vγ4Vδ4+, as we had found in the CIA model, although TCR junctional sequences were not examined to determine whether or not those with the particular junctional limitations seen in CIA predominated. When IL-23 pre-cultured αβ T cells or γδ T cells were highly purified prior to re-stimulation with APCs plus the IRBP peptide, both T cell types produced only low levels of IL-17. However, in both cases, IL-17 levels were greatly augmented by adding the other T cell type to the culture. Cell contact was required for this IL-17 augmentation, although it was not for IFNγ; instead, co-culturing the αβ and γδ T cells induced approximately equal amounts of IFNγ, whether or not cell–cell contact was possible. In contrast, highly purified γδ T cells cultured without αβ T cells not only produced little IL-17, but also produced only low levels of IFNγ. Consistently, T cells from TCRδ–/– mice, following immunization with the IRBP peptide emulsified in CFA, produced about four times more IL-17 in vitro if they were taken from mice that had first been reconstituted with γδ T cells. However, T cells from sham-reconstituted TCRδ–/– mice made approximately equal amounts of IFNγ as did those from γδ T cell reconstituted-TCRδ–/– mice. After stimulating T cells from these mice in vitro with APCs and IRBP peptide, they were also found to be able to transfer disease efficiently to naïve wild type mice, but only if they had first been reconstituted with γδ T cells before the initial immunization with IRBP peptide in CFA (Cui et al. 2009). The results from this study show that at least in some settings, high level production of IL-17 and of IFNγ depends on cross-talk between αβ and γδ T cells, including a cell contact-dependent type of cross-talk in the case of IL-17.

A response by IL-17-producing γδ T cells has also been recently reported in mice with experimental auto-immune encephalomyelitis (EAE) (Sutton et al. 2009), a model of MS. Here, IL-17-secreting γδ T cells were detectable in the brains of mice by day 7 after disease induction, with the numbers peaking on days 10–14. Although the absolute numbers of γδ T cells obtained from the brains were small, 50–80% of these produced IL-17 at the peak of the response. Moreover, most of the infiltrating cells expressed Vγ4. In the EAE model, disease is induced by subcutaneous injection of CFA emulsified with a myelin oligodendrocyte glycoprotein-derived (MOG) peptide, followed by two subsequent intra-peritoneal injections of pertussis toxin on days 0 and 2. Hence, it is possible that the responding γδ T cells are again the TCR-junctionally limited Vγ4Vδ4+ subset we discovered in mice with CIA, because an intradermal CFA injection alone is already sufficient to induce their response (unpublished observations). Interestingly, mice lacking γδ T cells (TCRδ–/– mice) had less severe EAE symptoms and produced substantially less IL-17, compared to wild type mice of the same background. Moreover, T cells from mice with EAE, expanded in vitro with the MOG peptide, can adoptively transfer disease into naïve hosts, but this is greatly inhibited if γδ T cells are first removed from the transferred population (Sutton et al. 2009). These findings show that in EAE, as in CIA, IL-17 producing γδ T cells act to increase both disease incidence and severity.

Despite the above three examples of IL-17-producing γδ T cells playing an exacerbating role in auto-immune disease, they can instead play a protective role in auto-immune diseases in which IL-17 does not appear to be pathologic. A recent publication by Bordon et al. (2009) illustrates this point. In this study, a mouse strain was generated that showed reduced susceptibility to DSS-induced colitis, a model of auto-immune colitis in which intestinal inflammation is induced by feeding the animals dextran sodium sulfate. This strain was knocked out for the gene encoding the D6 CC chemokine receptor, a molecule thought to be a non-signaling "decoy" receptor capable of binding several CC chemokines, which is able to scavenge and remove large amounts of inflammation-inducing chemokines. The D6 chemokine receptor is normally expressed abundantly by stromal intestinal cells and intestinal B cells. However, the D6 knockout mice, when treated with DSS, were found to produce increased amounts of certain pro-inflammatory cytokines, including TNFα, IFNγ, and IL-17A. The amounts of CC chemokines produced, and the types and numbers of infiltrating inflammatory cells, did not differ between D6 knockouts and wild type controls. When IL-17A activity was blocked in the D6 knockout mice by injection of a neutralizing anti-IL-17 monoclonal antibody, the severity of DSS-induced colitis increased in these mice. Intestinal γδ T cells were implicated as the source of the excess IL-17 being produced in the D6 knockout mice, because the knockouts showed a threefold higher percentage of IL-17-producing γδ T cells in the lamina propria of their colons compared to wild type, whereas other IL-17 producing cells were relatively rare and did not differ between wild type and the D6 knockout mice (Bordon et al. 2009). The findings of Bordon et al. are at odds with those previously reported for two other colitis models, in which IL-17 was instead found to promote intestinal inflammation (Elson et al. 2007; Hue et al. 2006). They are in agreement, however, with a previous finding for DSS-induced colitis, in which IL-17 was likewise shown to decrease intestinal inflammation (Ogawa et al. 2004), although this was thought to be due to the ability of IL-17 to decrease the production of other pro-inflammatory cytokines, including TNFα and IFNγ. Another study using IL-17 knockout mice presented evidence that in fact, IL-17A and IL-17F play different roles in DSS-induced colitis, and that whereas IL-17A reduces disease severity, IL-17F exacerbates it (Yang et al. 2008). This is consistent with findings from the Bordon et al. report, because they used an IL-17A-specific neutralizing antibody in their study and found a protective role for IL-17A. Bordon et al. point out that colitis is a complex disease process in which an inadequate inflammatory response at the induction of the disease process can lead to increased inflammation later on, because it can allow colonic bacteria to penetrate into the lamina propria. They suggested that D6 normally plays an anti-inflammatory role that not only suppresses IL-17 production, but also suppresses antibacterial or epithelial repair processes in the colon, and actually promotes the susceptibility of wild type mice to colitis when exposed to DSS. If so, it seems likely that instead of IL-17, the level of IL-22, which was not examined in the Bordon et al. study, might also be suppressed by D6 along with IL-17. IL-22 is often produced by T cells that secrete IL-17 (Spolski and Leonard 2009), including γδ T cells (Simonian et al. 2009; Sutton et al. 2009), augmented by IL-17 (Spolski and Leonard 2009), and has been shown to promote both the secretion of anti-bacterial peptides (Liang et al. 2006) and epithelial repair (Aujla and Kelis 2009). Findings from our laboratories in another disease model also revealed a critical protective role for IL-22 produced by γδ T cells, as discussed further below. Furthermore, IL-22 has been previously shown to play a role in protecting against inflammatory damage in both DSS-induced colitis and a T-cell mediated colitis model (Zenewicz et al. 2008). In any case, the results of Bordon et al. help us to understand how γδ T cells are important in maintaining the epithelium of the gut, as shown in earlier studies (Komano et al. 1995; Roberts et al. 1996), and suggest that their ability to produce IL-17 and/or associated cytokines is an important part of this process.

A recent publication concerning the NOD mouse model of type I diabetes presented evidence that IL-17-producing γδ T cells also play a protective role in this auto-immune disease model (Han et al. 2010). Here, a transfer model of diabetes was used to examine disease incidence and severity, using T cells from diabetic NOD mice adoptively transferred into NOD/SCID hosts. Although IL-17 production by splenocytes was increased almost fourfold in T cells cultured from diabetic mice compared to those from non-diabetic controls, IL-17 does not appear to affect the disease, because administering a neutralizing IL-17 monoclonal antibody in vivo (using an IL-17A-specific reagent) in these mice had no significant effect on the development of diabetes. However, IL-17-producing γδ T cells were found to be unexpectedly abundant in NOD mice, and represented the main source of IL-17. When γδ T cells from non-diabetic NODs were co-transferred along with diabetic T cells, they were found to markedly reduce both the ensuing disease incidence and severity in the NOD/SCID hosts. The IL-17 that the γδ T cells produced did not mediate this protective effect; rather, TGF-β, which is co-produced by these cells, was found to be critically important (Han et al. 2010). Hence, although IL-17 producing γδ T cells exacerbate some types of auto-immune disease, in others they can play a protective role, and this effect may be mediated by cytokines other than IL-17.

γδ T cells have been previously implicated in several auto-immune diseases in humans, including rheumatoid arthritis (Bucht et al. 1992; Keystone et al. 1991; Soderstrom et al. 1994; Chaouni et al. 1990; Holoshitz et al. 1989; Jacobs and Haynes 1992), MS (Shimonkevitz et al. 1993; Wucherpfennig et al. 1992), and type I diabetes (Gyarmati et al. 1999; Lang et al. 1991), but to date, no reports linking IL-17-producing γδ T cells to auto-immune disease in humans have been made. Apart from the very small study discussed above involving rheumatoid arthritis patients (Ito et al. 2009), this question appears largely to have not yet been examined. It seems likely that such associations exist, because a recent publication showed an expansion of IL-17-producing γδ T cells in patients with active tuberculosis (Peng et al. 2008), and in the normal control subjects for this study, more than half of the IL-17 producing T cells expressed a γδ TCR. Thus, IL-17-producing γδ T cells appear to be as common in humans as they are in mice, and probably will prove to be involved in human inflammatory diseases as well.

4 Exacerbation of Auto-immunity by IL-17-Producing γδ T Cells Results in Part from Promotion of the Development of T_H17 Cells

IL-17-producing γδ T cells may provide larger than expected effects in disease models because rather than just contributing additional IL-17 and other cytokines, they may act as early-responding inducers of T_H17 cells. In their study using cells derived from IRBP peptide-stimulated mice, Cui et al. (2009) presented some evidence by showing that maximum IL-17 levels were achieved only when γδ T cells and αβ T cells were cultured together and able to make contact with one another.

Sutton et al. (2009) also showed that supernatants from IL-17-producing γδ T cells stimulated with IL-1β and IL-23 contained IL-17 and IL-21, and that both substantially stimulate IL-17 production by T_H17 cells activated in vitro with anti-CD3 and anti-CD28. In support of this idea, we also found in an in vivo experiment that depletion of Vγ4+ IL-17-producing γδ T cells from mice during CIA induction reduces the percentage of CD4+ IL-17+ cells that subsequently develops in these mice (Huang et al. unpublished results). Thus, providing an early source of cytokines that augment T_H17 development may be important in the role of IL-17-producing γδ T cells.

5 Requirements for Activation by IL-17-Producing γδ T Cells

Most IL-17-producing γδ T cells appear to become biased towards production of this class of cytokines as a result of prior programming during their development in the thymus. For example, Shibata et al. (2008) showed that the mouse Vγ6Vδ1+ subset becomes predisposed to produce IL-17 during development in the fetal thymus, a process that depends upon IL-2. In addition, Jensen et al. presented evidence that γδ T cells developing in the adult thymus become biased to produce IL-17 if they fail to encounter ligands during thymic development. If ligands are present, they instead develop into IFNγ-secreting T cells (Jensen et al. 2008). Ribot et al. also presented evidence that γδ T cells in the thymus develop with a preference to produce either IL-17 or IFNγ, but not both. They concluded that the CD27 molecule controls this process, such that CD27+ γδ TCR+ thymocytes develop a bias to secrete IFNγ, whereas CD27- γδ TCR+ thymocytes develop a bias to secrete IL-17 (Ribot et al. 2009). Several papers now support the concept that such IL-17-biased peripheral γδ T cells can be activated to secrete IL-17 by cytokine stimulation alone and without need for prior activation through the TCR. This was first shown by Shibata et al. (2007), who demonstrated by intracellular cytokine staining that about 50% of peritoneal cells from naïve mice when cultured with recombinant IL-23 for 7 h, produce IL-17. Cheng et al. (2008) similarly found that γδ T cells purified from mice immunized with a retinal protein-derived peptide, when cultured with a mixture of IL-1β, IL-7, and IL-23, but not antigen, were stimulated to produce IL-17. Likewise, Ito et al. (2009) found that a mixture of IL-23 and IL-1β was sufficient to induce both proliferation and IL-17 secretion by γδ T cells, without a need for antigen. Additional stimulation through the TCR greatly augmented the response. Sutton et al. (2009) confirmed that γδ T cells can produce IL-17 after stimulation with IL-1β and IL-23 only, and they went on to show that IL-17-producing γδ T cells also produce IL-17F, IL-21, and IL-22, cytokines known to be co-produced along with IL-17 (IL-17A) by T_H17 cells. Martin et al. (2009) conducted a more in-depth comparison of IL-17 producing γδ T cells with T_H17 cells, and reported that the γδ T cells, like T_H17 cells, express CCR6. The aryl hydrocarbon receptor (AhR), which is known to stimulate both IL-17 and IL-22 secretion, was shown to be essential to induce

IL-22 production by these cells. Moreover, IL-17-producing γδ T cells were found in the Martin et al. study to express dectin-1, TLR1, and TLR2, plus stimulation with TLR2 ligands was shown to be sufficient to induce proliferation of the γδ T cells. Adding IL-23 greatly augmented their proliferation, and the highest proliferation was seen when IL-23 and ligands for TLR2 and TLR1 or dectin-1 were used simultaneously.

6 IL-17-Producing γδ T Cells Can Protect Against Inflammatory Damage

γδ T cells that produce IL-17 have rather unexpectedly been associated with protection *against* inflammatory damage in some instances (Braun et al. 2008; Han et al. 2010; Simonian et al. 2009). For example, in a murine bleomycin-induced lung injury model, despite the observation that the infiltrating γδ T cells were producing IL-17, mice lacking γδ T cells developed increased interstitial inflammation and fibrosis. The effect was attributed to the ability of the γδ T cells to promote epithelial repair, and epithelial cells were shown to replicate more slowly in this study in mice lacking γδ T cells compared to wild type mice (Braun et al. 2008). As already mentioned, in a NOD mouse type-I diabetes model, IL-17 producing γδ T cells were also associated with reduced inflammation, because mice that received adoptively transferred γδ T cells producing IL-17, as well as diabetogenic αβ T cells, developed less insulitis and pancreatitis. Here, the IL-17 appeared to play no role in the disease, but rather the protective effect was thought to be mediated by co-produced TGFβ. Recent results from our laboratories suggest that some IL-17-producing γδ T cells are also IL-22 producers, and that these mediate a protective effect that in a mouse model of hypersensitivity pneumonitis, act to reduce lung inflammation and ensuing fibrosis (Simonian et al. 2010). In this model, the responding γδ T cells all express the Vγ6Vδ1 TCR, and whereas many of these produce IL-17A, others instead express IL-17F (but not IL-17A) along with IL-22. IL-22 has already been shown to protect against inflammatory damage in several other disease models (Wahl et al. 2009; Zenewicz et al. 2007, 2008). The IL-22 appears to be critical for the protective role of these γδ T cells in hypersensitivity pneumonitis, because mice with defective aryl hydrocarbon (AhR) receptors that poorly produce IL-22, showed exacerbated inflammation and fibrosis compared to wild type, as did mice treated with a drug that blocks IL-22 secretion, compared to sham-treated controls. Moreover, reconstituting the AhR-defective or drug-treated mice with recombinant IL-22 largely restored the protective effect (Simonian et al. 2010).

It will be interesting to see whether in future studies, IL-17-producing γδ T cells that protect against inflammatory damage are also found to co-produce IL-22. Thus far, we have not been able to detect IL-22 production among the γδ T cells that respond during CIA, which instead increase inflammatory damage and disease incidence. As mentioned above, these γδ T cells express a different TCR than the one characterized in the hypersensitivity pneumonitis model, a junctionally-limited Vγ4Vδ4+ TCR, and it is possible that γδ T cell subsets having different TCRs also differ intrinsically in their cytokine bias. In the studies on auto-immune models described above, with the exception of our study on CIA, the TCRs of the responding IL-17-producing γδ T cells were not characterized. Although IL-23 can induce the production of both IL-17A and IL-22, the two also may be differentially regulated (Siegemund et al. 2009), and stimulation via a ligand of the AhR receptor seems to be the major pathway to activate IL-22 secretion (Veldhoen et al. 2008). IL-22 has also been associated with inflammatory damage, however, and in fact exacerbates skin lesions in psoriasis (Tohyama et al. 2009; Wolk et al. 2009), so it cannot be simply assumed that T cells producing IL-22 are protective.

7 Concluding Remarks

Several different studies now show that γδ T cells that secrete IL-17 can be stimulated to produce it by certain cytokines, without a need for antigens or deliberate stimulus through the TCR. IL-17-producing γδ T cells have been shown to have an exacerbating effect on some auto-immune disease models in mice, whereas in others, they have been shown to play a protective role. Those that play an exacerbating role may largely accomplish this by stimulating the subsequent development of T_H17 αβ T cells, as it has now been shown in several studies that cross-talk between αβ and γδ T cells appears to be an important process in the IL-17 response. Some evidence suggests that those which have a protective role may secrete IL-22 as well. Whether the γδ T cells are able to secrete IL-22, as well as IL-17A or IL-17F, may depend upon external signals they receive over the course of the disease, or this may be pre-determined during their development in the thymus. Whether TCR-defined γδ T cell subsets differ in this regard remains to be determined.

References

Afzali B, Mitchell P, Lechler RI, *et al.* (2010) Translational mini-review series on Th17 cells: Induction of interleukin-17 production by regulatory T cells. Clin Exp Immunol 159:120–30.

Aujla SJ, and Kelis JK (2009) IL-22: A critical mediator in mucosal host defense. J Mol Med 87:451–454

Bordon Y, Hansell CA, Sester DP, *et al.* (2009) The atypical chemokine receptor D6 contributes to the development of experimental colitis. J Immunol 182:5032–5040

Braun RK, Ferrick C, Neubauer P, *et al.* (2008) IL-17 producing γδ T cells are required for a controlled inflammatory response after bleomycin-induced lung injury. Inflammation 31:167–179

Bucht A, Soderstrom K, Hultman T, *et al.* (1992) T cell receptor diversity and activation markers in the Vδ1 subset of rheumatoid synovial fluid and peripheral blood T lymphocytes. Eur J Immunol 22:567–574

Chaouni I, Radal M, Simony-Lafontaine J, *et al.* (1990) Distribution of T-cell receptor-bearing lymphocytes in the synovial membrane from patients with rheumatoid arthritis. J Autoimmun 3:737–745

Cheng L, Cui Y, Shao H, *et al.* (2008) Mouse γδ T cells are capable of expressing MHC class II molecules, and of functioning as antigen-presenting cells. J Neuroimmunol 203:3–11
Cornelissen F, Mus AM, Asmawidjaja PS, *et al.* (2009) Interleukin-23 is critical for full-blown expression of a non-auto-immune destructive arthritis and regulates interleukin-17A and RORγt in γδ T cells. Arthri Res Ther 11:R194
Cui Y, Shao H, Lan C, *et al.* (2009) Major role of γδ T cells in the generation of IL-17 uveitogenic T cells. J Immunol 183:560–567
Curtis MM, and Way SS (2009) Interleukin-17 in host defense against bacterial, mycobacterial and fungal pathogens. Immunol 126:177–185
Elson CO, Cong Y, Weaver CT, *et al.* (2007) Monoclonal anti-interleukin 23 reverses active colitis in a T cell-mediated model in mice. Gastroenterol 132:2359–2370
Gyarmati J, Szekeres-Bartho J, Fischer B, *et al.* (1999) Fetal type lymphocytes in insulin dependent diabetes mellitus. Autoimmunity 30:63–69
Hamada S, Umemura M, Shiono T, *et al.* (2008) IL-17A produced by γδ T cells plays a critical role in innate immunity against *Listeria monocytogenes* infection in the liver. J Immunol 181:3456–3463
Han G, Wang R, Chen G, *et al.* (2010) Interleukin-17-producing γδ T cells protect NOD mice from type 1 diabetes through a mechanism involving transforming growth factor-β. Immunol 129:197–206
Hayes SM, Li L, and Love PE (2005) TCR signal strength influences αβ/γδ lineage fate. Immunity 22:583–593.
Holoshitz J, Koning F, Coligan JE, *et al.* (1989) Isolation of $CD4^-$ $CD8^-$ mycobacteria-reactive T lymphocyte clones from rheumatoid arthritis synovial fluid. Nature 339:226–229
Hue S, Ahern P, Buonocore S, *et al.* (2006) Interleukin-23 drives innate and T cell-mediated intestinal inflammation. J Exp Med 203:2473–2483
Huppert J, Closhen D, Croxford A, *et al.* (2010) Cellular mechanisms of IL-17-induced blood-brain barrier disruption. FASEB J 24:1023–34
Ito Y, Usui T, Kobayashi K, *et al.* (2009) γδ T cells are the predominant source of Interleukin-17 in affected joints in collagen-induced arthritis, but not in rheumatoid arthritis. Arthrit & Rheum 60:2294–2303
Jacobs MR, and Haynes BF (1992) Increase in TCR γδ lymphocytes in synovia from rheumatoid arthritis patients with active synovitis. J Clin Immunol 12:130–136
Jensen KD, Su X, Shin S, *et al.* (2008) Thymic selection determines γδ T cell effector fate: Antigen-naive cells make interleukin-17 and antigen-experienced cells make interferon-γ. Immunity 29:90–100
Keystone EC, Rittershaus C, Wood N, *et al.* (1991) Elevation of a γδ T cell subset in peripheral blood and synovial fluid of patients with rheumatoid arthritis. Clin Exp Immunol 84:78–82
Komano H, Fujiura Y, Kawaguchi M, *et al.* (1995) Homeostatic regulation of intestinal epithelia by intraepithelial γδ T cells. Proc Natl Acad Sci USA 92:6147–6151
Kotake S, Udagawa N, Takahashi N, *et al.* (1999) IL-17 in synovial fluids from patients with rheumatoid arthritis is a potent stimulator of osteoclastogenesis. J Clin Invest 103:1345–1352
Lang FP, Schatz DA, Pollock BH, *et al.* (1991) Increased T lymphocytes bearing the γδ T cell receptor in subjects at high risk for insulin dependent diabetes. J Autoimmun 4:925–933
Liang SC, Tan XY, Luxenberg DP, *et al.* (2006) Interleukin (IL)-22 and IL-17 are co-expressed by Th17 cells and cooperatively enhance expression of antimicrobial peptides. J Exp Med 203:2271–2279
Martin BH, K., Cua DJ, Stockinge B, *et al.* (2009) Interleukin-17-producing γδ T cells slectively expand in response to pathogen products and environmental signals. Immunity 131:321–330
Meeks KD, Sieve AN, Kolls JK, *et al.* (2009) IL-23 Is required for protection against systemic Infection with *Listeria monocytogenes*. J Immunol 183:8026–8034
O'Brien RL, Roark CL, and Born WK (2009) IL-17-producing γδ T cells. Eur J Immunol 39:662–666
Ogawa A, Andoh A, Araki Y, *et al.* (2004) Neutralization of interleukin-17 aggravates dextran sulfate sodium-induced colitis in mice. Clin Immunol 110:55–62
Peng MY, Wang ZH, Yao CY, *et al.* (2008) Interleukin 17-producing gamma delta T cells increased in patients with active pulmonary tuberculosis. Cell Mol Immunol 5:203–208
Ribot JC, deBarros A, Fang DJ, *et al.* (2009) CD27 is a thymic determinant of the balance between interferon-γ and interleukin 17-producing γδ T cell subsets. Nature Immunol 10: 427–436

Roark CL, French JD, Taylor MA, *et al.* (2007) Exacerbation of collagen-induced arthritis by oligoclonal, IL-17-producing γδ T cells. J Immunol 179:5576–5583
Roark CL, Simonian PL, Fontenot AP, *et al.* (2008) γδ T cells: An important source of IL-17. Curr Opin Immunol 20:353–357
Roberts SJ, Smith AL, West AB, *et al.* (1996) T-cell αβ⁺ and γδ⁺ deficient mice display abnormal but distinct phenotypes toward a natural, widespread infection of the intestinal epithelium. Proc Natl Acad Sci USA 93:11774–11779
Romani L, Fallarino F, De Luca A, *et al.* (2008) Defective tryptophan catabolism underlies inflammation in mouse chronic granulomatous disease. Nature 451:211–215
Sakaguchi N, Takahashi T, Hata H, *et al.* (2003) Altered thymic T-cell selection due to a mutation of the ZAP-70 gene causes auto-immune arthritis in mice. Nature 426:454–460
Shibata K, Yamada H, Hara H, *et al.* (2007) Resident Vδ1+ γδ T cells control early infiltration of neutrophils after *Escherichia coli* infection via IL-17 production. J Immunol 178:4466–4472
Shibata K, Yamada H, Nakamura R, *et al.* (2008) Identification of CD25+ γδ T cell as fetal thymus-derived naturally occurring IL-17 producers. J Immunol 181:5940–5947
Shimonkevitz R, Colburn C, Burnham JA, *et al.* (1993) Clonal expansions of activated γδ T cells in recent-onset multiple sclerosis. Proc Natl Acad Sci USA 90:923–927
Siegemund S, Schutze N, Schulz S, *et al.* (2009) Differential IL-23 requirement for IL-22 and IL-17A production during innate immunity against *Salmonella enterica* serovar Enteritidis. Int Immunol 21:555–565
Simonian PL, Roark CL, Wehrmann F, *et al.* (2009) IL-17A expressing T cells are essential for bacterial clearance in a murine model of hypersensitivity pneumonitis. J Immunol 182:6540–6549
Simonian PL, Wehrmann F, Roark CL, *et al.* (2010) γδ T cells protect against lung fibrosis via IL-22. J Exp Med 207:2239–53
Soderstrom K, Bucht A, Halapi E, *et al.* (1994) High expression of Vγ8 is a shared feature of human γδ T cells in the epithelium of the gut and in the inflamed synovial tissue. J Immunol 152:6017–6027
Spolski R, and Leonard WJ (2009) Cytokine mediators of Th17 function. Eur J Immunol 39:658–661
Sutton CE, Lalor SJ, Sweeney CM, *et al.* (2009) Interleukin-1 and IL-23 induce innate IL-17 production from γδ T cells, amplifying Th17 responses and auto-immunity. Immunity 31:331–341
Tohyama M, Hanakawa Y, Shirakata Y, *et al.* (2009) IL-17 and IL-22 mediate IL-20 subfamily cytokine production in cultured keratinocytes via increased IL-22 receptor expression. Eur J Immunol 39:2779–2788
Umemura M, Yahagi A, Hamada S, *et al.* (2007) IL-17-mediated regulation of innate and acquired immune response against pulmonary *Mycobacterium bovis* Bacille Calmette-Guerin infection. J Immunol 178:3786–3796
Veldhoen M, Hirota K, Westendorf AM, *et al.* (2008) The aryl hydrocarbon receptor links TH17-cell-mediated auto-immunity to environmental toxins. Nature 453:106–109
Wahl C, Wegenka UM, Leithauser F, *et al.* (2009) IL-22-dependent attenuation of T cell-dependent (ConA) hepatitis in herpes virus entry mediator deficiency. J Immunol 182:4521–4528
Wolk K, Witte E, Warszawska K, *et al.* (2009) The Th17 cytokine IL-22 induces IL-20 production in keratinocytes: A novel immunological cascade with potential relevance in psoriasis. Eur J Immunol 39:3570–3581
Wucherpfennig KW, Newcombe J, Li H, *et al.* (1992) γδ T-cell receptor repertoire in acute multiple sclerosis lesions. Proc Natl Acad Sci USA 89:4588–4592
Yang XO, Chang SH, Park H, *et al.* (2008) Regulation of inflammatory responses by IL-17F. J Exp Med 205:1063–1075
Zenewicz LA, Yancopoulos GD, Valenzuela DM, *et al.* (2007) Interleukin-22 but not interleukin-17 provides protection to hepatocytes during acute liver inflammation. Immunity 27:647–659
Zenewicz LA, Yancopoulos GD, Valenzuela DM, *et al.* (2008) Innate and adaptive interleukin-22 protects mice from inflammatory bowel disease. Immunity 29:947–957

Index

S. Jiang (ed.), *T_H17 Cells in Health and Disease*,
DOI 10.1007/978-1-4419-9371-7, © Springer Science+Business Media, LLC 2011

B

C

J

K

L

MIX
Papier aus verantwortungsvollen Quellen
Paper from responsible sources
FSC® C105338

If you have any concerns about our products,
you can contact us on
ProductSafety@springernature.com

In case Publisher is established outside the EU,
the EU authorized representative is:
Springer Nature Customer Service Center GmbH
Europaplatz 3, 69115 Heidelberg, Germany

Printed by Libri Plureos GmbH
in Hamburg, Germany